Contents

Invasive Wild Pigs in North America

Ecology, Impacts, and Management

Invasive Wild Pigs in North America

Ecology, Impacts, and Management

Edited by
Kurt C. VerCauteren, James C. Beasley,
Stephen S. Ditchkoff, John J. Mayer,
Gary J. Roloff, and Bronson K. Strickland

CRC Press is an imprint of the
Taylor & Francis Group, an **informa** business

Front Cover: Image of an adult male wild pig exhibiting the Eurasian wild boar phenotype. Because of widespread selective releases of hybrid or purebred animals exhibiting this phenotype, there has been a shift among wild pig populations in North America from animals that mostly resemble the wild-living domestic/feral phenotype to that of the partial or full Eurasian wild boar phenotype. (Image courtesy of ©shutterstock.com.)

CRC Press
Taylor & Francis Group
6000 Broken Sound Parkway NW, Suite 300
Boca Raton, FL 33487-2742

Printed on acid-free paper

International Standard Book Number-13: 978-0-367-86173-5 (Paperback)
International Standard Book Number-13: 978-1-138-03581-2 (Hardback)

Visit the Taylor & Francis Web site at
http://www.taylorandfrancis.com

and the CRC Press Web site at
http://www.crcpress.com

Foreword: The Dilemma with Wild Pigs in North America

This book, *Invasive Wild Pigs in North America: Ecology, Impacts, and Management*, introduces you to the villain of a horror story, describing biological and behavioral attributes that enhance the villain's capacity to be so good at being very bad. Like most villains, this one has its supporters, which complicates a solution. Unfortunately, the story is not fictional, it is real life. The authors introduce you to the villain, lead you through scenarios of destruction, and offer approaches and actions to combat it. The horror is real, no sector is left untouched; fields of crops destroyed, urban landscapes devastated, sensitive wildlife species pushed towards extinction, malaise among humans and livestock, entire ecosystems altered, and the list goes on. The depth of destruction endured by the time this story fully unfolds may well depend upon whether managers, scientists, and public entities incorporate information enclosed within this book to cooperatively stop this villain.

A few years ago, I was asked to develop a programmatic approach to address invasive wild pig issues across the United States. At the time, there was limited reliable information available in a concise location. My endeavors to learn more on the topic introduced me to most authors contributing to this book. I benefitted greatly through interactions with them, gleaning knowledge and insights. A few are the old guards who recognized early on an emerging problem and took steps to learn more and make others aware. All are determined in their diligent efforts to combat an increasing threat to agriculture, natural resources, property, and human health and safety, and are true subject matter experts within their respective fields.

Wild pigs are one of the world's worst invasive species. Or perhaps the best, if the ranking depends on them being good at being invasive. *Merriam-Webster Dictionary* defines "invasive" as "an organism that is not native to the place where found and tends to grow and spread easily, usually to the detriment of native species and ecosystems." Paraphrasing, a successful invasive species is capable of surviving when introduced to new environments, and then generally creates problems. Wild pigs not only survive but often thrive in new environments, and they always cause problems. Attributes that enable them to succeed include their intelligence and adaptability, a lack of natural predators, high fecundity, and an opportunistic, generalist omnivore diet.

The US Department of Agriculture has implemented a national response to reduce, and in some areas eliminate, the risks and damages inflicted by wild pigs to property, agriculture, natural resources, and human health. Serving as the lead federal entity, the US Department of Agriculture operates in a cooperative effort with other federal, state, tribal, and local entities that share a common interest in reducing or eliminating problems caused by wild pigs. Since environmental conditions and laws governing wild pigs vary considerably among states, our strategy is to provide resources and expertise at a national level, while allowing flexibility to manage operational activities from a more local perspective. Many of the book's authors were consulted and provided insight in the development of our national response. Here they expand upon and fully develop much of the information gleaned from those conversations.

The editors admit that there are a plethora of unanswered questions surrounding wild pigs and the authors point out gaps in knowledge and necessary research to fill those voids. This book is a comprehensive review of what is currently known and provides a combination of insights regarding the topic from true subject matter experts. I am truly grateful for the editors and authors alike who combined their efforts to create a repository of the current knowledge available on wild pigs. I hope their efforts serve to arm more people with this knowledge to combat this villain through direct management, support of science-informed state and federal regulations, and/or additional studies to further our understanding.

Dale L. Nolte
USDA/APHIS/WS/National Feral Swine Damage Management Program

Acknowledgments

KURT C. VERCAUTEREN:

As primary editor of this volume, I wholeheartedly appreciate my coeditors for all of their efforts and collegiality throughout the process. Jim, Jack, Steve, Gary, and Bronson all played major roles in bringing this project to fruition and I will miss our monthly calls. I thank the Taylor & Francis Group of CRC Press, especially Randy Brehm, for the opportunity to produce this book and all the guidance and patience along the way. I am indebted to the authors for their commitment; this book is the culmination and assemblage of their fantastic contributions. I gratefully acknowledge Michael Glow for assisting in many ways, most especially keeping us organized and attending to details. The effort was supported by the USDA/APHIS/WS/National Wildlife Research Center and National Feral Swine Damage Management Program. I also want to thank my most influential mentor, Scott Hygnstrom, for all the career and life lessons and experiences he has taught and shared with me. And most especially, huge hugs to my wife, Tammy, and daughters, Karlie and Grace, for supporting and putting up with me through this and all my other endeavors.

JAMES C. BEASLEY:

I am extremely grateful to Dr VerCauteren and the other coeditors for the opportunity to contribute to the production of this book. I thank all of my coeditors for their commitment and dedication throughout the development of the book, and also would like to thank all the chapter authors who contributed their time and expertise. My efforts were partially supported by the US Department of Energy under award # DE-EM0004391 to the University of Georgia Research Foundation. I also would like to thank my mentor, Gene Rhodes, for all he has done to guide and support me through my professional journey. And finally, I am forever grateful to my family and particularly my wife, Rochelle, and our two sons, Matthew and Michael, for their unwavering support of all of my endeavors.

STEPHEN S. DITCHKOFF:

I greatly appreciate the opportunity to contribute to the production of this book, and to work alongside my fellow coeditors. As described in the Introduction, I believe that this book will greatly enhance our success in managing wild pigs in North America, and to play a small part means a lot. I also would like to acknowledge the School of Forestry and Wildlife Sciences at Auburn University, whose support enabled me to contribute my time to the production of this book.

JOHN J. ("JACK") MAYER:

It has been an honor both to work with my coeditors and to be involved in the writing of this volume. The topic of wild pigs has grown enormously since I started doing research on these animals years ago; hopefully this book will provide an excellent starting place for researchers beginning on their own journeys toward better understanding and dealing with these animals. I would like to thank my two sons, J.J. and Loren, who constantly challenged their old man over the years with stimulating and thought-provoking questions about wild pigs. Lastly, I would like to thank my wife, Mary, the love of my life, who has put up with my interest in and fascination with wild pigs for many decades now, including helping to raise a number of wild piglets in our house. My time and contributions toward the completion of this book were supported by the US Department of Energy

Office of Environmental Management under Contract DE-AC09-08SR22470 to Savannah River Nuclear Solutions LLC.

GARY J. ROLOFF:

These types of products do not come easily, and I thank my coeditors for their commitment, hard work, and focus to complete this book. I especially acknowledge the leadership of Dr VerCauteren in keeping us on task. I truly appreciated our open lines of communication and willingness to receive and give constructive criticisms; this book is better because we all showed some humility towards a common goal. Personally, this was a highly rewarding product professionally as I learned more about pigs, established personal relationships with individuals recognized for their global expertise on wild pigs, and produced a product that will advance the management of pigs. I thank Michigan State University and the Michigan Department of Natural Resources-Wildlife Division, whose support allowed me to contribute time to this book.

BRONSON K. STRICKLAND:

Wild pig population expansion will be a challenge for wildlife biologists, natural resource managers, and landowners for decades to come. As time passes, and more research is generated, we continue to learn about the pervasive impacts of wild pigs on the landscape. To that end, the editors all agreed on the critical need of a book that compiled and synthesized information on wild pig ecology and management in North America, and we hope this book serves as a foundation to build upon for years to come. I am grateful to Dr VerCauteren for allowing me to contribute as an editor and author. I very much enjoyed the collaboration and discussion among this group of coeditors—it was a demanding but worthwhile endeavor. I want to acknowledge and thank the Mississippi State University Extension Service and the Forest and Wildlife Research Center for support during the development of this book. As always, I'm most grateful to my family, Kacey, Sadie, and Cia, for their continued love and support even when my work keeps me on a computer, or away from home.

Editors

PRIMARY EDITOR

Dr Kurt C. VerCauteren leads research on invasive wild pigs at the National Wildlife Research Center (NWRC) of USDA/APHIS/Wildlife Services. He obtained a BS degree in wildlife from the University of Wisconsin-Stevens Point and an MS and PhD in wildlife ecology from the University of Nebraska-Lincoln. In 1999, he became a Research Wildlife Biologist at NWRC, where he has focused primarily on damage and disease issues associated with deer, elk, wild pigs, and other mammals. Diseases he has focused on include bovine tuberculosis, chronic wasting disease, and rabies. He has authored or coauthored more than 190 peer-reviewed scientific publications and 23 book chapters. Kurt has served as an associate editor of the *Wildlife Society Bulletin* and *Prairie Naturalist.* Awards he has received include NWRC Research Scientist of the Year, NWRC Publication of the Year, and Wildlife Services Supervisor of the Year.

COEDITORS

Dr James C. Beasley is an Associate Professor at the Savannah River Ecology Lab and the Warnell School of Forestry and Natural Resources at the University of Georgia. His research program is focused on wild pig ecology and management, carnivore ecology and management, spatial ecology and population dynamics of wildlife, wildlife health and disease ecology, and scavenging ecology. In addition to his research in the United States, Jim is involved in numerous international or overseas research projects, with recent or current projects in Belarus, Japan, Hawaii, and Guam. Since 2014, Jim also has served as the International Atomic Energy Association's wildlife advisor to the Fukushima Prefecture Government in Japan in response to the nuclear accident that occurred there in 2011. Jim earned an AAS in Pre-Professional Forestry from Paul Smith's College, a BS in Wildlife Science from SUNY-Environmental Science and Forestry, and an MS and PhD in Wildlife Ecology from Purdue University, where he studied the spatial ecology and population dynamics of mesopredators. Over the last 12 years he has published over 100 peer reviewed research articles and book chapters, and coedited the book *Ecology and Management of Terrestrial Invasive Species in the United States.* His research has been featured in several hundred media outlets such as the *New York Times*, Animal Planet, CNN, *USA Today*, *National Geographic*, BBC News, and NPR. Jim currently serves as the research chair of the National Wild Pig Task Force research subcommittee, is a member of the South Carolina Wild Pig Task Force, an active member of The Wildlife Society (TWS), a Certified Wildlife Biologist with TWS, and an associate editor for *Human–Wildlife Interactions* and *Pest Management Science.* In his spare time Jim is an avid hunter and fisherman who enjoys traveling and spending time outdoors with his family.

Dr Stephen S. Ditchkoff has been a Professor at Auburn University in the School of Forestry and Wildlife Sciences since he was hired in 2001. He received his BS degree from Michigan State University in Fisheries and Wildlife, his MS from the University of Maine in Wildlife Ecology, and his PhD from Oklahoma State University in Wildlife Ecology. His research is focused on the ecology and management of large mammals, with white-tailed deer and wild pigs being the primary species he studies. He has authored or coauthored more than 80 peer-reviewed scientific articles and 3 book chapters, and has published more than 40 popular articles in outlets such as *Deer & Deer Hunting* and *Wildlife Trends.* In addition to his research responsibilities, he teaches both undergraduate and graduate courses in the wildlife program at Auburn University.

Dr John J. Mayer received both his BA in biology and PhD in zoology from the University of Connecticut. He is currently a research scientist and the Environmental Sciences & Biotechnology manager at the Savannah River National Laboratory in Aiken, South Carolina. Dr Mayer has been conducting research on wild pigs for 46 years. Although mostly focused on morphological work, it has also included research on wild pigs in the areas of systematics, behavior, population biology, reproductive biology, damage/impacts, and management/control techniques. He is the senior author of *Wild Pigs in the United States*. Dr Mayer's work with wild pigs has spanned three continents and included over 20,000 specimens examined/measured. He was also one of the National Geographic Society team of scientists who unearthed and examined the legendary, or perhaps infamous, "Hogzilla".

Dr Gary J. Roloff is a Professor in the Department of Fisheries and Wildlife at Michigan State University (MSU). Gary directs the Applied Forest and Wildlife Ecology Laboratory (AFWEL) at MSU, where he oversees several research projects including wild pig ecology and control, deer herbivory effects on regenerating northern hardwoods, effects of structural retention on wildlife in clearcut areas, and marten and snowshoe hare ecology, among others. Prior to working at MSU, Gary worked 11 years as a wildlife management specialist for Boise Cascade Corporation (Boise), an integrated forest and wood products company. While at Boise, Gary had responsibilities for projects in the Pacific Northwest, Minnesota, and the southeast United States. Gary's educational background includes a BS from the University of Wisconsin-Steven Point, an MS from Eastern Kentucky University, and a PhD from Michigan State University. Gary served the professional organization of wildlife biologists (The Wildlife Society) as the Michigan Chapter President, Treasurer of the Biological Diversity Working Group, Secretary and President of the North Central Section, and as an Associate Editor for the *Journal of Wildlife Management*.

Dr Bronson K. Strickland is the Extension Wildlife Specialist and St. John Family Professor of Wildlife Management at Mississippi State University. He received a bachelor's degree in Forest Resources from the University of Georgia, a master's degree from Texas A&M University-Kingsville, and a PhD from Mississippi State University. He and coauthors have published more than 60 peer-reviewed scientific articles, 1 book, 2 book chapters, and over 60 popular articles. Bronson's applied science in wildlife management is focused on white-tailed deer and the impacts of wild pigs. Bronson's educational outreach efforts include face-to-face seminars and workshops, websites, social media, and a podcast. He currently serves on the National and Mississippi Wild Pig Task Force and is a Certified Wildlife Biologist with The Wildlife Society and a Professional Member of The Boone & Crockett Club.

Contributors

Scott F. Beckerman
USDA/APHIS/Wildlife Services
Springfield, Illinois

Michael J. Bodenchuk
USDA/APHIS/Wildlife Services
San Antonio, Texas

Rafael Borroto-Páez
Cuban Zoological Society
Havana, Cuba

Raoul K. Boughton
Range Cattle Research and Education Center
Department of Wildlife Ecology and Conservation
University of Florida
Ona, Florida

Ryan K. Brook
Department of Animal and Poultry Science
College of Agriculture and Bioresources
University of Saskatchewan
Saskatoon, Canada

James C. Cathey
Natural Resources Institute
Texas A&M University
College Station, Texas

Joseph L. Corn
Southeastern Cooperative Wildlife Disease Study
College of Veterinary Medicine
University of Georgia
Athens, Georgia

Johanna Delagado-Acevedo
Department of Biological and Environmental Science
Texas A&M University-Commerce
Commerce, Texas

Robert M. Denkhaus
Fort Worth Nature Center and Refuge
Fort Worth, Texas

Doug R. Dufford
Illinois Department of Natural Resources
Lena, Illinois

Dwayne R. Etter
Michigan Department of Natural Resources
Lansing, Michigan

Bethany A. Friesenhahn
Ceasar Kleberg Wildlife Research Institute
Texas A&M-Kingsville
Kingsville, Texas

Joshua A. Gaskamp
Noble Research Institute
Ardmore, Oklahoma

Michael P. Glow
USDA/APHIS/Wildlife Services
National Wildlife Research Center
Fort Collins, Colorado

Steven M. Gray
Applied Forest and Wildlife Ecology Laboratory
Department of Fisheries and Wildlife
Michigan State University
East Lansing, Michigan

Steven C. Hess
Pacific Islands Ecosystem Research Center
US Geological Survey
Hawaii National Park, Hawaii

Billy Higginbotham
Agrilife Extension Service
Texas A&M University
Overton, Texas

Karmen M. Hollis-Etter
University of Michigan-Flint
Flint, Michigan

David A. Keiter
School of Natural Resources
University of Nebraska-Lincoln
Lincoln, Nebraska

John C. Kilgo
Southern Research Station
USDA/Forest Service
Ellenton, South Carolina

Stephanie Kramer-Schadt
Department of Ecological Dynamics
Leibniz Institute for Zoo and Wildlife Research
Berlin, Germany

Michael J. Lavelle
USDA/APHIS/Wildlife Services
National Wildlife Research Center
Fort Collins, Colorado

Jesse S. Lewis
College of Integrative Sciences and Arts
Arizona State University
Mesa, Arizona

Creighton M. Litton
Department of Natural Resources and Environmental Management
University of Hawaii-Mānoa
Honolulu, Hawaii

Michael T. Mengak
Warnell School of Forestry and Natural Resources
University of Georgia
Athens, Georgia

Craig A. Miller
Illinois Natural History Survey
Prairie Research Institute
University of Illinois
Champaign, Illinois

Ryan S. Miller
USDA/APHIS/Veterinary Services
Fort Collins, Colorado

Robert A. Montgomery
Department of Fisheries and Wildlife
Michigan State University
East Lansing, Michigan

Melissa Nichols
Michigan Department of Natural Resources
Lansing, Michigan

Dale L. Nolte
USDA/APHIS/Wildlife Services
National Feral Swine Damage Management Program
Fort Collins, Colorado

J. Alfonso Ortega-S
Ceasar Kleberg Wildlife Research Institute
Texas A&M University-Kingsville
Kingsville, Texas

Sylvia Ortmann
Department of Evolutionary Ecology
Leibniz Institute for Zoo and Wildlife Research
Berlin, Germany

Kim M. Pepin
USDA/APHIS/Wildlife Services
National Wildlife Research Center
Fort Collins, Colorado

Antoinette J. Piaggio
USDA/APHIS/Wildlife Services
National Wildlife Research Center
Fort Collins, Colorado

Peter E. Schlichting
College of Integrative Sciences and Arts
Arizona State University
Mesa, Arizona

Andrew L. Smith
Department of Wildlife, Fisheries and Aquaculture
College of Forest Resources
Mississippi State University
Mississippi State, Mississippi

Mark D. Smith
School of Forestry and Wildlife Sciences
Auburn University
Auburn, Alabama

Timothy J. Smyser
USDA/APHIS/Wildlife Services
National Wildlife Research Center
Fort Collins, Colorado

Nathan P. Snow
USDA/APHIS/Wildlife Services
National Wildlife Research Center
Fort Collins, Colorado

Milena Stillfried
Department of Ecological Dynamics
Leibniz Institute for Zoo and Wildlife Research
Berlin, Germany

Roberto Tamez-González
Centro de Investigaciones en Ciencias y Desarrollo de la Salud
Universidad Autónoma de Nuevo León
San Nicolás de los Garza, Nuevo León, Mexico

Ben S. Teton
Tejon Ranch Conservancy
Lebec, California

Jorge G. Villarreal-González
Consejo Estatal de Flora y Fauna de Nuevo Léon
Monterrey, Nuevo Léon, Mexico

Nathaniel H. Wehr
Department of Natural Resources and Environmental Management
University of Hawaii-Mānoa
Honolulu, Hawaii

Michael D. White
University of California-Berkley
Encinitas, California

Bradley E. Wilson
USDA/APHIS/Wildlife Services
Springfield, Illinois

Michael J. Yabsley
Warnell School of Forestry and Natural Resources
University of Georgia
Athens, Georgia

Stam M. Zervanos
Penn State University-Berks
Reading, Pennsylvania

1 Introduction

Kurt C. VerCauteren, John J. Mayer,
James C. Beasley, Stephen S. Ditchkoff,
Gary J. Roloff, and Bronson K. Strickland

CONTENTS

1.1 INTRODUCTION

If one wanted to design the perfect invasive animal species, i.e., a highly adaptable generalist (e.g., could live almost anywhere and eat almost anything), able to quickly increase population size and rapidly expand its range, difficult to control and manage, and causes extensive and diverse damage to natural and anthropogenic environments, one need look no further than wild pigs (*Sus scrofa*; Figure 1.1). As members of the species *S. scrofa*, wild pigs in North America share both a conspecific kinship with domestic pigs as well as folklore and mythology of native Eurasian wild boar of the Old World. In the scientific and public literature, individuals of the species are referred to as wild pigs, wild hogs, wild boar, feral swine, feral pigs, and other derivations of this theme. In this book, we call free-ranging suids "wild pigs" (Keiter et al. 2016). We use "wild boar" when discussing literature from the native range of *S. scrofa*, or when referring to recent introductions of pigs with European lineage.

The history of wild pigs in continental North America dates back to the 15th century (Mayer 2018), corresponding to initial European exploration of the mainland, and possibly the 13th century in the Pacific Basin with human colonization of the Hawaiian Archipelago (see Chapter 17). Despite recognition as a destructive species in North America as early as 1505, wild pigs persisted, thrived, and gradually expanded their range, mostly occurring in the southeastern United States (Mayer 2018). However, over the last few decades the distribution of wild pigs substantially expanded to include all major geographical regions of continental North America (Figure 1.2), largely due to human-facilitated movements. With the exception of successful eradication efforts in a few areas, wild pigs now occur at greater densities and across greater portions of the continent than ever before.

North America and its associated islands are not alone in dealing with ramifications of non-native wild pigs. Wild pigs have a long history of introduction and subsequent establishment throughout non-polar regions of the world. Hence, wild pigs are one of the widest-ranging mammals in the world, resulting in substantial impacts on ecosystems throughout their range (Melletti and Meijaard 2018). Even in their native range, wild boar populations have increased and individuals occur in historically unoccupied regions as a result of human activities such as deliberate releases for sport hunting, habitat alterations like reforestation, increased availability of certain agricultural crops such as corn, and changing climatic conditions to include milder winters and springs (Tack 2018).

Until the late 1980s, wild pigs successfully evaded focused attention of wildlife professionals. This occurred largely because range expansion and fostering of wild pig populations often happened secretly, thereby allowing wild pigs to infiltrate many North American ecosystems. At one time recently, wild pigs were found in 48 US states (Mayer and Beasley 2018), 6 Canadian provinces, at least 11 Mexican states (Mayer 2018), and numerous islands in the Caribbean and Pacific

FIGURE 1.1 Images of wild pigs: (a) depicts one of more pure Eurasian lineage, (b) is of typical, thin-haired wild pigs found in southern North America, (c) demonstrates characteristics of domestic pig pelage, and (d) shows typical pelage of young wild pigs. (Photos (a–c) by the US Department of Agriculture and (d) by J. Gaskamp. With permission.)

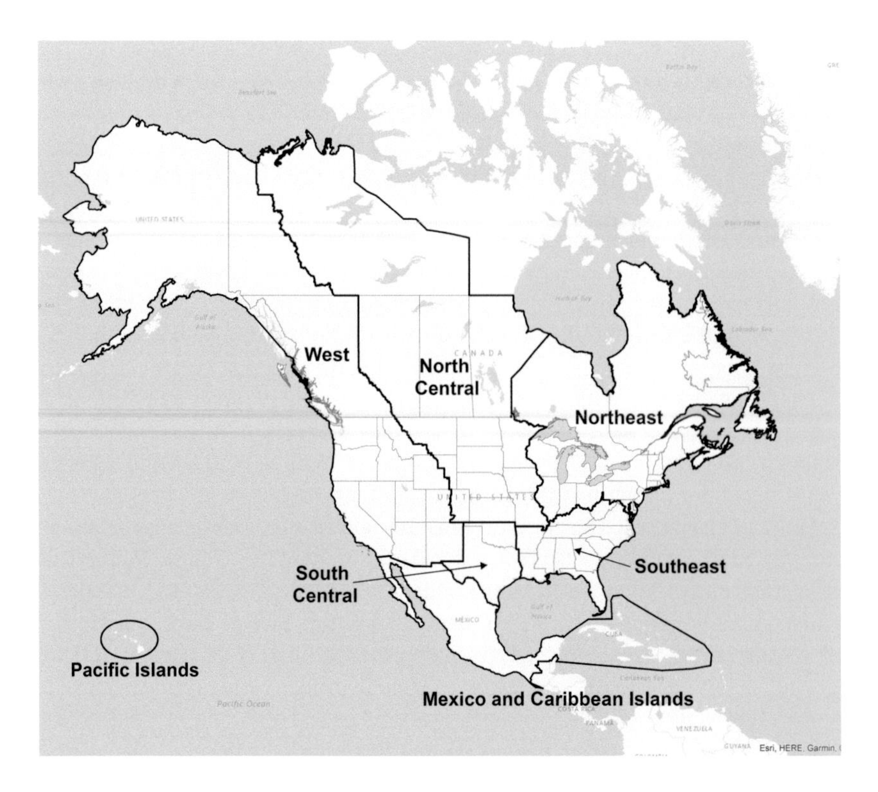

FIGURE 1.2 Map of North America depicting general boundaries of regional chapters.

FIGURE 1.3 Image (a) depicts typical damage to crop and (b) shows rooting damage to natural habitat. (Photos (a) by the US Department of Agriculture and (b) by J. Beasley. With permission.)

Basin (see Chapters 17, 18). These increases resulted in corresponding surges in the magnitude and scope of ecological and social damages being inflicted by wild pigs (Figure 1.3; see Chapter 7). Consequently, a multitude of far-reaching direct and indirect costs, some of which have yet to be realized, have been attributed to wild pigs.

Costs generally outweigh benefits of wild pigs, but this debate is contentious in some regions of North America and can add to management complexities. For example, in some parts of North America, substantial industries exist relative to wild pig hunting and trapping, often adding significant financial and political incentives to retain and perpetuate wild pigs in these areas. Because wild pigs yield economic benefits to individuals that trap or process them for human consumption or other purposes, segments of society can have polarized views towards wild pigs, uniquely complicating management of this invasive species. Further, wild pigs in North America are second only to white-tailed deer (*Odocoileus virginianus*) in numbers harvested annually by recreational hunters pursuing large game (Kaufman et al. 2004), and in some US states wild pigs are legally classified as a game species. A broad divergence of opinions toward wild pigs even exists among governmental agencies with respect to classification (e.g., livestock, game animal, invasive species), and resultant management authority (e.g., departments of agriculture if classed as livestock, wildlife departments if classified as game species; see Chapter 11). This variation among administrative oversight confounds continent-wide management of wild pigs.

Wildlife professionals, the non-professional public, and numerous policy makers fortunately realize that wild pig populations are inflicting irreversible harm to some ecosystems in North America. Accordingly, support and urgency for controlling wild pigs increased substantially, but the tasks of control and eradication are monumental endeavors. In fact, some argue wild pigs in parts of North

America (e.g., Florida) are naturalized (see Chapter 6) and substantial control or eradication is not feasible. Where wild pigs have existed for hundreds of years and local native species have adapted, management may more effectively focus on small-scale protection of critical resource areas (e.g., sea turtle nesting sites, unique ecological communities or areas). In contrast, newly occupied areas (e.g., northern latitudes of North America) where populations are relatively isolated hold greatest promise for eradication, likely warranting significant resource expenditures.

It is an exciting and challenging time to be a researcher or manager of wild pigs in North America. For researchers, the demand for applied results is high, with considerable knowledge gaps ranging from basic wild pig biology (e.g., reproductive rates, survival) and ecology (e.g., food habits, competition with native species), to development of new tools and strategies for control (e.g., toxicants; see Beasley et al. 2018). For managers, the integration of ecological, social, and economic factors affecting wild pig management poses substantial challenges, but impacts of wild pigs on natural and anthropogenic resources place an urgency on eliminating these animals where possible and reducing their densities and impacts elsewhere. Integrated management, innovative research, an informed and educated public, and effective policy and regulations against wild pigs will be imperative for successful management of this species. One objective of this book is to facilitate that process.

Our overarching goal with this book was to compile and synthesize current knowledge and understanding of wild pigs in North America, addressing all aspects of their life history, ecology, biology, strategies for management, and environmental and agricultural impacts. Information presented in this book builds upon previously published volumes focused on wild pigs in North America (e.g., Mayer and Brisbin 1991, reprinted 2008) and incorporates the large body of research and information produced in the last decade. It also updates and integrates important international resources (e.g., Tisdell 1982, Briedermann and Stöcker 2014, Melletti and Meijaard 2018) by including new literature or information relevant to North America. In total, 54 authors, all experts in their fields, from Canada, Mexico, the Pacific Islands, the continental United States, and elsewhere devoted valuable time to generate and summarize the extensive amount of information found within this book. We trust their efforts will be rewarded through improved understanding and management of wild pigs.

The book is structured in 3 broad sections, covering wild pig biology and ecology, techniques for management and research, and regional chapters. The first section includes an introduction and chapters related to taxonomy, spatial ecology, population dynamics, diseases and parasites, and the naturalized niche of wild pigs. The second section includes chapters on damage to resources, management, research methods, human dimensions and education, and policy and legislation. In the last section, we divided North America and associated islands into geographic areas, and each chapter provides greater region-specific information and detail (Figure 1.2). Wild pig researchers and practitioners provided detailed information and case studies specific to each region. The last section also includes a chapter on wild pigs at the wildland-urban interface, an especially challenging emerging issue. We conclude the book with our vision of the future of wild pigs in North America. We are confident that eradication is a reasonable goal in some locations (and perhaps regions) and believe that population control is achievable in the other areas, but both will require immense effort and political and social fortitude.

ACKNOWLEDGMENTS

Contributions from K.C.V. were supported by the USDA. Contributions from J.J.M. were supported by the US Department of Energy Office of Environmental Management under Contract DE-AC09-08SR22470 to Savannah River Nuclear Solutions LLC. Contributions from J.C.B. were partially funded by the US Department of Energy under award # DE-EM0004391 to the University of Georgia Research Foundation. Mention of commercial products or companies does not represent an endorsement by the US Government.

REFERENCES

Beasley, J. C., S. S. Ditchkoff, J. J. Mayer, M. D. Smith, and K. C. VerCauteren. 2018. Research priorities for managing invasive wild pigs in North America. *Journal of Wildlife Management* 82:674–681.

Briedermann, L., and B. Stöcker. 2014. *Schwarzwild*. Kosmos, Stuttgart, Germany.

Kaufman, K., R. Bowers, and N. Bowers. 2004. *Kaufman Focus Guide to Mammals of North America*. Houghton Mifflin, New York.

Keiter, D., J. J. Mayer, and J. C. Beasley. 2016. What's in a "common" name? A call for consistent terminology for referring to non-native *Sus scrofa*. *Wildlife Society Bulletin* 40:384–387.

Mayer, J. J. 2018. Introduced wild pigs in North America: History, problems and management. Pages 299–312 *in* M. Melletti and E. Meijaard, editors. *Ecology, Evolution and Management of Wild Pigs and Peccaries: Implications for Conservation*. Cambridge University Press.

Mayer, J. J., and J. C. Beasley. 2018. Wild pigs. Pages 219–248 *in* W. C. Pitt, J. C. Beasley, and G. W. Witmer, editors. *Ecology and Management of Terrestrial Vertebrate Invasive Species in the United States*. CRC Press, LLC, Taylor & Francis Group, Boca Raton, FL.

Mayer, J. J., and I. L. Brisbin, Jr. 1991. *Wild Pigs in the United States: Their History, Comparative Morphology, and Current Status*. The University of Georgia Press, Athens, GA, USA. (Reprinted in 2008.)

Melletti, M., and E. Meijaard. 2018. *Ecology, Evolution and Management of Wild Pigs and Peccaries: Implications for Conservation*. Cambridge University Press.

Tack, J. 2018. *Wild Boar (Sus scrofa) Populations in Europe: A Scientific Review of Population Trends and Implications for Management*. European Landowners' Organization, Brussels, Belgium.

Tisdell, C. A. 1982. *Wild Pigs: Environmental Pest or Economic Resource?* Pergamon Press, Sydney, Australia.

2 Wild Pig Taxonomy, Morphology, Genetics, and Physiology

John J. Mayer, Timothy J. Smyser, Antoinette J. Piaggio, and Stam M. Zervanos

CONTENTS

2.1 INTRODUCTION

True wild pigs (*Sus scrofa*) belonging to the mammalian Family Suidae are not indigenous to North America (Wilson and Reeder 2005). In fact, the only pig-like mammals native to the Western Hemisphere are the peccaries (Family Tayassuidae), and only 2 of the 3 extant peccary species have ranges that extend into North America (Sowls 2013). The presence of wild pigs belonging to the species *Sus scrofa* in the Nearctic zoogeographic realm is solely attributable to introductions by mankind (see also Section 2.7 on the presence of introduced warthogs (*Phacochoerus africanus*) in Texas). Such introductions were both intentional (e.g., Eurasian wild boar released as a new big game species) and accidental (e.g., escaped domestic pigs that have gone wild or feral; Mayer and Brisbin 2008, 2009).

Wild pigs in North America exhibit a broad spectrum of morphological, genetic, and physiological diversity. This diversity stems from widely varying taxonomic or ancestral origins of these animals. Free-ranging domestic pigs and introduced Eurasian wild boar comprised the initial types of wild *S. scrofa* established in North America. Hybridization between feral pigs and wild boar has diversified these populations; however, the ancestral makeup of these animals was far more complex than a simple combination of 2 wild pig types. For example, domestic pigs introduced into North America over the centuries have morphologically varied from archaic or primitive domestic stock, to derived colonial forms, and most recently to highly modified, selectively bred modern domestic breeds. Such changes within ancestral and founding stocks have shaped the variation currently seen in these animals across North America (Mayer 2018).

This chapter provides a taxonomic, morphological, genetic, and physiological baseline for wild pigs in North America. Aside from its importance to the basic science of these animals, the understanding of this information about wild pigs is also useful to the successful management of this invasive species, a discussion of which is provided at the end of this chapter.

2.2 TAXONOMY

All wild pigs found at present in North America belong to 1 species, *S. scrofa*, initially described by Linnaeus in 1758. The taxonomic hierarchy of this species is as follows: Class Mammalia, Order Artiodactyla, Suborder Suina, Family Suidae, Subfamily Suinae, Tribe Suini, and Genus *Sus* (Wilson and Reeder 2005, Groves 2008). The genus *Sus* contains 9 species divided into 2 groups: 1) *Sus scrofa* Group that includes *S. scrofa* (Eurasian wild boar), and *S. salvanius* (pygmy hog); and 2) *Sus verrucosus* Group that includes *S. verrucosus* (Javan warty pig), *S. bucculentus* (Indochina warty pig), *S. celebensis* (Sulawesi warty pig), *S. barbatus* (bearded pig), *S. ahoenobarbus* (Palawan pig), *S. cebifrons* (Visayan pig), and *S. philippensis* (Philippine pig; Groves 2008). Eurasian wild boar (*S. scrofa*) further contains 15 recognized subspecies (modified from Wilson and Reeder 2005, Groves 2008, Mayer and Brisbin 2008) as follows:

- *S. s. algira* Loche 1867:59 - North African wild boar; type locality country of Beni Sliman, Algeria.
- *S. s. attila* Thomas 1912:105 - Eastern European wild boar; type locality Kolozsvar, Transylvania, Romania.
- *S. s. cristatus* Wagner 1839:435 - Indian wild boar; type locality probably the Malabar Coast of India.
- *S. s. davidi* Groves 1981:37 - Southwest Asian wild boar; type locality Sind, Pakistan.
- *S. s. leucomystax* Temminck 1842:6 - Japanese wild boar; type locality Japan.
- *S. s. lybicus* Gray 1868:31 - Middle Eastern wild boar; type locality Xanthus, near Gunek, Turkey.
- *S. s. meridionalis* Forsyth Major 1882:119 - Mediterranean wild boar; type locality Sardinia.
- *S. s. moupinensis* Milne-Edwards 1871:93 - Chinese wild boar; type locality Moupin, Sze-Chwan, China.
- *S. s. nigripes* Blanford 1875:112 - Central Asian wild boar; type locality Tien Shan Mountains, Kashgar District, Sinkiang, China.
- *S. s. riukiuanus* Kuroda 1924:11 - Ryukyu wild boar; type locality Kabira, Ishigakijima, Ryukyu Islands.
- *S. s. scrofa* Linnaeus 1758:49 - Western European wild boar; type locality Germany.
- *S. s. sibiricus* Staffe 1922:51 - Mongolian wild boar; type locality Tunkinsk Mountains, southern Siberia.
- *S. s. taivanus* Swinhoe 1863:360 - Taiwanese wild boar; type locality Taiwan.
- *S. s. ussuricus* Heude 1888:54 - Siberian wild boar; type locality Ussuri Valley, eastern Siberia.
- *S. s. vittatus* Boie 1828:240 - Indonesian wild boar; type locality Sumatra.

With the exception of localized domestic pigs on Sulawesi and its offshore islands (which were selectively bred from Sulawesi warty pigs), the Eurasian wild boar is considered to be the wild ancestor to both the ancient and modern breeds of domestic pigs found globally (Groves 1981, 2008; Giuffra et al. 2000; Okomura et al. 2001; Alves et al. 2003; Larson et al. 2005, 2007; Wu et al. 2007). As such, Eurasian wild boar and domestic pigs are considered conspecifics. Aside from the debate over the use of scientific nomenclature for domestic animals (e.g., Groves 1971, Van Gelder 1979, Clutton-Brock 1981), the application of *S. scrofa* for Eurasian wild boar, domestic pigs, feral pigs, as well as hybrids between these forms is commonly accepted by most systematic zoologists (MacDonal and Frädrich 1991, Mayer and Brisbin 2008).

It has been generally agreed that Eurasian wild boar that have been introduced into North America primarily belonged to the subspecies *S. s. scrofa* (Bratton 1977). However, various wild boar introductions onto this continent may have in fact been represented by up to 7 of the aforementioned recognized subspecies found in the Old World (Mayer and Brisbin 2008, Mayer 2018). Further, since most current populations in North America consist of animals with hybridized wild boar and domestic or feral pig ancestries, the use of any sub-specific designation would be invalid from a nomenclatural perspective (International Commission on Zoological Nomenclature 1999). Therefore, *S. scrofa* would be the most accurate scientific name to use for identifying any of the introduced wild pigs in North America.

In contrast, based on anecdotal morphological evidence (i.e., reported facial warts) and the origin from Southeast Asia through Melanesia and finally into Polynesia, Groves (1983) postulated that wild pigs found in the Hawaiian Archipelago were likely either pure Sulawesi warty pigs or hybrids between *S. scrofa* and *S. celebensis* (Sulawesi warty pig). However, mitochondrial DNA and morphological analyses have shown that the pigs that were dispersed throughout the Polynesian subregion of the Pacific basin, including the Hawaiian Islands, belonged solely to the species *S. scrofa* (Larson et al. 2005, 2007; Linderholm et al. 2016).

The present-day combination of wild pigs in North America consists of pure feral pigs, pure Eurasian wild boar (e.g., in Michigan, New Hampshire, and Canada), and pigs with hybrid ancestry (Figure 2.1). Pure Eurasian wild boar and hybrids have resulted from the increasingly selective introduction of animals exhibiting all or some of the Eurasian wild boar phenotype for hunting purposes (J. J. Mayer, Savannah River National Laboratory, unpublished data). Here, we refer collectively to these animals as wild pigs. However, when appropriate for specific context, we use "wild boar," "feral pig," or "hybrid," consistent with Figure 2.1.

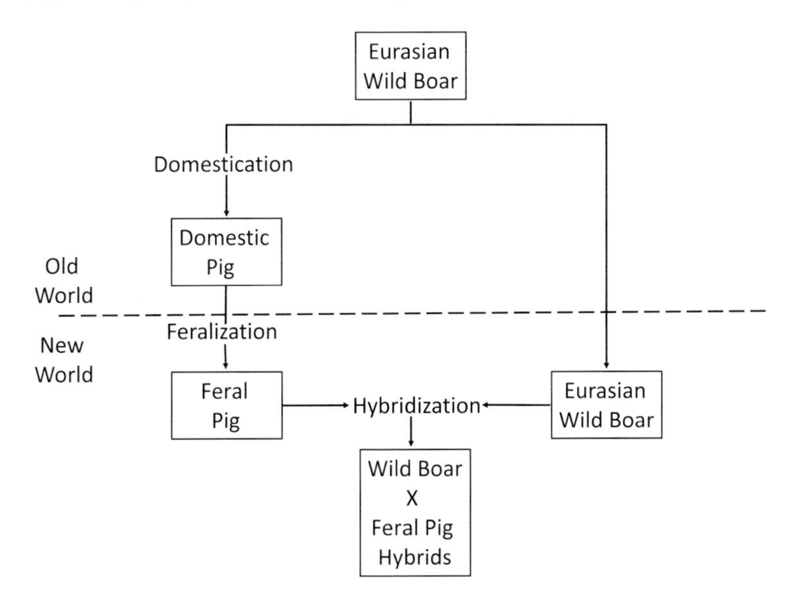

FIGURE 2.1 The phylogeny of wild pig populations found in North America at present. (Reproduced from Mayer and Brisbin 2009.)

2.3 MORPHOLOGY

With up to 7 wild pig subspecies represented in North America, combined with selective breeding for specific traits, these animals exhibit a wide variety of physical characteristics. This variation primarily involves differences in body mass, external dimensions, skull morphology, coat coloration patterns, and hair morphology (Mayer and Brisbin 2008, 2009).

In general, wild pigs in North America are medium- to large-sized (e.g., 70–100 kg) animals, with a barrel-like stout body (often with flattened sides), and a relatively long, pointed head supported by a short and thickset neck (Figure 2.2). The snout ends in a seemingly hairless, damp, disk-like plate (i.e., rhinarial pad) on which 2 external nostrils open anteriorly. The eyes are comparatively small (Figure 2.2). The ears are relatively large, broad structures, which taper to a point at the tip (Figure 2.2). The pinnae can vary from an erect to a drooped or lop-eared form. The legs are short and slender, and the front limbs constitute slightly less than one-half of the height at the shoulder (Figure 2.2). Each foot has 4 toes with the lateral toes being shorter and positioned higher up the limb than the central pair. All toes are dorsally covered by keratinized hooves supported by fleshy pads on the rear ventral surface of each digit. The tail is short in length, the conformation of which can vary from straight to curled (Mayer and Brisbin 2009).

FIGURE 2.2 Appearance and coat colorations of wild pigs found in North America at present.

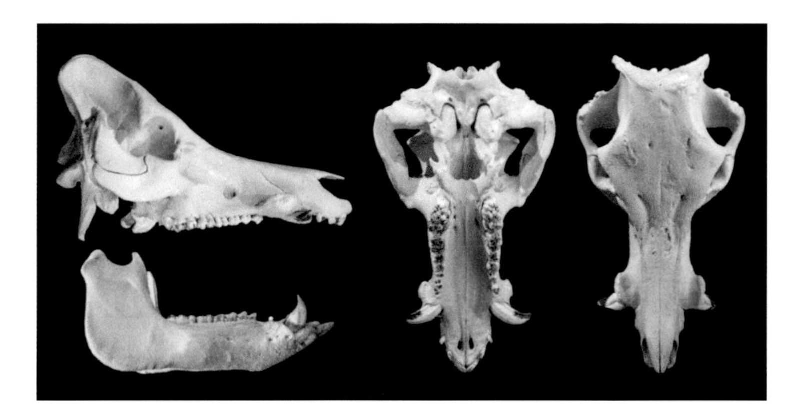

FIGURE 2.3 Lateral views of the cranium and lower jaw (left), and ventral (middle) and dorsal (right) views of the cranium of a wild pig skull, adult male, from the Savannah River Site, South Carolina. (Courtesy of L. Russo. With permission.)

The skull is elongate and roughly triangular in appearance from lateral, dorsal, and ventral views, tapering from the proximal to the distal ends (Figure 2.3). The rostral portion comprises 62–74% of the cranial length in adult males, which then decreases in females, from older to younger individuals, and from Eurasian wild boar to hybrids and finally to feral pigs (Figure 2.4). From a lateral perspective, the dorsal profile of the cranium varies from being flat or straight to deeply dished or concave (Mayer and Brisbin 2008, 2009). Cranial and mandibular measurements can be used to determine the taxonomic type of *S. scrofa* to which a pig belongs (i.e., Eurasian wild boar, feral pig, hybrid or domestic pig). The accuracy of such identifications increases with age and is higher in crania versus mandibles, and in males versus females of a comparable age (Mayer and Brisbin 1993, 2008; Figure 2.5). Summary data of 5 skull measurements from adult North American wild pigs are presented in Table 2.1.

Wild pigs have 44 permanent teeth with the dental formula: I 3/3, C 1/1, P 4/4, M 3/3. Young pigs have 28 deciduous teeth (di 3/3, dc 1/1, dp 3/3), with the deciduous set erupting first followed by the permanent set. Permanent dentition in wild pigs is fully erupted at about 3 years of age, at which time these animals are defined as adults (Mayer and Brisbin 2009). The permanent canine teeth are enlarged, taper toward the distal end of the crown, and are larger in males than in females (Figure 2.6). Based on the size and shape of the canines, the sex can be determined for pigs as young as 30–51 weeks of age. The permanent canines of males continuously grow throughout life (Mayer and Brisbin 1988; Figure 2.7). The molariform teeth are low-crowned bunodont or multi-cuspidate. The lower third molar is the longest tooth of the permanent set (Mayer and Brisbin 2009). The aging of wild pigs has used the pattern of tooth eruption and replacement (up to 3 years of age) and the incremental cementum lines, pulp cavity width, and wear in the teeth for aging animals older than 3 years of age (Mayer 2002).

The coat of wild pigs is coarse and bristly, and can vary from sparsely to densely haired with respect to coverage on any specific animal. Some individuals exhibit a well-developed mane along the neck, shoulder, and forward portion of the lower back. The tail is covered with hair, especially toward the tip. These animals have 3 types of hair including bristles or guard hairs, underfur, and vibrissae. The presence and size of these structures varies with the age and ancestral type of the individual pig, environmental parameters (e.g., temperature) and seasonally. Wild pig bristles average about 50 mm (2–149 mm) in length mid-dorsally and 32 mm (1–76 mm) laterally. The bristles can be solid colored or banded, of varying shades of black, red-brown or white. The underfur, also variable in color, has solid coloration along the entire hair shaft. The vibrissae are concentrated on the mental glands (i.e., a round, raised structure seen on the anterior end of the throat), and scattered in a more widespread pattern over the snout, muzzle, around the eyes, and on the upper and lower lips. Hair is present year-round, but is shed or molted from late April through June, which occurs in a sequential pattern of replacement over the entire body (Mayer and Brisbin 2009).

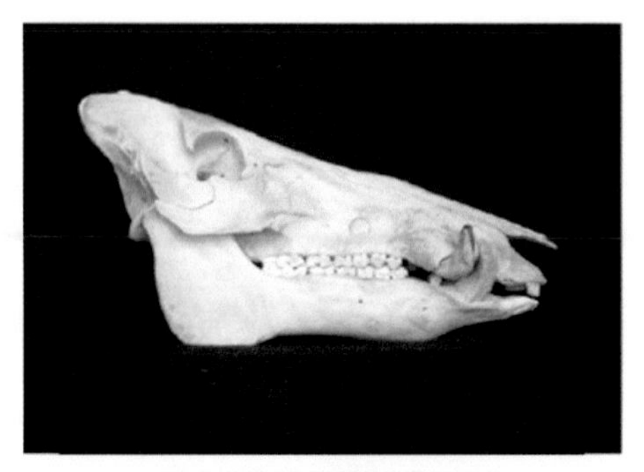

Eurasian Wild Boar

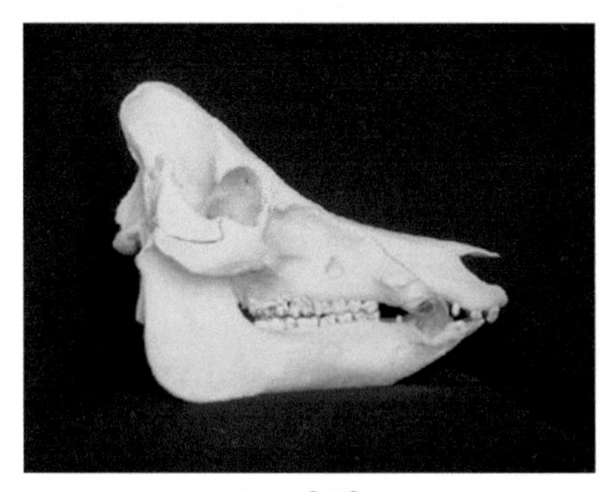

Feral Pig

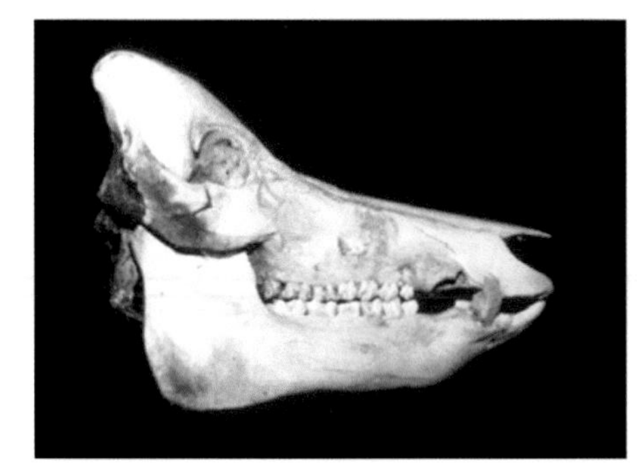

Hybrid

FIGURE 2.4 Lateral views of the taxonomic variation observed in wild pig skulls in North America. All of these specimens were adult (i.e., with fully erupted permanent dentition) males. (Adapted from Mayer and Brisbin 2009.)

The number and arrangement of the teats in wild pigs are comparable to that seen in domestic pigs. The location of the teats in both sexes extends from the thoracic area back to the inguinal region of the underside. Typically, the teats are arranged in pairs as follows: 1–2 inguinal pairs, 2–3 abdominal pairs, and 2 thoracic pairs. However, staggered arrangements or supernumerary teats occur that can result in odd numbers of total teats on any one individual. Overall, the numbers of teats in wild pigs range from 3 to 16. The number of teats varies taxonomically, increasing from Eurasian wild boar to hybrids to feral pigs (Mayer and Brisbin 2009).

Sexually mature males possess a thickened subcutaneous layer of tissue, commonly referred to as the "shield," which overlies the outermost muscles in the lateral shoulder region. This unique

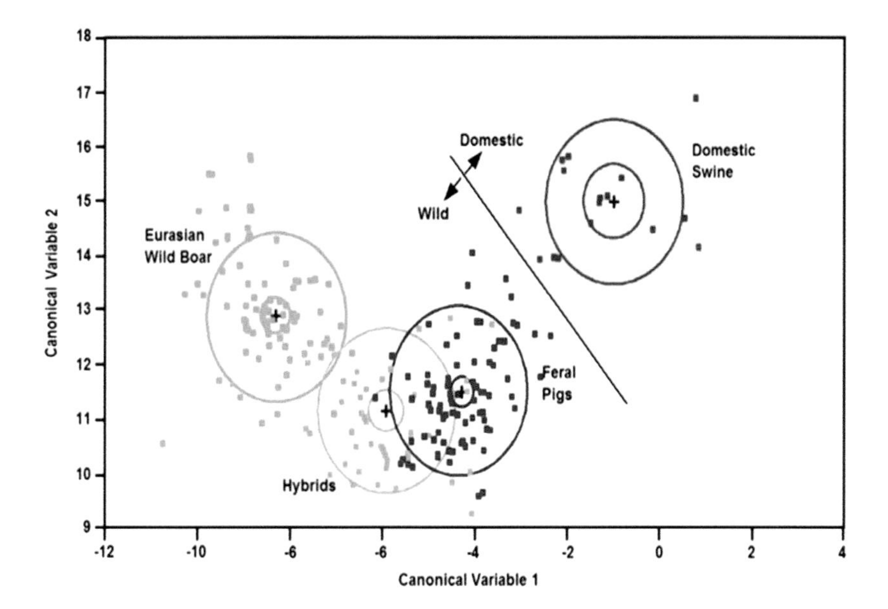

FIGURE 2.5 Graph of first 2 canonical variables showing the resolution of the 4 known target groups (i.e., wild boar, feral pigs, wild boar/feral pig hybrids, and domestic swine). Individual specimens are plotted as well as the 90% (inner circle) and 50% (outer circle) contours and the target group centroids (plus sign). The separation of domestic versus all wild target groups is also illustrated.

anatomical structure, a secondary sexual characteristic, serves a reported protective function for boars that potentially fight for breeding opportunities with sows in estrous. The shield initially develops as early as 9–12 months of age, and then increases in size with age, physical stature, and body mass (Mayer and Brisbin 2009).

TABLE 2.1
Summary of 5 Skull Measurements of 191 Adult Wild Pigs (i.e., with Fully Erupted Permanent Dentition) from Introduced North American Populations

Measurement (mm)	Sex	N	Mean	Observed range	SD
Condylobasal length	Female	71	282.6	233–351	21.5
	Male	120	303.8	246–388	26.3
Occipitonasal length	Female	71	285.7	235–371	25.8
	Male	120	308.2	249–409	29.6
Zygomatic breadth	Female	71	139.5	114–167	12.1
	Male	120	151.7	116–230	18.2
Mandibular length	Female	71	221.7	119–281	21.6
	Male	120	244.6	198–324	24.8
Posterior mandibular	Female	71	121.7	97–152	12.4
Width	Male	120	135.4	104–196	17.4

Note: Data were segregated by sex, and included a combined sample of wild boar, feral pigs, and hybrids. These measurements are defined in Mayer and Brisbin 2008.

Source: Adapted from Mayer and Brisbin (2009).

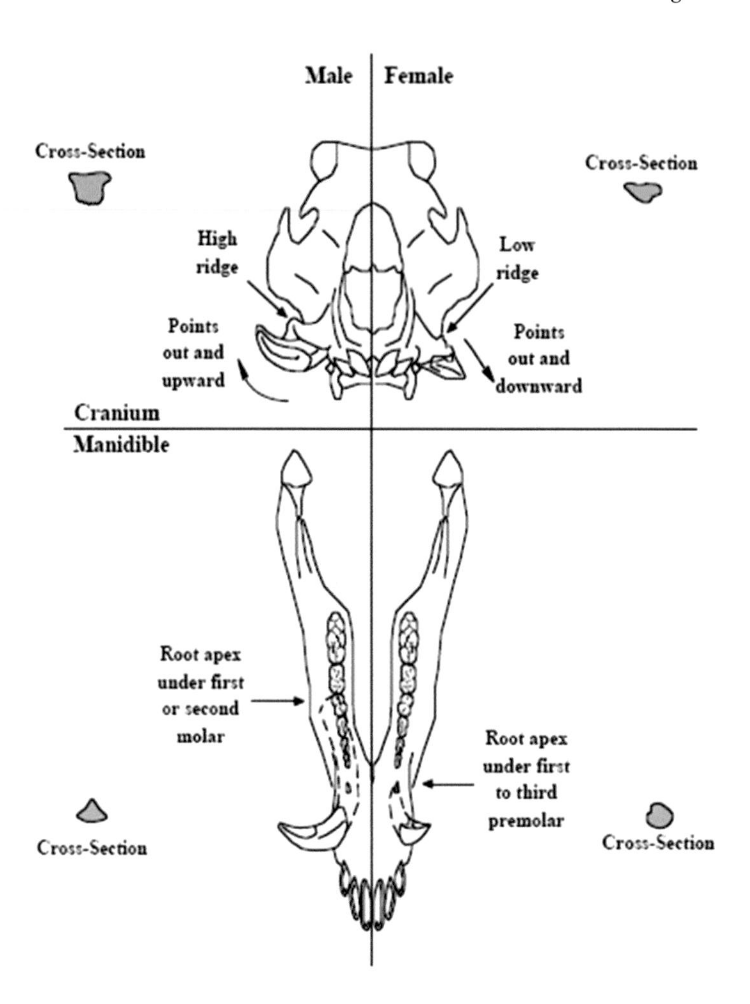

FIGURE 2.6 Appearance of the sexual dimorphism observed in the permanent canines of male and female wild pigs in North America. (Reproduced from Mayer and Brisbin 2009.)

Total body mass (i.e., intact weight with no internal organs removed) is a physical parameter that differs greatly among populations of wild pigs. Again, that variability relates back to wild pig types wthat formed the founding stocks of these different populations. Wild pigs are born at approximately 900 grams (range 494–1,620 grams), or at approximately 0.9% of their adult body mass. As these animals increase in age, females grow at a consistent and slightly faster rate toward their adult body mass than males. In contrast, males exhibit a greater absolute weight gain than females of the same age. This sexually dimorphic difference with males being larger is consistent for all age classes. Overall, males are about 1.2–1.4 times larger than females. The average adult body mass of wild pigs is approximately 85 kg. Comparable body masses for adults of each sex average 70–75 kg for females and 95–100 kg for males. Some insular populations of wild pigs are substantially smaller than this, with adult animals averaging around 40 kg. Growth in body mass among wild pigs continues until about the fifth year of life, after which body mass decreases with advancing age. Exceptional specimens have been reported to exceed 300 kg, all of which were males. Recent reports of harvested wild pigs exceeding 450 kg have turned out to be capture-reared males that either were released or escaped into the wild before being killed (Mayer and Brisbin 2009). Total body mass in wild pigs can be estimated by using regression equations from gutted weight, heart

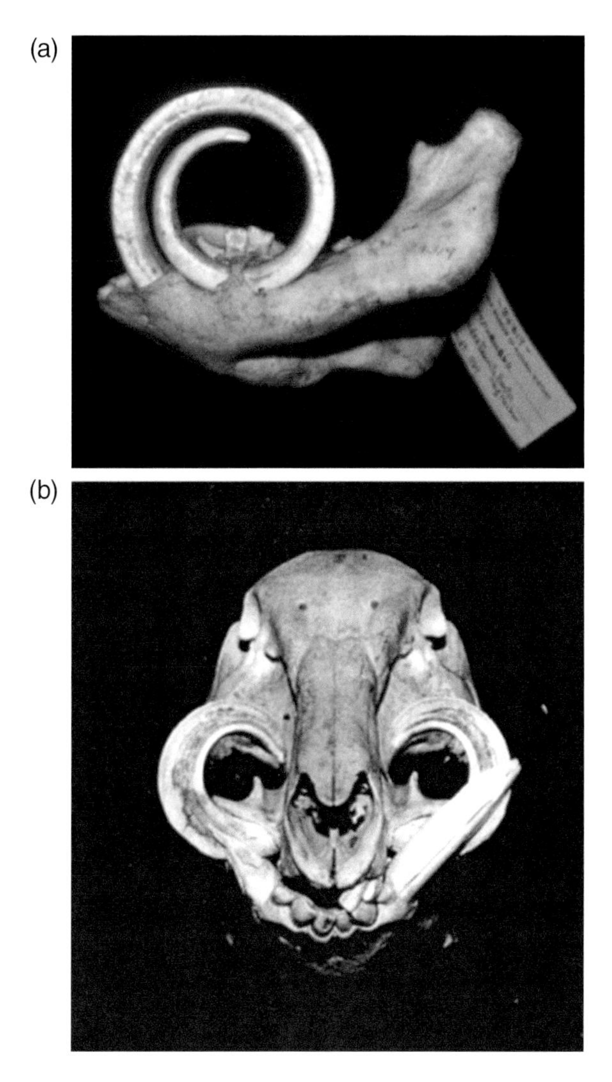

FIGURE 2.7 The permanent canines in male wild pigs are ever-growing. This is illustrated in both (a) the lower canine and (b) the upper canine. (Reproduced from Mayer and Brisbin 2009.)

girth, external body dimensions, and cranial measurements (Mayer and Brisbin 2009, J. J. Mayer, unpublished data).

Due to the aforementioned diverse taxonomic origins and similar variation seen in total body mass, the external dimensions in wild pigs are highly variable. Wild pigs are born on average at approximately 22–28% of the adult size for various dimensions. Summary statistics for 7 external body measurements of adult wild pigs are provided in Table 2.2. Males were on average 4.5–9.2% larger than females for the various body dimensions (Table 2.2). The most significant differences were seen in shoulder height followed by head or snout length (Table 2.2). As with body mass, mean differences in various external dimensions between the sexes increase with age. The growth of body length was slightly faster in males, and ceased at 3–5 years of age (Mayer and Brisbin 2009). Feral animals exhibit the most variability for each of the dimensions listed in Table 2.2. In general, Eurasian wild boars have proportionately shorter total lengths, head-body lengths and ear lengths and longer hind foot lengths, shoulder heights, and snout lengths than feral pigs. The hybrids are intermediate between the 2 parental types (Mayer and Brisbin 2008).

TABLE 2.2
Summary of 7 External Body Measurements of 358 Adult Wild Pigs (i.e., with Fully Erupted Permanent Dentition) from Introduced North American Populations

Measurement (mm)	Sex	N	Mean	Observed Range	SD	Difference (%)
Total length	F	174	1,594.4	1,045–1,910	143.0	5.5
	M	184	1,682.4	1,160–2,083	160.2	
Head-body length	F	174	1,305.3	840–1,580	116.4	5.8
	M	184	1,380.9	950–1,750	129.6	
Tail length	F	174	288.5	130–395	46.7	4.5
	M	184	301.6	90–410	60.5	
Hind foot length	F	174	269.8	185–333	23.2	5.6
	M	184	285.3	200–380	24.4	
Ear length	F	174	149.9	86–204	21.5	4.7
	M	183	157.2	92–250	24.4	
Shoulder height	F	174	698.4	515–1,000	70.3	9.2
	M	184	761.8	535–1,060	95.5	
Snout length	F	174	226.6	155–305	27.3	7.5
	M	184	244.4	160–368	31.7	

Note: Data were segregated by sex, and included a combined sample of wild boar, feral pigs, and hybrids. These measurements are defined in Mayer and Brisbin 2008.
Source: Reproduced from Mayer and Brisbin (2009).

Coat coloration observed among wild pigs includes solid, spotted or mottled, belted, grizzled patterns, and combinations of these. Light or dark points (i.e., the distal portions of the snout, ears, legs, and tail), and sharp contrast in dorsal and ventral colors also occurs (Figure 2.2). Most coloration patterns are expressed as uniform or mixed variations of the basic colors (i.e., black, red-brown or white). A striped juvenile pattern is also seen in neonates and young piglets in populations with partial or full wild boar ancestry. At between 4 to 6 months of age, this striped pattern transitions to a reddish coloration pattern, which then transitions to the adult grizzled/wild boar pattern at about 1 year of age (Mayer and Brisbin 2009; Figure 2.2).

Two uncommon morphological features observed in wild pigs are syndactylous (i.e., "mule-footed") hooves and neck wattles (Figure 2.8). The presence of these unusual structures is neither widespread nor frequently observed even where they are known to occur. In a few locations, both characters are found to occur within the same population (Mayer and Brisbin 2008).

2.4 GENETICS

2.4.1 Genetic Origins of *Sus scrofa*

Eurasian wild boar (Figure 2.1) diverged from other members of the genus *Sus* in Island Southeast Asia (also referred to as Maritime Southeast Asia—an area spanning Malay Peninsula, Sumatra, Java, Borneo, and the Philippines) between 3.5 and 5.3 million years ago (Frantz et al. 2013). Wild boar subsequently expanded out of Southeast Asia to colonize much of the Eurasian supercontinent. With the colonization of Eurasia, distinct lineages of wild boar in the east (Asia) and west (Europe) began to diverge. A variety of molecular tools including targeted resequencing of genes within the mitochondrial (mtDNA) genome, complete mtDNA sequencing, and whole genome sequences have been used to estimate the split between Asian and European wild boar lineages at 746,000 (Fernández et al. 2011) to 1 million years ago (Groenen et al. 2012).

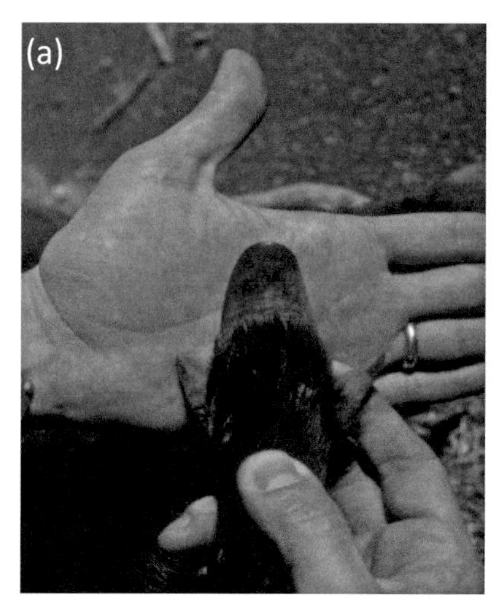

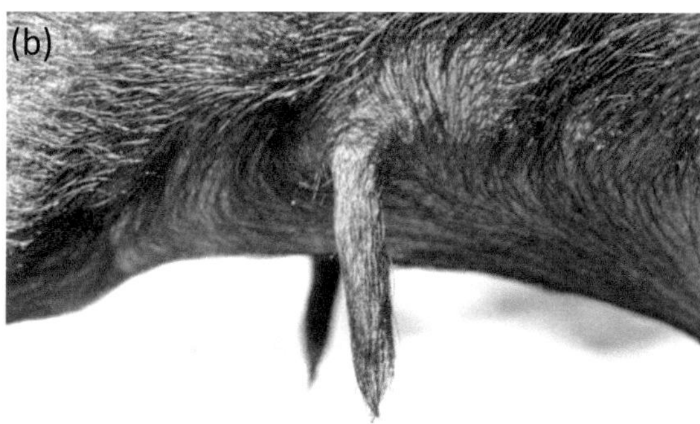

FIGURE 2.8 Wild pigs exhibiting (a) syndactylous or "mule-footed" condition of the middle toes and (b) neck waddles or wattles. (Reproduced from Mayer and Brisbin 2009.)

2.4.2 Domestication of Wild Pigs

Within these distinct genetic lineages, there is strong scientific evidence that *Sus scrofa* was independently domesticated approximately 9,000 years ago in both the Middle East (modern day Turkey) and China (Larson et al. 2005, 2010, 2011; Ramos-Onsins et al. 2014; Frantz et al. 2013). Specifically, zooarchaeological evidence from numerous sites in what is now eastern Turkey indicates this region served as the epicenter for western pig domestication (Larson et al. 2011, Ramos-Onsins et al. 2014). The brief appearance of Middle Eastern mtDNA haplotypes among ancient European pigs suggests that early domestic pigs were brought from the Middle East to Europe by humans (Larson et al. 2007, Ramos-Onsins et al. 2014). Subsequent introgression with sympatric wild boar populations served to replace Middle Eastern with European haplotypes (Larson et al. 2007, Ramos-Onsins et al. 2014). Genetic and zooarchaeological data suggest a similar process of reticulated domestication also occurred in what is modern day China. Archaeological evidence demonstrates that primary domestication occurred within the Mekong Valley and along the Yangtze River (Wu et al. 2007). Pigs in other regions across Asia possess mtDNA haplotypes uniquely found within local wild boar populations, yet a lack of archaeological evidence for primary domestication suggests a similar colonization pattern that started with domesticated pigs, first developed elsewhere, followed by interbreeding with sympatric wild boar populations (Larson et al. 2010). Thus, additional research is needed to determine if other locations in Asia, Southeast Asia, and the Indian subcontinent were sites of parallel *de novo* domestication or represent secondary domestication as observed in Europe (Ramos-Onsins et al. 2014). Regardless, the genetic record clearly demonstrates independent domestication of both Eastern and Western wild boar lineages and subsequent complex, reticulated domestication in which there was extensive and ongoing gene flow between early domestic herds and sympatric wild boar populations with movement of archaic pigs across broad geographic areas by humans (Frantz et al. 2013).

Despite the contemporaneous emergence of domestic pigs in many places throughout Eurasia, early Asian and European domestic pigs were under disparate selective pressures. In Asia early animal husbandry practices involved separating domestic pigs from wild boar, with domestic pigs generally raised in sties and reared on food scraps and refuse (White 2011). This selective regime produced a short-legged, round-bodied animal that was well adapted for confinement (White 2011). In contrast, early European swine were generally herded through forested habitats, with animals foraging on naturally available food items and seasonally fattened on mast resources (White 2011). Thus, European pigs were not only maintained under similar selective pressures as their wild progenitors, but genetic isolation was far from complete with wild boar and early domestic pigs routinely interbreeding (White 2011, Frantz et al. 2013, Ramírez et al. 2015). Such continued gene flow from wild boar into early European domestic breeds would have slowed the genetic and morphological divergence of pigs from wild boar.

During the 1600s and 1700s, growing human populations and decreasing forest availability in Europe created economic and ecological pressures for the intensification of pig husbandry, with pigs increasingly confined and raised on feed as in Asia (White 2011, Ramírez et al. 2015). The genetic and morphological divergence of European pigs from sympatric wild boar populations was greatly accelerated in the late 1700s and continuing into the 1800s with the crossbreeding of short, round-bodied pigs imported from China with the larger-bodied indigenous pig that had dominated Europe to that point (White 2011, Bosse et al. 2014, Ramírez et al. 2015). With the hybridization of genetic lineages that had been separated for >750,000 years, improved breeds (those breeds of mixed European and Asian origins; Larson et al. 2011) were well established by the mid-1800s and offered higher fecundity, faster weight gain, and greater feed conversion ratios (Bosse et al. 2014). Much of this crossbreeding consisted of pairing Chinese sows with European boars. Given that mtDNA is maternally inherited and there is no recombination in the mtDNA genome, many improved European breeds have a high frequency of Asian mtDNA haplotypes (e.g., >50% of the Large White breed have Asian mtDNA haplotypes; Giuffra et al. 2000, Bosse et al. 2014). Within the nuclear genome, the extent of Asian ancestry varies among improved breeds, reflecting the specific pattern of crossbreeding used in breed development (see also Ojeda et al. 2011). Using whole genome sequence, Groenen et al. (2012) estimated the Asian contribution to the genomes of common European breeds at 33% for Hampshire, 38% for Landrace, and 38% for Large White.

2.4.3 Genetic Origins of *Sus scrofa* Introduced to North America

Through this period of rapid change in husbandry practices and genetic characteristics, pigs were continuously arriving in the United States from Europe as a consequence of European exploration and colonization. The first introduction of *S. scrofa* to North America was associated with Christopher Columbus' second voyage in 1493 and unconfined release of domestic pigs onto several islands in the Caribbean (Towne and Wentworth 1950, Mayer and Brisbin 2008; see Chapter 18). Perhaps more notably, in 1539 the Spanish exploration party led by Hernando de Soto introduced pigs along their expedition route throughout Florida and the southeastern United States (see Chapter 16). Seeding pigs during exploration was a common practice of the time to establish a source of readily available meat for subsequent expeditions (Bianco et al. 2015). These early free-ranging pig populations were established in the United States prior to the introgression of Chinese breeds into European domestic lines or the strong genetic isolation between domestic pigs and wild boar. Thus, these early domestic pigs were likely only modestly differentiated, both genetically and morphologically, from European wild boar. With the progression of European colonization over subsequent centuries, pigs would continue to flow to the United States, however this coincided with ongoing Asian introgression and a period of active breed development in Europe. Colonists found abundant forests in North America to be conducive for free-range livestock practices, thus creating an environment in which contemporary breeds could genetically mix with established feral populations from earlier introductions. The historic husbandry practice of seasonally releasing pigs into forested habitats to fatten on fallen mast crops continued in the United States until the closure of open range, which varied state-by-state, but generally ceased by the 1960s. Thus, the genetic origins of wild pigs in the United States is complex, representing a mix of early European domestic pigs that closely resembled their wild ancestors augmented with improved breeds of mixed European and Asian ancestry.

The subsequent translocation of wild boar from Europe to the United States further complicated the genetic ancestry of wild pigs. Wild boar were first introduced to Corbin's Park, a private game preserve in New Hampshire, in 1889 from southwestern Germany (Mayer and Brisbin 2008). Since that time there have been numerous other wild boar introductions with variable genetic and morphological effects on associated wild pig populations. Perhaps the most notable wild boar introduction occurred at Hooper Bald, North Carolina, in 1912. Like Corbin's Park, these wild boar were introduced to a private game preserve for the establishment of a huntable population. Within 10 years, wild boar escaped from the preserve and colonized the mountains along the Tennessee and North Carolina border (Mayer and Brisbin 2008). The genetic impacts of the Hooper Bald introduction, however, were greatly magnified by the capture of descendants of introduced wild boar and subsequent release in numerous locations ranging from Florida to California (Mayer and Brisbin 2008; see Chapter 16).

2.4.4 Genetic Analyses of Wild Pigs in the United States

A number of recent studies have used molecular tools to both disentangle the complex history of wild pigs in the United States and elucidate contemporary processes shaping these populations. McCann et al. (2014) compared mtDNA control region sequences from wild pigs sampled across their introduced range in the United States to published reference samples from wild and domestic pigs sampled across the world. High mtDNA diversity among US wild pig populations demonstrates that a wide variety of source populations have contributed to contemporary wild pig populations. The most common haplotypes observed in the United States were associated with European domestic pigs and Eurasian wild boar (McCann et al. 2014). However, McCann et al. (2014) also observed Asian haplotypes among wild pig populations on the US mainland. While historic introductions of Asian pigs to the United States is possible, the detection of Asian haplotypes is more likely attributable to the contribution of improved breeds (European, Asian crossbreeds) to wild pig populations in the United States. In their description of spatial patterns of mtDNA diversity, McCann et al. (2014) documented a limited number of haplotypes widely distributed among long-established wild pig populations. These same haplotypes also appeared among newly emerging populations, suggesting that the translocation of wild pigs from long-established populations has contributed to the formation of new populations. However, McCann et al. (2014) also observed novel haplotypes among newly emerging populations that have not been observed elsewhere within the United States, thus indicating that releases from novel genetic sources have also contributed to recent expansion of wild pigs.

Although analysis of mtDNA is suitable for describing the diversity of sources that have contributed to US wild pig populations, this molecular tool lacks the resolution to resolve finer patterns of population structure that are important for describing patterns of both natural and artificial connectivity among populations and identifying source populations as new populations emerge. Accordingly, McCann et al. (2018) used a low-density single nucleotide polymorphism (SNP) array to describe patterns of population genetic structure at a continental scale. With 959 wild pigs sampled across the United States and genotyped with 88 SNP loci, McCann et al. (2018) observed disparate geographic patterns of genetic differentiation. Long-established populations largely formed cohesive genetic clusters that spanned broad geographic areas (e.g., associations throughout south-eastern states east of the Mississippi River, south-central states west of the Mississippi River). This observation conforms to the genetic pattern expected when natural processes of immigration and emigration among proximate populations (isolation by distance) govern genetic associations. In contrast, unique genetic signatures were associated with localized, often recently established, populations found in Indiana, Illinois, Arizona, and North Carolina. Such a pattern is a characteristic of anthropogenic introduction – the creation of new wild pig populations from various sources through either deliberate release or accidental escape in otherwise unoccupied landscapes. Upon establishment, localized and isolated populations experience founder effects, genetic drift, and novel selective pressure, and thus develop unique genetic signatures. For example, the population in Mohave County, Arizona is known to have been established prior to 1900 by domestic pigs that escaped from a nearby ranch, with no indication of recent immigration into that population (Mayer and Brisbin 2008, T. J. Smyser and A. J. Piaggio, USDA/APHIS/WS/National Wildlife Research Center, unpublished data). This isolated population displayed a strong genetic association with domestic pigs with effectively no genetic association with wild boar, a characteristic that is not found among other US wild pig populations (T. J. Smyser and A. J. Piaggio, unpublished data).

Additionally, McCann et al. (2018) observed genetic associations between geographically disjunct populations – a genetic pattern attributable to human-mediated movement as opposed to recent or ongoing gene flow. For example, populations in northern California were genetically associated with wild pigs in Great Smoky Mountains National Park (McCann et al. 2018), likely reflecting the well-documented collection of introduced wild boar from Hooper Bald, North Carolina and release in Monterey County, California between 1925 and 1926 (Mayer and Brisbin 2008). Despite the success of McCann et al. (2018) in describing genetic patterns at the continental-scale, observed substructure within the described genetic clusters suggested that wild pig populations may have been genetically partitioned at finer spatial scales than could be fully resolved with the available dataset (959 samples genotyped at 88 SNP loci).

Tabak et al. (2017) provided insights into the genetic processes that structure wild pig populations at fine spatial scales. Employing 43 microsatellite loci and intensive sampling (699 individuals) of wild pigs in California, Tabak et al. (2017) described population genetic structure across the state, quantified both natural and anthropogenic rates of genetic exchange among populations, and described societal drivers of wild pig movements by humans. Their results demonstrate that wild pig populations are strongly differentiated at fine spatial scale with low rates of natural dispersal among proximate populations. This finding corroborates social patterns observed among wild boar populations in Poland and Belarus in which microsatellite analyses revealed females were highly philopatric and matriarchal lineages formed long-term social groups (Podgórski et al. 2014). Despite limited natural connectivity among proximate populations, Tabak et al. (2017) documented the genetic signals of human-assisted migration connecting geographically disparate populations. Tabak et al. (2017) also identified a number of societal predictors for rates of human-assisted translocation, specifically demonstrating that the frequency of wild pig translocation increased as the social interest and opportunity for recreational hunting of wild pigs increased. Thus, genetic attributes of wild pigs at the population scale appeared to be shaped by the dichotomous pressures of philopatry, with minimal exchange of individuals among proximate populations, and long-distance anthropogenic translocation, the frequency of which was related to recreational pig hunting.

A parallel study in Florida conducted by Hernández et al. (2018) corroborated patterns observed in California (Tabak et al. 2017). Again, employing intensive sampling across the state (454 individuals) and a large panel of microsatellite markers (52 loci), Hernández et al. (2018) described patterns of population genetic structure and quantified rates of migration among populations. Similar to California, wild pig populations in Florida were shaped by both strong philopatry and ongoing long-distance translocation (Hernández et al. 2018). The state of Florida allows for the legal transport of live wild pigs to holding facilities, where pigs are sold for slaughter (see Chapter 8) or transferred to privately-owned high-fenced shooting operations. Analyses by Hernández et al. (2018) demonstrate that the presence of economic incentives for the active movement of wild pigs by private citizens created an environment in which holding facilities served as foci for frequent translocation. Both Tabak et al. (2017) and Hernández et al. (2018) demonstrate that humans introduced this destructive species and continue to exacerbate the problem by transporting them to vacant habitats.

Microsatellite markers, as used by both Tabak et al. (2017) and Hernández et al. (2018), are difficult to compare across studies without extensive calibration of allele calls. Given the economic importance of pigs, a high-density (HD) SNP array has been developed; this molecular tool provides high genetic resolution and universal allele calls, facilitating comparisons of genotypes across studies. Principal component analysis (SNP & Variation Suite, Golden Helix, Bozeman, Montana) was used to compare HD SNP genotypes of invasive wild pigs sampled throughout the United States to those of various domestic pig breeds and wild boar populations from the species' native range. This analysis demonstrated that invasive wild pigs in the United States overwhelmingly fall along a genetic continuum between Western domestic breeds (those developed in Europe and the United States from European stock) and European wild boar. With the limited sampling and analysis conducted to date, Asian genetic signatures are limited to Hawaii, Guam, and central Florida (T. J. Smyser and A. J. Piaggio, unpublished data; Figure 2.9). Recent evaluation of preliminary SNP data with genetic clustering algorithms (ADMIXTURE; Alexander et al. 2009) in Missouri corroborated the patterns observed by Tabak et al. (2017) and Hernández et al. (2018)—wild pig populations in Missouri demonstrate fine-scale genetic structuring with a subset of genetic clusters recently augmented from novel genetic sources (either from wild pigs collected outside the state or from captive herds).

The application of genetic resources is beginning to elucidate the patterns that have shaped and are shaping US wild pig populations. Pigs were first introduced to the US mainland from Europe at a time in which domestic pigs were only modestly differentiated from wild boar. Since the time of

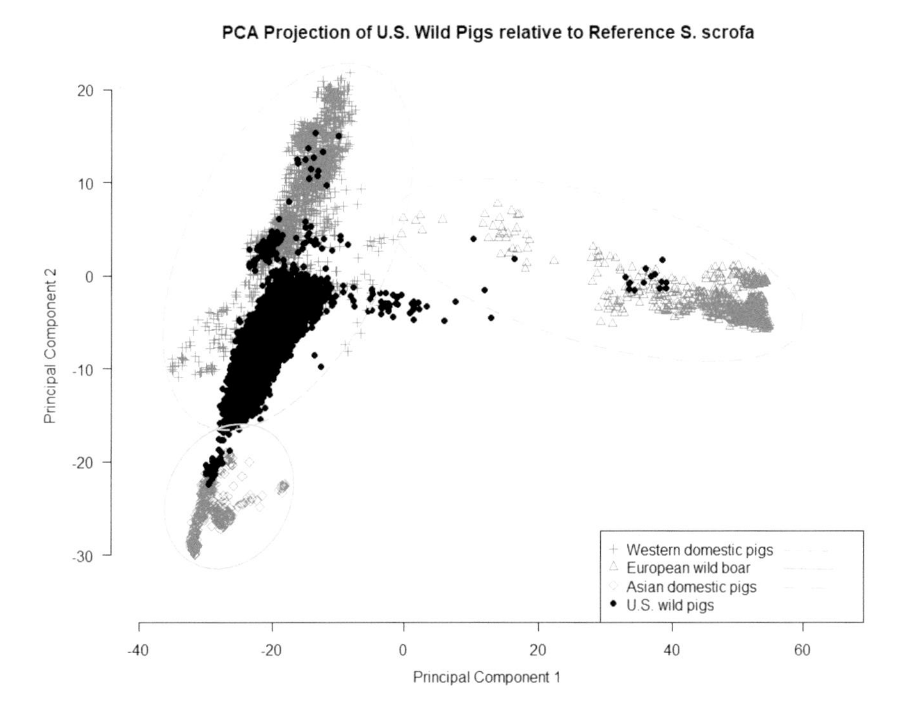

FIGURE 2.9 Reduction of HD SNP genotypes (29,375 loci) with principal component analysis to present wild pigs sampled throughout their invaded range in the United States relative to reference samples of domestic and wild pigs sampled throughout the world.

establishment, wild pig populations likely have been supplemented as a consequence of free-range livestock practices that continued over successive centuries through the early to mid-20th century, a timeframe over which genetic and phenotypic attributes of domestic pigs were rapidly diverging from wild boar. As demonstrated across multiple states with a variety of molecular tools, philopatric tendencies of wild pigs result in genetic populations that are naturally structured at fine spatial scales. However, these analyses have also demonstrated that translocation of wild pigs is an ongoing management challenge that will certainly impact control of this destructive, invasive species.

2.5 PHYSIOLOGY

This section focuses on basic ecological energetics as they apply to wild pig ecology. Because physiological information on wild pigs is limited, information from studies of domestic pigs and wild boar is included, with the understanding that there may be minor differences in some parameters. Much of this discussion is based on wild pig physiology information found in Zervanos (2009).

2.5.1 Energy Balance

The total energy budget for an animal is a balance between intake and utilization, where:

$$\text{Total forage energy intake} = \text{metabolism} + \text{production} + \text{excretory energy loss} \qquad (2.1)$$

Metabolism can further be divided into energy for maintenance (including basal metabolism and thermoregulation) and activity (mostly locomotion). Production includes energy for growth and reproduction. Excretory energy loss can be divided into urinary and fecal, however for this discussion urinary loss is excluded.

2.5.2 Forage Energy Intake

The efficiency of forage energy intake for an herbivore is based on the amount and quality of available forage consumed in a given period of time, less the energy involved in grazing (Canas et al. 2003). Because wild pigs are broad-spectrum feeders (see Chapter 3), it is difficult to get an accurate perspective on forage intake efficiency from wild populations (Oliver and Brisbin 1993). Additionally, there are physiological and landscape constraints on efficiency. For example, pigs lack sweat glands and can die if exposed to full or partial sun at temperatures above 23°C for extended periods of time (Porter and Gates 1969). Thus, foraging in open fields is constrained by time (Dexter 1999, Armstrong and Robertson 2000, Canas et al. 2003, Choquenot and Ruscoe 2003). The landscape also poses constraints. Choquenot and Ruscoe (2003) suggested that, because wild pigs require riverine woodlands to thermoregulate when radiant heat loads are high, their foraging efficiency declines as the distance of food resources to riverine woodlands increases. Refuge habitats, such as riverine woodlands, also play a role in reproduction, juvenile survival, and predator avoidance (including hunters). Finally, the efficiency of forage intake is also influenced by the quality and quantity of food resources in a given area. As would be expected, the higher the quality and quantity, the higher the efficiency of intake (Spitz and Janeau 1995, Dexter 1999, Armstrong and Robertson 2000, Canas et al. 2003).

Lab studies have demonstrated the effects of various environmental factors on food consumption. Seasonal changes in food consumption were observed in young captive wild pigs fed *ad libitum* over a 2-year period (Pepin and Mauget 1989). During both years, food intake was high during autumn and decreased during winter. In part, this may be related to the seasonal effects of photoperiod on food consumption. Weiler et al. (1996) found that wild boar increased food consumption with increasing day length, reached a maximum during decreasing day length, and reached lowest levels under short-day conditions. This seasonal change in food consumption may reflect photoperiod-induced changes in gonadal hormones in wild boar (Claus and Weiler 1994).

Ambient temperature may also influence food intake. Young domestic pigs maintained at 33°C decreased food consumption by 30% compared to those maintained at 23°C (Collin et al. 2001a, b). Campbell and Taverner (1988) found similar results with young pigs maintained at 14°C and 32°C. At high temperatures, reduced food intake would result in lower heat production associated with feeding and digestion, thus facilitating thermoregulation (Collin et al. 2001a).

2.5.3 Assimilation of Energy

Energy that is absorbed into the body during digestion is assimilated energy. It is equal to the forage energy intake less the fecal energy loss plus metabolic costs associated with the digestive and absorptive processes. The degree to which energy is assimilated into the animal is based on forage digestibility, and 1 important factor that determines digestibility is fiber content. In herbivores, microbes in the gut help digest some of the fiber; pigs use microbes in their cecum and colon to extract energy from a fibrous diet. van Wieren (2000) found that wild boar were better able to digest fibrous forages, such as acorns, than domestic pigs (although this may not apply to fiber content of commercial feeds; Hodgkinson et al. 2008). This adaptation would allow wild boars to extract more energy from natural forage. Whether this applies to other types of wild pigs (e.g., feral pigs and hybrids), still needs to be determined. Using commercial feeds, Zervanos et al. (1983) found no difference in assimilation rates between domestic and feral pigs, 87.8% and 85.1%, respectively. In addition, both groups exhibited significantly lower assimilation efficiency when water stressed. This suggests that the digestive processes between domestic and wild pigs are not significantly different.

2.5.4 Assimilated Energy Expenditure

Assimilated energy is the energy available for both metabolic and production expenditures.

$$\text{Assimilated energy} = \text{metabolic energy expenditure} + \text{production energy expenditure} \quad (2.2)$$

If assimilated energy decreases, then the expenditures must decrease and vice-versa. Metabolic energy expenditures include energy needed for basal metabolism (measured as the basal metabolic rate (BMR) or standard metabolic rate (SMR)), locomotion (skeletal muscle use) and thermoregulation. It is not possible to measure these expenditures separately for a free-ranging animal. However, estimating field metabolic rates (FMR) is possible. FMR collectively represents all metabolic expenditures, and to some extent the production expenditure, of a free-ranging animal. One way to measure FMR is to use isotopes of molecules associated with metabolism (Nagy 1975, 1987; Zervanos and Day 1977; McNab 2002). Using isotopes from field data, Nagy (1987) found that FMR for mammalian herbivores correlated strongly to body mass with a slope = 0.73. Thus, FMR can be estimated for a mammalian herbivore such as a wild pig by using a basic allometric equation:

$$y = axb \quad (2.3)$$

where: y = FMR in kJ/day; a = regression coefficient 5.95; x = mass in g; and b = slope (0.73). In addition, assuming an assimilation rate of 80% and substituting a coefficient of 0.58 in the above equation, feeding rate in g dry weight/day can be estimated (Nagy 1987). This equation can then be used to calculate energy and carrying capacities for wild pig populations.

For wild pigs, locomotion is a primary energy expenditure, particularly during grazing. The amount of grazing energy required varies based on forage availability/amount and nutritional quality. If forage is sparse and of low quality the locomotion cost associated with grazing and associated stress is high. Conversely, if forage density and quality is high, grazing energy drops substantially (Armstrong and Robertson 2000, Canas et al. 2003). Canas et al. (2003) estimated that wild boar at low forage availability of 2,000 mJ/ha had a cost of grazing that approached 130 kJ/kg body mass, and as forage availability increased to 16,000 mJ/ha the cost of grazing decreased to 30 kJ/kg.

Another way to estimate the cost of locomotion is to determine the rate of movement and apply an allometric equation developed by Taylor et al. (1982) for terrestrial animal locomotion. For mammals:

$$\text{Energy of Locomotion} = 10.7\ M_b^{-0.316} \times v_g + 6.03\ M_b^{-0.303} \tag{2.4}$$

where Energy of Locomotion is in watts/kg; M_b = body mass in kg; v_g = speed in m/sec. Using data from Ilse and Hellgren (1995a), this equation can be applied to estimate the energy of locomotion of wild pigs inhabiting rangeland of mixed honey mesquite (*Prosopis glandulosa*) and live-oak (*Quercus virginiana*) in Texas. They estimated that wild pigs moved 113.6 m/h (0.032 m/s). Assuming an average mass of 75 kg for an adult male (Ilse and Hellgren 1995b), the locomotion energy requirement would be 1.72 watts/kg.

Homeothermic animals must remain in thermal equilibrium with their environment to survive (i.e., thermal energy in must equal thermal energy out). Based on this thermal equilibrium principle, Porter and Gates (1969) estimated climate conditions for survival of domestic pigs. They concluded that at wind speeds of 100 cm/s, light-colored pigs could not survive full sun at ambient temperatures (Ta) above 23°C; partial shade at Ta above 35°C; and full shade at Ta above 40°C. Hair and bristle color make a noticeable difference. A dark colored pig, because of the increase in radiant heat absorption, would be more heat stressed in full or partial sun. Pigs lack functional pelage (i.e., a dense and insulative coat of hair). Under winter conditions in still air they could survive Ta of −27°C, but only −15°C at wind speeds of 1,000 cm/s (Porter and Gates 1969). Surviving at these temperatures would require significant increases in metabolic energy expenditure (Verstegen and Close 1994). Additionally, because pigs have a reduced ability to produce additional metabolic heat through non-shivering thermogenesis, newborn pigs particularly have difficulty surviving cold temperatures (Herpin et al. 2002). Wild pigs could compensate for this problem by farrowing and reproducing during warmer seasons.

Overall, it would seem that pigs are poorly adapted to withstand climate extremes. The range of temperatures for thermal comfort for domestic pigs appears to be narrow, between 23°C and 25°C (Quiniou et al. 2001). However, wild pigs compensate for their reduced thermoregulatory abilities by using behavioral adaptations, such as seeking shade, wallowing, and building communal nests (see Chapter 3). For example, Ilse and Hellgren (1995b) suggested that wild pigs in south Texas selected bedding sites with higher canopy cover than did the more desert-adapted collared peccaries (*Pecari tajacu*). Fernandez-Llario (2005) acknowledged the use of wallowing by wild boar for thermoregulation, in addition to serving a sexual function for mature males.

2.5.5 Production Energy Expenditure

2.5.5.1 Growth

During the growth phase of mammals, a high percentage of assimilated energy supports increasing body size (McNab 2002). Under ideal conditions, domestic pigs can channel as much as 32% of their assimilated energy into growth (fat and protein; Verstegen and Close 1994). However, this will vary based on how much of the assimilated energy is used for metabolic energy expenditures, such as thermoregulation and locomotion. For example, because of increased energy demands required to maintain thermal balance, both low and high temperatures have negative effects on overall growth of domestic pigs (Close and Stanier 1984, Christon 1988, Verstegen and Close 1994).

As expected, growth is maximized under optimal energy intake. This rarely occurs in the wild as seasonal, habitat, and climate conditions impact food availability. It has been well documented that declines in energy and nutritional values of forage result in deceased growth in wild boar (Wolkers et al. 1994, Spitz et al. 1998). There is also evidence that annual endogenous cycles in food consumption play a role in growth rates (Spitz et al. 1998). Increased food intake in autumn corresponded to increased growth of young female wild boars (Pepin and Mauget 1989). These endogenous feeding cycles are also linked to the annual reproductive cycle. Decreasing day length optimizes reproductive preparations during fall and winter in males and females. Thus, by fall, if food supply is not limited, wild pigs reach their maximum mass and begin gonadal development in preparation for mating (Pepin and Mauget 1989, Claus and Weiler 1994). In addition, survival is enhanced by hormonal changes related to seasonal changes in temperature and food supply and thus help regulate annual growth patterns. Seasonal changes in hormones, such as gonadal steroids and growth hormone (GH), appear to control the seasonal growth pattern of wild boars. With increasing day length GH levels rise and facilitate protein synthesis, particularly for muscle development. In winter, high GH levels help with fat catabolism, thus providing energy during periods of low food availability (Claus and Weiler 1994).

2.5.5.2 Reproduction

The energetic cost of reproduction in wild pigs is difficult to assess. Much of the cost is related to factors that are not easily determined, such as the type of mating behavior (e.g., defending territories, defending dominance, finding mates) and the costs of gestation, lactation, and rearing young (McNab 2002). One index of reproduction cost is litter size. Wild pigs produce litters that average between 5 to 8 young (Read and Harvey 1989, Mauget 1991, Taylor et al. 1998; see Chapter 4). Compared to other ungulates, these numbers are high and result in high reproductive energy costs. Litter size for wild pigs is influenced by female body mass and age, as larger females tend to have larger litters (Fernandez-Llario and Mateos-Quesada 1998). Taylor et al. (1998) found that older wild pigs in Texas had larger litter sizes than yearlings. Older females tend to be larger with higher energy stores (particularly fat) and thus are able to support more young during gestation and lactation. Environmental factors can also affect litter size (see Chapter 4). Frauendorf et al. (2016) found that wild boar litter size was positively correlated to higher summer temperature and rainfall. Higher summer temperatures and rainfall would result in higher forage energy intake; making more energy available for allocation to reproduction. Variations in annual mast production can also have a significant impact on litter size (Gamelon et al. 2013, 2017).

2.5.5.3 Ecological Energetics

Estimating energy budgets for free-ranging animals is also difficult. Armstrong and Robertson (2000) present an excellent summary of factors that affect energy budget estimates. These include cost of thermoregulation under various conditions of temperature, insulation, wind, and precipitation; cost of grazing under various forage densities, digestibility, terrain; and cost of sheltering behavior. Choquenot and Ruscoe (2003) presented a model for estimating population growth, as it relates to forage intake, of wild pigs in Australia by using the linear regression equation:

$$r = -0.193 + (0.007 \times IR) \tag{2.5}$$

where r is the population rate of increase and IR is the rate of pasture intake based on metabolic body weight (g/kg$^{0.75}$/day). They estimated that r would be maximum when IR was at 58 g/kg$^{0.75}$/day at a forage density of 1000 kg/ha. This would occur under ideal grazing conditions. The rate of pasture intake (IR) is directly influenced by factors affecting grazing efficiency, including pasture biomass, inter- and intra-specific competition for forage, and proximity of forage to refuge such as riverine woodlands used by pigs (Choquenot and Ruscoe 2003).

2.6 IMPLICATIONS FOR MANAGEMENT

The more known about the biology of an invasive species, the better the chances of being able to control or eliminate the organism, and this axiom absolutely applies to invasive wild pigs (Mayer and Beasley 2018). Along those lines, Campbell and Long (2009) noted that successful management of wild pigs depends upon programs that incorporate sound biological information. For example, some biological parameters (e.g., litter sizes; Mayer and Brisbin 2009) important to management vary taxonomically. If these data are lacking for a localized wild pig population with known taxonomy, then inferring parameters from other populations that are taxonomically similar can help inform management. Hence, understanding taxonomic, morphological, genetic, and physiological characteristics of wild pigs in North America can support successful management of these invasive animals.

Using the most accurate taxonomic identification for wild pigs when developing laws or regulations for managing these animals is critical for rules to be successful. Use of inaccurate or unclear nomenclature for common names can be problematic (Keiter et al. 2016). For example, to develop legislation specifically applying to "feral swine" (e.g., as was done early on in Michigan) or "Russian or European boar" (e.g., as was done early on in Colorado) can lead to ineffective laws that are difficult to enforce. Both federal and state case law establishes that "feral pigs" are not "wild boar" and vice versa (J. J. Mayer, unpublished data). As such, a law prohibiting private ownership of "feral pigs" could successfully be litigated if the plaintiff wanted to establish a high-fenced shooting operation for pure Eurasian wild boar. However, if taxonomic status of a wild pig population is in question, it is important to identify these animals as accurately as possible. Perhaps the simplest solution would be to use a general term for wild *S. scrofa* in laws or regulations (e.g., wild pig, wild hog, or wild swine), or to define these animals with an all-encompassing identification (e.g., "feral swine is used to refer collectively to all free-ranging swine (*Sus scrofa*) belonging to the family Suidae" or "feral swine (also called wild pigs; *Sus scrofa*)"). With that said, morphological or genetic analyses can be used to make more rigorous taxonomic determinations of a local wild pig population.

Knowledge regarding wild pig morphology can play a role in management. For example, the sex of wild pigs can be determined via canine tooth morphology, perhaps obtained from analyzing photographs or from skeletal remains. Wild pigs can also be aged (i.e., for survival data) using dental morphology and characteristics from skeletal remains. If measurements of total body mass are not available (e.g., for human health risk assessment of contaminants, age class estimation, body condition monitoring), total body mass of wild pigs can be derived from several morphological dimensions. Because morphology in these animals is variable, awareness of basic phenotypic composition of local populations facilitates population monitoring, such as identifying new illegal introductions (e.g., in the Great Smoky Mountains National Park, W. H. Stiver, National Park Service, personal communication, or the Loxahatchee National Wildlife Refuge, J. J. Mayer, unpublished data). Knowing the morphological dimensions of local wild pigs, which can be highly variable (especially between island and mainland populations), can serve to improve management tools such as traps (e.g., tailoring box trap or corral trap door size to size of local animals).

In addition to supporting taxonomy, genetic analyses permit reconstructing and monitoring wild pig dispersal and colonization (McCann et al. 2014). Most importantly, this allows identification of new human-assisted translocations versus range expansion via dispersal (Spencer et al. 2005, Morelle et al. 2016). Understanding pathways of illegal introductions and natural range expansion through dispersal is informative to land and resource managers trying to mitigate or prevent further invasions (Hampton et al. 2004, Hernández et al. 2018). Additionally, knowing historic translocation patterns might enable identification and halting of future illegal introductions between known source and recipient populations or areas. Simply being able to confirm that illegal translocations are occurring is useful information (Spencer et al. 2005, Tabak et al. 2017). Such molecular techniques can also be used to identify hybridization events between different types of pigs (e.g., recently released/escaped domestic pigs and long-established wild pigs; Spencer et al. 2005, Morelle et al. 2016). The ongoing exchange of animals between domestic and wild environs has implications for disease transmission. Genetic analysis would provide information regarding connectivity and interbreeding among populations (Hernández et al. 2018). Such analyses could further serve to identify areas for eradication isolated from areas subjected to natural reinvasion by dispersal (Tabak et al. 2017). Genetic analyses can also be used to identify individual animals in support of studies estimating population size and density (Fickel and Hohmann 2006, Ebert et al. 2012).

Lastly, knowing physiological parameters that affect how wild pigs survive, behave, and reproduce provides context for mechanistically understanding potential limiting effects of forage availability, habitat, climate, and seasonal changes. Knowing these limitations is critical in applying the principles of ecological energetics to effective management of wild pigs.

2.7 INTRODUCED POPULATION OF COMMON WARTHOGS IN SOUTH TEXAS

In 2013, 20 common warthogs escaped into the wild from a high-fenced shooting operation on a private ranch near the Chaparral Wildlife Management Area (CWMA) located in La Salle and Dimmit Counties in southern Texas (S. Lange, CWMA Manager, personal communication). It is believed that these pigs burrowed under the high fence to escape (Holder 2015). Given the subsequent widespread reported distribution of these animals (i.e., ~50–65 km apart), it is also possible that warthogs have been escaping or been released by other ranches in this area (McDonald 2015, Sams 2015). The first warthogs observed on the CWMA, which is also surrounded by a high fence, occurred in September of 2014 during an aerial survey of the deer population (Holder 2015, Tompkins 2015). That animal, a large male warthog, apparently had also burrowed under the CWMA's high fence (Tompkins 2015). Between 2014 and 2016, a total of 12 warthogs were harvested by hunters on the CWMA and 6 others were live trapped and euthanized by CWMA personnel (Holder 2015, Tompkins 2015). In 2017, USDA/APHIS/Wildlife Services removed a 3–4 year old male warthog during aerial operations on the CWMA with a hole in its ear consistent with an ear tag, potentially suggesting additional releases or escapes are occurring (M. J. Bodenchuk, USDA/APHIS/Wildlife Services, personal communication).

This Texas warthog population appears to be small but reproducing and growing (Tompkins 2015). Between 2014 and 2015, observations have been made on the CWMA of lactating/nursing females, a female with 4 piglets, and various juvenile animals (Holder 2015, McDonald 2015, Tompkins 2015). In addition, observations of juvenile pigs were made on the CWMA in 2017 (M. J. Bodenchuk, personal communication). Most observations of these animals have consisted of either singles or pairs, but one group of 6 warthogs has been reported (Holder 2015, Sams 2015).

How these animals will negatively impact both natural and anthropogenic environments in this region of southern Texas is just beginning to be understood (Tompkins 2015). The warthog's burrowing behavior has the potential to damage ranch fence lines (McDonald 2015), allowing livestock to escape/wander, giving access to predators, and resulting in costly repairs. The first road-killed warthog in the area was observed in 2017 (S. Lange, personal communication). Samples from an adult male warthog collected on the CWMA tested negative for classical swine fever, pseudorabies, and swine brucellosis (B. T. Mesenbrink, USDA/AHPIS/Wildlife Services, personal communication).

At present, warthogs on the CWMA appear to be declining, but these animals are still being observed on the private lands surrounding the WMA. These animals appear to be dispersing along the drainage corridor/basin associated with the Nueces River system and have recently been found in LaSalle, Dimmit, McMullen and Duval Counties (S. Lange, personal communication). Warthogs have been reported as far north as Woodward and as far south as Encinal, both in LaSalle County (Sams 2015).

ACKNOWLEDGMENTS

The authors thank G.J. Roloff for his constructive editorial comments and suggestions that improved an early draft of this chapter. Contributions from T.J.S. and A.J.P. were supported by the USDA. Contributions from J.J.M. were supported by the US Department of Energy Office of Environmental Management under Contract DE-AC09-08SR22470 to Savannah River Nuclear Solutions LLC. Mention of commercial products or companies does not represent an endorsement by the US Government.

REFERENCES

Alexander, D. H., J. Novembre, and K. Lange. 2009. Fast model-based estimation of ancestry in unrelated individuals. *Genome Research* 19:1655–1664.

Alves, E., C. Óvilo, M. C. Rodríguez, and L. Silió. 2003. Mitochondrial DNA sequence variation and phylogenetic relationships among Iberian pigs and other domestic and wild pig populations. *Animal Genetics* 34:319–324.

Armstrong, H. M., and A. Robertson. 2000. Energetics of free-ranging large herbivores: When should costs affect foraging behavior. *Canadian Journal of Zoology* 78:1604–1615.

Bianco, E., H. W. Soto, L. Vargas, and M. Pérez-Enciso. 2015. The chimerical genome of Isla del Coco feral pigs (Costa Rica), an isolated population since 1793 but with remarkable levels of diversity. *Molecular Ecology* 24:2364–2378.

Bosse, M., H. J. Megens, O. Madsen, L. A. F. Frantz, Y. Paudel, R. P. M. A. Crooijmans, and M. A. M. Groenen. 2014. Untangling the hybrid nature of modern pig genomes: A mosaic derived from biogeographically distinct and highly divergent *Sus scrofa* populations. *Molecular Ecology* 23:4089–4102.

Bratton, S. P. 1977. Wild hogs in the United States—Origin and nomenclature. Pages 1–4 *in* G. W. Wood, editor. *Research and Management of Wild Hog Populations*. Belle Baruch Forest Science Institute of Clemson University, Georgetown, SC.

Campbell, R. G., and M. R. Taverner. 1988. Relationship between energy intake and protein and energy metabolism, growth, and body composition of pigs kept at 14 or 32°C from 9 to 20 kg. *Livestock Production Science* 18:289–303.

Campbell, T. A., and D. B. Long. 2009. Wild pig damage and damage management in forested ecosystems. *Forest Ecology and Management* 257:2319–2326.

Canas, C. R., R. A. Quiroz, C. Leon-Velarde, A. Posadas, and J. Osorio. 2003. Quantifying energy dissipation by grazing animals in harsh environments. *Journal of Theoretical Biology* 225:351–359.

Choquenot, D., and W. A. Ruscoe. 2003. Landscape complementation and food of large herbivores: Habitat-related constraints on the foraging efficiency of wild pigs. *Journal of Animal Ecology* 72:14–26.

Christon, R. 1988. The effect of tropical ambient temperature on growth and metabolism in pigs. *Journal of Animal Science* 66:3112–3123.

Claus, R., and U. Weiler. 1994. Endocrine regulation of growth and metabolism in the pig: A review. *Livestock Production Science* 37:245–260.

Close, W. H., and M. W. Stanier. 1984. Effects of plane of nutrition and environmental temperature in the growth and development of the early weaned piglet 2. Energy metabolism. *Animal Production* 38:221–231.

Clutton-Brock, J. 1981. *Domesticated Animals from Early Times*. University of Texas Press, Austin, TX.

Collin, A., J. van Milgen, and J. Le Dividich. 2001a. Modelling the effect of high, constant temperature on food intake in young growing pigs. *Animal Science* 72:519–527.

Collin, A., J. van Milgen, S. Dubois, and J. Noblet. 2001b. Effect of high temperature and feeding level on energy utilization in piglets. *Journal of Animal Science* 79:1849–1857.

Dexter, N. 1999. The influence of pasture distribution and temperature on habitat selection by feral pigs in a semi-arid environment. *Wildlife Research* 25:547–559.

Ebert, C., F. Knauer, B. Spielberger, B. Thiele, and U. Hohmann. 2012. Estimating wild boar *Sus scrofa* population size using faecal DNA and capture-recapture modeling. *Wildlife Biology* 18:142–152.

Fernández, A. I., E. Alves, C. Óvilo, M. C. Rodríguez, and L. Silio. 2011. Divergence time estimates of East Asian and European pigs based on multiple near complete mitochondrial DNA sequences. *Animal Genetics* 42:86–88.

Fernandez-Llario, P. 2005. The sexual function of wallowing in male wild boar (*Sus scrofa*). *Journal of Ethology* 23:9–14.

Fernandez-Llario, P., and P. Mateos-Quesada. 1998. Body size and reproductive parameters in the wild boar *Sus scrofa*. *Acta Theriologica* 43:439–444.

Fickel, J., and U. Hohmann. 2006. A methodological approach for non-invasive sampling for population size estimates in wild boars (*Sus scrofa*). *European Journal of Wildlife Research* 52:28–33.

Frantz, L. A. F., J. G. Schraiber, O. Madsen, H. J. Megens, M. Bosse, Y. Paudel, G. Semiadi, et al. 2013. Genome sequencing reveals fine scale diversification and reticulation history during speciation in *Sus*. *Genome Biology* 14:R107.

Frauendorf, M., F. Gethoffer, U. Siebert, and O. Keuling. 2016. The influence of environmental and physiological factors on the litter size of wild boar (*Sus scrofa*) in an agriculture dominated area in Germany. *Science of the Total Environment* 541:877–882.

Gamelon, M., M. Douhard, E. Baubet, O. Gimenez, S. Brandt, and J. M. Gailard. 2013. Fluctuating food resources influence developmental plasticity in wild boar. *Biology Letters* 9:1–4.

Gamelon, M., S. Focardi, E. Baubet, S. Brandt, B. Franzetti, F. Ronchi, S. Venner, et al. 2017. Reproductive allocation in pulse-resource environments: A comparative study in two populations of wild boars. *Oecologia* 183:1065–1076.

Giuffra, E., J. M. H. Kijas, V. Amarger, Ö. Carlborg, J. T. Jeon, and L. Andersson. 2000. The origin of the domestic pig: Independent domestication and subsequent introgression. *Genetics* 154:1785–1791.

Groenen, M. A. M., A. L. Archibald, H. Uenishi, C. K. Tuggle, Y. Takeuchi, M. F. Rothschild, C. Rogel-Gaillard, et al. 2012. Analyses of pig genomes provide insight into porcine demography and evolution. *Nature* 491:393–398.

Groves, C. 1981. *Ancestors for the pigs: Taxonomy and phylogeny of the genus Sus*. Technical Bulletin No. 3. Department of Prehistory, Research School of Pacific Studies, Australian National University, Canberra, Australia.

Groves, C. 2008. Current views on taxonomy and zoogeography of the genus *Sus*. Pages 15–29 *in* U. Albarella, K. Dobney, A. Ervynck, and P. Rowley-Conwy, editors. *Pigs and Humans: 10,000 Years of Interaction*. Oxford University Press, Oxford, UK.

Groves, C. P. 1971. Request for a declaration modifying Article 1 so as to exclude names proposed for domestic animals from zoological nomenclature. *Bulletin of Zoological Nomenclature* 27:269–272.

Groves, C. P. 1983. Pigs east of the Wallace Line. *Journal de la Société des Océanistes* 77:105–119.

Hampton, J. O., P. B. S. Spencer, D. L. Alpers, L. E. Twigg, A. P. Woolnough, J. Doust, T. Higgs, and J. Pluske. 2004. Molecular techniques, wildlife management and the importance of genetic population structure and dispersal: A case study with feral pigs. *Journal of Applied Ecology* 41:735–743.

Hernández, F. A., B. M. Parker, C. L. Pylant, T. J. Smyser, A. J. Piaggio, S. L. Lance, M. P. Milleson, J. D. Austin, and S. Wisely. 2018. Invasion ecology of wild pigs (*Sus scrofa*) in Florida, USA: The role of humans in the expansion and colonization of an invasive wild ungulate. *Biological Invasion* 20:1865–1880.

Herpin, P., M. Damon, and J. Le Dividich. 2002. Development of thermoregulation and neonatal survival in pigs. *Livestock Production Science* 78:25–45.

Hodgkinson, S. M., M. Schmidt, and N. Ulloa. 2008. Comparison of the digestible energy content of maize, oats, and alfalfa between European wild boar (*Sus scrofa* L.) and Landrace x Large White pig (*Sus scrofa domesticus*). *Animal Feed Science and Technology* 144:167–173.

Holder, K. 2015. Warthogs loose in south Texas. *The Devine News*. June 10. <http://www.devinenews.com/contentitem/397482/1707/warthogs-loose-in-south-texas>. Accessed 6 May 2016.

Ilse, L. M., and E. C. Hellgren. 1995a. Resource partitioning by sympatric populations of collared peccaries and feral hogs in southern Texas. *Journal of Mammalogy* 76:784–799.

Ilse, L. M., and E. C. Hellgren. 1995b. Spatial use and group dynamics of sympatric collared peccaries and feral hogs in southern Texas. *Journal of Mammalogy* 76:993–1002.

International Commission on Zoological Nomenclature. 1999. *International Code of Zoological Nomenclature*. Fourth Edition. The International Trust for Zoological Nomenclature, London, UK.

Keiter, D. A., J. J. Mayer, and J. C. Beasley. 2016. What's in a "common" name? A call for consistent terminology for non-native *Sus scrofa*. *Wildlife Society Bulletin* 40:384–387.

Larson, G., T. Cucchi, and K. Dobney. 2011. Genetic aspects of pig domestication. Pages 14–37 *in* M. F. Rothschild and A. Ruvinsky, editors. *The Genetics of the Pigs*, Second Edition. CAB International, Cambridge, MA.

Larson, G., T. Cucchi, M. Fujita, E. Matisoo-Smith, J. Robins, A. Anderson, B. Rolett, et al. 2007. Phylogeny and ancient DNA of *Sus* provides insights into neolithic expansion in Island Southeast Asia and Oceania. *Proceedings of the National Academy of Sciences of the United States of America* 104:4834–4839.

Larson, G., K. Dobney, U. Albarella, M. Fang, E. Matisoo-Smith, J. Robins, S. Lowden, et al. 2005. Worldwide phylogeography of wild boar reveals multiple centers of pig domestication. *Science* 307:1618–1621.

Larson, G., R. Liu, X. Zhao, J. Yuan, D. Fuller, L. Barton, K. Dobney, et al. 2010. Patterns of East Asian pig domestication, migration, and turnover revealed by modern and ancient DNA. *Proceedings of the National Academy of Science of the United States of America* 107:7686–7691.

Linderholm, A., D. Spencer, V. Battista, L. Frantz, R. Barnett, R. C. Fleischer, H. F. James, et al. 2016. A novel MC1R allele for black coat colour reveals the Polynesian ancestry and hybridization patterns of Hawaiian feral pigs. *Royal Society Open Science* 3:160304. dx.doi.org/10.1098/rsos.160304.

MacDonal, A. A., and H. Frädrich. 1991. Pigs and peccaries: What are they? Pages 7–19 *in* R. H. Barrett and F. Spitz, editors. *Biology of Suidae*. Imprimerie des Escartons, Briancon, France.

Mauget, R. 1991. Reproductive biology of the wild Suidae. Pages 49–64 *in* R. H. Barrett and F. Spitz, editors. *Biology of Suidae*. Imprimerie des Escartons, Briancon, France.

Mayer, J. J. 2002. *A Simple Field Technique for Age Determination of Adult Wild Pigs: Environmental Information Document*. WSRC-RP-2002-00635. Westinghouse Savannah River Company, Aiken, SC.

Mayer, J. J. 2018. Introduced wild pigs in North America: History, problems and management. Pages 881–921 *in* M. Melletti and E. Meijaard, editors. *Ecology, Evolution and Management of Wild Pigs and Peccaries: Implications for Conservation*. Cambridge University Press, Cambridge, UK.

Mayer, J. J., and I. L. Brisbin, Jr. 1988. Sex identification of *Sus scrofa* based on canine morphology. *Journal of Mammalogy* 69:408–412.

Mayer, J. J., and I. L. Brisbin, Jr. 1993. Distinguishing feral hogs from introduced wild boar and their hybrids: A review of past and present efforts. Pages 28–49 *in* C. W. Hanselka and J. F. Cadenhead, editors. *Feral Swine: A Compendium for Resource Managers*. Texas Agricultural Extension Service, Kerrville, TX.

Mayer, J. J., and I. L. Brisbin, Jr. 2008. *Wild Pigs in the United States: Their History, Comparative Morphology, and Current Status*. Second Edition. The University of Georgia Press, Athens, GA.

Mayer, J. J., and I. L. Brisbin, Jr., editors. 2009. *Wild Pigs: Biology, Damage, Control Techniques and Management*. SRNL-RP-2009-00869. Savannah River National Laboratory, Aiken, SC.

Mayer, J. J., and J. C. Beasley. 2018. Wild pigs. Pages 219–248 *in* W. C. Pitt, J. C. Beasley, and G. W. Witmer, editors. *Ecology and Management of Terrestrial Vertebrate Invasive Species in the United States*. CRC Press, LLC, Taylor & Francis Group, Boca Raton, FL.

McCann, B. E., M. J. Malek, R. A. Newman, B. S. Schmit, S. R. Swafford, R. A. Sweitzer, and R. B. Simmons. 2014. Mitochondrial diversity supports multiple origins for invasive pigs. *Journal of Wildlife Management* 78:202–213.

McCann, B. E., T. J. Smyser, B. S. Schmit, R. A. Newman, A. J. Piaggio, M. J. Malek, S. R. Swafford, R. A. Sweitzer, and R. B. Simmons. 2018. Molecular population structure for feral swine in the United States. *Journal of Wildlife Management* 82:821–832.

McDonald, R. C. 2015. Warthogs spotted loose in south Texas. *Wilson County News*. June 24. <http://www.wilsoncountynews.com/article.php?id=66670>. Accessed 6 May 2016.

McNab, B. K. 2002. *The Physiological Ecology of Vertebrates*. Cornell University Press, Ithaca, NY.

Morelle, K., J. Fattebert, C. Mengal, and P. Lejeune. 2016. Invading or recolonizing? Patterns and drivers of wild boar population expansion into Belgian agroecosystems. *Agriculture, Ecosystems & Environment* 222:267–275.

Nagy, K. A. 1975. Water and energy budgets of free-living animals: Measurements using isotopically labelled water. Pages 227–245 *in* N. F. Hadley, editor. *Environmental Physiology of Desert Organisms*. Dowden, Hutchinson, and Ross, Stroudsburg, PA.

Nagy, K. A. 1987. Field metabolic rate and food requirement scaling in mammals and birds. *Ecological Monographs* 57:111–128.

Ojeda, A., S. E. Ramos-Onsins, D. Marletta, L. S. Huang, J. M. Folch, and M. Pérez-Enciso. 2011. Evolutionary study of a potential selection target region in the pig. *Heredity* 106:330–338.

Okomura, N., Y. Kurosawa, E. Kobayashi, R. Watanobe, N. Ishiguro, H. Yasue, and T. Mitsuhashi. 2001. Genetic relationship amongst the major non-coding regions of mitochondrial DNAs in wild boars and serval breeds of domesticated pigs. *Animal Genetics* 32:139–147.

Oliver, W. L. R., and I. L. Brisbin, Jr. 1993. Introduced and feral pigs: Problems, policy, and priorities. Pages 179–191 *in* W. R. Oliver, editor. *Pigs, Peccaries, and Hippos*. International Union for Conservation of Nature and Natural Resources (IUCN). Gland, Switzerland.

Pepin, D., and R. Mauget. 1989. The effect of planes of nutrition on growth and attainment of puberty in female wild boars raised in captivity. *Animal Reproduction Science* 20:71–77.

Podgórski, T., D. Lusseau, M. Scandura, L. Sönnichsen, and B. Jędrzejewska. 2014. Long-lasting, kin-directed female interactions in a spatially structured wild boar social network. *PLOS ONE* 9:e99875.

Porter, W. G., and D. M. Gates. 1969. Thermodynamic equilibria of animals with environment. *Ecological Monographs* 39:245–270.

Quiniou, N., J. Noblet, J. van Milgen, and S. Dubois. 2001. Modelling heat production and energy balance in group-housed growing pigs exposed to cold or hot ambient temperatures. *British Journal of Nutrition* 85:97–106.

Ramírez, O., W. Burgos-Paz, E. Casas, M. Ballester, E. Bianco, I. Olalde, G. Santpere, et al. 2015. Genome data from a sixteenth century pig illuminate modern breed relationships. *Heredity* 114:175–184.

Ramos-Onsins, S. E., W. Burgos-Paz, A. Manunza, and M. Amills. 2014. Mining the pig genome to investigate the domestication process. *Heredity* 113:471–484.

Read, A. F., and P. H. Harvey. 1989. Life history differences among the eutherian radiations. *Journal of Zoology (London)* 219:329–353.

Sams, L. 2015. Warthogs on the loose in Texas—African hogs likely escapees from ranches that import exotics. *Texas Hunting*. May 9. <https://www.lsonews.com/warthogs-loose-texas-african-hogs-likely-escapees-ranches-importexotics/>. Accessed 6 May 2016.

Sowls, L. K. 2013. *Javelinas and Other Peccaries: Their Biology, Management, and Use*. Second Edition. Texas A&M University Press, College Station, TX.

Spencer, P. B. S., and J. O. Hampton. 2005. Illegal translocation and genetic structure of feral pigs in Western Australia. *Journal of Wildlife Management* 69:377–384.

Spitz, F., and G. Janeau. 1995. Daily selection of habitat in wild boar (*Sus scrofa*). *Zoology* 237:423–434.

Spitz, F., G. Valet, and I. L. Brisbin, Jr. 1998. Variations in body mass of wild boars from southern France. *Journal of Mammalogy* 79:251–259.

Tabak, M. A., A. J. Piaggio, R. S. Miller, R. A. Sweitzer, and H. B. Ernest. 2017. Anthropogenic factors predict movement of an invasive species. *Ecosphere* 8:e01844.

Taylor, C. R., N. C. Heglund, and G. M. O. Maloiy. 1982. Energetics and mechanics of terrestrial locomotion: Metabolic energy consumption as a function of speed and body size in birds and mammals. *Journal of Experimental Biology* 97:1–21.

Taylor, R. B., E. C. Hellgren, T. M. Gabor, and L. M. Ilse. 1998. Reproduction of feral pigs in southern Texas. *Journal of Mammalogy* 79:1325–1331.

Tompkins, S. 2015. Warthog invasion grows in south Texas. *San Antonio Express News*. December 10. <http://www.expressnews.com/sports/outdoors/article/Warthog-invasion-grows-in-South-Texas-6689406.php>. Accessed 6 May 2016.

Towne, C. W., and E. N. Wentworth. 1950. *Pigs from Cave to Cornbelt*. University of Oklahoma Press, Norman, OK.

Van Gelder, R. G. 1979. Comments on request for a declaration modifying Article 1 so as to exclude names proposed for domestic animals from zoological nomenclature. Z.N. (S.) 1935. *Bulletin of Zoological Nomenclature* 36:5–9.

van Wieren, S. E. 2000. Digestibility and voluntary intake of roughages by wild boar and Meishan pigs. *Animal Science* 71:149–156.

Verstegen, M. W. A., and W. H. Close. 1994. The environment and the growing pigs. Page 333 *in* D. J. A. Cole, J. Wiseman, and M. A. Varley, editors. *Principles of Pig Science*. Nottingham University Press, Nottingham, UK.

Weiler, U., R. Claus, M. Dehnhard, and S. Hofacker. 1996. Influence of the photoperiod and a light reverse program on metabolically active hormones and food intake in domestic pigs compared with a wild boar. *Canadian Journal of Animal Science* 76:531–539.

White, S. 2011. From globalized pig breeds to capitalist pigs: A study in animal cultures and evolutionary history. *Environmental History* 16:94–120.

Wilson, D. E., and D. M. Reeder, editors. 2005. *Mammal Species of the World: A Taxonomic and Geographic Reference*. Third Edition. Johns Hopkins University Press, Baltimore, MD.

Wolkers, J., T. Wensing, J. T. Schonewille, and A. T. Van-Klooser. 1994. Undernutrition in relation to changed tissue composition in wild boar (*Sus scrofa*). *Comparative Biochemistry and Physiology (A)* 108:623–628.

Wu, G. S., Y. G. Yao, K. X. Qu, Z. L. Ding, H. Li, M. G. Palanichamy, Z. Y. Duan, N. Li, Y. S. Chen, and Y. P. Zhang. 2007. Population phylogenomic analysis of mitochondrial DNA in wild boars and domestic pigs revealed multiple domestication events in East Asia. *Genome Biology* 8:R245.

Zervanos, S. M. 2009. Wild pig physiological ecology. Pages 145–156 *in* J. J. Mayer and I. L. Brisbin, editors. *Wild Pigs: Biology, Damage, Control Techniques and Management*. SRNL-RP-2009-00869, Savannah River National Laboratory, Aiken, SC.

Zervanos, S. M., and J. L. Day, 1977. Water and energy requirements of captive and free-living collared peccaries (*Tayassu tajacu*). *Journal of Wildlife Management* 41:527–532.

Zervanos, S. M., W. D. McCort, and H. B. Graves, 1983. Salt and water balance of feral versus domestic Hampshire hogs. *Physiological Zoology* 56:67–77.

3 Wild Pig Spatial Ecology and Behavior

Steven M. Gray, Gary J. Roloff, Robert A. Montgomery, James C. Beasley, and Kim M. Pepin

CONTENTS

3.1 INTRODUCTION

The behaviors and spatial ecology of invasive wild pigs (*Sus scrofa*) in modern North America reflect the capacity for wildlife to evolutionarily adapt, develop complex behaviors honed from interactions with the environment, survive, and in some cases thrive in increasingly human-dominated landscapes. Biologically, wild pigs reach sexual maturity at a young age, are capable of farrowing large litters, breed year-round (Comer and Mayer 2009), and have high survivability in the wild (see Chapter 4). Behaviorally, wild pigs exhibit substantial plasticity and can readily adjust life history strategies in response to their surroundings. For example, in urban landscapes wild pigs increase nocturnal activity to minimize their interactions with humans (Hanson and Karstad 1959, Podgórski et al. 2013). Furthermore, as dietary generalists (Senior et al. 2016) wild pigs can readily adapt to a variety of environments (Baskin and Danell 2003) and these attributes help make wild pigs successful invaders in novel environments.

The spatial ecology of wild pigs is highly varied across their North American range. Wild pigs are capable of traversing long distances in short amounts of time (Lapidge et al. 2004) and can move fluidly (and often undetected) through a variety of landscapes. Depending on geographic location, wild pigs can be found utilizing a diversity of vegetation types and features. The underlying motivations associated with space use in wild pigs include interactions among available forage, cover, breeding condition, social structure, and their need to thermoregulate. Therefore, spatial and temporal patterns of landscape use for wild pigs illustrate highly variable and complex life-history strategies, making efforts to describe scalable principles (i.e., across space and time) on wild pig behaviors challenging.

In this chapter, we summarize information on the behaviors and spatial ecology of wild pigs in North America. Although our focus is on free-ranging pigs, we include relevant research on captive animals and from other wild pig populations worldwide, both native and non-native, to provide a thorough synthesis of the primary topics concerning behaviors and the spatial ecology of wild pigs. Consistent with other chapters in this book, we refer to pigs within and outside of their native ranges as wild boar and wild pigs, respectively.

3.2 BEHAVIORS

3.2.1 Social Organization

Most of what is known about the social organization of wild pigs stems from research on native and invasive populations occurring outside of North America. Wild pigs are a highly social species with matrilineal group structure often consisting of multiple generations of females and their young, referred to as sounders (Spitz 1986, Kaminski et al. 2005). After birthing, daughters will primarily remain with their mother. However, in some instances, yearling females will leave to form sister groups and pursue reproductive opportunities (Kaminski et al. 2005). Genetic and spatial structure analyses of wild boar social groups confirmed that individuals closely associated in space were often inter-related (Poteaux et al. 2009). Yet social groups are flexible and their composition can fluctuate over time (Ilse and Hellgren 1995a). In fact, female groups maintain a fission-fusion society, where individuals form sub-groups that readily merge and separate, thereby exchanging individuals (Gabor et al. 1999), although the mechanisms regulating the fission-fusion process of social groups are poorly understood. Adult males are mostly solitary but younger males are sometimes observed in bachelor groups (Braza and Alvarez 1989, Fernández-Llario et al. 1996). Furthermore, structure of wild boar social groups apparently affects individual willingness to incur risk (Focardi et al. 2015). Adult males, yearlings, and piglets will enter potentially risky areas first, allowing for a foraging advantage, with adult and sub-adult females being more risk averse (Focardi et al. 2015). Additionally, larger social groups had longer effective foraging times than smaller groups (Focardi et al. 2015). In North America, the social organization of wild pigs is comparable to what has been observed in native populations, but anecdotal evidence suggests variability among regions (e.g., larger groups in areas of the southwest United States), but the mechanisms regulating group structure in North America remain poorly understood.

3.2.2 Territoriality

Territoriality has been observed in several different contexts for wild pigs, however this remains a topic in need of further inquiry. Males are often solitary and grapple with other males for access to resources and mating opportunities (Graves 1984, Mayer 2009). For females, increased territoriality has been observed during the farrowing period, when a female will actively defend a nesting area from predators and conspecifics (Graves 1984, Fernández-Llario 2004). Territoriality has also been observed among female social groups, presumably to ensure access to resources. In Fort Benning, Georgia, individual females within a sounder exhibited extensive home range overlap whereas each respective sounder often occupied an exclusive area (Sparklin et al. 2009). However, the precise mechanisms influencing territoriality remain unclear, and variation in this behavior may correspond to geographic location, density, resource availability, and climate. We caution that results from Sparklin et al. (2009) were based on a limited sample, but their findings provide initial insights into territorial behavior among sounders. This phenomenon was also observed in Texas, where it was hypothesized that territoriality was necessary due to poor habitat quality, high population density, and limited access to resources that resulted in increased competition among sounders (Gabor et al. 1999). Furthermore, it has been suggested that territoriality may be best observed via temporal partitioning among sounders, rather than spatial avoidance. For example, sounders may exhibit spatial overlap yet rarely encounter one another due to temporal partitioning of overlapping home ranges (J. C. Beasley, Savannah River Ecology Laboratory, University of Georgia, personal communication).

3.2.3 Rooting

Wild boar are known to adjust foraging strategies to take advantage of seasonally available food resources (Sandom et al. 2013). One of the primary modes of foraging is through rooting, where individuals will upturn soil using their spade-like snouts in search of subterranean food. Rooting can vary in intensity, from sifting through surface leaf-litter to deeply excavating large areas (Figure 3.1). Wild pigs use their superior sense of smell and sensitivity of their snout to explore and understand the areas they occupy.

Where wild pigs choose to root can vary seasonally, again based on food availability (Baron 1982, Sandom et al. 2013). Wild pigs tend to root in saturated areas (e.g., swamps and lowlands), presumably because the soil structure is more pliable (Mitchell and Mayer 1997, Welander 2000). There seems to be a tendency to root in these areas during the summer or in geographic regions that feature a wet season, but rooting has also been noted during times of low mast production (Graves 1984). Across landscapes, rooting tends to occur at higher frequency during winter and spring, likely due to the decreased availability of aboveground food (Wilson 2004, Sandom et al. 2013). During these seasons, rooting will commonly occur along ridgelines and in areas where mast is plentiful (Stegeman 1938, Barrett 1982). During years of poor acorn crops, wild pigs will shift their activity to exploit alternative food sources. For example in the Great Smoky Mountains National Park, wild pigs were observed to shift their activity to beech (*Fagus* spp.) forests where intensive rooting was observed (Bratton et al. 1982). Wild pigs will also revisit previously disturbed areas rather than root a new site (Groot Bruinderink and Hazebroek 1996, Parkes et al. 2015), presumably related to the quantity and quality of forage at these locations. In a coastal dune system on St. Catherine's Island, Georgia, wild pigs returned to previously rooted sites approximately every 5 years (Oldfield and Evans 2016).

3.2.4 Diet

Wild pigs are omnivorous and opportunistic (Stegeman 1938, Sweeney et al. 2003), a strategy that allows them to utilize a variety of food resources. In fact, the ability of this species to adapt and

FIGURE 3.1 Forest floor rooting in the central Lower Peninsula of Michigan detailing (a) low-intensity and (b) high-intensity rooting events. (Photos by S. Gray. With permission.)

opportunistically exploit resources is one of the reasons why wild pigs are such successful invaders and are found in so many areas worldwide (Senior et al. 2016). Furthermore, the diet of wild pigs fluctuates by season and spatial location (Table 3.1). Throughout both their native and invasive ranges, wild pig diets principally rely upon plant matter (Schley and Roper 2003, Ballari and Barrios-García 2014). Wild pigs eat subterranean foods such as roots and bulbs (Giménez-Anaya et al. 2008, Schlichting et al. 2015), but consumption of these items is more common between fall and early summer when other food sources are scarce (Dardaillon 1987). When available, acorns appear to be a common food item (Sjarmidi et al. 1996, Elston and Hewitt 2010), as wild pigs will scour forest stands and ridgelines where hard mast is abundant (Stegeman 1938, Barrett 1982). A review of diets throughout the native range indicated that mast, such as acorns and similar energy-rich food items, were a staple in most wild boar diets (Schley and Roper 2003). Wild pig diets tend to be low in energy, high in fiber, and often seasonally deficient in protein (Baber and Coblentz 1987). To offset protein deficiency, wild pigs may rely heavily on animal matter (Wilcox and Van Vuren 2009).

TABLE 3.1
Percent Composition of Material by Volume in Wild Pig Diets in the United States

		Percent Composition of Diet				
Region	**Location**	**Plant**	**Fungi**	**Animal**	**Other**	**References**
South	Great Smoky Mountains National Park, NC/TN	84.6–99.3	0.0–18.6	0.7–10.8	Trace	Ackerman et al. (1978)
South	Davis Mountains, TX	97.5	—	2.5	—	Adkins et al. (2006)
South	Horn Island, Gulf Islands National Seashore, MS	77.5	3.8	18.8	—	Baron (1979)
West	Dye Creek Ranch, Tehama County, CA	98.4	—	1.6	—	Barrett (1978)
West	Kipahulu Valley, Maui, HI	93.2	—	6.8	—	Diong (1982)
South	Yturria Ranch, Willacy County, TX	95.2	—	4.8	—	Everitt and Alaniz (1980)
West	Island of Hawaii, HI	93.0–96.0	—	4.0–7.0	—	Giffin (1970)
West	Mendocino County, CA	71.5	—	28.5	—	Grover (1983)
South	Tellico Wildlife Management Area, TN	89.4	—	6.4	4.2	Henry and Conley (1972)
South	Great Smoky Mountains National Park, NC and TN	98.0	—	2.0	—	Howe et al. (1981)
West	Monterey, San Luis Obispo, and San Benito Counties, CA	80.9–94.2	—	5.8–19.1	—	Pine and Gerdes (1973)
South	Great Smoky Mountains National Park, NC and TN	99.1	—	0.3	0.2	Scott and Pelton (1975)
South	Cottle, Dickens, Foard, King, and Motley Counties, TX	98.4[a]	—	1.7[a]	—	Schlichting et al. (2015)
South	Aransas National Wildlife Refuge, TX	75.0–83.9	1.7–14.6	16.1–25.0	—	Springer (1975)
South	Chaparral Wildlife Management Area, TX	93.0	6.7	—	0.3	Taylor and Hellgren (1997)
South	Hobcaw Barony, SC	83.0–97.4	0.8–11.7	1.3–5.6	—	Wood and Roark (1980)

[a] Diets reported as percentages using point frame sampling techniques described in Chamrad and Box (1964).
Source: Adapted from Ditchkoff and Mayer (2009).

Compared to plant matter, animal matter makes up a considerably smaller proportion of wild pig diets (Table 3.1). However, the occurrence of animal matter in wild boar diets is common (Irizar et al. 2004, Giménez-Anaya et al. 2008). Wild pigs will consume invertebrates, birds, herptiles, small and large mammals, and carrion. In fact, wild pigs have been shown to be efficient members of the scavenging community (Turner et al. 2017). One of the most common animal components consumed by wild pigs is earthworms (Hanson and Karstad 1959, Challies 1975). Wild pigs appear to exploit animal food resources based on availability. For example, in Spain, wild boar consumed birds more frequently during the molting period when juvenile birds were both accessible and easily preyed upon (Giménez-Anaya et al. 2008). In the southeastern United States, wild pigs were observed targeting herpetofauna when they were vulnerable during the breeding season (Jolley et al. 2010). Overall, animal protein may play an important role in the growth and survival of wild pigs (Schley and Roper 2003), and an absence of these proteins could degrade body condition (Klaa 1992). We note that while wild pigs will also consume fungi, in most seasons this makes up a small proportion of the overall diet (Fournier-Chambrillon et al. 1995, Baubet et al. 2004), but we caution that the presence of fungi is difficult to assess via dietary studies (Schley and Roper 2003) unless using molecular methods. It is also worth noting that wild boar are notorious for their ability to locate truffles, which aids in truffle dispersal and survival (Piattoni et al. 2014). According to a review of wild pig diets in the native and invasive ranges, it appears that animal matter and fungi are consumed in greater quantities in the invasive range (Ballari and Barrios-García 2014).

In landscapes where agriculture is abundant, wild pigs may heavily use this resource. Their use of crop fields has been documented mainly during the summer and fall months (Sparklin et al. 2009, Schlichting et al. 2015), after fields have been recently planted or when crops are mature. For example, in the plains of Texas, crops were an abundant dietary item throughout the year with peak consumption occurring in the fall (>60% of the total diet; Schlichting et al. 2015). In an agroecosystem of Spain, wild boar primarily fed on crops, with ~77% of the diet by volume consisting of corn (Herrero et al. 2006). Taller crops such as cereal, wheat, and corn are favored during the late summer months as these crops provide both food and shelter (Łabudzki et al. 2009). When crops are available, some wild pig populations will exclusively feed and reside in agricultural fields. In an assessment of damage caused by wild boar in the native range, frequently damaged crop types included oats, rye, and potatoes with a distinct preference for oats (Andrzejewski and Jezierski 1978), while other studies found high amounts of maize, wheat, barley, and alfalfa in wild pig diets (Herrero et al. 2006, Gentle et al. 2015). However, it is important to note that some of these dietary relationships may strictly be driven by the availability of crops in the landscape.

3.2.5 Wallowing

Wallowing, where an individual will cover its body in mud, is a common behavior exhibited by wild pigs. The primary function of this behavior is to reduce body temperature and protect against potentially harmful insects and parasites (Eisenberg and Lockhart 1972, Frädrich 1974, Diong 1982). Wallows can be naturally inundated shallow depressions or may be created via rooting (Figure 3.2). Wallows tend to be oval in shape and frequently occur along trails, streams, and areas of slow moving or standing water (Belden and Pelton 1976). Moreover, wild pigs seek wallows in areas that are cool, wet, and shaded (Stegeman 1938). These areas can often be identified by trampled and muddy vegetation near the wallowing site (Stegeman 1938). Wallows are used most frequently during the summer (Crouch 1983). However, in some instances wallows are used year-round, even in cold weather months, and may be visited repeatedly by individuals or groups (Stegeman 1938). In Spain, mature males primarily used wallows during the rutting period (October-February), which suggests wallowing might also serve a sexual function (Fernández-Llario 2005). The duration of wallowing is variable but individuals will often remain in wallows longer when temperatures are warm (Graves 1984). In fact, during times of the year when temperatures are high, wild pigs will spend the entire diurnal period wallowing before moving to shade in the early evening (Graves 1984).

FIGURE 3.2 (a) Wild boar wallow found in the Chernobyl Exclusion Zone, Belarus, and (b) wild pig wallow in South Carolina. (Photos by J. Beasley. With permission.)

3.2.6 Rubbing and Tusking

Rubbing and tusking behaviors are performed as a method of self-grooming to remove mud, hair, and ectoparasites (Conley et al. 1972, Graves 1984). It has also been suggested that these behaviors might serve as a method for marking territory (Conley et al. 1972) but additional research is needed. Rubbing is often performed near wallows where an individual will remove excess mud using the base of a nearby tree. Rubbing is conducted on a variety of objects as long as they are sturdy and upright, but preference has been shown for smaller diameter (<15 cm) pines (Stegeman 1938). Wild pigs have also been observed rubbing against creosote-soaked poles (Campbell and Long 2009) and, in some cases, this behavior is vigorous enough to cause damage (Figure 3.3).

Similar in function to rubbing, tusking is when an individual will use its tusks to scrape a tree causing it to release sap or pitch, which is then rubbed against to serve as a deterrent to ticks and lice (Graves 1984). Tusking can vary from a single slash to prolonged and repeated tusking events (Graves 1984), which can result in tree mortality in extreme cases (Stegeman 1938). Of the few studies on tusking, preferred trees are pine and hemlock (Graves 1984), and access to pitch is considered to be one of the main drivers of this behavior (Stegeman 1938).

FIGURE 3.3 Damage to a creosote-soaked pole due to vigorous rubbing by wild pigs in Mississippi. (Photo by C. Gibson. With permission.)

3.2.7 Bedding

Wild pigs will opportunistically bed in shallow depressions or construct beds for resting (Figure 3.4). Beds share similarities to farrowing nests and can be difficult to differentiate (Figure 3.4); however, resting beds tend to be smaller in size (Mayer et al. 2002), and are often less structurally complex (Frädrich 1974). Resting beds will often be rooted to expose mineral soil, and sometimes are lined with plant material, although this is more common in beds occupied during cold-weather months (Conley et al. 1972, Mayer et al. 2002). Beds are frequently placed along ridgelines and in areas with dense cover (Stegeman 1938, S. M. Gray, Applied Forest and Wildlife Ecology Laboratory, Michigan State University, personal observation). These areas are often heavily shaded and a single bed may be occupied by one or more individuals (Hanson and Karstad 1959). When resting, wild pigs will remain in close physical contact with others sharing the same bed (Hediger 1950, 1955). Many times a bed will be used more than once and some individuals will use the same bed repeatedly (Frädrich 1974, S. M. Gray, personal observation).

3.2.8 Scent Marking

Wild pigs have several glands located on their body that they use for scent marking. The most studied glands are the carpal glands that occur in a series along the posterior of the forefeet (Farnesi et al. 1999, Bacchetta et al. 2007). These glands are found in both males and females, and are often larger in mature individuals (Heise-Pavlov et al. 2005, Bacchetta et al. 2007). Carpal glands likely function as a means of intra-sounder communication and serve as a method for delineating territories and facilitating reproduction (Graves 1984, Bacchetta et al. 2007). Furthermore, it may be possible to utilize carpal gland secretions to affect wild pig behavior. In captive trials, study subjects were attracted to and investigated carpal gland secretions collected from unknown individuals

FIGURE 3.4 Examples of (a) spring loafing bed and (b) farrowing nest in the eastern Upper Peninsula of Michigan and (c) farrowing nest in South Carolina. (Photos (a) and (b) by M. Haen and (c) by J. Beasley. With permission.)

(B. S. Schmit, USDA/APHIS/WS/National Wildlife Research Center, personal communication). However, additional research on the function of carpal glands is necessary.

There are also several glands that occur on the upper lip of wild pigs that are concentrated around the tusks. Use of these glands primarily occurs when an individual is tusking and they are rubbed against a tree or associated object. Wild pigs also mark areas by generating saliva, which helps release the scent associated with these glands (Diong 1982). These submaxillary glands appear to serve a sexual function, as their removal hinders the ability of males to induce the mating stance in females and these individuals consequently had lower libido (Perry et al. 1980). The preputial gland is found in males near the distal end of the penis (Mayer 2009). Secretions from this gland emit a scent called "muskone" (Barrett 1978), a foul-smelling odor that is commonly associated with males (Mayer 2009). Marking via this gland occurs during urination and has also been observed during carpal marking (Mayer 2009). Excretions from the preputial gland are believed to play a role in mating and dominance, as dominant males have been found to scent mark more frequently with this gland than subordinate males after being in close proximity to multiple females (Mayer and Brisbin 1986). Preorbital glands are found below the eye. Overall, there is limited information on this gland and its associated function, however it may play a role in reproduction. Females will use these glands to mark objects, primarily trees, during the time leading up to estrus (Meynhardt 1982). Wild pigs have several other glands but the purpose of these glands in scent marking remains largely unknown (Groves and Giles 1989).

3.2.9 Vigilance

Wild pigs will exhibit vigilance behaviors to detect and escape from potential predators or sources of anthropogenic disturbance. Research on vigilance behavior has primarily been conducted on wild boar populations in their native range. When assessing vigilance behavior, Quenette and Desportes (1992) observed an oscillating pattern of long and short scanning events, rather than a random scanning pattern. The temporal pattern of this behavior varied by individual, and increasing the number of conspecifics lowered overall vigilance in individuals (Quenette and Desportes 1992). Similarly in Poland, Podgórski et al. (2016) noted that wild boar spent a small proportion of time exhibiting vigilance behaviors. The authors further noticed that vigilance varied seasonally, where wild boar were less vigilant during seasons featuring limited access to food resources (Podgórski et al. 2016). In both wild and captive studies, collective vigilance was found to decrease with increasing group size (Quenette and Gerard 1992, Podgórski et al. 2016). Additionally, Focardi et al. (2015) found the presence of conspecifics improved foraging efficiency in adult males since other individuals frequently resided near the periphery of the group and shared the burden of vigilance. Due to a lack of natural predators in North America, it can be argued that vigilance of wild pigs may substantially differ from what has been observed in the native range. Thus, further research is needed to better understand the role of vigilance in wild pigs inhabiting North America.

3.3 SPATIAL ECOLOGY

3.3.1 Home Range

Multiple studies on wild pig home ranges exist throughout North America (Table 3.2). Wild pigs exhibit notable variation in home range sizes among geographic regions, with these differences oftentimes attributed to biotic (e.g., vegetation types, pig densities, human activity, sex, age) and abiotic (e.g., distribution of water or roads, climate, weather, season) factors. Additionally, it is noteworthy that different methods used to calculate home ranges vary in strengths and weaknesses and thereby, will yield variable estimates (Kernohan et al. 2001). This makes comparisons of home ranges among studies that use different estimators difficult. A second issue is that studies of home ranges are often conducted over different time scales or at different times during the year,

TABLE 3.2
List of Home Range Sizes Reported for Wild Pigs in the United States Organized by Study, Sex, Region, and Method

Region	Location	Home Range Size (km²) Female (*n*)	Male (*n*)	Method[a]	Reference
South	Davis County, TX	28.3 (6)	35.0 (7)	95% MCP	Adkins and Harveson (2007)
South	Davis County, TX	43.4 (6)	58.7 (7)	95% AK	Adkins and Harveson (2007)
West	Santa Catalina Island, CA	1.5 (4)	2.4 (6)	95% utilization contour	Baber and Coblentz (1986)
West	Santa Catalina Island, CA	0.7 (4)	1.4 (6)	MCP	Baber and Coblentz (1986)
West	Dye Creek Ranch, CA	10.0–25.0	50.0+	–	Barrett (1978)
South	Kleberg and San Patricio Counties, TX	1.3–5.5 (4–7)[b]	1.3–5.5 (5–9)[b]	95% FK	Campbell et al. (2010)
South	San Patricio County, TX	0.7–1.4 (10)[c]		95% FK	Campbell et al. (2012)
South	Savannah River Plant, SC	7.9 (4)	14.0 (3)	MCP	Crouch (1983)
West	Kipahula Valley, Maui, HI	1.1 (4)	2.0 (5)	MCP	Diong (1982)
Central	Southern MO	4.7 (7)[d]	3.8 (6)[d]	MKDE	Fischer et al. (2016)
South	Congaree National Park, SC	1.9 (9)	2.2 (7)	95% FK	Friebel and Jodice (2009)
South	Congaree National Park, SC	1.4 (9)	1.5 (7)	100% MCP	Friebel and Jodice (2009)
South	Lowndes County, AL	4.0 (6)	4.0 (5)	95% AK	Gaston et al. (2008)
North	Central Lower Peninsula, MI	1.3 (6)	3.0 (2)	BBMM	S. M. Gray, unpublished data
West	Dye Creek Ranch, CA	1.5 (3)	—	Modified MCP[e]	Grover (1983)
South	Grenada County, MS	3.8–15.1 (1–4)	2.1–8.1 (3–6)	MCP	Hayes et al. (2009)
South	Grenada County, MS	3.2–15.0 (1–4)	2.9–8.1 (3–6)	95% AK	Hayes et al. (2009)
South	Savannah River Plant, SC	5.7 (5)	13.8 (6)	MCP	Hughes (1985)
South	Savannah River Plant, SC	4.4 (1)	5.3 (4)	Modified MCP[e]	Kurz and Marchinton (1972)
South	Savannah River Plant, SC	2.5 (2)	4.0 (2)	MCP	J. J. Mayer, unpublished data
South	TXU Corporation's Big Brown Mine, TX	6.5 (10)	15.8 (6)	MCP	Mersinger and Silvy (2007)
South	Kent County, TX	23.9 (11)	52.2 (3)	MCP	Schlichting et al. (2016)
South	Great Smoky Mountain National Park, NC/TN	2.7–3.5 (4)	3.8–3.9 (9)	MCP	Singer et al. (1981)
South	Fort Benning Military Reservation, AL/GA	—	2.0–3.7 (4–5)[f]	95% K	Sparklin et al. (2009)
West	Santa Cruz Island, CA	1.0 (5)	1.4 (5)	–	Sterner (1990)
South	Hobcaw Barony, SC	1.8 (3)	2.3 (3)	MCP	Wood and Brenneman (1980)
South	Southern TX	4.6–17.2 (5–6)	7.7–19.4 (5–10)	95% K	Wyckoff et al. (2012)

[a] Abbreviations for commonly used methods to estimate home range: BBMM = Brownian Bridge Movement Models, MCP = Minimum Convex Polygon (100%), 95% MCP = 95% Minimum Convex Polygons, 95% AK = 95% Adaptive Kernel, 95% FK = 95% Fixed Kernel, 95% K = 95% Kernel, MKDE = Movement-based Kernel Density Estimator.

[b] Mean home ranges pooled across sex and study area.

[c] Mean home ranges of control population pooled by sex.

[d] Mean home ranges of control population.

[e] Modified version of minimum convex polygon using ¼ of an individual's range to test for outlying points (Harvey and Barbour 1965).

[f] Home range estimates for controls. Sample size corresponds to sounders rather than individuals. Use of individual and multiple animals per sounder in some home range estimates.

Source: Adapted from Mayer (2009).

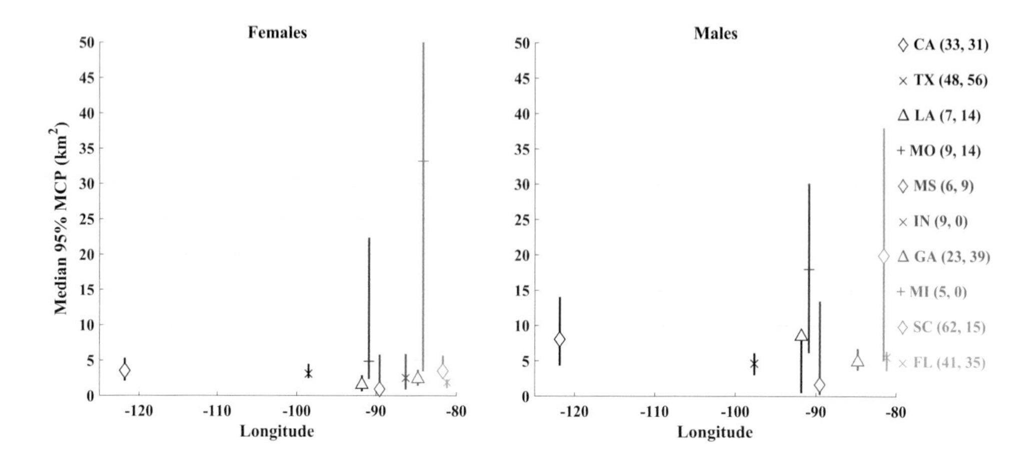

FIGURE 3.5 95% Minimum convex polygon (MCP) home range sizes from 456 pigs occurring in 10 states depicted by state and sex. Points represent median values across individual pigs in multiple studies within each state. Error bars are 95% confidence intervals of the median. Sample sizes are shown in brackets next to the state name (Female, Male). We excluded 5 individuals which had home range sizes that were outliers (≫100 km²). These included 4 males from Michigan, Louisiana, and South Carolina and one female from Michigan. These individuals did not have stationary home range movements, which violates the assumptions of most methods for estimating home range size.

introducing additional challenges in comparing home range estimates across studies since wild pigs may alter their movements depending on season (Kay et al. 2017). Telemetry studies are also further complicated by the difficulty in equipping and maintaining collars attached to wild pigs, as these animals grow rapidly and their behaviors (e.g., rubbing, wallowing) can hinder collar performance and lead to potential malfunction (see Chapter 9).

Home ranges for wild pigs in the United States have previously been summarized by Mayer (2009) and Schlichting et al. (2016), as well as in this chapter (Figure 3.5; Table 3.2). To date, individual radio telemetry studies have generally been based on <10 individuals (by sex) and used 95% minimum convex polygon (MCP) or kernel methods to estimate home range sizes (Table 3.2). One exception is Kay et al. (2017) that reanalyzed global positioning system (GPS) data from 226 wild pigs from 14 different studies across the southern United States to compare home range size estimators (MCP versus autocorrelated kernel density estimator (AKDE); Figure 3.5) and analyze determinants of home range size. AKDE is a novel kernel approach that accounts for temporal autocorrelation that is inherent in GPS data (Fleming et al. 2015). On average, wild pig monthly MCP home ranges in the southern United States are 3.4 km² (sd = 4.6 km²) with an overall home range size of 6.1 km² (sd = 7.8 km²; Kay et al. 2017). In comparison, average overall wild pig home ranges using AKDE were 12.4 km² (sd = 21.0 km²). Seasonal variations in home range sizes are well documented, with most differences attributed to changes in food (Calenge et al. 2002, Hayes et al. 2009) or water availability (Kay et al. 2017). For example, Singer et al. (1981) found a 3-fold increase in home range sizes of wild pigs during winters following a mast failure, compared to winters following good mast years. Similarly, in Mississippi, home ranges were larger during the dry season than in the wet season when food was more abundant (Hayes et al. 2009). Indeed, telemetry evidence suggests sounders will exclude other groups from core areas in environments with more dense wild pig populations (Sparklin et al. 2009), but this pattern may not be consistent for individual animals (Baber and Coblentz 1986, Boitani et al. 1994). Generally, home range size tends to non-linearly decrease as density of conspecifics increases (Saunders and McLeod 1999), with apparent density-dependent effects occurring when wild pig densities were <9 pigs/km² (Figure 3.6; Table 3.3). However, exceptions to this pattern exist (e.g., agricultural landscapes of Belgium; Morelle and Lejeune 2014), likely related to variation in nutritional carrying capacity of the occupied areas.

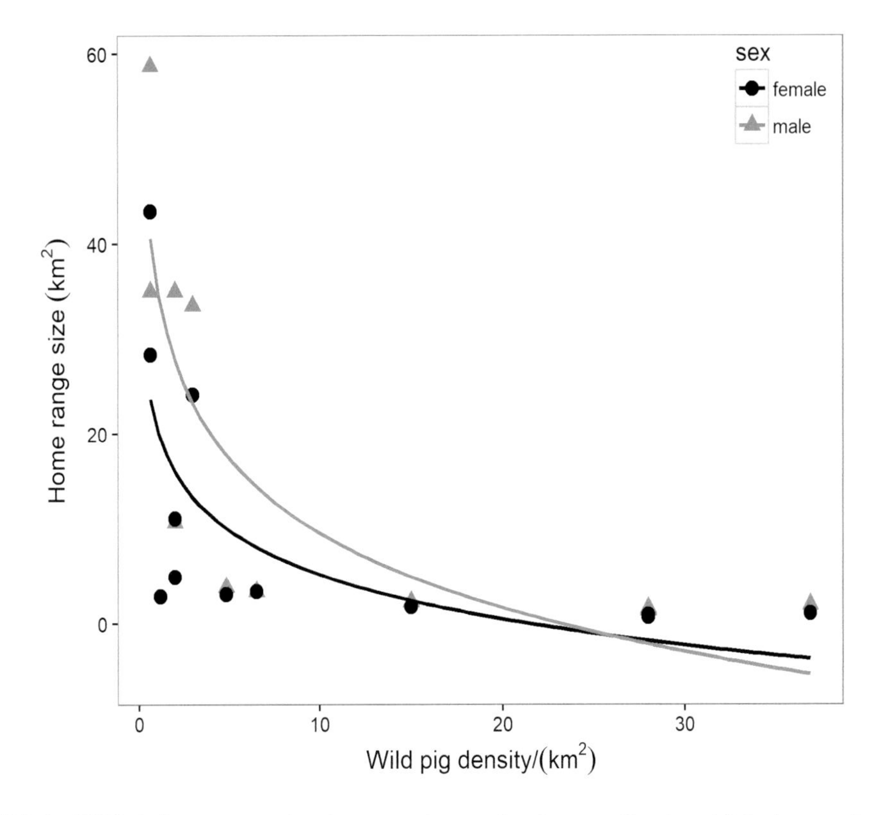

FIGURE 3.6 Wild pig home range sizes in comparison to density according to published research.

Wild pig home range sizes also vary by sex. Generally, home ranges of males are 3.5–5 km^2 larger than females in North America (Kay et al. 2017), although most of this information stems from research in the southern United States. These trends have also been observed in the invasive populations of Australia (Caley 1997, Morini et al. 2014). Differences in home range sizes between sexes may be related to differences in body mass (Saunders and McLeod 1999), but behavior also plays a role. This pattern is especially true during the breeding season, when males will travel long distances to secure mating opportunities. However, it is worth noting that differences in home range sizes between sexes are not always evident (South Carolina: Wood and Brenneman 1980; Italy: Massei et al. 1997).

For wild pigs, younger individuals appear to use larger home ranges than adults, likely related to younger individuals exploring and establishing territories (Alabama: Gaston et al. 2008; Germany: Keuling et al. 2008a). This was particularly noticeable in females (New Zealand: McIlroy 1989) and in younger individuals during the hunting season (France: Calenge et al. 2002). Younger wild boar tend to flush more easily compared to adults, potentially contributing to larger home ranges for younger animals when subjected to hunting pressure (Maillard et al. 1996).

The shape of wild pig home ranges varies, but has generally been described as circular to elongate, depending on topography, human activity, and the distribution of food and other pigs, among other factors (Singer et al. 1981, McIlroy 1989). Wild pig home range shapes can vary by season, with summer home ranges being more circular than winter (Singer et al. 1981), but these patterns may differ depending on the analytical method being used. Furthermore, natural and anthropogenic features such as roads, rivers, and railways can also contribute to the overall shape of wild pig home ranges.

Wild boar home ranges typically consist of a high-use (core) area, which is used for resting, and peripheral areas that are primarily used for feeding (Boitani et al. 1994). Core areas are often small

TABLE 3.3
Description of Density Calculation Methods, Reported Densities, Home Range Sizes by Sex, and Home Range Estimators Used in Published Research on Wild Pigs

Density		Home Range Size (km²)			
Calculation method	Avg. Pigs/ km²	Male	Female	Estimator	Reference
Capture-recapture techniques, computed using program CAPTURE by dividing estimate of population size by area of use[a]	28	1.4	0.7	MCP	Saunders and McLeod (1999)
Calculated using removal methods, observation, comparison between communities, and program CAPTURE[b]	28	1.6	0.9	MCP	Saunders and McLeod (1999)
Unknown	37	2	1.1	MCP	Saunders and McLeod (1999)
Unknown	15	2.3	1.8	MCP	Saunders and McLeod (1999)
Unknown	2	10.7	4.9	MCP	Saunders and McLeod (1999)
Combination of mark-recapture, observation, and the ratio method[c]	3	33.5	24.1	MCP	Saunders and McLeod (1999)
Unknown	2	35	11.1	MCP	Saunders and McLeod (1999)
Unknown	4.85[d]	3.85	3.1	MCP	Singer et al. (1981)
Based on number of individuals captured/total area[e]	0.65	35	28.3	95% MCP	Adkins and Harveson (2007)
Based on number of individuals captured/total area[e]	0.65	58.7	43.4	95% AK	Adkins and Harveson (2007)
Unknown	6.5	3.4	3.4	95% AK	Campbell et al. (2010)
Program MARK abundance estimate/avg. effective sampling area[f]	1.2	—	2.85	95% K	Sparklin et al. (2009)

[a] Method found in Baber and Coblentz (1986).
[b] Method found in Coblentz and Baber (1987).
[c] Method found in Caley (1993).
[d] Value found in Singer et al. (1981).
[e] Method found in Ilse and Hellgren (1995b).
[f] Method and value found in Hanson (2006).

and provide dense cover, with some potential variation among sexes. For example, core areas of males were more heterogeneous and used less frequently compared to females in Italy (Boitani et al. 1994). Males tend to use portions of their home ranges more widely while female movements are more concentrated (Wood and Brenneman 1980, Boitani et al. 1992).

Most wild pig populations worldwide are subjected to population control and hunting, and researchers have documented the ways in which these events have impacted home ranges. In response to aerial gunning in Texas, Campbell et al. (2010) did not observe differences in home range sizes before and after control events. Similarly, researchers in Missouri found that wild pig home range centroids did not shift following pursuit by dogs, hunters, helicopters, or after trap-and-release events, although core home range area shrunk and diurnal movement distances increased, suggesting some fluctuations in space use in response to harassment (Fischer et al. 2016). However, individual pig response to hunting pressure can be mixed. After drive hunts in Germany, some individuals increased while others reduced the size of their resting ranges (Sodeikat and Pohlmeyer 2007). Although it is generally accepted that wild boar avoid human activity, they will move into human-dominated areas to exploit anthropogenic food sources and agriculture when other forage is scarce (Bieber and Ruf 2005, Cahill et al. 2012).

3.3.2 Habitat Use and Selection

Habitat use is broadly concerned with the distribution of individuals throughout landscapes (Hutto 1985). Habitat selection is the behavioral response exhibited by individuals where a particular area is selected in favor of other areas (Hutto 1985, Block and Brennan 1993). In wild pigs, habitat use and selection can widely vary since this species is a generalist and has the ability to thrive in a diversity of environments. As such, wild pigs primarily use habitats in relation to their availability in the landscape (Ilse and Hellgren 1995b, Gabor et al. 2001), but seasonal differences exist (Baber and Coblentz 1986, Keuling et al. 2009b). However, trends in resource selection by wild pigs will become clearer as research is conducted using GPS data and robust analytical techniques at multiple spatial and temporal scales.

As with most animals, the primary determinants of wild pig distributions are the availability of water, food, and cover (McIlroy 1989). Wild pigs will respond to fluctuations in these components by spatially adjusting their habitat selection to capitalize upon seasonally available resources (Germany: Keuling et al. 2009b; Texas: Campbell and Long 2010). In many cases, wild pigs select areas that satisfy all of these life history requirements. In the Sierra Foothills of California, wild pigs selected oak thickets as both ample cover and acorns for food were available (Barrett 1982). In central Alabama, wild pigs selected wetlands during the summer months as these areas provided food, water, and cover, as well as quality areas for farrowing (Gaston et al. 2008). In Tennessee, wild pigs used oak-pine slopes during the winter because these areas offered thermoregulatory cover and acorns (Singer et al. 1981). During the summer months, however, individuals moved extensively between multiple forested habitats to meet food, cover, and thermoregulatory requirements (Singer et al. 1981). In South Carolina, changes in monthly habitat use were mainly related to food availability (Kurz and Marchinton 1972). Wild pigs on Santa Cruz Island, California, moved from canyon bottom to ridgelines due to weather and food availability (Van Vuren 1984). In an area of their native range in Eurasia, wild boar selected European beech (*Fagus sylvatica*) and hornbeam (*Carpinus betulus*) forests during the winter since these habitats offered forage and cover (Fonseca 2007). These habitat selection patterns also hold true in agricultural areas, where wild boar will utilize wheat, cereal, and corn during the summer months when these crops are mature and provide both food and shelter (Łabudzki et al. 2009).

Wild pigs have an affinity for water and associated vegetation types (e.g., riparian areas, marshes, swamps). Overall, wild pigs are poor thermoregulators and require access to water to cool down during periods of high temperatures. This affinity for water is especially pronounced in arid areas where open water can be scarce. For example, in east and south Texas, open water and saturated soils were important habitat components affecting the distribution of wild pigs (Mersinger and Silvy 2007, Cooper and Sieckenius 2016). Wild pigs in the southeast United States also selected habitats closely associated with water, which can be exacerbated by drought conditions (Friebel and Jodice 2009). In response to high temperatures on Santa Catalina Island, California wild pigs selected canyon bottoms that offered a cool environment and increased access to water (Baber and Coblentz 1986), patterns that were attributed to a physiological need for water. The influence of water on spatial distribution of wild pigs is most pronounced during drier seasons. For example, in southern Sweden wild boar preferred areas with abundant water in all seasons except during springtime when water was generally present throughout the landscape (Thurfjell et al. 2009).

Because wild pigs seek areas with adequate food, cover, and water, selection of areas associated with water like wet scrub-shrub, bottomland forest, and emergent marsh, is likely related to the dense vegetation and productivity of these areas. Wild pigs used the margins of fresh and brackish marshes in all seasons of the year in coastal South Carolina, as these areas featured an abundance of roots and tubers (Wood and Brenneman 1980). Other studies have also observed intense rooting in wetter habitats that feature a rich herbaceous understory (Bratton et al. 1982). In the central Lower Peninsula of Michigan, wild pigs will often bed in areas that provide thick vegetative cover during the daytime, such as cattail marshes where extensive rooting has been observed along the edges of these habitats (S. M. Gray, personal observation). In east Texas, wild pigs selected riparian corridors that had dense vegetation, thus serving as visual cover and aiding in thermoregulation (Mersinger and Silvy 2007). Preference for riparian habitat has also been supported by research on vehicle collisions with wild pigs, where collisions occurred more frequently when in proximity to streams and associated bottomland vegetation (Beasley et al. 2014). Selection for riparian habitats has also been observed in wild pigs outside of North America, most notably in Australia (Caley 1997, Dexter 1998).

Scale is an important consideration in habitat use and selection studies, as this will ultimately influence inferences (Johnson 1980). For example, in the Chihuahuan Desert of Texas, population level habitat selection (second-order; *sensu* Johnson 1980) indicated selection for open-canopied evergreen woodlands by wild pigs (Adkins and Harveson 2007). However, individual-level selection was consistent with generalist behaviors that showed minimal preference for any particular habitat type (Adkins and Harveson 2007). Given this study occurred in a desert ecosystem, these findings highlight the importance of the scale of analysis and that wild pigs may display stronger selection in harsher climates. Overall, additional research is needed to better understand how wild pigs in North America select resources across multiple scales.

3.3.3 Activity and Movement

Over most of their range, wild pigs are generally considered to be nocturnal (Keuling et al. 2008b, Campbell and Long 2010). This is broadly considered to represent a shift in behavior due to human interactions, where anthropogenic disturbance has caused wild pigs to adopt a predominately nocturnal activity pattern (Stegeman 1938, Keuling et al. 2008b). Indeed, in areas farther from human settlements wild boar have shown increased daytime activity (Podgórski et al. 2013). Similar to other aspects of wild pig spatial ecology, activity patterns are influenced by changes in food, cover, and temperature (Van Vuren 1984, Campbell and Long 2010). In landscapes with thick vegetative cover, wild pigs will be more active during the daytime (Stegeman 1938). During warm summer days, wild pigs will bed during the day and become more active at night. This pattern has also been observed in captive wild pigs (Blasetti et al. 1988). Conversely, during colder winter days, wild pigs increase activity during the afternoon (Kurz and Marchinton 1972, Van Vuren 1984). Seasonally, wild pigs also alter their activity in response to temperature (Van Vuren 1984, Blasetti et al. 1988, Campbell and Long 2010). For example, in North America wild pigs reduced daily movements in relation to temperature extremes while monthly and overall movement decreased in conjunction with rising mean temperatures (Kay et al. 2017).

Compared to other aspects of wild pig spatial ecology, research on the movement ecology of this species is limited (Morelle et al. 2014). Wild pig movement has predominantly been described as general wandering over short distances, particularly in landscapes with suitable food, cover, and minimal human disturbance. However, wild pigs are also capable of large-scale movements where individuals traverse long distances over short periods of time. For example, genetic studies in Australia indicated that wild pigs moved >100 km over several days to breed (Lapidge et al. 2004). Similarly, experimentally translocated sows in South Carolina made extensive movements post-release in an apparent attempt to locate their previous home range, with average daily movement rates and home range sizes that were significantly larger than those of non-translocated pigs (J. C. Beasley, unpublished manuscript). In this scenario, the magnitude of difference between groups was moderated by the quality of the surrounding resources they were introduced to, with wild pigs introduced into predominantly pine habitats exhibiting more extensive movements than those introduced into bottomland hardwood habitats (J. C. Beasley, unpublished manuscript). These extensive movements were maintained for ~20–30 days prior to establishment of stable home ranges, at which point most individuals appeared to have assimilated within new sounders (J. C. Beasley, unpublished manuscript). The potential for large-scale movements in wild pigs has also been observed in populations that migrate. For example, some individual wild boar were recorded moving distances >250 km during the migration period in Poland (Andrzejewski and Jezierski 1978).

Movement metrics such as maximum distance or average hourly distance moved in a day or longer can also be informative for understanding wild pig spatial ecology and guiding management. In general, wild pigs are fairly sedentary in their movements during a given day (average = 0.35 km, maximum = 0.8 km) or month (average = 0.42 km, maximum = 2.1 km; Kay et al. 2017). Average hourly movement and maximum movement distances are affected by weather, season, geographic region, sex/age, and distance to water sources (Kay et al. 2017). In general, wild pigs in drier ecoregions (e.g., Southern Texas Plains and Ozark Highlands) have above average movement rates across all scales, while wild pigs in mesic ecoregions (e.g., Mississippi Alluvial Plain and Southern Coastal Plain) have below-average movement rates. Interestingly, the shape of the relationship between some of these factors and movement changes depending on the temporal scale of the analysis (i.e., daily, monthly, or overall time). For example, the maximum distance moved by males can be up to 1.3 km farther than females at the monthly scale, whereas at the daily scale sub-adult males show similar maximum daily movement rates to sub-adult females. Similarly at the monthly scale, maximum daily movements increase dramatically for individuals whose home range centroids were farther from a water source, but this relationship was not apparent at the daily scale. The most striking difference based on temporal scale of the data is the relationship with meteorological variables. At the daily scale, the relationship between meteorological variables and movement tends to be concave, with movement being greatest at intermediate values of temperature and pressure, whereas at the monthly scale, the relationship was linear, with movement rates decreasing as temperature and pressure increased (Kay et al. 2017). This indicates that meteorological patterns could be used as indicators for the likelihood or frequency that wild pigs may visit bait piles placed near their home range. Specifically, they are less likely to visit if temperature and pressure are outside their optimal movement ranges. These observations came from the southern United States and we caution that movement relationships from more northern areas may differ, presenting an opportunity for future research.

The configuration of wild pig movements is largely driven by the surrounding landscape. Wild pigs exhibit a relatively high degree of site fidelity (e.g., bedding or resting sites; Oliveira-Santos et al. 2016). When undisturbed, wild pigs are methodical in their movements, regularly using the same trails (Hanson and Karstad 1959, Frädrich 1974), even those that were forged by conspecifics or other species (Graves 1984). Although wild pigs can move freely in a variety of environments, some studies indicate that certain manmade and natural features influence movement. For example, in agricultural areas wild boar prefer to move along vegetation edges and will utilize narrow landscape features (Thurfjell et al. 2009), presumably for protective cover. In south Texas, paved roads with prolonged traffic deterred movements of wild pigs (Wyckoff et al. 2012). Similarly in Australia, rivers, roads, and railways were a barrier to wild pig movements (Saunders and Kay 1991).

Some studies on wild pig movements have focused on responses to management and control activities. The responses of wild pigs to control activities are variable and ultimately depend on control method and repeated exposure. In general, wild boar reduce their activity after being subjected to hunting pressure (Sodeikat and Pohlmeyer 2007, van Doormaal et al. 2015). However, this can change with repeated hunting events (Fischer et al. 2016). If cover is dense, wild pigs may remain stationary when encountered by hunters. But individuals will flush from their hiding location, often in a random direction, once they feel imminently threatened (Marini et al. 2008). Also, flight distances tend to be small (<300 m), with many recorded distances being <25 m (Marini et al. 2008). After fleeing, individuals selected vegetation that offered abundant cover and forage; they also reduced their overall movement (Thurfjell et al. 2013). During drive hunts wild boar tended to flee, while individuals often hid and remained stationary in response to still hunting (Thurfjell et al. 2013). Furthermore, in response to drive hunts, wild boar used larger resting ranges, were reluctant to return to the same resting sites, and individuals subjected to the greatest amount of hunting pressure displayed the most erratic movements (Scillitani et al. 2009). The influence of other hunting techniques, such as aerial gunning, has elicited various responses by wild pigs. During aerial gunning in Texas, movement of wild pigs increased and ≥50% of individuals displayed movements outside of their documented home ranges but returned to their initial home range soon after aerial gunning ceased (Campbell et al. 2010). Other aerial removal events also reported minimal impact on movements of surviving wild pigs (Saunders and Bryant 1988, Dexter 1996), but individual responses can vary.

3.3.4 Natal Dispersal

Natal dispersal (hereafter dispersal) refers to the movement of individuals between their natal range and the location of first breeding (Howard 1960). Individually, dispersal has important implications for reproduction and survival (Réale et al. 2000, Cote et al. 2009), while contributing to gene flow among populations (Bowler and Benton 2005, Clobert et al. 2009). In wild boar, dispersal occurs most often in males but has also been observed in females (Keuling et al. 2009a, Prévot and Licoppe 2013). As males mature, they often become solitary and disperse from social groups in search of mating opportunities. Dispersal in females is less common, but is likely triggered by the same process. In general, dispersal distances are primarily greater in males than in females (males = 3.80–4.90 km, females = 1.60–2.49 km; Keuling et al. 2009a, Prévot and Licoppe 2013). It is worth noting that information pertaining to dispersal in wild pigs in North America is limited and merits additional study.

The age when most dispersal events occur varies by sex and location. In general, the maximum dispersal rate coincides with the age at which individuals reach sexual maturity (Truvé and Lemel 2003). Variation in the timing of dispersal has been noted in North America. In the southern United States, dispersal was observed in individuals between 5 and 10 months of age (Crouch 1983, Hughes 1985). In Texas dispersal appears to occur later, where researchers reported a peak in male dispersal at 16 months of age, when individuals reached 40–50 kg (Gabor et al. 1999). In Sweden, dispersal in males began at ~10 months of age, peaked at 13 months, and declined at 16 months (Truvé and Lemel 2003). For females, dispersal occurred when individuals were between 7 and 11 months of age (Truvé and Lemel 2003). When dispersal occurs, female siblings may disperse together, thus lowering individual fitness costs (Kaminski et al. 2005). This phenomenon has also been observed in males, where dispersing individuals will form bachelor groups (Gabor et al. 1999). In some instances, younger males and females have been observed in a single group, and are believed to be siblings that have dispersed together (S. M. Gray, personal observation; J. C. Beasley, personal observation). Kaminski et al. (2005) noted that decisions by females to remain or leave the natal group was primarily determined by apparent fitness costs and benefits. For example, individuals born to younger mothers can be nutritionally deficient compared to young from more experienced mothers, making it more beneficial to remain in the group longer (Kaminski et al. 2005). Conversely, the ratio of mothers to daughters within a sounder may influence the likelihood of dispersal. In social groups where female yearlings outnumbered the adults, the probability of dispersal was high (Kaminski et al. 2005).

Other mechanisms also influence dispersal in wild pigs. In terms of density, research in Sweden determined that the density of conspecifics was negatively correlated with wild boar dispersal (Truvé et al. 2004). In southwest Australia, researchers reported that heavier animals often dispersed farthest to access breeding opportunities and that dispersal events (3–11 km) occurring over short periods of time were also observed in sows (Hampton et al. 2004). Furthermore, landscape features can affect the spatial configuration of dispersal in wild pigs. For instance, in France, animals dispersed towards natural areas rather than agriculture, often following the edges of open water (Dardaillon and Beugnon 1987). The dispersal process for wild pigs differs from some other large ungulates. The presence of rivers, roads, and railways was less influential on dispersal events in wild boar than in red deer (*Cervus elaphus*) in Belgium (Prévot and Licoppe 2013).

3.4 APPLICATIONS AND MANAGEMENT

Wild pigs are often described as intelligent animals, capable of exploiting a variety of habitats and resources. Given their expansive range, both in North America and abroad, there is need to understand the intricacies of the spatial ecology and behavior of wild pigs in these locations, as they are often quite different. To date, many methods have been developed to better manage this species using information gathered from behavioral and spatial ecology research. One of the most notable advances is the use of GPS telemetry devices to obtain detailed spatiotemporal information on wild pigs. A technique that explicitly uses this information is called the *Judas pig* approach. This technique exploits the social tendencies of wild pigs, and uses GPS/telemetry information from a tagged animal to locate other individuals in the landscape. This has been an effective method during removal efforts conducted in Australia (McIlroy and Gifford 1997) as well as in the United States, most notably California (Wilcox et al. 2004, Macdonald and Walker 2008). However, the Judas pig method is most effective when used in concert with other control methods or when wild pig populations persist at low levels. Thus, the Judas pig technique is especially useful in newly colonized areas that often have populations at low-densities (McIlroy and Gifford 1997), or when surviving individuals flee into previously unoccupied areas making them difficult to relocate. Similarly, despite the role of illegal translocations in the rapid spread of wild pig populations across North America (see Chapter 2), there is limited information available on the movement behavior of translocated pigs. Such data are essential to the development of management strategies in the event of a known release. For example, within a few weeks translocated sows often moved several kilometers from the release site in South Carolina, although the distance moved decreased with increasing habitat quality at the release site (J. C. Beasley, unpublished manuscript), suggesting localized management around suspected release locations is likely to be ineffective. Similarly, detailed spatial and temporal information can provide further insight into population connectivity and potential source-sink dynamics in response to management activities. Understanding how individuals move among populations and refill gaps in the landscape could be crucial in informing cost-effective and efficient control strategies at the landscape scale.

Another important application in understanding spatial behavior is for optimizing deployment of management tools such as traps or toxicants. Knowledge of factors driving the spatial ecology of pigs can inform how, when, and where to deploy traps or pharmaceutical baits in a manner that pigs are most likely to visit the attractant quickly and frequently. Similarly, for deployment of any toxicant, an understanding of group dynamics (i.e., dominance, interactions) during feeding is critical. Lastly, it is important to understand contact behavior among wild pigs to assess disease risk, and to test how management could affect disease risk through behavioral and spatial changes in how pigs come in contact with others in the area (see Pepin et al. 2016). Because wild pigs can be vectors of several potentially harmful zoonoses, there is a need to know which control methods can be carried out to limit this potential for disease spread (Pepin and VerCauteren 2016). Currently, dispersal of wild pigs in North America is poorly understood and remains a topic in need of further inquiry. Dispersal has important implications for managers as this can lead to disease spread and can further hinder removal efforts. In North America, wild pigs are believed to regularly disperse, especially juveniles/sub-adults when a new litter is born. However, much is left to be learned as to the ultimate function and consequences of this behavior from an ecological and management standpoint. Similarly, social structure and territoriality of wild pigs have important implications for managers, yet few studies have investigated these topics in North America. Additional research exploring group composition and social interactions within sounders as well as territorial interactions among sounders may inform and improve management efforts.

3.5 RESEARCH NEEDS

As pointed out throughout this chapter, there are several areas where research is lacking on the spatial ecology and behavior of wild pigs in North America. First, in terms of behavior, there remain many opportunities to explore wild pig communication and the function of several glands used in scent-marking. Research to date is limited and a holistic understanding of the strategies used by wild pigs may aid in the development of an effective attractant to be used in management and control. Second, the majority of home range studies on wild pigs have estimated home ranges using MCP methods. Only recently have studies begun employing more informative estimators and appropriately accounted for the temporal autocorrelation that is inherent in GPS data, especially when collected frequently (i.e., hourly). The use of statistically rigorous estimators could provide less biased and more detailed estimates of home range. For example, AKDE has the advantage of incorporating an underlying movement process that predicts space mechanistically from the observed locations, thus learning from patterns in the data to estimate the overall home range (Fleming et al. 2015). However, care must be taken when applying these methods because adjustments must be made when natal dispersal or long-distance translocations are included in the dataset (i.e., movement that is non-stationary). A study by Kay et al. (2017) determined that MCP methods are an accurate metric for calculating home range size in most circumstances due to the limited movements exhibited by wild pigs. However, process-based models will be important for understanding other characteristics of home range and movement (e.g., behavior; see Hanks et al. 2015). Third, research on the movement of wild pigs in North America is limited. In fact, the movement ecology of this species is considerably understudied in comparison to other ungulates (Morelle et al. 2014). Most research on wild pig movement has occurred in the native and international invasive ranges, making the application of inferences to populations in North America limited. Recent advances in state-space (Patterson et al. 2008) and other behaviorally based models have increased the potential and applicability for movement studies on wild pigs. Fourth, most research on wild pigs has occurred in areas where animals are fairly common and often occur at high densities (e.g., Texas, Georgia, South Carolina). Continued range expansion and translocation of wild pigs into areas of northern United States and Canada merit additional research. These areas feature newly established populations and are currently at low densities in areas that are ecologically distinct based on climate and vegetation in comparison to the southern United States. These smaller populations are important for providing insight into how the spatial ecology and behavior of wild pigs differ during invasion as well as in new climates and habitat conditions. It is also important to understand how these populations respond to management activities as populations are removed and groups are potentially fractured. Additionally, spatial research on wild pigs in urban and residential areas is nonexistent in North America (see Chapter 19). Interactions between humans and wild pigs will likely increase moving forward; therefore, understanding how animals use and move within human-dominated landscapes can inform control and management efforts in these areas.

ACKNOWLEDGMENTS

We thank B.K. Strickland and S.S. Ditchkoff for thoughtful reviews, edits, and suggestions. Contributions from S.M.G., G.J.R., and R.A.M. were supported by Michigan State University and the Michigan Department of Natural Resources. Contributions from J.C.B. were partially funded by the US Department of Energy under award # DE-EM0004391 to the University of Georgia Research Foundation. Contributions from K.M.P. were supported by the USDA. Mention of commercial products or companies does not reflect an endorsement by the US Government.

REFERENCES

Ackerman, B., M. Harmon, and F. Singer. 1978. Part II. Seasonal food habits of European wild boar. Pages 94–137 *in* F. Singer, editor. *Studies of European Wild Boar in the Great Smoky Mountains National Park: 1st Annual Report; A Report for the Superintendent.* Uplands Field Research Laboratory, Great Smoky Mountains National Park, Gatlinburg, TN.

Adkins, R. N., and L. A. Harveson. 2007. Demographic and spatial characteristics of feral hogs in the Chihuahuan Desert, Texas. *Human-Wildlife Conflicts* 1:152–160.

Adkins, R. N., L. A. Harveson, and C. A. Jones. 2006. Summer diets of feral hogs in the Davis Mountains, Texas. *The Southwestern Naturalist* 51:578–580.

Andrzejewski, R., and W. Jezierski. 1978. Management of a wild boar population and its effects on commercial land. *Acta Theriologica* 23:309–339.

Baber, D. W., and B. E. Coblentz. 1986. Density, home range, habitat use, and reproduction in feral pigs on Santa Catalina Island. *Journal of Mammalogy* 67:512–525.

Baber, D. W., and B. E. Coblentz. 1987. Diet, nutrition, and conception in feral pigs on Santa Catalina Island. *The Journal of Wildlife Management* 51:306–317.

Bacchetta, R., P. Mantecca, L. Lattuada, F. Quaglia, G. Vailati, and M. Apollonio. 2007. The carpal gland in wild swine: Functional evaluations. *Italian Journal of Zoology* 74:7–12.

Ballari, S. A., and M. N. Barrios-García. 2014. A review of wild boar *Sus scrofa* diet and factors affecting food selection in native and introduced ranges. *Mammal Review* 44:124–134.

Baron, J. 1982. Effects of feral hogs (*Sus scrofa*) on the vegetation of Horn Island, Mississippi. *American Midland Naturalist* 107:202–205.

Baron, J. S. 1979. Vegetation damage by feral hogs on Horn Island, Gulf Islands National Seashore, Mississippi. Thesis, University of Wisconsin, Madison, WI.

Barrett, R. 1978. The feral hog at Dye Creek Ranch, California. *California Agriculture* 46:283–355.

Barrett, R. 1982. Habitat preferences of feral hogs, deer, and cattle on a Sierra foothill range. *Journal of Range Management* 35:342–346.

Baskin, L., and K. Danell. 2003. *Ecology of Ungulates: A Handbook of Species in Eastern Europe and Northern and Central Asia*. Springer-Verlag, Berlin, Heidelberg, Germany.

Baubet, E., C. Bonenfant, and S. Brandt. 2004. Diet of the wild boar in the French Alps. *Galemys* 16:99–111.

Beasley, J. C., T. E. Grazia, P. E. Johns, and J. J. Mayer. 2014. Habitats associated with vehicle collisions with wild pigs. *Wildlife Research* 40:654–660.

Belden, R., and M. Pelton. 1976. Wallows of the European wild hog in the mountains of east Tennessee. *Journal of Tenessee Academy of Science* 51:91–93.

Bieber, C., and T. Ruf. 2005. Population dynamics in wild boar *Sus scrofa*: Ecology, elasticity of growth rate and implications for the management of pulsed resource consumers. *Journal of Applied Ecology* 42:1203–1213.

Blasetti, A., L. Boitani, M. Riviello, and E. Visalberghi. 1988. Activity budgets and use of enclosed space by wild boars (*Sus scrofa*) in captivity. *Zoo Biology* 7:69–79.

Block, W., and L. Brennan. 1993. The habitat concept in ornithology: Theory and applications. *Current Ornithology* 11:35–91.

Boitani, L., L. Mattei, P. Morini, and B. Zagarese. 1992. Experimental release of captivity reared wild boar (*Sus scrofa*). Pages 413–417 *in* F. Spitz, G. Janeau, G. Gonzalez, and S. Aulagnier, editors. *Ongulés/Ungulates 91*. Institute Recherche Grand Mammiferes, Paris-Toulouse, France.

Boitani, L., L. Mattei, D. Nonis, and F. Corsi. 1994. Spatial and activity patterns of wild boars in Tuscany, Italy. *Journal of Mammalogy* 75:600–612.

Bowler, D. E., and T. G. Benton. 2005. Causes and consequences of animal dispersal strategies: Relating individual behaviour to spatial dynamics. *Biological Reviews* 80:205–225.

Bratton, S. P., M. E. Harmon, and P. S. White. 1982. Patterns of European wild boar rooting in the western Great Smoky Mountains. *Castanea* 47:230–242.

Braza, F., and F. Alvarez. 1989. Utilisation de l'habitat et organisation sociale du sanglier (*Sus scrofa* L.) à Doñana (Sud-Ouest de l'Espagne). *Canadian Journal of Zoology* 67:2047–2051.

Cahill, S., F. Llimona, L. Cabañeros, and F. Calomardo. 2012. Characteristics of wild boar (*Sus scrofa*) habituation to urban areas in the Collserola Natural Park (Barcelona) and comparison with other locations. *Animal Biodiversity and Conservation* 35:221–233.

Calenge, C., D. Maillard, J. Vassant, and S. Brandt. 2002. Summer and hunting season home ranges of wild boar (*Sus scrofa*) in two habitats in France. *Game and Wildlife Science* 19:281–301.

Caley, P. 1993. Population dynamics of feral pigs (*Sus scrofa*) in a tropical riverine habitat complex. *Wildlife Research* 20:625–636.

Caley, P. 1997. Movements, activity patterns and habitat use of feral pigs (*Sus scrofa*) in a tropical habitat. *Wildlife Research* 24:77–87.

Campbell, T. A., and D. B. Long. 2009. Feral swine damage and damage management in forested ecosystems. *Forest Ecology and Management* 257:2319–2326.

Campbell, T. A., and D. B. Long. 2010. Activity patterns of wild boars (*Sus scrofa*) in southern Texas. *The Southwestern Naturalist* 55:564–567.

Campbell, T. A., D. B. Long, M. J. Lavelle, B. R. Leland, T. L. Blankenship, and K. C. VerCauteren. 2012. Impact of baiting on feral swine behavior in the presence of culling activities. *Preventive Veterinary Medicine* 104:249–257.

Campbell, T. A., D. B. Long, and B. R. Leland. 2010. Feral swine behavior relative to aerial gunning in southern Texas. *Journal of Wildlife Management* 74:337–341.

Challies, C. N. 1975. Feral pigs (*Sus scrofa*) on Auckland Island: Status, and effects on vegetation and nesting sea birds. *New Zealand Journal of Zoology* 2:479–490.

Chamrad, A. D., and T. W. Box. 1964. A point frame for sampling rumen contents. *Journal of Wildlife Management* 28:473–477.

Clobert, J., L. Galliard, J. Cote, S. Meylan, and M. Massot. 2009. Informed dispersal, heterogeneity in animal dispersal syndromes and the dynamics of spatially structured populations. *Ecology Letters* 12:197–209.

Coblentz, B. E., and D. W. Baber. 1987. Biology and control of feral pigs on Isla Santiago, Galapagos, Ecuador. *Journal of Applied Ecology* 24:403–418.

Comer, C. E., and J. J. Mayer. 2009. Wild pig reproductive biology. Pages 51–75 *in* J. J. Mayer and I. L. Brisbin, editors. *Wild Pigs: Biology, Damage, Control Techniques, and Management*. SRNL-RP-2009-00869, Savannah River National Laboratory, Aiken, SC.

Conley, R., V. Henry, and G. Matschke. 1972. *Final Report for the European Hog Research Project W-34*. Tennessee Game and Fish Commission, Nashville, TN.

Cooper, S. M., and S. S. Sieckenius. 2016. Habitat selection of wild pigs and northern bobwhites in shrub-dominated rangeland. *Southeastern Naturalist* 15:382–393.

Cote, J., A. Dreiss, and J. Clobert. 2009. Social personality trait and fitness. *Proceedings of the Royal Society of London B: Biological Sciences* 275:2851–2858.

Crouch, L. C. 1983. Movements of and habitat utilization by feral hogs at the Savannah River Plant, South Carolina. Dissertation, Clemson University, Clemson, SC.

Dardaillon, M. 1987. Seasonal feeding habits of the wild boar in a Mediterranean wetland, the Camargue (southern France). *Acta Theriologica* 32:389–401.

Dardaillon, M., and G. Beugnon. 1987. The influence of some environmental characteristics on the movements of wild boar *Sus scrofa*. *Biology of Behaviour* 12:82–92.

Dexter, N. 1996. The effect of an intensive shooting exercise from a helicopter on the behaviour of surviving feral pigs. *Wildlife Research* 23:435–441.

Dexter, N. 1998. The influence of pasture distribution and temperature on habitat selection by feral pigs in a semi-arid environment. *Wildlife Research* 25:547–559.

Diong, C. 1982. Population biology and management of the feral pig (*Sus scrofa* L.) in Kipahulu Valley, Maui. Dissertation, University of Hawaii, Honolulu, HI.

Ditchkoff, S. S., and J. J. Mayer. 2009. Wild pig food habits. Pages 105–143 *in* J. J. Mayer and I. L. Brisbin, editors. *Wild Pigs: Biology, Damage, Control Techniques, and Management*. SRNL-RP-2009-00869, Savannah River National Laboratory, Aiken, SC.

Eisenberg, J. F., and M. Lockhart. 1972. An ecological reconnaissance of Wilpattu National Park, Ceylon. *Smithsonian Contributions to Zoology* 101:1–118.

Elston, J. J., and D. G. Hewitt. 2010. Intake of mast by wildlife in Texas and the potential for competition with wild boars. *The Southwestern Naturalist* 55:57–66.

Everitt, J., and M. Alaniz. 1980. Fall and winter diets of feral pigs in south Texas. *Journal of Range Management* 33:126–129.

Farnesi, R., D. Vagnetti, B. Santarella, and S. Tei. 1999. Morphological and ultrastructural study of carpal organ in adult female wild swine. *Anatomia, Histologia, Embryologia* 28:31–38.

Fernández-Llario, P. 2004. Environmental correlates of nest site selection by wild boar *Sus scrofa*. *Acta Theriologica* 49:383–392.

Fernández-Llario, P. 2005. The sexual function of wallowing in male wild boar (*Sus scrofa*). *Journal of Ethology* 23:9–14.

Fernández-Llario, P., J. Carranza, and S. Hidalgo de Trucios. 1996. Social organization of the wild boar (*Sus scrofa*) in Doñana National Park. *Miscellania Zoologica* 19:9–18.

Fischer, J. W., D. McMurtry, C. R. Blass, W. D. Walter, J. Beringer, and K. C. VerCauteren. 2016. Effects of simulated removal activities on movements and space use of feral swine. *European Journal of Wildlife Research* 62:285–292.

Fleming, C. H., W. F. Fagan, T. Mueller, K. A. Olson, P. Leimgruber, and J. M. Calabrese. 2015. Rigorous home range estimation with movement data: A new autocorrelated kernel density estimator. *Ecology* 96:1182–1188.

Focardi, S., F. Morimando, S. Capriotti, A. Ahmed, and P. Genov. 2015. Cooperation improves the access of wild boars (*Sus scrofa*) to food sources. *Behavioural Processes* 121:80–86.

Fonseca, C. 2007. Winter habitat selection by wild boar *Sus scrofa* in southeastern Poland. *European Journal of Wildlife Research* 54:361–366.

Fournier-Chambrillon, C., D. Maillard, and P. Fournier. 1995. Diet of the wild boar (*Sus scrofa* L.) inhabiting the Montpellier garrigue. *Journal of Mountain Ecology* 3:174–179.

Frädrich, H. 1974. A comparison of behavior in the Suidae. Pages 133–143 *in* V. Geist and F. R. Walther, editors. *The Behavior of Ungulates and Its Relation to Management*. IUCN, Morges, Switzerland.
Friebel, B. A., and P. G. Jodice. 2009. Home range and habitat use of feral hogs in Congaree National Park, South Carolina. *Human-Wildlife Interactions* 3:49–63.
Gabor, T. M., E. C. Hellgren, R. A. Bussche, and N. J. Silvy. 1999. Demography, sociospatial behaviour and genetics of feral pigs (*Sus scrofa*) in a semi-arid environment. *Journal of Zoology* 247:311–322.
Gabor, T. M., E. C. Hellgren, and N. J. Silvy. 2001. Multi-scale habitat partitioning in sympatric suiforms. *The Journal of Wildlife Management* 65:99–110.
Gaston, W., J. B. Armstrong, W. Arjo, and H. L. Stribling. 2008. Home range and habitat use of feral hogs (*Sus scrofa*) on Lowndes County WMA, Alabama. National Conference on Feral Hogs, St. Louis, MO.
Gentle, M., J. Speed, and D. Marshall. 2015. Consumption of crops by feral pigs (*Sus scrofa*) in a fragmented agricultural landscape. *Australian Mammalogy* 37:194–200.
Giffin, J. 1970. *Feral Game Mammal Survey: Feral Pig Survey* (Hawaii), Project No. W-5-R–21, Job No. 46 (21). State of Hawaii, Division of Fish and Game, Honolulu, HI.
Giménez-Anaya, A., J. Herrero, C. Rosell, S. Couto, and A. García-Serrano. 2008. Food habits of wild boars (*Sus scrofa*) in a Mediterranean coastal wetland. *Wetlands* 28:197–203.
Graves, H. 1984. Behavior and ecology of wild and feral swine (*Sus scrofa*). *Journal of Animal Science* 58:482–492.
Groot Bruinderink, G. W. T. A., and E. Hazebroek. 1996. Wild boar (*Sus scrofa scrofa* L.) rooting and forest regeneration on podzolic soils in the Netherlands. *Forest Ecology and Management* 88:71–80.
Grover, A. M. 1983. The home range, habitat utilization, group behavior, and food habits of the feral hog (*Sus scrofa*) in northern California. Dissertation, California State University, Sacramento, CA.
Groves, C., and J. Giles. 1989. Suidae. Pages 1044–1049 *in* D. Walton and B. Richardson, editors. *Fauna of Australia*. Australian Government Publishing Service, Canberra, Australia.
Hampton, J., J. R. Pluske, and P. B. Spencer. 2004. A preliminary genetic study of the social biology of feral pigs in south-western Australia and the implications for management. *Wildlife Research* 31:375–381.
Hanks, E. M., M. B. Hooten, and M. W. Alldredge. 2015. Continuous-time discrete-space models for animal movement. *The Annals of Applied Statistics* 9:145–165.
Hanson, L. 2006. Demography of feral pig populations at Fort Benning, Georgia. Thesis, Auburn University, Auburn, AL.
Hanson, R., and L. Karstad. 1959. Feral swine in the southeastern United States. *Journal of Wildlife Management* 23:64–74.
Harvey, M. J., and R. W. Barbour. 1965. Home range of *Microtus ochrogaster* as determined by a modified minimum area method. *Journal of Mammalogy* 46:398–402.
Hayes, R., S. Riffell, R. Minnis, and B. Holder. 2009. Survival and habitat use of feral hogs in Mississippi. *Southeastern Naturalist* 8:411–426.
Hediger, H. 1950. *Wild Animals in Capitivity: An Outline of the Biology of Zoological Gardens* (G. Sircom, Trans. 1964 ed.). Dover Publishing, New York.
Hediger, H. 1955. *The Psychology and Behaviour of Animals in Zoos and Circuses* (G. Sircom, Trans. 1968 ed.). Dover Publishing, New York.
Heise-Pavlov, S., P. Heise-Pavlov, and A. Bradley. 2005. Carpal glands in feral pigs (*Sus domesticus*) in tropical lowland rainforest in north-east Queensland, Australia. *Journal of Zoology* 266:73–80.
Henry, V. G., and R. H. Conley. 1972. Fall foods of European wild hogs in the southern Appalachians. *Journal of Wildlife Management* 36:854–860.
Herrero, J., A. García-Serrano, S. Couto, V. M. Ortuño, and R. García-González. 2006. Diet of wild boar *Sus scrofa* L. and crop damage in an intensive agroecosystem. *European Journal of Wildlife Research* 52:245–250.
Howard, W. E. 1960. Innate and environmental dispersal of individual vertebrates. *American Midland Naturalist* 63:152–161.
Howe, T. D., F. J. Singer, and B. B. Ackerman. 1981. Forage relationships of European wild boar invading northern hardwood forest. *Journal of Wildlife Management* 45:748–754.
Hughes, T. W. 1985. Home range, habitat utilization, and pig survival of feral swine on the Savannah River Plant. Thesis, Clemson University, Clemson, SC.
Hutto, R. L. 1985. Habitat selection by nonbreeding, migratory land birds. Pages 455–476 *in* M. L. Cody, editor. *Habitat Selection in Birds*. Academic Press, New York.
Ilse, L. M., and E. C. Hellgren. 1995a. Spatial use and group dynamics of sympatric collared peccaries and feral hogs in southern Texas. *Journal of Mammalogy* 76:993–1002.
Ilse, L. M., and E. C. Hellgren. 1995b. Resource partitioning in sympatric populations of collared peccaries and feral hogs in southern Texas. *Journal of Mammalogy* 76:784–799.

Irizar, I., N. A. Laskurain, and J. Herrero. 2004. Wild boar frugivory in the Atlantic Basque Country. *Galemys* 16:125–134.

Johnson, D. H. 1980. The comparison of usage and availability measurements for evaluating resource preference. *Ecology* 61:65–71.

Jolley, D. B., S. S. Ditchkoff, B. D. Sparklin, L. B. Hanson, M. S. Mitchell, and J. B. Grand. 2010. Estimate of herpetofauna depredation by a population of wild pigs. *Journal of Mammalogy* 91:519–524.

Kaminski, G., S. Brandt, E. Baubet, and C. Baudoin. 2005. Life-history patterns in female wild boars (*Sus scrofa*): Mother–daughter postweaning associations. *Canadian Journal of Zoology* 83:474–480.

Kay, S. L., J. W. Fischer, A. J. Monaghan, J. C. Beasley, R. Boughton, T. A. Campbell, S. M. Cooper, et al. 2017. Quantifying drivers of wild pig movement across multiple spatial and temporal scales. *Movement Ecology* 5:14.

Kernohan, B. J., R. A. Gitzen, and J. J. Millspaugh. 2001. Analysis of animal space use and movements. Pages 125–166 *in* J. J. Millspaugh and J. M. Marzluff, editors. *Radio Tracking and Animal Populations*. Academic Press, San Diego, CA.

Keuling, O., N. Stier, and M. Roth. 2008a. Annual and seasonal space use of different age classes of female wild boar *Sus scrofa* L. *European Journal of Wildlife Research* 54:403–412.

Keuling, O., N. Stier, and M. Roth. 2008b. How does hunting influence activity and spatial usage in wild boar *Sus scrofa* L.? *European Journal of Wildlife Research* 54:729–737.

Keuling, O., K. Lauterbach, N. Stier, and M. Roth. 2009a. Hunter feedback of individually marked wild boar *Sus scrofa* L.: Dispersal and efficiency of hunting in northeastern Germany. *European Journal of Wildlife Research* 56:159–167.

Keuling, O., N. Stier, and M. Roth. 2009b. Commuting, shifting or remaining?: Differential spatial utilisation patterns of wild boar *Sus scrofa* L. in forest and field crops during summer. *Mammalian Biology—Zeitschrift für Säugetierkunde* 74:145–152.

Klaa, K. 1992. The diet of wild boar (*Sus scrofa* L.) in the National Park of Chrea (Algeria). *Ongulés/Ungulates* 91:403–407.

Kurz, J. C., and R. L. Marchinton. 1972. Radiotelemetry studies of feral hogs in South Carolina. *Journal of Wildlife Management* 36:1240–1248.

Łabudzki, L., G. Górecki, J. Skubis, and M. Wlazełko. 2009. Forest habitats use by wild boar in the Zielonka Game Investigation Centre. *Acta Scientiarum Polonorum-Silvarum Colendarum Ratio et Industria Lignaria* 8:51–57.

Lapidge, S. J., B. Cowled, and M. Smith. 2004. Ecology, genetics and socio-biology: Practical tools in the design of target-specific feral pig baits and baiting procedures. *Proceedings of the Vertebrate Pest Conference* 21:317–322.

Macdonald, N., and K. Walker. 2008. *A New Approach for Ungulate Eradication: A Case Study for Success*. Prohunt Inc., Ventura, CA.

Maillard, D., P. Fournier, and C. Fournier-Chambrillon. 1996. Influence of food availability and hunting on wild boar (*Sus scrofa* L.) home range size in Mediterranean habitat. *Proceedings of the Wild Boar Symposium* 24–27.3.1996 in Sopron, Hungary, 69–81.

Marini, F., B. Franzetti, A. Calabrese, S. Cappellini, and S. Focardi. 2008. Response to human presence during nocturnal line transect surveys in fallow deer (*Dama dama*) and wild boar (*Sus scrofa*). *European Journal of Wildlife Research* 55:107–115.

Massei, G., P. Genov, B. Staines, and M. Gorman. 1997. Factors influencing home range and activity of wild boar (*Sus scrofa*) in a Mediterranean coastal area. *Journal of Zoology* 242:411–423.

Mayer, J. J. 2009. Wild pig behavior. Pages 77–104 *in* J. J. Mayer and I. L. Brisbin, editors. *Wild Pigs: Biology, Damage, Control Techniques, and Management*. SRNL-RP-2009-00869, Savannah River National Laboratory, Aiken, SC.

Mayer, J. J., and I. L. Brisbin. 1986. A note on the scent-marking behavior of two captive-reared feral boars. *Applied Animal Behaviour Science* 16:85–90.

Mayer, J. J., F. D. Martin, and I. L. Brisbin. 2002. Characteristics of wild pig farrowing nests and beds in the upper Coastal Plain of South Carolina. *Applied Animal Behaviour Science* 78:1–17.

McIlroy, J. 1989. Aspects of the ecology of feral pigs (*Sus scrofa*) in the Murchison area, New Zealand. *New Zealand Journal of Ecology* 12:11–22.

McIlroy, J., and E. Gifford. 1997. The 'Judas' pig technique: A method that could enhance control programmes against feral pigs, *Sus scrofa*. *Wildlife Research* 24:483–491.

Mersinger, R. C., and N. J. Silvy. 2007. Range size, habitat use, and dial activity of feral hogs on reclaimed surface-mined lands in east Texas. *Human-Wildlife Conflicts* 1:161–167.

Meynhardt, H. 1982. *Schwarzwild-Report: Mein Leben unter Wildschweinen*. Verlag J. Neumann, Leipzig, East Germany.

Mitchell, J., and R. Mayer. 1997. Diggings by feral pigs within the Wet Tropics World Heritage Area of north Queensland. *Wildlife Research* 24:591–601.

Morelle, K., F. Lehaire, and P. Lejeune. 2014. Is wild boar heading towards movement ecology? A review of trends and gaps. *Wildlife Biology* 20:196–205.

Morelle, K., and P. Lejeune. 2014. Seasonal variations of wild boar *Sus scrofa* distribution in agricultural landscapes: A species distribution modelling approach. *European Journal of Wildlife Research* 61:45–56.

Morini, P., L. Boitani, L. Mattei, and B. Zagarese. 2014. Space use by pen-raised wild boars (*Sus scrofa*) released in Tuscany (Central Italy)-II: Home range. *Journal of Mountain Ecology* 3:112–116.

Oldfield, C. A., and J. P. Evans. 2016. Twelve years of repeated wild hog activity promotes population maintenance of an invasive clonal plant in a coastal dune ecosystem. *Ecology and Evolution* 6:2569–2578.

Oliveira-Santos, L. G., J. D. Forester, U. Piovezan, W. M. Tomas, and F. A. Fernandez. 2016. Incorporating animal spatial memory in step selection functions. *Journal of Animal Ecology* 85:516–524.

Parkes, J. P., T. A. Easdale, W. M. Williamson, and D. M. Forsyth. 2015. Causes and consequences of ground disturbance by feral pigs (*Sus scrofa*) in a lowland New Zealand conifer-angiosperm forest. *New Zealand Journal of Ecology* 39:34–42.

Patterson, T. A., L. Thomas, C. Wilcox, O. Ovaskainen, and J. Matthiopoulos. 2008. State–space models of individual animal movement. *Trends in Ecology & Evolution* 23:87–94.

Pepin, K. M., A. J. Davis, J. C. Beasley, R. Boughton, T. Campbell, S. M. Cooper, W. Gaston, S. Hartley, J. C. Kilgo, and S. M. Wisely. 2016. Contact heterogeneities in feral swine: Implications for disease management and future research. *Ecosphere* 7:e01230.

Pepin, K. M., and K. C. VerCauteren. 2016. Disease-emergence dynamics and control in a socially-structured wildlife species. *Scientific Reports* 6:25150.

Perry, G. C., R. L. S. Patterson, H. J. H. MacFie, and C. G. Stinson. 1980. Pig courtship behaviour: Pheromonal property of androstene steroids in male submaxillary secretion. *Animal Science* 31:191–199.

Piattoni, F., A. Amicucci, M. Iotti, F. Ori, V. Stocchi, and A. Zambonelli. 2014. Viability and morphology of *Tuber aestivum* spores after passage through the gut of *Sus scrofa*. *Fungal Ecology* 9:52–60.

Pine, D., and G. Gerdes. 1973. Wild pigs in Monterey County, California. *California Fish and Game* 59:126–137.

Podgórski, T., G. Baś, B. Jędrzejewska, L. Sönnichsen, S. Śnieżko, W. Jędrzejewski, and H. Okarma. 2013. Spatiotemporal behavioral plasticity of wild boar (*Sus scrofa*) under contrasting conditions of human pressure: Primeval forest and metropolitan area. *Journal of Mammalogy* 94:109–119.

Podgórski, T., S. de Jong, J. W. Bubnicki, D. P. J. Kuijper, M. Churski, and B. Jędrzejewska. 2016. Drivers of synchronized vigilance in wild boar groups. *Behavioral Ecology* 27:1097–1103.

Poteaux, C., E. Baubet, G. Kaminski, S. Brandt, F. S. Dobson, and C. Baudoin. 2009. Socio-genetic structure and mating system of a wild boar population. *Journal of Zoology* 278:116–125.

Prévot, C., and A. Licoppe. 2013. Comparing red deer (*Cervus elaphus* L.) and wild boar (*Sus scrofa* L.) dispersal patterns in southern Belgium. *European Journal of Wildlife Research* 59:795–803.

Quenette, P., and J. Desportes. 1992. Temporal and sequential structure of vigilance behavior of wild boars (*Sus scrofa*). *Journal of Mammalogy* 73:535–540.

Quenette, P., and J. Gerard. 1992. From individual to collective vigilance in wild boar (*Sus scrofa*). *Canadian Journal of Zoology* 70:1632–1635.

Réale, D., B. Y. Gallant, M. Leblanc, and M. Festa-Bianchet. 2000. Consistency of temperament in bighorn ewes and correlates with behaviour and life history. *Animal Behaviour* 60:589–597.

Sandom, C. J., J. Hughes, and D. W. Macdonald. 2013. Rewilding the Scottish Highlands: Do wild boar, *Sus scrofa*, use a suitable foraging strategy to be effective ecosystem engineers? *Restoration Ecology* 21:336–343.

Saunders, G., and H. Bryant. 1988. The evaluation of a feral pig eradication program during a simulated exotic disease outbreak. *Wildlife Research* 15:73–81.

Saunders, G., and B. Kay. 1991. Movements of feral pigs (*Sus scrofa*) at Sunny Corner, New South Wales. *Wildlife Research* 18:49–61.

Saunders, G., and S. McLeod. 1999. Predicting home range size from the body mass or population densities of feral pigs, *Sus scrofa* (Artiodactyla: Suidae). *Australian Journal of Ecology* 24:538–543.

Schley, L., and T. J. Roper. 2003. Diet of wild boar *Sus scrofa* in western Europe, with particular reference to consumption of agricultural crops. *Mammal Review* 33:43–56.

Schlichting, P. E., S. R. Fritts, J. J. Mayer, P. S. Gipson, and C. B. Dabbert. 2016. Determinants of variation in home range of wild pigs. *Wildlife Society Bulletin* 40:487–493.

Schlichting, P. E., C. L. Richardson, B. Chandler, P. S. Gipson, J. J. Mayer, and C. B. Dabbert. 2015. Wild pig (*Sus scrofa*) reproduction and diet in the Rolling Plains of Texas. *The Southwestern Naturalist* 60:321–326.

Scillitani, L., A. Monaco, and S. Toso. 2009. Do intensive drive hunts affect wild boar (*Sus scrofa*) spatial behaviour in Italy? Some evidences and management implications. *European Journal of Wildlife Research* 56:307–318.

Scott, C., and M. Pelton. 1975. Seasonal food habits of the European wild hog in the Great Smoky Mountains National Park. *Proceedings of the Southeastern Association of Game and Fish Commissioners* 29:585–593.

Senior, A. M., C. E. Grueber, G. Machovsky-Capuska, S. J. Simpson, and D. Raubenheimer. 2016. Macronutritional consequences of food generalism in an invasive mammal, the wild boar. *Mammalian Biology—Zeitschrift für Säugetierkunde* 81:523–526.

Singer, F. J., D. K. Otto, A. R. Tipton, and C. P. Hable. 1981. Home ranges, movements, and habitat use of European wild boar in Tennessee. *Journal of Wildlife Management* 45:343–353.

Sjarmidi, A., G. Valet, and F. Spitz. 1996. Autumn frugivory in wild boar (*Sus scrofa*). Pages 213–224 *in* M. Mathias, M. Santos-Reis, G. Amori, R. Libois, A. Mitchell-Jones, and M. Saint Girons, editors. European mammals. *Proceedings of the I European Congress of Mammalogy*. Lisboa, Portugal.

Sodeikat, G., and K. Pohlmeyer. 2007. Impact of drive hunts on daytime resting site areas of wild boar family groups (*Sus scrofa* L.). *Wildlife Biology in Practice* 3:28–38.

Sparklin, B. D., M. S. Mitchell, L. B. Hanson, D. B. Jolley, and S. S. Ditchkoff. 2009. Territoriality of feral pigs in a highly persecuted population on Fort Benning, Georgia. *Journal of Wildlife Management* 73:497–502.

Spitz, F. 1986. Current state of knowledge of wild boar biology. *Pig News and Information* 7:171–175.

Springer, M. D. 1975. Food habits of wild hogs on the Texas Gulf Coast. Thesis, Texas A&M University, College Station, TX.

Stegeman, L. C. 1938. The European wild boar in the Cherokee National Forest, Tennessee. *Journal of Mammalogy* 19:279–290.

Sterner, J. D. 1990. Population characteristics, home range, and habitat use of feral pigs on Santa Cruz Island, California. Thesis, University of California, Berkeley, CA.

Sweeney, J., J. Sweeney, and S. Sweeney. 2003. Feral Hog (*Sus scrofa*). Pages 1164–1179 *in* G. Feldhammer, B. Thompson, and J. Chapman, editors. *Wild Mammals of North America: Biology, Management, and Conservation*. The Johns Hopkins University Press, Baltimore, MD.

Taylor, R. B., and E. C. Hellgren. 1997. Diet of feral hogs in the western South Texas Plains. *The Southwestern Naturalist* 42:33–39.

Thurfjell, H., J. P. Ball, P.-A. Åhlén, P. Kornacher, H. Dettki, and K. Sjöberg. 2009. Habitat use and spatial patterns of wild boar *Sus scrofa* (L.): Agricultural fields and edges. *European Journal of Wildlife Research* 55:517–523.

Thurfjell, H., G. Spong, and G. Ericsson. 2013. Effects of hunting on wild boar *Sus scrofa* behaviour. *Wildlife Biology* 19:87–93.

Truvé, J., and J. Lemel. 2003. Timing and distance of natal dispersal for wild boar *Sus scrofa* in Sweden. *Wildlife Biology* 9:51–57.

Truvé, J., J. Lemel, and B. Söderberg. 2004. Dispersal in relation to population density in wild boar (*Sus scrofa*). *Galemys* 16:75–82.

Turner, K. L., E. F. Abernethy, L. M. Conner, O. E. Rhodes, and J. C. Beasley. 2017. Abiotic and biotic factors modulate carrion fate and vertebrate scavenging communities. *Ecology* 98:2413–2424.

van Doormaal, N., H. Ohashi, S. Koike, and K. Kaji. 2015. Influence of human activities on the activity patterns of Japanese sika deer (*Cervus nippon*) and wild boar (*Sus scrofa*) in central Japan. *European Journal of Wildlife Research* 61:517–527.

Van Vuren, D. 1984. Diurnal activity and habitat use by feral pigs on Santa Cruz Island, California. *California Fish and Game* 70:140–144.

Welander, J. 2000. Spatial and temporal dynamics of wild boar (*Sus scrofa*) rooting in a mosaic landscape. *Journal of Zoology* 252:263–271.

Wilcox, J., E. T. Aschehoug, C. A. Scott, and D. H. Van Vuren. 2004. A test of the Judas technique as a method for eradicating feral pigs. *Transactions of the Western Section of the Wildlife Society* 40:120–126.

Wilcox, J. T., and D. H. Van Vuren. 2009. Wild pigs as predators in oak woodlands of California. *Journal of Mammalogy* 90:114–118.

Wilson, C. J. 2004. Rooting damage to farmland in Dorset, southern England, caused by feral wild boar *Sus scrofa*. *Mammal Review* 34:331–335.

Wood, G. W., and R. E. Brenneman. 1980. Feral hog movements and habitat use in coastal South Carolina. *Journal of Wildlife Management* 44:420–427.

Wood, G. W., and D. N. Roark. 1980. Food habits of feral hogs in coastal South Carolina. *Journal of Wildlife Management* 44:506–511.

Wyckoff, A. C., S. E. Henke, T. A. Campbell, D. G. Hewitt, and K. C. VerCauteren. 2012. Movement and habitat use of feral swine near domestic swine facilities. *Wildlife Society Bulletin* 36:130–138.

4 Wild Pig Population Dynamics

Nathan P. Snow, Ryan S. Miller,
James C. Beasley, and Kim M. Pepin

CONTENTS

4.1 INTRODUCTION

Understanding population dynamics of wild pigs (*Sus scrofa*) is an exercise in determining how and why populations increase, decrease, and fluctuate throughout time and space. Many factors are responsible for differences and fluctuations in population sizes of wild pigs, including birth synchronization, climate, pulsing of resources, landscape structure, and disease, among others (Lozan 1995, Bieber and Ruf 2005, Acevedo et al. 2006). While these factors may cause drastic reductions in populations of wild pigs periodically, high reproductive potential offers resiliency to recover, sustain, and flourish under a wide variety of conditions (Mayer 2009). Because of their high reproductive success, wild pigs are proving to be one of the most challenging invasive species to manage throughout North America and the world (Lowe et al. 2000), and knowledge of population dynamics is critical for making effective management decisions.

The capacity of the landscape to support wild pigs throughout North America is unknown, but predicted to be quite high (Lewis et al. 2017) because of favorable climatic conditions coupled with heterogeneous landscapes containing mixtures of agriculture and cover. Favorable habitat conditions and ability of wild pigs to expand geographically (e.g., expanded 6.5–12.6 km northward per year during 1982–2012 in the United States; Snow et al. 2017a) demonstrates the importance of gaining a better understanding of wild pig population dynamics in North America.

In this chapter we present what is currently known about population dynamics of wild pigs in North America, including population demographics, survival and longevity, and reproduction. We then examine current techniques used to estimate population densities of wild pigs, and the different modeling approaches available for understanding population dynamics. Finally, we combine these concepts into a discussion of how understanding and estimating population dynamics can inform management of wild pigs, and what gaps of knowledge still exist for improving management in North America.

4.1.1 Population Dynamics in Brief

Studying the life cycle of wild pigs requires translation of information from individual animals to the population. Individual wild pigs experience varying rates of birth, growth, maturation, fertility, and mortality. Collectively these rates, often referred to as vital rates, describe the progression of individuals through a life cycle (Caswell 2014). The dynamics of populations, including wild pigs, are directly determined by these vital rates. Population dynamics can often be investigated using just 2 critical vital rates, survival and birth, both of which can fluctuate with age of animals, season, landscape, available resources, population genetics, and other factors. These vital rates determine population growth rate, how populations recover from perturbations such as harvest or disease, and are fundamental knowledge for successful management. Thus, quantifying these vital rates can provide important insight into population status and serve as the basis for predicting harvest rates necessary to achieve a particular management objective (Hone 1999).

The concept of *carrying capacity* provides a useful theory for discussing population dynamics of wild pigs. This theory suggests that vital rates for populations of wild animals trend toward capacity of the landscape to support those populations, and that densities fluctuate around this capacity over time. In reality, understanding the relationships between carrying capacity and vital rates for wild pigs is challenging because of perpetual variation (i.e., seasonally or annually) in food resources, climatic conditions, landscape alterations, and other factors that affect the population dynamics of wild pigs over the short-term (e.g., hunting, predators, new population introductions).

4.2 POPULATION DEMOGRAPHICS

Despite the importance of age-specific vital rates (e.g., fecundity and survival) to the understanding of population dynamics, empirical data are limited for wild pigs in their non-native ranges. Furthermore, much available information has been derived from trapping or hunting records, a source of data that may have inherent biases due to hunter preference or variability in susceptibility to harvest as a function of animal sex or age. As a result, elucidation of vital and demographic rates, and development of robust population models that account for the unique social structure and behavioral ecology of pigs is an active area of ongoing research (Keiter et al. 2017a).

4.2.1 Sex Ratio

Information on wild pig sex ratio is generally localized, and restricted to reports of fetal counts or data collected from culled populations. In both wild pigs and wild boar, fetal sex ratios typically do not differ statistically from parity. However, the majority of studies reported a slight male bias (Mayer 2009) and few studies have explicitly tested or controlled for factors potentially contributing to variability in fetal sex ratios (but see Servanty et al. 2007). Sex-allocation theory posits females should invest more energy in production of males under favorable environmental conditions, particularly for polygynous species where males have greater variance in individual fitness than females, because during good conditions males should be able to mate with more females (Trivers and

Willard 1973). Given that wild pigs are polygamous and opportunistic generalists, male-biased fetal sex ratios could be expected in pigs, especially in years following high mast production. However, the magnitude of energy allocation to produce males during favorable environmental conditions may be constrained in species like pigs that are polytocous (have multiple offspring per litter) and have high reproductive output (Byers and Moodie 1990).

For wild boar, neither female age, weight, nor mast production appear to influence fetal sex ratios (Servanty et al. 2007). However, fetal sex ratios can vary with litter size, with a male bias observed for average litter sizes (4–6 individuals) and a female-bias in large litters (e.g., 8 individuals; Servanty et al. 2007). Individuals weaned in larger litters may be of poorer condition due to increased competition for resources, suggesting a reproductive advantage for females in larger litters because females sexually mature more quickly than males (Ahmad et al. 1995, Mayer 2009). These reports provide evidence that wild boar exhibit fetal sex-ratio patterns that support the Trivers Willard hypothesis (Trivers and Willard 1973, Servanty et al. 2007), as well as the Byers and Moodie (1990) hypothesis that male-biased fetal sex ratios might only occur in wild pigs with small to moderate litter sizes.

It is currently unknown whether fetal sex ratios of wild pigs vary with litter size within their non-native ranges. Thus, prior studies failing to detect differences in fetal sex ratio in wild pigs may be confounded if differences in sex ratio exist as a function of litter size. However, males can reach puberty more quickly within their non-native range compared to wild boar, and litter sizes in wild pigs are generally larger than those of wild boar due to the influence of domestication (Mayer and Brisbin 2009). Therefore, deviations from Servanty et al. (2007) could be expected within non-native range and warrant further exploration.

Similar to fetal sex ratios, the proportion of post-natal male and female pigs in a population generally approximates a 1:1 ratio, although some deviations from this pattern have been reported (Mayer 2009). For example, both female-biased (e.g., Ahmad et al. 1995, Fernández-Llario and Mateos-Quesada 2005) and male-biased (Diong 1973, Boitani et al. 1995) populations have been reported throughout much of their native range. Similarly, both male- and female-biased populations have been observed in non-native range, including the United States (Giles 1980, Peine and Farmer 1990, Gabor et al. 1999, Schuyler et al. 2002, Lapidge et al. 2004). However, these observations should be interpreted with caution as most post-natal data on sex ratios have been based solely on harvest and trapping records, which can be biased towards 1 sex (Mayer 2009). In fact, there is widespread evidence that wild boar sex ratios vary among age classes within a population, as populations appear to become more female-dominated with age (Andrzejewski and Jezierski 1978, Dardaillon 1989, Garzon-Heydt 1992, Massolo and Della Stella 2006, Gamelon et al. 2014, Albrycht et al. 2016). It is currently unknown whether a similar pattern in favor of females occurs with age in wild pigs throughout non-native range.

4.2.2 Age Structure

Similar to our knowledge of sex ratios, much of the available data on age structure of pig populations comes from trapping and hunting records from non-native (e.g., Pine and Gerdes 1973, Barrett 1978, Henry and Conley 1978, Diong 1983, Clarke et al. 1992, Mayer 2009), and native ranges (e.g., Fernández-Llario and Mateos-Quesada 2005, Massolo and Della Stella 2006, Merta et al. 2014). Within North America, most of these studies were conducted >30 years ago. Since then, extensive range expansion and growth of wild pig populations has occurred (McClure et al. 2015, Snow et al. 2017a), and coupled with changes in environmental and societal conditions, updated information on population age structure of wild pigs in different geographies would be valuable.

Although variability in age composition among populations occurs, no distinct differences are apparent between non-native and native ranges of wild pigs. Rather, differences in data collection methods, harvest pressure, and environmental conditions likely determine population age structure. Typical for large mammals, the age structure of wild pigs and wild boar typically is biased towards younger age classes, with generally 40–70% of a population comprised of individuals ≤1 year of age (Mayer 2009).

Population age structure can be influenced by harvest pressure because hunters often target older and larger individuals (e.g., Ditchkoff et al. 2017). Intensive harvest pressure not only has the potential to directly shift age or sex composition of populations, but can elicit a compensatory reproductive response through the production of larger litters, increased breeding by both yearlings and adults, and altered timing of birth (Servanty et al. 2011, Gamelon et al. 2012). Thus, the proportion of young individuals within a population may be exacerbated in areas of intensive harvest pressure (Servanty et al. 2011). Instituting a bounty program in attempt to reduce a population of wild pigs proved counterproductive, in part, because of this compensatory response (Ditchkoff et al. 2017).

One substantial limitation of many studies assessing demography of wild pig populations has been collection of unbiased data. For example, records from hunter harvests may be biased toward wild pigs of older age. Thus, remote cameras have become a routine tool in wild pig management programs and could easily be adapted to more explicitly collect unbiased demographic data (Maselli et al. 2014). Assuming no differences in detectability between sexes or age classes, remote cameras represent a potential useful tool for further exploration of demographic parameters and social dynamics in wild pigs.

4.3 SURVIVAL AND LONGEVITY

Studies investigating survival of wild pigs in North America are relatively limited compared to other ungulate species, with only 5 studies published since 2000 (Adkins and Harveston 2007, Gaston 2008, Hanson et al. 2009, Hayes et al. 2009, Keiter et al. 2017b). Further, studies reporting survival are limited to populations within the historic non-native range (i.e., from introduction through the 1980s) in Alabama, California, Georgia, Mississippi, Tennessee, and Texas. From these limited reports, survival varies greatly within and across age classes with no consistent patterns, and does not differ from wild boar (Miller 2017; Figure 4.1).

Average annual survival for all ages of wild pigs is 0.56 (standard deviation = 0.26) and can be as great as 0.91 (Pine and Gerdes 1973). Studies for wild pigs in Australia indicated that mortality declines with age class, with animals <1 year of age having the greatest mortality rates (Masters 1979, Giles 1980, Saunders 1988). However, increased mortality in younger age classes is not consistently observed for wild pigs in North America, and there are no observed differences in age-specific survival probability in some populations (Gaston 2008, Miller 2017). Furthermore, no known-fate studies have successfully quantified survival of neonate wild pigs despite the importance of neonate survival to population models (Keiter et al. 2017b). Two studies reported greater survival for pigs <1 year of age (not including neonate piglets) when compared to adults (Barrett 1971, Gaston 2008). Studies investigating longevity indicate wild pigs commonly reach 8 years old in North America, with typical maximum lifespans estimated to be 10 years of age (Conley et al. 1972, Barrett 1978). Studies of wild boar and wild pigs in Australia indicated that individuals may reach 12–14 years of age (Jezierski 1977, Dzięciołowski et al. 1992).

Anthropogenic mortality from recreational hunting or population control is thought to be the greatest source of mortality among wild pigs and wild boar (Okarma et al. 1995, Sweeney et al. 2003). However, few studies have explicitly investigated cause-specific mortality with rigor. One study of wild boar in Sweden attributed 94% of mortality to hunting, 4% to vehicle collisions, and 2% to natural causes (Lemel 1999). However, studies distinguishing natural and anthropogenic mortality are not available for North American wild pig populations. There is some evidence that mortality varies among seasons (Hayes et al. 2009) with 1 study indicating that mortality may be greatest during post-farrowing periods (Henry and Conley 1978). However, to date no studies in North America have explicitly investigated seasonal differences in survival across multiple years with the intent of determining both trends and potential mechanisms (e.g., environmental conditions) important for survival. This paucity of survival studies differs from other large game species

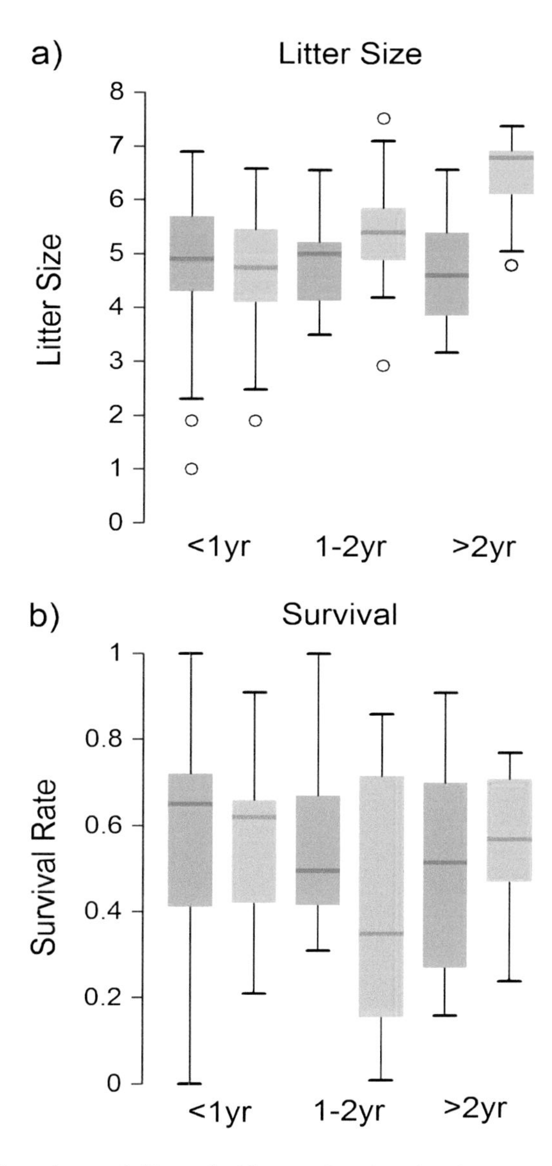

FIGURE 4.1 Fetal (a) litter sizes and (b) survival by age class for wild pigs in North America (yellow shading) and wild boar in native range (blue shading). Solid line indicates median, box is the upper and lower quartiles, and whiskers note the minimum and maximum litter size. (Adapted from Miller 2017.)

in North America and may occur due to wild pigs being invasive with the majority of research resources focused on damage and control rather than basic biology of the species.

4.4 REPRODUCTION

4.4.1 AGE-SPECIFIC BREEDING PATTERNS

Fecundity of wild pigs under favorable conditions can be extremely high (Briedermann 1990, Ahmad et al. 1995, Fernández-Llario and Carranza 2000, Fernández-Llario and Mateos-Quesada 2005). Females may start to reproduce within their first year. The onset of female reproduction primarily depends on body weight and available nutrition, such that females with optimal forage reproduce earlier (Briedermann 1990, Ahmad et al. 1995). Female wild boar do not typically reproduce until they have reached ≥30 kg (Fernández-Llario and Mateos-Quesada 1998, Fernández-Llario et al. 1999), although females of ~20 kg have occasionally bred (Bieber and Ruf 2005). In non-native range, age of sexual maturity is variable, documented to range from 3 to 12 months (Giles 1980, Mayer and Brisbin 2009). There are currently no studies in North America relating age of first breeding to weight of females; however, several studies have reported age. First breeding has been documented to occur at 5–8 months of age in the Great Smoky Mountains National Park (Johnson et al. 1982) and as early as 6 months in South Carolina (Wood and Brenneman 1977). On Santa Catalina Island in California, the majority (83%) of pigs did not breed until after 1 year of age, with a small proportion of females (4%) breeding at 5–7 months of age (Baber and Coblentz 1986). One study in California found that age of first breeding corresponded with availability of forage resources, with breeding occurring earlier under abundant resources (6–8 months) and later under scarce resources (8–10 months; Barrett 1978).

The number of animals breeding in each age class varies greatly among studies by both time of year and environmental conditions, but tends to increase with age. For females <1 year of age, this has been reported to be as high as 75% (Hanson et al. 2009). Among 5 studies of wild pigs in Georgia, Florida, and Texas, the percentage of animals under 1 year of age that were pregnant ranged from 31% to 75% (Belden and Frankenberger 1979, 1990; Taylor et al. 1998, Gabor et al. 1999, Ditchkoff et al. 2012). Percentage of adult females pregnant also varied greatly, ranging from 12% (Henry 1966) to 100% (Barrett 1971). Most studies found that pregnancy rates for adult females ranged from 92% (Henry 1966) to 50% (Singer and Ackerman 1981). An important limitation of these studies is that most did not sample pigs throughout all seasons or over multiple years, so reported variations in pregnancy rates are likely influenced by seasonal environmental conditions that vary among study sites and years.

4.4.2 Litter Size

Information on litter size for wild pigs in North America primarily comes from California, Florida, Georgia, South Carolina, North Carolina, Tennessee, and Texas (Table 4.1), with only 5 studies conducted since 2000. One study reported fetal litter sizes for a small number of pigs in Illinois (McCann 2003). Based on these 27 studies, mean live fetuses per sow was 5.3 (95% CI = 4.8–5.7). Individual fetus counts ranged from 1 to 12, with one account of 16 fetuses (Hanson and Karstad 1959).

Older wild pig sows (>1 year old) in North America tend to have larger litters compared to wild boar (Mayer and Brisbin 2009, Miller 2017, R. S. Miller, USDA/APHIS/Veterinary Services, unpublished data; Figure 4.1). Differences in litter sizes between wild boar and North American wild pigs could result from differences in ovulatory rate rather than differences in fetal survival or uterine capacity (Hagen and Kephart 1980, Hagen et al. 1980). This has been proposed as an artifact of hybridization with domestic pigs; however, the underlying biological mechanisms remain unstudied for wild pigs (Mayer and Brisbin 2009). Embryonic losses measured by differences between number of corpora lutea and fetuses range from 25% to 34% (Barrett 1978, Sweeney 1979, Baber and Coblentz 1986, Taylor et al. 1998). In addition to embryonic losses, stillborn and post-natal losses between 8% and 38% have been reported (Henry 1966, Baber and Coblentz 1986). Average fetal litter sizes in pigs less than 1 year of age tend to be smaller than adults (Figure 4.1); however, there is a wide range from 2.9 to 6.6 piglets (Sweeney et al. 1979, Singer and Ackerman 1981). Two studies observed that litter size in wild pigs declined beyond 3–5 years of age (Barrett 1978, Baber and Coblentz 1986). In contrast, increases in litter size with sow age and sow body size have been commonly reported for wild boar, although the degree of increase with age varies greatly among studies (Briedermann 1990, Fernández-Llario and Mateos-Quesada 1998). Seasonal and annual availability of forage has been proposed as a

TABLE 4.1
Mean Fetal Litter Size for Wild Pigs in North America by State and Location

State	Location[a]	Mean Fetal Litter Size	Citation
California	Dye Creek Ranch	5.6	Barrett (1971)
California	Dye Creek Ranch	6.1	Barrett (1978)
California	Monterey County	4.2	Pine and Gerdes (1969)
California	Monterey County	5.0	Pine and Gerdes (1973)
California	Santa Catalina Island	5.0	Baber and Coblentz (1986)
Florida	Fisheating Creek WR	6.5	Belden and Frankenberger (1990)
Florida	Levy County	5.5	Belden and Frankenberger (1979)
Florida	Merritt Island NWR	6.8	Strand (1980)
Georgia	Fort Benning	5.6	Hanson et al. (2009)
Georgia	Fort Benning	5.9	Ditchkoff et al. (2012)
Georgia	Ossabaw Island	4.9	Sweeney et al. (1979)
Georgia	Ossabaw Island	5.1	Graves (1984)
Illinois		4.5	McCann (2003)
South Carolina	Hobcaw Barony	4.7	Sweeney (1979)
South Carolina	Hobcaw Barony	5.3	Wood and Brenneman (1977)
South Carolina	Savannah River Site	5.7	Mayer and Brisbin (2009)
South Carolina	Savannah River Site	7.4	Sweeney et al. (1979)
South Carolina	Savannah River Site	8.4	Sweeney (1979)
Tennessee	Tellico WMA	4.7	Conley et al. (1972)
Tennessee	Tellico WMA	4.9	Henry (1966)
Tennessee/North Carolina	Great Smoky Mountains NP	3.0	Duncan (1974)
Tennessee/North Carolina	Great Smoky Mountains NP	3.0	Johnson et al. (1982)
Tennessee/North Carolina	Great Smoky Mountains NP	4.8	Singer and Ackerman (1981)
Tennessee/North Carolina	Great Smoky Mountains NP	4.8	Singer et al. (1978)
Texas	Aransas NWR	4.2	Springer (1977)
Texas	Central and southern Texas	5.4	Delgado-Acevedo et al. (2011)
Texas	Chaparral WMA	4.7	Gabor et al. (1999)
Texas	Gulf Coast prairies	5.2	Taylor et al. (1998)
Texas	South Texas plains	5.4	Taylor et al. (1998)

[a] WR = Wildlife Refuge, NWR = National Wildlife Refuge, WMA = Wildlife Management Area, NP = National Park.

mechanism contributing to variation in litter sizes with age and weight (Bieber and Ruf 2005, Mayer and Brisbin 2009, Miller 2017).

4.4.3 Temporal Patterns in Breeding

Wild pigs are polyestrous and physiologically capable of producing more than 1 litter in a year. Studies reporting the number of litters per year are limited in North America. Frequency of littering to date has not been the subject of explicit investigation although it is important for understanding population dynamics. Multiple litters per year have been reported in California (Baber and Coblentz 1986), Great Smoky Mountain National Park (Conley et al. 1972, Johnson et al. 1982), and Texas (Springer 1977, Taylor et al. 1998; Table 4.2). Across these studies, wild pig sows produced 1.4 (95% CI = 0.8–2.0) litters per year on average. Wild pigs in Tennessee did not produce multiple litters per year in 1 study (Henry 1966); however, another study in Tennessee reported it was common for wild sows to breed within a month of farrowing, but that conception was rare and produced small

TABLE 4.2
Litters per Year for Wild Pigs in North America by State and Age Class

State	Age	Litters Per Year	Citation
California	All ages	1	Pine and Gerdes (1973)
California	All ages	2.0	Barrett 1978)
California	All ages	0.86	Baber and Coblentz (1986)
Georgia	0–1yr	1	Hanson et al. (2009)
Georgia	>1yr	2.28	Hanson et al. (2009)
Georgia	>1yr	2.9	Hanson et al. (2009)
Texas	0–1yr	0.49	Taylor et al. (1998)
Texas	1–2yr	0.85	Taylor et al. (1998)
Texas	>3yr	1.57	Taylor et al. (1998)

litters (Conley et al. 1972). In California on Santa Catalina Island, adult sows produced an average of 0.86 litters per year (Baber and Coblentz 1986), while Barrett (1978) observed that wild sows at Dye Creek Ranch in California produced an average of 2 litters per year, but Pine and Gerdes (1973) did not observe multiple litters per year in Monterey, California.

Farrowing of wild pigs in North America occurs year round; however, some seasonal increases have qualitatively been reported. Timing of farrowing is likely linked to seasonal availability of forage (Mayer and Brisbin 2009). Most studies qualitatively report an increase in farrowing activity in spring (April-May; Pine and Gerdes 1969, 1973; Springer 1977; Singer et al. 1978; Johnson et al. 1982) or fall (September-November; Pine and Gerdes 1969, 1973; Barrett 1978; Baber and Coblentz 1986). Populations occurring in Florida (Belden and Frankenberger 1990) and Texas (Taylor et al. 1998) demonstrated peaks in farrowing during winter months. However, this is not exclusive to southern populations as peaks in farrowing in December and January have been reported in Tennessee (Henry 1966, Singer et al. 1978). Two pulses in farrowing in south-central Texas during 2017 were observed during January-February and again in June-July (N. P. Snow, USDA/APHIS/WS/National Wildlife Research Center, personal observation). Farrowing during the summer months appears to be least common with only a few studies reporting peaks in farrowing during June and July in California (Henry 1966, Barrett 1978, Singer et al. 1978, Baber and Coblentz 1986) and Texas (Taylor et al. 1998).

Studies in North America that observed multiple peaks in farrowing could result from mature females having multiple litters per year or lack of synchrony in breeding among individuals within a population. This pattern generally differs from wild boar that tend to have a single, synchronous peak of breeding that occurs in late winter through late spring (Figure 4.2). Data from North America (i.e., California, Florida, South Carolina, Texas) indicate no common temporal peaks in proportion of animals farrowing (Briedermann 1990, Fernández-Llario and Carranza 2000, Gethöffer 2005, R. S. Miller, unpublished data; Figure 4.2). Current studies are limited in several ways. Most studies do not report empirically the number of animals farrowing over time and only provide a qualitative accounting of temporal breeding activity. Studies that monitor individual animals over time to determine how often farrowing occurs for an individual are not available, making it impossible to determine if multiple litters in a year are common or the observed patterns simply reflect a lack of synchrony in breeding. Additionally, all available North American studies have investigated wild pig reproduction in more temperate regions and to date there are no studies describing reproduction available for populations occurring above 38° latitude or the arid regions of the southwestern deserts of the United States and Mexico.

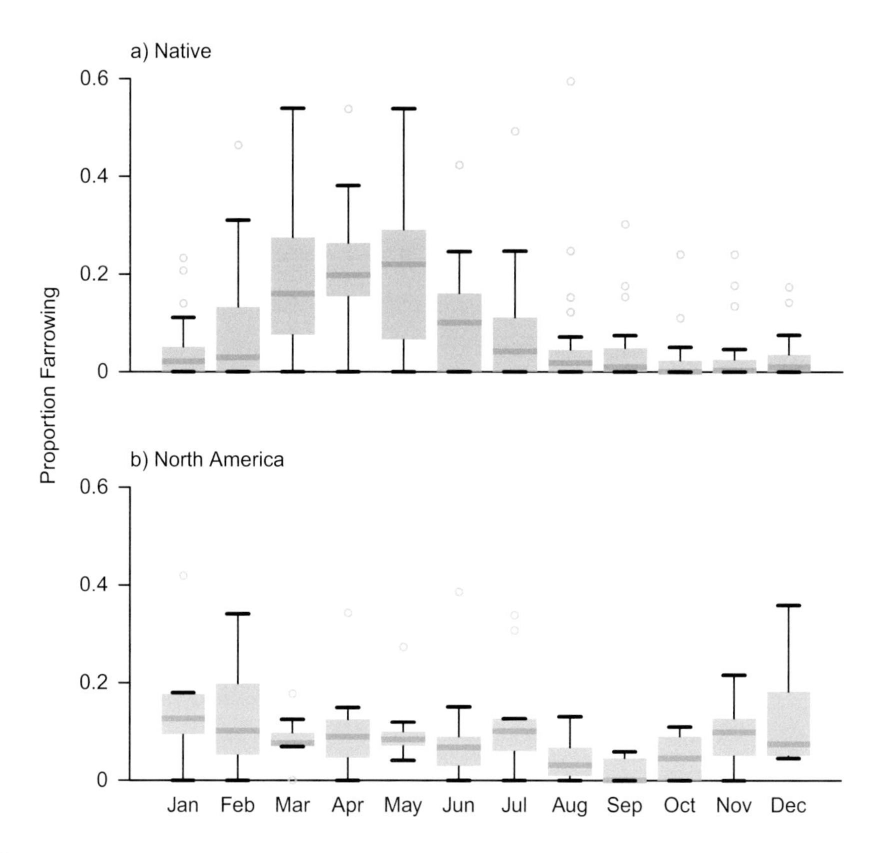

FIGURE 4.2 Proportion of wild sows farrowing on (a) native and (b) non-native (North American) ranges. Solid line indicates median, box is the upper and lower quartiles, and whiskers note the minimum and maximum litter size. (R.S. Miller, unpublished data.)

4.4.4 Environmental Influences on Reproduction

There is a well-established relationship between domestic and wild pig fecundity and environmental resources (i.e., food availability). Increases in nutrition prior to ovulation increase both ovulation rates and implantation in domestic pigs (Soede et al. 2011), and has been exploited as production practice in piggeries (Hazeleger et al. 2005). On native range in Europe, wild boar are consumers of pulsed resources with subsequent reproduction varying in response to availability of those resources. For instance, litter size and increased reproductive activity of wild boar was associated with availability of crops (Frauendorf et al. 2016) and hard mast (i.e., acorns and other nuts; Servanty et al. 2009, Vetter et al. 2015). Massei et al. (1996) found that both proportion of females breeding and litter size depended on annual production of olives and acorns in Mediterranean wild boar. Both cultivated land and hard mast forests have been associated with increased growth rates of European wild pig populations (Osada et al. 2015, Salinas et al. 2015, Vetter et al. 2015); however, crops and naturally available mast are expected to influence populations differently. Cultivated land provides a constant forage resource at the expense of reduced habitat (Morelle and Lejeune 2015), while hard mast forests provide a forage resource that can vary annually, but also provide security cover and thermoregulation (Canu et al. 2015).

Climatic conditions that increase forage resources have also been shown to affect productivity in wild pigs. This has been most clearly linked to precipitation with drought conditions associated with reduced post-natal litter sizes in wild boar (Fernández-Llario and Carranza 2000), and increased percentage of pregnant sows during years with increased rainfall (Fernández-Llario and Mateos-Quesada 2005).

Wild pigs in a variety of habitats in North America have shown reproductive responses to forage conditions similar to those observed for wild boar in native range. In Tennessee, Matschke (1964) attributed reproductive failure in wild pig populations to multiple years of oak mast failure. Examination of ovaries suggested that the primary effect was severely reduced ovulation rates. Greater ovulation rates and increased numbers of piglets per female have been documented during good mast years in Tennessee (Johnson et al. 1982) and North Carolina (Henry 1966). Similar effects on pregnancy rates were reported in 2 California populations in response to varying acorn production (Pine and Gerdes 1973, Baber and Coblentz 1986). Agricultural production influences ovulation rates and litter sizes in California and Georgia. In California, Barrett (1978) reported larger litters from females that used agricultural pastures compared to females that only consumed natural forage. These differences were suspected to result from increased conception and implantation rates for pigs with access to pastures. Similarly, increased ovulation rates were reported on a coastal island in Georgia where females utilized cultivated grape leaves as forage (Warren and Ford 1997).

4.4.5 Domestic Introgression

Wild pig populations in North America are often composed of hybrids between wild boar and domestic pigs (Mayer and Brisbin 2008; see Chapter 2). This hybridization may result in genotypes that produce larger litters, potentially facilitating invasion success. While currently unstudied in North America, domestic introgression (i.e., hybridization between domestic and wild pigs) has been observed to increase litter size in European wild boar (Fulgione et al. 2016). Recent studies found genetic introgression from domestic pigs into central and northern wild boar populations is more recent and more common than expected (Frantz et al. 2013, Goedbloed et al. 2013). Similarly, wild boar litter size has been observed to increase from an average of 3–4 piglets in the south (Fernández-Llario and Mateos-Quesada 1998, Bywater et al. 2010, Fonseca et al. 2011) to an average of 6 piglets in regions of central and northern Europe with domestic introgression (Gethöffer et al. 2007). Frantz et al. (2013) proposed that this observed increase in litter size may be contributing to dramatic increases in wild boar abundance over the last 20 years in northern Europe (Massei et al. 2015, Morelle et al. 2016).

While currently unreported for North America, domestic introgression likely contributes to increased litter size and may in part be responsible for observed increases in litter sizes for North American wild pigs relative to European wild pigs (Miller 2017). We estimate that average litter size of wild pigs in North America is 5.3, but individual studies have documented 8.4 (Table 4.1), compared to average litter size of 4.75 for wild boar (Bywater et al. 2010). Variation in litter size due to introgression can be influenced by type of domestic pig breed. For example, the American Yorkshire, the most common domestic pig breed in the United States, averages 13 piglets with 16–20 piglets common (Pond and Mersmann 2001). Studies directly investigating domestic introgression in North American wild pig populations are currently unavailable.

4.5 ESTIMATION OF POPULATION DENSITY

4.5.1 Role of Monitoring Densities

When eradication of wild pigs from an area is attempted, it is important to monitor changes in abundance as control efforts are implemented (Cruz et al. 2005, Morrison et al. 2007, McCann and Garcelon 2008, Davis et al. 2018). Just a few missed animals during an eradication attempt can result in a re-established population given high reproductive potential and population growth rates of wild pigs. Thus, population monitoring is critical to declare whether eradication has been achieved and to avoid removing control resources before the effort is completed (Cruz et al. 2005, 2009; Morrison et al. 2007). Similarly, for populations where the objective is reduced abundance (as opposed to eradication), monitoring wild pigs pre- and post-control can reveal the efficacy of control efforts (Davis et al. 2016). In addition, monitoring how wild pig populations change relative to different control techniques and effort helps improve effectiveness, recognizing that effectiveness tends to decline as abundance declines (Cruz et al. 2009).

4.5.2 Wild Pig Densities in North America

Most available information on wild pig densities in North America comes from discrete studies, providing limited perspective of how densities vary through space and time. To date, there have been no long-term studies of wild pig densities in North America, which are critical for understanding factors affecting reproductive success, estimating important demographic parameters and their relationships to environmental conditions, and validating predictive models of population dynamics. Nonetheless, because a substantial number of independent studies across a diverse range of habitats have been conducted, we have reasonably informed knowledge of maximum density wild pigs may occur at in North America (see pages 169–172 in Mayer and Brisbin 2009).

Not all estimates of density are directly comparable. For example, density estimates compiled in previous publications (Mayer and Brisbin 2009) tended to be substantially greater than estimates from more recent studies conducted in traditionally high-density areas (range from 0.3 to 5.1 pigs/km^2; Table 4.3). These discrepancies could partly be due to error in assessing area of the study site (Keiter et al. 2017a), which can likely be more accurately estimated in studies using GPS flight tracks from aerial gunning work, but less accurately for other techniques such as mark–recapture using traps (Davis et al. 2017). The error in estimating area can be particularly great if it is assumed to approximate the area of a trapping or camera grid if the grid does not allow for individuals to be detected at multiple traps (i.e., area could be largely overestimated in this case). In some management applications, trapping grids are not feasible and area searched must be estimated on a trap-by-trap or camera-by-camera basis to convert estimates of abundance to density (Davis et al. 2017). Few studies have attempted to quantify distance at which pigs are attracted to bait (e.g., Campbell et al. 2012, Davis et al. 2017), which likely varies regionally due to age, sex, behavior, weather, and landscape. Snow and VerCauteren (2019) identified that bait sites placed within 1 km of where females lived (1.25 km for males) had ≥50% probability of daily visitation by each sex.

Here we update a previous summary of reported densities (see pages 169–172 in Mayer and Brisbin 2009) to incorporate several recent studies using conventional estimation methods (Table 4.3). Reported densities on continental North America (e.g., up to 27 pigs/km^2) tend to be less than associated islands (e.g., up to ~50 pigs/km^2 on Hawaii), which are consistent with global estimates of densities of wild pigs on islands (Lewis et al. 2017). Variation in densities were associated with differing environmental conditions related to forage availability such as agriculture and precipitation during the wet and dry seasons and have been used to estimate potential wild pig densities in North America (Lewis et al. 2017; Figure 4.3).

4.6 MODELS OF POPULATION DYNAMICS

4.6.1 Role of Dynamic Population Models in Management

For decades ecologists have recognized the importance of considering how management affects wild pig population dynamics to help identify optimal control strategies (Hone 2012). Dynamic population models (i.e., models that consider changes in abundance over space and/or time) can be an efficient means of improving our knowledge for managing wild pig populations and damage because of their ability to examine a variety of ecological factors simultaneously (Hone 2012). These types of models have been used extensively to manage game species in North America and have applications to invasive species management. For wild pigs, these models have been used to:

1) Determine removal intensity needed to cause a desired reduction in wild pig abundance (Servanty et al. 2011, Gamelon et al. 2012, Hone 2012), determine effectiveness of aerial gunning as a function of population density (Hone 1990), and determine targeted population densities that reduce damage (Hone 2002, Krull et al. 2016).

TABLE 4.3
Recent (Since 2008) Estimates of Wild Pig Densities in the United States by Site, Year of the Study, and Method Used to Estimate Density

Site, state; year	Estimated (Adult Pigs/km²)	References	Methods
Savannah River Site, SC; 2014	0.9–1.3 (bottomland[c]) 0.9–1.7 (mixed habitat[d]) 0.7–1.2 (upland[e])	Keiter et al. (2017a)	[a]Lincoln-Peterson using biomarking
Savannah River Site, SC; 2014	1.05–1.35 (bottomland) 0.9–1.7 (mixed habitat) 0.8–1.3 (upland)	Keiter et al. (2017a)	[a]Lincoln-Peterson using camera data and unique natural marks
Savannah River Site, SC; 2014	1.0–3.5 (bottomland) 1.0–3.4 (mixed habitat) 1.0–3.4 (upland)	Keiter et al. (2017a)	[a]Camera-based Spatially Explicit Capture-Recapture method using unique natural marks
Savannah River Site, SC; 2014	1.1–3.9 (bottomland) 1.1–3.9 (mixed habitat) 1.1–3.9 (upland)	Keiter et al. (2017a)	[a]Trap-based Spatially Explicit Capture-Recapture method using unique marks
Savannah River Site, SC; 2014	1.2–1.9 (bottomland) 0.3–1.2 (mixed habitat) 0.4–1.7 (upland)	Keiter et al. (2017a)	[a]Removal model applied to trapping grid data
Savannah River Site, SC; 2014	1.2–2.5 (bottomland) 3.8–4.3 (mixed habitat) 1.6–2.1 (upland)	A.J. Piaggio[f], unpublished data	[a]Genetic Capture–Mark–Recapture using fecal sampling along transects
Baylor and Wilbarger Counties, TX; 2015	0.9 (Site 1) 1.3 (Site 2) 2.8 (Site 3)	Davis et al. (2017)	[b]Counts by aerial gunning and flight tracks to calculate area searched
Wichita Mountains National Wildlife Refuge, OK; 2015–2016	1.1–1.3	A.J. Davis[f], unpublished data	Removal model with aerial gunning data and flight tracks to calculate area searched
Colberson, Reeves, Loving, and Ward (Site 1); Burnet (Site 2), and Matagorde (Site 3) counties, TX; 2016	0.3–0.5 (Site 1) 1.9–3.0 (Site 2) 3.1–5.1 (Site 3)	A.J. Davis, unpublished data	Removal model with aerial gunning data and flight tracks to calculate area searched
Fort Benning Military Reservation, GA; 2004	3.7–10.0	Hanson et al. (2008)	Capture–Mark–Recapture for open populations using trapping for marking, cameras for recaptures and survival estimation

[a] Estimates of adult population; pre-dispersal aged individuals were excluded.
[b] Method for estimating abundance was not modern but method for calculating area searched was.
[c] Bottomland refers to swamp and riparian bottomland habitat with hardwoods.
[d] Mixed habitats are dominated by upland pines, but include some bottomlands.
[e] Upland refers to upland pines habitat.
[f] USDA/APHIS/WS/National Wildlife Research Center.

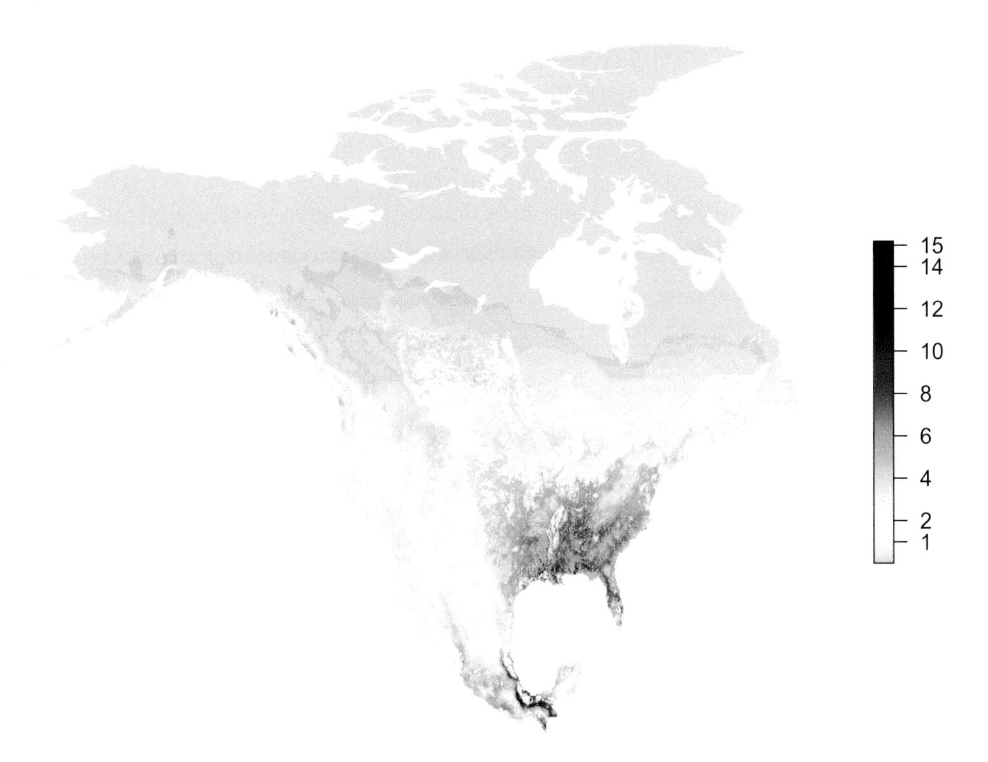

FIGURE 4.3 The predicted population density of wild pigs (individuals/km^2) in North America based on biotic and abiotic environmental factors that are important for wild pig density (empirically derived from 129 populations of wild pigs). Biotic factors included agriculture, vegetation cover, and large carnivore richness while abiotic factors were precipitation and potential evapotranspiration. (Adapted from Lewis et al. 2017.)

2) Determine changes in population growth rates due to environmental factors (Dexter and McLeod 2015, Salinas et al. 2015), and geographic spread (Snow et al. 2017b).
3) Identify optimal strategies for trap placement and baiting (Choquenot et al. 1993), and assess potential effects of incorporating nonlethal tools into control programs (Burton et al. 2013, Pepin et al. 2017a).
4) Predict how toxicants should be distributed on a landscape for effective population reduction (Hone 1992, Snow et al. 2017b, Lavelle et al. 2018, N. P. Snow, unpublished data).
5) Estimate demographic parameters while accounting for environmental variation and movement (Levy et al. 2016).
6) Assess which demographic processes are most important for managers to target (Mellish et al. 2014, Miller 2017, M. A. Tabak, USDA/APHIS/Veterinary Services, unpublished data), and effects of immigration on control efficiency (Pepin et al. 2017b).
7) Choose optimal spatial (McMahon et al. 2010, Pepin et al. 2017b) or temporal population control strategies (Pepin et al. 2017b).
8) Plan for strategies that meet particular management objectives within the cost constraints of particular situations (McMahon et al. 2010, Beeton et al. 2015).

There is substantial variation in vital rates for wild pigs across North America, owing to the diversity of environmental conditions that occur across such a large geographic extent (Mayer and Brisbin 2009). For this reason, dynamic population models are especially useful for interpreting variation in the effects of control on population densities of wild pigs, and understanding the uncertainty associated with these interpretations. By examining multiple sets of demographic and environmental conditions, population models also can provide confidence for implementing a particular strategy under particular conditions, and inform managers of risks associated with particular management decisions. However, models are only as good as their inputs, and thus the utility of results is limited by the ecological concepts underlying model structure and field data used to inform the model. Thus, previous studies of wild pig demographic processes are useful for predicting population dynamics, but are also fundamental for structuring modeling tools based on wild pig ecology. To maximize practical application of dynamic population models, it is essential that managers and researchers continue to work together to identify appropriate objectives and collect appropriate data.

4.6.2 Modeling Approaches

A variety of population modeling approaches have been applied to wild pig management problems, ranging in complexity from: deterministic population-level models (Dexter and McLeod 2015, Krull et al. 2016) and matrix projection models based on vital rates (Servanty et al. 2011, Gamelon et al. 2012, Davis et al. 2016, Miller 2017, M. A. Tabak, unpublished data), to spatial, individual-based models (Burton et al. 2013, Salinas et al. 2015, Pepin et al. 2017a, b). Deterministic population-level models typically have the advantage of being easily described by a set of equations and are not computationally intense. To add more realistic biology, population-level models can be implemented by explicitly modeling multiple subgroups (e.g., by age or sex) with separate vital rates, as part of the entire population. However, as population-level models express ecological processes at the population level, differences among animals are overlooked and detailed spatial ecology is difficult to capture.

Spatial, individual-based models of population dynamics allow researchers to incorporate the greatest level of detail, such as differences among animals in reproductive output, survivorship, movement rates, capture rates, and spatial location on the landscape. These models provide a powerful means of assessing management effects when ecological processes and management actions vary widely among individual animals. Individual-based models are most efficient when applied to small populations of animals because computational intensity increases dramatically in larger populations (>10,000). A current deficiency with all types of dynamic population models for wild pig management, especially in North America, is lack of user-friendly interfaces (e.g., McMahon et al. 2010). Future research on developing simple platforms for using population models, and education of managers on how to use them efficiently, will help to inform more cost-effective control of wild pigs throughout North America. Population models will also inform the probability of success from these control activities.

4.7 USING POPULATION MODELS TO INFORM MANAGEMENT

A common question asked by wild pig managers is "What proportion of the population do I need to remove to control or eradicate pigs in my area?". Several population modeling studies tailored to wild pigs have provided insights into this question. Based on published studies throughout the world, Mayer (2009) estimated that a 50–60% annual rate of removal for wild pigs is required to keep the population stable and not increasing. Hunting of wild pigs for sport and subsistence is only estimated to remove 10–30% of the population annually (Mayer 2009); therefore, other control and eradication programs are needed to reduce or eliminate wild pigs from an area. Other options such as helicopter shooting is highly effective at removing wild pigs when their populations are dense, but this approach is extremely ineffective when densities are low (Choquenot et al. 1999). Timing of removals is also important. For example, an intensive removal of 66% of a population during the first 3-months of a control effort greatly reduced subsequent births of wild pigs and resulted in more effective control than a slower-paced removal strategy (Anderson and Stone 1993).

Modeling efforts with data from Santa Cruz Island, California using an age-structured Leslie matrix model (Klinger et al. 2011) suggested that 45–65% of the population needed to be removed annually to cause population declines for a population at carrying capacity, whereas annual removal rates could be as low as 20% to cause decline once mortality was additive when the population was well below carrying capacity. In the same study, the authors determined that >70% of animals would need to be removed annually to have a high probability of eradication. In another study based on population density and control data from Hawaii, the authors regressed annual change in abundance against proportion of the population removed to derive removal rates that caused population growth rates <1 (Hess et al. 2006). The authors found that >41.3% of the population needed to be removed annually to cause a population decline. A similar result (i.e., 40% population removal) was obtained using an individual-based model of wild pig data from Tennessee (Salinas et al. 2015). These lower harvest requirements are likely more accurate than those from Santa Cruz Island because the individual-based model implemented a daily time scale for population dynamics and harvesting, thus accounting for integrated effects of harvesting on demographics throughout the year.

Other recent work using an individual-based model with demographic data from South Carolina showed that annual removal rates as low as 20% can reduce a population to half its carrying capacity within 5 years (Pepin et al. 2017b). This study indicated that removal of wild pigs should be spatially prioritized, by dividing the area into a set of management zones where each zone is sequentially prioritized for elimination, and temporally prioritized so that removal is additive rather than compensatory (Figure 4.4). This work found that annual removal rates between 20% and 40% can lead to population reductions of 40–90% within 5 years (Figure 4.4). These results corroborate results from a spatially explicit model based on an Australian wild pig population, which found that harvesting

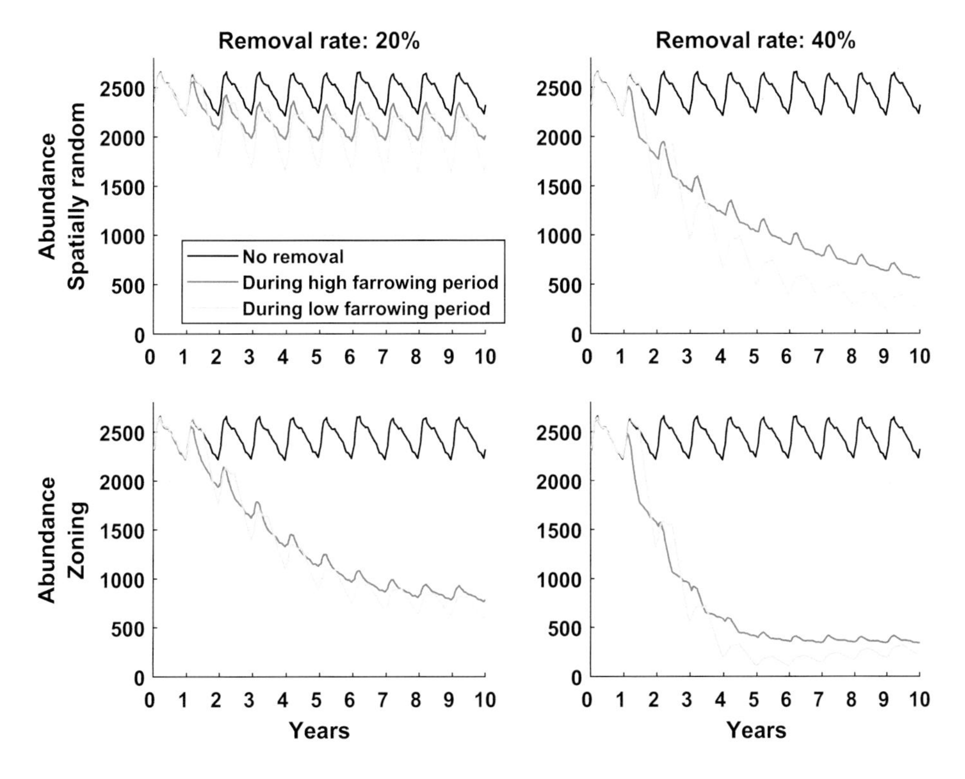

FIGURE 4.4 Population trajectories from an individual-based model using wild pig data from South Carolina. Different conditions for spatial (top versus bottom plots) and temporal prioritization (lines within the plots) of removal are shown for 2 annual harvesting rates (right versus left plots). For spatially random removals, individuals were selected at random across space. For zoning, animals on the east side were prioritized—including new immigrants—such that the population was cleared from east to west. Density-dependent immigration from surrounding, uncontrolled areas occurred freely (based on natal dispersal and dispersal to find free space). The landscape was heterogeneous with high- and low-density patches. Each line is the mean of 30 replicate simulations. (Adapted from Pepin et al. 2017a.)

rates of 9–17% could cause a 25% population reduction in 5 years (McMahon et al. 2010). Together these studies show that annual harvesting rates below 40% can control wild pig populations if control activities are tactically implemented, and the intensity that meets a particular management objective will not only depend on vital rates but also on the spatial and temporal patterns by which pigs are removed (Pepin et al. 2017b).

Throughout much of North America eradication of wild pigs is not feasible in the foreseeable future, thus managers are focused on sustainable control rather than eradication. In these situations, controlled areas likely face continuous recolonization pressure from surrounding, uncontrolled populations (Bodenchuk 2014). A population model of wild pigs in Australia showed that wild pigs from uncontrolled areas were drawn into controlled areas at high rates (Dexter and McLeod 2015). Therefore, control at local scales was more successful when this density-dependent immigration was weak. However, at the landscape level, control efforts were effective when density-dependent immigration (i.e., from natural dispersal) was strong because surrounding wild pigs are being drawn into control areas and removed. When density-dependent immigration was high, annual harvesting rates of up to 50% caused an immediate reduction in abundance of up to 30% (Figure 4.5), but abundance may be reduced in the long-term (Pepin et al. 2017a). In the absence of immigration, the same control intensity led to >95% population reduction within 5 years. In this latter work, immigration was assumed to occur as soon as space became available in the controlled area, by natal dispersal or individuals from the uncontrolled area. A better understanding of natural and human-induced immigration rates and factors determining them is needed for more accurate prediction of optimal control strategies in landscapes with uncontrolled zones.

Wild pig populations may require different management approaches relative to similarly sized ungulates (Servanty et al. 2011) because wild pigs have fast population turnover rates (Focardi et al. 2008). Matrix projection models based on vital rates from populations in Europe have suggested that population growth rates are most sensitive to juvenile survival (Servanty et al. 2011). In another study based on wild boar vital rates and hunter harvest rates, hunting focused on medium sized females was found to be most efficient for controlling the population to a target size (Gamelon et al. 2012). Similarly, an age-structured population model based on demographic data from Texas found that recruitment rates (i.e., piglet survival and number of litters per year) were most influential in determining population growth rate (Mellish et al. 2014). The same study also estimated population growth rate of pigs in Texas to be 32% annually, which suggested that the population could quintuple in 5 years (Mellish et al. 2014).

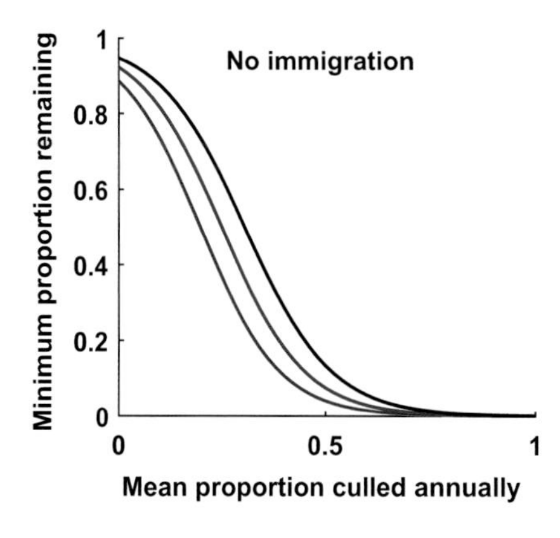

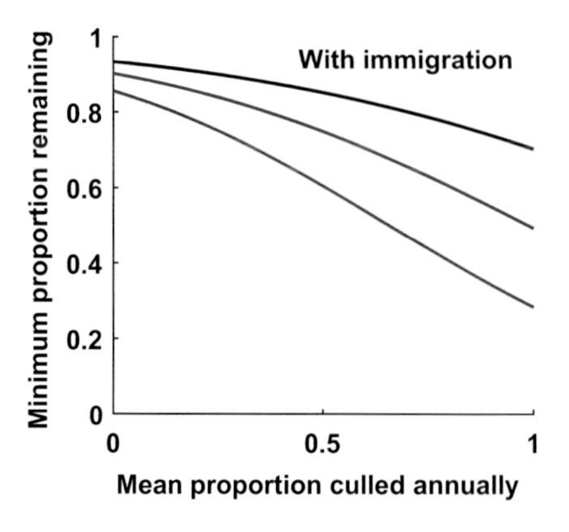

FIGURE 4.5 Predicted proportion of the wild pig population remaining after 5 years at different culling intensities in the absence (left panel) and presence (right panel) of immigration from a neighboring, uncontrolled population. Colors indicate a range of intrinsic fecundities (Blue: field-based measures from South Carolina (Mayer and Brisbin 2009); Red: 50% of blue; Black: 300% of blue). Lines are predictions from generalized linear models fit to simulated data (points). (Adapted from Pepin et al. 2017a.)

Sensitivity analysis of matrix projection models for wild pig populations in California, Georgia, Tennessee, and Texas indicated that variation in age-specific vital rates for wild pigs can have differing effects on population growth rates (Miller 2017). These models found that for these populations, wild pigs <1 year of age have the greatest effect on population growth rates (Figure 4.6). Survival of animals <1 year of age had the largest contribution (40.6%) to variation in population growth rate while survival of animals >1 year of age did not significantly contribute to population growth rates (Figure 4.6). Also, fecundity of wild pigs <1 year of age and 1–2 years of age contributed the most (18.4%) to variation in population growth, while wild pigs >2 years old contributed little to variation in population growth (Figure 4.6). These results indicate that there may be significant age-specific differences in the effect of vital rates on population growth. In addition, for these populations, recruitment and participation of the youngest age class (i.e., <1 year old) in reproduction had the greatest influence on population growth. The contribution of age-specific vital rates, specifically fecundity, to population growth appear most sensitive to the presence of crop agriculture and mast producing forests (Miller 2017, M. A. Tabak, unpublished data). Current studies suggest that wild pigs <1 year of age contribute most to population growth rates when forage resources are high and adults contribute more when forage conditions are poor (Bieber and Ruf 2005, M. A. Tabak, unpublished data) but this may differ between native and non-native populations (Miller 2017).

Contraceptives are currently not available in North America for wild pigs, but population models have been useful for evaluating the potential value of contraceptives for nonlethal population reduction (Burton et al. 2013, Pepin et al. 2017a). An individual-based model using demographic data from Fort Benning, Georgia found that a combined strategy of sterilant baits and lethal control was more effective than either method on its own (Burton et al. 2013). A recent study suggested that potential benefits of using a sterilant in addition to lethal control depend strongly on fecundity, removal rates, and immigration from uncontrolled areas (Pepin et al. 2017a; Figure 4.7). Fertility control could be especially beneficial when immigration from uncontrolled populations or fecundity is too high to be controlled by lethal control alone (Pepin et al. 2017a). One mechanism by

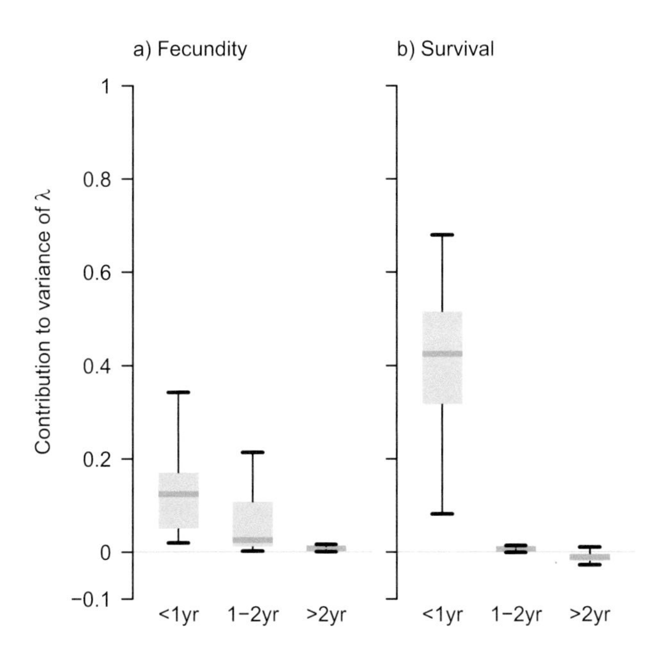

FIGURE 4.6 Contribution of age-specific (a) fecundity and (b) survival to variation in population growth rate (λ) for wild pig populations in California, Georgia, Tennessee, and Texas. Solid line indicates median, box is the upper and lower quartiles, and whiskers note the minimum and maximum litter size. (Adapted from Miller 2017.)

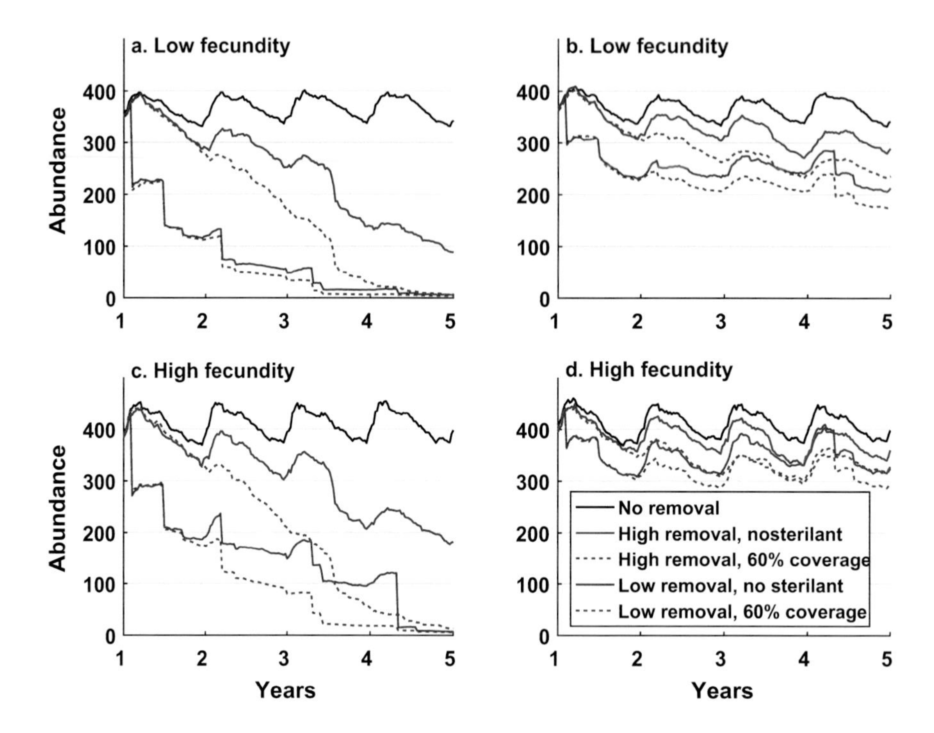

FIGURE 4.7 Population trajectories from an individual-based simulation model using data from South Carolina. Different conditions for fecundity (a. & b.—low; c. & d.—high) and whether immigration from uncontrolled populations occurs (b. & d.) or not (a. & c.). Within each plot, 2 harvesting trajectories (red—high versus blue—low) based on removal counts from USDA/APHIS/Wildlife Services operations in Texas are compared (solid lines). The dotted lines show population trajectories when 60% of the females are sterilized annually in addition to the removal treatments. We assumed that individuals were selected at random, sterilized individuals could be removed, and the sterilant had a 2-year life span. Even with these inefficiencies, substantial gains in population reduction could be obtained depending on fecundity, harvesting rate, and immigration conditions. Gains from fertility control were proportionately larger when harvesting had smaller effects relative to when harvesting had larger effects. (Adapted from Pepin et al. 2017a.)

which fertility control could be especially helpful in populations with immigration is through the *placeholder effect*, where sterilized individuals occupy space that might otherwise be occupied by fecund individuals (Gese and Terletzky 2015), thus maintaining lower abundances. Future research should investigate potential cost-effectiveness of using this type of combined strategy in populations subject to immigration.

Ultimately, attempts to reduce populations of wild pigs are undertaken to protect human and domestic animal health and reduce damage to natural and anthropogenic resources. However, relationships between population control, damage reduction, and disease risks, as well as control costs, are poorly understood throughout North America, and globally (e.g., Hone 1994). There is evidence that differing densities of wild pigs and differing age-sex compositions of wild pig populations inflict different intensities of damage (Hone 1994, Krull et al. 2016, A. J. Davis, USDA/APHIS/WS/National Widlife Research Center, unpublished data) and may also influence differences in disease transmission risk (Hone 2002). For instance, a study in Australia found that large proportions of the wild pig population must be removed to have substantial reduction in rooting damage (Cowled and Garner 2008). In contrast, a study in New Zealand found the relationship to be exponential, such that only a small proportion of pigs needed to be removed to decrease damage substantially (Krull et al. 2016). Results of the latter study were potentially related to the population hovering at carrying capacity, which forced wild pigs to seek below-ground resources via rooting rather than just foraging for above ground resources. A recent study in North America showed that cost-effectiveness of control strongly depends on relationships between pig density and damage levels (A. J. Davis, unpublished data). For example, when the relationship is logarithmic such that many pigs need to be removed to affect damage levels, then it is more cost-effective to drastically reduce the population even though the control expenses are high. In contrast, if the relationship is exponential such that removing a small number of pigs reduces damage levels dramatically, then it is most cost-effective to only reduce the population by a small amount, thus expending fewer resources on control efforts. More research relating pig densities to control costs and damage reduction, specifically reductions in disease risk, is needed to predict cost-effectiveness for managers in particular conditions.

4.8 KNOWLEDGE GAPS IN NORTH AMERICA

Although substantial progress in our understanding of wild pig population dynamics has occurred in the last few decades, surprisingly, several key attributes of vital rates remain poorly understood, especially in comparison with other large ungulates in North America. These data gaps stem from a historic lack of funding for pig research by many state and federal agencies. The limited knowledge of wild pig ecology is further exacerbated by their secretive nature, complex social structure, unique genetic composition due to hybridization between domestic pigs and wild boar, and varying management approaches and public perception of pigs across their range in North America. Consequently, many methodological and analytical approaches commonly used for other ungulate species have proven inadequate for pigs, requiring modification or development of new tools to aid in research and control of this important invasive species (Campbell et al. 2006, 2010, 2012; Bevins et al. 2014).

Ironically, the basic reproductive biology of *Sus scrofa* is better studied than it is for most other vertebrates due to their domestic importance as livestock and for biomedical industries (e.g., Pond and Houpt 1978, Mendl et al. 1992, Pond and Mersmann 2001). However, there are critical gaps in specific knowledge of reproduction and survival in wild pigs. For example, while it is generally understood that wild pigs are capable of producing multiple litters per year, the extent to which this occurs within individuals, timing of reproduction, and spatio-temporal variability in these parameters or how biotic and abiotic environmental factors contribute to this variability are poorly understood for wild pigs in North America. Similarly, few studies have quantified survival rates in wild pigs, and to date no studies have successfully determined survival rates of neonate pigs; data essential to development of robust population models (Keiter et al. 2017a). Likewise, accurate estimates of density are needed to both develop appropriate control strategies and determine efficacy of management. However, while densities of wild pigs and wild boar have been reported extensively (Mayer 2009), these estimates have been derived using inconsistent methodologies and in many cases have failed to adequately account for animal movement (but see Keiter et al. 2017a). Given the distinct social structure and behavior of wild pigs, there is growing need for development and evaluation of analytical approaches that are optimized for estimating wild pig densities (Keiter et al. 2017a). Ultimately, modeling efforts are limited by extent and quality of empirical data from which they can be developed. Developing long-term studies that measure population dynamics as a function of environmental conditions and control efforts are much needed to form predictive models of population dynamics and obtain more robust estimates of vital rates. Thus, there remains a critical need for both empirical data on age-specific vital rates collected under various biotic and abiotic scenarios, as well as advancements in modeling techniques that incorporate diverse behaviors and environmental conditions across which wild pig populations are found.

Due to many uncertainties and multiple geographic scales of management objectives for wild pigs in North America, adaptive management (i.e., structured, iterative decision making) would be helpful for guiding decisions on wild pig management. Adaptive management models incorporate an understanding of population dynamics of the system to identify optimal management strategies while accounting for uncertainties in management responses (Williams 1996). Future research to address this gap will need to consider funding complexity and multiple stakeholder values as well as cost-effectiveness of different harvesting strategies in space and time. Although this challenge sounds immense, a formal mechanism for making decisions while considering potentially conflicting interests and highly uncertain wild pig ecology would be extremely valuable for informed population management planning throughout their non-native ranges (Brondum et al. 2017).

4.9 CONCLUSIONS

Understanding population dynamics of wild pigs throughout North America provides valuable information for reducing, eliminating, or constricting invasive populations. Wild pig populations are predicted to have enormous unrealized potential in North America. They are more fecund than wild pigs in their native ranges because of domestic introgression. Advances in population modeling approaches are useful for examining ever-changing relationships of vital rates (e.g., survival and fecundity) based on age of animals, seasons, landscapes, available resources, population genetics, and other factors. These models are providing important suggestions regarding intensity and rate of removal of wild pigs to keep populations from expanding, and possibly declining. However, application of these models could be widespread if more empirical data were generated to inform them. Specifically, there is a critical need for empirical data on age-specific vital rates, carrying capacities, and densities of wild pigs collected under various biotic and abiotic scenarios, particularly throughout the expanding range of wild pigs in North America.

ACKNOWLEDGMENTS

We thank G.J. Roloff, S.S. Ditchkoff, and K.C. VerCauteren for their comments on this chapter. Contributions from N.P.S., R.S.M., and K.M.P. were supported by the USDA. Contributions from J.C.B. were partially funded by the US Department of Energy under award # DE-EM0004391 to the University of Georgia Research Foundation. Mention of commercial products or companies does not represent an endorsement by the US Government.

REFERENCES

Acevedo, P., M. A. Escudero, R. Muñoz, and C. Gortázar. 2006. Factors affecting wild boar abundance across an environmental gradient in Spain. *Acta Theriologica* 51:327–336.

Adkins, R. N., and L. A. Harveston. 2007. Demographic and spatial characteristics of feral hogs in the Chihuahuan Desert, Texas. *Human-Wildlife Conflicts* 1:152–160.

Ahmad, E., J. E. Brooks, I. Hussain, and M. H. Khan. 1995. Reproduction in Eurasian wild boar in central Punjab, Pakistan. *Acta Theriologica* 40:163–173.

Albrycht, M., D. Merta, J. Bobek, and S. Ulejczyk. 2016. The demographic pattern of wild boars (*Sus scrofa*) inhabiting fragmented forest in north-eastern Poland. *Baltic Forestry* 22:251–258.

Anderson, S. J., and C. P. Stone. 1993. Snaring to control feral pigs *Sus scrofa* in a remote Hawaiian rain forest. *Biological Conservation* 63:195–201.

Andrzejewski, R., and W. Jezierski. 1978. Management of a wild boar population and its effects on commercial land. *Acta Theriologica* 23:309–339.

Baber, D. W., and B. E. Coblentz. 1986. Density, home range, habitat use, and reproduction in feral pigs on Santa Catalina Island. *Journal of Mammalogy* 67:512–525.

Barrett, R. 1978. The feral hog at Dye Creek Ranch, California. *California Agriculture* 46:283–355.

Barrett, R. H. 1971. Ecology of the feral hog in Tehama County, California. Dissertation, University of California, Berkeley, CA.

Beeton, N. J., C. R. McMahon, G. J. Williamson, J. Potts, J. Bloomer, M. N. Bester, L. K. Forbes, and C. N. Johnson. 2015. Using the spatial population abundance dynamics engine for conservation management. *Methods in Ecology and Evolution* 6:1407–1416.

Belden, R. C., and W. B. Frankenberger. 1979. Brunswick hog study. Final performance report, P-R Project W-41-R. Florida Fresh Water Fish and Game Commission Wildlife Research Laboratory, Gainesville, FL.

Belden, R. C., and W. B. Frankenberger. 1990. Biology of a feral hog population in south central Florida. *Proceedings of the Annual Conference of the Southeastern Association of Fish & Wildlife Agencies* 44:231–249.

Bevins, S. N., K. Pedersen, M. W. Lutman, T. Gidlewski, and T. J. Deliberto. 2014. Consequences associated with the recent range expansion of nonnative feral swine. *BioScience* 64:291–299.

Bieber, C., and T. Ruf. 2005. Population dynamics in wild boar *Sus scrofa*: Ecology, elasticity of growth rate and implications for the management of pulsed resource consumers. *Journal of Applied Ecology* 42:1203–1213.

Bodenchuk, M. 2014. Method-specific costs of feral swine removal in a large metapopulation: The Texas experience. *Proceedings of the Vertebrate Pest Conference* 26:269–271.

Boitani, L., P. Trapanese, L. Mattei, and D. Nonis. 1995. Demography of a wild boar (*Sus scrofa*, L.) population in Tuscany, Italy. *Gibier faune sauvage* 12:109–132.

Briedermann, L. 1990. *Schwarzwild.* Second Edition. VEB Deutscher Landwirtschaftsverlag., Berlin, Gerrmany.

Brondum, M. C., Z. A. Collier, C. S. Luke, B. L. Goatcher, and I. Linkov. 2017. Selection of invasive wild pig countermeasures using multicriteria decision analysis. *Science of the Total Environment* 574: 1164–1173.

Burton, J. L., J. D. Westervelt, and S. S. Ditchkoff. 2013. *Simulation of Wild Pig Control via Hunting and Contraceptives.* ERDC Innovative Solutions, Champaign, IL.

Byers, J. A., and J. D. Moodie. 1990. Sex-specific maternal investment in pronghorn, and the question of a limit on differential provisioning in ungulates. *Behavioral Ecology and Sociobiology* 26:157–164.

Bywater, K. A., M. Apollonio, N. Cappai, and P. A. Stephens. 2010. Litter size and latitude in a large mammal: The wild boar *Sus scrofa*. *Mammal Review* 40:212–220.

Campbell, T. A., S. J. Lapidge, and D. B. Long. 2006. Using baits to deliver pharmaceuticals to feral swine in southern Texas. *Wildlife Society Bulletin* 34:1184–1189.

Campbell, T. A., D. B. Long, M. J. Lavelle, B. R. Leland, T. L. Blankenship, and K. C. VerCauteren. 2012. Impact of baiting on feral swine behavior in the presence of culling activities. *Preventive Veterinary Medicine* 104:249–257.

Campbell, T. A., D. B. Long, and B. R. Leland. 2010. Feral swine behavior relative to aerial gunning in southern Texas. *The Journal of Wildlife Management* 74:337–341.

Canu, A., M. Scandura, E. Merli, R. Chirichella, E. Bottero, F. Chianucci, A. Cutini, and M. Apollonio. 2015. Reproductive phenology and conception synchrony in a natural wild boar population. *Hystrix, the Italian Journal of Mammalogy* 26:77–84.

Caswell, H. 2014. *Matrix Population Models.* John Wiley & Sons, Ltd, Wiley StatsRef: Statistics Reference Online. https://doi.org/10.1002/9781118445112.stat07481

Choquenot, D., J. Hone, and G. Saunders. 1999. Using aspects of predator-prey theory to evaluate helicopter shooting for feral pig control. *Wildlife Research* 26:251–261.

Choquenot, D., R. J. Kilgour, and B. S. Lukins. 1993. An evaluation of feral pig trapping. *Wildlife Research* 20:15–22.

Clarke, C., R. Dzieciolowski, D. Batcheler, and C. Frampton. 1992. A comparison of tooth eruption and wear and dental cementum techniques in age determination of New Zealand feral pigs. *Wildlife Research* 19:769–777.

Conley, R., V. Henry, and G. Matschke. 1972. *Final Report for the European Hog Research Project W-34.* Tennessee Game and Fish Commission, Nashville, TN.

Cowled, B., and G. Garner. 2008. A review of geospatial and ecological factors affecting disease spread in wild pigs: Considerations for models of foot-and-mouth disease spread. *Preventive Veterinary Medicine* 87:197–212.

Cruz, F., V. Carrion, K. J. Campbell, C. Lavoie, and C. J. Donlan. 2009. Bio-economics of large-scale eradication of feral goats from Santiago Island, Galápagos. *Journal of Wildlife Management* 73:191–200.

Cruz, F., C. J. Donlan, K. Campbell, and V. Carrion. 2005. Conservation action in the Galapagos: Feral pig (*Sus scrofa*) eradication from Santiago Island. *Biological Conservation* 121:473–478.

Dardaillon, M. 1989. Age-class influences on feeding choices of free-ranging wild boars (*Sus scrofa*). *Canadian Journal of Zoology* 67:2792–2796.

Davis, A. J., M. B. Hooten, R. S. Miller, M. L. Farnsworth, J. Lewis, M. Moxcey, and K. M. Pepin. 2016. Inferring invasive species abundance using removal data from management actions. *Ecological Applications* 26:2339–2346.

Davis, A. J., B. Leland, M. Bodenchuk, K. C. VerCauteren, and K. M. Pepin. 2017. Estimating population density for disease risk assessment: The importance of understanding the area of influence of traps using wild pigs as an example. *Preventative Veterinary Medicine* 141:33–37.

Davis, A. J., R. McCreary, J. Psiropoulos, G. Brennan, T. Cox, A. Partin, and K. M. Pepin. 2018. Quantifying site-level usage and certainty of absence for an invasive species through occupancy analysis of camera-trap data. *Biological Invasions* 0:877–890.

Delgado-Acevedo, J., A. Zamorano, R. W. DeYoung, T. A. Campbell, D. G. Hewitt, and D. B. Long. 2011. Promiscuous mating in feral pigs (*Sus scrofa*) from Texas, USA. *Wildlife Research* 37:539–546.

Dexter, N., and S. R. McLeod. 2015. Modeling ecological traps for the control of feral pigs. *Ecology and Evolution* 5:2036–2047.

Diong, C. 1973. Studies of the Malayan wild pig in Perak and Johore. *Malayan Nature Journal* 26:120–151.

Diong, C. H. 1983. *Population Biology and Management of the Feral Pig (Sus scrofa L.) in Kipahulu Valley, Maui*. University of Hawaii, Honolulu, HI.

Ditchkoff, S. S., R. W. Holtfreter, and B. L. Williams. 2017. Effectiveness of a bounty program for reducing wild pig densities. *Wildlife Society Bulletin* 41:548–555.

Ditchkoff, S. S., D. B. Jolley, B. D. Sparklin, L. B. Hanson, M. S. Mitchell, and J. B. Grand. 2012. Reproduction in a population of wild pigs (*Sus scrofa*) subjected to lethal control. *The Journal of Wildlife Management* 76:1235–1240.

Duncan, R. W. 1974. Reproductive biology of the European wild hog (*Sus scrofa*) in the Great Smoky Mountains National Park. Thesis, University of Tennessee, Knoxville, TN.

Dzięciołowski, R. M., C. M. H. Clarke, and C. M. Frampton. 1992. Reproductive characteristics of feral pigs in New Zealand. *Acta Theriologica* 37:259–270.

Fernández-Llario, P., and J. Carranza. 2000. Reproductive performance of the wild boar in a Mediterranean ecosystem under drought conditions. *Ethology Ecology & Evolution* 12:335–343.

Fernández-Llario, P., J. Carranza, and P. Mateos-Quesada. 1999. Sex allocation in a polygynous mammal with large litters: The wild boar. *Animal Behaviour* 58:1079–1084.

Fernández-Llario, P., and P. Mateos-Quesada. 1998. Body size and reproductive parameters in the wild boar *Sus scrofa*. *Acta Theriologica* 43:439–444.

Fernández-Llario, P., and P. Mateos-Quesada. 2005. Influence of rainfall on the breeding biology of wild boar (*Sus scrofa*) in a Mediterranean ecosystem. *Folia Zoologica* 54:240–248.

Focardi, S., J. M. Gaillard, F. Ronchi, and S. Rossi. 2008. Survival of wild boars in a variable environment: Unexpected life-history variation in an unusual ungulate. *Journal of Mammology* 89:1113–1123.

Fonseca, C., A. A. Da Silva, J. Alves, J. Vingada, and A. M. Soares. 2011. Reproductive performance of wild boar females in Portugal. *European Journal of Wildlife Research* 57:363–371.

Frantz, A. C., F. E. Zachos, J. Kirschning, S. Cellina, S. Bertouille, Z. Mamuris, E. A. Koutsogiannouli, and T. Burke. 2013. Genetic evidence for introgression between domestic pigs and wild boars (*Sus scrofa*) in Belgium and Luxembourg: A comparative approach with multiple marker systems. *Biological Journal of the Linnean Society* 110:104–115.

Frauendorf, M., F. Gethöffer, U. Siebert, and O. Keuling. 2016. The influence of environmental and physiological factors on the litter size of wild boar (*Sus scrofa*) in an agriculture dominated area in Germany. *Science of the Total Environment* 541:877–882.

Fulgione, D., D. Rippa, M. Buglione, M. Trapanese, S. Petrelli, and V. Maselli. 2016. Unexpected but welcome. Artificially selected traits may increase fitness in wild boar. *Evolutionary Applications* 9:769–776.

Gabor, T. M., E. C. Hellgren, R. A. Bussche, and N. J. Silvy. 1999. Demography, sociospatial behaviour and genetics of feral pigs (*Sus scrofa*) in a semi-arid environment. *Journal of Zoology* 247:311–322.

Gamelon, M., S. Focardi, J. M. Gaillard, O. Gimenez, C. Bonenfant, B. Franzetti, R. Choquet, F. Ronchi, E. Baubet, and J. F. Lemaître. 2014. Do age-specific survival patterns of wild boar fit current evolutionary theories of senescence? *Evolution* 68:3636–3643.

Gamelon, M., J. M. Gaillard, S. Servanty, O. Gimenez, C. Toigo, E. Baubet, F. Klein, and J. D. Lebreton. 2012. Making use of harvest information to examine alternative management scenarios: A body weight-structured model for wild boar. *Journal of Applied Ecology* 49:833–841.

Garzon-Heydt, P. 1992. Study of a population of wild boar *Sus scrofa* castilianus Thomas, 1912 in Spain, based on hunting data. Pages 489–492 *in* B. Bobek, K. Perzanowski, and W. Regelin, editors. *Global Trends in Wildlife Management Transactions of the 18th International Union of Game Biologists, Krakow, 1987.* Swiat Press, Krakow-Warszawa, Poland.

Gaston, W. 2008. Feral pig (*Sus scrofa*) survival, home range, and habitat use at Lowndes County Wildlife Management Area, Alabama. Thesis, Auburn University, Auburn, AL.

Gese, E. M., and P. A. Terletzky. 2015. Using the "placeholder" concept to reduce genetic introgression of an endangered carnivore. *Biological Conservation* 192:11–19.

Gethöffer, F. 2005. Reproduktionsparameter und Saisonalität der Fortpflanzung des Wildschweins (*Sus scrofa*) in drei Untersuchungsgebieten Deutschlands. Dissertation, University of Veterinary Medicine, Hanover, Germany.

Gethöffer, F., G. Sodeikat, and K. Pohlmeyer. 2007. Reproductive parameters of wild boar (*Sus scrofa*) in three different parts of Germany. *European Journal of Wildlife Research* 53:287–297.

Giles, J. 1980. The ecology of feral pigs in western New South Wales. PhD Dissertation, University of Sydney, New South Whales, Australia.

Goedbloed, D. J., H. J. Megens, P. van Hooft, J. M. Herrero-Medrano, W. Lutz, P. Alexandri, R. Crooijmans, M. Groenen, S. E. van Wieren, R. C. Ydenberg, and H. H. T. Prins. 2013. Genome-wide single nucleotide polymorphism analysis reveals recent genetic introgression from domestic pigs into Northwest European wild boar populations. *Molecular Ecology* 22:856–866.

Graves, H. 1984. Behavior and ecology of wild and feral swine (*Sus scrofa*). *Journal of Animal Science* 58:482–492.

Hagen, D., and K. Kephart. 1980. Reproduction in domestic and feral swine. I. Comparison of ovulatory rate and litter size. *Biology of Reproduction* 22:550–552.

Hagen, D., K. Kephart, and P. Wangsness. 1980. Reproduction in domestic and feral swine. II. Interrelationships between fetal size and spacing and litter size. *Biology of Reproduction* 23:929–934.

Hanson, L. B., J. B. Grand, M. S. Mitchell, D. B. Jolley, B. D. Sparklin, and S. S. Ditchkoff. 2008. Change-in-ratio density estimator for feral pigs is less biased than closed mark–recapture estimates. *Wildlife Research* 35:695–699.

Hanson, L. B., M. S. Mitchell, J. B. Grand, D. B. Jolley, B. D. Sparklin, and S. S. Ditchkoff. 2009. Effect of experimental manipulation on survival and recruitment of feral pigs. *Wildlife Research* 36: 185–191.

Hanson, R., and L. Karstad. 1959. Feral swine in the southeastern United States. *The Journal of Wildlife Management* 23:64–74.

Hayes, R., S. Riffell, R. Minnis, and B. Holder. 2009. Survival and habitat use of feral hogs in Mississippi. *Southeastern Naturalist* 8:411–426.

Hazeleger, W., N. Soede, and B. Kemp. 2005. The effect of feeding strategy during the pre-follicular phase on subsequent follicular development in the pig. *Domestic Animal Endocrinology* 29:362–370.

Henry, V. 1966. European wild hog hunting season recommendations based on reproductive data. *Southeastern Association of Game and Fish Commissioners* 20:139–145.

Henry, V. G., and R. H. Conley. 1978. Survival and mortality of European wild hogs. *The Annual Conference of the Southeastern Association of Fish & Wildlife Agencies* 32:93–99.

Hess, S. C., J. J. Jeffrey, D. L. Ball, and L. Babich. 2006. Efficacy of feral pig removals at Hakalau Forest National Wildlife Refuge. *Transactions of the Western Section of the Wildlife Society* 42:53–67.

Hone, J. 1990. Predator prey theory and feral pig control, with emphasis on evaluation of shooting from a helicopter. *Australian Wildlife Research* 17:123–130.

Hone, J. 1992. Modelling of poisoning for vertebrate pest control, with emphasis on poisoning feral pigs. *Ecological Modelling* 62:311–327.

Hone, J. 1994. *Analysis of Vertebrate Pest Control*. Cambridge University Press, Cambridge, UK.

Hone, J. 1999. On rate of increase (r): Patterns of variation in Australian mammals and the implications for wildlife management. *Journal of Applied Ecology* 36:709–718.

Hone, J. 2002. Feral pigs in Namadgi National Park, Australia: Dynamics, impacts and management. *Biological Conservation* 105:231–242.

Hone, J. 2012. *Applied Population and Community Ecology: The Case of Feral Pigs in Australia*. Wiley-Blackwell, Oxford, UK.

Jezierski, W. 1977. Longevity and mortality rate in a population of wild boar. *Acta Theriologica* 22:337–348.

Johnson, K. G., R. W. Duncan, and M. R. Pelton. 1982. Reproductive biology of European wild hogs in the Great Smokey Mountains National Park. *Proceedings of the Annual Conference of the Southeastern Fish and Wildlife Agencies* 36:552–564.

Keiter, D. A., A. J. Davis, O. E. Rhodes, F. L. Cunningham, J. C. Kilgo, K. M. Pepin, and J. C. Beasley. 2017a. Effects of scale of movement, detection probability, and true population density on common methods of estimating population density. *Scientific Reports* 7:9446.

Keiter, D. A., J. C. Kilgo, M. A. Vukovich, F. L. Cunningham, and J. C. Beasley. 2017b. Development of known-fate survival monitoring techniques for juvenile wild pigs (*Sus scrofa*). *Wildlife Research* 44:165–173.

Klinger, R., J. Conti, J. K. Gibson, S. M. Ostoja, and E. Aumack. 2011. What does it take to eradicate a feral pig population? Pages 78–86 *in* C. R. Veitch, M. N. Clout, and D. R. Towns, editors. *Island Invasions, Eradication and Management.* IUCN.

Krull, C. R., M. C. Stanley, B. R. Burns, D. Choquenot, and T. R. Etherington. 2016. Reducing wildlife damage with cost-effective management programmes. *Plos One* 11:15.

Lapidge, S. J., B. Cowled, and M. Smith. 2004. Ecology, genetics and socio-biology: Practical tools in the design of target-specific feral pig baits and baiting procedures. *Proceedings of the Vertebrate Pest Conference* 21:317–322.

Lavelle, M. J., N. P. Snow, J. M. Halseth, E. H. VanNatta, H. N. Sanders, and K. C. VerCauteren. 2018. Evaluation of movement behaviors to inform toxic baiting strategies for invasive wild pigs (*Sus scrofa*). *Pest Management Science* 74:2504–2510.

Lemel, J. 1999. *Populationstillväxt, dynamik och spridning hos vildsvinet, Sus scrofa, i mellersta Sverige: Slutrapport.* Svenska Jägareförbundet.

Levy, B., C. Collins, S. Lenhart, M. Madden, J. L. Corn, R. A. Salinas, and W. H. Stiver. 2016. A metapopulation model for feral hogs in Great Smokey Mountain National Park. *Natural Resource Modeling* 29:71–97.

Lewis, J. S., M. L. Farnsworth, C. L. Burdett, D. M. Theobald, M. Gray, and R. S. Miller. 2017. Biotic and abiotic factors predicting the global distribution and population density of an invasive large mammal. *Scientific Reports* 7:44152.

Lowe, S., M. Browne, S. Boudjelas, and M. De Poorter. 2000. 100 of the world's worst invasive alien species: A selection from the global invasive species database. The Invasive Species Specialist Group (ISSG) a specialist group of the Species Survival Commission (SSC) of the World Conservation Union (IUCN).

Lozan, M. 1995. Factors that limit the number of wild boars (*Sus scrofa* L.) in the Republic of Moldova. *Journal of Mountain Ecology* 3:211.

Maselli, V., D. Rippa, G. Russo, R. Ligrone, O. Soppelsa, B. D'Aniello, P. Raia, and D. Fulgione. 2014. Wild boars' social structure in the Mediterranean habitat. *Italian Journal of Zoology* 81:610–617.

Massei, G., P. V. Genov, and B. W. Staines. 1996. Diet, food availability and reproduction of wild boar in a Mediterranean coastal area. *Acta Theriologica* 41:307–320.

Massei, G., J. Kindberg, A. Licoppe, D. Gačić, N. Šprem, J. Kamler, E. Baubet, U. Hohmann, A. Monaco, and J. Ozoliņš. 2015. Wild boar populations up, numbers of hunters down? A review of trends and implications for Europe. *Pest Management Science* 71:492–500.

Massolo, A., and R. M. Della Stella. 2006. Population structure variations of wild boar *Sus scrofa* in central Italy. *Italian Journal of Zoology* 73:137–144.

Masters, K. 1979. Feral pigs in the south-west of Western Australia. Final Report to Feral Pig Committee. Agriculture Protection Board and Department of Conservation and Land Management, Western Australia.

Matschke, G. H. 1964. The influence of oak mast on European wild hog reproduction. *Proceedings Annual Conference Southeast Association of Game and Fish Commission* 18:35–39.

Mayer, J. J. 2009. Wild pig population biology. Pages 157–191 *in* J. Mayer, and I. L. Brisbin Jr., editors. *Wild Pigs: Biology, Damage, Control Techniques and Management.* Savannah River National Laboratory, Aiken, SC.

Mayer, J. J., and I. L. Brisbin. 2008. *Wild Pigs in the United States: Their History, Comparative Morphology, and Current Status.* University of Georgia Press, Athens, GA.

Mayer, J. J., and I. L. Brisbin, editors. 2009. *Wild Pigs: Biology, Damage, Control Techniques and Management.* Savannah River National Laboratory, Aiken, SC.

McCann, B. 2003. *The Feral Hog in Illinois.* Thesis, Southern Illinois University, Carbondale, IL.

McCann, B. E., and D. K. Garcelon. 2008. Eradication of feral pigs from Pinnacles National Monument. *Journal of Wildlife Management* 72:1287–1295.

McClure, M. L., C. L. Burdett, M. L. Farnsworth, M. W. Lutman, D. M. Theobald, P. D. Riggs, D. A. Grear, and R. S. Miller. 2015. Modeling and mapping the probability of occurrence of invasive wild pigs across the contiguous United States. *PloS One* 10:e0133771.

McMahon, C. R., B. W. Brook, N. Collier, and C. J. A. Bradshaw. 2010. Spatially explicit spreadsheet modelling for optimising the efficiency of reducing invasive animal density. *Methods in Ecology and Evolution* 1:53–68.

Mellish, J. M., A. Sumrall, T. A. Campbell, B. A. Collier, W. H. Neill, B. Higginbotham, and R. R. Lopez. 2014. Simulating Potential Population Growth of Wild Pig, *Sus scrofa*, in Texas. *Southeastern Naturalist* 13:367–376.

Mendl, M., A. J. Zanella, and D. M. Broom. 1992. Physiological and reproductive correlates of behavioural strategies in female domestic pigs. *Animal Behaviour* 44:1107–1121.
Merta, D., P. Mocala, M. Pomykacz, and W. Frackowiak. 2014. Autumn-winter diet and fat reserves of wild boars (*Sus scrofa*) inhabiting forest and forest-farmland environment in south-western Poland. *Folia Zoologica* 63:95–102.
Miller, R. S. 2017. Interaction among societal and biological drivers of policy at the wildlife-agricultural interface. Dissertation, Colorado State University, Fort Collins, CO.
Morelle, K., J. Fattebert, C. Mengal, and P. Lejeune. 2016. Invading or recolonizing? Patterns and drivers of wild boar population expansion into Belgian agroecosystems. *Agriculture, Ecosystems & Environment* 222:267–275.
Morelle, K., and P. Lejeune. 2015. Seasonal variations of wild boar *Sus scrofa* distribution in agricultural landscapes: A species distribution modelling approach. *European Journal of Wildlife Research* 61:45–56.
Morrison, S. A., N. Macdonald, K. Walker, L. Lozier, and M. R. Shaw. 2007. Facing the dilemma at eradication's end: Uncertainty of absence and the Lazarus effect. *Frontiers in Ecology and the Environment* 5:271–276.
Okarma, H., B. Jędrzejewska, W. Jędrzejewski, Z. A. Krasiński, and L. Miłkowski. 1995. The roles of predation, snow cover, acorn crop, and man-related factors on ungulate mortality in Białowieża Primeval Forest, Poland. *Acta Theriologica* 40:197–217.
Osada, Y., T. Kuriyama, M. Asada, H. Yokomizo, and T. Miyashita. 2015. Exploring the drivers of wildlife population dynamics from insufficient data by Bayesian model averaging. *Population Ecology* 57:485–493.
Peine, J. D., and J. A. Farmer. 1990. Wild hog management program at Great Smoky Mountains National Park. *Proceedings of the Fourteenth Vertebrate Pest Conference* 1990:221–227.
Pepin, K. M., A. J. Davis, F. L. Cunningham, K. C. VerCauteren, and D. C. Eckery. 2017a. Potential effects of incorporating fertility control into typical culling regimes in wild pig populations. *PloS One* 12:e0183441.
Pepin, K. M., A. J. Davis, and K. C. VerCauteren. 2017b. Efficiency of different spatial and temporal strategies for reducing vertebrate pest populations. *Ecological Modelling* 365:106–118.
Pine, D., and G. Gerdes. 1973. Wild pigs in Monterey County, California. *California Fish and Game* 59:126–137.
Pine, D. S., and G. L. Gerdes. 1969. Wild pig study in Monterey County, California. California Department of Fish and Game, Region 3, San Francisco, CA.
Pond, W. G., and K. A. Houpt. 1978. *The Biology of the Pig*. Cornell University Press, Ithaca, NY.
Pond, W. G., and H. J. Mersmann. 2001. *Biology of the Domestic Pig*. Comstock Pub. Associates, Cornell University Press.
Salinas, R. A., W. H. Stiver, J. L. Corn, S. Lenhart, C. Collins, M. Madden, K. C. VerCauteren, et al. 2015. An individual-based model for feral hogs in Great Smoky Mountains National Park. *Natural Resource Modeling* 28:18–36.
Saunders, G. R. 1988. The ecology and management of feral pigs in New South Wales. Thesis, Macquarie University, Sydney, New South Whales, Australia.
Schuyler, P., D. Garcelon, and S. Escover. 2002. Eradication of feral pigs (*Sus scrofa*) on Santa Catalina Island, CA. Turning the tide: The eradication of invasive species. *IUCN SSC Invasive Species Specialist Group*. IUCN, Gland, Switzerland:274–286.
Servanty, S., J. M. Gaillard, D. Allainé, S. Brandt, and E. Baubet. 2007. Litter size and fetal sex ratio adjustment in a highly polytocous species: The wild boar. *Behavioral Ecology* 18:427–432.
Servanty, S., J. M. Gaillard, F. Ronchi, S. Focardi, E. Baubet, and O. Gimenez. 2011. Influence of harvesting pressure on demographic tactics: Implications for wildlife management. *Journal of Applied Ecology* 48:835–843.
Servanty, S., G. Jean-Michel, T. Carole, B. Serge, and B. Eric. 2009. Pulsed resources and climate-induced variation in the reproductive traits of wild boar under high hunting pressure. *Journal of Animal Ecology* 78:1278–1290.
Singer, F. J., and B. B. Ackerman. 1981. Food availability, reproduction, and condition of European wild boar in Great Smoky Mountains National Park. US Department of Interior, National Park Service, Southeast Region [and] Uplands Field Research Laboratory, Great Smoky Mountains National Park.
Singer, F., B. Ackerman, M. Harmon, and A. Tipton. 1978. Studies of European wild boar in the Great Smoky Mountains National Park; Part I. Census, trapping and population biology of European wild boar–1977. US Department of the Interior, National Park Service, Southeast Regional Office. Report for the Superintendent 1477.

Snow, N. P., J. A. Foster, J. C. Kinsey, S. T. Humphrys, L. D. Staples, D. G. Hewitt, and K. C. VerCauteren. 2017b. Development of toxic bait to control invasive wild pigs and reduce damage. *Wildlife Society Bulletin* 41:256–263.

Snow, N. P., M. A. Jarzyna, and K. C. VerCauteren. 2017a. Interpreting and predicting the spread of invasive wild pigs. *Journal of Applied Ecology* 54:2022–2032.

Snow, N. P., and K. C. VerCauteren. 2019. Movement responses inform effectiveness and consequences of baiting wild pigs for population control. *Crop Protection* 124:104835.

Soede, N. M., P. Langendijk, and B. Kemp. 2011. Reproductive cycles in pigs. *Animal Reproduction Science* 124:251–258.

Springer, M. 1977. Ecologic and economic aspects of wild hogs in Texas. Pages 37–46 *in* G. W. Wood, editor. *Research and Management of Wild Hog Populations*: *Proceedings of a Symposium*. Belle W. Baruch Forest Science Institute, Clemson University, Clemson, SC.

Strand, D. K. 1980. Reproductive ecology and behavior of the Florida feral hog (*Sus scrofa*). Thesis, Florida Institute of Technology, Melbourne, FL.

Sweeney, J. R. 1979. Ovarian activity in feral swine. *Bulletin of the South Carolina Academy of Science* 41, 74 pp.

Sweeney, J. M., J. R. Sweeney, and E. E. Provost. 1979. Reproductive biology of a feral hog population. *Journal of Wildlife Management* 43:555–559.

Sweeney, J., J. Sweeney, and S. Sweeney. 2003. Feral hog (*Sus scrofa*). *in* G. Feldhammer, B. Thompson, and J. Chapman, editors. *Wild Mammals of North America: Biology, Management, and Conservation*. The Johns Hopkins University Press, Baltimore, MD.

Taylor, R. B., E. C. Hellgren, T. M. Gabor, and L. M. Ilse. 1998. Reproduction of feral pigs in southern Texas. *Journal of Mammalogy* 79:1325–1331.

Trivers, R. L., and D. E. Willard. 1973. Natural selection of parental ability to vary the sex ratio of offspring. *Science* 179:90–92.

Vetter, S. G., T. Ruf, C. Bieber, and W. Arnold. 2015. What is a mild winter? Regional differences in within-species responses to climate change. *PloS One* 10:e0132178.

Warren, R. J., and C. R. Ford. 1997. Diets, nutrition, and reproduction of feral hogs on Cumberland Island, Georgia. *Proceedings Annual Conference Southeast Association of Fish and Wildlife Agencies* 51:285–296.

Williams, B. K. 1996. Adaptive optimaization and the harvest of biological populations. *Mathematical Biosciences* 136:1–20.

Wood, G. W., and R. E. Brenneman. 1977. Research and management of feral hogs on Hobcaw Barony. Pages 23–35 *in* G. W. Wood, editor. *Research and Management of Wild Hog Populations*. Belle Baruch Forest Science Institute of Clemson University, Georgetown, SC.

5 Diseases and Parasites That Impact Wild Pigs and Species They Contact

Joseph L. Corn and Michael J. Yabsley

CONTENTS

5.1 INTRODUCTION

Wild pigs (*Sus scrofa*) presumably are susceptible to all diseases of domestic swine, although few of these pathogens cause clinical disease in wild pigs. Thus, understanding disease dynamics in wild pigs is of importance because of their ability to serve as reserviors of disease agents, and to transmit them to wildlife, livestock, and humans. The World Organization for Animal Health (OIE), established in 1924, addresses worldwide issues associated with animal diseases. A recent review identified that wild pigs can be a host for 34 OIE-listed pathogens that can cause diseases in livestock, poultry, wildlife, and humans, with most being of concern to bovids (Miller et al. 2017). Most information on diseases in wild pigs in the United States is from surveys for evidence of selected disease agents. Data from other countries in North America and Caribbean and Pacific Islands are limited due to lack of surveillance. In this chapter, we discuss several viral, bacterial, and parasitic disease agents in detail, and provide tables summarizing other disease agents detected in wild pigs, including some with little or no published information available. Disease agents discussed in detail are ones that received significant attention through research and/or surveys. We also discuss management issues, foreign animal diseases, role of humans in spread of diseases in wild pigs, and trade issues.

5.2 BACTERIAL DISEASES

5.2.1 Swine Brucellosis

5.2.1.1 Etiology

Swine brucellosis is caused by a gram-negative bacteria of the species *Brucella suis*. There are 5 biovars, or physiologically unique strains, within the species of *B. suis*. Biovars 1 and 3 occur worldwide, and principal hosts are domestic and wild pigs. Biovar 2 occurs in swine and European hare (*Lepus europaeus*) in Europe; biovar 4 in reindeer and caribou (*Rangifer tarandus*) in Arctic regions; and biovar 5 in rodents in what was previously known as the Soviet Union (Bishop and Bosman 1994, MacMillan et al. 2006). Domestic swine have been experimentally infected with several *Brucella* species including *B. suis* biovar 4, *B. abortus*, *B. melitensis*, *B. ovis*, and *B. neotomae*, but in each case there was no invasion of the genital tract, no transmission between pigs, and the bacteria were found only in lymph nodes draining the infection sites (MacMillan et al. 2006).

5.2.1.2 Prevalence and Distribution

Antibodies against *Brucella* spp. occur in wild pig populations across most of their US range (Table 5.1; Figure 5.1). Serologic tests for antibodies are specific to *Brucella* genus, but not species. The Southeastern Cooperative Wildlife Disease Study summarized published and unpublished data collected in the United States from 1955 to 2004 in an unpublished report. Antibodies against *Brucella* spp. were detected in 6% (986 of 15,916) of wild pigs tested nationwide during this time, as well as in 9 of 15 states where testing had occurred. Average prevalence of seropositive wild pigs in counties where seropositive animals occurred was 8%. Pedersen et al. (2012) reported 3.5% (159 of 4,479) of wild pigs seropositive for *Brucella* spp. in 13 of 35 states during 2009–2010. *Brucella suis* biovar 1 is most common in wild pigs in the United States, and has been cultured from wild pigs in numerous southeastern states (Table 5.1). However, there are also rare isolations of other species including *B. abortus* strain 19, a vaccine strain previously used in cattle, from wild pigs in Texas (Corn et al. 1986); *B. abortus* biovar 1, vaccine strain *B. abortus* strain 19, and vaccine strain *B. abortus* RB51 from a single population of wild pigs in South Carolina (Stoffregen et al. 2007); and *B. suis* biovar 3 from wild pigs in Hawaii (Pedersen et al. 2014).

It is likely that *B. suis* in wild pigs is more widespread than testing has documented. Factors affecting occurrence of *B. suis* in wild pig populations include length of time a localized population has been established in the area, disease status of source animals in the initial introduction, disease status of animals in subsequent introductions, and potential for previous contact with infected domestic swine (factors reviewed in Corn et al. 2009).

5.2.1.3 Transmission

Transmission of *B. suis* is through direct contact via alimentary and reproductive tracts (Bishop and Bosman 1994, MacMillan et al. 2006). *Brucella suis* may be spread among domestic and wild pigs through coitus, and through consumption of aborted fetuses and other contaminated tissues.

5.2.1.4 Clinical Signs

Classical signs of brucellosis in domestic swine include abortion, infertility, orchitis (inflammation of testicles), posterior paralysis, and lameness. Clinical signs may be transient and mortality is uncommon (MacMillan et al. 2006). Reports on clinical signs in wild pigs are rare. *Brucella suis* biovar 1 was recovered from 9 wild pigs necropsied as part of a survey in Florida (Becker et al. 1978). Of these 9, 1 boar had an abscess in the seminal vesicle that was culture positive, and

TABLE 5.1
Bacterial Pathogens Detected in Wild Pigs in the United States and Selected Other Countries

Agent	Test	Year	US State or Country (With Province if Available)	No. Positive/No. Tested	References
Actinobacillus pleuropneumoniae	Serology	2011–2012	United States	113/199	Baroch et al. (2015)
	Serology	2009–2014	Canada (SK)	20/20	McGregor et al. (2015)
	Serology	2015	Guam (Yigo and Santa Rita)	41/44	Cleveland et al. (2017)
Bartonella spp.	PCR	2007–2009	NC	15/76	Beard et al. (2011)
Brucella suis biovar 1	Culture	1977–1978	FL	9/9	Becker et al. (1978)
	Culture	<1995	FL	2	Ewalt et al. (1997)
	PCR	2015–2017	GA	19/87 (of feces)	Lama and Bachoon (2018)
	Culture	1974	SC	1/2	Wood et al. (1976)
	Culture	2002–2003	SC	55/80	Stoffregen et al. (2007)
	Culture	1985	TX	4/100	Corn et al. (1986)
	Culture	2005	TX	2/40	Musser et al. (2013)
	Culture	2015	TX	49/376	Pedersen et al. (2017c)
	Culture	1979–1980	Southeast	3/292	Zygmont et al. (1982)
	Culture	2010–2012	United States	20/183	Pedersen et al. (2014)
Brucella suis biovar 3	Culture	2010–2012	United States	1/183	Pedersen et al. (2014)
Brucella abortus biovar 1	Culture	2002–2003	SC	21/80	Stoffregen et al. (2007)
	Culture	2011	TX	1/12	Higgins et al. (2012)
Brucella abortus strain 19	Culture	2002–2003	SC	8/80	Stoffregen et al. (2007)
	Culture	1985	TX	1/100	Corn et al. (1986)
Brucella abortus strain RB51	Culture	2002–2003	SC	6/80	Stoffregen et al. (2007)
Brucella spp.	Serology	1977–1989	CA	23/611	Drew et al. (1992)
	Serology	1981–1983	CA	21/136	Clark et al. (1983)
	Serology	1994–1995	CA	14/462	Sweitzer et al. (1996)
	Serology	1974–1989	FL	238/1,015	Van der Leek et al. (1993a)
	Serology	1977–1978	FL	50/95	Becker et al. (1978)
	Serology	1956	GA	2/30	Hanson and Karstad (1959)
	Serology	1970–1971	HI	21/268	Griffin (1972)
	Serology	2008–2011	MS	16/499	Jack et al. (2012)
	Serology	1993–2005	MO	1/321	Hartin et al. (2007)
	Serology	1990	NC/TN	0/108	New et al. (1994)
	Serology	2007–2010	NC	9/98	Sandfoss et al. (2012)
	Serology	2009–2015	OH	2/138	Linares et al. (2018)
	Serology	1996	OK	0/120	Saliki et al. (1998)

(*Continued*)

TABLE 5.1 (CONTINUED)
Bacterial Pathogens Detected in Wild Pigs in the United States and Selected Other Countries

Agent	Test	Year	US State or Country (With Province if Available)	No. Positive/No. Tested	References
	Serology	2010–2011	OK	2/282	Gaskamp et al. (2016)
	Serology	1974–1975	SC	46/255	Wood et al. (1976)
	Serology	1987–1988	SC	76/569	Wood et al. (1992)
	Serology	1999	SC	100/227	Gresham et al. (2002)
	Serology	2002–2003	SC	39/80	Stoffregen et al. (2007)
	Serology	2006	SC	7/50	Corn et al. (2009)
	Serology	1975–1976	TX	1/1	Randhawa et al. (1977)
	Serology	1985	TX	17/124	Corn et al. (1986)
	Serology	2004–2006	TX	41/368	Wyckoff et al. (2009)
	Serology	2005	TX	4/40	Musser et al. (2013)
	Serology	2006–2007	TX	5/409	Campbell et al. (2008)
	Serology	2015	TX	37/376	Pedersen et al. (2017c)
	Serology	1979–1980	Southeast	21/292	Zygmont et al. (1982)
	Serology	2009–2010	United States	159/4,479	Pedersen et al. (2012)
	Serology	2015	Guam (Yigo and Santa Rita)	1/46	Cleveland et al. (2017)
	Serology	N/A	Mexico (Baja CA Sur)	2/15	Perez-Rivera et al. (2017)
	Culture	2006	CA	12/30	Jay-Russel et al. (2012)
Campylobacter spp.	Culture	2007–2009	NC	7/161	Thakur et al. (2011)
Clostridium difficile	Culture	ca 2007	TX	1/1	Ramlachan et al. (2007)
Clostridium hathewayi	Serology	1981–1983	CA	67/135	Clark et al. (1983)
Coxiella burnetti	Serology	1975–1976	TX	1/1	Randhawa et al. (1977)
	PCR	2015–2017	GA	24/87	Lama and Bachoon (2018)
Escherichia coli (virulent strains)	Culture	2006	CA	2/40	Jay et al. (2007)
E. coli O157:H7	Serology	1993–2005	MO	0/321	Hartin et al. (2007)
Francisella tularensis	Serology	ca 2011	TX	50%, 15%	Texas Tech University (2011)
	Serology	2011–2012	United States	130/199	Baroch et al. (2015)
Lawsonia intracellularis	Serology	2015	Guam (Yigo and Santa Rita)	41/44	Cleveland et al. (2017)
	Serology	1981–1983	CA	111/130	Clark et al. (1983)
Mycobacterium avium	Culture/ Histology	1980	HI	12/61	Essey et al. (1981)
Mycobacterium bovis	Culture/ Histology	1983	HI	1/68	Essey et al. (1983)
	Serology	2009–2015	OH	0/13	Linares et al. (2018)
	Serology	2007–2015	United States	1/2,725	Pedersen et al. (2017d)
	Serology	N/A	AR, LA, OK, TX	16/50	Baker et al. (2011)

(Continued)

TABLE 5.1 (CONTINUED)
Bacterial Pathogens Detected in Wild Pigs in the United States and Selected Other Countries

Agent	Test	Year	US State or Country (With Province if Available)	No. Positive/No. Tested	References
Mycoplasma hyopneumoniae	Serology	2011–2012	United States	32/199	Baroch et al. (2015)
	Culture	2007–2009	NC	8/161	Thakur et al. (2011)
Salmonella spp.	Culture	2013–2015	TX	194/442	Cummings et al. (2016)
	Serology	2011–2012	United States	80/199	Baroch et al. (2015)
	Serology	N/A	Mexico (Baja CA Sur)	0/15	Perez-Rivera et al. (2017)
	Culture	2009–2015	OH	21/138	Linares et al. (2018)
Staphylococcus aureus (methicillin susceptible)	PCR	2011–2012	AL, AR, FL, GA, HI, and LA	34/37	Baroch et al. (2015)
Streptococcus suis	Serology	1974–1993	CA	137/738	Smith (1994)
Yersinia pestis	Serology	1981–1983	CA	9/59	Clark et al. (1983)
	Serology	1984	CA	17/23	Nelson et al. (1985)
	Serology	1994	CA	5/69	Sweitzer et al. (1996)

pus was present in the uterus of a non-pregnant sow, but cultures from the uterus were contaminated. Zygmont et al. (1982) cultured *B. suis* biovar 1 from 2 wild pigs from Florida and 1 from Louisiana. A sow <8 months old was in poor condition and had purulent metritis (pus producing inflammation of the uterus) and ulcerative dermatitis of the hind limbs, 1 boar was in poor condition, and 1 was in fair condition but no other necropsy results were noted. *Brucella suis* biovar 1 also was isolated from 4 wild pigs necropsied during a survey in Texas (Corn et al. 1986). Of these 4, 1 boar was in good condition, 1 sow was pregnant and in good condition but with vaginal discharge, 1 sow was in fair condition and pregnant, and 1 sow was in good condition and not pregnant.

5.2.1.5 Management Concerns: Wildlife Health

As swine are the reservoir for *B. suis* biovar 1, wild pigs would serve as a potential source for infection of other wildlife; however, documented cases of *B. suis* transmitting from wild pigs to other wildlife are lacking and hence impacts are unknown.

5.2.1.6 Management Concerns: Domestic Animal Health

Wild pigs are the reservoir for *B. suis* biovar 1 in the United States, as domestic swine are considered free of brucellosis. The presence of *B. suis* in wild pigs represents a risk for infection of domestic swine, cattle, horses, and dogs, and documented cases of transmission exist. Over half (26 of 41) of domestic swine herds infected with *B. suis* in Texas between 2003 and 2008 had possible or definite wild pig exposure (Black and Corn 2008). Show pigs in a vaccinated herd in Georgia became infected with *B. suis* as a result of inadequate biosecurity and wild pig exposure (Coats and Black 2005). *Brucella suis* from a Michigan transitional swine herd (i.e., captive wild pigs or domestic pigs likely to be exposed to wild pigs; Corn et al. 2009) was most closely related to a source from wild pigs in the southeastern United States (Black 2009). Cattle naturally infected with *B. suis* biovar 1 were reported in Florida (Ewalt et al. 1997) and Texas (Black 2009,

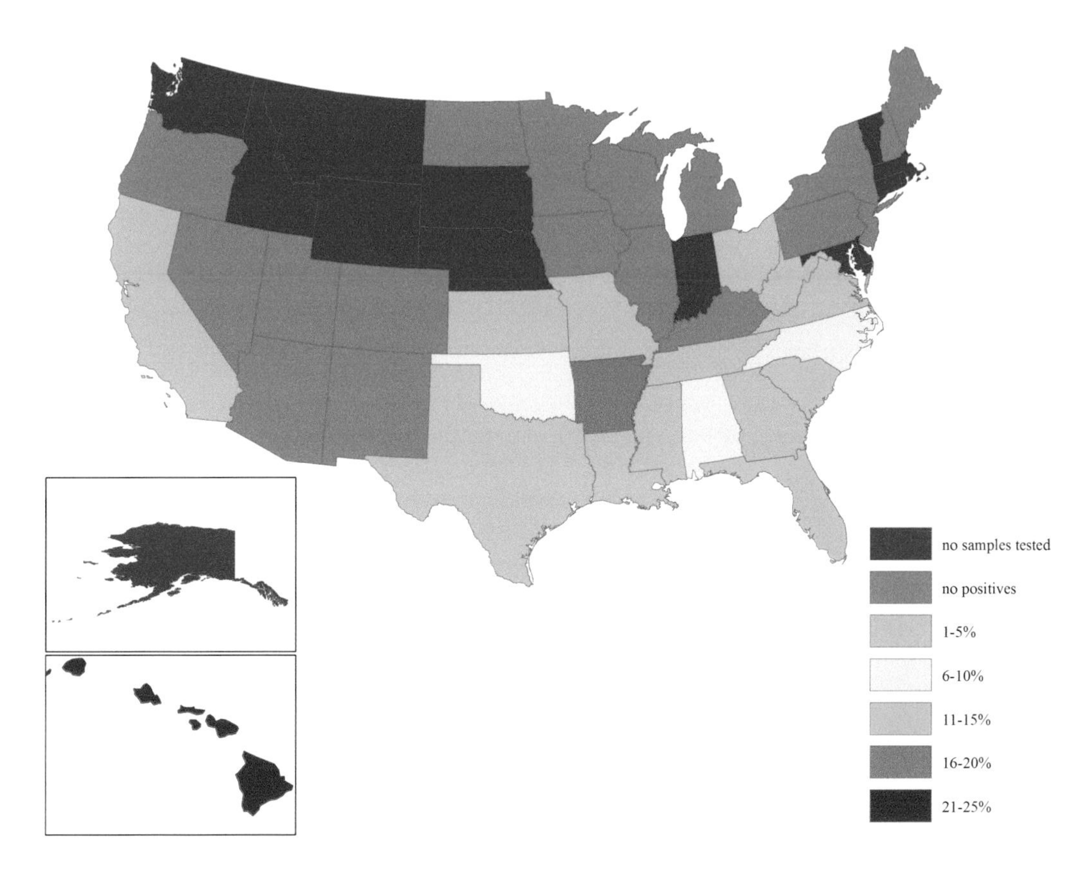

FIGURE 5.1 General prevalence of antibodies to *Brucella* in wild pigs in the United States from 2011 to 2017. (Reproduced from the US Department of Agriculture, unpublished data.)

Musser et al. 2013), and infected wild pigs were present in both areas. As infections of *B. suis* biovar 1 are widespread in wild pigs but rare and non-contagious in cattle, it is presumed that wild pigs were the source of these infections (Ewalt et al. 1997). The first record of brucellosis in a horse in the United States was isolation of *B. suis* from an aborted fetus in 1924 (McNutt and Murray 1924). In a survey of dogs collected by veterinarians servicing kennels and pet dogs in Georgia, all dogs seropositive for *Brucella* spp. were dogs with a history of hunting wild pigs (Ramamoorthy et al. 2011).

5.2.1.7 Management Concerns: Public Health

Brucella suis biovars 1 and 3 are highly pathogenic to humans and infection by biovar 1 has occurred in humans in the United States as a result of contact with wild pigs (Woldemeskel 2013). Infection in these cases occurred while field dressing or butchering wild pigs. Bigler et al. (1977) reported that 22% (6 of 27) of human cases of brucellosis during 1974–1975 were associated with hunting wild pigs in Florida, and Starnes et al. (2004) reported 2 cases associated with hunting wild pigs in South Carolina. Additionally, 3 cases of brucellosis in 2007–2008, 1 each in Florida, Pennsylvania, and South Carolina, were attributed to exposure during field dressing and butchering wild pig carcasses while hunting in Florida (Centers for Disease Control and Prevention 2009). Simoes and Justino (2013) and Franco-Paredes et al. (2017) reported additional cases in wild pig hunters in Florida and Georgia. The Florida Department of Health (2012) reported that most cases of brucellosis in Florida were from hunters that had contact with blood or other fluid and tissues from wild pigs.

5.2.2 Leptospirosis

5.2.2.1 Etiology

Leptospirosis is caused by *Leptospira interrogans*, a gram-negative spirochete bacterium. There are over 300 serovars or distinguishable strains classified based on unique cell surface antigens. Some serovars are more adapted to occur in certain hosts, (i.e., maintenance hosts) that maintain the organism in nature and rarely develop clinical disease. In the United States, most common serovars detected in domestic swine are Bratislava, Grippotyphosa, and Pomona, although reports of other serotypes are rare. Serovars Grippotyphosa, Pomona, and Canicola have been isolated from wild pigs (Forrester 1992).

5.2.2.2 Prevalence and Distribution

Most studies on prevalence of leptospirosis have been based on detection of antibodies in which the titer level is important; titers ≥100 are suggestive of exposure whereas higher titers (e.g., ≥800) are indicative of active or recent infections (Table 5.2). Reports of antibodies against *Leptospira* occur across the range of wild pigs in the United States, and in most tested populations outside the United States (Table 5.2). Few United States studies investigated presence of leptospires in kidneys, which is what indicates a shedding risk. One of the larger studies, Pedersen et al. (2017a), detected bacteria in kidneys of 3.4% (23 of 677) of wild pigs from numerous states, including Mississippi which had the highest prevalence at 26% (13 of 53). *Leptospira* was also isolated from kidney samples from 20% (1 of 5) of wild pigs from southern Georgia, but the isolate was not typed (Gorman et al. 1962).

5.2.2.3 Transmission

Transmission of *Leptospira* is generally via contact with urine from an infected animal or through ingestion of bacteria in contaminated water; however, the bacteria can enter through cuts, abrasions, or mucous membranes. Transmission by these routes is always associated with wet environments as the bacteria must remain moist. Additional routes of transmission can be in utero infection and through ingestion or inhalation of bacteria in placenta or fetal tissues.

5.2.2.4 Clinical Signs

Reports of mortality in wild pigs do not exist, and domestic pigs rarely die due to *Leptospira* infections. Domestic pigs may suffer abortions, poor conception rates, anorexia, icterus (jaundice), and may display weakness and an inability to stand or walk. Detection of lesions is rare in adult wild pigs but scattered foci of interstitial nephritis (inflammation of tubules within kidneys) may be observed. Piglets may have hepatic necrosis (liver failure), intraperitoneal fluid, and disseminated interstitial nephritis. In a study in Germany, moderate to severe chronic lymphoplasmacytic interstitial inflammation was twice as likely to occur in seropositive wild pigs, and 3 of 10 pigs with these lesions had leptospires detected by silver staining, 2 of which were confirmed with polymerase chain reaction (PCR; Jansen et al. 2007).

5.2.2.5 Management Concerns: Wildlife Health

The *Leptospira* serovars found in pigs can also infect a wide range of mammals. Serologic studies indicate exposure is common and many of these wildlife species can shed bacteria, but disease is rarely detected. Although sporadic clinical cases are reported in wildlife (e.g., pinnipeds, red fox (*Vulpes vulpes*), and fox squirrel (*Sciurus niger*); Kingscote, 1986, Cameron et al. 2008, Dirsmith et al. 2013), experimental studies in certain species, such as white-tailed deer (*Odocoileus virginianus*) and raccoons (*Procyon lotor*), suggest disease is possible and may be underreported (Reilly 1970, Shotts 1981).

5.2.2.6 Management Concerns: Domestic Animal Health

Leptospirosis is a concern for many species of domestic animals including dogs, horses, cattle, sheep, and goats. Although less often diagnosed, domestic cats can also become infected and shed

TABLE 5.2

Prevalence of *Leptospira interrogans* Serovars in Wild Pigs in the United States and Selected Countries Based on Microagglutination Assay

Country	State/Region (for United States)	Positive Titer[a]	N[b]	All Serotypes[c]	Bratislava	Canicola	Grippotyphosa	Hardjo	Icterohemorrhagiae	Pomona	Reference
United States	CA	N/A[d]	136	118	N/A	N/A	N/A	N/A	N/A	N/A	Clark et al. (1983)
	CA (Santa Cruz Island)	100	223	32	12	3	16	0	15	2	Blumenshine et al. (2009)
	FL[e]	N/G[f]	39	9	0	6	1	0	2	0	Forrester (1992)
	FL[g]	100	324	107	58	2	15	N/T[h]	37	4	Chatfield et al. (2013)
	HI	100	804	272	92	N/T	N/T	N/T	113	1	Buchholz et al. (2016)
	OK	100	117	52	34	1	8	20	4	27	Saliki et al. (1998)
	TN/NC[i]	100	108	48	39	39	40	43	39	43	New et al. (1994)
	TX[j]	100	120	N/G	18	0	0	0	21	3	Corn et al. (1986)
	TX	100	376	184	117	3	16	4	54	112	Pedersen et al. (2017b)
	29 states, 2012–2014	100	642	340	197	62	60	32	165	87	Pedersen et al. (2017a)
	28 states, 2007–2011	200	2055	269	106	12	75	32	59	124	Pedersen et al. (2015)
Guam	Yigo and Santa Rita	100	46	11	6	0	0	1	6	2	Cleveland et al. (2017)
Mexico[k]	Baja California Sur	100	15	4	N/G	0	N/G	0	N/G	0	Perez-Rivera et al. (2017)

[a] Cutoff titer researchers used to determine a positive result.
[b] Total # of wild pigs sampled.
[c] Total # of wild pigs seropositive for 1 or more serotypes. This number will not always be the sum of the 6 serovars shown because some individual pigs had antibodies to 2 or more serovars and some investigators may have tested pigs for other serovars not shown.
[d] Not applicable (these researchers used an ELISA) .
[e] 1 pig had antibodies to serovar Autumnalis.
[f] Not given.
[g] Antibodies were detected to serovars Australis (25 pigs), Autumnalis (37), Caelledoni (3), Cynopteri (6), Djasaiman (4), Mankarso (3), Georgia (4), Alexi (4), Pyrogenes (3). All pigs were negative for serovars Ballum, Bataviae, Borincana, Javnica, Wolffi, and Tarassovi.
[h] Not tested.
[i] Great Smoky Mountains National Park.
[j] 0 pigs had antibodies to serovar Autumnalis while 1 had antibodies to serovar Tarassovi.
[k] Antibodies to serovars Bratslava, Grippotyphosa, Icterohemorrhagiae, Tarassovi, and Pyrogenes were detected but exact numbers are not known.

bacteria. Vaccination decreases infection risk in some species. Ellis (2015) provides a review on animal leptospirosis.

5.2.2.7 Management Concerns: Public Health

Leptospirosis is an important zoonotic disease, especially in tropical and subtropical countries, where people are in close contact with animal urine, infected tissues, or water contaminated with animal urine. It can cause fever, chills, intense headaches, nausea, vomiting, muscle aches and in severe cases, kidney disease and pulmonary hemorrhage. Additional information on clinical disease, diagnosis, and epidemiology have been reviewed (McBride et al. 2005, Costa et al. 2015).

In general, prevalence and titers of serovar Icterohaemorrhagiae, a common zoonotic serovar in pigs, were low, suggesting pigs do not serve as an important reservoir. Serovars Pomona and Bratislava are zoonotic and common in wild pigs. Historically, these serovars are less common in people as other serovars; however, there are increasing numbers of human cases with serogroup Australis that includes Bratislava (Saliki et al. 1998, Pedersen et al. 2015, Bucholz et al. 2016, Pedersen et al. 2017b).

5.3 VIRAL DISEASES

5.3.1 Pseudorabies

5.3.1.1 Etiology

Pseudorabies (Aujeszky's disease) is caused by suid herpesvirus 1, commonly referred to as pseudorabies virus (PRV). This virus is in the family Herpesviridae, subfamily Alphaherpesvirus. There is 1 serotype, but multiple strains (Kluge et al. 1992, Müller et al. 2011).

5.3.1.2 Prevalence and Distribution

Antibodies against PRV have been detected in wild pig populations across most of their range in the United States (Table 5.3; Figure 5.2). Antibodies against PRV were detected in 28% (4,276 of 15,399) of wild pigs tested nationwide during 1955–2004 based on published and unpublished data (Southeastern Cooperative Wildlife Disease Study, unpublished data), and in 10 of 16 states where testing occurred. Average prevalence of seropositive wild pigs in counties known to contain seropositive animals was 30%. Pedersen et al. (2013) reported results of serologic surveys conducted by the United States Department of Agriculture during 2009–2012. Of 8,498 wild pigs sampled, 18% were antibody positive within 25 of 35 states.

Corn et al. (2004) found that PRV persists in wild pig populations over time. Surveys conducted at 10 sites from 13 to 22 years after a previous detection of seropositive animals determined that seropositive wild pigs occurred in all 10 populations. Pseudorabies occurred in at least 6 of 10 populations for over 20 years. Pedersen et al. (2013) also found that PRV persisted in counties with previous positive test results for wild pigs, but not in all counties, and suggested the negative results were likely due to detectability (e.g., inadequate sample size relative to prevalence). However, Hernández et al. (2018) found that 7% (14 of 212) of seronegative wild pigs were PCR positive, suggesting that serologic testing alone can underestimate prevalence.

Corn et al. (2009) surveyed for evidence of PRV exposure in wild pigs in areas with high densities of transitional domestic pig operations (captive wild pigs or domestic pigs with potential exposure to wild pigs) in South Carolina and commercial production pig facilities (intensively managed and lacking potential for direct exposure to wild pigs) in North Carolina. Wild pigs tested seropositive for PRV (20.0%) in South Carolina, but PRV was not detected in wild pigs in North Carolina. Wild pigs were present in South Carolina prior to PRV eradication efforts in domestic pigs, whereas populations in North Carolina were established more recently (Southeastern Cooperative Wildlife Disease Study 2004, Corn and Jordan 2017) and may not have been present until after eradication of PRV from domestic pigs in the area (Corn et al. 2009).

TABLE 5.3
Viral Pathogens Detected in Wild Pigs in the United States and Selected Other Countries

Agent	Test	Year	US State or Country (With Province if Available)	No. Positive/ No. Tested	References
Pseudorabies virus (suid herpesvirus 1)	Serology	1981–1983	CA	4/135	Clark et al. (1983)
	Serology	1980–1989	FL	579/1,662	Van der Leek et al. (1993b)
	Serology	2014–2016	FL	224/436	Hernández et al. (2018)
	PCR	2014–2016	FL	38/549	Hernández et al. (2018)
	Serology	1981–1986	GA	55/661	Pirtle et al. (1989)
	Serology	2008–2011	MS	37/499	Jack et al. (2012)
	Serology	1993–2005	MO	0/321	Hartin et al. (2007)
	Serology	2007	NE	1 (n/a)	Wilson et al. (2009)
	Serology	2009–2012	NH	1/34	Musante et al. (2014)
	Serology	2001–2007	NC, TN	16/497	Cavendish et al. (2008)
	Serology	2009–2015	OH	0/139	Linares et al. (2018)
	Serology	1996	OK	0/120	Salki et al. (1998)
	Serology	2010–2011	OK	68/282	Gaskamp et al. (2016)
	Serology	1987–1988	SC	55/569	Wood et al. (1992)
	Serology	2002	SC	138/227	Gresham et al. (2002)
	Serology	2006	SC	10/50	Corn et al. (2009)
	Serology	1985	TX	46/124	Corn et al. (1986)
	Serology	2004–2006	TX	109/369	Wyckoff et al. (2009)
	Serology	2006–2007	TX	145/409	Campbell et al. (2008)
	Serology	2010	TX	73/279	Campbell et al. (2011)
	Serology	1979–1983	Southeastern United States	93/423	Nettles and Erickson (1984)
	Serology	2001–2002	Southeastern United States	38/100	Corn et al. (2004)
	Serology	2009–2012	United States	1,529/8,498	Pedersen et al. (2013)
	Serology	2015	Guam (Yigo and Santa Rita)	29/45	Cleveland et al. (2017)
	Serology	N/A[a]	Mexico (Baja California Sur)	0/15	Perez-Rivera et al. (2017)
Porcine Reproductive and Respiratory Syndrome	Serology	2007, 2010	HI (Hawaii)	1/52	Stephenson et al. (2015)
	Serology	2007, 2010	HI (Oahu)	19/292	Stephenson et al. (2015)
	Serology	2006–2007	NC	1/120	Corn et al. (2009)
	Serology	2009–2015	OH	0/72	Linares et al. (2018)
	Serology	1996	OK	2/117	Saliki et al. (1998)
	Serology	2010–2011	OK	1/282	Gaskamp et al. (2016)
	Serology	2004–2006	TX	2/137	Wyckoff et al. (2009)
	Serology	2006–2007	TX	3/409	Campbell et al. (2008)
	Serology	2010	TX	2/134	Campbell et al. (2011)
	Serology	N/A	AR, LA, OK, TX	1/50	Baker et al. (2011)
	Serology	2011–2012	United States	4/162	Baroch et al. (2015)

(*Continued*)

TABLE 5.3 (CONTINUED)
Viral Pathogens Detected in Wild Pigs in the United States and Selected Other Countries

Agent	Test	Year	US State or Country (With Province if Available)	No. Positive/ No. Tested	References
	Serology	2009–2014	Canada (SK)	0/22	McGregor et al. (2015)
	Serology	2015	Guam (Yigo and Santa Rita)	6/46	Cleveland et al. (2017)
	Serology	N/A	Mexico (Baja California Sur)	0/15	Perez-Rivera et al. (2017)
Porcine circovirus type 2	Serology	2007, 2010	HI (Hawaii)	21/52	Stephenson et al. (2015)
	Serology	2007, 2010	HI (Oahu)	25/292	Stephenson et al. (2015)
	Serology	2006–2007	NC	86/120	Corn et al. (2009)
	Serology	2007–2010	NC	53/90	Sandfoss et al. (2012)
	Serology	2006	SC	29/49	Corn et al. (2009)
	Serology	N/A	AR, LA, OK, TX	21/50	Baker et al.(2011)
	Serology	2011–2012	United States	41/199	Baroch et al. (2015)
	PCR	2009–2014	Canada (SK)	0/58	McGregor et al. (2015)
	Serology	2015	Guam (Yigo and Santa Rita)	16/44	Cleveland et al. (2017)
Torque teno virus	PCR	2009–2014	Canada (SK)	1/8	McGregor et al. (2015)
Influenza A viruses					
H1N1	Serology	1993–1994	KS	15/20	Gipson et al. (1999)
H1N1	Serology	1996	OK	13/117	Saliki et al. (1998)
H1N1	Serology	2009–2014	Canada (SK)	0/22	McGregor et al. (2015)
H1N1	Serology	N/A	Mexico (Baja California Sur)	7/15	Perez-Rivera et al. (2017)
H3N2	Serology	2005–2006	CA	5/94	Hall et al. (2008)
H3N2	Serology	2005–2006	MS	1/99	Hall et al. (2008)
H3N2	Serology	2005–2006	TX	57/472	Hall et al. (2008)
H3N2	Serology	2009–2014	Canada (SK)	0/22	McGregor et al. (2015)
H3N2	Serology	N/A	Mexico (Baja California Sur)	9/15	Perez-Rivera et al. (2017)
H1N1	Serology	2005–2006	TX	4/472	Hall et al. (2008)
H3N2 + H1N1	Serology	2005–2006	TX	7/472	Hall et al. (2008)
HuH1N1	Serology	2006–2007	NC	87/119	Corn et al. (2009)
rH1N1	Serology	2006–2007	NC	8/119	Corn et al. (2009)
H1N1	Serology	2006–2007	NC	17/119	Corn et al. (2009)
H3N2	Serology	2006–2007	NC	56/119	Corn et al. (2009)
SIV	Serology	2007–2010	HI (Hawaii)	1/52	Stephenson et al. (2015)
SIV	Serology	2007–2010	HI (Oahu)	25/292	Stephenson et al. (2015)
H1	Serology	N/A	AR, LA, OK, TX	1/50	Baker et al. (2011)
H3	Serology	N/A	AR, LA, OK, TX	20/50	Baker et al. (2011)
pH1N1	Serology	2011	TX	2/2	Clavijo et al. (2013)
H3N2, H1N1	Serology	2011–2012	United States	182/1989	Feng et al. (2014)
Influenza A general	Serology	2010	TX	6/102	Campbell et al. (2011)
Influenza A general	Serology	2015	TX	53/376	Pedersen et al. (2017d)

(Continued)

TABLE 5.3 (CONTINUED)
Viral Pathogens Detected in Wild Pigs in the United States and Selected Other Countries

Agent	Test	Year	US State or Country (With Province if Available)	No. Positive/ No. Tested	References
Influenza A general	Serology	2015	Guam (Yigo and Santa Rita)	0/47	Cleveland et al. (2017)
Swine influenza	PCR	2011–2012	United States	9/1983	Feng et al. (2014)
West Nile Virus	Serology	2001–2004	FL, GA, TX	50/222	Gibbs et al. (2006)
Venezuelan equine encephalitis	Serology	1963–1973	TX	31/51	Smart et al. (1975)
Eastern equine encephalitis	Serology	1992–1992	GA	62/376	Elvinger et al. (1996)
Vesicular exanthema of swine	Serology	1973	CA	17/49	Smith and Latham (1978)
San Miguel sea lion virus	Serology	1973	CA	13/49	Smith and Latham (1978)
Porcine parvovirus	Serology	1993–1994	KS	14/20	Gipson et al. (1999)
	Serology	1990	NC/TN	15/108	New et al. (1994)
	Serology	1996	OK	20/117	Saliki et al. (1998)
	Serology	2010	TX	130/137	Campbell et al. (2011)
	Serology	2015	Guam (Yigo and Santa Rita)	16/46	Cleveland et al. (2017)
Vesicular stomatitis virus	Serology	ca 1958	GA	46/58	Hanson and Karstad (1958)
	Serology	1981–1982	GA (Ossabaw Island)	84/158	Fletcher et al. 1985)
	Serology	1982, 1983	GA (Ossabaw Island)	~12% and 60%	Stallknecht et al. 1985)
	Serology	1984, 1985	GA (Ossabaw Island)	~32% and 36%	Stallknecht et al. (1987)
	Serology	1990	GA (Ossabaw Island)	58/243	Stallknecht et al. (1993)
	Virus isolation	1990	GA (Ossabaw Island)	5/54	Stallknecht et al. (1993)
	Serology	1996	OK	0/120	Saliki et al. (1998)
	Serology	1979–1985	Southeastern United States	75/941	Stallknecht et al. (1986)
Infectious bovinerhinotracheitis	Serology	1993–1997	GA (Ossabaw Island)	~73/208	Killmaster et al. (2011)
Hepatitis E virus	Serology	2009–2015	OH	1/32	Linares et al. (2018)
	Virus isolation	ca 1987	TX	1	Crandell et al. (1987)
Porcine rotavirus	Serology	N/A	AR, FL, GA, NC, SC, TX	9/306	Dong et al. (2011)
	Serology	1990	NC/TN	0/108	New et al. (1994)
(TGE)/(PRCV)[b]	Serology	N/A	FL	0262	Woods et al. (1990)

(*Continued*)

TABLE 5.3 (CONTINUED)
Viral Pathogens Detected in Wild Pigs in the United States and Selected Other Countries

Agent	Test	Year	US State or Country (With Province if Available)	No. Positive/ No. Tested	References
	Serology	N/A	GA (Ossabaw Island)	0/184	Woods et al. (1990)
	Serology	1996	OK	0/120	Saliki et al. (1998)
	Serology	N/A	TX	0/114	Woods et al. (1990)
	Serology	2009–2014	Canada (SK)	0/22	McGregor et al. (2015)
PEDV[c]	Serology	2009–2015	OH	1/45	Linares et al. (2018)
	Serology	2015	Guam (Yigo and Santa Rita)	1/44	Cleveland et al. (2017)
Swine enterovirus	Serology	1993–1994	KS	13/20	Gipson et al. (1999)

[a] Not available.
[b] Transmissible gastroenteritis/Porcine respiratory coronavirus.
[c] Porcine epidemic diarrhea virus.

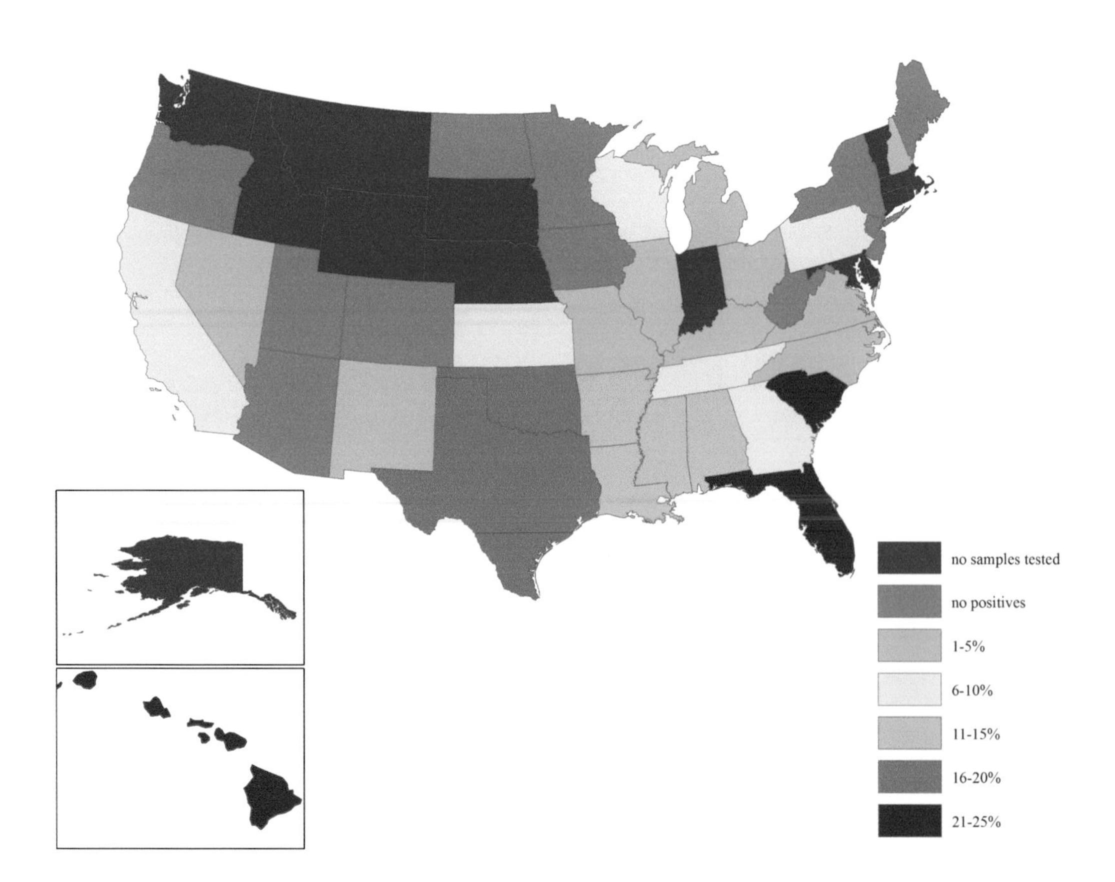

FIGURE 5.2 General prevalence of antibodies to pseudorabies virus in wild pigs in the United States from 2011 to 2017. (Reproduced from the US Department of Agriculture, unpublished data.)

Additionally, PRV may have occurred in wild pigs from South Carolina as a result of association between wild pigs and infected transitional pigs; compared to North Carolina, where biosecurity measures at commercial facilities would have prevented such contact. There are numerous factors affecting occurrence and prevalence of PRV in wild pig populations including length of time the population has been present, disease status of source animals in the initial introduction, disease status of animals in subsequent introductions, and potential for previous contact with infected domestic swine (reviewed in Corn et al. 2009). It is likely that distribution of PRV in wild pigs is more widespread than testing has documented.

5.3.1.3 Transmission

Pseudorabies transmission in domestic swine is primarily by direct contact, but also occurs via inhalation, venereal transmission, and orally via ingestion of contaminated material (Gustafson 1986). Pseudorabies may be isolated from nasal secretions, tonsils, and genital organs (Hahn et al. 1997), and Hernández et al. (2018) detected PRV in blood, nasal swabs, oral swabs, and vaginal swabs by PCR of presumably chronically infected wild pigs in Florida. Documented transmission in wild pigs has been via the respiratory route, consumption of animals that had died of acute infection (Hahn et al. 1997), and through venereal contact (Romero et al. 1997, 2001). Hahn et al. (1997) found that consumption of tissues from acutely infected animals resulted in transmission of virus to wild pigs, but consumption of tissues from latently infected animals did not. Romero et al. (1997, 2001) found that PRV indigenous to wild pigs is primarily transmitted to wild or domestic swine via the venereal route, but in a modeling experiment Smith (2012) found that venereal transmission alone might not account for high PRV seroprevalence found in wild pigs.

5.3.1.4 Clinical Signs

Clinical disease associated with PRV infection in free-ranging wild pigs has not been reported. Experimental infection of young wild boar with a virus isolated from a European domestic pig did not result in clinical signs (Tozzini et al. 1982); however, Müller et al. (2001) found that virulence in wild boar was strain dependent. Clinical signs in wild pigs from the United States experimentally infected by Hahn et al. (1997), using several wild pig-derived strains of PRV, were mild to nonexistent, although mortality occurred among wild pigs <4 weeks of age. These wild pigs developed subclinical, latent infections. Wild pigs experimentally infected using a domestic swine strain (Funkhauser virus) exhibited fever, central nervous system signs, and death, depending on dosage (Hahn et al. 1997). Clinical signs of PRV infection in domestic swine depend on virus strain, infectious dose, and animal age, ranging from respiratory and neurologic signs in young pigs to respiratory signs and abortion in adult swine (Kluge et al. 1992).

5.3.1.5 Management Concerns: Wildlife Health

As the only reservoir of PRV in the United States, wild pigs serve as a risk and source of infection for other wildlife, especially scavengers and predators, including threatened and endangered species. Worldwide, infections have been reported in an endangered Florida panther (*Felis concolor coryi*; Glass et al. 1994), Norway rats (*Rattus norvegicus*; Maes et al. 1979), raccoons (Kirkpatrick et al. 1980), a roe deer (*Capreolus capreolus*; Nikolitsch 1954), black bear (*Ursus americanus*; Pirtle et al. 1986, Schultze et al. 1986), captive brown bears (*Ursus arctos*; Zanin et al. 1997), captive coyotes (*Canis latrans*; Raymond et al. 1997), farmed mink (*Mustela vison*; Konrad and Blazek 1958), and farmed red fox (Ljubashenko et al. 1958). Additionally, the range of Louisiana black bear (*U. americanus luteolus*), which was delisted from the federal List of Endangered and Threatened Wildlife in 2016 (Federal Register 2016), overlaps with PRV-infected wild pigs. Experimental infections have been reported in numerous other species of mammals and some birds (Galloway 1938, Shahan et al. 1947, Trainer and Karstad 1963, Gustafson 1986, Stallknecht and Howerth 2001) but not poikilothermic (cold-blooded) animals (Beran 1992).

Clinical signs in wildlife include anorexia, depression, hypersalivation, paralysis, convulsions, lethargy, incoordination, shaking, nervous excitability, and coma, but infections typically are fatal in species other than swine (Trainer and Karstad 1963, Wright and Thawley 1980, Schultze et al. 1986, Zanin et al. 1997, Raymond et al. 1997). For example, reports of mortality due to PRV infection in Florida panthers have been documented (Glass et al. 1994). However, raccoons survived an experimental infection with PRV in which direct transmission was documented (Platt et al. 1983).

5.3.1.6 Management Concerns: Domestic Animal Health

The US Department of Agriculture, National Pork Producers Council, and state animal health agencies began an eradication program for PRV in 1989 (King 1990). All 50 states were granted Stage V (Free) status by 2004 (Anderson and Leafstedt 2004, US Department of Agriculture 2008), but isolated detections of PRV continued (Anderson and Leafstedt 2005, Engle and Snelson 2006, Snelson 2010). Wild pigs currently serve as the reservoir for PRV in the United States, and the potential for transmission of PRV from wild pigs into domestic swine populations is of serious concern to the domestic swine industry and animal health agencies. Increasing distribution of wild pigs in the United States resulted in increased opportunities for transmission of PRV from wild to domestic swine (Corn et al. 2005, Bevins et al. 2014), and Wyckoff et al. (2009) documented contacts between wild pigs and domestic pigs in Texas. Fatal infections in dogs used to hunt wild pigs also have been documented (Cramer et al. 2011), and additional unpublished reports exist from other states including Arkansas and Florida.

5.3.1.7 Management Concerns: Public Health

Illnesses in humans suspected from PRV infection have been reported, but none confirmed. In these cases, pruritus (intense itching), fever, sweating, weakness, and/or possible central nervous system involvement were reported (Mravak et al. 1987).

5.3.2 Porcine Reproductive and Respiratory Syndrome

5.3.2.1 Etiology, Prevalence, and Distribution

Porcine reproductive and respiratory syndrome (PRRS) was first observed during 1987–1988 in domestic swine herds in North Carolina, Iowa, and Minnesota (Keffaber 1989), and then in Western Europe in 1990–1991 (Wensvoort et al. 1991). The PRRS virus (PRRSV) is a positive-sense single-stranded RNA virus of the order Nidovirales and family Arteriviridae. The virus was first isolated in the Netherlands in 1990 (Wensvoort et al. 1991) and identified as the Lelystad virus. Shortly after, a similar strain was isolated in the United States (Collins et al. 1992) and identified as VR-2332. The North American and European isolates of PRRSV induce similar clinical symptoms but represent 2 distinct viral genotypes that differ in virulence, genetic, and antigenic properties (Nelsen et al. 1993, 1999). Antibodies against PRRSV occur in low numbers of wild pigs across the United States (Table 5.3).

5.3.2.2 Transmission

In domestic pigs, infection most likely occurs through oronasal, fecal, and urinary routes (Rossow et al. 1994, Albina 1997). Aerial transmission occurs over distances up to 20 km, however this infection route is probably more important over shorter distances (i.e., less than 3 km; Albina 1997, Le Potier et al. 1997, Kristensen et al. 2004). Semen from infected intact and vasectomized boars carries PRRSV (Gradil et al. 1996, Rossow 1998). Pregnant sows exposed to PRRSV can transmit the virus to fetuses in utero (McCaw 1995). Zimmerman et al. (1997) reported that some waterfowl harbored and shed PRRSV following experimental exposure. Although infection was subclinical, feces of infected birds remained infectious for at least 24 days. No other species have been found susceptible to experimental infection (Albina 1997).

5.3.2.3 Clinical Signs

Clinical signs depend on sex, age, pregnancy status, trimester of gestation, and co-infection with other diseases or bacteria (Rossow 1998, Animal Health Australia 2004). Clinical signs in infected gilts or sows range from none to lethargy, fever, agalactia (absence of milk flow), pneumonia, anorexia, subcutaneous and hind limb edema, blue or red discoloration of the ears and vulva, a post-weaning delayed return to estrus, and in some cases, death (Hopper et al. 1992, Terpstra et al. 1992, Done and Paton 1995). Infection of pregnant gilts or sows in the third-trimester may result in late-term abortion or premature farrowing with mummified, stillborn, and partially autolyzed fetuses (Loula 1991, Hopper et al. 1992, Terpstra et al. 1992). Clinical signs in neonatal pigs classically include labored or rapid breathing, anorexia, post-injection bleeding, rough hair coats, shaking, diarrhea, cutaneous erythema, fever, blue discoloration of the ears, eyelid edema, conjunctivitis, and periocular edema. Mortality approaching 100% may occur (Loula 1991, Anonymous 1992, Hopper et al. 1992, Cooper et al. 1995). Clinical signs in weaned pigs may include lethargy, fever, pneumonia, failure to thrive, and an increase in mortality from co-infection with bacteria or other viruses (Loula 1991, Terpstra 1992, Rossow et al. 1994). Concurrent infection with PRRSV and bacteria or other viruses may be clinically significant, based on observed increases in postweaning mortality (Hopper et al. 1992, Done and Paton 1995). Subclinical signs are common in recently weaned pigs (Hopper et al. 1992). Infection in unbred gilts or sows, boars, and finishing pigs may manifest as a transient fever and inappetence (Hopper et al. 1992, Gradil et al. 1996). Some infected boars may display loss of libido while most show no clinical signs (Gordon 1992; Prieto et al. 1996a, 1996b).

The PRRSV enters hosts through mucosal surfaces, and replicates in local macrophages. From there, PRRSV distributes to regional lymphoid tissues with subsequent systemic distribution to macrophages in multiple tissues (Duan et al. 1997, Rossow 1998). Gross lesions generally occur in respiratory and lymphoid systems, and are usually more remarkable in juvenile pigs (Benfield et al. 1999). In general, PRRS manifests as a reproductive disease in adult sows, and a respiratory disease in juvenile pigs.

5.3.2.4 Management Concerns

Porcine Reproductive and Respiratory Syndrome is not zoonotic or known to affect animals other than swine. Infection of some avian species occurred experimentally, but no current data suggest that infection in other species occurs naturally. Porcine Reproductive and Respiratory Syndrome is endemic in domestic pig populations throughout most of the world. Its impacts primarily relate to production losses in the domestic swine industry. The extent and severity of these losses vary, as manifestation of the disease varies from farm to farm.

5.3.3 Porcine Circovirus Type 2 (PCV2)

5.3.3.1 Etiology, Prevalence, and Distribution

Initially described as postweaning multisystemic wasting syndrome (PMWS), Porcine circovirus type 2 (PCV2) was first recognized in domestic pigs in Canada in 1991 (Allan et al. 1998, Clark 1997, Meehan et al. 1998). Porcine circovirus diseases (PCVD), as this group of swine diseases is now termed, including PMWS (Segalés 2012), have since been documented in swine herds worldwide (Chae 2004, Opriessnig et al. 2007, Grau-Roma et al. 2011, Segalés et al. 2013). Circoviruses may infect birds, mammals, and fish (Delwart and Li, 2012), but PCV2 is the only circovirus known to be pathogenic in mammals (Franzo et al. 2015). Four PCV2 genotypes are known (Franzo et al. 2015). Antibodies against PCV2 occur in wild pigs across the United States (Table 5.3).

5.3.3.2 Transmission

Previous studies on domestic swine indicated that PCV2 transmission occurs by direct contact via oronasal, fecal, and urinary routes (Magar et al. 2000, Bolin et al. 2001, Shibata et al. 2003, Caprioli et al. 2006). González-Barrio et al. (2015) found PCV2 DNA in oronasal and rectal sections but not in genital secretions in wild boar and concluded that oronasal secretions were the predominant means of spread because of greater shedding rates. Additionally, transmission via exposure to contaminated feed has been reported (Ghebremariam and Gruys 2005). These studies suggest that PCV2 shed from infected individuals and can persist in the environment.

5.3.3.3 Clinical Signs

A number of porcine circovirus diseases or porcine circovirus associated diseases (PCVAD) are now associated with PCV2. These diseases include systemic disease (PCV2-SD), characterized by wasting, pallor (paleness), respiratory distress, diarrhea, and sometimes icterus (jaundice); lung disease (PCV2-LD), with respiratory distress; enteric disease (PCV2-ED), with diarrhea; reproductive disease (PCV2-RD), with later term abortions, still births, and mummification; and porcine dermatitis and nephropathy syndrome (PDNS), in which swine are clinically anorexic and depressed (Segalés 2012).

5.3.3.4 Management Concerns

Porcine circovirus type 2 is not zoonotic or known to affect animals other than swine. Given the high prevalence of animals seropositive for antibodies against PCV2 in surveys of wild pigs, these animals may serve as a source of infection for domestic swine.

5.3.4 Influenza

5.3.4.1 Etiology, Prevalence, and Distribution

Influenza viruses are members of the family Orthomyxoviridae, and are classified as Influenza A, B, or C (Mahy 1997). Influenza A viruses cause epidemics in humans and domestic swine and are commonly isolated from a wide range of wild birds (Olsen et al. 2006). Birds that live in aquatic environments, specifically Anseriformes and Charadriiformes, are major reservoirs for these viruses, but Influenza A viruses infect a wide range of species worldwide (Webster et al. 1992). Antibodies to H1N1 and H3N2 influenza virus subtypes have been detected in wild pigs across the United States (Table 5.3), and these are the predominant subtypes circulating in domestic swine in the United States (Vincent et al. 2008).

5.3.4.2 Transmission

Transmission of swine influenza viruses is via direct contact including nose-to-nose contact and aerosol spread (Easterday 1986, Vincent et al. 2008).

5.3.4.3 Clinical Signs

Influenza virus infections in swine are an acute respiratory disease characterized by fever, inactivity, decreased feeding, respiratory distress, coughing, sneezing, conjunctivitis, and nasal discharge. Severity varies, but the incubation period is 1–3 days with sudden onset of disease, and recovery 4–78 days after onset (Vincent et al. 2008).

5.3.4.4 Management Concerns: Wildlife and Domestic Animal Health

Aquatic birds are the source of all Influenza A viruses, and domestic swine are reservoirs of H1N1 and H3N2 Influenza A viruses. These viruses are found in domestic pigs worldwide and are a primary cause of respiratory disease in domestic swine. Influenza viruses have significant implications for domestic animals and public health, with domestic swine susceptible to both avian and human Influenza A viruses (Brown 2000). Domestic swine are considered "mixing vessels" for Influenza A viruses, as infection may result in re-assortment of viruses. Domestic swine provide an intermediate step in a pathway of virus movement from wild birds to humans (Hinshaw et al. 1981). Sun et al. (2015) demonstrated that these viruses are also highly infectious and transmissible in wild pigs.

Interchange of disease agents between domestic and wild pigs is discussed in the brucellosis and PRV sections of this chapter and applies to influenza viruses. Corn et al. (2009) found that wild pigs near commercial swine premises in North Carolina had a high prevalence of antibodies against several influenza viruses that were circulating in domestic swine at that time (Table 5.3). Per this example, wild pigs near domestic swine or domestic poultry premises may be susceptible to infection by influenza viruses present in these domestic animals.

5.3.4.5 Management Concerns: Public Health

A number of H1N1 and H3N2 viruses recovered from wild pigs have also been found in humans (Corn et al. 2009, Clavijo et al. 2013, Feng et al. 2014). Among viruses recovered from wild pigs is the pandemic H1N1 influenza virus (pH1N1) (Clavijo et al. 2013). This virus was first detected in humans in 2009 (Dawood et al. 2009), and by 2010 reported in over 200 countries. Shrestha et al. (2011) estimated that in the United States alone, during 2009–2010, there were 60.8 million cases, 274,304 hospitalizations, and 12,469 deaths. This virus also was reported in domestic swine from at least 20 countries (Njabo et al. 2012), and apparently can be transmitted from pig to pig, pig to human, and human to pig (Nfon et al. 2011). Transmission from wild pigs to humans has not been reported.

5.3.5 Other Pathogens Detected

Serologic surveys detected other disease agents in wild pigs, however, beyond evidence of exposure in a small number of surveys, data on infection by these agents in wild pigs is lacking. Wild pigs may serve as a reservoir for these other disease agents, may serve to transmit some of these agents to wildlife, domestic animals, and/or humans, and may provide for environmental contamination. In other cases, wild pig exposure may simply reflect presence of the organism in the environment or be indicative of spillover from domestic animals without further transmission. Tables 5.1 and 5.3 provide summaries of additional viral and bacterial disease agents detected by serology or culture. These tables are primarily limited to data from wild pigs in the United States because few reports of pathogens in wild pigs exist from Canada, Mexico, or the Caribbean or Pacific Islands, but this is due to a significant lack of surveillance in these areas.

5.4 PARASITIC DISEASES

Wild pigs are susceptible to parasites that infect domestic pigs, with numerous ectoparasites and >50 species of internal parasites reported. Below are details on some of the more important parasites of wild pigs, where importance is defined as resulting in clinical disease or because the agent is zoonotic. Table 5.4 provides a summary of additional parasites of wild pigs in the United States.

5.4.1 Toxoplasmosis

5.4.1.1 Etiology, Prevalence, and Distribution

Toxoplasmosis, caused by the protozoan parasite *Toxoplasma gondii*, is an important zoonosis considered a leading cause of foodborne illness in the United States. This parasite occurs nearly worldwide in a wide range of felid definitive hosts, and numerous mammalian and avian intermediate hosts. Prevalence of the parasite varies considerably by region and host, although prevalence is generally greatest in carnivores and omnivores. Wild pig exposure in the United States is widespread, but prevalence varies from nearly 0% on an isolated island where felines were absent to almost 50% (Table 5.5). A comprehensive review of toxoplasmosis in people and animals has been published (Tenter et al. 2000).

TABLE 5.4
Parasites Reported from Wild Pigs in the United States and Other Countries That Are Not Included in Text

Parasite	Zoonotic?	US State(s) (or Country with Provinces) with Reports	Transmission Route	Management Concerns
Protozoans				
Eimeria spp.[a]	No	CA, FL	Direct	No concern
Isospora suis	No	FL	Direct	No concern
Intestinal coccidia	No	United States (AL, AR, FL, GA, SC, TN, TX, VA), Canada (SK)	Direct	No concern
Sarcocystis miescheriana (syn. *S. suicanis*)	No	AL, AZ, AR, FL, LA, MT, OK, TX	Ingestion of oocysts from canid definitive hosts	No concern
Sarcocystis spp.	No	AL, AR, CA, GA, KY, LA, MS, NC, SC, TN, TX, VA, WV	Ingestion of oocysts	No concern
Trypanosoma cruzi	Yes	TX	Reduviid vector	No concern
Cryptosporidium spp.	Some strains	CA, TX	Direct	May facilitate contamination of water sources for other animals
Giardia spp.	Some strains	CA, TX	Direct	May facilitate contamination of water sources for other animals
Neospora caninum	No	AL, AZ, AR, FL, GA, HI, IL, IN, IA, KY, LA, MI, MS, MO, NC, NJ, NM, OH, OK, SC, TN, TX	Ingestion of oocysts from canid definitive hosts	Can cause abortion in cattle, pigs may contribute to maintenance of the parasite in nature
Balatidium coli	Yes	FL	Direct	No concern
Nematodes				
Stephanurus dentatus	No	United States (AL, AR, FL, GA, HI, LA, MS, NC, SC, TN, TX, VA, WV), Canada (NS), Guam (Yigo and Santa Rita)	Direct ingestion or ingestion of earthworm paratenic hosts, percutaneous	See text
Ascaris suum	Yes	FL, GA, KS, LA, NC, SC, TN, TX, VA, WV	Direct	See text
Trichuris suis	Yes	United States (GA, HI, KS, LA, NC, SC, TN, WV), Canada (SK)	Direct	No concern
Gongylonema pulchrum or *Gongylonema* sp.	Yes	AL, AR, FL, GA, HI, LA, NC, SC, TN, TX, WV	Ingestion of coprophagous insects (e.g., beetles and cockroaches)	Wide-host range, no importance to health

(Continued)

TABLE 5.4 (CONTINUED)
Parasites Reported from Wild Pigs in the United States and Other Countries That Are Not Included in Text

Parasite	Zoonotic?	US State(s) (or Country with Provinces) with Reports	Transmission Route	Management Concerns
Hyostrongylus rubidus	No	AR, FL, GA, HI, TX	Direct	Can cause gastritis when worm burdens are high
Ascarops strongylina	No	AL, AR, FL, GA, HI, LA, NC, SC, TN, TX, VA, WV	Ingestion of beetle intermediate host	Can cause gastritis when worm burdens are high
Physocephalus sexalatus	No	AR, FL, GA, HI, LA, NC, SC, TN, TX, WV	Ingestion of beetle intermediate host	Can cause gastritis when worm burdens are high
Oesophagostomum spp.[b]	No	AR, FL, GA, HI, LA, SC, TX	Direct	No concern
Ceratospira ophthalmica	No	HI	Vector borne	Rare, no concern
Strongyloides ransomi	No	FL, GA, LA, MS, TX, VA	Percutaneous, transmammary	Cause of mortality in domestic piglets so may be concern for young wild pigs
Aonchotheca (=Capillaria) putorii	No	SC, TN	Direct	No concern
Haemonchus contortus	No		Direct	Rare, no concern to swine; this parasite can cause disease in a wide range of ungulates
Physaloptera sp.	No		Beetle and cockroach intermediate hosts	Rare, no concern
Trichostrongylus axei	No	FL	Direct	No concern
Globocephalus spp.	No	AL, AR, FL, GA, HI, LA, MS, NC, SC, TN, TX, VA, WV	Direct	Very common, no reported disease
Metastrongylus spp.[c]	No	United States (AL, AR, CA, FL, GA, HI, KS, LA, MS, OH, SC, TN, TX, VA, WV), Canada (SK), Guam (Yigo and Santa Rita)	Direct or through ingestion of earthworm intermediate hosts	Heavy infections can cause pneumonia especially if pigs have malnutrition or are stressed
Trematodes				
Alaria mesocercariae	Yes	OK	Direct penetration of skin by cercaria or ingestion of paratenic hosts with mesocercariae	No concern

(*Continued*)

TABLE 5.4 (CONTINUED)
Parasites Reported from Wild Pigs in the United States and Other Countries That Are Not Included in Text

Parasite	Zoonotic?	US State(s) (or Country with Provinces) with Reports	Transmission Route	Management Concerns
Fascioloides magna	No	TX	Ingestion of metacercaria from plant materials; aquatic snail intermediate host	Aberrant hosts, no eggs passed in feces, rare
Fasciola hepatica	Yes	AL, GA	Ingestion of metacercaria from plant materials; aquatic snail intermediate host	Few likely involved in maintenance of parasite in nature, rare
Brachylaima virginianum	No	GA, TN, WV	Ingestion of metacercaria	Rare, no concern
Fibricola sp.	No	GA	Ingestion of metacercaria	Rare, no concern
Pseudospelotrema (= *Pseudospelotrematoides*) sp.	No	GA	Ingestion of metacercaria	Rare, no concern
Paragonimus kellictoti	Yes	AL, GA	Ingestion of metacercaria in crayfish intermediate hosts	Rare, no concern
Cestodes				
Taenia hydatigena cystercercoids	No	HI	Ingestion of eggs from canid/felid definitive hosts	Rare, no concern
Spirometra pleurocercoids	Yes	FL, TX	Ingestion of infected copepods	Risk to people if eaten undercooked
Acanthocephalans				
Macracanthorhychus hirudinaceus	Yes	AL, AR, FL, GA, LA, SC	Ingestion of beetle intermediate host	Can perforate the intestine of domestic swine but this has not been reported in wild pigs
Lice				
Haematophinus suis	No	United States (AL, AR, CA, FL, GA, HI, KS, LA, SC, TN), Guam (Yigo and Santa Rita)	Direct, environment	Can cause dermatitis and poor growth in domestic pigs but has not been reported in wild pigs; can transmit swinepox virus and hog cholera virus
Mites				
Demodex sp.	No	CA (in a fecal sample)	Direct	No concern

(*Continued*)

TABLE 5.4 (CONTINUED)
Parasites Reported from Wild Pigs in the United States and Other Countries That Are Not Included in Text

Parasite	Zoonotic?	US State(s) (or Country with Provinces) with Reports	Transmission Route	Management Concerns
Demodex phylloides		AL, AR, FL, GA, NC, SC, VA, WV	Direct	Pathogenic to domestic pigs, no lesions noted in wild pigs
Sarcoptes scabiei	Yes	FL, HI	Direct	Pathogenic to domestic pigs, low mite loads and no lesions in wild pigs
Fleas				
Pulex irritans	N/A	CA, HI	Direct, contact with other hosts, environment	No concern
Ticks				
Amblyomma americanum	TBD[d]	AL, AR, FL, GA, SC, TX, VA	Environment	Transmits pathogens
Amblyomma auricularium	N/A	FL	Environment	No concern
Amblyomma maculatum	TBD	AR, GA, FL, MS, SC, TX	Environment	Transmits pathogens
Amblyomma mixtum (=*A. cajennense*)	N/A	TX	Environment	Transmits pathogens
Amblyomma tenellum	N/A	TX	Environment	No concern
Amblyomma breviscutatum	N/A	Guam (Yigo and Santa Rita)	Environment	No concern
Ixodes scapularis	TBD	FL, LA, SC, TX	Environment	Transmits pathogens
Ixodes pacificus	TBD	CA		Transmits pathogens
Dermacentor variabilis	TBD	FL, GA, SC, TN, TX	Environment	Transmits pathogens
Dermacentor albipictus		NH, TX	Environment	No concern
Dermacentor halli	N/A	TX	Environment	No concern

[a] Including *Eimeria porci*, *E. debliecki*, *E. neodebliecki*, *E. scabra*, *E. suis*, *E cerdonis*, *E. polita*, *E. perminuta*, and *E. spinose*.
[b] Including *Oesophagostomum brevicaudum*, *O. dentatum*, and *O. quadrispinulatum*.
[c] Including *Metastrongylus apri*, *M. salmi*, and *M. pudendotectus*.
[d] Tick-borne disease—these tick species are known to transmit pathogens to people or domestic animals.

Sources: Barbero et al. (1959), Becklund (1962), Alicata (1964), Coombs and Springer (1974), Foreyt et al. (1975), Prestwood et al. (1975), Smith and Hawkes (1978), Smith et al. (1982), Greiner et al. (1984), Pence et al. (1988), McKenzie and Davidson (1989), Forrester (1992), Atwill et al. (1997), Gray et al. (1999), Allan et al. (2001), Shender et al. (2002), Bevins et al. (2013), Sanders et al. (2013), Calero-Bernal et al. (2015), McGregor et al. (2015), Cerqueira-Cézar et al. (2016), Comeaux et al. (2016), Corn et al. (2016), Rodriguez-Rivera et al. (2016), Cleveland et al. (2017), Johnson et al. (2017), Linares et al. (2018), Merrill et al. (2018).

TABLE 5.5
Prevalence of *Toxoplasma gondii* in Wild Pigs in the United States and Other Selected Countries

Country	US State(s) (or Country with Provinces)	Assay	Year(s)	No. Infected/ No. Tested	References
United States	CA (Humboldt Co.)	Serology	1975–1976	1/7	Franti et al. (1976)
	CA	Serology	1981–1983	17/135	Clark et al. (1983)
	CA	Serology	1994	8/69	Sweitzer et al. (1996)
	CA (Santa Cruz Island, cat free)	Serology	2005	0/256	Blumenshine et al. (2009)
	FL	Serology	1977–1979	14/457	Forrester (1992)
	FL	Serology	N/A	12/457	Burridge et al. (1979)
	GA	Serology	N/A	0/8	Walton and Walls (1964)
	GA (Ossabaw Island, mostly cat-free island)	Serology	1992–1994	11/1,264	Dubey et al. (1997)
	NC	Serology	2007–2009	23/83	Sandfoss et al. (2011)
	OH	Serology	2009–2015	11/76	Linares et al. (2018)
	SC	Serology	1993	55/149	Dideprich et al. (1996)
	SC	Serology	1999	89/181	Gresham et al. (2002)
	TN/NC (Great Smoky Mountains National Park)	Serology	1990	33/108	Dideprich et al. (1996)
	TX	Serology	2015	30/376	Pedersen et al. (2017b)
	WV	Isolation	1979–1980	1 or 2/6	Smith et al. (1982)
	26 states	Serology	2006–2010	576/3,247[a]	Hill et al. (2014)
	South			325/2,055	
	Midwest			147/699	
	West			102/474	
	Northeast			2/21	
Canada	Saskatchewan	Serology	2009–2014	0/20	McGregor et al. (2015)
Guam	Yigo and Santa Rita	Serology	2015	5/47	Cleveland et al. (2017)

[a] Prevalence in some individual states was as high as 43%.

5.4.1.2 Transmission

The parasite is maintained in a complex life cycle of felid definitive hosts and numerous species of intermediate hosts. Within the felid host, the parasite undergoes sexual replication in intestinal epithelial cells where they form oocysts. These oocysts pass unsporulated and develop to the infective stage in 24–48 hours. The oocysts are environmentally hardy and can survive for months under ideal conditions (cool, moist, no sunlight exposure). Once intermediate hosts (including people) ingest oocysts, the parasites leave the gastrointestinal tract and reproduce by asexual replication in various tissues where they form cysts filled with zoites. If these cysts are ingested by another intermediate host, the zoites can cause infection, undergo additional replication, and form new cysts.

Because pigs scavenge, infection can occur through ingestion of other intermediate hosts or by ingestion of oocysts from the environment. Pigs may also become infected through vertical transmission (pregnant sows to offspring). No significant difference in prevalence has been found between males and females, but prevalence tends to be greater in females (Dubey et al. 1997, Gresham et al. 2002). Also, prevalence in wild pigs increased with age as expected for a chronic infection (Dubey et al. 1997, Sandfoss et al. 2011).

5.4.1.3 Clinical Signs and Pathology

Most omnivore and carnivore intermediate hosts, including wild pigs, do not typically develop clinical signs including lesions. For those hosts that develop disease, they are often found dead. Should clinical signs be detected they will relate to specific organs in which damage has occurred (e.g., difficulty breathing with lung lesions, neurologic signs due to brain damage). Animals may also be lethargic. Abortion occurs in some species due to toxoplasmosis. At necropsy, enlarged spleens and lung and muscle lesions may be evident, although many animals may appear grossly normal. Histologically, animals have tissue necrosis and inflammation, neither of which are symptomatic for *T. gondii* so the presence of protozoans must be established. If intralesional protozoa are detected, they must be identified using sequence analysis, immunohistochemistry, or in situ hybridization because several related parasites such as *Sarcocystis* and *Neospora* can cause similar lesions.

5.4.1.4 Management Concerns: Wildlife and Domestic Animal Health

Although sporadic reports of clinical disease due to *T. gondii* infections exist in domestic pigs, there are no reported cases of disease in wild pigs. Because transmission of *T. gondii* can be by carnivory/scavenging, prevalence is generally greater in carnivore and omnivore species, although they rarely develop clinical disease. However, even in these species, disease may occur in young animals or those that are immunosuppressed. For example, raccoons, foxes, and coyotes are commonly asymptomatically infected with *T. gondii*, but can develop severe disease when they become infected with canine distemper virus. In wildlife species that evolved in areas where felids are absent (e.g., marsupials in Australia, prosimians in Madagascar, birds in Hawaii), animals can develop severe disease due to *T. gondii* infections. Domestic cats and dogs rarely develop clinical disease. The parasite also occurs in a variety of small and large ruminants, but disease is rare.

5.4.1.5 Management Concerns: Public Health

In the United States, domestic pig products historically were the primary routes of transmission of *T. gondii* to people. Campaigns to encourage proper cooking of pork and improved sanitation (exclusion of rodents, cooking of food fed to pigs) led to a significant decrease in *T. gondii* in commercial swine operations in the United States. However, *T. gondii* remains a concern outside the United States (Dubey 2009). Free-range and organic farming conditions can also increase risks of infection for pigs (Dubey 2009). As noted above, numerous species can serve as intermediate hosts for the parasite, and ingestion of any undercooked products from these hosts can result in infection. Although classified as a foodborne illness, people can also become infected with *T. gondii* via other routes such as direct ingestion of oocysts while gardening or handling felid feces, drinking water contaminated with oocysts, and organ transplants. Although most infected people experience no clinical signs, some may develop eye disease leading to vision loss, immunocompromised individuals may develop disseminated disease or encephalitis (inflammation of the heart or brain), and infection of fetuses can lead to birth defects or death.

5.4.2 Trichinosis

5.4.2.1 Etiology

Trichinosis is caused by nematodes in the genus *Trichinella*. These parasites occur in a wide range of mammal, reptile, and bird hosts and many species are zoonotic. There are numerous species of *Trichinella*, with many not formally described and named. These parasites are grouped into: 1) those that encyst in tissues, and 2) those that do not. The most important species for human health worldwide is *T. spiralis*, as it is the predominant species found in pigs. Other zoonotic species in North America include *T. nativa*, *Trichinella* sp. T6, *T. murrelli*, and *T. pseudospiralis*.

TABLE 5.6
Prevalence of *Trichinella* spp. in Wild Pigs in the United States and Selected Other Countries

Country	State or Province	Detection Method	Year(s)	No. Infected/ No. Tested	References
United States	CA	Serology	1994	2/69	Sweitzer et al. (1996)
	CA (Santa Cruz Island)	Serology	2005	0/256	Blumenshine et al. (2009)
	FL	Serology	1988–1989	14/179	Forrester (1992)
		Digestion	1982–1983	1/26	Forrester et al. (1985)
	NC	Serology	2007–2009	11/83	Sandfoss et al. (2011)
	OH	Serology	2009–2015	4/69	Linares et a. (2018)
	SC	Serology	1999	70/179	Gresham et al. (2002)
	TX	Digestion	1985	0/100	Corn et al. (1986)
	TX	Digestion	1998–1999	0/5	Pozio et al. (2001)
	TX	Serology	2015	13/376	Pedersen et al. (2017b)
	Southeastern United States	Digestion	1979–1980	0/187	Smith et al. (1982)
	26 states	Serology	2006–2010	98/3,247	Hill et al. (2014)
	South			87/2,055[a]	
	Midwest			147/699	
	West			1/474	
	Northeast			1/21	
Canada	ON, QC, MB, and SK	Serology and Digestion	1994	0/391	Gajadhar et al. (1997)
	SK	Digestion	2009–2014	0/22	McGregor et al. (2015)
Guam	Yigo and Santa Rita	Serology	2015	0/47	Cleveland et al. (2017)

[a] Prevalence varied by state (Georgia 13%, Virginia 10%).

5.4.2.2 Prevalence and Distribution

Trichinella spiralis occurs in domestic pigs worldwide and, in some regions, is common in wild pigs. Prevalence in wild pigs in the United States varies by detection method (serology vs. digestion of tissues), but in general it is found in most states in the Southeast and Midwest with prevalence being much less in the West and Northeast regions (Table 5.6).

Other *Trichinella* species of importance in North America include *T. nativa* and *Trichinella* sp. T6, which occur in various carnivores and omnivores in the Holarctic regions of North America. However, these 2 species are undocumented in wild pigs to date. Another widespread and zoonotic species in the United States and southern Canada is *T. murrelli*, a common parasite of many carnivore hosts. Although *T. murrelli* is undocumented in wild pigs in North America, there is a possible report in a wild boar from Iran (Kia et al. 2009).

Only 1 non-encapsulating species occurs in North America, *T. pseudospiralis*. Although this species has a broad mammalian and avian host range worldwide, in the United States it has only been detected in a black vulture (*Coragyps atratus*) from Alabama, Florida panthers from Florida, and a wild pig from Texas (Lindsay et al. 1995, Gamble et al. 2005, Reichard et al. 2015). The *T. pseudospiralis* infection in the wild pig in 2001 was found during a routine inspection and the pig had a worm burden of 143.5 larvae per gram of diaphragm muscle (Gamble et al. 2005). *Trichinella pseudospiralis* is likely more widespread in wild pigs as it has also been reported from wild boar in Europe.

5.4.2.3 Transmission

Adult *Trichinella* species live in intestines where they mate and females release larvae. These larvae circulate throughout the body and either encapsulate in various muscles (for those that encapsulate) or remain free in various muscles. When larvae are ingested by another host they mature to adults in the intestines and the life cycle continues. Thus, *Trichinella* transmission is via carnivory or scavenging so, prevalence is greatest in carnivore and omnivore hosts. Once infected, hosts can maintain viable parasites for months or years. Humans are dead-end hosts.

5.4.2.4 Clinical Signs and Pathology

There are no reports of disease in wild pigs or other wild hosts of *Trichinella* spp., only humans develop disease (discussed below).

5.4.2.5 Management Concerns: Public Health

The only concern with this group of parasites is human health. Similar to *T. gondii*, historically, pork products were the primary source of human infection with *Trichinella* spp., termed trichinosis. However, efforts to eradicate this parasite from commercial pig operations in the United States have led to a significant decrease in prevalence, and currently the greatest source of infection is ingestion of undercooked meat from wild game. In 2011, 2 Iowa hunters who had consumed a wild pig experienced *T. spiralis* infection (Holzbauer et al. 2014). Interviews of the 2 hunters indicated they processed the meat without gloves and consumed food and drink during processing. Outside of the United States, the parasite remains a significant concern in commercial pigs.

The disease is rarely fatal and severity of clinical signs is related to number of larvae present in tissues (i.e., high numbers of ingested larvae generally result in more severe disease). After infection, the parasite develops within the intestine, possibly resulting in abdominal pain and diarrhea. After larvae are produced they penetrate the intestinal wall, enter the blood stream, and enter muscle cells. During this stage, patients may develop myalgia (muscle pain), fever and fluid accumulation, especially under the eyes. On rare occasions the disease may be severe and include myocarditis or encephalitis. Global epidemiology and the public health burden of trichinosis have been reviewed (Gottstein et al. 2009, Murrell and Pozio 2011).

5.4.3 Ascarids

5.4.3.1 Etiology, Prevalence, and Distribution

Ascaris suum is a large ascarid worm that resides in the small intestine of pigs. This parasite is found worldwide in domestic and wild pigs (Table 5.4), and prevalence is generally high (e.g., 67% of 241 wild pigs were positive in Florida; Forrester 1992).

5.4.3.2 Transmission

The life cycle is direct. Eggs passed in feces of pigs are initially undeveloped, but develop to the infectious stage in approximately 2 weeks. Transmission occurs when these masked eggs are ingested, hatch, and the larvae migrate to liver and lungs. Most pigs pass large numbers of eggs per gram of feces so environments can become contaminated quickly.

5.4.3.3 Clinical Signs and Pathology

Domestic pigs can develop severe disease from *A. suum* infections and it is a cause of economic losses. Adult worms in the intestine can reduce growth rate of young pigs but the most severe disease is associated with larvae migrating into the liver and lungs. Liver lesions consist of white spots that result in liver condemnation. Further, large numbers of larvae can cause pulmonary edema resulting in difficulty breathing, weight loss, and lethargy. Clinical signs and lesions are rarely noted in wild pigs; however, when present they must be distinguished from those caused by the kidney worm, *Stephanurus dentatus*.

5.4.3.4 Management Concerns

Pigs are the primary host for this parasite. Only rare reports of *A. suum* exist for sheep and cattle, both of which may develop disease due to larvae migrating through the lungs. *Ascaris suum* can cause visceral larva migrans in humans and infections generally occur in pig farmers or those in close contact with pigs. *Ascaris suum* is closely related to the human ascarid, *A. lumbricoides*, and some researchers consider them a single species. However, there are genetic differences and hybrids appear to be sterile, suggesting that they are reproductively isolated (Søe et al. 2016).

5.4.4 Kidney Worms

5.4.4.1 Etiology, Prevalence, and Distribution

The kidney worm is a large nematode, *Stephanurus dentatus*. This parasite is widespread in domestic and wild pigs (Table 5.4) and is most common in tropical and subtropical areas. However, it occurs in other areas (e.g., Canada) due to movement of infected pigs.

5.4.4.2 Transmission

Eggs pass in the urine of infected pigs. The young parasite undergoes some development, the egg then hatches, and larvae undergo further development in the environment to become an infective larval stage. Transmission can occur by ingestion of larvae, penetration of the skin by larvae, or by ingesting intermediate paratenic host earthworms that had previously ingested larvae.

5.4.4.3 Clinical Signs and Pathology

Disease is caused by migration of larvae through the liver and other organs. Gross lesions can be similar to those of *A. suum* larva migration. Affected pigs may have a loss of appetite, lose body condition, and have hindquarter paralysis. Large numbers of parasites can cause death. The adults form cysts in fat surrounding the kidney and these cysts can extend to surrounding areas of the abdominal cavity.

5.4.4.4 Management Concerns

Pigs are the only known host. This parasite is economically important and can cause severe losses of domestic pigs. No known public health concerns exist, but lesions in domestic and wild pigs can be dramatic and noticed by hunters.

5.5 SELECTED FOREIGN ANIMAL DISEASES

African swine fever (ASF), classical swine fever (CSF), and foot-and-mouth disease (FMD) were among the OIE-listed diseases affecting pigs in 2017 (World Organization for Animal Health 2017). These diseases impact domestic pigs in many parts of the world, and can infect wild pigs, but none of these diseases currently occur in North America.

5.5.1 African Swine Fever

African swine fever virus is a complex DNA virus and the only member of the family Asfarviridae (Dixon et al. 2005). Clinical signs in domestic pigs, wild boar, and wild pigs are similar but wild African Suidae develop subclinical and asymptomatic infections and act as reservoirs (Penrith and Vosloo 2009). African swine fever virus affects all age groups of swine; strains are classified as high, moderate, or low virulence with disease in peracute, acute, subacute, and chronic forms (Sánchez-Vizcaíno et al. 2015). Peracute and acute ASF mortality is up to 100%. Peracute ASF does not result in lesions while acute ASF includes fever, erythema, or cyanosis (reddish or bluish discoloration of the skin) of the skin, and functional failure of internal organs. Subacute ASF results in similar but milder lesions and 30–70% mortality. Clinical signs in chronic ASF are nonspecific and there is no vaccine for control or prevention (Gallardo et al. 2015, Sánchez-Vizcaíno et al. 2015).

African swine fever was first described after introduction of European domestic swine to Kenya, and later in other countries of sub-Saharan Africa (Penrith and Vosloo 2009). The first large epidemic of ASF outside of Africa occurred during 1960–1995 and affected countries in Europe, as well as Cuba, Dominican Republic, Haiti, and Brazil in the Americas. A second epidemic began in 2007 when ASF occurred in Georgia, and is ongoing with spread to Armenia, Azerbaijan, the Russian Federation, and more recently to the Ukraine and Belarus and to Lithuania, Poland, Latvia, and Estonia in the Continental European Union (Bosch et al. 2016). In August 2018, ASF was reported in northeastern China and, as of September 2019, has spread throughout Southeast Asia including Mongolia, Viet Nam, Cambodia, Democratic People's Republic of Korea, Lao People Democratic Republic, Myanmar, the Philippines, and the Republic of Korea (Food and Agricultural Organization of the United Nations 2019).

African swine fever virus is transmitted via direct contact between infected and susceptible pigs, contact with contaminated fomites, consumption of meat from infected pigs, and via ticks of the genus *Ornithodoros* (Penrith and Vosloo 2009). In Africa, warthogs (*Phacochoerus* spp.), bushpigs (*Potamochoerus larvatus*), and soft ticks in the genus *Ornithodoros* are the natural hosts and vectors of ASF virus (Jori et al. 2013); however, most of the recent outbreaks in Africa have been a result of movement of infected pigs or pig products (Penrith and Vosloo 2009). In the ongoing epidemic involving countries in the European Union (EU), Eastern Europe, and Russia, domestic pigs are the major hosts in Russia, whereas wild boar are of most concern in the EU (Gallardo et al. 2015, Bosch et al. 2016). As wild pigs from the United States are susceptible to ASF virus (McVicar et al. 1981), they would be at risk for infection and could potentially serve as hosts if ASF virus was introduced into the United States.

5.5.2 Classical Swine Fever

Classical swine fever is caused by an RNA virus in the genus *Pestivirus*, family Flaviviridae (Lindenbach et al. 2007). The clinical course is identified as acute, chronic, or late onset; up to 90% of domestic swine with acute infections die, while 100% of those with chronic or latent infections succumb to the disease (Moennig 2015). Transmission of CSF virus is by direct contact with infected animals or via infected feces, carcasses, and contaminated food including garbage (Moennig 2015). The United States was declared free of CSF in 1978, but CSF is still one of the most important viral diseases affecting domestic swine worldwide and is endemic in most of the world where domestic swine are kept except Australia, Canada, the EU, and the United States (Edwards et al. 2000, Moennig 2015). Several European countries have had outbreaks of CSF in wild boar in recent years, and infected wild pigs have been reported in other regions including India (2016) and South Korea (2015). The EU enacted control measures for CSF in domestic swine and wild boar (Council of the European Union 2001), including oral vaccination in wild boar (Moennig 2015), as wild boar are a source of infection for domestic swine (Fritzemeier et al. 2000). In Germany during 1993–1997, 60% (55 of 92) of outbreaks of CSF in domestic swine were a result of contact with wild boar (Fritzemeier et al. 2000).

Wild pigs in the United States are susceptible to CSF virus infection (Brugh et al. 1964), and prior to its eradication in the United States, CSF was documented in wild pigs (Shaw 1941, Hanson and Karstad 1959). A nationwide survey for CSF in wild pigs in the United States conducted from 1979 to 1987 did not detect CSF (Nettles et al. 1989), and Sandfoss et al. (2012) did not detect CSF in wild pigs in North Carolina during 2007 to 2010. Classical swine fever virus was used in attempts to eliminate wild pigs from Santa Cruz and Santa Rosa Islands, California. Wild pigs in Santa Cruz were intentionally infected 3 times; once just prior to 1944 (Wheeler 1944), once in 1950, and once in the early to mid-1950s (Nettles et al. 1989). Wild pigs in Santa Rosa were infected twice; once in

1949 and again in either 1952 or 1953 (Nettles et al. 1989). After each introduction, large numbers of wild pigs died, but within a few years the wild pig populations rebounded and evidence of CSF disappeared (Nettles et al. 1989). Surveys conducted in wild pigs from both islands in 1987 did not detect CSF (Nettles et al. 1989).

Based on studies done in Europe, the course of an outbreak of CSF in wild pigs in the United States would depend on the number of susceptible animals in a given population (Moennig 2015). Rossi et al. (2005) found during an outbreak in France that CSF dies out in populations of 2000 or less within a year, but in larger populations it persists and may become endemic for years. Artois et al. (2002) estimated that CSF would perpetuate in wild boar in Europe at a threshold of 200 susceptible pigs in an area of 220 km^2. The goal of CSF control in Europe is to reduce the affected susceptible population to the threshold level (Artois et al. 2002, Moennig 2015). The threshold level may be the same for wild pigs in the United States, and so the course of an outbreak of CSF in wild pigs in the United States would depend on the population involved. Where wild pigs are present during an outbreak in domestic swine, transmission could occur if they had direct or indirect contact with infected domestic swine. Subsequent transmission to additional pigs would depend on the density of wild pigs in the area, the size of their population, and if wild pigs in the affected population had access to additional domestic swine premises.

5.5.3 Foot-and-Mouth Disease

Foot-and-mouth disease is caused by an RNA virus in the genus *Aphthovirus*, family Picornaviridae (Belsham 1993); 7 serotypes and multiple subtypes exist (Grubman and Baxt 2004). Clinical signs vary depending on virulence of the subtype and species infected but are characterized by an acute febrile reaction and development of vesicular lesions on the mouth, nares, muzzle, feet, and teats (Alexandersen et al. 2003). The host range includes all cloven-hooved domestic animals and wildlife. Foot-and-mouth disease is highly contagious, and transmission is via direct or indirect contact with infected animals, contaminated fomites or, a contaminated environment (Alexandersen et al. 2001, 2003).

Foot-and-mouth disease is considered to be the most economically significant disease of livestock worldwide (Knight-Jones et al. 2016). Globally, countries are identified as FMD-free without vaccination, FMD-free with vaccination, and FMD endemic. FMD is endemic in parts of Africa and Asia and some countries in South America. Outbreaks continue to occur in parts of Africa and Asia (Knight-Jones et al. 2016).

Maintenance and spread of FMD by wildlife are a major obstacle to control in some endemic regions (Knight-Jones et al. 2016). Although a wide range of cloven-hooved wildlife are susceptible, only Cape buffalo (*Syncerus caffer*) in Africa play a role in long-term maintenance of FMD virus (Weaver et al. 2013). Wild pigs have been shown experimentally to be susceptible to infection and to transmit FMD virus (Mohamed et al. 2011), and infected wild boar were detected in an outbreak in FMD-free Bulgaria in 2010–2011 (Alexandrov et al. 2013). Clusters of seropositive wild boar were found in Bulgaria (Alexandrov et al. 2013), and seropositive wild boar also were found in FMD-endemic regions in Turkey (Khomenko et al. 2012), but a modeling study found that wild boar in the area would not maintain FMD in the region without the presence of infected domestic animals (Breithaupt et al. 2012). Corn et al. (2009) postulated that aerosol spread of swine influenza from domestic swine to wild pigs predicted potential aerosol spread of FMD in similar situations. Subsequent spread among wild pigs and to domestic swine would depend on factors including the density and distribution of wild pigs and their access to domestic swine. Although it does not appear that wild pigs would maintain FMD long-term, spread through wild pig populations and to domestic swine in limited time and geographic areas could be significant. Another possible concern is increased co-occurrence of wild pigs and farmed cervids that are also susceptible to FMD (Miller et al. 2017).

5.6 ECONOMICS AND INTERNATIONAL TRADE

Ongoing economic impacts resulting from diseases in wild pigs include livestock producer costs for biosecurity, management of disease outbreaks such as PRV and *B. suis*, reduced productivity, and other health costs, veterinary costs, and diagnostic tests required for interstate movement of pigs. Estimates of these costs for the United States are not available. Presence of PRV, *B. suis*, and other diseases currently endemic in wild pigs in the United States has minimal effect on international trade; as with interstate movement, producers exporting live commercial swine are required by some countries (e.g., Canada) to test for diseases only found in wild or transitional pigs such as PRV and *B. suis*. However, economic and trade impacts resulting from introduction of a disease, specifically FMD, ASF or CSF, would be significant, but are specific for each disease. An introduction of FMD would affect all producers, with severe impacts even if the disease was not in commercial swine. The United States would lose its FMD-free status until proven otherwise. In some cases, the United States has negotiated for regionalization (i.e., Highly Pathogenic Avian Influenza), but overall a finding of FMD would be severely detrimental to the United States.

The Terrestrial Animal Health Code (Terrestrial Code), developed by the OIE, sets standards for both animal health and welfare as well as veterinary public health worldwide. The Terrestrial Code addresses importation of domestic and wild pigs or wild boar, including live animals, semen, embryos, fresh meat, meat products, products of animal origin such as for pharmaceutical use or for trophies from wild pigs; products for use in animal feeding and agricultural; or industrial use, bristles, litter, and manure. Notifiable diseases are those that require prompt reporting to health officials once identified and necessitate immediate action to prevent and control an outbreak. The ASF or CSF status of a country, zone, or compartment of a country depends on several factors: 1) whether ASF or CSF is notifiable, 2) if there is an ongoing awareness program to encourage reporting, 3) the knowledge that the Veterinary Authority has of domestic and wild pigs and their habitat of wild pigs, and 4) the presence or absence of ASF or CSF in wild pigs (World Organization for Animal Health 2016a, b). Veterinary Authority is the government entity of the respective country responsible for implementation of animal health and welfare measures (World Organization for Animal Health 2016a). Appropriate surveillance for CSF in both domestic and captive wild pigs should take into account natural and artificial boundaries, ecology, and the risk of disease spread (World Organization for Animal Health 2016b).

5.7 HUMAN INTERACTIONS: MOVEMENT BY PEOPLE

All animals may be hosts to various bacteria, viruses, fungi, helminths, protozoa, ticks, and other arthropods, and should be viewed as "biological packages" (Davidson and Nettles 1992). Translocation of wildlife may result in introduction of pathogens to new environments and establishment of new hosts for pathogens already present in the environment, and in both cases may impact health of introduced animals as well as wildlife, domestic animals and humans. Efforts to restore wildlife populations should always consider potential health issues (Davidson and Nettles 1992, Corn and Nettles 2001, Gaydos and Corn 2001). Although wild pigs are not translocated or reintroduced as part of state or federal wildlife agency wildlife restoration efforts in the United States, wild pigs are moved and introduced by private persons into areas for the purpose of hunting (see Chapter 11). Efforts to maintain health of wild pigs are not of consideration here, but as wild pigs are trapped and moved from one location or state to another without consideration for health issues or other negative impacts, the pathogens they carry are moved with them. As wild pigs are not native to North America, all infections in wild pigs either are the result of the movement of pathogens in wild pigs from one geographic area to another, or of wild pigs serving as a new host for an already established pathogen.

Because collecting specimens in a timely manner from a newly established wild pig population is difficult, there are few documented examples of the introduction of pathogens into new geographic areas via introduction of wild pigs. For example, in 2008, wild pigs transported in a trailer to a high-fenced shooting operation in Colorado were confiscated and subsequently tested positive for PRV (Colorado Parks & Wildlife 2008). Wilson et al. (2009) reported occurrence of PRV in wild pigs after trapping and illegal transport from Texas to Nebraska. While being held in Nebraska, some of these animals escaped and 1 of 2 subsequently collected was seropositive for PRV.

5.8 CONCLUSIONS

Wild pigs can harbor an array of pathogens and serve as reservoirs for disease, threatening native wildlife populations, the livestock industry, and human health and safety. Contamination of food and water destined for human consumption by wild pigs is becoming increasingly common. While wild pigs are an inherently destructive species (see Chapter 7), the spread of pathogens, including potential spread of foreign animal diseases (e.g., ASF, CSF, FMD) into North America, is particularly alarming as it could have severe implications for livestock production, ultimately affecting consumers and economies on a regional or even global scale. When humans contribute to unregulated movements of wild pigs across landscapes, unexpected consequences surrounding disease transmission are likely. Although endemic diseases of concern may be challenging enough to manage without the presence of wild pigs, spread of wild pigs susceptible to infection into such an environment can only exacerbate the situation. Increasing biosecurity measures to prevent transmission of disease from wild pigs to domestic livestock and food and water resources intended for human use, preventing the translocation of wild pigs, and increased management and surveillance efforts focused on populations of wild pigs are essential steps towards maintaining the health of wildlife, livestock, and humans.

REFERENCES

Albina, E. 1997. Porcine reproductive and respiratory syndrome: Ten years of experience (1986–1996) with this undesirable viral infection. *Veterinary Research* 28:305–352.

Alexandersen, S., M. B. Oleksiewicz, and A. I. Donaldson. 2001. The early pathogenesis of foot-and-mouth disease in pigs infected by contact: A quantitative time course study using TaqMan RT-PCT. *Journal of General Virology* 82:747–755.

Alexandersen, S., Z. Zhang, A. I. Donaldson, and A. J. M. Garland. 2003. The pathogenesis and diagnosis of foot-and-mouth disease. *Journal of Comparative Pathology* 129:1–36.

Alexandrov, T., D. Stefanov, P. Kamenov, A. Miteva, S. Khomenko, K. Sumption, H. Meyer-Gerbaulet, and K. Depner. 2013. Surveillance of foot-and-mouth disease (FMD) in susceptible wildlife and domestic ungulates in southeast of Bulgaria following a FMD case in wild boar. *Veterinary Microbiology* 166:84–90.

Alicata, J. E. 1964. *Parasitic Infections of Man and Animals in Hawaii.* Hawaii Agricultural Experimental Station, Technical Bulletin No. 61. 138 pp.

Allan, G. M., F. McNeilly, S. Kennedy, B. Daft, E. G. Clarke, J. A. Ellis, D. M. Haines, B. M. Meehan, and B. M. Adair. 1998. Isolation of porcine circovirus-like viruses from pigs with a wasting disease in the USA and Europe. *Journal of Veterinary Diagnostic Investigation* 10:3–10.

Allan, S. A., L. A. Simmons, and M. J. Burridge. 2001. Ixodid ticks on white-tailed deer and feral swine in Florida. *Journal of Vector Ecology* 26:93–102.

Anderson, P. L., and J. W. Leafstedt. 2004. Report of the Committee on Pseudorabies. *Proceedings of the United States Animal Health Association* 108:480–489.

Anderson, P. L., and J. W. Leafstedt. 2005. Report of the Committee on Pseudorabies. *Proceedings of the United States Animal Health Association* 109:534–537.

Animal Health Australia. 2004. *Disease Strategy: Porcine Reproductive and Respiratory Syndrome (Version 3.0).* Australian Veterinary Emergency Plan (AUSTVETPLAN), Edition 3, Primary Industries Ministerial Council, Canberra, ACT, 54 pp.

Anonymous. 1992. Porcine reproductive and respiratory syndrome (PRRS or blue-eared pig disease). *Veterinary Record* 130:87–89.

Artois, M., K. R. Depner, V. Guberti, J. Hars, S. Rossi, and D. Rutili. 2002. Classical swine fever (hog cholera) in wild boar in Europe. *Office International Des Epizooties Revue Scientifique et Technique* 21:287–303.

Atwill, E. R., R. A. Sweitzer, M. G. Pereira, I. A. Gardner, D. Van Vuren, and W. M. Boyce. 1997. Prevalence of and associated risk factors for shedding *Cryptosporidium parvum* oocysts and *Giardia* cysts within feral pig populations in California. *Applied and Environmental Microbiology* 63:3946–3949.

Baker, S. R., K. M. O'Neil, M. R. Gramer, and S. A. Dee. 2011. Estimates of the seroprevalence of production-limiting diseases in wild pigs. *Veterinary Record* 168:564.

Barbero, B. B., L. H. Karstad, and J. R. Shepperson. 1959. Studies on helminth parasitism of feral swine in Georgia. *Bulletin of the Georgia Academy of Sciences* 17:94–104.

Baroch, J. A., C. A. Gagnon, S. Lacouture, and M. Gottschalk. 2015. Exposure of feral swine (*Sus scrofa*) in the United States to selected pathogens. *The Canadian Journal of Veterinary Research* 79:74–78.

Beard, A. W., R. G. Maggi, S. Kennedy-Stoskopf, N. A. Cherry, M. R. Sandfoss, C. S. DePerno, and E. B. Breitschwerdt. 2011. *Bartonella* spp. in feral pigs, southeastern United States. *Emerging Infectious Diseases* 17:893–895.

Becker, H. N., R. C. Belden, T. Breault, M. J. Burridge, W. B. Frankenberger, and P. Nicoletti. 1978. Brucellosis in feral swine in Florida. *Journal of the American Veterinary Medical Association* 173:1181–1182.

Becklund, W. W. 1962. Occurrence of a larval trematode (Diplostomidae) in a larval cestode (Diphyllobothriidae) from *Sus scrofa* in Florida. *Journal of Parasitology* 48:286.

Belsham, G. J. 1993. Distinctive features of foot-and-mouth disease virus, a member of the picornavirus family: Aspects of virus protein synthesis, protein processing and structure. *Progress in Biophysics and Molecular Biology* 60:241–260.

Benfield, D. A., J. E. Collins, S. A. Dee, P. G. Halbur, H. S. Joo, K. M. Lager, W. L. Mengeling, M. P. Murtaugh, K. D. Rossow, G. W. Stevenson, and J. J. Zimmerman. 1999. Porcine reproductive and respiratory syndrome. Pages 201–232 *in* B. E. Straw, S. D'Allare, W. L. Mengeling, and D. J. Taylor, editors. *Diseases of Swine*. Ames: Iowa State University Press.

Beran, G. W. 1992. Transmission of Aujeszky's disease virus. Pages 93–112 *in* R. B. Morrison, editor. *First International Symposium on Eradication of Pseudorabies (Aujeszky's) Virus*. St. Paul: University of Minnesota, College of Veterinary Medicine.

Bevins, S., E. Blizzard, L. Bazan, and P. Whitley. 2013. *Neospora caninum* exposure in overlapping populations of coyotes (*Canis latrans*) and feral swine (*Sus scrofa*). *Journal of Wildlife Diseases* 49:1028–1032.

Bevins, S. N., K. Pedersen, M. W. Lutman, T. Gidlewski, and T. J. Deliberto. 2014. Consequences associated with the recent range expansion of nonnative feral swine. *BioScience* 64:291–299.

Bigler, W. J., G. L. Hoff, W. H. Hemmert, J. A. Tomas, and H. T. Janowski. 1977. Trends of brucellosis in Florida: An epidemiologic review. *American Journal of Epidemiology* 105:245–251.

Bishop, G. C., and P. P. Bosman. 1994. *Brucella suis* infection. Pages 1053–1066 *in* J. A. W. Coetzer, G. R. Thomson, and R. C. Tustin, editors. *Infectious Diseases of Livestock with Special Referenced to South Africa*. Volume 2. Oxford: Oxford University Press.

Black, C. 2009. Report of the Feral Swine Subcommittee on Brucellosis and Pseudorabies. *Proceedings of the Annual Meeting of the United Sates Animal Health Association* 113:219–220.

Black, C., and J. Corn. 2008. Report of the Feral Swine Subcommittee on Brucellosis and Pseudorabies. *Proceedings of the Annual Meeting of the United Sates Animal Health Association* 112:167–171.

Blumenshine, K. M., H. Kinde, and S. Patton. 2009. Biometric and disease surveillance of an insular population of feral pigs on Santa Cruz Island, California. *Proceedings of the California Islands Symposium* 7:387–402.

Bolin, S. R., W. C. Stoffregen, G. P. Nayar, and A. L. Hamel. 2001. Postweaning multisystemic wasting syndrome induced after experimental inoculation of cesarean-derived, colostrum deprived piglets with type 2 porcine circovirus. *Journal of Veterinary Diagnostic Investigation* 13:185–194.

Bosch, J., A. Rodrígues, I. Iglesias, M. J. Muñoz, C. Jurado, J. M. Sánchez-Vizcaíno, and A. Torre. 2016. Update on the risk of introduction of African swine fever by wild boar into disease-free European Union countries. *Transboundary and Emerging Diseases* 64:1424–1432.

Breithaupt, A., K. Depner, B. Haas, T. Alexandrov, L. Polihronova, G. Georgiev, H. Meyer-Gerbaulet, and M. Beer. 2012. Experimental infection of wild boar and domestic pigs with a foot and mouth disease virus strain detected in the southeast of Bulgaria in December of 2010. *Veterinary Microbiology* 159:33–39.

Brown, I. H. 2000. The epidemiology and evolution of influenza viruses in pigs. *Veterinary Microbiology* 74:29–46.

Brugh, M., J. W. Foster, and F. A. Hayes. 1964. Studies on the comparative susceptibility of wild European and domestic swine to hog cholera. *American Journal of Veterinary Research* 107:1124–1127.

Buchholz, A. E., A. R. Katz, R. Galloway, R. A. Stoddard, and S. M. Goldstein. 2016. Feral swine *Leptospira* seroprevalence survey in Hawaii, USA, 2007–2009. *Zoonoses and Public Health* 63:584–587.

Burridge, M. J., W. J. Bigler, D. J. Forrester, and J. M. Hennemann. 1979. Serologic survey for *Toxoplasma gondii* in wild animals in Florida. *Journal of the American Veterinary Medical Association* 175:964–967.

Calero-Bernal, R., S. K. Verma, S. Oliveira, Y. Yang, B. M. Rosenthal, and J. P. Dubey. 2015. In the United States, negligible rates of zoonotic sarcocystosis occur in feral swine that, by contrast, frequently harbour infections with *Sarcocystis miescheriana*, a related parasite contracted from canids. *Parasitology* 142:549–556.

Cameron, C. E., R. L. Zuerner, S. Raverty, K. M. Colegrove, S. A. Norman, D. M. Lambourn, S. J. Jeffries, and F. M. Gulland. 2008. Detection of pathogenic *Leptospira* bacteria in pinniped populations via PCR and identification of a source of transmission for zoonotic leptospirosis in the marine environment. *Journal of Clinical Microbiology* 46:1728–1733.

Campbell, T. A., D. B. Long, L. R. Bazan, B. V. Thomsen, S. Robbe-Austerman, R. B. Davey, L. A. Soliz, S. R. Swafford, and K. C. VerCauteren. 2011. Absence of *Mycobacterium bovis* in feral swine (*Sus scrofa*) from the southern Texas border region. *Journal of Wildlife Diseases* 47:974–978.

Campbell, T. A., R. W. DeYoung, E. M. Wehland, L. I. Grassman, D. B. Long, and J. Delgado-Acevedo. 2008. Feral swine exposure to selected viral and bacterial pathogens in southern Texas. *Journal of Swine Health and Production* 16:312–315.

Caprioli, A., F. McNeilly, I. McNair, P. Lagan-Tregaskis, J. Ellis, S. Krakowka, J. Mckillen, F. Ostanello, and G. Allan. 2006. PCR detection of porcine circovirus type 2 (PCV2) DNA in blood, tonsillar and faecal swabs from experimentally infected pigs. *Research in Veterinary Science* 81:287–292.

Cavendish, T., W. Stiver, and E. K. Delozier. 2008. Disease surveillance of wild hogs in Great Smoky Mountains National Park: A focus on pseudorabies. National Conference on Feral Hogs. Paper 7. St. Louis, MO.

Centers for Disease Control and Prevention. 2009. *Brucella suis* infection associated with feral swine hunting: Three states, 2007–2008. *Morbidity and Mortality Weekly Report* 58:618–621.

Cerqueira-Cézar, C. K., K. Pedersen, R. Calero-Bernal, O. C. Kwok, I. Villena, and J. P. Dubey. 2016. Seroprevalence of *Neospora caninum* in feral swine (*Sus scrofa*) in the United States. *Veterinary Parasitology* 226:35–37.

Chae, C. 2004. Postweaning multisystemic wasting syndrome: A review of aetiology, diagnosis and pathology. *Veterinary Journal* 168:41–49.

Chatfield, J., M. Milleson, R. Stoddard, D. M. Bui, and R. Galloway. 2013. Serosurvey of leptospirosis in feral host (*Sus scrofa*) in Florida. *Journal of Zoo and Wildlife Medicine* 44:404–407.

Clark, E. G. 1997. Post-weaning multisystemic wasting syndrome. *Proceedings of the American Association of Swine Practitioners Annual Meeting* 28:499–501.

Clark, R. K., D. A. Jessup, D. W. Hird, R. Ruppanner, and M. E. Meyer. 1983. Serologic survey of California wild hogs for antibodies against selected zoonotic disease agents. *Journal of the American Veterinary Medical Association* 183:1248–1251.

Clavijo, A., A. Nikooienejad, M. S. Esfahani, R. P. Metz, S. Schwartz, E. Atashpaz-Gargari, T. J. Deliberto, M. W. Lutman, K. Pederson, L. R. Bazan, L. G. Koster, M. Jenkins-Moore, S. L. Swenson, M. Zhang, T. Beckham, C. D. Johnson, and M. Bounpheng. 2013. Identification and analysis of the first 2009 pandemic H1N1 influenza virus from US feral swine. *Zoonoses and Public Health* 60:327–335.

Cleveland, C. A., A. DeNicola, J. P. Dubey, D. E. Hill, R. D. Berghaus, and M. J. Yabsley. 2017. Survey for selected pathogens in wild pigs (*Sus scrofa*) from Guam, Marianna Islands, USA. *Veterinary Microbiology* 205:22–25.

Coats, M., and C. Black. 2005. Report of the Feral Swine Subcommittee on Brucellosis and Pseudorabies. *Proceedings of the Annual Meeting of the United Sates Animal Health Association* 109:249–251.

Collins, J. E., D. A. Benfield, W. T. Christianson, L. Harris, J. C. Hennings, D. P. Shaw, S. M. Goyal, S. McCullough, R. B. Morrison, H. S. Joo, D. Gorcyca, and D. Chladek. 1992. Isolation of swine infertility and respiratory syndrome virus (isolate ATCC VR-2332) in North America and experimental reproduction of the disease in gnotobiotic pigs. *Journal of Veterinary Diagnostic Investigation* 4:117.

Colorado Parks & Wildlife. 2008. CPW News Release. <cpw.state.co.us/aboutus/Pages/News-Release-Archive-Details.aspx?NewsID=3782>.

Comeaux, J. M., R. Curtis-Robles, B. C. Lewis, K. J. Cummings, B. T. Mesenbrink, B. R. Leland, M. J. Bodenchuk, and S. A. Hamer. 2016. Survey of feral swine (*Sus scrofa*) infection with the agent of Chagas Disease (*Trypanosoma cruzi*) in Texas, 2013–14. *Journal of Wildlife Diseases* 52:627–630.

Coombs, D. W., and M. D. Springer. 1974. Parasites of feral pig x European wild boar hybrids in southern Texas. *Journal of Wildlife Diseases* 10:436–441.

Cooper, V. L., A. R. Doster, R. A. Hesse, and N. B. Harris. 1995. Porcine reproductive and respiratory syndrome: NEB-1 PRRSV infection did not potentiate bacterial pathogens. *Journal of Veterinary Diagnostic Investigation* 7:313–320.

Corn, J. L., J. C. Cumbee, R. Barfoot, and G. A. Erickson. 2009. Pathogen exposure in feral swine populations geographically associated with high densities of transitional swine premises and commercial swine production. *Journal of Wildlife Diseases* 45:713–721.

Corn, J. L., J. C. Cumbee, B. A. Chandler, D. E. Stallknecht, and J. R. Fischer. 2005. Implications of feral swine expansion: Expansion of feral swine in the United States and potential implications for domestic swine. *Proceedings of the United States Animal Health Association* 109:295–297.

Corn, J. L., R. Duhaime, J. Alfred, J. Mertins, B. Leland, R. Sramek, J. Moczygema, and D. Shaw. 2016. Survey for ticks on feral swine within a cattle fever tick–infested landscape in Texas, USA. *Systematic and Applied Acarology* 21:1564–1570.

Corn, J. L., and T. R. Jordan. 2017. Development of the National Feral Swine Map, 1982–2016. *Wildlife Society Bulletin* 41:758–763.

Corn, J. L., and V. F. Nettles. 2001. Health protocol for translocation of free-ranging elk. *Journal of Wildlife Diseases* 37:413–426.

Corn, J. L., D. E. Stallknecht, N. M. Mechlin, M. P. Luttrell, and J. R. Fischer. 2004. Persistence of pseudorabies virus in feral swine populations. *Journal of Wildlife Diseases* 40:307–310.

Corn, J. L., P. K. Swiderek, B. O. Blackburn, G. A. Erickson, A. B. Thiermann, and V. F. Nettles. 1986. Survey of selected diseases in wild swine in Texas. *Journal of the American Veterinary Medical Association* 189:1029–1032.

Costa, F., J. E. Hagan, J. Calcagno, M. Kane, P. Torgerson, M. S. Martinez-Silveira, C. Stein, B. Abela-Riddler, and A. I. Ko. 2015. Global morbidity and mortality of leptospirosis: A systematic review. *PLoS Neglected Tropical Diseases* 9:e0003898.

Council of the European Union. 2001. Council Directive 2001/89/EC of 23 October 2001 on community measures for the control of classical swine fever. <http://eur-lex.europa.eu/legal-content/EN/TXT/?uri=CELEX%3A32001L0089>. Accessed 29 Sept 2019.

Cramer, S. D., G. A. Campbell, B. L. Njaa, S. E. Morgan, S. K. Smith II, W. R. McLin IV, B. W. Brodersen, A. G. Wise, G. Scherba, I. M. Langohr, and R. K. Maes. 2011. Pseudorabies virus infection in Oklahoma hunting dogs. *Journal of Veterinary Diagnostic Investigation* 23:915–923.

Crandell, R. A., W. L. Schwartz, and D. B. Lawhorn. 1987. Isolation of infectious rhinotracheitis virus from a latently infected feral pig. *Veterinary Microbiology* 14:191–195.

Cummings, K. J., L. D. Rodriguez-Rivera, M. K. Grigar, S. C. Rankin, B. T. Mesenbrink, B. R. Leland, and M. J. Bodenchuk. 2016. Prevalence and characterization of *Salmonella* isolated from feral pigs throughout Texas. *Zoonoses and Public Health* 63:436–441.

Davidson, W. R., and V. F. Nettles. 1992. Relocation of wildlife: Identifying and evaluating disease risks. *Transactions of the North American Wildlife and Natural Resources Conference* 57:466–473.

Dawood, F. S., S. Jain, L. Finelli, M. W. Shaw, S. Lindstrom, R. J. Garten, L. V. Gubareva, et al. 2009. Emergence of a novel swine-origin influenza A (H1N1) virus in humans. *New England Journal of Medicine* 360:2605–2615.

Delwart, E., and L. Li. 2012. Rapidly expanding genetic diversity and host range of the *Circoviridae* viral family and other rep encoding small circular ssDNA genomes. *Virus Research* 164:114–121.

Dideprich, V., J. C. New, G. P. Noblet, and S. Patton. 1996. Serologic survey of *Toxoplasma gondii* antibodies in free-ranging wild hogs (*Sus scrofa*) from the Great Smoky Mountains National Park and from sites in South Carolina. *Journal of Eukaryotic Microbiology* 43:122S.

Dirsmith, K., K. VanDalen, T. Fry, B. Charles, K. VerCauteren, and C. Duncan. 2013. Leptospirosis in fox squirrels (*Sciurus niger*) of Larimer County, Colorado, USA. *Journal of Wildlife Diseases* 49:641–645.

Dixon, L. K., J. M. Escribano, C. Martins, D. L. Rock, M. L. Salas, and P. J. Wilkinson. 2005. Asfarviridae. *Proceedings of the Report of the International Committee on Taxonomy of Viruses* 8:135–143.

Done, S. H., and D. J. Paton. 1995. Porcine reproductive and respiratory syndrome: Clinical disease, pathology and immunosuppression. *Veterinary Record* 136:32–35.

Dong, C., J. Meng, X. Dai, J.-H. Liang, A. R. Feagins, X.-J. Meng, N. M. Belfiore, et al. 2011. Restricted enzooticity of hepatitis E virus genotypes 1 to 4 in the United States. *Journal of Clinical Microbiology* 49:4164–4172.

Drew, M. L., D. A. Jessup, A. B. Burr, and C. E. Franti. 1992. Serologic survey for brucellosis in feral swine, wild ruminants, and black bear of California, 1977 to 1989. *Journal of Wildlife Diseases* 28:355–363.

Duan, X., H. J. Nauwynck, and M. B. Pensaert. 1997. Virus quantification and identification of cellular targets in the lungs and lymphoid tissues of pigs at different time intervals after inoculation with porcine reproductive and respiratory syndrome virus (PRRSV). *Veterinary Microbiology* 56:9–19.

Dubey, J. P. 2009. Toxoplasmosis in pigs: The last 20 years. *Veterinary Parasitology* 164:89–103.

Dubey, J. P., E. A. Rollor, K. Smith, O. C. Kwok, and P. Thulliez. 1997. Low seroprevalence of *Toxoplasma gondii* in feral pigs from a remote island lacking cats. *Journal of Parasitology* 83:839–841.

Easterday, B. C. 1986. Swine influenza. Pages 244–254 *in* A. D. Leman, B.Straw, R. D. Glock, W. L. Mengeling, R. H. C. Penny, and E. Scholl, editors. *Diseases of Swine*. Ames: Iowa State University Press.

Edwards, S., A. Fukusho, P. C. Lefèvre, A. Lipowski, Z. Pejsak, P. Roehe, and J. Westergaard. 2000. Classical swine fever: The global situation. *Veterinary Microbiology* 73:103–119.

Ellis, W. A. 2015. Animal leptospirosis. *Current Topics in Microbiology and Immunology* 387:99–137.

Elvinger, F., C. A. Baldwin, A. D. Liggett, K. N. Tang, and D. E. Stallknecht. 1996. Prevalence of exposure to eastern equine encephalomyelitis virus in domestic and feral swine in Georgia. *Journal of Veterinary Diagnostic Investigation* 8:481–484.

Engle, M., and H. Snelson. 2006. Report of the committee on transmissible diseases of swine. *Proceedings of the United States Animal Health Association* 110:668–672.

Essey, M. A., R. L. Payne, E. M. Himes, and D. Luchsinger. 1981. Bovine tuberculosis surveys of axis deer and feral swine on the Hawaiian island of Molokai. *Proceedings of the United States Animal Health Association* 85:538–549.

Essey, M. A., D. E. Stallknecht, E. M. Himes, and S. K. Harris. 1983. Follow-up survey of feral swine for *Mycobacterium bovis* infection on the Hawaiian island of Molokai. *Proceedings of the United States Animal Health Association* 87:589–595.

Ewalt, D. R., J. B. Payeur, J. C. Rhyan, and P. L. Geer. 1997. *Brucella suis* biovar 1 in naturally infected cattle: A bacteriological, serological, and histological study. *Journal of Veterinary Diagnostic Investigations* 9:417–420.

Federal Register. 2016. Endangered and threatened wildlife and plants; removal of the Louisiana black bear from the federal list of endangered and threatened wildlife and removal of similarity-of-appearance protections for the American Black Bear. *Federal Register 50 CFR Part 1781*, No. 48:13124–13171.

Feng, Z., J. A. Baroch, L.P. Long, Y. Xu, F. L. Cunningham, K. Pedersen, M. W. Lutman, B. S. Schmit, A. S. Bowman, T. J. DeLiberto, and X. F. Wan. 2014. Influenza A subtype H3 viruses in feral swine, United States, 2011–2012. *Emerging Infectious Diseases* 20:843–846.

Fletcher, W. O., D. E. Stallknecht, and E. W. Jenney. 1985. Serologic surveillance for vesicular stomatitis virus on Ossabaw Island, Georgia. *Journal of Wildlife Diseases* 21:100–104.

Florida Department of Health. 2012. Brucellosis guide to surveillance and investigation. Florida Department of Health. http://www.floridahealth.gov/diseases-and-conditions/disease-reporting-and-management/disease-reporting-and-surveillance/_documents/gsi-brucellosis.pdf. Accessed 1 Nov 2018.

Food and Agricultural Organization of the United Nations. 2019. ASF situation in Asia update. http://www.fao.org/ag/againfo/programmes/en/empres/ASF/situation_update.html. Accessed 29 Sept 2019.

Foreyt, W. J., A. C. Todd, and K. Foreyt. 1975. *Fascioloides magna* (Bassi, 1875) in feral swine from southern Texas. *Journal of Wildlife Diseases* 11:554–559.

Forrester, D. J. 1992. *Parasites and Diseases of Wild Mammals of Florida*. Gainesville: University Press of Florida.

Forrester, D. J., J. A. Conti, and R. C. Belden. 1985. Parasites of the Florida panther (*Felis concolor coryi*). *Proceedings of the Helminthological Society of Washington* 52:96–97.

Franco-Paredes, C., D. Chastain, P. Taylor, S. Stocking, and B. Sellers. 2017. Boar hunting and brucellosis caused by *Brucella suis*. *Travel Medicine and Infectious Disease* 16:18–22.

Franti, C. E., H. P. Riemann, D. E. Behymer, D. Suther, J. A. Howarth, and R. Ruppanner. 1976. Prevalence of *Toxoplasma gondii* antibodies in wild and domestic animals in northern California. *Journal of the American Veterinary Medical Association* 169:901–906.

Franzo, G., M. Cortey, A. M. M. G. de Castro, U. Piovezan, M. P. J. Szabo, M. Drigo, J. Segalés, and L .J. Richtzenhain. 2015. Genetic characterization of porcine circovirus type 2 (PCV2) strains from feral pigs in the Brazilian Pantanal: An opportunity to reconstruct the history of PCV2 evolution. *Veterinary Microbiology* 178:158–162.

Fritzemeier, J., J. Teuffert, I. Greiser-Wilke, C. Staubach, H. Schlüter, and V. Moennig. 2000. Epidemiology of classical swine fever in Germany in the 1990s. *Veterinary Microbiology* 77: 29–41.

Gajadhar, A. A., J. R. Bisaillon, and G. D. Appleyard. 1997. Status of *Trichinella spiralis* in domestic swine and wild boar in Canada. *Canadian Journal of Veterinary Research* 61:256–259.

Gallardo, M. C., A. T. Reoyo, J. Fernández-Pinero, I. Iglesias, M. J. Muñoz, and M. L. Arias. 2015. African swine fever: A global view of the current challenge. *Porcine Health Management* 1:21.

Galloway, I. A. 1938. Aujeszky's disease. *Veterinary Record* 50:745–763.

Gamble, H. R., E. Pozio, J. R. Lichtenfels, D. S. Zarlenga, and D. E. Hill. 2005. *Trichinella pseudospiralis* from a wild pig in Texas. *Veterinary Parasitology* 132:147–150.

Gaskamp, J. A., K. L. Gee, T. A. Campbell, N. J. Silvy, and S. L. Webb. 2016. Pseudorabies virus and *Brucella abortus* from an expanding wild pig (*Sus scrofa*) population in southern Oklahoma, USA. *Journal of Wildlife Diseases* 52:383–386.

Gaydos, J. K., and J. L. Corn. 2001. Health aspects of large mammal restoration. Pages 149–162 *in* D. S. Maehr, R. F. Noss, and J. L. Larkin, editors. *Large Mammal Restoration*. Washington: Island Press.

Ghebremariam, M. K., and E. Gruys. 2005. Postweaning multisystemic wasting syndrome in pigs with particular emphasis on the causative agent, the mode of transmission, the diagnostic tools and the control measures. *Veterinary Quarterly* 27:105–116.

Gibbs, S. E. J., N. L. Marlenee, J. Romines, D. Kavanaugh, J. L. Corn, and D. E. Stallknecht. 2006. Antibodies to West Nile virus in feral swine from Florida, Georgia, and Texas, USA. *Vector-Borne and Zoonotic Diseases* 6:261–265.

Gipson, P. S., J. K. Veatch, R. S. Matlack, and D. P. Jones. 1999. Health status of a recently discovered population of feral swine in Kansas. *Journal of Wildlife Diseases* 35:624–627.

Glass, C. M., R. G. McLean, J. B. Katz, D. S. Maehr, C. B. Cropp, L. J. Kirk, A. J. McKeirnan, and J. F. Evermann. 1994. Isolation of pseudorabies (Aujeszky's Disease) virus from a Florida Panther. *Journal of Wildlife Diseases* 30:180–184.

González-Barrio, D., M. P. Martín-Hernando, and F. Ruiz-Fons. 2015. Shedding patterns of endemic Eurasian wild boar (*Sus scrofa*) pathogens. *Research in Veterinary Science* 102:206–211.

Gordon, S. C. 1992. Effects of blued-eared disease on a breeding and fattening unit. *Veterinary Record* 130:513–515.

Gorman, G. W., S. McKeever, and R. D. Grimes. 1962. Leptosirosis in wild mammals from southwestern Georgia. *American Journal of Tropical Medicine and Hygiene* 11:518–524.

Gottstein, B., E. Pozio, and K. Nöckler. 2009. Epidemiology, diagnosis, treatment, and control of trichinellosis. *Clinical Microbiology Reviews* 22:127–145.

Gradil, C., C. Dubuc, and M. D. Eaglesome. 1996. Porcine reproductive and respiratory syndrome virus: Seminal transmission. *Veterinary Record* 138:521–522.

Grau-Roma, L., L. Fraile, and J. Segalés. 2011. Recent advances in the epidemiology, diagnosis and control of diseases caused by porcine circovirus type 2. *Veterinary Journal* 187:23–32.

Gray, M. L., F. Rogers, S. Little, M. Puette, D. Ambrose, and E. P. Hoberg. 1999. Sparganosis in feral hogs (*Sus scrofa*) from Florida. *Journal of the American Veterinary Medical Association* 215:204–208.

Greiner, E. C., P. P. Humphrey, R. C. Belden, W. B. Frankenberger, D. H. Austin, and E. P. Gibbs. 1984. Ixodid ticks on feral swine in Florida. *Journal of Wildlife Diseases* 20:114–119.

Gresham, C. S., C. A. Gresham, M. J. Duffy, C. T. Faulkner, and S. Patton. 2002. Increased prevalence of *Brucella suis* and pseudorabies virus antibodies in adults of an isolated feral swine population in coastal South Carolina. *Journal of Wildlife Diseases* 38:653–656.

Griffin, J. 1972. Ecology of the feral pig on the Island of Hawaii, 1968–1972. Final Report Pittman-Robertson Project #W–15-3, Study #11, State of Hawaii, Department of Land and Natural Resources, Division of Fish and Game.

Grubman, M. J., and B. Baxt. 2004. Foot-and-mouth disease. *Clinical Microbiology Reviews* 17:465–493.

Gustafson, D. P. 1986. Pseudorabies. Pages 274–289 *in* A. D. Leman, B. Straw, R. D. Glock, W. L. Mengeling, R. H. C. Penny, and E. Scholl, editors. *Diseases of Swine*. Ames: Iowa State University Press.

Hahn, E. C., G. R. Page, P. S. Hahn, K. D. Gillis, C. Romero, J. A. Annelli, and E. P. J. Gibbs. 1997. Mechanisms of transmission of Aujeszky's disease virus originating from feral swine in the USA. *Veterinary Microbiology* 55:123–130.

Hall, J. S., R. B. Minnis, T. A. Campbell, S. Barras, R. W. DeYoung, K. Pabilonia, M. L. Avery, H. Sullivan, L. Clark, and R. G. McLean. 2008. Influenza exposure in United States feral swine populations. *Journal of Wildlife Diseases* 44:362–368.

Hanson, R. P. and L. Karstad. 1958. Feral swine as a reservoir of vesicular stomatitis virus in southeastern United States. *Proceedings of the United States Livestock Sanitary Association* 62:309–315.

Hanson, R. P., and L. Karstad. 1959. Feral swine in the southeastern United States. *Journal of Wildlife Management* 23:64–74.

Hartin, R. E., M. R. Ryan, and T. A. Campbell. 2007. Distribution and disease prevalence of feral hogs in Missouri. *Human–Wildlife Interactions* 1:186–191.

Hernández, F. A., K. A. Sayler, C. Bounds, M. P. Milleson, A. N. Carr, and S. W. Wisely. 2018. Evidence of pseudorabies virus shedding in feral swine (*Sus scrofa*) populations of Florida, USA. *Journal of Wildlife Diseases* 54:45–53.

Higgins, J., T. Stuber, C. Quance, W. H. Edwards, R. V. Tiller, T. Linfield, J. Rhyan, A. Berte, and B. Harris. 2012. Molecular epidemiology of *Brucella abortus* isolates from cattle, elk, and bison in the United States, 1998–2011. *Applied and Environmental Microbiology* 78:3674–3684.

Hill, D. E., J. P. Dubey, J. A. Baroch, S. R. Swafford, V. F. Fournet, D. Hawkins-Cooper, D. G. Pyburn, et al. 2014. Surveillance of feral swine for *Trichinella* spp. and *Toxoplasma gondii* in the USA and host-related factors associated with infection. *Veterinary Parasitology* 205:653–665.

Hinshaw, V. S., R. G. Webster, B. C. Easterday, and W. J. Bean Jr. 1981. Replication of avian influenza A viruses in mammals. *Infection and Immunity* 34:354–361.

Holzbauer, S. M., W. A. Agger, R. L. Hall, G. M. Johnson, D. Schmitt, A. Garvey, H. S. Bishop, K. Rivera, M. E. de Almeida, D. Hill, B. E. Stromberg, R. Lynfield, and K. E. Smith. 2014. Outbreak of *Trichinella spiralis* infections associated with a wild boar hunted at a game farm in Iowa. *Clinical Infectious Disease* 59:1750–1756.

Hopper, S. A., M. E. C. White, and W. Twiddy. 1992. An outbreak of blue-eared pig disease (porcine reproductive and respiratory syndrome) in four pig herds in Great Britain. *Veterinary Record* 131:140–144.

Jack, S. W., J. C. Cumbee Jr., and D. C. Godwin. 2012. Serologic evidence of *Brucella* and pseudorabies in Mississippi feral swine. *Human-Wildlife Interactions* 6:89–93.

Jansen, A., E. Luge, B. Guerra, P. Wittschen, A. D. Gruber, C. Loddenkemper, T. Schneider, M. Lierz, D. Ehlert, B. Appel, K. Stark, and K. Nöckler. 2007. Leptospirosis in urban wild boars, Berlin, Germany. *Emerging Infectious Diseases* 13:739–742.

Jay, M. T., M. Cooley, D. Carychao, G. W. Wiscomb, R. A. Sweitzer, L. Crawford-Miksza, J. A. Farrar, D. K. Lau, J. O'Connell, A. Millington, R. V. Asmundson, E. R. Atwill, and R. E. Mandrell. 2007. *Escherichia coli* O157:H7 in feral swine near spinach fields and cattle, central California coast. *Emerging Infectious Diseases* 13:1908–1911.

Jay-Russel, M. T., A. Bates, L. Harden, W. G. Miller, and R. E. Mandrell. 2012. Isolation of *Campylobacter* from feral swine (*Sus scrofa*) on the ranch associated with the 2006 *Escherichia coli* O157:H7 spinach outbreak investigation in California. *Zoonoses and Public Health* 59:314–319.

Johnson, E. M., Y. Nagamori, R. A. Duncan-Decocq, P. N. Whitley, A. Ramachandran, and M. V. Reichard. 2017. Prevalence of *Alaria* infection in companion animals in north central Oklahoma from 2006 through 2015 and detection in wildlife. *Journal of the American Veterinary Medical Association* 250:881–886.

Jori, F., L. Vial, M. L. Penrith, R. Pérez-Sánchez, E. Etter, E. Albina, V. Michaud, and F. Roger. 2013. Review of the sylvatic cycle of African swine fever in sub-Saharan Africa and the Indian Ocean. *Virus Research* 173:212–227.

Keffaber, K. K. 1989. Reproductive failure of unknown etiology. *American Association of Swine Practitioners Newsletter* 1:1–9.

Khomenko, S., T. Alexandrov, N. Bulut, S. Aktas, and K. Sumption. 2012. Surveillance for FMD in wild boar in 2011–2012: Results from Bulgaria and Turkey. EuFMD Open session 2012, 29–31 October 2012, Jerez, Spain.

Kia, E. B., A. R. Meamar, F. Zahabiun, and H. Mirhendi. 2009. The first occurrence of *Trichinella murrelli* in wild boar in Iran and a review of Iranian trichinellosis. *Journal of Helminthology* 83:399–402.

Killmaster, L. F., D. E. Stallknecht, E. W. Howerth, J. K. Moulton, P. F. Smith, and D. G. Mead. 2011. Apparent disappearance of vesicular stomatitis New Jersey virus from Ossabaw Island, Georgia. *Vector-Borne and Zoonotic Diseases* 11:559–565.

King, L. J. 1990. Pseudorabies: A new model for disease eradication. *Proceedings of the United States Animal Health Association* 94:352–354.

Kingscote, B. F. 1986. Leptospirosis in red foxes in Ontario. *Journal of Wildlife Diseases* 22:475–478.

Kirkpatrick, C. M., C. L. Kanitz, and S. M. McCrocklin. 1980. Possible role of wild mammals in the transmission of pseudorabies to swine. *Journal of Wildlife Diseases* 16:601–614.

Kluge, J. P., G. W. Beran, H. T. Hill, and K. B. Platt. 1992. Pseudorabies (Aujeszky's disease). Pages 312–323 *in* A. D. Leman, B. E. Straw, W. L. Mengeling, S. D'Allaire, and D. J. Taylor, editors. *Diseases of Swine*. Ames: Iowa State University Press.

Knight-Jones, T. J. D., L. Robinson, B. Charleston, L. L. Rodriguez, C. G. Gay, K. J. Sumption, and W. Vosloo. 2016. Global foot-and-mouth disease research update and gap analysis: 2—epidemiology, wildlife, and economics. *Transboundary and Emerging Diseases* 63 (Suppl. 1):14–29.

Konrad, J., and K. Blazek. 1958. Aujeszky's disease in mink. *Veterinary Bulletin* 29:307.

Kristensen, C. S, A. Botner, H. Takai, J. P. Nielsen, and S. E. Jorsal. 2004. Experimental airborne transmission of PRRS virus. *Veterinary Microbiology* 99:197–202.

Lama, J. K., and D. S. Bachoon. 2018. Detection of *Brucella suis*, *Campylobacter jejuni*, and *Escherichia coli* strains in feral pig (*Sus scrofa*) communities of Georgia. *Vector Borne and Zoonotic Diseases* 18:350–355.

Le Potier, M. F., P. Blanquefort, E. Morvan, and E. Albina. 1997. Results of a control programme for the porcine reproductive and respiratory syndrome in the French 'Pays de la Loire' region. *Veterinary Microbiology* 55:355–360.

Linares, M., C. Hicks, A. S. Bowman, A. Hoet, and J. W. Stull. 2018. Infectious agents in feral swine in Ohio, USA (2009–2015): A low but evolving risk to agriculture and public health. *Veterinary and Animal Science* 6:81–85. doi:10.1016/j.vas.2018.06.002.

Lindenbach, B. D., H. J. Thiel, and C. M. Rice. 2007. Flaviviridae: The viruses and their replication. Pages 1101–1152 *in* D. M. Knipe and P. M. Howley, editors. *Fields Virology*. Philadelphia: Lippincott-Raven Publishing.

Lindsay, D. S., D. S. Zarlenga, H. R. Gamble, F. al-Yaman, P. C. Smith, and B. L. Blagburn. 1995. Isolation and characterization of *Trichinella pseudospiralis* Garkavi, 1972 from a black vulture (*Coragyps atratus*). *Journal of Parasitology* 81:920–923.

Ljubashenko, S. Y., A. F. Tyulpanova, and V. M. Grishin. 1958. Aujeszky's disease in mink, arctic fox, and silver fox. *Veterinary Bulletin* 34:244, Abstract 1386.

Loula, T. 1991. Mystery pig disease. *Agri-Practice* 12:23–34.

MacMillan, A. P., H. Schleicher, J. Korslund, and W. Stoffregen. 2006. Brucellosis. Pages 603–612 *in* B. E. Straw, J. J. Zimmerman, S. D'Allaire, and D. J. Taylor, editors. *Disease of Swine*. Ames: Blackwell Publishing.

Maes, R. K., C. L. Kanitz, and D. P. Gustafson. 1979. Pseudorabies virus infections in wild and laboratory rats. *American Journal of Veterinary Research* 40:393–396.

Magar, R., R. Larochelle, S. Thibault, and L. Lamontagne. 2000. Experimental transmission of porcine circovirus type 2 (PCV2) in weaned pigs: A sequential study. *Journal of Comparative Pathology* 123:258–269.

Mahy, B. 1997. Influenza A virus (FLUA). Pages 170–171 *in* B. W. J. Mahy, editor. *A Dictionary of Virology*. San Diego: Academic Press.

McBride, A. J., D. A. Athanazio, M. G. Reis, and A. I. Ko. 2005. Leptospirosis. *Current Opinion in Infectious Diseases* 18:376–386.

McCaw, M. 1995. PRRS control: Whole herd management concepts and research update. *Proceedings of the North Carolina Health Hogs Seminar*, pp. 57–64.

McGregor, G. F., M. Gottschalk, D. I. Godson, W. Wilkins, and T. K. Bollinger. 2015. Disease risks associated with free-ranging wild boar in Saskatchewan. *Canadian Veterinary Journal* 56:839–844.

McKenzie, M. E., and W. R. Davidson. 1989. Helminth parasites of intermingling axis deer, wild swine and domestic cattle from the island of Molokai, Hawaii. *Journal of Wildlife Diseases* 25:252–257.

McNutt, S. H., and C. Murray. 1924. Bacterium abortion (Bang) isolated from the fetus of an aborting mare. *Journal of the American Veterinary Medical Association* 97:576–580.

McVicar, J. W., C. A. Mebus, H. N. Becker, R. C. Belden, and E. P. J. Gibbs. 1981. Induced African swine fever in feral pigs. *Journal of the American Veterinary Medical Association* 179:441–446.

Meehan, B. M., F. McNeilly, D. Todd, S. Kennedy, V. A. Jewhurst, J. A. Ellis, L. E. Hassard, E. G. Clark, D. M. Haines, and G. M. Allen. 1998. Characterization of novel circovirus DNAs associated with wasting syndromes in pigs. *Journal of General Virology* 79:2171–2179.

Merrill M. M., R. K. Boughton, C. C. Lord, K. A. Sayler, B. Wight, W. M. Anderson, and S. M. Wisely. 2018. Wild pigs as sentinels for hard ticks: A case study from south-central Florida. *International Journal for Parasitology. Parasites and Wildlife* 7:161–170.

Miller, R. S., S. J. Sweeney, C. Slootmaker, D. A. Grear, P. A. Di Salvo, D. Kiser, and S. A. Shwiff. 2017. Cross-species transmission potential between wild pigs, livestock, poultry, wildlife, and humans: Implications for disease risk management in North America. *Scientific Reports* 7:7821.

Moennig, V. 2015. The control of classical swine fever in wild boar. *Frontiers in Microbiology* 6:1211.

Mohamed, F., S. Swafford, H. Petrowski, A. Bracht, B. Schmit, A. Fabian, J. M. Pacheco, E. Hartwig, M. Berninger, C. Carillo, G. Mayr, K. Moran, D. Kavanaugh, H. Leibrecht, W. White, and S. Metwally. 2011. Foot-and-mouth disease feral swine: Susceptibility and transmission. *Transboundary and Emerging Diseases* 58:358–371.

Mravak, S., U. Bienzle, H. Feldmeier, H. Hampl, and K. O. Habermehl. 1987. Pseudorabies in man. *Lancet* 1:501–502.

Müller, T., E. C. Hahn, F. Tottewitz, M. Kramer, B. G. Klupp, T. C. Mettenleiter, and C. Freuling. 2011. Pseudorabies virus in wild swine: A global perspective. *Archives of Virology* 156:1691–1705.

Müller, T. F., J. Teuffert, R. Zellmer, and F. J. Conraths. 2001. Experimental infection of European wild boars and domestic pigs with pseudorabies viruses with differing virulence. *American Journal of Veterinary Research* 62:252–258.

Murrell, K. D. and E. Pozio. 2011. Worldwide occurrence and impact of human trichinellosis, 1986–2009. *Emerging Infectious Diseases* 17:2194–2202.

Musante, A. R., K. Pedersen, and P. Hall. 2014. First reports of pseudorabies and winter ticks (*Dermacentor albipictus*) associated with an emerging feral swine (*Sus scrofa*) population in New Hampshire. *Journal of Wildlife Diseases* 50:121–124.

Musser, J. M. B., A. L. Schwartz, I. Srinath, and K. A. Waldrup. 2013. Use of serology and bacterial culture to determine prevalence of *Brucella* spp. in feral swine (*Sus scrofa*) in proximity to a beef cattle herd positive for *Brucella suis* and *Brucella abortus*. *Journal of Wildlife Diseases* 49:215–220.

Nelsen, C., M. Murtaugh, and K. Faaberg. 1999. Porcine reproductive and respiratory syndrome virus comparison: Divergent evolution on two continents. *Journal of Virology* 73:270–280.

Nelson, E. A., J. Christopher-Hennings, T. Drew, W. Wensvoort, J. E. Collins, and D. A. Benfield. 1993. Differentiation of US and European isolates of porcine reproductive and respiratory syndrome virus by monoclonal antibodies. *Journal of Clinical Microbiology* 31:3184–3189.

Nelson, J. H., R. H. Decker, A. M. Barnes, B. C. Nelson, T. J. Quan, A. R. Gillogly, and G. S. Phillips. 1985. Plague surveillance using wild boar and wild carnivore sentinels. *Journal of Environmental Health* 47:306–309.

Nettles, V. F., J. L. Corn, G. A. Erickson, and D. A. Jessup. 1989. A survey of wild swine in the United States for evidence of hog cholera. *Journal of Wildlife Diseases* 25:61–65.

Nettles, V. F., and G. A. Erickson. 1984. Pseudorabies in wild swine. *Proceedings of the United States Animals Health Association* 88:505–506.

New, J. C. Jr., K. Delozier, C. E. Barton, P. J. Morris, and L. N. D. Potgieter. 1994. A serologic survey of selected viral and bacterial diseases of European wild hogs, Great Smoky Mountains National Park, USA. *Journal of Wildlife Diseases* 30:103–106.

Nfon, C. K., Y. Berhane, T. Hisanaga, S. Zhang, K. Handel, H. Kehler, O. Labrecque, et al. 2011. Characterization of H1N1 swine influenza viruses circulating in Canadian pigs in 2009. *Journal of Virology* 85:8667–8679.

Nikolitsch, M. 1954. Die Aujeszkysch Krankheit beim Reh. *Wierner Tierarztliche Monatsschrift* 41:603–605.

Njabo, K. Y., T. L. Fuller, A. Chasar, J. P. Pollinger, G. Cattoli, C. Terregino, I. Monne, J. M. Reynes, R. Nyouom, and T. B. Smith. 2012. Pandemic A/H1N1/2009 influenza virus in swine, Cameroon, 2010. *Veterinary Microbiology* 156:189–192.

Olsen, B., V. J. Munster, A. Wallensten, J. Waldenström, A. D. Osterhaus, and R. A. Fouchier. 2006. Global patterns of influenza A virus in wild birds. *Science* 312:384–388.

Opriessnig, T., X.-J. Meng, and P. G. Halbur. 2007. Porcine circovirus type 2-associated disease: Update on current terminology, clinical manifestations, pathogenesis, diagnosis, and intervention strategies. *Journal of Veterinary Diagnostic Investigation* 19:591–615.

Pedersen, K., T. D. Anderson, S. N. Bevins, K. L. Pabilonia, P. N. Whitley, D. R. Virchow, and T. Gidlewski. 2017a. Evidence of leptospirosis in the kidneys and serum of feral swine (*Sus scrofa*) in the United States. *Epidemiology and Infection* 145:87–94.

Pedersen, K., N. E. Bauer, S. Rodgers, L. R. Bazan, B. T. Mesenbrink and T. Gidlewski. 2017b. Antibodies to various zoonotic pathogens detected in feral swine (*Sus scrofa*) at abattoirs in Texas, USA. *Journal of Food Protection* 80:1239–1242.

Pedersen, K., N. E. Bauer, S. Olsen, A. M. Arenas-Gamboa, A. C. Henry, T. D. Sibley, and T. Gidlewski. 2017c. Identification of *Brucella* spp. in feral swine (*Sus scrofa*) at abattoirs in Texas, USA. *Zoonoses and Public Health* 64:647–54.

Pedersen, K., S. N. Bevins, J. A. Baroch, J. C. Cumbee, S. C. Chandler, B. S. Woodruff, T. T. Bigelow, and T. J. DeLiberto. 2013. Pseudorabies in feral swine in the United States, 2009–2012. *Journal of Wildlife Diseases* 49:709–713.

Pedersen, K., R. S. Miller, T. D. Anderson, K. L. Pabilonia, J. R. Lewis, R. L. Mihalco, C. Gortazar, and T. Gidlewski. 2017d. Limited antibody evidence of exposure to *Mycobacterium bovis* in feral swine (*Sus scrofa*) in the USA. *Journal of Wildlife Diseases* 53:30–36.

Pedersen, K., S. N. Bevins, B. S. Schmit, M. W. Lutman, M. P. Milleson, C. T. Turnage, T. T. Bigelow, and T. J. DeLiberto. 2012. Apparent prevalence of swine brucellosis in feral swine in the United States. *Human-Wildlife Interactions* 6:38–47.

Pedersen, K., K. L. Pabilonia, T. D. Anderson, S. N. Bevins, C. R. Hicks, J. M. Kloft, and T. J. Deliberto. 2015. Widespread detection of antibodies to *Leptospira* in feral swine in the United States. *Epidemiology and Infection* 143:2131–2136.

Pedersen, K., C. R. Quance, S. Robbe-Austerman, A. J. Piaggio, S. N. Bevins, S. M. Goldstein, W. D. Gaston, and T. J. DeLiberto. 2014. Identification of *Brucella suis* from feral swine in selected states in the USA. *Journal of Wildlife Diseases* 50:171–179.

Pence, D. B., R. J. Warren, and C. R. Ford. 1988. Visceral helminth communities of an insular population of feral swine. *Journal of Wildlife Diseases* 24:105–112.

Penrith, M. L., and W. Vosloo. 2009. Review of African swine fever: Transmission, spread and control. *Journal of the South African Veterinary Association* 80:58–62.

Perez-Rivera C. M., M. S. Lopez, G. A. Franco, and R. C. Napoles. 2017. Detection of antibodies against pathogens in feral and domestic pigs (*Sus scrofa*) of the Sierra la Laguna Biosphere Reserve, Mexico. *Veterinaria Mexico* 4: doi:10.21753/vmoa.4.1.378

Pirtle, E. C., M. E. Roelke, and J. Brady. 1986. Antibodies against pseudorabies virus in the serum of a Florida black bear cub. *Journal of the American Veterinary Medical Association* 189:1164.

Pirtle, E. C., J. M. Sacks, V. F. Nettles, and E. A. Rollor. 1989. Prevalence and transmission of pseudorabies virus in an isolated population of feral swine. *Journal of Wildlife Diseases* 25:605–607.

Platt, K. B., D. L. Graham, and R. A. Faaborg. 1983. Pseudorabies: Experimental studies in raccoons with different virus strains. *Journal of Wildlife Diseases* 19:297–301.

Pozio, E., D. B. Pence, G. La Rosa, A. Casulli, and S. E. Henke. 2001. *Trichinella* infection in wildlife of the southwestern United States. *Journal of Parasitology* 87:1208–1210.

Prestwood, A. K., F. E. Kellogg, S. R. Pursglove, and F. A. Hayes. 1975. Helminth parasitisms among intermingling insular populations of white-tailed deer, feral cattle, and feral swine. *Journal of the American Veterinary Medical Association* 166:787–789.

Prieto, C., R. Sanchez, S. Martin-Rillo, P. Suarez, I. Simarro, A. Solana, and J. M. Castro. 1996a. Exposure of gilts in early gestation to porcine reproductive and respiratory syndrome virus. *Veterinary Record* 138:536–539.

Prieto, C., P. Suarez, J. M. Bautista, R. Sanchez, S. M. Rillo, I. Simarro, A. Solana, and J. M. Castro. 1996b. Semen changes in boars after experimental infection with porcine reproductive and respiratory syndrome (PRRS) virus. *Theriogenology* 45:383–395.

Ramamoorthy, S., M. Woldemeskel, A. Ligett, R. Snider, R. Cobb, and S. Rajeev. 2011. *Brucella suis* infection in dogs, Georgia, USA. *Emerging Infectious Diseases* 17:2386–2387.

Ramlachan, N., R. C. Anderson, K. Andrews, G. Laban, and D. J. Nisbet. 2007. Characterization of an antibiotic resistant *Clostridium hathewayi* strain from a continuous-flow exclusion chemostat culture derived from the cecal contents of a feral pig. *Anaerobe* 13:153–160.

Randhawa, A. A., V. P. Kelly, and E. F. Baker Jr. 1977. Agglutinins to *Coxiella burnetii* and *Brucella* spp, with particular reference to *Brucella canis*, in wild animals of southern Texas. *Journal of the American Veterinary Medical Association* 171:939–942.

Raymond, J. T., R. G. Gillespie, M. Woodruff, and E. B. Janovitz. 1997. Pseudorabies in captive coyotes. *Journal of Wildlife Diseases* 33:916–918.

Reichard, M. V., M. Criffield, J. E. Thomas, J. M. Paritte, M. Cunningham, D. Onorato, K. Logan, M. Interisano, G. Marucci, and E. Pozio. 2015. High prevalence of *Trichinella pseudospiralis* in Florida panthers (*Puma concolor coryi*). *Parasites and Vectors* 8:67.

Reilly, J. R. 1970. The susceptibility of five species of wild animals to experimental infection with *Leptospira grippotyphosa*. *Journal of Wildlife Diseases* 6:289–294.

Rodriguez-Rivera, L. D., K. J. Cummings, I. McNeely, J. S. Suchodolski, A. V. Scorza, M. R. Lappin, B. T. Mesenbrink, B. R. Leland, and M. J. Bodenchuk. 2016. Prevalence and diversity of *Cryptosporidium* and *Giardia* identified among feral pigs in Texas. *Vector Borne and Zoonotic Diseases* 16:765–768.

Romero, C. H., P. Meade, J. Santagata, K. Gillis, G. Lollis, E. C. Hahn, and E. P. J. Gibbs. 1997. Genital infection and transmission of pseudorabies virus in feral swine in Florida, USA. *Veterinary Microbiology* 55:131–139.

Romero, C. H., P. N. Meade, J. E. Shultz, H. Y. Chung, E. P. Gibbs, E. C. Hahn, and G. Lollis. 2001. Venereal transmission of pseudorabies viruses indigenous to feral swine. *Journal of Wildlife Diseases* 37:289–296.

Rossi, S., E. Fromont, D. Pontier, C. Crucière, J. Hars, J. Barrat, X. Pacholek, and M. Artois. 2005. Incidence and persistence of classical swine fever in free-ranging wild boar (*Sus scrofa*). *Epidemiology and Infection* 133:559–568.

Rossow, K. D. 1998. Review Article: Porcine reproductive and respiratory syndrome. *Veterinary Pathology* 35:1–20.

Rossow, K. D., E. M. Bautista, S. M. Goyal, T. W. Molitor, M. P. Murtaugh, R. B. Morrison, D. A. Benfield, and J. E. Collins. 1994. Experimental porcine reproductive and respiratory syndrome virus infection in one-, four-, and 10-week-old pigs. *Journal of Veterinary Diagnostic Investigation* 6:3–12.

Saliki, J. T., S. J. Rodgers, and G. Eskew. 1998. Serosurvey of selected viral and bacterial diseases in wild swine from Oklahoma. *Journal of Wildlife Diseases* 34:834–838.

Sánchez-Vizcaíno, J. M., L. Mur, J. C. Gomez-Villamandos, and L. Carrasco. 2015. An update on the epidemiology and pathology of African swine fever. *Journal of Comparative Pathology* 152:9–21.

Sanders, D. M., A. L. Schuster, P. W. McCardle, O. F. Strey, T. L. Blankenship, and P. D. Teel. 2013. Ixodid ticks associated with feral swine in Texas. *Journal of Vector Ecology* 38:361–373.

Sandfoss, M., C. DePerno, S. Patton, J. Flowers, and S. Kennedy-Stoskopf. 2011. Prevalence of antibody to *Toxoplasma gondii* and *Trichinella* spp. in feral pigs (*Sus scrofa*) of eastern North Carolina. *Journal of Wildlife Diseases* 47:338–343.

Sandfoss, M. R., C. S. DePerno, C. W. Betsill, M. B. Palamar, G. Erickson, and S. Kennedy-Stoskopf. 2012. A serosurvey for *Brucella suis*, classical swine fever virus, porcine circovirus type 2, and pseudorabies virus in feral swine (*Sus scrofa*) of eastern North Carolina. *Journal of Wildlife Diseases* 48:462–466.

Schultze, A. E., R. K. Maes, and D. C. Taylor. 1986. Pseudorabies and volvulus in a black bear. *Journal of the American Veterinary Medical Association* 189:1165–1186.

Segalés, J. 2012. Porcine circovirus type 2 (PCV2) infections: Clinical signs, pathology and laboratory diagnosis. *Virus Research* 164:10–19.

Segalés, J., T. Kekarainen, and M. Cortey. 2013. The natural history of *Porcine circovirus* type 2: From an inoffensive virus to a devastating swine disease? *Veterinary Microbiology* 165:13–20.

Shahan, M. S., R. L. Knudson, H. R. Seibold, and C. N. Dale. 1947. Aujeszky's disease (pseudorabies): A review, with notes on two strains of the virus. *North American Veterinarian* 28:440–449.

Shaw, A. C. 1941. The European wild hog in America. *Transactions of the North American Wildlife Conference* 5:436–441.

Shender L. A., R. G. Botzler, and T. L. George. 2002. Analysis of serum and whole blood values in relation to helminth and ectoparasite infections of feral pigs in Texas. *Journal of Wildlife Diseases* 38:385–394.

Shibata, I., Y. Okuda, S. Yazawa, M. Ono, T. Sasaki, M. Itagaki, N. Nakajima, Y. Okabe, and I. Hidejima. 2003. PCR detection of porcine circovirus type 2 DNA in whole blood, serum, oropharyngeal swab, nasal swab, and feces from experimentally infected pigs and field cases. *Journal of Veterinary Medical Science* 65:405–408.

Shotts, E. B. 1981. Leptospirosis. Pages 138–147 *in* W. R. Davidson, F. A. Hayes, V. F. Nettles, and F. E. Kellogg, editors. *Diseases and Parasites of White-Tailed Deer*. Tallahassee: Tall Timbers Research Station.

Shrestha, S. S., D. L. Swerdlow, R. H. Borse, V. S. Prabhu, L. Finelli, C. Y. Atkins, K. Owusu-Edusei, B. Bell, P. S. Mead, M. Biggerstaff, L. Brammer, H. Davidson, D. Jernigan, M. A. Jhung, L. A. Kamimoto, T. L. Merlin, M. Nowell, S. C. Redd, C. Reed, A. Schuchat, and M. I. Meltzer. 2011. Estimating the burden of 2009 pandemic influenza A (H1N1) in the United States (April 2009–April 2010). *Clinical Infectious Diseases* 52(Supplement 1):S75–82.

Simoes, E. M., and J. D. Justino. 2013. Brucellosis infection in a feral swine hunter. *Nurse Practitioner* 38:49–53.

Smith, C. R. 1994. Wild carnivores as plague indicators in California: A cooperative interagency disease surveillance program. *Proceedings of the Vertebrate Pest Conference* 16:192–199.

Smith, G. 2012. Preferential sexual transmission of pseudorabies virus in feral swine populations may not account for observed seroprevalence in the USA. *Preventive Veterinary Medicine* 103:145–156.

Smith, A. W., and A. B. Latham. 1978. Prevalence of vesicular exanthema of swine antibodies among feral mammals associated with the southern California coastal zones. *American Journal of Veterinary Research* 39:291–296.

Smith, H. J. and B. Hawkes. 1978. Kidney worm infection in feral pigs in Canada with transmission to domestic swine. *Canadian Veterinary Journal* 19:40–43.

Smith, H. M., W. R. Davidson, V. F. Nettles, and R. R. Gerrish. 1982. Parasitisms among wild swine in southeastern United States. *Journal of the American Veterinary Medical Association* 181:1281–1284.

Smart, D. L., D. O. Trainer, and T. M. Yuill. 1975. Serologic evidence of Venezuelan equine encephalitis in some wild and domestic populations of southern Texas. *Journal of Wildlife Diseases* 11:195–200.

Snelson, H. 2010. Report of the Committee on Transmissible Diseases of Swine. *Proceedings of the United States Animal Health Association* 114:538–544.

Søe, M. J., C. M. Kapel, and P. Nejsum. 2016. *Ascaris* from humans and pigs appear to be reproductively isolated species. *PLoS Neglected and Tropical Diseases* 10:e0004855.

Southeastern Cooperative Wildlife Disease Study. 2004. National feral swine map, 2004. University of Georgia, Athens.

Stallknecht, D. E., W. O. Fletcher, G. A. Erickson, and V. F. Nettles. 1987. Antibodies to vesicular stomatitis New Jersey type virus in wild and domestic sentinel swine. *American Journal of Epidemiology* 125:1058–1065.

Stallknecht, D. E., and E. W. Howerth. 2001. Pseudorabies (Aujeszky's Disease). Pages 164–170 *in* E. S. Williams and I. K. Barker, editors. *Infectious Diseases of Wild Mammals*. Ames: Iowa State University Press.

Stallknecht, D. E., D. M. Kavanaugh, J. L. Corn, K. A. Eernisse, J. A. Comer, and V. F. Nettles. 1993. Feral swine as a potential amplifying host for vesicular stomatitis virus New Jersey serotype on Ossabaw Island, Georgia. *Journal of Wildlife Diseases* 29:377–383.

Stallknecht, D. E., V. F. Nettles, G. A. Erickson, and D. A. Jessup. 1986. Antibodies to vesicular stomatitis virus in populations of feral swine in the United States. *Journal of Wildlife Diseases* 22:320–325.

Stallknecht, D. E., V. F. Nettles, W. O. Fletcher, and G. A. Erickson. 1985. Enzootic vesicular stomatitis New Jersey type in an insular feral swine population. *American Journal of Epidemiology* 122:876–883.

Starnes, C. T., R. Talwani, J. A. Horvath, W. A. Duffus, and C. S. Bryan. 2004. Brucellosis in two hunt club members in South Carolina. *Journal of the South Carolina Medical Association* 100:113–115.

Stephenson, R. J., B. R. Trible, Y. Wang, M. A. Kerrigan, S. M. Goldstein, and R. R. R. Rowland. 2015. Multiplex serology for common viral infections in feral pigs (*Sus scrofa*) in Hawaii between 2007 and 2010. *Journal of Wildlife Diseases* 51:239–243.

Stoffregen, W. C., S. C. Olsen, C. J. Wheeler, B. J. Bricker, M. V. Palmer, A. E. Jensen, S. M. Halling, and D. P. Alt. 2007. Diagnostic characterization of a feral swine herd enzootically infected with *Brucella*. *Journal of Veterinary Diagnostic Investigations* 19:227–237.

Sun, H., F. L. Cunningham, J. Harris, Y. Xu, L-P Long, K. Hanson-Door, J. A. Baroch, P. Fioranelli, M. W. Lutman, T. Li, K. Pederson, B. S. Schmit, J. Cooley, X. Lin, R. G. Jarman, T. J. DeLiberto, and X. F. Wan. 2015. Dynamics of virus shedding and antibody responses in influenza A virus-infected feral swine. *Journal of General Virology* 96:2569–2578.

Sweitzer, R. A., I. A. Gardner, B. J. Gonzales, D. van Vuren, and W. M. Boyce. 1996. Population densities and disease surveys of wild pigs in the coast ranges of central and northern California. *Proceeding of the Vertebrate Pest Conference* 17:75–82.

Tenter, A. M., A. R. Heckeroth, and L. M. Weiss. 2000. *Toxoplasma gondii*: From animals to humans. *International Journal for Parasitology* 30:1217–1258.

Terpstra, C., G. Wensvoort, and L. van Leengoed. 1992. Persistence of Lelystad virus in herds affected by porcine epidemic abortion and respiratory syndrome. *Proceedings of the 12th International Pig Veterinary Society Congress*. The Hague, The Netherlands 12:118.

Texas Tech University. 2011. Researchers warn of tularemia in Texas feral hogs. News Release 24 January 2011, Article ID 572701.

Thakur, S., M. Sandfoss, S. Kennedy-Stoskopf, and C. S. DePerno. 2011. Detection of *Clostridium difficile* and *Salmonella* in feral swine population in North Carolina. *Journal of Wildlife Diseases* 47:774–776.

Tozzini, F., A. Poli, and G. D. Croce. 1982. Experimental infection of European wild swine (*Sus scrofa* L.) with pseudorabies virus. *Journal of Wildlife Diseases* 18:425–428.

Trainer, D., and L. Karstad. 1963. Experimental pseudorabies in some wild North American mammals. *Zoonoses Research* 2:135–151.

US Department of Agriculture. 2008. Pseudorabies (Aujeszky's disease) and its eradication: A review of the US experience. United States Department of Agriculture/Animal and Plant Health Inspection Service, Technical Bulletin No. 1923, 231 pp.

Van der Leek, M. L., H. N. Becker, P. Humphrey, C. L. Adams, R. C. Belden, W. B. Frankenberger, and P. L. Nicoletti. 1993a. Prevalence of *Brucella* sp. antibodies in feral swine in Florida. *Journal of Wildlife Diseases* 29:410–415.

Van der Leek, M. L., H. N. Becker, E. C. Pirtle, P. Humphrey, C. L. Adams, B. P. All, G. A. Erickson, R. C. Belden, W. B. Frankenberger, and E. P. J. Gibbs. 1993b. Prevalence of pseudorabies (Aujeszky's disease) virus antibodies in feral swine in Florida. *Journal of Wildlife Diseases* 29:403–409.

Vincent, A. L., W. Ma, K. M. Lager, B. H. Janke, and J. A. Richt. 2008. Swine influenza viruses: A North American perspective. *Advances in Virus Research* 72:127–154.

Walton, B. C., and K. W. Walls. 1964. Prevalence of toxoplasmosis in wild animals from Fort Stewart, Georgia, as indicated by serological tests and mouse inoculation. *American Journal of Tropical Medicine and Hygiene* 13:530–533.

Weaver, G. V., J. Domenech, A. R. Thiermann, and W. B. Karesh. 2013. Foot and mouth disease: A look from the wild side. *Journal of Wildlife Diseases* 49:759–785.

Webster, R. G., W. J. Bean, O. T. Gorman, T. M. Chambers, and Y. Kawaoka. 1992. Evolution and ecology of influenza A viruses. *Microbiological Reviews* 56:152–179.

Wensvoort, G., C. Terpstra, J. M. A. Pol, E. A. ter Laak, M. Bloemrad, E. P. Dekluyver, C. Kragtan, L. Van Buiten, A. Den Bensten, F. Wagenaar, J. M. Boehkuijsen, P. L. Moonen, T. Zestra, E. A. De Boer, H. J. Tibben, M. F. De Jong, P Van Veld, G. J. R. Groenland, J. A. van Gennped, M. T. Voets, J. H. M. Verheijden, and J. Braamskamp. 1991. Mystery swine disease in the Netherlands: The isolation of Lelystad virus. *Veterinary Quarterly* 13:121–130.

Wheeler, S. A. 1944. California's little known Channel Islands. *United States Naval Institute Proceedings* 70:257–270.

Wilson, S., A. R. Doster, J. D. Hoffman, and S. E. Hygnstrom. 2009. First record of pseudorabies in feral swine in Nebraska. *Journal of Wildlife Diseases* 45:874–876.

Woldemeskel, M. 2013. Zoonosis due to *Brucella suis* with special reference to infection in dogs (Carnivores): A brief review. *Open Journal of Veterinary Medicine* 3:213–221.

Wood, G. W., J. B. Hendricks, and D. E. Goodman. 1976. Brucellosis in feral swine. *Journal of Wildlife Diseases* 12:579–582.

Wood, G. W., L. A. Woodward, D. C. Mathews, and J. R. Sweeney. 1992. Feral hog control efforts on a coastal South Carolina plantation. *Proceedings of the Annual Conference Southeastern Association of Fish and Wildlife Agencies* 46:167–178.

Woods, R. D., E. C. Pirtle, J. M. Sacks, and E. P. J. Gibbs. 1990. Serologic survey for transmissible gastroenteritis virus neutralizing antibodies in selected feral and domestic swine sera in the southern United States. *Journal of Wildlife Diseases* 26:420–422.

World Organization for Animal Health. 2016a. African swine fever. Terrestrial Animal Health Code. <http://www.oie.int/index.php?id=169&L=0&htmfile=chapitre_asf.htm>. Accessed 1 Nov 2018.

World Organization for Animal Health. 2016b. Infection with classical swine fever virus. Terrestrial Animal Health Code. http://www.oie.int/index.php?id=169&L=0&htmfile =chapitre_csf.htm. Accessed 1 Nov 2018.

World Organization for Animal Health. 2017. OIE-listed diseases, infections and infestations in force in 2017. <www.oie.int/en/animal-health-in-the-world/oie-listed-diseases-2017/>. Accessed 14 Mar 2017.

Wright, J. C., and D. G. Thawley. 1980. Role of the raccoon in the transmission of pseudorabies: A field and laboratory investigation. *American Journal of Veterinary Research* 41:581–583.

Wyckoff, A. C., S. E. Henke, T. A. Campbell, D. G. Hewitt, and K. C. VerCauteren. 2009. Feral swine contact with domestic swine: A serologic survey and assessment of potential for disease transmission. *Journal of Wildlife Diseases* 45:422–429.

Zanin, E., I. Capua, C. Casaccia, A. Zuin, and A. Moresco. 1997. Isolation and characterization of Aujeszky's disease virus in capture brown bears from Italy. *Journal of Wildlife Diseases* 33:632–634.

Zimmerman, J. J., K. J. Yoon, E. C. Pirtle, R. W. Wills, T. J. Sanderson, and M. J. McGinley. 1997. Studies of porcine reproductive and respiratory syndrome (PRRS) virus infection in avian species. *Veterinary Microbiology* 55:329–336.

Zygmont, S. M., V. F. Nettles, E. B. Shotts, W. A. Carmen, and B. O. Blackburn. 1982. Brucellosis in wild swine: A serologic and bacteriologic survey in the southeastern United States and Hawaii. *Journal of the American Veterinary Medical Association* 181:1285–1287.

6 The Naturalized Niche of Wild Pigs in North America

Peter E. Schlichting, James C. Beasley, and Kurt C. VerCauteren

CONTENTS

6.1 INTRODUCTION

Since domesticated roughly 9,000 years ago (Epstein and Bichard 1984; see Chapter 2), pigs (*Sus scrofa*) have had a close economic and cultural relationship with humans. As a result, pigs have accompanied humans throughout history as they have expanded into new areas. Pigs were considered such a valuable food resource that Polynesians transported them in long canoes across the Pacific Ocean (Mayer and Brisbin 2008), and pigs were among the most common domestic livestock in European explorations and colonies in continental North America (Towne and Wentworth 1950). Traditional pig husbandry practices allowed pigs to free range with periodic roundups for slaughter, resulting in the establishment of local wild populations (Mayer and Brisbin 2008, Mayer and Beasley 2017). In some areas of North America, these populations have survived into modern times (Figure 6.1) and have been present on the landscape for several hundred years. An argument could be made that pigs in these long-term populations have become a naturalized part of the ecosystem and culture.

Naturalization of invasive species is a difficult topic to conceptualize and definitions vary by scientific field, leading to problematic and ambiguous terminology. According to Colautti and MacIsaac (2004), "naturalized" has been applied to several stages of the invasion process, ranging from introduced species that are established in natural environments to those that have a negative effect on habitats. Additionally, naturalization has been defined by Richardson et al. (2000) as "starting when abiotic and biotic barriers to survival are surmounted and when various barriers to regular reproduction are overcome." In many regions of the Western hemisphere, introduced, invasive wild pigs fit all these definitions of naturalization and have for many generations.

Ecologists have invested considerable effort to address mechanisms that determine the naturalization of species and how this affects niche space of native species. Charles Darwin's naturalization hypothesis (Darwin 1859) states that species are more likely to become naturalized in the absence of closely related species because there is niche space left unoccupied. The theory assumes closely related species exploit similar niches, leading to competition, and that these species are also susceptible to similar predators and pathogens. Thus, the shared niche space with native species makes establishment of invasive species less likely. Empirical tests of Darwin's naturalization

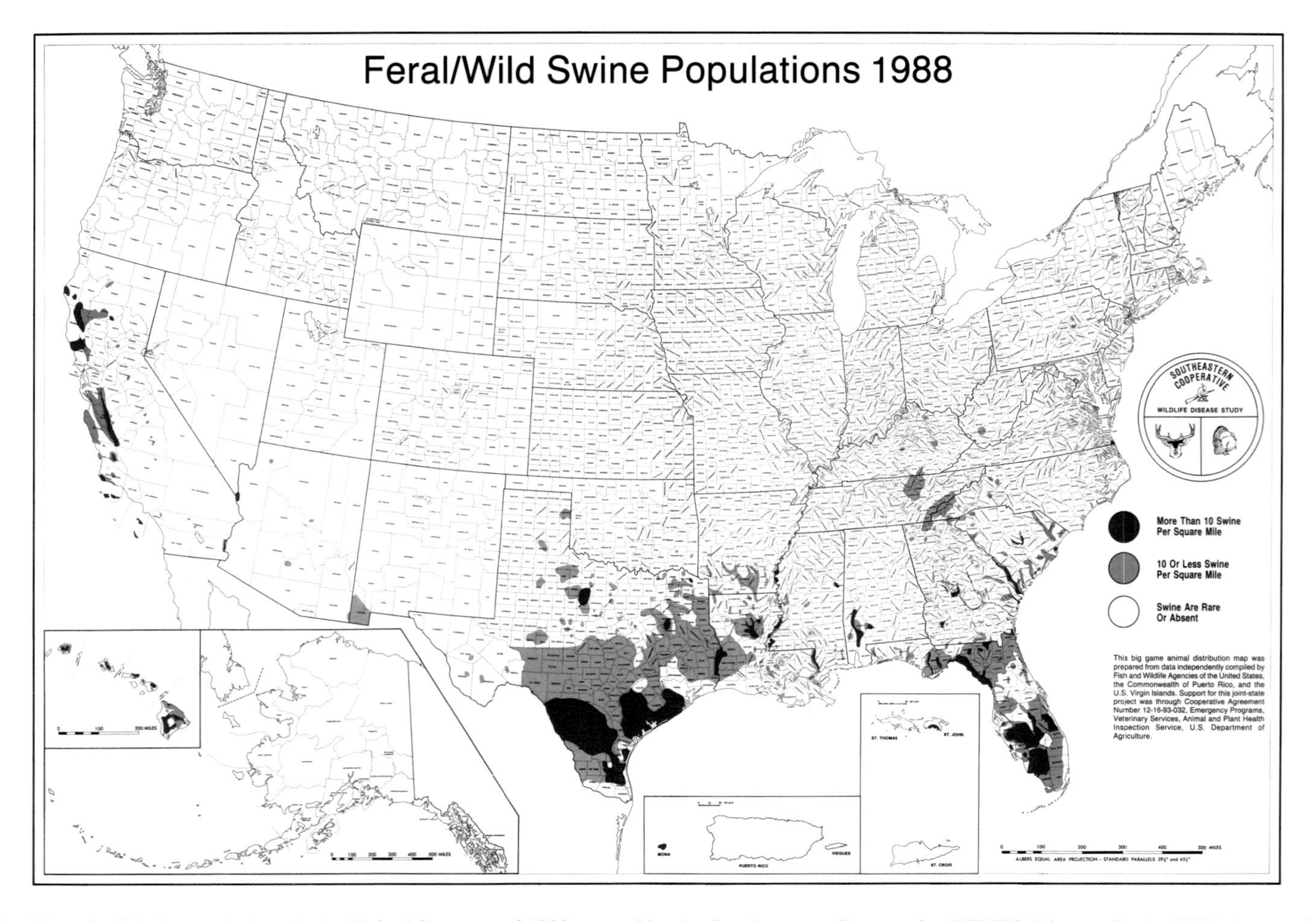

FIGURE 6.1 Map of wild pig populations in the United States as of 1988 created by the Southeastern Cooperative Wildlife Disease Study. Wild pig populations documented here represent some of the oldest and most established populations in North America including southern Texas and Florida, California, and Hawaii. Other populations throughout the Southeast are smaller and more isolated and many are concentrated along riverine corridors and in protected areas (e.g., Great Smoky Mountain National Park).

hypothesis have been best described by the distribution and invasiveness of plant species (e.g., Duncan and Williams 2002, Fargione et al. 2003, Pearson et al. 2012, Carboni et al. 2013), but attempts to verify this hypothesis in vertebrates have been met with limited success (Ricciardi and Mottiar 2006). Wild pigs were historically absent from the Western hemisphere and, having no direct analog across most of their introduced range, may be considered to be filling empty niche space. If differences in niche space exist between wild pigs and native species, diversity could be maintained if native species are limited more by intraspecific competition than by competition with wild pigs (Chesson 2000), despite the realization that wild pigs do clearly have negative impacts on some native species and environments.

The goal of this chapter is to discuss the concept of naturalization relative to wild pigs in the Western hemisphere, and examine the mechanisms behind this phenomenon. We discuss how wild pig niches interact with those of native species, their place in food web dynamics, and their role as ecological engineers. We also evaluate their cultural significance and how wild pigs have become integrated into modern society, despite the suite of impacts caused by this destructive invasive species. We focus our review on the roles wild pigs are playing in native systems and do not address their negative financial, agricultural, or ecological impacts, as these topics are covered extensively in other chapters (e.g., Chapters 5, 7, 10).

6.2 NICHE INTERACTIONS WITH NATIVE SPECIES

Where wild pigs have become integrated into native systems, they did so by competing at some level with native species, filling niche space unoccupied by native species, or filling niches vacated by extinct or locally extirpated native species. Wild pigs overlap in dietary niche space with many native species throughout their introduced range due to their generalist diet (Ditchkoff and Mayer 2009, Ballari and Barrios-García 2014). Species groups thought to compete with wild pigs for food include cervids (Wood and Barrett 1979, Everitt and Alaniz 1980, Wood and Roark 1980, Taylor and Hellgren 1997), mesocarnivores (Turner et al. 2017), gallinaceous birds (Wood and Barrett 1979), rodents (Wood and Barrett 1979, Focardi et al. 2000), and native herbivores (Sweitzer and Van Vuren 2002). Competition between wild pigs and native species may be most intense for hard mast (e.g., acorns), a critical food resource and driver of population dynamics for many native species (Bieber and Ruf 2005). Wild pigs are voracious consumers of mast and can limit seedling establishment via seed predation, leading to potential long-term decline in hard mast availability via decreased recruitment (Siemann et al. 2009, Sanguinetti and Kitzberger 2010). Wild pigs focus their habitat use in areas with mast species, and mast failure can lead to shifts in space use, diet, reproduction, and survival (Singer et al. 1981, Massei et al. 1997; see Chapters 3, 4). Increased competition, especially during mast failure events could have substantive impacts on native vertebrate and invertebrate species who also rely on mast resources (Yarrow 1987, Shea and Chesson 2002). The complexity of dietary interactions between pigs and native species warrants more investigation to not only document overlap, but to also illuminate potential competition between species and the effect that competition is having on those species.

The impacts of wild pigs should be most evident in species that extensively overlap in niche space. The only pig-like mammals in the Western hemisphere are the peccaries (family Tayassuidae), although they employ different life history strategies and are not closely related phylogenetically to wild pigs (Theimer and Keim 1998). Desbiez et al. (2009) investigated niche overlap and partitioning among wild pigs, white-lipped peccaries (*Tayassu pecari*), and collared peccaries (*Pecari tajacu*) in the Brazilian Pantanal where wild pigs have been present for around 200 years and have not driven either peccary species to local extinction, ruling out strict competitive exclusion. Fecal analyses indicated dietary overlap was higher between the native peccary species than between either peccary species and wild pigs. Habitat use also varied among species, with wild pigs spending a greater proportion of time in lowland areas and around water locations while avoiding upland and forested habitats, which were preferred by peccaries (Desbiez et al. 2009). Another study of niche overlap found dietary overlap was much greater between wild pigs and white-lipped peccaries than between wild pigs and collared peccaries (Galetti et al. 2015). Additionally, native peccaries shifted their foraging when wild pigs were present, possibly to avoid direct interaction and competition. Although wild pigs and peccaries currently coexist in the Pantanal in South America, anthropogenic changes to the system such as the creation of water holes for domestic livestock and increased deforestation could have unintended effects on niche partitioning among these species by increasing habitat suitability for wild pigs (Desbiez et al. 2009, Galetti et al. 2015).

Niche partitioning between wild pigs and collared peccaries has also been documented in Texas (Ilse and Hellgren 1995a, b). Wild pigs have been present in coastal Texas since the 1500s (Towne and Wentworth 1950; see Chapters 2, 15), and current population estimates for the state are greater than 2.6 million individuals (Timmons et al. 2012, Lewis et al. 2019). As in the Pantanal, wild pigs and peccaries coexist under current ecological conditions and differ in their diet and habitat use, suggesting niche overlap may be limited except in spring when both species use the same abundant resources (Ilse and Hellgren 1995a). However, collared peccary group size and densities were found to be lower in areas of high wild pig density, suggesting peccaries may have been displaced by wild pigs (Ilse and Hellgren 1995b, Gabor and Hellgren 2000).

Wild pigs and peccaries have several important differences that suggest they should be able to differentiate niche space and coexist. Cranial morphology of wild pigs facilitates more efficient rooting and a more powerful bite than peccaries, allowing them to exploit a greater range of food resources (Sicuro and Oliveira 2002). Wild pigs and peccaries also have divergent digestive tracks, allowing peccaries to focus foraging on succulents and reduce their dependence on water (Langer 1978, 1979; Carl and Brown 1986). These behavioral and physiological differences appear to limit the degree of direct competition between wild pigs and their closest analogs in the Western hemisphere, allowing both species to coexist. However, anthropogenic changes to the environment and increasing wild pig populations may change the dynamics of this coexistence in the future.

Wild pigs may also become naturalized into invaded systems by filling niche space left vacant by extirpated species. Many extinct or locally extirpated species had substantive effects on other species through the creation and maintenance of habitat, the movement of nutrients, and on competitive interactions (Catling 2001). For example, black bears (*Ursus americanus*) and brown bears (*Ursus arctos*) were once widespread throughout North America and their regional removal or population suppression has implications for disturbance regimes, nutrient dynamics, and seed dispersal. Brown bear digging for edible roots has been shown to increase the fitness of several plants (Edge et al. 1990, Tardiff and Stanford 1998), and this foraging strategy can create large areas of disturbance. Bears also are important seed dispersers (especially of fleshy fruits) as they can excrete thousands of seeds in a single scat (Willson 1993, Willson and Gende 2004, Enders and Vander Wall 2012), and can transport seeds up to 3000 m from their source (Patten 1993). The ecological consequences of the extirpation of bears from the majority of their native range (Hall 1981, Catling 2001) might be mitigated by the addition of wild pigs to the landscape, which can occupy similar niche space. Through rooting behavior, wild pigs can modify plant and soil dynamics, especially in areas with high densities of wild pigs (Bueno et al. 2013, Boughton and Boughton 2014). It has been argued that wild pig rooting in California has functionally replaced the role brown bears once held in oak savannah communities (Sweitzer and Van Vuren 2002; Figure 6.2). Wild pigs are documented fecal seed dispersers in Australia and Europe (Grice 1996, Lynes and Campbell 2000, Schmidt et al. 2004), and undoubtedly contribute to plant dispersal throughout their invasive range. Epizoochory (dispersal of seeds by attaching to their fur) also has been documented (Schmidt et al. 2004). With home ranges up to 34 km^2 (Adkins and Harveson 2007) and maximum daily movements between 6.8 and 12.9 km (Spitz and Janeau 1990, Podgórski et al. 2013, Kay et al. 2017; see Chapter 3), wild pigs have great potential to facilitate the movement of plant species in much the same way as bears. Thus, wild pigs may partially fulfill this ecological function in areas that now support fewer or no bears.

FIGURE 6.2 Extensive rooting damage by wild pigs in an oak savanna woodland of southern California. (Photo by B. Teton. With permission.)

6.3 FOOD WEB DYNAMICS OF WILD PIGS

Wild pigs have also become naturalized into the food webs as a predator, prey item, and scavenger. Predation and scavenging by pigs could be a stabilizing influence on ecosystems because they increase the connectivity of food webs (Wilson and Wolkovich 2011), even though it may be an additive form of mortality for prey species. Although their diet is primarily comprised of plant material, wild pigs eat animal material year round, consuming all life stages of various invertebrates, fish, amphibians, reptiles, birds, and mammals (Ditchkoff and Mayer 2009, see Chapter 3). The occurrence of animal material is ubiquitous across diet studies, but potentially may be underrepresented due to rapid digestion. A large proportion of animal consumption comes through rooting, with invertebrates being the most common animal material found in dietary studies on wild pigs. Although consumption of a wide variety of invertebrates have been documented (Ditchkoff and Mayer 2009), annelids are a particularly important food resource for wild pigs and can constitute up to 90% of stomach contents in some instances (Hanson and Karstad 1959, Wood and Roark 1980, Diong 1982).

Predation of vertebrates is less common than invertebrates, but has the potential to impact species of economic and conservation concern. Most predation by wild pigs is opportunistic and the result of novel circumstances, concentrated prey, or random discovery while foraging. Wild pigs are documented predators of reptiles and amphibians and are especially impactful as nest predators or when prey is congregated for breeding. For example, a single wild pig consumed 49 eastern spadefoot toads (*Scaphiopus holbrookii*) when the toads were breeding in the southeastern United States (Jolley et al. 2010). Additionally, wild pigs are a predator of turtle nests (Fordham et al. 2006, Whytlaw et al. 2013), and Elsey et al. (2012) reported over half of landowners surveyed reported American alligator (*Alligator mississippiensis*) nests were damaged by wild pigs. Predation of larger game may be opportunistic and the result of novel circumstances or defenseless prey. Wild pigs have been documented actively killing and feeding on a Rio Grande turkey (*Meleagris gallopavo*) after they were feeding alongside each other moments before (Ditchkoff and Mayer 2009). Predation of white-tailed deer (*Odocoileus virginianus*; Springer 1975, Jolley 2007) and domestic neonates (Pavlov and Hone 1982) also has been recorded. Given their generalist diet and opportunistic foraging strategy, wild pigs depredate ground-nesting bird nests (Tolleson et al. 1993, Sanders 2017) and nest depredation is typically the most important factor limiting population growth rate (Vangilder and Kurzejeski 1995). Population-level effects of wild pigs on most vertebrate prey species are unlikely to occur through predation alone due to their opportunistic foraging behavior and the limited amount of overall diet comprised of animal material (Ditchkoff and Mayer 2009). However, predation by wild pigs in conjunction with native predators could result in additive mortality for specific groups such as ground-nesting birds and reptiles, as well as species already threatened (Lever 1985).

Throughout their invasive range wild pigs may also serve as an important prey item for native carnivores. There are many accounts of wild pigs being preyed upon by native predators including American alligators, some raptors, and many medium to large carnivores, yet predation is usually limited to piglets with limited effects on wild pig population dynamics (reviewed in Mayer 2009). Those species where predation has led to observable changes in either predator populations or wild pig populations are discussed here. Coyotes (*Canis latrans*) are a widespread and common canid throughout much of North America and now overlap in distribution with wild pigs throughout their introduced range. Coyotes prey on piglets and juvenile wild pigs (Keiter et al. 2017) but are assumed to be unable to regularly prey upon adults. Coyotes have shown a numerical response to increases in wild pig numbers (Holling 1966, Stevens 1996), suggesting the presence of wild pigs may be beneficial for coyote populations. Wild pigs also have been implicated in the establishment of golden eagles (*Aquila chrysaetos*) on the Channel Islands of southern California. The presence of a highly productive ungulate allowed eagles to nest on the islands due to increased food availability (Roemer et al. 2001). Once wild pigs were eradicated from most islands, nesting eagles switched prey items and nearly drove the indigenous island fox (*Urocyon littoralis*) to extinction via predation (Roemer et al. 2001).

Larger mammalian predators have the potential to influence wild pig populations because of their ability to take adults as well as juveniles. Black bears can actively prey on all age classes of wild pigs (Stegeman 1938) and are known predators of domestic pigs (Horstman and Gunson 1982). Although anecdotal evidence suggests wild pigs can mortally wound bears during agonistic encounters (reviewed in Mayer 2009), they likely are a valuable source of protein for bears where the 2 species overlap. Mountain lions or panthers (*Puma concolor*) are another large carnivore species that may benefit from the presence of wild pigs in areas where the 2 species overlap, including western North America, South America, and in the southern US (Florida). Mountain lions can catch and kill both juvenile and adult wild pigs (Bruce 1941), and Maehr et al. (1990) reported wild pigs constituted 22.7–58.7% of biomass consumed by mountain lions over a 12-year period. Mountain lion diets are plastic (Iriarte et al. 1990) and predation on wild pigs may be relative to the abundance of alternative prey species in an area (Hopkins 1991). Wild pigs can be an important component of large carnivore diets where their ranges overlap and their high reproductive rate makes them resilient to predation, which could benefit endangered carnivores such as Florida panthers, red wolves (*Canis rufus*), and Mexican wolves (*Canis lupus*). However, wild pigs are carriers of diseases such as pseudorabies that can negatively impact carnivore species (Höfle et al. 2004; see Chapter 5). How wild pig populations influence native predator species is not well understood and deserves further investigation, especially because wild pig interactions with native predators will continue to increase if their ranges are allowed to expand.

Wild pigs also play a pivotal role within food webs both as a scavenger and as a scavenged item by native species. Coyotes, red fox (*Vulpes vulpes*), gray fox (*Urocyon cinereoargenteus*), turkey vultures (*Cathartes aura*), and black vultures (*Coragyps atratus*) all regularly scavenge wild pig carcasses, and the presence of pig tissue in their diet is usually attributed to scavenging (Turner et al. 2017; Figure 6.3). Coyote scavenging of wild pigs can most commonly be tied to hunter harvest or lethal control measures (Schrecengost et al. 2008), and coyotes may actively defend carcass piles (Turner et al. 2017). Wild pigs are also important scavengers of carrion (DeVault and Rhodes 2002, Turner et al. 2017) and carrion can make up a substantial portion of their diet (Thomson

FIGURE 6.3 Wild pigs feature prominently in invaded systems as both a food source for scavengers and a scavenger of carrion. Typical scavengers of wild pig carcasses include (a) coyotes (*Canis latrans*) and (b) turkey vultures (*Cathartes aura*) where their ranges overlap. Wild pigs regularly consume carrion including (c) rodents (*Rattus* sp., tail seen in photo) and (d) white-tailed deer (*Odocoileus virginianus*) used in scavenging trials. (Photos (a–c) by K. Turner and J. Beasley and (d) by M. Vukovich. With permission.)

and Challies 1988). Wild pigs have been documented consuming carrion of rodents, lagomorphs, reptiles, carnivores, and large vertebrates such as white-tailed deer (DeVault and Rhodes 2002, Schley and Roper 2003, Abernethy et al. 2016, Turner et al. 2017; Figure 6.3). Wild pig scavenging is opportunistic and, although they are documented to consume carrion across studies, the magnitude of this impact on native scavengers is unknown and warrants further investigation. Interestingly, Turner et al. (2017) failed to record any instances of cannibalism across 96 scavenging trials of wild pig carcasses, although wild pigs did investigate these carcasses and pigs have been identified in stomach contents in other studies (Taylor and Hellgren 1997), suggesting cannibalism of conspecifics may occur.

6.4 WILD PIGS AS ECOSYSTEM ENGINEERS

Ecosystem engineers are species that "directly or in-directly modulate the availability of resources to other species, by causing physical state changes in biotic or abiotic materials. In so doing they modify, maintain, and/or create habitat" (Jones et al. 1994). This term excludes trophic and competitive interactions discussed above and instead focuses on the physical manipulation of habitat that changes the abundance and distribution of other species. Wild pigs can have extensive impacts on their environment (see Chapter 7), including geomorphic changes that creates habitat for other species (Barrios-Garcia and Ballari 2012, Bevins et al. 2014, Keiter and Beasley 2017), and thus are often considered ecosystem engineers.

In particular, rooting and wallowing by wild pigs can cause significant physical changes to soils, plant communities, and water quality, which can directly influence resource availability for other species (Figure 6.2). Rooting disrupts soil structure by mixing soil horizons and accelerates the decomposition of organic matter, boosting nutrient mobility and soil aeration (Bratton 1975, Singer et al. 1984). Rooting has also been shown to decrease the availability of soil nitrogen but increase the availability of nitrate and ammonia, indicating alteration in nitrogen transformation processes (Siemann et al. 2009, Singer et al. 1984). However, other studies have found no difference in soil properties between rooted and unrooted plots (Moody and Jones 2000, Cushman et al. 2004, Tierney and Cushman 2006), suggesting the effects of wild pigs on soil properties may be site dependent and need further study. Changes to plant communities from rooting may be negative or positive and can become evident in several ways including a reduction in plant cover (Singer et al. 1984), and changes in diversity and composition (Arrington et al. 1999). Rooting can promote plant growth in geophyte species (bulbs and rhizomes), potentially creating a positive feedback loop by attracting more pigs to the site (Palacio et al. 2013). Rooting increased diversity and species richness in a marshy wetland environment while decreasing broadleaf plant cover (Arrington et al. 1999). Conversely, the introduction of invasive plant species by wild pig rooting is well documented (Singer et al. 1984, Aplet et al. 1991, Cushman et al. 2004, Tiernay and Cushman 2006, Siemann et al. 2009), and particularly concerning if invasive plants cause permanent changes to community composition and structure. Rooting and wallowing can also influence water quality and aquatic communities. Pig presence has been shown to increase pathogens, nitrate concentrations, acidity, and turbidity in streams and wetlands (Singer et al. 1984, Kaller and Kelso 2006, Doupe et al. 2010). Aquatic plant and animal communities can also be affected with changes in invertebrates (Kaller and Kelso 2006), macrophytes (Doupe et al. 2010), and species richness (Arrington et al. 1999).

Disturbance and habitat manipulation by wild pigs varies with soil type, climate, vegetation composition and structure, and population density, resulting in both positive and negative effects. Sensitive habitat types (Nilsson and Grelsson 1995) such as islands, wetlands, and alpine environments are particularly susceptible to degradation by wild pigs and their impact is well characterized in the literature (Genov et al. 2017). Less fragile habitat types may be able to sustain repeated disturbance by wild pigs without losses in biodiversity or ecosystem function, but the magnitude of wild pig disturbance is dependent upon wild pig density. At low densities wild pig impacts may be beneficial for maintaining habitat and species diversity by maintaining an intermediate disturbance regime (Kotanen 1995, Arrington et al. 1999), but excessive degradation and impacts likely occur at high densities. The length of time wild pigs are present in an area and continually modifying soil and vegetation could have important implications for how pigs are naturalizing into invaded systems. Most wild pig research to date in their invasive range has been limited to short-term (2–3 year) studies, which may not be sufficient to elucidate the long-term impacts of wild pigs. We reiterate the calls of Genov et al. (2017) for long-term studies of the ecological impacts of this ecosystem engineer throughout their invasive range. We also strongly suggest density estimates are integrated into study designs to more fully understand the impacts wild pigs have on native ecosystems and species (see Chapter 9). Inclusion of density estimates should help identify causal mechanisms and relationships between wild pigs and other species throughout their invasive range.

6.5 WILD PIGS IN SOCIETY

In conjunction with their effects on native systems and communities, the long-term persistence of wild pig populations has implications for a naturalized role in human culture. Wild boar have long held a prominent position in cultures across their native range and featured heavily in Indo-European mythologies where they tend to symbolize strength, valor, fertility, and wildness (Sztych 2014). Hunting wild boar was generally a noble pursuit and its popularity led to the local extirpation of wild boar from many areas of Europe. Pigs have also been an important social component in other cultures including the Far East where a system of pigsty-privies was developed for pigs to convert human waste into fertilizer (Nemeth 1987). While wild pigs have yet to become fully assimilated into the cultural landscapes of all invaded areas, they are beginning to become "naturalized" into some cultures and have even become a source of pride. For example, the University of Arkansas-Fayetteville mascot is the Razorback and is depicted on their logo (Figure 6.4). The University of Arkansas also keeps a live wild pig, Tusk IV, which travels to select sporting events and makes public appearances to promote the university. Further integration into modern society is

FIGURE 6.4 Symbols of wild pig integration into the culture of North America including the logo of the University of Arkansas, Piggly Wiggly grocery stores, and an advertisement for the 7th Annual Wild Hog Festival in Temple, Oklahoma.

expected because wild pig populations historically limited to rural or natural areas have begun to invade urban areas (Cahill et al. 2012). This phenomenon will greatly expand the visibility of wild pigs for the general public and could lead to increased interest in wild pig management.

Within their invasive range, wild pigs have become most engrained in the culture of the southeastern United States. For centuries domestic pigs served as a backbone of southern farms and played a central role in the economy and culture (Bass 1995). Southern states contained 2/3 of the nation's pigs (~1 million animals) in 1860 and hog drives provided pork to many northeastern cities (Hilliard 2014). Free ranging during the spring and summer, annual fall hog roundups and harvesting were a community event and a "time of genuine joy" for the community (Bass 1995). Consumption of pork was so ubiquitous at the time that a Georgian physician, Dr John S. Wilson, humored that "hogs" lard is the very oil that moves the machinery of life." While pigs at that time may not have been considered fully wild pigs, these free-ranging practices allowed for the establishment of wild populations throughout the South and no distinction was made between the 2 forms at the time. The establishment of stock laws in the 20th century reduced the number of free-ranging animals, but not an appetite for pork nor its place in culture (Bass 1995). Grocery stores, such as Piggly Wiggly (Figure 6.4), marketed pork and county and state fairs emerged as a continuation of crop and animal agriculture celebration and harvest. Wild pigs have recently become the focus of local festivals throughout southeastern North America and Texas with crafts, music, and food (Figure 6.4). Most offer pig-catching contests from piglets to 45 kg using wild pigs trapped within the area. Perhaps taking the place of annual fall roundups, these festivals further entwine wild pigs into the cultural fabric.

The popularity of wild pig hunting has increased in recent times and has moved from a means of subsistence to a recreational pursuit. Unsurprisingly, the culture of wild pig hunting is strongly developed in certain areas, especially the southern United States, and its cultural traditions are entrenched (Golding 2011). Interest in wild pig hunting accelerated after the killing of "Hogzilla," a wild pig purported to be ~360 kg, and the airing of television programs, including "Hogs Gone Wild" that highlighted wild pig hunting for sport. A simple internet search for "guided hog hunting" returns thousands of hits, providing a relatively cheap alternative to native game like white-tailed deer. In some areas wild pig hunting has surpassed deer hunting in popularity (Tolleson et al. 1995), especially where there are unlimited bag limits and year-round hunting opportunities. Wild pigs have the distinction of being considered both a game and pest species, with their value to stakeholders being highly variable (Zivin et al. 2000; see Chapter 11). The presence of wild pigs has increasingly become an economic benefit for some, largely due to hunting fees generated on private lands (Bevins et al. 2014). Consequently, wild pig hunters and those earning income from these hunters have been identified as a primary cause of wild pig introductions (Tabak et al. 2017). As long as wild pigs remain a valuable resource, it is likely humans will continue to transport and establish pig populations in the absence of laws and enforcement sufficient to discourage this behavior (see Chapter 11).

In addition to subsistence and recreational hunting, wild pigs have become an economic incentive in parts of their introduced range. The sale of captured wild pigs to processing plants has provided income to local trappers in both Australia (O'Brien 1987) and the United States (Timmons et al. 2012; see Chapter 8). In some states wild pigs can be sold dead or alive to meat companies, for as much as US$0.88kg, and companies even organize hunting tournaments with prizes and buy carcasses (Wild Boar Meat Company 2018). Wild pig meat is even becoming popular with chefs where they can buy meat from processing plants (Associated Press 2017). Income generated from these activities may provide incentives to maintain wild pig populations for future revenue, thereby perpetuating populations and spread of wild pigs. Economic incentives have helped shape policy decisions on wild pigs as well. When use of a warfarin-based toxicant was considered in Texas, trappers, hunters, and meat processors formed an unlikely alliance with animal rights organizations to influence policymakers to ban its use.

6.6 CONCLUSIONS

The idea of the naturalized niche is complicated to conceptualize for wild pigs due to the myriad of impacts that wild pigs have on native systems, the variety of species and habitat types they interact with, and variation in the time since introduction. The naturalization of wild pigs in the Western hemisphere certainly increases competition with many native species. Although wild pig populations may reduce the density of native species, there is little evidence to suggest that wild pig presence is driving species to localized extinction via competition and wild pigs may be filling empty niche space in many areas. Naturalized wild pig niches are likely more impactful on food webs and disturbance regimes than currently understood and we hope to see future research that investigates these topics in recently established and historic populations to understand the role of time since invasion. The influence of invasive species in sensitive habitats such as islands or wetlands may be overridingly negative, but more resilient communities may sustain wild pig presence with fewer negative impacts. The impacts of wild pigs are also tightly linked to the density of that population and we suggest future research should make concerted efforts to estimate density and relate it to diversity losses and damage estimates. Additionally, the role of wild pigs in culture where they are invasive is not well documented in recent times, and the introgression of wild pigs into modern culture deserves further investigation outside of consumptive and commercial endeavors (see Chapter 10).

Despite increased abundance and continued range expansion (Bevins et al. 2014, Snow et al. 2017) across the invasive range of wild pigs, natural resource managers still take great efforts to educate the public about their negative impacts, making it unlikely that they will ever enjoy the revered status of wild boar. Currently, most stakeholders have a negative opinion of wild pigs in both long-naturalized and recently established populations (Adams et al. 2005, Harper et al. 2016). However, the complete eradication of wild pigs is unlikely except in select circumstances such as islands and fenced properties (Cruz et al. 2005, McCann and Garcelon 2008), as well as small and recently established populations (e.g., New York State). In most areas, therefore, the naturalization of wild pigs should continue to be a focus of management actions and scientific research.

ACKNOWLEDGMENTS

We thank M.P. Glow and J. Smith for their contributions to this chapter and K. Turner, M. Vukovitch, and B. Teton for the use of their pictures. Lastly, we acknowledge the University of Arkansas and

Piggly Wiggly for the use of their logos. Contributions from K.C.V. were supported by the USDA. Contributions from J.C.B. were partially funded by the US Department of Energy under award # DE-EM0004391 to the University of Georgia Research Foundation. Mention of commercial products or companies does not represent an endorsement by the US Government.

REFERENCES

Abernethy, E. F., K. L. Turner, J. C. Beasley, T. L. DeVault, W. C. Pitt, and O. E. Rhodes. 2016. Carcasses of invasive species are predominantly utilized by invasive scavengers in an island ecosystem. *Ecosphere* 7:e01496.doi:10.1002/ecs2.1496.

Adams, C. E., B. J. Higginbotham, D. Rollins, R. B. Taylor, R. Skiles, M. Mapston, and S. Turman. 2005. Regional perspectives and opportunities for feral hog management in Texas. *Wildlife Society Bulletin* 33:1312–1320.

Adkins, R. N., and L. A. Harveson. 2007. Demographic and spatial characteristics of feral hogs in the Chihuahuan Desert, Texas. *Human-Wildlife Conflicts* 1:152–160.

Aplet, G. H., S. J. Anderson, and C. P. Stone. 1991. Association between feral pig disturbance and the composition of some alien plant assemblages in Hawaii Volcanoes National Park. *Vegetatio* 95:55–62.

Arrington, D. A., L. A. Toth, and J. W. Koebel. 1999. Effects of rooting by feral hogs *Sus scrofa* L. on the structure of a floodplain vegetation assemblage. *Wetlands* 19:535–544.

Associated Press. 2017. Feral hogs becoming popular with Louisiana chefs. Outdoor News. <https://www.outdoornews.com/2017/04/28/feral-hogs-becoming-popular-louisiana-chefs/>. Accessed 1 July 2018.

Ballari, S. A. and M. N. Barrios-García. 2014. A review of wild boar *Sus scrofa* diet and factors affecting food selection in native and introduced ranges. *Mammal Review* 44:124–134.

Barrios-Garcia, M. N., and S. A. Ballari. 2012. Impact of wild boar (*Sus scrofa*) in its introduced and native range: A review. *Biological Invasions* 14:2283–2300.

Bass, S. J. 1995. "How'bout a hand for the hog": The enduring nature of the swine as a cultural symbol in the South. *Southern Cultures* 1:301–320.

Bevins, S. N., K. Pedersen, M. W. Lutman, T. Gidlewski, and T. J. Deliberto. 2014. Consequences associated with the recent range expansion of nonnative feral swine. *BioScience* 64:291–299.

Bieber, C., and T. Ruf. 2005. Population dynamics in wild boar *Sus scrofa*: Ecology, elasticity of growth rate and implications for the management of pulsed resource consumers. *Journal of Applied Ecology* 42:1203–1213.

Boughton, E. H., and R. K. Boughton. 2014. Modification by an invasive ecosystem engineer shifts a wet prairie to a monotypic stand. *Biological Invasions* 16:2105–2114.

Bratton, S. P. 1975. The effect of the European wild boar, *Sus scrofa*, on gray beech forest in the Great Smoky Mountains. *Ecology* 56:1356–1366.

Bruce, J. 1941. Mostly about the habits of the wild boar. *California Conservationist* 6:14–21.

Bueno, C. G., J. Azorín, D. Gómez-García, C. L. Alados, and D. Badía. 2013. Occurrence and intensity of wild boar disturbances, effects on the physical and chemical soil properties of alpine grasslands. *Plant and Soil* 373:243–256.

Cahill, S., F. Llimona, L. Cabañeros, and F. Calomardo. 2012. Characteristics of wild boar (*Sus scrofa*) habituation to urban areas in the Collserola Natural Park (Barcelona) and comparison with other locations. *Animal Biodiversity and Conservation* 35:221–233.

Carboni, M., T. Münkemüller, L. Gallien, S. Lavergne, A. Acosta, and W. Thuiller. 2013. Darwin's naturalization hypothesis: Scale matters in coastal plant communities. *Ecography* 36:560–568.

Carl, G. R., and R. D. Brown. 1986. Comparative digestive efficiency and feed intake of the collared peccary. *Southwestern Naturalist* 31:79–85.

Catling, P. M. 2001. Extinction and the importance of history and dependence in conservation. *Biodiversity* 2:2–14.

Chesson, P. 2000. Mechanisms of maintenance of species diversity. *Annual Review of Ecology and Systematics* 31:343–366.

Colautti, R. I., and H. J. MacIsaac. 2004. A neutral terminology to define "invasive" species. *Diversity and Distributions* 10:135–141.

Cruz, F., C. J. Donlan, K. Campbell, and V. Carrion. 2005. Conservation action in the Galapagos: Feral pig (*Sus scrofa*) eradication from Santiago Island. *Biological Conservation* 121:473–478.

Cushman, J. H., T. A. Tierney, and J. M. Hinds. 2004. Variable effects of feral pig disturbances on native and exotic plants in a California grassland. *Ecological Applications* 14:1746–1756.

Darwin, C. 1859. *On the Origin of the Species by Means of Natural Selection: or, the Preservation of Favoured Races in the Struggle for Life*. J. Murray, London, UK.

Desbiez, A. L. J., S. A. Santos, A. Keuroghlian, and R. E. Bodmer. 2009. Niche partitioning among white-lipped peccaries (*Tayassu pecari*), collared peccaries (*Pecari tajacu*), and feral pigs (*Sus scrofa*). *Journal of Mammalogy* 90:119–128.

DeVault, T. L., and O. E. Rhodes. 2002. Identification of vertebrate scavengers of small mammal carcasses in a forested landscape. *Acta Theriologica* 47:185–192.

Diong, C. H. 1982. Population biology and management of the feral pig (*Sus scrofa* L.) in Kipahula Valley, Maui. Dissertation, University of Hawaii, Honolulu, HI.

Ditchkoff, S. S., and J. J. Mayer. 2009. Wild pig food habits. Pages 105–143 *in* J. J. Mayer and I. L. Brisbin, Jr., editors. *Wild Pigs: Biology, Damage, Control Techniques and Management*. SRNL-RP-2009-00869. Savannah River National Laboratory, Aiken, SC.

Doupe, R. G., J. Mitchell, M. J. Knott, A. M. Davis, and A. J. Lymbery. 2010. Efficacy of exclusion fencing to protect ephemeral floodplain lagoon habitats from feral pigs (*Sus scrofa*). *Wetlands Ecology and Management* 18:69–78.

Duncan, R. P. and P. A. Williams. 2002. Ecology: Darwin's naturalization hypothesis challenged. *Nature* 417:608–609.

Edge, W. D., C. Les-Marcum, and S. L. Olson-Edge. 1990. Distribution and grizzly bear, *Ursus arctos*, use of yellow sweetvetch, *Hedysarum sulphurescens*, in northwestern Montana (USA) and southeastern British Columbia (Canada). *Canadian Field-Naturalist* 104:435–438.

Elsey, R. M., E. C. Mouton Jr., and N. Kinler. 2012. Effects of feral swine (*Sus scrofa*) on alligator (*Alligator mississippiensis*) nests in Louisiana. *Southeastern Naturalist* 11:205–218.

Enders, M. S., and S. B. Vander Wall. 2012. Black bears *Ursus americanus* are effective seed dispersers, with a little help from their friends. *Oikos* 121:589–596.

Epstein, J., and M. Bichard. 1984. Pig. Pages 145–162 *in* I. L. Mason, editor. *Evolution of Domesticated Animals*. Longman, London, UK and New York.

Everitt, J. H., and M. A. Alaniz. 1980. Fall and winter diets of feral pigs in south Texas. *Journal of Rangeland Management* 33:126–129.

Fargione, J., C. S. Brown, and D. Tilman. 2003. Community assembly and invasion: An experimental test of neutral versus niche processes. *Proceedings of the National Academy of Sciences* 100:8916–8920.

Focardi, S., D. Capizzi, and D. Monetti. 2000. Competition for acorns among wild boar (*Sus scrofa*) and small mammals in a Mediterranean woodland. *Journal of Zoology* 250:329–334.

Fordham, D., A. Georges, B. Corey, and B. W. Brook. 2006. Feral pig predation threatens the indigenous harvest and local persistence of snake-necked turtles in northern Australia. *Biological Conservation* 133:379–388.

Gabor, T. M., and E. C. Hellgren. 2000. Variation in peccary populations: Landscape composition or competition by an invader? *Ecology* 81:2509–2524.

Galetti, M., H. Camargo, T. Siqueira, A. Keuroghlian, C. I. Donatti, M. L. S. P. Jorge, F. Pedrosa, C. Z. Kanda, and M. C. Ribeiro. 2015. Diet overlap and foraging activity between feral pigs and native peccaries in the Pantanal. *PloS One* 10: p.e0141459.

Genov, P. V., S. Focardi, F. Morimando, L. Scilliani, and A. Ahmed. 2017. Ecological impact of wild boar in natural ecosystems. Pages 404–419 *in* M. Melletti and E. Meijaard, editors. *Ecology, Conservation and Management of Wild Pigs and Peccaries*. Cambridge University Press, Cambridge, UK.

Golding, M. 2011. Panther tract: Wild boar hunting in the Mississippi Delta. University Press of Mississippi, Jackson, MS.

Grice, A. C. 1996. Seed production, dispersal and germination in *Cryptostegia grandiflora* and *Ziziphus mauritiana*, two invasive shrubs in tropical woodlands of northern Australia. *Austral Ecology* 21:324–331.

Hall, E. R. 1981. *The Mammals of North America*. Second Edition. John Wiley and Sons, New York.

Hanson, R. P., and L. Karstad. 1959. Feral swine in the southeastern United States. *The Journal of Wildlife Management* 23:64–74.

Harper, E. E., C. A. Miller, J. J. Vaske, M. T. Mengak, and S. Bruno. 2016. Stakeholder attitudes and beliefs toward wild pigs in Georgia and Illinois. *Wildlife Society Bulletin* 40:269–273.

Hilliard, S. B. 2014. *Hog Meat and Hoecake: Food Supply in the Old South, 1840–1860*. University of Georgia Press, Athens, GA.

Höfle, U., J. Vicente, I. G. Fernández de Mera, D. Villanúa, P. Acevedo, F. Ruiz-Fons, and C. Gortázar. 2004. Health risks in game production: The wild boar. *Galemys* 16:197–206.

Holling, C. S. 1966. The functional response of invertebrate predators to prey density. *The Memoirs of the Entomological Society of Canada* 98:5–86.

Hopkins, R. A. 1991. Ecology of the puma in the Diablo Range, California. Dissertation, University of California, Berkeley, CA.

Horstman, L. P., and J. R. Gunson. 1982. Black bear predation on livestock in Alberta. *Wildlife Society Bulletin* 10:34–39.

Ilse, L. M., and E. C. Hellgren. 1995a. Resource partitioning in sympatric populations of collared peccaries and feral hogs in southern Texas. *Journal of Mammalogy* 76:784–799.

Ilse, L. M., and E. C. Hellgren. 1995b. Spatial use and group dynamics of sympatric collared peccaries and feral hogs in southern Texas. *Journal of Mammalogy* 76:993–1002.

Iriarte, J. A., W. L. Franklin, W. E., Johnson, and K. H. Redford. 1990. Biogeographic variation of food habits and body size of the America puma. *Oecologia* 85:185–190.

Jolley, D. B. 2007. Reproduction and herpetofauna depredation of feral pigs at Fort Benning, Georgia. Thesis, Auburn University, Auburn, AL.

Jolley, D. B., S. S. Ditchkoff, B. D. Sparklin, L. B. Hanson, M. S. Mitchell, and J. B. Grand. 2010. Estimate of herpetofauna depredation by a population of wild pigs. *Journal of Mammalogy* 91:519–524.

Jones, C. G., J. H. Lawton, and M. Shachak. 1994. Organisms as ecosystem engineers. *Oikos* 69:373–386.

Kaller, M. D., and W. E. Kelso. 2006. Swine activity alters invertebrate and microbial communities in a coastal plain watershed. *The American Midland Naturalist* 156:163–177.

Kay, S. L., J. W. Fischer, A. J. Monaghan, J. C. Beasley, R. Boughton, T. A. Campbell, S. M. Cooper, et al. 2017. Quantifying drivers of wild pig movement across multiple spatial and temporal scales. *Movement Ecology* 5:14.

Keiter, D. A., and J. C. Beasley. 2017. Hog heaven? Challenges of managing introduced wild pigs in natural areas. *Natural Areas Journal* 37:6–16.

Keiter, D. A., J. C. Kilgo, M. A. Vukovich, F. L. Cunningham, and J. C. Beasley. 2017. Development of known-fate survival monitoring techniques for juvenile wild pigs (*Sus scrofa*). *Wildlife Research* 44:165–173.

Kotanen, P. M. 1995. Responses of vegetation to a changing regime of disturbance: Effects of feral pigs in a Californian coastal prairie. *Ecography* 18:190–199.

Langer, P. 1978. Anatomy of the stomach of the collared peccary, *Dicotyles tajacu* (L., 1758) (Artiodactyla: Mammalia). *Zeitschrift fur Saugetierkunde* 43:42–59.

Langer, P. 1979. Adaptational significance of the forestomach of the collared peccary, *Dicotyles tajacu* (L. 1758) (Mammalia: Artiodactyla). *Mammalia* 43:235–246.

Lever, C. 1985. *Naturalized Mammals of the World*. Longman, London, UK.

Lewis, J. S., J. L. Corn, J .J. Mayer, T. R. Jordan, M. L. Farnsworth, C. L. Burdett, K. C. VerCauteren, S. S. Sweeney, and R. S. Miller. 2019. Historical, current, and potential population size estimates of invasive wild pigs (*Sus scrofa*) in the United States. *Biological Invasions* doi:10.1007/s10530–019–01983–1.

Lynes, B. C., and S. D. Campbell. 2000. Germination and viability of mesquite (*Prosopis pallida*) seed following ingestion and excretion by feral pigs (*Sus scrofa*). *Tropical Grasslands* 34:125–128.

Maehr, D. S., R. C. Belden, E. D. Land, and L. Wilkins. 1990. Food habits of panthers in southwest Florida. *Journal of Wildlife Management* 54:420–423.

Massei, G., P. V. Genov, B. W. Staines, and M. L. Gorman. 1997. Factors influencing home range and activity of wild boar (*Sus scrofa*) in a Mediterranean coastal area. *Journal of Zoology* 242:411–423.

Mayer, J. J. 2009. Natural predators of wild pigs in the United States. Pages 193–204 *in* J. J. Mayer and I. L. Brisbin, Jr., editors. *Wild Pigs: Biology, Damage, Control Techniques, and Management*. SRNL-RP-2009-00869. Savannah River National Laboratory, Aiken, SC.

Mayer, J. J., and J. C. Beasley. 2017. Wild pigs. Pages 221–250 *in* W. C. Pitt, J. C. Beasley, and G. W. Witmer, editors. *Ecology and Management of Terrestrial Vertebrate Invasive Species in the United States*. CRC Press, Boca Raton, FL

Mayer, J. J., and I. L. Brisbin. 2008. *Wild Pigs in the United States: Their History, Comparative Morphology, and Current Status*. The University of Georgia Press, Athens, GA.

McCann, B. E., and D. K. Garcelon. 2008. Eradication of feral pigs from Pinnacles National Monument. *Journal of Wildlife Management* 72:1287–1295.

Moody, A., and J. A. Jones. 2000. Soil response to canopy position and feral pig disturbance beneath *Quercus agrifolia* on Santa Cruz Island, California. *Applied Soil Ecology* 14:269–281.

Nemeth, D. J. 1987. *The Architecture of Ideology: Neo-Confucian Imprinting on Cheju Island, Korea*. University of California Press, Berkeley, CA.

Nilsson, C., and G. Grelsson. 1995. The fragility of ecosystems: A review. *Journal of Applied Ecology* 32:677–692.

O'Brien, P. H. 1987. Socio-economic and biological impact of the feral pig in New South Wales: An overview and alternative management plan. *The Rangeland Journal* 9:96–101.

Palacio, S., C. G. Bueno, J. Azorín, M. Maestro, and D. Gómez-García. 2013. Wild-boar disturbance increases nutrient and C stores of geophytes in subalpine grasslands. *American Journal of Botany* 100:1790–1799.

Patten, L. A. 1993. Seed dispersal patterns generated by brown bears (*Ursus arctos*) in southeastern Alaska. Thesis, Washington State University, Pullman, WA.

Pavlov, P. M., and J. Hone. 1982. The behavior of feral pigs, *Sus scrofa*, in flocks of lambing ewes. *Australian Wildlife Research* 9:101–109.

Pearson, D. E., Y. K. Ortega, and S. J. Sears. 2012. Darwin's naturalization hypothesis up-close: Intermountain grassland invaders differ morphologically and phenologically from native community dominants. *Biological Invasions* 14:901–913.

Podgórski, T., G. Baś, B. Jędrzejewska, L. Sönnichsen, S. Śnieżko, W. Jędrzejewski, and H. Okarma. 2013. Spatiotemporal behavioral plasticity of wild boar (*Sus scrofa*) under contrasting conditions of human pressure: Primeval forest and metropolitan area. *Journal of Mammalogy* 94:109–119.

Ricciardi, A., and M. Mottiar. 2006. Does Darwin's naturalization hypothesis explain fish invasions? *Biological Invasions* 8:1403–1407.

Richardson, D. M., P. Pyšek, M. Rejmácek, M. G. Barbour, F. D. Panetta, and C. J. West. 2000. Naturalization and invasion of alien plants: Concepts and definitions. *Diversity and Distributions* 6:93–107.

Roemer, G. W., T. J. Coonan, D. K. Garcelon, J. Bascompte, and L. Laughrin. 2001. Feral pigs facilitate hyperpredation by golden eagles and indirectly cause the decline of the island fox. *Animal Conservation* 4:307–318.

Sanders, H. N. 2017. Impacts of invasive wild pigs on wild turkey reproductive success. Thesis, Texas A&M University, College Station, TX.

Sanguinetti, J., and T. Kitzberger. 2010. Factors controlling seed predation by rodents and non-native *Sus scrofa* in *Araucaria araucana* forests: Potential effects on seedling establishment. *Biological Invasions* 12:689–706.

Schley, L., and T. J. Roper. 2003. Diet of wild boar *Sus scrofa* in western Europe, with particular reference to consumption of agricultural crops. *Mammal Review* 33:43–56.

Schmidt, M., K. Sommer, W. U. Kriebitzsch, H. Ellenberg, and G. von Oheimb. 2004. Dispersal of vascular plants by game in northern Germany. Part I: Roe deer (*Capreolus capreolus*) and wild boar (*Sus scrofa*). *European Journal of Forest Research* 123:167–176.

Schrecengost, J. D., J. C. Kilgo, D. Mallard, H. S. Ray, and K. V. Miller. 2008. Seasonal food habits of the coyote in the South Carolina coastal plain. *Southeastern Naturalist* 7:135–144.

Shea, K., and P. Chesson. 2002. Community ecology theory as a framework for biological invasions. *Trends in Ecology and Evolution* 17:170–176.

Sicuro, F. L., and L. F. B. Oliveira. 2002. Coexistence of peccaries and feral hogs in the Brazilian Pantanal wetland: An ecomorphological view. *Journal of Mammalogy* 83:207–217.

Siemann, E., J. A. Carrillo, C. A. Gabler, R. Zipp, and W. E. Rogers. 2009. Experimental test of the impacts of feral hogs on forest dynamics and processes in the southeastern US. *Forest Ecology and Management* 258:546–553.

Singer, F. J., D. K. Otto, A. R. Tipton, and C. P. Hable. 1981. Home ranges, movements, and habitat use of European wild boar in Tennessee. *Journal of Wildlife Management* 45:343–353.

Singer, F. J., W. T. Swank, and E. E. C. Clebsch. 1984. Effects of wild pig rooting in a deciduous forest. *Journal of Wildlife Management* 48:464–473.

Snow, N. P., M. A. Jarzyna, and K. C. VerCauteren. 2017. Interpreting and predicting the spread of invasive wild pigs. *Journal of Applied Ecology* 54:2022–2032.

Spitz, F., and G. Janeau. 1990. Spatial strategies: An attempt to classify daily movements of wild boar. *Acta Theriologica* 35:129–149.

Springer, M. D. 1975. Food habits of wild hogs on the Texas Gulf Coast. Thesis, Texas A&M University, College Station, TX.

Stegeman, L. R. C. 1938. The European wild boar in the Cherokee National Forest, Tennessee. *Journal of Mammalogy* 19:279–290.

Stevens, R. L. 1996. *The Feral Hog in Oklahoma*. Samuel Roberts Noble Foundation, Ardmore, OK.

Sweitzer, R. A., and D. H. Van Vuren. 2002. Rooting and foraging effects of wild pigs on tree regeneration and acorn survival in California's oak woodland ecosystems. US Department of Agriculture Forest Service General Technical Report 219–231.

Sztych, D. 2014. Kulturotwórcza rola dzika. *Wiadomości Zootechniczne* 52:142–154.

Tabak, M. A., A. J. Piaggio, R. S. Miller, R. A. Sweitzer, and H. B. Ernest. 2017. Anthropogenic factors predict movement of an invasive species. *Ecosphere* 6:e01844.

Tardiff, S. E., and J. A. Stanford.1998. Grizzly bear digging: Effects on subalpine meadow plants in relation to mineral nitrogen availability. *Ecology* 79:2219–2228.

Taylor, R. B., and E. C. Hellgren. 1997. Diet of feral hogs in the western South Texas Plains. *Southwestern Naturalist* 42:33–39.

Theimer, T. C., and P. Keim. 1998. Phylogenetic relationships of peccaries based on mitochondrial cytochrome b DNA sequences. *Journal of Mammalogy* 79:566–572.

Thomson, C., and C. N. Challies. 1988. Diet of feral pigs in the podocarp-tawa forests of the Urewera Ranges. *New Zealand Journal of Ecology* 11:73–78.

Tierney, T. A., and J. H. Cushman. 2006. Temporal changes in native and exotic vegetation and soil characteristics following disturbances by feral pigs in a California grassland. *Biological Invasions* 8:1073–1089.

Timmons, J. B., B. Higginbotham, R. Lopez, J. C. Cathey, J. Mellish, J. Griffin, A. Sumrall, and K. Skow. 2012. *Feral Hog Population Growth, Density and Harvest in Texas*. Texas A&M AgriLIFE, SP-472. Texas A&M University, College Station, TX.

Tolleson, D., D. Rollins, W. Pinchak, M. Ivy, and A. Hierman. 1993. Impact of feral hogs on ground-nesting gamebirds. Pages 76–83 *in* C. W. Hanselka and J. F. Cadenhead, editors. *Feral Swine: A Compendium for Resource Managers*. Texas Agricultural Extension Service, Kerrville, TX.

Tolleson, D. R., W. E. Pinchak, D. Rollins, and L. J. Hunt. 1995. Feral hogs in the Rolling Plains of Texas: Perspectives, problems, and potential. Pages 124–128 *in Wildlife Damage Management, Internet Center for Great Plains Wildlife Damage Control Workshop Proceedings*. University of Nebraska, Lincoln, NE.

Towne, C. W., and E. N. Wentworth. 1950. *Pigs from Cave to Corn Belt*. University of Oklahoma Press, Norman, OK.

Turner, K. L., E. F. Abernethy, L. M. Conner, O. E. Rhodes, and J. C. Beasley. 2017. Abiotic and biotic factors modulate carrion fate and vertebrate scavenging communities. *Ecology* 98:2413–2424.

Vangilder, L. D., and E. W. Kurzejeski. 1995. Population ecology of the eastern wild turkey in northern Missouri. *Wildlife Monographs* 130:1–50.

Whytlaw, P. A., W. Edwards, and B. C. Congdon. 2013. Marine turtle nest depredation by feral pigs (*Sus scrofa*) on the Western Cape York Peninsula, Australia: Implications for management. *Wildlife Research* 40:377–384.

Wild Boar Meat Company. 2018. Tournaments. <https://wildboarmeats.com/tournaments/>. Accessed 30 June 2018.

Willson, M. F. 1993. Mammals as seed-dispersal mutualists in North America. *Oikos* 67:159–176.

Willson, M. F., and S. M. Gende. 2004. Seed dispersal by brown bears, *Ursus arctos*, in southeastern Alaska. *The Canadian Field-Naturalist* 118:499–503.

Wilson, E. E., and E. M. Wolkovich. 2011. Scavenging: How carnivores and carrion structure communities. *Trends in Ecology and Evolution* 26:129–135.

Wood, G. W., and R. H. Barrett. 1979. Status of wild pigs in the United States. *Wildlife Society Bulletin* 7:237–246.

Wood, G. W., and D. N. Roark. 1980. Food habits of feral hogs in coastal South Carolina. *Journal of Wildlife Management* 44:506–511.

Yarrow, G. K. 1987. The potential for interspecific resource competition between white-tailed deer and feral hogs in the post oak savannah region of Texas. Dissertation, Stephen F. Austin University, Nacogdoches, TX.

Zivin, J., B. M. Huethand, and D. Zilberman. 2000. Managing a multiple-use resource: The case of feral pig management in California rangeland. *Journal of Environmental Economics and Management* 39:189–204.

7 Wild Pig Damage to Resources

Bronson K. Strickland, Mark D. Smith, and Andrew L. Smith

CONTENTS

7.1 INTRODUCTION

From a wildlife ecologist's perspective, wild pigs (*Sus scrofa*) are truly incredible animals. Few large mammals show plasticity necessary to maintain relatively high survival and reproductive rates among variable landscapes found throughout North America. From conifer forests of British Columbia to southern reaches of the Florida Everglades, and most places in between, wild pigs manage to adapt, exploit their local environment, and often persist. While populations of most native species remain relatively limited in many of these locales (based on resources available), wild pigs seem to remain untethered by such constraints and often thrive. Aside from occasional assistance provided by humans, wild pigs accomplish this feat through occupying what is, in many respects, a vacant niche (see Chapter 6). This allows them to exploit their surroundings and communities that dwell within, often proving invariant to factors that typically limit populations. This resourcefulness and utility is one of the primary factors making wild pigs such successful invaders; however, this success occurs at the expense of native wildlife, host environments, and human-focused land uses such as agriculture and forestry.

Wherever wild pigs exist on landscapes, damage tends to occur, often having implications on a wide range of variables and functions, of which many are cumulative. Damage may occur through direct consumption of plant and animal biomass via various forms of competition for resources (i.e., space, cover, forage/food) with native wildlife, and degradation of those ecosystems, communities, and services in areas inhabited by wild pigs. However, these negative impacts may also occur through more indirect processes frequently compromising the integrity of local environments and resident populations of other animals via disease and parasite transmission, reductions in water quality due to bacterial contamination and/or increased sedimentation and nutrient load, sediment loss, and through the propagation of additional invasive species, to name a few. Per unit area, a finite amount of food, water, and cover exists to support North America's native wildlife populations.

When wild pigs enter these systems—systems in which they did not evolve—associated wildlife and functions they collectively provide become vulnerable to a host of degenerations.

Due to facultative and opportunistic tendencies, wild pig damage is not exclusive to North America's natural landscapes. Wild pigs also elicit a variety of economic damages, such as impacts to agricultural commodities (e.g., row crops, forestry, livestock), infrastructure, and numerous health and public safety concerns (e.g., wild pig–vehicle collisions). However, what may be surprising to some is that North Americans have had to reconcile and cope with these damages for hundreds of years. During the 1600s, wild pigs caused similar damages and controversy among colonists. Domesticated pigs were a vital commodity to early colonial life where they were free-ranging—a practice that quickly escalated into conflict, ecologically and socially, among colonists and frequently with Native Americans. Wild pig depredation of wheat and corn, gardens, and grain bins were commonplace, with any reasonable solution amiss due to personal property clauses and ideals of the era.

As free-ranging personal property (i.e., livestock), pigs managed to root their way through early settler and Native American resources, and through the newly established social fabric holding these 2 factions at peace. Ultimately, and after several legal battles in Connecticut, Massachusetts, and Rhode Island General Courts, the final and well-received English conclusion was to drive all free-ranging pigs >8 km away from various townships and colonies (Conover and Conover 1987). This resulted in a shift from depredation of colonists' commodities to that of neighboring Native American settlements. Moreover, as pigs were considered personal property (versus depredating wildlife), people were prohibited from killing them to prevent or mitigate damage based on colonial law—laws Native Americans were expected to follow. Furthermore, if wild pigs were trapped, Native Americans were expected to feed and care for the animal(s) until the owner could claim his stock (Conover and Conover 1987). Unsurprisingly, disagreement and conflict quickly escalated among the factions, but the double-standard ethics of early English law were far from over. In 1637, Connecticut's General Court ruled that any Native American settlement in proximity to damage would be the sole entity responsible and liable for the damages of pigs, and under no circumstances would the nearest English settlement be held at fault (Conover and Conover 1987). The ultimate result of these conflicts, among others of the time (i.e., English law imposed on Native Americans), was King Phillip's War, which ended the lives of over 600 colonists and 3,000 natives (Conover 2007).

A couple hundred years later during the industrial era of the 1800s, free-ranging pigs continued to catalyze conflict even in densely populated areas such as New York City; representing what is likely one of the first cases of urban-wild pig conflict in the New World. Many paintings from the era commonly feature pigs roaming the streets of New York and rooting alongside Broadway (Alonso and Recarte 2008). Such accounts of daily life during this period showed free-roaming pigs as a simple and ubiquitous reality, as elegant ladies frequently crossed the streets of Broadway in the company of loose pigs and dogs. One writer, Ted Steinberg, described how Europeans viewed America at the time. He stated that when Europeans visualized "the Continent" in the 19th century, they typically thought of Native Americans—these strange and eccentric new tribes of indigenous people unknown on their continent. In contrast, when they pictured American cities, their imagination filled with pigs and unsanitary municipalities. Clearly, wild pigs loomed large in European images of the US urban scene (Alonso and Recarte 2008).

What change occurred that lead to abrupt intolerance of pigs? This fundamental shift from socially accepting a presence of free-roaming pigs within urban communities, to complete removal and prevention of what, for so long, was a common practice. Was it economic or ecological losses causing people to view wild pigs differently? Perhaps this shift in tolerance aligned with a "moral restoration"—an effort to remove the public stain of uncivilized and unsanitary conditions that plagued the image of city life in the early United States (Alonso and Recarte 2008). While it was likely a combination of these negative consequences, most importantly, a shift did occur and free-ranging pigs soon disappeared (Alonso and Recarte 2008), an accomplishment completed with a small fraction of the knowledge and technology we have access to today.

These less-discussed, historical accounts in North America demonstrate the ability of wild pigs to adapt to most environments, and their propensity to affect various plant, animal, and human populations that coexist with them; including those in urban areas (see Chapter 19). However we currently perceive threats of invasive wild pigs, historical parallels prove impacts are far from recent. While wild pigs are unlikely to be seen, or tolerated, roaming the streets of New York City today—and their impacts are unlikely to facilitate war among citizens in modern society—they are still controversial, divisive, and ecologically and economically detrimental as they were over 400 years ago. In fact, as civilization and populations of wild pigs continue to expand, it is likely their negative impacts will also. This chapter will review those primary types of impacts typically associated with wild pigs, while summarizing and discussing some of the other, perhaps less documented or considered, forms of damage.

7.2 DAMAGE TO NATIVE FLORA AND FAUNA

While a variety of factors contribute to the nature of wild pig damage, morphological characteristics uniquely position wild pigs to cause damage like no other North American vertebrate. Physical features such as a low center of gravity, strong neck, and various skull biometrics (e.g., eye location, snout, jaw structure; see Chapter 2) are beneficial for a generalist omnivore typically restricted to subterranean and herbaceous understory strata of terrestrial ecosystems. This phenotype gives wild pigs the ability to root up and forage upon a variety of resources normally inaccessible to most native North American ungulate species. These morphological adaptations, coupled with access to unexploited, subterranean resources has potentially given wild pigs a competitive advantage over other native species. Currently, this niche space is unoccupied in many North American ecosystems (see Chapter 6), allowing wild pigs to capitalize, and often thrive in many locales. As such, wild pig populations that occur on landscapes for a long time will necessarily, and perhaps irreversibly, affect local flora and fauna, as evidenced by their negative impacts on rare, threatened, endangered, and species of concern in Florida and other Gulf states (Thompson 1977, Engeman et al. 2004, West et al. 2009). Wild pigs currently co-occur with >55 imperiled species in southeastern North America (McClure et al. 2018), where their populations are currently most prolific and historically persistent (see Chapter 16).

7.2.1 Competitive Interactions

Wild pigs are recognized as one of the greatest *vertebrate* modifiers of natural plant communities in existence (Bratton 1975, Wood and Barrett 1979, Stone and Keith 1987), and are frequently considered to be ecosystem engineers (Crooks 2002, Bevins et al. 2014, Genov et al. 2018). Their diet is predominately comprised of plant materials (Seward et al. 2004, Mayer and Brisbin 2009, Ballari and Barrios-Garcia 2014), despite being classified as a generalist and omnivore (see Chapter 3). Although wild pigs likely occupy some unexploited niche spaces across North America, niche overlap with native wildlife is inevitable. Overlap among diets exist among many native species in North America, but negative implications of overlaps with wild pigs are particularly evident. Wild pigs seek energy-rich foods as their monogastric stomachs are less capable of extracting nutrients from cellulose (Massei and Genov 2004) compared to North American ruminants (Mayer and Brisbin 2009), and may cause wild pigs to forage for greater amounts of time daily (Seward et al. 2004). Because food resources are finite and ebb and flow over time, resource competition is one of many considerations when attempting to manage native wildlife populations coexisting with wild pigs. Consequently, understanding and mitigating competition within these invaded systems is one of the fundamental challenges facing biologists and managers, and is rarely a comprehensive process.

Direct and indirect consequences of competition between wild pigs and native wildlife may manifest in several forms. Manifestations occur through interactions known as exploitative, interference, and apparent competition. *Exploitative competition* is indirect and simply results from wild pigs finding and consuming resources faster than competitors (Pianka 1983). For example, hard and soft mast crops are finite and seasonal, and numerous native wildlife species depend on mast availability. Wild pigs readily consume hard mast, and even seek out acorns (*Quercus* spp.) cached by small mammals (Focardi et al. 2000). Pulsed availability of hard mast is linked to population dynamics of wild boar in native (Cutini et al. 2013) and introduced ranges (Johnson et al. 1982, Baber and Coblentz 1986; see Chapter 4). Yarrow (1987) documented a 46% seasonal diet overlap between wild pigs and white-tailed deer (*Odocoileus virginianus*) during fall (correlated with mast production) in the southeastern United States, with 25–28% overlap during winter, spring, and summer. Wild pigs eat 3–5% of their body mass daily (Bodenchuk 2008), thus as pig populations grow, exploitative competition with native wildlife is inevitable. Furthermore, burgeoning wild pig populations may only leave a small amount of viable vegetation intact for fertilization and reseeding. This could subsequently lead to fewer resources for wildlife that depend on understory vegetation (including mast) for either food or cover (Sweitzer and Van Vuren 2002). While some dietary studies focused primarily on deer (Yarrow 1987, Yarrow and Kroll 1989, Taylor and Hellgren 1997), others showed that overlap also exists among wild pigs and North American raccoons (*Procyon lotor*), opossums (*Didelphis virginiana*; Stegeman 1938), wild turkey (*Melagris* spp.), squirrel (*Sciurus* spp.), and black bear (*Ursus americanus*) diets (Wood and Barrett 1979). Dietary overlap also occurs between wild pigs and collared peccaries (*Pecari tajacu*), cattle (*Bos taurus*), striped skunks (*Mephitis mephitis*), red fox (*Vulpes vulpes*), gray fox (*Urocyon cinereoargenteus*), bobcats (*Lynx rufus*), muskrats (*Ondatra zibethicus*), nutria (*Myocastor coypus*), eastern cottontail and swamp rabbits (*Sylvilagus* spp.), hawks (Accipitridae), owls (Strigidae, Tytonidae), sandhill cranes (*Grus canadensis*), and waterfowl (Conley et al. 1972; Bratton 1974a, b; Thompson 1977; Baron 1979; Everitt and Alaniz 1980; Sweeney et al. 2003). Moreover, as wild pigs are opportunistic, they also scavenge, competing with coyotes (*Canis latrans*), vultures (Cathartidae), and eagles (Buteoninae) for carrion (Mayer and Beasley 2018).

Direct competition for food is one of the more obvious detrimental impacts of wild pigs on North American wildlife, but wild pigs also compete for other resources such as space and cover. Not only do wild pigs compete on the individual level, their high reproductive capacity (Melletti and Meijaard 2018) may allow them to outcompete native wildlife at population and community levels, through overcrowding and displacement; a characteristic common among many invasive species (Vitousek et al. 1996, Elton 2000). Although not well documented for wild pigs, *interference competition* occurs when one species physically excludes another from a resource, typically using aggressive or territorial behavior (Pianka 1983). For wild pigs, a sounder (or individual) may physically prevent native wildlife species from using a common food or water source, or perhaps areas of cover such as quality bedding and nesting habitat. Tolleson et al. (1995) reported that white-tailed deer avoid feeders, food plots, or natural foraging areas utilized by wild pigs. In addition, Taylor and Hellgren (1997) reported that wild pigs excluded deer from feeding on acorns, and a study by Keever (2014) suggested deer spatially and temporally partition use of forage resources to avoid wild pigs. In Australia, Natusch et al. (2017) documented short- and long-term negative effects of wild pigs on native birds. Presence of wild pigs on the landscape was associated with a simultaneous reduction in bird numbers (possibly interference competition), but, following departure of wild pigs, bird numbers increased in response to foraging opportunities availed by wild pig rooting and disturbed soil. After birds had exhausted the exposed food resources, bird numbers again declined. In Mississippi, Ivey et al. (2019) documented a 26% decline of vertebrate species occupying forest patches simultaneously occupied by wild pigs. Cause for the decline was unknown, but could represent interference competition for limited resources. What once may have been quality habitat prior to colonization of wild pigs was reduced in quality as herbaceous understory was rooted, consumed or trampled, denuded of mast, and colonized by wild pigs and additional invasive or non-native plants which were unpalatable and undesirable to existing native wildlife (see below; West et al. 2009).

Indirectly limiting native species can also occur through a process called *apparent competition*. This arises when a shared predator, disease, or pathogen increases because of species A, and indirectly affects species B (Holt 1977, Muller and Godfray 1997). Wild pigs exhibited this relationship on the California Channel Islands. Abundance of wild pigs promoted establishment of mainland golden eagle (*Aquila chrysaetos*) on the islands, leading to a restructuring of the trophic hierarchy on the island. Once on the islands, golden eagles also preyed upon island fox (*Urocyon littoralis*) causing population decline, and simultaneously diminishing competition between island fox and island spotted skunk (*Spilogale gracilis amphiala*), causing release of the skunk population (Roemer et al. 2002). In this case, wild pigs indirectly caused decline of island fox by promoting establishment of a common predator, golden eagles.

7.2.2 Predation

Wild pig morphology, coupled with a highly plastic and opportunistic diet, allows access to a wide variety of animal prey (McClure et al. 2018). Generally, nearly all animals spending at least a portion of their lives on the ground are candidates for predation by wild pigs (Appendix 7.1). This includes terrestrial, ground dwelling small mammals (Wilcox and Van Vuren 2009), nesting vertebrates (McClure et al. 2018) and eggs they may produce – particularly, but not exclusive to, herpetofauna (Jolley et al. 2010), ground-nesting birds (Tolleson et al. 1993, Rollins and Carroll 2001, Schaefer 2004), and fossorial invertebrates (e.g., arthropods, annelids, mollusks; Barrios-Garcia and Ballari 2012). While diet composition and food selection of wild pigs varies geographically and seasonally, animal matter varies from 0.3 to 30% of their diet in native and introduced ranges (Mayer and Brisbin 2009, Ballari and Barrios-Garcia 2014, McClure et al. 2018; Figure 7.1). Interestingly,

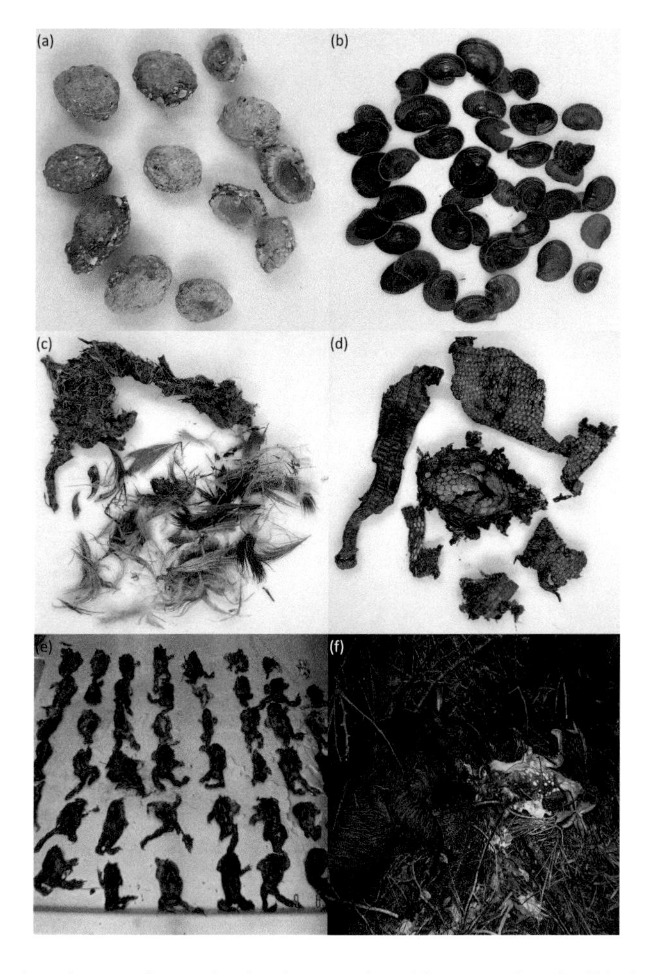

FIGURE 7.1 Examples of vegetative and animal matter found in stomachs of wild pigs in North America: (a) hickory mast, (b) freshwater mussels, (c) depredated or scavenged avian species, (d) depredated or scavenged armadillo, (e) numerous eastern spadefoot toads from a single pig, and (f) presumed opportunistic depredation of a white-tailed deer fawn. (Photos (a-d) by N. Hinson and (e, f) by D. Jolley. With permission.)

a comparative review of diets in native and introduced ranges revealed that wild pigs in introduced areas typically consume greater amounts of animal matter (Ballari and Barrios-Garcia 2014).

Regardless of proportion of animal matter in diets, it is widely accepted that wild pigs function as both predators and scavengers, capitalizing on various prey and carrion when opportunities occur (Ballari and Barrios-Garcia 2014). Whereas foraging style of wild pigs has been classified primarily as opportunistic (Mayer and Brisbin 2009), wild pigs also have capacity for spatial memory and can even exploit knowledge of conspecifics regarding food locations (Held et al. 2000). Sanders (2017) conducted an experiment to determine if wild pigs would modify their foraging strategy based on timing and density of simulated wild turkey nests, but found no change in search strategy. It is likely the search behavior of wild pigs varies among geographic scales, and the spatial and temporal distribution of resources. If resource availability demonstrates a spatial or temporal pattern wild pigs may develop a "search image" for the resource over time, a phenomena that commonly occurs in predators, giving them an advantage at locating specific prey more efficiently based on visual and olfactory cues and patterns (Ishii and Shimada 2010). The ability to modify search behavior could have wide implications, particularly regarding threats facing endangered species where they co-occur with wild pigs (Figure 7.2). For example, in both natural and introduced ranges, wild pigs typically occur in proximity to various water resources (e.g., swamps, marshes, streams, rivers), where impacts to numerous aquatic species inevitably occur. A recent risk summary by McClure et al. (2018) showed that aquatic crustaceans may be most vulnerable to wild pig foraging, as mean range overlaps ~86% with wild pigs. Mollusks are also vulnerable, with 18 of 19 identified by McClure et al. (2018) as co-occurring with wild pigs.

Not only do wild pigs effect such populations through direct predation, they may also elicit impacts indirectly through predation of other species within the community. For instance, wild pigs not only affect ground-nesting birds directly through nest depredation (Rollins and Carroll 2001), but also by competitively reducing local invertebrate abundances available to native species (Carpio et al. 2014) through their exhaustive and competitive foraging behavior (i.e., exploitative competition; Persson 1985, Polis et al. 1989).

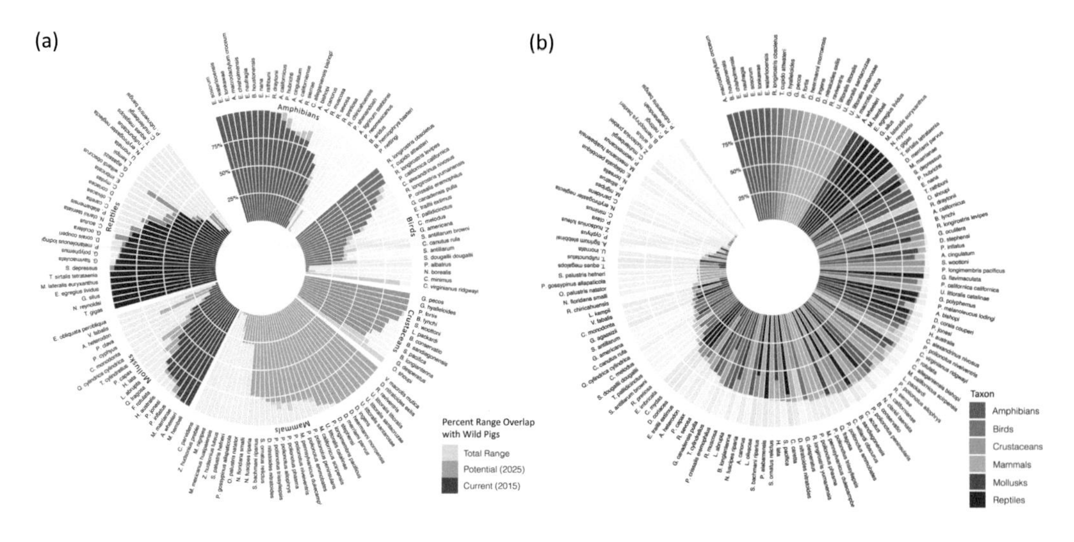

FIGURE 7.2 Most threatened and endangered species in the contiguous United States expected to be susceptible to wild pig impacts have extensive range overlap with wild pigs that is projected to increase by 2025. Proportion of each susceptible species' range currently occupied by wild pigs and estimated to potentially be occupied by 2025, displayed by (a) taxonomic group and (b) decreasing proportion of the range that is currently occupied. (Reproduced from McClure et al. 2018.)

7.2.3 Communal Impacts

Wild pigs perform a variety of functions (e.g., ecosystem engineers, disturbers, crop pests, frugivores, predators, competitors, depredators of seed banks, plant dispersers) depending on ecosystems or habitats in which their impacts occur (Ballari and Barrios-Garcia 2014; Figure 7.3). In addition to plant and animal impacts that occur from daily wild pig foraging activities, common impacts also include reductions in local species diversity and alterations in distribution and availability of local food resources (Kaller and Kelso 2006, Campbell and Long 2009, Mayer and Brisbin 2009, Melletti and Meijaard 2018). Wild pigs are one of the greatest vertebrate modifiers of natural plant communities (Bratton 1975, Wood and Barrett 1979, Stone and Keith 1987). As such, they are also considered ecosystem engineers (Hone 2002, Bevins et al. 2014, Genov et al. 2018), as they disproportionately affect species diversity, composition, and other ecological metrics in areas where they occur (Crooks 2002). This can be dichotomized into a host of well-documented positive or negative physical changes within the biotic and abiotic environment; which may have cumulative and cascading effects at the ecosystem level (Barrios-Garcia and Ballari 2012, Genov et al. 2018). In the floodplains of Florida, for example, a positive relationship was found between wild pig rooting behavior and plant species richness (Arrington et al. 1999), supporting similar results from deciduous forests of Sweden (Brunet et al. 2016). The study showed domination of control wet prairie plots by grasses and small forbs, whereas rooted wet prairie plots included grasses, forbs, several broadleaf species, and a larger compliment of sedges. Wild pig rooting behavior may also function as a surrogate to fire disturbance in some fire-suppressed regions of North America, such as California (Kotanen 1995). Concomitantly, wild pig rooting disturbance was beneficial to native wildlife that depended on establishment of early successional vegetation (Everitt and Alaniz 1980). This outcome was predicted by Genov et al. (2018) based on the intermediate disturbance hypothesis (Connell 1978) where intermediate levels of disturbance, such as grazing and rooting, may improve biodiversity and productivity of some plant species. While no distinction was implied between impacts in introduced versus native ranges of wild pigs, plant community responses are likely to be species specific (Barrios-Garcia and Ballari 2012) and depend on plant and soil characteristics, plant and herbivore interactions, among other plant-related factors.

While positive ecological effects from wild pigs exist within introduced ranges of North America, the ultimate question is whether positive impacts outweigh negative consequences (i.e., net gain). The answer likely varies geographically and politically; however, it is collectively understood among the scientific community that negative impacts (within their introduced range) far outweigh perceived benefits from wild pigs (ecologically or economically; Mayer and Brisbin 2009). For example, Engeman et al. (2003, 2004) estimated that wild pig rooting caused US$1.2–4.0M worth of damage annually to unique basin wetland systems and adjacent uplands in Savannas Preserve State Park in Florida. Not only does wild pig rooting behavior result in efficient removal of a wide range of organisms from various soil strata, it also disturbs the soil surface, in some areas extensively, facilitating propagation of rapidly invading plant species (Mungall 2001, Cushman et al. 2004, Boughton and Boughton 2014). These species are typically non-native, invasive, and less palatable or desirable as a nutritional resource by native wildlife (Siemann et al. 2009, West et al. 2009). Moreover, due to the prolific and colonizing nature of these non-native invasive plants, they typically outcompete native flora for sunlight and space, often resulting in monotypic patches of little to no ecological value to native wildlife which evolved in these systems. Barrios-Garcia and Simberloff (2013) conducted an experiment to determine how these introduced mammals may facilitate plant invasions and found establishment of non-native plant seedlings was twice the level found in non-rooted areas. Similarly, a 7-year study in Big Thicket National Preserve, a heavily forested ecosystem found in southeast Texas, revealed that diversity of understory woody plants increased with exclusion of wild pigs, and all large-fruiting species (e.g., oaks, hickories (*Carya* spp.)) preferred by wild pigs and native ungulates positively responded (Siemann et al. 2009). Moreover, wild pig rooting decreased plant cover and increased area of bare soil, apparently facilitating greater abundance of invasive Chinese tallow

References: Predation of native wildlif — Ditchkoff and Mayer 2009, Refer to Appendix 7.1 • Aquatics — Kaller and Kelso 2003, 2006 • Depredation of native flora — Ditchkoff and Mayer 2009 • Propagation of invasive species — Siemann et al, 2009 •Competition with native wildlife-Yarrow 1987 • Impacts to soil health — Singer et al. 1984 • Disruption of plant and animal diversity — Barrios-Garcia and ballari 2012

FIGURE 7.3 Wild pig damage extends far beyond negative impacts to agricultural crops. Wild pigs compete with native wildlife for food and space, disrupt propagation and colonization of native flora, disrupt geochemical processes, and facilitate cascading impacts within aquatic systems.

tree (*Sapium sebiferum*). Hone (2002) showed ground rooting by wild pigs reduced species richness on 2 sites, based on extent of rooting plant richness declined to 0% in extensively rooted areas. Similarly, in the Lower Peninsula of Michigan, wild pig rooting was associated with less coverage of native herbaceous plants and decreased plant diversity, but greater colonization of non-native plants did not occur (Gray et al. In Review).

Elsewhere in the world, results indicate that wild pigs are effective vectors of exotic plants through means other than soil disturbance. In central Israel (part of wild pig native range), study results showed that up to 91% of seedlings excreted and dispersed by wild pigs were exotic. Furthermore, 59% of seedlings dispersed via epizoochory (dispersal by fur) were also exotic (Dovrat et al. 2012). However, wild pigs may also promote colonization of native weed species. Boughton and Boughton (2014) showed, for example, that rooted plots shifted from bunchgrass dominated wet prairie to a monotypic patch of bloodroot (*Lachnanthes caroliana*). These results, combined with others, indicate that wild pigs have a variety of communal impacts throughout a given ecosystem, likely influencing future diversity (floral and faunal) and composition of plant understories and eventually overstories (Siemann et al. 2009). A cascading consequence to this degree of change in vegetation communities, and concurrent impacts on plant succession, would not only be a potential reduction in the plant diversity of an area, but also the diversity of fauna that depends on the services and functions provided by those species.

Wild pig occurrence and activity also results in a variety of geophysical and chemical effects both above- and below-ground. Rooting and wallowing modifies soil chemistry and nutrient cycling by mixing A1 and A2 soil horizons (Mayer and Brisbin 2009), accelerates erosion (Sierra 2001) and leaching of minerals and micronutrients (i.e., Ca, P, Zn, CU, Mg) from the forest floor (Singer et al. 1984), and affects soil compaction and water infiltration (Mungall 2001; Figure 7.3). Moreover, these behaviors may suppress or, at the very least, interfere with decomposition cycles and other microbial functions through widespread reductions of earthworms and other fossorial biota (Tisdell 1982) from foraging. Such impacts can have cascading, long-term effects throughout communities and ecosystems by influencing variability of soil fertility and function (Genov et al. 2018). The ability to alter species composition, reduce biodiversity, and modify species distributions throughout an ecosystem support why wild pigs are recognized as a significant engineer (Crooks 2002, Genov et al. 2018). Fundamental negative consequences of wild pig presence in an area clearly indicate necessity for current management, coupled with future efforts focused on monitoring and quantifying impacts.

Another critical component to consider and account for is the spatial scale at which damage occurs. Plant diversity and richness may increase on a microhabitat scale in areas rooted by wild pigs in Florida wetlands (Arrington et al. 1999), whereas consequences of rooting (or other damage) at larger spatial scales, may vary. For instance, in forests and other terrestrial systems, wild pig impacts may be isolated and relatively confined; however, in aquatic systems impacts may permeate throughout entire watersheds in areas of high wild pig densities. Wild pig rooting is widely implicated in declines in water quality related to increased erosion and sedimentation, spread of pathogens, and increased nutrient loading, but also in alterations of resident functional groups (i.e., grazers, collectors, scrapers). Kaller and Kelso (2006) showed that positive associations existed between gastropod grazers (snails) and wild pigs, as snails were likely attracted to an increase in organic matter (i.e., feces, detritus) resulting from wild pig activity. Although not entirely understood, results from Kaller and Kelso (2006) also suggested a total community shift occurred in riparian habitats they measured due to wild pig activity. Prior to wild pig exposure, riparian systems were dominated by a diverse group of insect taxa of collector and scraper functional groups, but post-pig activity these aquatic systems became dominated by gastropod grazers. Ramifications of a community shift within watersheds are of great importance. Decomposition of organic matter in such systems largely depend on various functional groups that process the materials. Ultimately, nutrient deposition, uptake, and changes to those energy cycles may affect macro- and microbiota abundance and diversity of entire systems, increase substrate accumulation of biofilms (Kaller and Kelso 2003, 2006), and alter littoral vegetation of the system, thus influencing various fauna (aquatic and terrestrial) that might subsequently use these systems. Additionally, presence of wild pigs within watersheds affects native wildlife through algal blooms, oxygen depletion, sediment loading (Mayer and Brisbin 2009), and even fish kills (Mapston 2004). The combined activities of rooting, wallowing, and defecating from wild pigs within aquatic systems can ultimately compromise health and longevity of such sensitive ecosystems through habitat degeneration, and may pose immediate threats to endemic aquatic and terrestrial life.

7.2.4 Forest Damage

Damage to forests is common where wild pigs occur in North America. Activity of wild pigs inhibits germination, and thus regeneration (natural and artificial) of numerous fruiting species, including conifer and deciduous taxa. In California, consumption of acorns by wild pigs reduced acorn availability for germination and limited forest regeneration (Sweitzer and Van Vuren 2002), while rooting modified forest structure and composition in Hawaii (Ralph and Maxwell 1984) and elsewhere (Ickes et al. 2003). In Spain, the number of holm oak (*Quercus ilex*) seedlings was ≤50% in areas rooted by wild boar, causing a change in recruitment rate and spatial distribution of this species (Gómez and Hódar 2008). Exploitative or competitive consumption of hard mast such as acorns, hickory nuts, tupelo (*Nyssa* spp.) seeds and soft mast such as wild grapes (*Vitis riparia*), muscadine (*Vitis rotundifolia*), dewberries, and others (e.g., *Rubus* spp.) may in itself significantly impact forest under- and overstory composition (Henry and Conley 1972, Wood and Roark 1980, Howe et al. 1981). In addition, pillaging of acorns and other seed caches deposited by tree squirrels (*Sciurus* spp.) and other frugivores and granivores may negatively impact regeneration of these fruiting species. Small mammals play an important role through secondary seed dispersal for some fruiting species (Debussche and Isenmann 1989, Vander Wall et al. 2005). Small mammal proclivity to deposit caches in preparation for overwintering could result in keystone species impacts, because of their disproportionate influence on seed fates (Steele et al. 2005). Wild pigs seek out and extract seed hoards deposited by small mammals (Focardi et al. 2000), not only directly competing with native frugivores and granivores, but also directly inhibiting germination of unrecovered caches. Combined with consumption of acorns and other seeds readily available on the forest floor, wild pigs potentially have significant impacts on germination and thus, regeneration of various forest species.

While many forest-dwelling species are responsible for seed dispersal through consumption and excretion (endozoochory), the role of wild pigs as seed dispersers is poorly understood in introduced ranges. However, a consensus of studies showed wild pigs function more as seed predators (Barrios-Garcia and Ballari 2012) by either partially or fully damaging the seed, resulting in seed mortality (Rudge 1976, Lott et al. 1995, Gómez et al. 2003, Siemann et al. 2009), in both native and introduced ranges. Herrero et al. (2006) further noted wild pigs were not important seed dispersers because consumed seeds were too large to avoid damage during digestion. Other research demonstrated that wild pigs are capable of successfully dispersing (through excretion) viable seeds of non-native plants such as mesquite (*Prosopsis pallida*), rubber vine (*Cryptostegia grandiflora*), Chinese date (*Ziziphus mauritiana*), and possibly Chinese tallow (Grice 1996, Lynes and Campbell 2000, Siemann et al. 2009). While consumption of mast is well documented through study of stomach contents in the introduced range, subsequent dispersal of seeds through epizoochory and endozoochory is less understood. Barrios-Garcia and Ballari (2012) suggested the combination of endozoochory and epizoochory may be critical for understanding relationships between rooting disturbance and presence of non-native plant species.

Although wild pig impacts to forests typically occur through direct consumption of seedlings, damages also occur through other activities. Wild pig wallowing and other non-consumptive behaviors such as rubbing, nesting, and bedding (Mayer and Brisbin 2009) may also affect tree health, quality, and other developmental processes (e.g., establishment, nutrient uptake). Mengak

(2015) reported timber was the 4th most frequently damaged crop for 30.5% of survey respondents, whereas both Poudyal et al. (2017) and Tanger et al. (2015) reported nearly US$1.5M/year in loss of timber due to wild pig damage in Tennessee and Louisiana, respectively. However, the type of forest-related damage was not clear. Negative impacts to forests have wide-ranging implications to local assemblages of vegetation (succession), nutrient cycles, as well as arrangement and distribution of species. While primarily documented in hardwood ecosystems (Bratton 1975, Bratton et al. 1982, Mayer et al. 2000, Sweitzer and Van Vuren 2002), forest damage issues are not restricted to beech (*Fagus* spp.), oak, and hickory forests of the North American interior. Restoration of longleaf pine (*Pinus palustris*) along coastal regions of the southeastern United States encountered many issues with wild pigs depredating seedlings soon after planting, as well as through depredation during the juvenile grass stage (Wood and Brenneman 1977, Lipscomb 1989, Campbell and Long 2009). During that period, wild pigs destroyed young longleaf pine seedlings at the rate of 6/minute/animal, and sustained damage of as many as 400 to 1,000 seedlings/day (Hopkins 1948, Wakely 1954). Impacts such as these may additionally limit other populations that are endemic to longleaf pine ecosystems, including some that are threatened or endangered, or considered species of concern. Examples include red-cockaded woodpeckers (*Leuconotopicus borealis*), gopher tortoises (*Gopherus polyphemus*), and other herpetofauna, as well as various carnivorous plants (*Sarracenia* spp.), which occur in the more mesic reaches of longleaf pine habitats (i.e., associated with pitcher plant bogs).

7.3 AGRICULTURAL IMPACTS

Wild pigs have been a continuous source of damage to various commodities since their arrival to the New World in colonial times (Conover 2007, Mayer and Brisbin 2008). In current markets, and considering contemporary scales of agricultural production in the United States (i.e., crop, livestock, forest resources), presence of wild pigs can result in significant and direct financial loss to agricultural producers and landowners. Types of agricultural damage vary regionally across the range of wild pigs and depend on crop type (e.g., cotton, corn, sorghum) and wild pig densities. Severity of damage varies depending on local abundance and distributions of wild pigs, but also by season (i.e., resource availability), landscape context and configuration (e.g., access), crop phenology, and active population management techniques. Damage such as direct consumption and trampling of various crops (e.g., corn, soybeans, peanuts, small grains), coupled with rooting of pastures and hay fields, in addition to subsequent damage to machinery (Engeman et al. 2007), are some of the most common forms of agricultural damage experienced by producers. Damage by wild pigs is indiscriminate between high-volume commercial growers or small-scale specialty crop producers, and may ultimately result in negative impacts to local, rural economies (Poudyal et al. 2017).

7.3.1 Identification and Quantification of Damage

Most knowledge of crop loss and forest damage from wild pigs in North America evolves from direct observations of damage and anecdotal reports, self-reporting surveys of producers (e.g., Mengak 2015, Anderson et al. 2016), and from food habit studies (e.g., Schley and Roper 2003, Ditchkoff and Mayer 2009), each of which are subject to limitations and assumptions. For example, food habit studies using stomach contents may not accurately represent importance of some food items by only quantifying amount and frequency of a particular food item. This approach potentially ignores nutritional importance of that food item. Whereas environmental impacts of site-specific agricultural and forestry damage can be reasonably quantified, reliable estimates of damage at state or national levels are more difficult to obtain. Reliable estimates of wild pig numbers do not exist on state or national scales making damage estimates dubious at best. For example, a widely cited estimate of damage nationally (Pimental 2007), suggested agriculture damage and associated wild pig control in the United States was US$1.5B annually, based on a crudely estimated national population of 5 million wild pigs (causing an estimated US$300 in damage/individual). While this national estimate seems reasonable, Timmons et al. (2012) estimated that wild pig populations in Texas numbered 1.8–3.4 million pigs, with 79% of the state containing suitable habitat. When simulating potential growth of wild pig populations in Texas, Mellish et al. (2014) suggested statewide population sizes could reach 3.6 to 16.9 million. Coupled with damage estimates of US$389.70/pig annually in areas with intensive agriculture (B. Higginbotham, Texas A&M AgriLife Research and Extension Center at Overton, unpublished report), nearly US$1.5B/year in damage could occur in Texas alone. However, caution must be exercised when projecting estimates of damages caused by wild pigs, particularly those which are themselves projections, of either population sizes or of future population growth. Moving forward, as new population metrics are developed and refined, and as models are improved (see Chapter 9), such methods to estimate and quantify damage will help support the demand for accurately assessing these impacts; in addition to measuring the success of damage management and mitigation.

Surveys by mail, phone, or online sent to stakeholder groups are one of the most frequently used methods to identify and quantify damage caused by wild pigs. Several of these surveys occurred in the past decade (see Chapter 10). Self-reporting estimates of damage are inherently susceptible to reporting bias and/or inaccuracy of estimates; however, they do offer a likely approximation of financial loss incurred by producers and landowners due to wild pig damage and in some cases provide insights into the diversity of damage experienced. Clearly, these estimates rely on ability of those taking the survey to accurately identify wild pig damage from other damage sources, then enumerate that damage into dollars, and accurately report those damages on the survey. Tegt and Strickland (2018) randomly selected a subsample of phone survey participants in Mississippi for face-to-face, onsite interviews to determine accuracy of reported damage cost estimates. They found that landowners perceived damage estimates <US$0.50/ha greater than the amount calculated onsite, resulting in a discrepancy of US$7M for the statewide damage estimate. In recent years, several surveys of landowners, producers, hunters, and general public provided a wealth of knowledge not only on types of damage wild pigs may cause to agricultural and forestry resources, but also reasonable estimates of financial losses incurred by landowners or producers. These estimates of agriculture and forestry damage have proven useful for guiding policy development, allocating resources for abatement, and targeting of educational materials.

7.3.2 Factors Influencing Damage

Surprisingly, few studies in North America identified factors influencing the process of agriculture and forestry damage by wild pigs. Whereas the European scientific literature is replete with studies examining spatiotemporal aspects of wild pig damage to crops (Herrero et al. 2006, Morelle and Lejeune 2015), linkages between population density and levels of damage (Bleier et al. 2012, Frackowiak et al. 2012, Bobek et al. 2017), annual and seasonal shifts in resource use (Keuling et al. 2008, 2009), or landscape structure and composition (Lombardini et al. 2017), these types of studies are currently lacking for North America. For example, Bobek et al. (2017) reported that damage to agricultural lands in Europe increased with increasing density of wild boar, but number of damage cases declined dramatically as distance from forest-farmland border increased. Moreover, they noted seasonal shifts in diets of wild boar from underground parts of herbs and grasses in meadow and grassland habitats in summer to that of higher energy cereal seeds in fall. Thurfjell et al. (2009) reported similar findings in Sweden in that most wild boar damage in agricultural fields occurred along field edges, with most damage occurring within 54 m from adjacent forest edges. Damage to smaller fields may occur throughout the entire field because all portions are close to a forest edge (Figure 7.4). Wild boar avoided agricultural fields for most parts of the year but less so during summer as these fields provided cover and food resources (Thurfjell et al. 2009). While examining patterns in crop damage over a 10-year period in Luxemborg, Schley et al. (2008) reported damage to grasslands (meadows and pastures; 51% of cases) was greater than that of crops (31% of cases).

FIGURE 7.4 Spatial patterns of wild pig crop damage in corn fields in the Mississippi Alluvial Valley during the 2016 growing season. White dots are verified wild pig damage locations. Although most crop damage occurs on field borders, spatial extent can be much greater depending on field size, food availability, and wild pig population density. (Photos by D. Foster. With permission.)

However, wild pigs actively sought maize over other available crops such as wheat, barley, and other cereal grains, potentially due to cover value afforded by this crop. Space use may even differ among age classes of wild boar (Keuling et al. 2008) and landscape composition (Keuling et al. 2009). In Germany, Keuling et al. (2009) categorized wild boar into 3 distinct behavioral types (i.e., field sows, commuters, forest sows) based upon movement patterns relative to availability of land cover types. How well these behavioral idiosyncrasies and patterns of damage by wild boar in Europe relate to non-native wild pigs in North America is currently unknown.

While these aspects of crop depredation are well studied for species such as white-tailed deer (DeVault et al. 2007) throughout North America, scant information exists for wild pigs and continues to be an area of much needed research. Only recently, Paolini et al. (2018) reported season-specific differences in resource selection by wild pigs in the Mississippi Alluvial Valley and noted that use of specific crop fields often depended on availability and proximity of other cover types, mainly wetlands. However, wild pigs responded to changes in crop phenology and selected corn crops later in the growing season toward harvest, whereas rice and soybean crops had greater probability of selection in the early growing season. When examining corn, cotton, and peanut fields within 10 days of initial planting in Alabama, Engeman et al. (2018) reported no damage to fields receiving protection through professional control operations, whereas 22% of those fields not protected received damage. Temporal availability of agricultural resources apparently leads to changes in spatial distributions of wild pigs across landscapes similar to observations outside of North America. However, further research into field and landscape scale resource selection, and hence damage, will likely increase as technologies such as unmanned aerial systems (UAS) incorporating remote sensors coupled with automated image processing gain widespread use (Samiappan et al. 2018) for assessing on-ground disturbances in crop fields by wild pigs (see Chapter 9).

7.3.3 Damage to Crops

Most commercially produced crops (e.g., corn, soybean, wheat) and pastureland in North America are subject to damage by wild pigs at some point during the growing and harvest seasons (Figure 7.5). For example, Mengak (2012) reported producers in 41 counties in southwest Georgia experienced wild pig damage to peanut and corn crops most frequently followed by cotton, but also reported substantial damage to wheat, wildlife food plots, pecans, hay/pasture, blackberries, grain sorghum,

FIGURE 7.5 Wild pig crop damage can occur most anytime throughout the growing season. Corn and soybean fields in the Mississippi Alluvial Valley are often depredated at the time of planting, and again once the crop matures and grain is produced. (Photos by D. Foster. With permission.)

oats, and other specialty crops. Y. Zhang (Auburn University, unpublished report) estimated wild pig damage to crops in Alabama to be approximately US$30M/year. Likewise, Anderson et al. (2016) estimated wild pigs caused US$190M/year in damage to corn, soybeans, wheat, rice, peanuts, and sorghum in 11 southeastern states. In Tennessee, Poudyal et al. (2017) reported US$26.2M in damage due to wild pigs with most (US$13.8M) associated with damage to crops (mean crop loss/respondent of US$2,630). Damage may be due to direct consumption of crops at various stages of development from seeding to harvest, trampling of crops, or rooting of seed beds; largely depending on crop type. Damage to these crops may be so substantial that producers change farming practices and types of crops planted to those of lesser value that may not incur damage from wild pigs (Mengak 2015, Poudyal et al. 2017). For example, at the field level, Engeman et al. (2018) estimated up to US$15,779 in lost yield in 1 peanut field in Alabama where damage by wild pigs amounted to 54% of the crop field area, because no control measures were undertaken to reduce damage at planting. However, damage and estimated values of potential crop loss/field vary substantially among fields and crop types, with cotton fields receiving the least amount of damage. Excluding the extreme example noted above, on average only 1–2% of field area was damaged by wild pigs (Engeman et al. 2018), similar to that reported in Anderson et al. (2016) for Alabama. Regardless, a significant amount of damage may occur among multiple crop fields within a farm, subsequently reducing whole-farm profitability.

7.3.4 Damage to Pastures/Hayfields

Wild pigs commonly damage pastures and hayfields through rooting when searching for food items, potentially resulting in significant loss of forage, owner/employee time to repair damage, and damage to machinery such as tractor axles or disk blades while trying to repair the damage (Figure 7.6). In many cases where damage is severe, farm operators simply operate around damaged areas within the field. Tegt and Strickland (2018) reported Mississippi landowners estimated costs to repair field damages were greater than costs associated with crop loss. Tanger et al. (2015) reported damage in 2013 to pastureland at US$2.3M in Louisiana, with loss of hay valued at US$9.9M, while producers spent about US$2.5M in re-disking ground rooted by wild pigs. Moreover, rooting activity may result in shifts in grass species composition over time (Bankovich et al. 2016) resulting in loss of value or shift in succession patterns in grassland communities through favoring non-native invasive species (Spatz and Mueller-Dombois 1972, Cushman et al. 2004).

7.3.5 Damage to Livestock

Although plant material composes much of wild pig diets (Ditchkoff and Mayer 2009), animal material, including that of common livestock animals, has also been found. While young of domestic sheep (*Ovis aries*), goat (*Capra hircus*), and cattle have been found in wild pig diets (Seward et al. 2004, Ditchkoff and Mayer 2009), frequency and severity of predation on livestock has not been well documented in North America. Furthermore, whether these were actually predation events or scavenging of carrion (i.e., stillborn fetuses, deceased adults) is unclear. In Australia, Choquenot et al. (1997) found that lamb predation increased with wild pig density, in addition to twin lambs being threatened by predation 5–6 times more than singletons. Currently, economic impacts of livestock

FIGURE 7.6 Rooting and wallowing behavior of wild pigs not only destroys crops, but often damages infrastructure of the field. In this example, heavy machinery will be needed to level and smooth soil before this area can be used for pasture or growing crops. (Photo by C. Gibson. With permission.)

predation by wild pigs in North America remains unknown. Rollins (1993) reported 33% of county agriculture agents listed wild pig predation on sheep and goats as problems within their counties, whereas Barrett and Birmingham (1994) reported nearly 1,473 sheep and goats killed in Texas and California by wild pigs. Some of these impacts may be underestimated because wild pig predation and subsequent consumption of carrion could easily be mistaken as low productivity in herds and flocks rather than predation. In contrast, wild pig predation may be confused with coyote predation, as signs of predation from both species appear similar, so severity of wild pig predation on livestock is difficult to discern and measure (Seward et al. 2004).

In a recent survey of landowners in Tennessee, Poudyal et al. (2017) found that 17% of respondents reported wild pigs accessed stored feed or grain and accessed housing for livestock, with potential implications for disease transmission. Wild pigs are carriers of numerous diseases and parasites that may affect livestock (Witmer et al. 2003, Corn et al. 2004, Hartin et al. 2007, Wyckoff et al. 2009; see Chapter 5) and humans (Davidson 2006, Meng et al. 2009), and they may also serve as agents for bioterrorism (Ditchkoff and West 2007). Swine brucellosis (*Brucella suis*), foot-and-mouth disease (*Aphthae epizooticae*), pseudorabies (*Suid herpesvirus 1*; PRV), trichinosis (*Trichinella* spp.), tuberculosis (*Mycobacterium tuberculosis*), vesicular stomatitis (*Vesiculovirus* sp.), and classical swine fever (*Pestivirus* sp.) are among some of the many diseases that are of significant concern to livestock producers, and to the industry as a whole. For example, Corn et al. (2004) reported prevalence of PRV antibodies in 38% of wild pigs (n = 100) collected from 10 sites in the southeastern United States. Likewise, Wyckoff et al. (2009) reported prevalence of PRV and swine brucellosis antibodies in 30% and 11% of wild pigs tested in southern Texas, respectively. However, Hartin et al. (2007) found only 0.3% antibody presence against swine brucellosis and did not detect PRV or classical swine fever antibodies in 321 wild pigs tested in Missouri. Although confinement operations may reduce exposure risk of domestic pigs to swine brucellosis and PRV, fenced or corralled domestic pigs may experience greater risk of exposure. Wyckoff et al. (2009) reported that 19% (7 of 37) of radio-marked wild pigs made contact with domestic pigs in Texas, and found greater visitation of wild pigs to pens containing domestic pigs than empty pens. Likewise, Cooper et al. (2010) reported that GPS collared wild pigs and free-range cattle frequented the same location (within 20 m) at water sources and irrigated forage fields along roads, posing threats of direct and indirect contact that may serve as an avenue for spread of diseases.

While diseases can be managed in domestic livestock through vaccinations, barriers against direct contact, and other animal husbandry practices, these are not viable options for wild pigs. As such, wild pig populations may serve as persistent disease reservoirs requiring continuous surveillance of affected wild pig populations and local livestock health. Moreover, given difficulties of eradication of wild pig populations on a landscape scale, wild pigs may serve as effective bioterrorism agents via introduction of foreign animal diseases (FADs), resulting in significant implications to the livestock industry. Paarlberg et al. (2002) modeled implications of an outbreak of foot-and-mouth disease and estimated US farm losses ranging between US$13.6–20.8B during the first year after an outbreak, mainly due to loss of exports and a consumer shift to poultry products. Likewise, outbreaks of classical swine fever in the commercial pork industry would result in an estimated US$2.6–4.1B in profit and animal loss over a 5-year (20-quarter) period (Paarlberg et al. 2009). In addition to livestock industry losses, a FAD outbreak would likely have devastating impacts on recreationally important species such as white-tailed deer and other free-ranging big game animals.

7.3.6 Other Farm-Related Damage

Other farm-related damage caused by wild pigs includes damage to machinery when repairing areas of rooting, damage to fencing, and other farm structures such as livestock feeders, watering units, streams and ponds, and damage to yards or gardens. Such irritants may stymy productivity and complicate day-to-day operations on both large- and small-scale farms. Poudyal et al. (2017) reported approximately US$750,000/year in farm loss due to wild pigs for damage to stock ponds and tanks, farm equipment, fences, and stored commodities in Tennessee. Likewise, Tanger et al. (2015) reported approximately US$1.5M in these same categories in Louisiana in 2013.

7.4 IMPACTS TO URBAN AND RURAL ENVIRONMENTS

Although wild pigs typically cause damage in rural and remote landscapes, a variety of other impacts also occur in anthropogenic landscapes (see Chapter 19). As wild pigs are opportunistic, damage to residential property such as yard rooting and wallowing, depredation of home-garden crops, and trampling of ornamental flowerbeds are common in neighborhoods that coexist with high wild pig densities. Wild pigs also damage irrigation systems (for access to water) and floodgates (Tisdell 1982), other miscellaneous items such as equipment, yard, and patio furniture (Mayer and Brisbin 2009), and fencing (Mapston 2004). Damage to fencing is of particular significance, as it can facilitate additional hazards, such as allowing predators or other pest species access to excluded areas, in addition to enabling livestock or confined pets to escape (Mapston 2004, Mayer and Brisbin 2009); an event that could precipitate further residential damages from animals running at-large. Another type of damage with potentially catastrophic consequences is rooting along levees that contain river floodwaters. During a record flooding event in the Mississippi Alluvial Valley in 2011, wild pig rooting on levees caused great concern among producers, residents, and engineers who feared that the rooting might compromise structural integrity of the levee and facilitate a breach (Figure 7.7).

While most wild pig impacts result directly from foraging behaviors (Mayer and Brisbin 2009), indirect impacts also arise in areas where they coexist with high densities of human populations. These include vehicle collisions, which pose a significant risk and could potentially lead to human fatalities or other physical harm. Previous estimates relating to economic impacts from wild pig–vehicle collisions have suggested a cost of >US$30M annually (Mayer and Johns 2007). However, as populations of both humans and pigs increased and expanded, revised estimates project a potential cost of >US$181M. This estimate was derived from a national estimate of wild pig populations in the United States (6.2 million animals; Mayer 2014) multiplied by annual percent population attrition from vehicle collisions (0.8%; Mayer and Johns 2007), multiplied by number of vehicle

FIGURE 7.7 Wild pig rooting compromises the structural integrity of levees that contain Mississippi River floodwaters. (Photo by P. Nimrod. With permission.)

collisions/animal involved (i.e., 10% of collisions involve 2 or more wild pigs) as 0.84 (Mayer and Johns 2007). Lastly, that value is multiplied by the mean damage cost-estimate/collision with deer or other wild-pig-sized mammals in the United States of US$4,341 (State Farm Insurance 2018).

Attacks on humans and domestic pets, however infrequent, also occur due to encroaching populations of wild pigs into more urban areas. While the majority of these incidents are typically a result of provoking animals by threatening, wounding, or cornering (Barrett 1971, Rappaport 1984, Westinghouse Savannah River Company 1994), cases of unprovoked attacks are documented (Bowie 2004). Such attacks can result in serious injury to both humans and pets through bite and puncture wounds, as well as lacerations, which may lead to infection and require medical attention (Barss and Ennis 1988, Gubler 1992, Hatake et al. 1995). Although rare, human fatalities occur as a direct result of wild pig attacks (Bryan 1937, Hatake et al. 1995). Ultimately, as wild pig populations continue to increase in size and range, and in some areas become more habituated to coexisting in both urban and rural environments, probability of such events are likely to increase.

7.5 CONCLUSIONS

As the human population in North America continues to grow and expand, conflicts between wild pigs and people will correspondingly increase. Agricultural producers (whether row crop, livestock, or forestry) are tasked with generating greater yields on less area. As a result, crop depredation and infrastructure damage will be even more impactful, further necessitating need for advancements in research related to reducing crop damage. Along with this need to quantify and lessen impacts on agricultural resources, research emphasis needs to be placed on relating wild pig occurrence to the impacts they have on the environment and native flora and fauna. More long-term studies are needed that measure environmental impacts to forests, soils, and aquatic resources, including biotic communities within these systems. Improved sensors and computing power should enable researchers to use UAS (Samiappan et al. 2018) and satellite technology to better quantify and monitor damage. These data could further (and more accurately) assist researchers and biologists with prescribing management responses to mitigate and alleviate impacts caused by wild pigs. Additionally, collective efforts of research, management, and outreach that more accurately quantify impacts to wildlife and economic damage will be imperative to more clearly highlight the extent and scope of damage caused by wild pigs. These results must then be effectively communicated to the public and state/federal officials so that legislative action becomes not only an economic priority, but an ecological and social priority as well (Smith 2018; see Chapters 10, 11).

ACKNOWLEDGMENTS

We thank the Mississippi State University Extension Service, Mississippi State University Department of Wildlife, Fisheries, and Aquaculture, Alabama Cooperative Extension System, and School of Forestry and Wildlife Sciences at Auburn University for providing the authors time and resources to research and develop this chapter. We thank editors and reviewers J.J. Mayer, G.J. Roloff, K.C. VerCauteren, and M.P. Glow for constructive comments.

REFERENCES

Ackerman, B. B., M. E. Harmon, and F. J. Singer. 1978. Part II. Seasonal food habits of European wild boar – 1977. Pages 94–137 *in* F. J. Singer, editor. *Studies of European Wild Boar in the Great Smoky Mountains National Park: 1st Annual Report; A Report for the Superintendent*. Uplands Field Research Laboratory, Great Smoky Mountains National Park, Gatlinburg, TN.

Alonso, E., and A. Recarte. 2008. Pigs in New York City; a study on 19th century urban "sanitation." Instituto Franklin-UAH, Madrid, Spain. <https://www.institutofranklin.net/sites/default/files/fckeditor/CS%20Pigs%20in%20New%20York.pdf>. Accessed 5 May 2018.

Anderson, A., C. Slootmaker, E. Harper, J. Holderieath, and S. A. Shwiff. 2016. Economic estimates of feral swine damage and control in 11 US states. *Crop Protection* 89:89–94.

Arrington, D. A., L. A. Toth, and J. W. Koebel. 1999. Effects of rooting by feral hogs *Sus scrofa* L. on the structure of a flood plain vegetation assemblage. *Wetlands* 19:535–544.

Baber, D. W., and B. E. Coblentz. 1986. Density, home range, habitat use, and reproduction in feral pigs on Santa Catalina Island. *Journal of Mammalogy* 67:512–525.

Baber, D. W., and B. E. Coblentz. 1987. Diet, nutrition and conception in feral pigs on Santa Catalina Island. *Journal of Wildlife Management* 51:306–317.

Ballari, S. A., and M. N. Barrios-Garcia. 2014. A review of wild boar *Sus scrofa* diet and factors affecting food selection in native and introduced ranges. *Mammal Review* 44:124–134.

Bankovich, B., E. Boughton, R. Boughton, M. L. Avery, and S. M. Wisely. 2016. Plant community shifts caused by feral swine rooting devalue Florida rangeland. *Agriculture, Ecosystems & Environment* 220:45–54.

Baron, J. S. 1979. Vegetation damage by feral hogs on Horn Island, Gulf Islands National Seashore, Mississippi. Thesis, University of Wisconsin, Madison, WI.

Barrett, R. H. 1971. Ecology of the feral hog in Tehama County, California. Dissertation, University of California, Berkeley, CA.

Barrett, R. H. 1978. The feral hog on the Dye Creek Ranch, California. *Hilgardia* 46:283–355.

Barrett, R. H., and G. H. Birmingham. 1994. Wild pigs. Pages D65–D70 *in* S. Hygnstrom, R. Timm, and G. Larsen, editors. *Prevention and Control of Wildlife Damage*. Cooperative Extension Service, University of Nebraska, Lincoln, NE.

Barrios-Garcia, M. N., and S. A. Ballari. 2012. Impact of wild boar (*Sus scrofa*) in its introduced and native range: A review. *Biological Invasions* 14:2283–2300.

Barrios-Garcia, M. N., and D. Simberloff. 2013. Linking the pattern to the mechanism: How an introduced mammal facilitates plant invasions. *Austral Ecology* 38:884–890.

Barss, P., and S. Ennis. 1988. Injuries caused by pigs in Papua New Guinea. *Medical Journal of Australia* 149:649–656.

Beach, R. 1993. Depredation problems involving feral hogs. Pages 67–75 *in* C. W. Hanselka and J. F. Cadenhead, editors. *Feral Swine: A Compendium for Resource Managers*. Texas Agricultural Extension Service, Kerrville, TX.

Bevins, S. N., K. Pedersen, M. W. Lutman, T. Gidlewski, and T. J. Deliberto. 2014. Consequences associated with the recent range expansion of nonnative feral swine. *BioScience* 64:291–299.

Bleier, N., R. Lehoczki, D. Ujvary, L. Szemethy, and S. Csanyi. 2012. Relationships between wild ungulates density and crop damage in Hungary. *Acta Theriologica* 57:351–359.

Bobek, B., J. Furtek, J. Bobek, D. Merta, and M. Wojciuch-Ploskonka. 2017. Spatio-temporal characteristics of crop damage caused by wild boar in north-eastern Poland. *Crop Protection* 93:106–112.

Bodenchuk, M. J. 2008. Feral hog management: Tying performance measures to resources protected. Pages 1–4 *in* S. M. Vantassel, editor. *2008 National Conference on Feral Hogs*. University of Nebraska, Lincoln, NE.

Boughton, E. H., and R. K. Boughton. 2014. Modification by an invasive ecosystem engineer shifts a wet prairie to a monotypic stand. *Biological Invasions* 16:2105–2114.

Bowie, C. 2004. Dog saves Edgefield man from wild hog attack. *The Edgefield Advertiser*, August 18:1.

Bratton, S. P. 1974a. The effect of the European wild boar (*Sus scrofa*) on the high-elevation vernal flora in Great Smoky Mountains National Park. *Bulletin of the Torrey Botanical Club* 101:198–206.

Bratton, S. P. 1974b. An integrated ecological approach to the management of European wild boar (*Sus scrofa*) in GRSM. NPS-SER Management Report No. 3. Uplands Field Research Laboratory, Great Smoky Mountains National Park, Gatlinburg, TN.

Bratton, S. P. 1975. The effect of the European wild boar, *Sus scrofa*, on gray beech forest in the Great Smoky Mountains. *Ecology* 56:1356–1366.

Bratton, S. P., M. E. Harmon, and P. S. White. 1982. Patterns of European wild boar rooting in the western Great Smoky Mountains. *Castanea* 47:230–242.

Brunet, J., P. O. Hedwall, E. Holmstrom, and E. Wahlgren. 2016. Disturbance of the herbaceous layer after invasion of a eutrophic temperate forest by wild boar. *Nordic Journal of Botany* 34:120–128.

Bryan, L. W. 1937. Wild pigs in Hawaii. *Paradise Pacific* 49:31–32.

Campbell, T. A., and D. B. Long. 2009. Feral swine damage and damage management in forested ecosystems. *Forest Ecology and Management* 257:2319–2326.

Carpio, A. J., J. Castro-López, J. Guerrero-Casado, L. Ruiz-Aizpurua, J. Vicente, and F. S. Tortosa. 2014. Effect of wild ungulate density on invertebrates in a Mediterranean ecosystem. *Animal Biodiversity and Conservation* 37:115–125.

Choquenot, D., B. Lukins, and G. Curran. 1997. Assessing lamb predation by feral pigs in Australia's semi-arid rangelands. *Journal of Applied Ecology* 34:1445–1454.

Conley, R. H., V. G. Henry, and G. H. Matschke. 1972. *Final Report for the European Hog Research Project W-34*. Tennessee Game and Fish Commission, Nashville, TN.

Connell, J. H. 1978. Diversity in tropical rain forests and coral reefs. *Science* 199:1302–1310.

Conover, D. O., and M. R. Conover. 1987. Wildlife management in colonial Connecticut and New Haven during their first century: 1636–1736. *Transactions of the Northeast Section of the Wildlife Society* 44:1–7.

Conover, M. R. 2007. America's first feral hog war. *Human-Wildlife Conflicts* 1:129–131.

Cooper, S. M., H. M. Scott, G. R. de la Garza, A. L. Deck, and J. C. Cathey. 2010. Distribution and interspecies contact of feral swine and cattle on rangeland in south Texas: Implications for disease transmission. *Journal of Wildlife Diseases* 46:152–164.

Corn, J. L., D. E. Stallknecht, N. M. Mechlin, M. P. Luttrell, and J. R. Fischer. 2004. Persistence of psuedorabies virus in feral swine populations. *Journal of Wildlife Diseases* 40:307–310.

Crooks, J. A. 2002. Characterizing ecosystem-level consequences of biological invasions: The role of ecosystem engineers. *Oikos* 97:153–166.

Cushman, J. H., T. A. Tierney, and J. M. Hinds. 2004. Variable effects of feral pig disturbances on native and exotic plants in a California grassland. *Ecological Applications* 14:1746–1756.

Cutini, A., F. Chianucci, R. Chirichella, E. Donaggio, L. Mattioli, and M. Apollonio. 2013. Mast seeding in deciduous forests of northern Apennines (Italy) and its influence on wild boar population dynamics. *Annals of Forest Science* 70:493–502.

Davidson, W. R., editor. 2006. Wild swine. Pages 105–134 *in Field Manual of Wildlife Diseases in the Southeastern United States*. Third Edition. Southeastern Cooperative Wildlife Disease Study, Athens, GA.

Debussche, M., and P. Isenmann. 1989. Fleshy fruit characters and the choices of bird and mammal seed dispersers in a Mediterranean region. *Oikos* 56:327–338.

de Nevers, G. 1993. What is feral hog damage? Pages 9–10 *in* W. Tietje and R. Barrett, editors. *The Wild Pig in California Oak Woodland: Ecology and Economics*. University of California, Berkeley, CA.

DeVault, T. L., J. C. Beasley, L. A. Humberg, B. J. MacGowan, M. I. Retamosa, and O. E. Rhodes, Jr. 2007. Intrafield patterns of wildlife damage to corn and soybeans in northern Indiana. *Human-Wildlife Conflicts* 1:205–2013.

DeVault, T. L., and O. E. Rhodes, Jr. 2002. Identification of vertebrate scavengers of small mammal carcasses in a forested landscape. *Acta Theriologica* 47:185–192.

Diaz, A. 2008. *Managing the Feral Hog Menace on Matagorda Island National Wildlife Refuge. Fish & Wildlife Journal*. US Fish and Wildlife Service, Department of the Interior, Washington, DC.

Ditchkoff, S. S., and J. J. Mayer. 2009. Wild pig food habits. Pages 99–138 *in* J. J. Mayer and I. L. Brisbin, Jr., editors. *Wild Pigs: Biology, Damage, Control Techniques, and Management*. SRNL-RP-2009-00869, Savannah River National Laboratory, Aiken, SC.

Ditchkoff, S. S., and B. C. West. 2007. Ecology and management of feral hogs. *Human-Wildlife Conflicts* 1:149–151.

Dovrat, G., A. Perevolotsky, and G. Ne'eman. 2012. Wild boar as seed dispersal agents of exotic plants from agricultural lands to conservation areas. *Journal of Arid Environments* 78:49–54.

Elton, C. S. 2000. *The Ecology of Invasions by Animals and Plants*. University of Chicago Press, Chicago, IL.

Engeman, R. M., H. T. Smith, R. Severson, M. A. Severson, S. A. Shwiff, B. Constantin, and D. Griffin. 2004. The amount and economic cost of feral swine damage to the last remnant of a basin marsh system in Florida. *Journal for Nature Conservation* 12:143–147.

Engeman, R. M., H. T. Smith, S. A. Shwiff, B. Constantin, J. Woolard, M. Nelson, and D. Griffin. 2003. Prevalence and economic value of feral swine damage to native habitat in three Florida state parks. *Environmental Conservation* 30:319–324.

Engeman, R. M., J. Terry, L. R. Stephens, and K. S. Gruver. 2018. Prevalence and amount of feral swine damage to three crops at planting. *Crop Protection* 112:252–256.

Engeman, R. M., J. Wolard, H. T. Smith, J. Bourassa, B. U. Constantin, and D. Griffin. 2007. An extraordinary patch of feral hog damage in Florida before and after initiating hog removal. *Human-Wildlife Conflicts* 1:271–275.

Everitt, J. H., and M. A. Alaniz. 1980. Fall and winter diets of feral pigs in south Texas. *Journal of Range Management* 33:126–129.

Focardi, S., D. Capizzi, and D. Monetti. 2000. Competition for acorns among wild boar (*Sus scrofa*) and small mammals in a Mediterranean woodland. *Journal of Zoology* 250:329–334.

Frackowiak, W., S. Gorczyca, D. Merta, and M. Wojciuch-Ploskonka. 2012. Factors affecting the level of damage by wild boar in farmland in north-eastern Poland. *Pest Management Science* 69:362–366.
Genov, P. V., S. Focardi, F. Morimando, L. Scillitani, and A. Ahmed. 2018. Ecological impact of wild boar in natural ecosystems. Pages 404–419 *in* M. Melletti and R. Meijaard, editors. *Ecology, Conservation and Management of Wild Pigs and Peccaries*. Cambridge University Press, New York.
Gómez, J. M., D. Garcia, and R. Zamora. 2003. Impact of vertebrate acorn- and seedling-predators on a Mediterranean *Quercus pyrenaica* forest. *Forest Ecology and Management* 180:125–134.
Gómez, J. M., and J. A. Hódar. 2008. Wild boars (*Sus scrofa*) affect the recruitment rate and spatial distribution of holm oak (*Quercus ilex*). *Forest Ecology and Management* 256:1384–1389.
Gray, S. M., G. J. Roloff, D. B. Kramer, D. R. Etter, K. C. VerCauteren, and R. A. Montgomery. In Review. Low-density feral swine population impacts in vegetation and forest soils in northern Michigan, USA. *Journal of Wildlife Management*.
Grice, A. C. 1996. Seed production, dispersal and germination in *Cryptostegia grandiflora* and *Ziziphus mauritiana*, two invasive shrubs in tropical woodlands of northern Australia. *Australian Journal of Ecology* 21:324–331.
Grover, A. M. 1983. The home range, habitat utilization, group behavior and food habits of the feral hog (*Sus scrofa*) in northern California. Thesis, California State University, Sacramento, CA.
Gubler, J. G. H. 1992. Septic arthritis of the knee induced by *Pasteurella multocida* and *Bacteroides fragilis* following an attack by a wild boar. *Journal of Wilderness Medicine* 3:288–291.
Hanson, R. P., and L. Karstad. 1959. Feral swine in the southeastern United States. *Journal of Wildlife Management* 23:64–74.
Hartin, R. E., M. R. Ryan, and T. A. Campbell. 2007. Distribution and disease prevalence of feral hogs in Missouri. *Human-Wildlife Conflicts* 1:186–191.
Hatake, K., Taniguchi, T., Negoro, M. Ouchi, H., Minami, T., and S. Hishida. 1995. A case of death of a woman attacked by a wild boar. *Research and Practice in Forensic Medicine* 38:275–277.
Hayes, R. B., N. B. Marsh, and G. A. Bishop. 1996. Sea turtle nest depredation by a feral hog: A learned behavior. Pages 129–134 *in* J. A. Keinath, D. E. Barnard, J. A. Musick, and B. A. Bell, editors. *Proceedings of the Fifteenth Annual Symposium on Sea Turtle Biology and Conservation*. NOAA Technical Memorandum MNFS-SEFSC 387, Hilton Head Island, SC.
Held, S., M. Mendl, C. Devereux, and R. W. Byrne. 2000. Social tactics of pigs in a competitive foraging task: The "informed forager" paradigm. *Animal Behaviour* 59:569–576.
Hellgren, E. C., and A. M. Holzem. 1992. *Feral Hog Food Habits in the Western Rio Grande Plains*. Texas Parks and Wildlife Department, Austin, TX.
Henry, V. G. 1969. Predation on dummy nests of ground-nesting birds in the southern Appalachians. *Journal of Wildlife Management* 33:169–172.
Henry, V. G., and R. H. Conley. 1972. Fall foods of European wild hogs in the southern Appalachians. *Journal of Wildlife Management* 36:854–860.
Herrero, J., A. Garcia-Serrano, S. Couto, V. M. Ortuno, and R. Garcia-Gonzalez. 2006. Diet of wild boar *Sus scrofa* L. and crop damage in an intensive agroecosystem. *European Journal of Wildlife Research* 52:245–250.
Holt, R. D. 1977. Predation, apparent competition, and the structure of prey communities. *Theoretical Population Biology* 12:197–229.
Hone, J. 2002. Feral pigs in Namadgi National Park, Australia: Dynamics, impacts and management. *Biological Conservation* 105:231–242.
Hopkins, W. 1948. Hogs or logs – longleaf pine seedlings and range hogs won't grow together. *Naval Stores Review* 57(43):12–13.
Howe, T. D., F. J. Singer, and B. B. Ackerman. 1981. Forage relationships of European wild boar invading northern hardwood forest. *Journal of Wildlife Management* 45:748–754.
Ickes, K., S. J. Dewalt, and S. C. Thomas. 2003. Resprouting of woody saplings following stem snap by wild pigs in a Malaysian rain forest. *Journal of Ecology* 91:222–233.
Ilse, L. M., and E. C. Hellgren. 1995. Resource partitioning by sympatric populations of collared peccaries and feral hogs in southern Texas. *Journal of Mammalogy* 76:784–799.
Ishii, Y., and M. Shimada. 2010. The effect of learning and search images on predator-prey interactions. *Population Ecology* 52:27–35.
Ivey, M., M. Colvin, B. K. Strickland, and M. Lashley. 2019. Reduced vertebrate diversity independent of spatial scale following feral swine invasions. *Ecology and Evolution*. 9:7761–7767.

Johnson, K. G., R. W. Duncan, and M. R. Pelton. 1982. Reproductive biology of European wild hogs in the Great Smoky Mountains National Park. *Proceedings of the Annual Conference of the Southeastern Fish and Wildlife Agencies* 36:552–564.

Jolley, D. B. 2007. Reproduction and herpetofauna depredation of feral pigs at Fort Benning, Georgia. Thesis, Auburn University, Auburn, AL.

Jolley, D. B., S. S. Ditchkoff, B. D. Sparklin, L. B. Hanson, M. S. Mitchell, and J. B. Grand. 2010. Estimate of herpetofauna depredation by a population of wild pigs. *Journal of Mammalogy* 91:519–524.

Kaller, M. D., and W. E. Kelso. 2003. Effects of feral swine on water quality in a coastal bottomland stream. *Proceedings of the Annual Conference of the Southeastern Association of Fish and Wildlife Agencies* 57:291–298.

Kaller, M. D., and W. E. Kelso. 2006. Swine activity alters invertebrate and microbial communities in a Coastal Plain watershed. *American Midland Naturalist* 156:163–178.

Keever, A. C. 2014. Use of N-mixture models for estimating white-tailed deer populations and impacts of predator removal and interspecific competition. Thesis, Auburn University, Auburn, AL.

Keuling, O., N. Stier, and M. Roth. 2008. Annual and seasonal space use of different age classes of female wild boar *Sus scrofa* L. *European Journal of Wildlife Research* 54:403–412.

Keuling, O., N. Stier, and M. Roth. 2009. Commuting, shifting or remaining? Different spatial utilization patterns of wild boar *Sus scrofa* L. in forest and field crops during summer. *Mammalian Biology* 74:145–152.

Kotanen, P. M. 1995. Responses of vegetation to a changing regime of disturbance: Effects of feral pigs in a Californian coastal prairie. *Ecography* 18:190–199.

Kroll, J. C. 1986. Interspecific competition between feral hogs and white-tailed deer in the post oak savannah region of Texas. Federal Aid Project W-109-R-8 Job. No. 44, Final Performance Report. Texas Parks and Wildlife Department, Austin, TX.

Lipscomb, D. J. 1989. Impacts of feral hogs on longleaf pine regeneration. *Southern Journal of Applied Forestry* 13:177–181.

Littauer, G. A. 1993. Control techniques for feral hogs. Pages 139–148 *in* C. W. Hanselka and J. F. Cadenhead, editors. *Feral Swine: A Compendium for Resource Managers*. Texas Agricultural Extension Service, Kerrville, TX.

Loggins, R. E., J. T. Wilcox, D. H. V. Vuren, and R. A. Sweitzer. 2002. Seasonal diet of wild pigs in oak woodlands of the central coast region of California. *California Fish and Game* 88:28–34.

Lombardini, M., A. Meriggi, and A. Fozzi. 2017. Factors influencing wild boar damage to agricultural crops in Sardinia (Italy). *Current Zooloogy* 63:507–514.

Lott, R. H., G. N. Harrington, A. K. Irvine, and S. McIntyre. 1995. Density-dependent seed predation and plant dispersion of the tropical palm *Normanbya normanbyi*. *Biotropica* 27:87–95.

Lynes, B., and S. D. Campbell. 2000. Germination and viability of mesquite (*Prosopis pallida*) seed following ingestion and excretion by feral pigs (*Sus scrofa*). *Tropical Grasslands* 34:125–128.

Mapston, M. E. 2004. Feral hogs in Texas. Document No. B-6149 5–04. Wildlife Services, Texas Cooperative Extension, Texas A&M University, College Station, TX.

Massei, G., and P. V. Genov. 2004. The environmental impact of wild boar. *Galemys* 16:135–145.

Mayer, J. J. 2014. Estimation of the number of wild pigs found in the United States. Savannah River National Laboratory, SNRL-STI-2014-00292. Aiken, SC.

Mayer, J. J., and J. C. Beasley. 2018. Wild pigs. Pages 219–248 *in* W. C. Pitt, J. C. Beasley, and G. W. Witmer, editors. *Ecology and Management of Terrestrial Vertebrate Invasive Species in the United States*. CRC Press, Boca Raton, FL.

Mayer, J. J., and I. L. Brisbin, Jr. 1995. Feral swine and their role in the conservation of global livestock genetic diversity. Pages 175–179 *in* R. D. Crawford, E. E. Lister, and J. T. Buckley, editors. *Proceedings of the Third Global Conference on Conservation of Domestic Animal Genetic Resources*. Rare Breeds International, Warwickshire, England.

Mayer, J. J., and I. L. Brisbin, Jr. 2008. *Wild Pigs in the United States: Their History, Comparative Morphology, and Current Status*. The University of Georgia Press, Athens, GA.

Mayer, J. J., and I. L. Brisbin, Jr., editors. 2009. *Wild Pigs: Biology, Damage, Control Techniques and Management*. Savannah River National Laboratory, SRNL-RP-2009-00869, Aiken, SC.

Mayer, J. J., and P. E. Johns. 2007. Characterization of wild pig-vehicle collisions. *Proceedings of the Wildlife Damage Management Conference* 12:175–187.

Mayer, J. J., E. A. Nelson, and L. D. Wike. 2000. Selective depredation of planted hardwood seedlings by wild pigs in a wetland restoration area. *Ecological Engineering* 15:S79–S85.

Mazzotti, F. J., and L. A. Brandt. 1994. Ecology of the American alligator in a seasonally fluctuating environment. Pages 485–506 *in* S. M. Davis and J. C. Ogden, editors. *Everglades: The Ecosystem and Its Restoration*. CRC Press, Boca Raton, FL.

McClure, M. L., C. L. Burdett, M. L. Farnsworth, S. J. Sweeney, and R. S. Miller. 2018. A globally-distributed alien invasive species poses risk to United States imperiled species. *Scientific Reports* 8:5331.

McIlhenny, E. A. 1976. *The Alligator's Life History*. Society for the Study of Amphibians and Reptiles. Lawrence, KS.

Melletti, M., and E. Meijaard, editors. 2018. *Ecology, Conservation and Management of Wild Pigs and Peccaries*. Cambridge University Press, New York.

Mellish, J. M., A. Sumrall, T. A. Campbell, B. A. Collier, W. H. Neill, B. Higginbotham, and R. R. Lopez. 2014. Simulating potential population growth of wild pig, *Sus scrofa*, in Texas. *Southeastern Naturalist* 13:367–376.

Meng, X. J., D. S. Lindsay, and N. Sriranganathan. 2009. Wild boars as sources of infectious diseases in livestock and humans. *Philosophical Transactions of the Royal Society B* 364:2697–2707.

Mengak, M. T. 2012. Georgia Wild Pig Survey Final Report. University of Georgia Cooperative Extension Wildlife Management Series Publication WMS-12-12, Athens, GA.

Mengak, M. T. 2015. Georgia Wild Pig Survey Final Report. Warnell School of Forestry and Natural Resources Outreach Publication No.16-23, Athens, GA.

Morelle, K., and P. Lejeune. 2015. Seasonal variation of wild boar *Sus scrofa* distribution in agricultural landscapes: A species distribution modelling approach. *European Journal of Wildlife Research* 61:45–56.

Muller, C. B., and H. C. J. Godfray. 1997. Apparent competition between two aphid species. *Journal of Animal Ecology* 66:57–64.

Mungall, E. C. 2001. Exotics. Pages 736–764 *in* S. Demarais and P. R. Krausman, editors. *Ecology and Management of Large Mammals in North America*. Prentice Hall, Upper Saddle River, NJ.

National Fish and Wildlife Laboratory. 1980. *Selected Vertebrate Endangered Species of the Seacoast of the United States*. FWS/OBS-80/01. Biological Services Program, US Fish and Wildlife Service, Department of the Interior, Washington, DC.

Natusch, D. J. D., M. Mayer, J. A. Lyons, and R. Shrine. 2017. Interspecific interactions between feral pigs and native birds reveal both positive and negative effects. *Austral Ecology* 42:479–485.

Neill, W. T. 1971. *The Last of the Ruling Reptiles: Alligators, Crocodiles, and Their Kin*. Columbia University Press, New York.

NMFS and USFWS (National Marine Fisheries Service and US Fish and Wildlife Service). 1991. *Recovery Plan for US Population of Atlantic Green Turtle*. National Marine Fisheries Service, Washington, DC.

Paarlberg, P. L., A. Hillberg Seitzinger, J. G. Lee, and K. H. Mathews, Jr. 2009. Supply reductions, export restrictions, and expectations for hog returns in a potential classical swine fever outbreak in the United States. *Journal of Swine Health and Production* 17:155–163.

Paarlberg, P. L., J. G. Lee, and A. H. Seitzinger. 2002. Potential revenue impact of an outbreak of foot and mouth disease in the United States. *Journal of the American Veterinary Medical Association* 220:988–992.

Paolini, K. E., B. K. Strickland, J. L. Tegt, K. C. VerCauteren, and G. M. Street. 2018. Seasonal variation in preference dictates space use in an invasive generalist. *PLoS ONE* 13: e0199078.

Persson, L. 1985. Asymmetrical competition: Are larger animals more competitively superior? *The American Naturalist* 126:261–266.

Pianka, E. R. 1983. *Evolutionary Ecology*. Third Edition. Harper & Row, Inc., New York.

Pimental, D. 2007. Environmental and economic costs of vertebrate species invasions into the United States. Pages 2–8 *in* G. W. Witmer, W. C. Pitt, and K. A. Fagerstone, editors. *Managing Vertebrate Invasive Species: Proceedings of an International Symposium*. US Department of Agriculture/Animal and Plant Health Inspection Service/Wildlife Services/National Wildlife Research Center, Ft. Collins, CO.

Pine, D. S., and G. L. Gerdes. 1973. Wild pigs in Monterey County, California. *California Fish and Game* 59:126–137.

Polis, G. A., C. A. Myers, and R. D. Holt. 1989. The ecology and evolution of intraguild predation: Potential competitors that eat each other. *Annual Review of Ecology and Systematics* 20:297–330.

Poudyal, N. C., C. Caplenor, O. Joshi, C. Maldonado, L. I. Muller, and C. Yoest. 2017. Characterizing the economic value and impacts of wild pig damage on a rural economy. *Human Dimensions of Wildlife* 22:538–549.

Ralph, C. J., and B. D. Maxwell. 1984. Relative effects of human and feral hog disturbance on a wet forest in Hawaii. *Biological Conservation* 30:291–303.

Rappaport, R. A. 1984. *Pigs for Ancestors: Ritual in the Ecology of a New Guinea People*. Second Edition. Waveland Press, Long Grove, IL.

Roemer, G. W., C. J. Donlan, and F. Courchamp. 2002. Golden eagles, feral pigs, and insular carnivores: How exotic species turn native predators into prey. *Proceedings of the National Academy of Sciences* 99:791–796.

Rollins, D. 1993. Statewide attitude survey on feral hogs in Texas. Pages 1–8 *in* C. W. Hanselka and J. F. Cadenhead, editors. *Feral Swine: A Compendium for Resource Managers*. Texas Agricultural Extension Service, Kerrville, TX.

Rollins, D., and J. P. Carroll. 2001. Impacts of predation on northern bobwhite and scaled quail. *Wildlife Society Bulletin* 29:39–51.

Rudge, M. 1976. A note on the food of feral pigs (*Sus scrofa*) of Auckland Island. *Proceedings of the New Zealand Ecological Society* 23:83–84.

Rutledge, A. 1970. *The Woods and Wild Things I Remember*. R. L. Bryan Company, Columbia, SC.

Samiappan, S., J. M. Prince Czarnecki, H. Foster, B. K. Strickland, J. L. Tegt, and R. J. Moorhead. 2018. Quantifying damage from wild pigs with small unmanned aerial systems. *Wildlife Society Bulletin* 42:304–309.

Sanders, H. N. 2017. Impacts of invasive wild pigs on wild turkey reproductive success. Thesis, Texas A&M University, Kingsville, TX.

Schaefer, J. 2004. Video monitoring of shrub-nests reveals nest predators. *Bird Study* 51:170–177.

Schley, L., M. Dufrene, A. Krier, and A. C. Frantz. 2008. Patterns of crop damage by wild boar (*Sus scrofa*) in Luxembourg over a 10–year period. *European Journal of Wildlife Research* 54:589–599.

Schley, L., and T. J. Roper. 2003. Diet of wild boar *Sus scrofa* in western Europe, with particular reference to the consumption of agricultural crops. *Mammal Review* 33:43–56.

Scott, C. D., and M. R. Pelton. 1975. Seasonal food habits of the European wild hog in the Great Smoky Mountains National Park. *Proceedings of the Annual Conference of the Southeastern Association of Fish & Wildlife Agencies* 29:585–593.

Seward, N. W., K. C. VerCauteren, G. W. Witmer, and R. M. Engeman. 2004. Feral swine impacts on agriculture and the environment. *Sheep and Goat Research Journal* 19:34–40.

Siemann, E., J. A., Carrillo, C. A. Gabler, R. Zipp, W. E. Rogers. 2009. Experimental test of the impacts of feral hogs on forest dynamics and processes in the southeastern US. *Forest Ecology and Management* 258:546–553.

Sierra, C. 2001. The feral pig (*Sus scrofa*, Suidae) in Cocos Island, Costa Rica: Rooting, soil alterations and erosion. *Revista de Biologia Tropical* 49:1159–1170.

Singer, F. J., W. T. Swank, and E. E. C. Clebsch. 1984. The effects of wild pig rooting in a deciduous forest. *Journal of Wildlife Management* 48:464–473.

Smith, A. L. 2018. Investigating the effectiveness of wild pig policy and legislation in the US. Thesis. Mississippi State University, Mississippi State, MS.

Spatz, G., and D. Mueller-Dombois. 1972. *Succession Patterns After Pig Digging in Grassland Communities of Mauno Loa, Hawaii*. International Biological Program Report 15. Island Ecosystems IRP.

Springer, M. D. 1975. Food habits of wild hogs on the Texas Gulf Coast. Thesis, Texas A&M University, College Station, TX.

State Farm Insurance. 2018. *How Likely Are You to Have a Deer Collision?* State Farm Insurance Company, Bloomington, IL. <https://www.statefarm.com/simple-insights/auto-and-vehicles/how-likely-are-you-to-have-a-deer-collision>. Accessed 6 Nov 2018.

Steele, M. A., L. A. Wauters, and K. W. Larsen. 2005. Selection, predation and dispersal of seeds by tree squirrels in temperate and boreal forests: Are tree squirrels keystone granivores. Pages 205–220 *in* P. M. Forget, J. E. Lambert, P. E. Hulme, and S. B. Vander Wall, editors. *Seed Fate: Predation. Dispersal and Seedling Establishment*. CABI Publishing, Cambridge, MA.

Stegeman, L. C. 1938. The European wild boar in the Cherokee National Forest, Tennessee. *Journal of Mammalogy* 19:279–290.

Stone, C. P., and J. O. Keith. 1987. Control of feral ungulates and small mammals in Hawaii's National Parks: Research and management strategies. Pages 277–287 *in* C. G. G. Richards and T. Y. Ku, editors. *Control of Mammal Pests*. Taylor and Francis, London, UK.

Sweeney, J. R., J. M. Sweeney, and S. W. Sweeney. 2003. Feral hog, *Sus scrofa*. Pages 1164–1179 *in* G. A. Feldhammer, B. C. Thompson, and J. A. Chapman, editors. *Wild Mammals of North America: Biology, Management, and Conservation*. The Johns Hopkins University Press, Baltimore, MD.

Sweitzer, R. A., and D. H. Van Vuren. 2002. Rooting and foraging effects of wild pigs on tree regeneration and acorn survival in California's oak woodland ecosystems. USDA Forest Service General Technical Report PSW-GTR-184.

Tanger, S. M., K. M. Guidry, and H. Nui. 2015. Monetary estimates of wild hog damage to agricultural producers in Louisiana. *Journal of the National Association of County Agricultural Agents* 8(2) December 2015. <https://www.nacaa.com/journal/index.php?jid=553>. Accessed 6 Aug 2018.

Taylor, R. B., and E. C. Hellgren. 1997. Diet of feral hogs in western South Texas Plains. *The Southwest Naturalist* 43:33–39.

Tegt, J. L., and B. K. Strickland. 2018. A comprehensive assessment of wild hog damage to Mississippi agriculture. Final Report, Land, Water, and Timber Resources Board. Jackson, MS.

Thompson, R. L. 1977. Feral hogs on national wildlife refuges. Pages 11–15 *in* G. W. Wood, editor. *Research and Management of Wild Hog Populations*. Belle Baruch Forest Science Institute of Clemson University, Georgetown, SC.

Thurfjell, H., J. P. Ball, P. Ahlen, P. Kornacher, H. Dettki, and K. Sjoberg. 2009. Habitat use and spatial patterns of wild boar *Sus scrofa* (L.): Agricultural fields and edges. *European Journal of Wildlife Research* 55:517–523.

Timmons, J. B, B. Higginbotham, R. Lopez, J. C. Cathey, J. Mellish, J. Griffin, A. Sumrall, and K. Skow. 2012. *Feral Hog Population Growth, Density, and Harvest in Texas*. Texas Agrilife Extension Service Publication SP-472. Texas A&M University, College Station, TX.

Tisdell, C. A. 1982. *Wild Pigs: Environmental Pest or Economic Resource?* Pergamon Press, New York.

Tolleson, D., D. Rollins, W. Pinchak, M. Ivy, and A. Hierman. 1993. Impact of feral hogs on ground-nesting gamebirds. Pages 76–83 *in* C. W. Hanselka and J. F. Cadenhead, editors. *Feral Swine: A Compendium for Resource Managers*. Texas Agricultural Extension Service, Kerrville, TX.

Tolleson, D. R., W. E. Pinchak, D. Rollins, and L. J. Hunt. 1995. Feral hogs in the rolling plains of Texas: Perspectives, problems, and potential. *Great Plains Wildlife Damage Control Conference* 12:124–128.

Towne, C. W., and E. N. Wentworth. 1950. *Pigs from Cave to Corn Belt*. University of Oklahoma Press, Norman, OK.

Vander Wall, S. B., K. M. Kuhn, and M. J. Beck. 2005. Seed removal, seed predation, and secondary dispersal. *Ecology* 86:801–806.

Vitousek, P. M., C. M. D'Antonio, L. L. Loope, and R. Westbrooks. 1996. Biological invasions as global environmental change. *American Scientist* 84:468–478.

Wakely, P. C. 1954. Planting the Southern pine. *Forest Service Agricultural Monograph* 18:1–233.

West, B. C., A. L. Cooper, and J. B. Armstrong. 2009. Managing wild pigs: A technical guide. *Human-Wildlife Interactions Monograph* 1:1–55.

Westinghouse Savannah River Company. 1994. Savannah River Site Deer Control Activities (U) – 1993. WSRC-IM-90-51. Westinghouse Savannah River Company, Savannah River Site, Aiken, SC.

Wilcox, J. T., and D. H. Van Vuren. 2009. Wild pigs as predators in oak woodlands of California. *Journal of Mammalogy* 90:114–118.

Witmer, G. W., R. B. Sanders, A. C. Taft. 2003. Feral swine – are they a disease threat to livestock in the United States? *Proceedings of the Wildlife Damage Management Conference* 10:316–325.

Wood, G. W., and R. H. Barrett. 1979. Status of wild pigs in the United States. *Wildlife Society Bulletin* 36:237–246.

Wood, G. W., and D. N. Roark. 1980. Food habits of feral hogs in coastal South Carolina. *Journal of Wildlife Management* 44:506–511.

Wood, G. W., and R. E. Brenneman. 1977. Research and management of feral hogs on Hobcaw Barony. Pages 23–35 *in* G. W. Wood, editor. *Research and Management of Wild Hog Populations*. Belle Baruch Forest Science Institute of Clemson University, Georgetown, SC.

Wyckoff, A. C., S. E. Henke, T. A. Campbell, D. G. Hewitt, and K. C. VerCauteren. 2009. Feral swine contact with domestic swine: A serologic survey and assessment of potential for disease transmission. *Journal of Wildlife Management* 45:422–429.

Yarrow, G. K. 1987. The potential for interspecific resource competition between white-tailed deer and the feral hogs in the post oak savannah region of Texas. Dissertation, Stephen F. Austin University, Nacodoches, TX.

Yarrow, G. K., and J. C. Kroll. 1989. Coexistence of white-tailed deer and feral hogs: Management implications (abstract only). *Southeast Deer Study Group* 12:13–14.

Zengel, S. A., and W. H. Conner. 2008. Could wild pigs impact water quality and aquatic biota in floodplain wetland and stream habitats at Congaree National Park, South Carolina? *Proceedings of the 2008 South Carolina Water Resources Conference*, Charleston, SC.

APPENDIX 7.1
Listing of Published Reports of Animals Consumed by Wild Pigs in the Continental United States

Taxonomic Group or Scientific Name (Common Name)	Part(s) Consumed	State(s) Reported From	Reference(s)
Annelida			
Hirudinidae (leech)	Entire organism	TX	Springer (1975)
Oligochaeta (earthworms)	Entire organism	CA, NC, TN	Ackerman et al. (1978), de Nevers (1993)
Lumbricidae (earthworms)	Entire organism	SC	Wood and Roark (1980)
Annelida sp. (earthworm)	Entire organism	CA, NC, TN, TX	Pine and Gerdes (1973), Scott and Pelton (1975), Springer (1975), Barrett (1978)
Lumbricoides sp. (earthworms)	Entire organism	GA	Hanson and Karstad (1959)
Lumbricus sp. (earthworms)	Entire organism	CA, TN TX	Conley et al. (1972), Henry and Conley (1972), Kroll (1986), Loggins et al. (2002)
Pheretima diffringens (earthworms)	Entire organism	GA	Hanson and Karstad (1959)
Arthropoda			
Arthropoda (arthropods - general)	Entire organism	TX	Ilse and Hellgren (1995)
Arachnida			
Aranea (spider)	Entire organism	CA	Grover (1983)
Ixodoidea (tick)	Entire organism	TX	Springer (1975)
Crustacea			
Decapoda (crayfish)	Entire organism	NC, TN	Ackerman et al. (1978)
Astacidae (crayfish)	Entire organism	TN	Conley et al. (1972), Henry and Conley (1972)
Cambarus sp. (crayfish)	Entire organism	TX	Springer (1975)
Brachyura (crab)	Entire organism	MS	Baron (1979)
Callinectes sapidus (blue crab)	Entire organism	MS	Baron (1979)
Insecta			
Insecta (insects - general)	Larvae/adults/ entire organism	CA, MS, TN	Stegeman (1938), Barrett (1978), Baron (1979), Grover (1983), Baber and Coblentz (1987)
Alleculidae (comb-clawed beetles)	Larvae/entire organism	TN	Conley et al. (1972), Henry and Conley (1972)
Bibionidae (march flies)	Larvae/entire organism	NC, TN, TX	Conley et al. (1972), Henry and Conley (1972), Springer (1975), Ackerman et al. (1978)
Caelifera (grasshopper)	Entire organism	CA, TX	Springer (1975), Grover (1983)
Brachystola magna (lubber grasshopper)	Entire organism	GA	Hanson and Karstad (1959)
Calliphoridae (blow flies)	Larvae/entire organism	TN, TX	Conley et al. (1972), Henry and Conley (1972), Springer (1975)
Carabidae (ground beetles)	Entire organism	NC, TN	Ackerman et al. (1978), Howe et al. (1981)
Coleoptera (beetles - general)	Adult and larvae/ entire organism	CA, MS, NC, SC, TN	Conley et al. (1972), Henry and Conley (1972), Scott and Pelton (1975), Baron (1979), Wood and Roark (1980), Grover (1983)

(Continued)

APPENDIX 7.1 (CONTINUED)
Listing of Published Reports of Animals Consumed by Wild Pigs in the Continental United States

Taxonomic Group or Scientific Name (Common Name)	Part(s) Consumed	State(s) Reported From	Reference(s)
Corydalidae (hellgrammites)	Larvae/entire organism	NC, TN	Ackerman et al. (1978)
Curculionidae (true or snout weevils)	Larvae/entire organism	NC, TN, TX	Conley et al. (1972), Henry and Conley (1972), Springer (1975), Ackerman et al. (1978)
Diptera (flies)	Larvae/adult/ entire organism	NC, TN, TX	Conley et al. (1972), Henry and Conley (1972), Scott and Pelton (1975), Springer (1975), Howe et al. (1981)
Dolichopodidae (long-legged flies)	Entire organism	NC, TN	Ackerman et al. (1978)
Elateridae (click beetles)	Larvae/entire organism	TN	Conley et al. (1972), Henry and Conley (1972)
Empididae (dance flies)	Entire organism	NC, TN	Ackerman et al. (1978)
Formicidae (ants - general)	Entire organism	NC, TN	Ackerman et al. (1978)
Geometridae (measuring worms)	Larvae/entire organism	TN	Conley et al. (1972), Henry and Conley (1972)
Helicoverpa zea (corn earworm)	Larvae/entire organism	NC, TN	Ackerman et al. (1978)
Lepidoptera (moths and butterflies)	Larvae, pupae/ entire organism	TN, TX	Conley et al. (1972), Henry and Conley (1972), Springer (1975), Taylor and Hellgren (1997)
Meloidae (blister beetles)	Entire organism	NC, TN	Ackerman et al. (1978)
Mutillidae (mutillid wasps)	Adult/entire organism	TN	Conley et al. (1972), Henry and Conley (1972)
Noctuidae (miller moths)	Larvae/entire organism	TN	Conley et al. (1972), Henry and Conley (1972)
Odonata (dragonflies and damsel flies)	Larvae/entire organism	TN, TX	Conley et al. (1972), Henry and Conley (1972), Springer (1975)
Phasmidae (walking sticks, stick insects)	Entire organism	NC, TN	Ackerman et al. (1978)
Phengodidae (glowworms)	Larvae/entire organism	TN	Conley et al. (1972), Henry and Conley (1972)
Rhagionidae (snipe flies)	Larvae/entire organism	TN	Conley et al. (1972), Henry and Conley (1972)
Scarabacidae (lamellicron beetles)	Larvae/entire organism	SC, TN, TX	Conley et al. (1972), Henry and Conley (1972), Springer (1975), Wood and Roark (1980)
Sialidae (alderflies)	Larvae/entire organism	TN	Conley et al. (1972), Henry and Conley (1972)
Sphingidae (hawkmoths)	Larvae/entire organism	NC, TN	Ackerman et al. (1978)
Tabanidae (deer flies)	Larvae/entire organism	NC, TN	Conley et al. (1972), Henry and Conley (1972), Ackerman et al. (1978)
Tenebrionidae (darkling beetles)	Larvae/entire organism	NC, TN	Conley et al. (1972), Henry and Conley (1972), Ackerman et al. (1978)

(*Continued*)

APPENDIX 7.1 (CONTINUED)
Listing of Published Reports of Animals Consumed by Wild Pigs in the Continental United States

Taxonomic Group or Scientific Name (Common Name)	Part(s) Consumed	State(s) Reported From	Reference(s)
Tipulidae (crane flies)	Larvae/entire organism	NC, TN, TX	Conley et al. (1972), Henry and Conley (1972), Scott and Pelton (1975), Springer (1975), Ackerman et al. (1978)
Trupaneidae (fruit flies)	Larvae/entire organism	TN	Conley et al. (1972), Henry and Conley 1972
Myriapoda			
Diplopoda (millipedes)	Adult/entire organism	TN	Conley et al. (1972), Henry and Conley (1972)
Juliformia (millipede)	Entire organism	NC, TN	Ackerman et al. (1978)
Polydesmoidae (millipedes)	Entire organism	NC, TN	Ackerman et al. (1978)
Chilopoda (centipedes)	Entire organism	CA, NC, TN, TX	Conley et al. (1972), Henry and Conley (1972), Scott and Pelton (1975), Springer (1975), de Nevers (1993)
Geophilidae (centipedes)	Entire organism	SC	Wood and Roark (1980)
Scolopendromorpha (centipedes)	Entire organism	NC, TN	Ackerman et al. (1978)
Scolopendridae (centipedes)	Entire organism	SC	Wood and Roark (1980)
Gastropoda (snails)	Entire organism	NC, TN	Ackerman et al. (1978)
Polygridae (land snails)	Entire organism	NC, TN	Howe et al. (1981)
Pelecypoda (clams and mussels)	Entire organism	SC	Wood and Roark (1980)
Elliptio icterina (variable spike mussel)	Entire organism	SC	Zengel and Conner (2008)
Uniomerus carolinianus (Florida pondhorn)	Entire organism	SC	Zengel and Conner (2008)
Nematode			
Nematoda (roundworms)	Entire organism	MS, NC, TN	Ackerman et al. (1978), Baron (1979)
Vertebrates			
Fish			
Osteichthyes (fish - general)	Carrion	MS	Baron (1979)
Gambusia sp. (topminnows)	Entire organism	TX	Springer (1975)
Amphibians			
Anura (frogs - general)	Entire organism	GA, TN	Stegeman (1938), Hanson and Karstad (1959)
Hyla sp. (tree frog)	Entire organism	GA	Jolly et al. (2010)
Pseudacris regilla (Pacific treefrog)	Entire organism	CA	Wilcox and Van Vuren (2009)
Rana berlandieri (Rio Grande leopard frog)	Entire organism	TX	Springer (1975)
Rana pipiens shenocephala (southern leopard frog)	Entire organism	SC	Wood and Roark (1980)
Rana sylvatica (wood frog)	Eggs	TN	Conley et al. (1972)
Scaphiopus holbrooki (spadefoot toad)	Entire organism	GA	Jolly et al. (2010)
Caudata (salamanders - general)	Entire organism	NC, TN	Scott and Pelton (1975), Ackerman et al. (1978)
Ambystoma maculatum (spotted salamander)	Entire organism	SC	Sweeney et al. (2003)

(Continued)

APPENDIX 7.1 (CONTINUED)
Listing of Published Reports of Animals Consumed by Wild Pigs in the Continental United States

Taxonomic Group or Scientific Name (Common Name)	Part(s) Consumed	State(s) Reported From	Reference(s)
Desmognathus ochrophaeus carolinensis (Carolina mountain dusky salamander)	Entire organism	NC, TN	Ackerman et al. (1978)
Eurycea bislineata wilderae (Blue Ridge two-lined salamander)	Entire organism	NC, TN	Ackerman et al. (1978)
Leurognathus marmoratus intermedius (shovel-nosed salamander)	Entire organism	NC, TN	Ackerman et al. (1978)
Plethodontidae (lungless salamanaders)	Entire organism	NC, TN	Howe et al. (1981)
Plethodon cinereus cinereus (red-backed salamander)	Entire organism	NC, TN	Ackerman et al. (1978)
Plethodon glutinosus (slimy salamander)	Entire organism	SC	Sweeney et al. (2003)
Plethodon jordani jordani (Jordan's salamander)	Entire organism	NC, TN	Ackerman et al. (1978)
Plethodon wrighti (pygmy salamander)	Entire organism	NC, TN	Ackerman et al. (1978)
Pseudotriton ruber schencki (black-chinned red salamander)	Entire organism	NC, TN	Ackerman et al. (1978)
Reptiles			
Alligator mississippiensis (American alligator)	Eggs	FL, LA	Neill 1971, McIlhenny (1976), Mazzotti and Brandt (1994)
Testudines (turtle - general)	Eggs, hatchlings, entire organism	GA, NC, TN	Hanson and Karstad (1959), Ackerman et al. (1978)
Caretta caretta (loggerhead sea turtle)	Eggs and hatchlings	FL, GA	Thompson (1977), Mayer and Brisbin (1995), Hayes et al. (1996)
Chelonia mydas (Atlantic green turtle)	Eggs and hatchlings	FL	National Marine Fisheries Service and US Fish and Wildlife Service (1991)
Gopherus berlandieri (Texas tortoise)	Entire organism	TX	Taylor and Hellgren (1997)
Lepidochelys kempii (Kemp's Ridley sea turtle)	Eggs	TX	Diaz (2008)
Terrapene carolina carolina (eastern box turtle)	Shell fragment	NC, TN	Ackerman et al. (1978)
Lacertilia (lizard - general)	Entire organism	TN	Stegeman (1938)
Anolis carolinensis (green anole)	Entire organism	GA, SC	Wood and Roark (1980), Jolly et al. (2010)
Cnemidophorus sexlineatus (prairie racerunner lizard)	Entire organism	TX	Taylor and Hellgren (1997)
Gerrhonotus multicarinatus (alligator lizard)	Entire organism	CA	de Nevers (1993)
Ophisaurus ventralis (eastern glass lizard)	Entire organism	SC	Wood and Roark (1980)

(Continued)

APPENDIX 7.1 (CONTINUED)
Listing of Published Reports of Animals Consumed by Wild Pigs in the Continental United States

Taxonomic Group or Scientific Name (Common Name)	Part(s) Consumed	State(s) Reported From	Reference(s)
Phrynosoma cornutum (Texas horned lizard)	Entire organism	TX	Taylor and Hellgren (1997)
Sceloporus occidentalis (western fence lizard)	Entire organism	CA	Grover (1983)
Sceloporus undulatus (eastern fence lizard)	Entire organism	GA	Jolly et al. (2010)
Xantusia riversiana (island night lizard)	Entire organism	CA	National Fish and Wildlife Laboratory (1980)
Serpentes (snakes - general)	Entire organism	GA, MS, NC, SC, TN, TX	Hanson and Karstad (1959), Sweeney et al. (2003), Scott and Pelton (1975), Springer (1975), Ackerman et al. (1978), Baron (1979)
Coluber constrictor (eastern racer)	Entire organism	CA	Wilcox and Van Vuren (2009)
Contia tenuis (sharp-tailed snake)	Entire organism	CA	Wilcox and Van Vuren (2009)
Crotalus sp. (rattlesnake)	Entire organism	CA	Grover (1983)
Crotalus viridis (western rattlesnake)	Entire organism	CA	Wilcox and Van Vuren (2009)
Diadophis punctatus (ringneck snake)	Entire organism	CA	de Nevers (1993)
Storeria occipitomaculata (red-bellied snake)	Entire organism	GA	Jolly et al. (2010)
Thomnophis sirtalis (eastern garter snake)	Entire organism	SC	Wood and Roark (1980)
Birds			
Aves (birds - general)	Entire organism, carrion	MS, NC, TN, TX	Scott and Pelton (1975), Springer (1975), Ackerman et al. (1978), Baron (1979)
Bonasa umbellus (ruffed grouse)	Eggs	TN	Henry (1969)
Callipepla californica (California quail)	Entire organism	CA	Wilcox and Van Vuren (2009)
Colinus virginianus (northern bobwhite)	Eggs	TX	Tolleson et al. (1993)
Geococcyx californianus (roadrunner)	Entire organism	TX	Taylor and Hellgren (1997)
Melanerpes formicivorus (acorn woodpecker)	Entire organism	CA	Wilcox and Van Vuren (2009)
Meleagris gallopavo (wild turkey)	Eggs and entire organism	SC, TN, TX	Henry (1969), J. J. Mayer, Savannah River National Laboratory, unpublished data
Phalaenoptilus nuttallii (common poorwill)	Carrion	CA	Wilcox and Van Vuren (2009)
Pipilo crissalis (California towhee)	Entire organism	CA	Wilcox and Van Vuren (2009)
Richmondena cardinalis (cardinal)	Entire organism	TX	Taylor and Hellgren (1997)
Thryomanes bewickii (Bewick's wren)	Entire organism	CA	Wilcox and Van Vuren (2009)
Zenaidura macroura (mourning dove)	Carrion	TX	Taylor and Hellgren (1997)

(*Continued*)

APPENDIX 7.1 (CONTINUED)
Listing of Published Reports of Animals Consumed by Wild Pigs in the Continental United States

Taxonomic Group or Scientific Name (Common Name)	Part(s) Consumed	State(s) Reported From	Reference(s)
Mammals			
Mammalia (mammals - general)	Entire organism, carrion	CA, NC, TN	Scott and Pelton (1975), Ackerman et al. (1978), Grover (1983)
Didelphis virginiana (opossum)	Carrion	TX	Taylor and Hellgren (1997)
Sorex sp. (shrew)	Entire organism	NC, TN	Ackerman et al. (1978)
Sorex trowbridgii (Trowbridge's shrew)	Entire organism	CA	Wilcox and Van Vuren (2009)
Scapanus latimanus (broad-footed mole)	Entire organism	CA	Wilcox and Van Vuren (2009)
Dasypus novemcinctus (nine-banded armadillo)	Entire organism	TX	Kroll (1986)
Sylvilagus bachmani (brush rabbit)	Carrion	CA	Wilcox and Van Vuren (2009)
Sylvilagus floridanus (eastern cottontail)	Carrion	TX	Taylor and Hellgren (1997)
Rodentia (rodents - general)	Entire organism, carrion	GA, SC, TX	Hanson and Karstad (1959), Springer (1975), DeVault and Rhodes (2002)
Citellus beecheyi (ground squirrel)	Carrion	CA	Pine and Gerdes (1973)
Microtus californicus (California vole)	Entire organism	CA	Wilcox and Van Vuren (2009)
Microtus sp. (voles)	Entire organism	CA	Loggins et al. (2002)
Ondatra zibethicus (muskrat)	Entire organism	MS	Hanson and Karstad (1959)
Neotoma fuscipes (dusky-footed woodrat)	Entire organism	CA	Wilcox and Van Vuren (2009)
Peromyscus sp. (deer mice)	Entire organism	NC, SC, TN, TX	Ackerman et al. (1978), Wood and Roark (1980), Taylor and Hellgren (1997)
Peromyscus maniculatus (deer mouse)	Entire organism	CA	Wilcox and Van Vuren (2009)
Peromyscus truei (piñon mouse)	Entire organism	CA	Wilcox and Van Vuren (2009)
Reithrodontomys megalotis (western harvest mouse)	Entire organism	CA	Wilcox and Van Vuren (2009)
Spermophilus beecheyi (California ground squirrel)	Entire organism	CA	Loggins et al. (2002), Wilcox and Van Vuren (2009)
Tamias striatus (eastern chipmunk)	Entire organism	TN	Stegeman (1938)
Thomomys bottae (Botta's pocket gopher)	Carrion	CA	Pine and Gerdes (1973), Wilcox and Van Vuren (2009)
Sus scrofa (wild pig)	Carrion	GA, TX	Hanson and Karstad (1959), Hellgren and Holzem (1992), Taylor and Hellgren (1997)
Axis axis (chital or axis deer)	Carrion	TX	J. J. Mayer, unpublished data
Bos taurus (domestic cattle)	Carrion, calves/ entire organism	CA, FL, TX	Towne and Wentworth (1950), Springer (1975), Barrett (1978), J. J. Mayer, unpublished data
Capra hircus (domestic goat)	Carrion, kids/ entire organism	TX	Mayer and Brisbin (2008), Beach (1993), Littauer (1993)

(*Continued*)

APPENDIX 7.1 (CONTINUED)
Listing of Published Reports of Animals Consumed by Wild Pigs in the Continental United States

Taxonomic Group or Scientific Name (Common Name)	Part(s) Consumed	State(s) Reported From	Reference(s)
Odocoileous hemionus (mule/ black-tailed deer)	Carrion	CA	Pine and Gerdes (1973), Grover (1983), Loggins et al. (2002), Wilcox and Van Vuren (2009)
Odocoileous virginianus (white-tailed deer)	Carrion, fawns/ entire organism	SC, TX	Rutledge (1970), Springer (1975), Hellgren and Holzem (1992), Beach (1993), Taylor and Hellgren (1997), Jolley 2007, J. J. Mayer, unpublished data
Ovis aries (domestic sheep)	Carrion, lambs/ entire organism	CA, SC, TX	Rutledge (1970), Barrett (1978), Beach (1993), Littauer (1993)

Source: Adapted from Ditchkoff and Mayer (2009).

8 Management of Wild Pigs

Stephen S. Ditchkoff and Michael J. Bodenchuk

CONTENTS

8.1 INTRODUCTION

In its earliest form in North America, "wildlife management" was described by Aldo Leopold (1933) in his book *Game Management* as "the art of making land produce sustained annual crops of wild game for recreational use." Today, however, wildlife management has taken on a broader definition. The term management in a wildlife context currently refers to intended actions and goals directed towards a population or species of interest. Usually, management actions directly or indirectly (e.g., habitat modification) manipulate wildlife populations toward some goal that is beneficial to either the animal or society. Management actions include modification of habitats, influence or control of human actions or public interest groups, or manipulation of individuals or populations of a species of interest.

With regards to manipulation of wildlife populations, there tend to be 3 primary categories into which most, if not all, wildlife management programs can be assigned: 1) management of wildlife populations for consumptive use (game management), where populations are manipulated to enhance either the total numbers or quality of individuals for harvest by hunters or trappers, 2) efforts to reduce declines of, or increase growth of populations of species that are of conservation concern, and 3) intentional reduction or eradication of populations of pest, exotic, or invasive species. White-tailed deer (*Odocoileus virginianus*) are a classic example of game management today, where manipulation of population quality, density, age structure, and sex ratio is achieved through habitat enhancement, selective harvest, and legislation intended to enhance the hunting experience (Hansen 2011, Jacobson et al. 2011). A popular management success story for species of conservation concern is population recovery of the bald eagle (*Haliaeetus leucocephalus*). After recognizing that impaired reproductive success due to pesticides was causing population declines, legislative actions aimed at reducing environmental concentrations of these toxins and management actions designed to enhance breeding success and survival resulted in one of the most successful conservation stories in US history (Saalfeld et al. 2009). Lastly, an example of management for control of exotic invasive species is the Burmese python (*Python bivittatus*) in southern Florida, where population reduction and eradication of this species is a top priority due to negative impacts on native mammalian communities (Reed et al. 2012).

Wild pigs (*Sus scrofa*) are a highly destructive invasive species for which population control should be a top management priority (management category #3 from above). Wild pigs cause damage to floral and faunal communities, contribute to degradation of water quality, damage agricultural production, spread diseases, and negatively impact other aspects of human society (Ditchkoff and West 2007; see Chapter 7). This has led to the belief that wild pigs are 1 of the 10 most important invasive species warranting management consideration, and 1 of the top 2 most important invasive terrestrial vertebrates (the other is the Burmese python) in North America (North American Invasive Species Network 2015). Pimental (2007) estimated that wild pigs cause US$1.5B in agricultural damage and control costs in the United States annually and Adams et al. (2005) reported the average farmer in Texas spent over US$2,500 on control efforts for wild pigs. More recently, Anderson et al. (2016) estimated that annual agricultural losses from wild pigs across 10 states exceeded US$190M. Due to economic and environmental impacts of wild pigs, efforts to reduce damage by wild pigs are commonplace and estimated to be economically substantial.

Management of wild pigs generally leads towards reduction or eradication of populations, or actions resulting in decreases in damage caused by wild pigs. Population reduction or eradication primarily includes lethal methodologies (except contraception), whereas efforts to reduce damage caused by wild pigs may consist of lethal or nonlethal population control, as well as exclusionary, diversionary, or deterrent methods. This chapter discusses tools and methodologies currently available to manage wild pig populations and wild pig damage in North America, particularly (but not exclusively) on tools and methodologies supported by scientific data. Finally, strategic application of available tools and methodologies are explored and specific case studies from successful wild pig control programs are described.

8.2 SUCCESS OF A MANAGEMENT PROGRAM

The success of any wildlife management program depends upon development of a sound plan, and management programs for wild pigs are no different. Sound plans include careful design and well-defined goals and benchmarks. Although multiple actions theoretically suitable for successful completion of the management objectives likely exist, some actions are more suitable due to social pressures, political constraints, biological obstacles, or economic hurdles. Thus, failure to develop a plan using a situational and objective-driven approach usually leads to an unsuccessful management program.

A successful management plan needs a starting point: identification of the problem. Without clearly defining the problem at the outset, pathways to a solution are obscure. For example, a landowner may want to significantly reduce wild pig numbers on their property. The simple solution is to recommend some form of lethal control that will reduce local wild pig density. If the landowner is a farmer that is experiencing significant crop damage due to wild pigs, then that recommended action may be suitable. However, if the landowner is having problems with wild pigs getting into deer feeders, then exclusion (e.g., fencing) from feeders may be a better solution. Defining the problem should always be the first step in development of a management plan, thereby allowing more objective assessment of possible actions.

Once the problem has been clearly identified, the next step is to identify objectives. Objectives describe desired outcomes of the program, and the best objectives are defined in measurable terms. For example, possible objectives for a management program designed to reduce crop damage by wild pigs could be: 1) reduce the wild pig population by 75%, 2) completely eliminate all wild pigs on the property, 3) reduce the wild pig population to a level such that crop damage is reduced by 75%, or 4) reduce the wild pig population such that crop damage is reduced by 75% and maintained at that level over time. Each of these objectives is a possible solution to the problem; however, they all may require different management actions or combinations of actions. In addition to an objective that defines the level of wild pig density or damage reduction, there may be additional objectives that establish economic constraints, define a time frame in which the program must be completed, or set restrictions on actions that can be taken to solve the problem. By specifically defining the objectives, the pathway from starting to ending point becomes clearer.

When the problem is clearly identified and objectives have been defined, possible management actions suitable for the plan become readily apparent. In our previous example, the objective was to reduce crop damage by 75%; hence, a short-term but intense lethal control program may be suitable to achieve the objective. However, if the objective was to reduce crop damage by 75% and maintain it at that level indefinitely, a short-term program will likely not be successful because the program will require continued monitoring of the population and occasional follow-up removal operations. Without question, these simplistic examples can easily be addressed with common sense. However, wild pig management programs often are far more complex than addressing needs of a single landowner. Rather, management programs may span counties or states and require coordination of multiple agencies and thousands of individual landowners. Conversely, management programs may be on public land where constraints of public perception force a departure from preferred actions and a reliance upon more creative solutions.

Ultimately, success of any management program will be a function of whether management objectives are satisfied. However, it is important to understand that objectives can vary widely depending on the management program. In some cases, complete eradication of a population is the ultimate objective, but eradication may not be an achievable objective in other situations. Rather, population reduction and/or reduction of damage by wild pigs may be the objective of a management program. Bomford and O'Brien (1995) developed a list of 6 criteria that must be met for eradication to be a feasible management option: 1) rate of removal must exceed rate of increase of the population, 2) immigration must not be possible, 3) all reproductive animals must be at risk of removal, 4) animals must be detectable at low densities, 5) cost:benefit analysis must favor eradication over continued control, and 6) the social and political environments should be amenable to eradication. Both Santa Cruz Island, California (Parkes et al. 2010) and Pinnacles National Monument in California (McCann and Garcelon 2008) satisfied these criteria and thus eradication was a feasible objective. However, eradication is not a viable objective in most areas because at least one of these criteria cannot be satisfied. In these situations, more realistic objectives will usually focus on population reduction or reduction of damage. The wild pig management program at Great Smoky Mountains National Park (GSMNP) is such an example. Despite US National Park Service (NPS) policies that mandated control or eradication of invasive species and a 50-plus year concerted effort on the part of NPS personnel to reduce or eradicate populations of wild pigs (Peine and Farmer 1990, Stiver and Delozier 2009), they continue to persist within the park. Unfortunately, the tools and technology available for controlling wild pigs today are not sufficient for wild pig eradication in an area such as GSMNP, and thus, criteria #3 of Bomford and O'Brien's (1995) list cannot be satisfied at this time. Until more advanced wild pig control technologies become available, a more realistic objective of population and damage reduction is likely the only alternative. However, control efforts in the park have demonstrated results in terms of limited population growth and reduced impact on the landscape (Peine and Farmer 1990, Salinas et al. 2015, Levy et al. 2016). The GSMNP case study provides an excellent example of the need to clearly define management objectives. In terms of the NPS mandate for eradication of invasive species, the management program has fallen well short; but in terms of control, the management program has been successful. While somewhat semantic, the ability to report "success" of a management program has important implications for the sociopolitical climate surrounding an issue, and for this reason it is important that the original stated objectives are realistic.

8.3 MANAGEMENT OPTIONS

No single methodology for wild pig control is suitable for all situations. Biological conditions, social attitudes, and political constraints have forced wildlife management professionals to develop a diverse array of tools for controlling wild pigs, usually implemented in an adaptive and integrative form. Here we describe these tools and provide supporting documentation from the scientific literature where possible.

8.3.1 Trapping

Use of traps to capture wild pigs in North America predates damage management efforts. Domesticated pigs in North America historically were free roaming and subsequently captured for food and to mark for ownership in the fall through trapping (Mayer and Beasley 2018). Dobie (1929), using John Young's (b. 1856) words, reported on the use of traps in Texas in the late 1800s: "Sometimes the settlers made traps to catch the hogs. A trap was nothing but a picket pen with a door swung from the top. The door opened inward when pushed against and fell shut as soon as force against it was released. It could not open outward. Many of the hogs were used to trying to get inside the wood-fenced cornfields. Some corn would be sprinkled around the pen up to the gate. The gate was not solid and the hog to be trapped could see the profusion of corn under the gate, or door, and inside the pen. He would nose against the door, the door would open inward, and, head down, he would enter. Hard on his heels would follow another hog. Even if the door swung shut the trapped hogs would be too engaged – for a while – in eating to notice their imprisonment. Any hogs outside would be frantic to join in the feast and would very likely find the door. Thus several hogs might be trapped at once, among them great-tusked outlaw boars."

In its simplest form, trapping has changed little over the last several centuries. Traps still basically consist of walls, a door, and sometimes a mechanism that causes the door to close (Hamrick et al. 2011, Higginbotham 2014; see Chapter 9). Where the original corral traps of the 1800s were constructed of wood, modern traps are constructed of galvanized wire livestock panels, or panels specifically made for wild pig trapping (Littauer 1993, Higginbotham 2014). Minimum height of trap panels is generally accepted as 1.5 m (Littauer 1993), unless a top is added to the trap. However, many wildlife professionals prefer panels ~2 m high to minimize pigs escaping over the top. The 2 trap types most commonly used are corral traps (Figure 8.1) and box traps (Figure 8.2). Both effectively capture wild pigs, but corral traps are generally preferred due to greater size and ability to capture more animals at one time (Williams et al. 2011a). Corral traps are generally circular or "teardrop shaped" but may also be "football shaped" with a gate at each end (Hamrick et al. 2011). Most commonly used without a top, corral traps allow non-target captures (e.g., deer, raccoons (*Procyon lotor*), turkey (*Meleagris* spp.)) to escape. Open top traps should avoid corners where pigs can concentrate and escape over the top by climbing on one another. Choquenot et al. (1993) indicated that trapping with corral traps could lead to population reductions of 80–90% in 2 weeks, and that females were more susceptible to trapping. However, most trapping programs in North America have experienced less success, suggesting that sustained effort over longer periods is necessary to achieve success of this magnitude. Bodenchuk (2014) reported on cost efficiency of trapping in Texas where 585 wild pigs were removed. Average cost/pig removed was US$46.95, and the main costs were travel and labor. These data included significant efforts to revisit properties following successful trapping to identify when wild pigs returned to the site.

Just as divergent opinions exist among trappers on size and shape of traps, wire spacing within panels is a debated topic. Common "livestock panels" have 10-cm × 10-cm spacing on the panel (Higginbotham 2014), but this will allow small pigs to escape. "Horse panels" with 5-cm × 10-cm spacing will hold the smallest piglets, but panels this size weigh twice as much as livestock panels and cost more. Panels with graduated spacing that have small gaps close to the ground and larger gaps higher on the panel, such as Jager Pro's 18–60™ Trap Panel, strike a compromise between

FIGURE 8.1 Corral trap with remote activated drop gate. Corral-style traps are generally preferred over other trap designs due to their greater size and ability to capture more animals at one time. Corral traps that utilize a remote activated drop gate allow managers greater flexibility regarding trap activation. (Photo by S. Zenas. With permission.)

FIGURE 8.2 The standard design of a box trap commonly deployed to capture wild pigs. (Photo by R. DeYoung. With permission.)

panel weight and the ability to hold small pigs. However, piglets occasionally congregate in the corners and climb over one another to larger spaces above where escape is possible.

While opinions vary regarding the ideal dimensions for a corral trap, the minimum recommended size consists of 3 galvanized panels 6 m in length. Tied end-to-end and set in a circle, the diameter of this trap, with a gate installed would be approximately 6 m. Most professional trappers would agree that traps this small would exclude trap-wary wild pigs and a 4- or 5-panel trap would be preferred (Higginbotham 2014). Panels for corral traps should overlap 1 mesh-width (10 cm) and stand upright on the ground (Higginbotham 2014). Panels are secured with smooth wire ties to standard T-posts driven around the perimeter of the trap: T-posts should be located on the outside of the panels to maximize trap strength. The number of posts necessary to secure the trap may vary with soil type and presence of rocks, but even in sandy soil 1 T-post/meter of circumference is usually adequate (G. Silvers, A to Z Wildlife Control and Consulting, personal communication). Panel traps are a type of corral trap constructed of preassembled components designed for easy assembly in less time or with fewer personnel. Panel traps have the advantage of not requiring T-posts to secure them (although use of T-posts improves strength of the trap considerably) and components are pre-fabricated to fit securely together. However, component sections of panel traps individually weigh more than livestock panels of similar length and access to the trap site by truck and trailer is usually a prerequisite for use.

Box traps are smaller than corral traps and are primarily used to capture individual wild pigs, or in urban and suburban settings. A welded steel frame (usually made from angle iron) with galvanized wire panels welded in place, is constructed with a gate opening at one end. Common dimensions for a box trap are 1.22 m × 1.22 m × 4.44 m, allowing the trap to be transported in a standard truck bed. While many commercially produced box traps have wire on all sides, including the floor of the trap, it is advisable to cover the floor with soil to reduce neophobia in pigs. Williams et al. (2011a) reported that corral-style traps caught 4 times more pigs than box-style traps, suggesting corral traps are generally preferred over box traps. However, Long and Campbell (2012) indicated that use of rooter gates on box traps increases capture rates as opposed to side-swing gates.

Drop traps, such as the Boar Buster™, use the design of a panel trap without placing the components on the ground (Gaskamp 2015). Assembled sections are suspended approximately 1 m above the ground on established posts and pigs have 360° access to bait placed at the center of the trap. When activated, the trap drops to the ground, capturing pigs within the trap frame. Falling traps eliminate "gate shyness" as a barrier to capture. Proponents suggest that pigs do not seem to notice the trap above their heads until it is triggered, although other researchers and managers have noted some wild pigs are wary of drop traps and unwilling to enter under them. Drop nets, long used for wildlife capture (Jedrzejewski and Kamler 2004), have been successfully utilized for wild pig capture (Gaskamp and Gee 2011), and have been reported to have greater capture success than corral traps (Gaskamp 2012). However, J. R. Sandoval (Texas AgriLife Extension, personal communication) noted numerous wild pig escapes when large sounders were involved in capture events as some pigs were close to the edge of the net when activated. Drop nets were 10.5 m × 10.5 m with 10-cm mesh. A light weight net, as opposed to metal traps, make drop nets highly portable and an important tool when trapping is necessary in remote areas with only foot-access.

A variety of gate designs exist and each has pros and cons. In general, gates fall into the broad categories of passive (i.e., "no gate" design; Higginbotham 2014), guillotine, saloon, and rooter styles. Passive gates are constructed by overlapping the 2 ends of galvanized trap panels in a snail-shell circular fashion so that wild pigs can enter the trap simply by pushing through the opening. During the prebaiting phase, the ends are apart to allow free access into and out of the trap. Once pigs are accustomed to entering freely, the inside panel is set to where it rests against the inside of the first panel. Pigs accustomed to feeding in the trap push through the gap and the spring tension of the panel snaps it back after the pig enters. Once inside, pigs cannot force the panels apart. Guillotine gates are set above the trap opening and, when activated, drop straight down within a track in the gate support frame (Figure 8.3). Lightweight guillotine gates may have a latching

FIGURE 8.3 Corral trap utilizing a guillotine- or drop-style door. Guillotine-style gates can be triggered in a variety of ways ranging from animal-activated root sticks to human-activated remote control. (Photo by C. Jaworowski. With permission.)

system when closed to prevent trapped pigs from lifting them with their snout. Saloon gates are hinged on the side and swing closed when activated. Usually spring loaded, saloon gates may also require a latch system to prevent captured pigs from forcing them open (Figure 8.4). Rooter gates are generally constructed of square, tubular steel with 3 side-by-side sections in a 1-m-wide gate (Mapston 2010; Figure 8.5). During prebaiting the gates are wired open. Rooter gates are activated once pigs are inside the trap, but are designed with the same principle described by Dobie (1929) above. Once activated, wild pigs may still force their way into the trap through the gate. Camera data in Texas indicated multiple captures after gate activation (R. M. Smith, USDA/APHIS/Wildlife Services, personal communication). However, Smith et al. (2014) reported that while rooter gates captured a few more pigs than other gate types, their increased cost and complexity likely did not

FIGURE 8.4 Corral trap utilizing a saloon-style trap door. Saloon gates are hinged on the side, and swing closed when activated. An additional latch may be required to prevent captured pigs from forcing the gate open and escaping. (Photo by the National Feral Swine Damage Management Program. With permission.)

FIGURE 8.5 Partially constructed corral trap utilizing a rooter-style trap door. The design of the rooter door allows additional pigs to be captured after the gate has been activated. However, it is possible for captured pigs to escape when pigs outside the trap are forcing their way inside. (Photo by the National Feral Swine Damage Management Program. With permission.)

outweigh benefits accrued from increased captures. Smith et al. (2014) also noted wild pigs escaped traps when others forced their way into the trap through the rooter gate.

Gate dimensions are an important consideration for most trappers. Smaller gates weigh and cost less, but are potentially less effective than larger/wider gates. Higginbotham (2014) noted that wider gates might reduce time necessary for pigs to acclimate to the trap and enter freely. However, Metcalf et al. (2014) did not find a significant difference in catch rates based on gate widths of 0.9–1.8 m. A 1-m wide gate is generally the minimum acceptable width, and gates up to 2.5-m wide have been commercially produced. Wide gates may reduce gate shyness, but also add to cost and weight of the trap and present logistical issues getting the gate into remote locations.

Regardless of gate design, the gate must be activated to capture animals, and gates may be activated by either humans or the animals themselves. Higginbotham (2014) listed 8 different categories of activation, including human activated, no trigger, tripwire trigger, pressure plate trigger, trough trigger, rooter stick trigger, bucket trigger, and tire trigger. A tripwire trigger involves a string (such as braided fishing line) set to release the gate by pulling a pin or removing a prop, once an animal bumps into it while feeding. Use of a tripwire trigger is less costly than human activation systems, but is subject to activation by non-target wildlife and does not selectively allow the entire sounder to enter the trap. Many other designs are employed to reduce non-target activation and increase the number of animals captured in a single event. The rooter stick trigger (Figure 8.6) across a pile of bait tends to reduce activation by species other than pigs, which knock the stick from its place when rooting in the bait pile; however, raccoons, Virginia opossums (*Didelphis virginiana*), and black bears (*Ursus americanus*) will trip root stick triggers. A tire trigger involves bait placed inside a tire attached to the activation string. While other animals may enter the trap and eat bait, adult pigs will eventually move the tire to access additional bait, thereby activating the gate.

Human-activated gates range from a simple string tied to a gate with a person hiding in a blind, to line-of-sight remote activation, to remote monitoring and activation via computer or cellular device. The latter systems allow trappers to review images of pigs in the trap in near-real time and make decisions about activation based on number of pigs present, employee workload, and weather. Continual monitoring is available and these systems reduce personnel and mileage requirements when compared to conventional trapping tools. Cellular contracts are necessary and some contracts include external monitoring by personnel outside of the trapper's organization. While these trigger

FIGURE 8.6 Root stick trigger mechanism. Root stick triggers activate the trap door when animals feeding inside the trap knock the stick from its place. Non-target animals are capable of activating the trap using this method. (Photo by C. Jaworowski. With permission.)

systems can be effective, they are often not practical for most landowners due to cost: initial costs for remote activated gates are US$2,000–5,000/system, and there is an additional monthly cost for cellular service. Additionally, rural or remote areas where use of these traps may be desirable often have unreliable cellular coverage, so these trap types may not be suitable for use in all areas.

Ancillary equipment for trapping includes lure or bait holders, remote feeders, and trail cameras. Bait holders can be made inexpensively and allow pigs to self-feed with minimal interference by non-target wildlife. For example, a heavy-duty plastic 189-liter drum with 2-cm holes drilled 10 cm above the bottom can be wired to a post inside the trap and filled with small bait such as whole, shelled corn. Adding water to the barrel up to the holes will allow corn at the bottom to ferment. Once wild pigs become accustomed to the trap, they will push the barrel around the post, dispensing small amounts of corn. This prevents consumption of large amounts of bait as well as reducing non-target attraction and the need to revisit the site daily to rebait (R. Sramek, USDA/APHIS/Wildlife Services, personal communication). Lure holders can be made from 10-cm PVC pipe, capped on one side and sealed with a removable cap on the other. Holes drilled into the pipe will allow an olfactory attractant to disperse while protecting it from pigs and other wildlife (J. R. Sandoval, personal communication).

Wide varieties of baits are used to capture wild pigs and response to baits may vary seasonally (Campbell and Long 2008, Lavelle et al. 2017). Whole corn is probably the most common bait for trapping wild pigs due to its availability and attractiveness. Many trappers choose to alter whole corn by soaking it in water and allowing it to ferment in the belief that this will increase attractiveness (fermented corn is also used to decrease attractiveness to non-target wildlife). However, Williams et al. (2011b) reported no difference in attractiveness (time to first detection) between dry, fermented, or mixed (dry and fermented) corn bait. Further, pigs in their study fed for longer at dry corn sites than at either fermented or mixed corn bait sites. Dried fish, "soured" livestock feed, reclaimed distiller grains, and surplus produce have all been used to capture wild pigs. During periods when food is readily available (e.g., periods of mast production, supplemental deer bait, ripening agricultural crops), attracting wild pigs to trap sites may be difficult. Alternatively, baiting during periods of food scarcity will increase attractiveness of the bait (McIlroy et al. 1993). During late winter, after crops are harvested and pigs require more food due to cold temperatures, bait may also be more readily accepted. McIlroy and Gifford (2005) suggested that use of estrous sows as lures in traps may be effective in situations where population density is extremely low and remaining animals have become trap shy or difficult to cull with other methods. In contrast, Choquenot et al. (1993) reported that use of estrous females as lures was not effective when trapping.

Location of traps is critical to successful trapping. Higginbotham (2014) noted that novice trappers tend to set traps at the location of damage, which is often in a crop field. However, placing the trap between damage sites and bedding locations tends to be more effective. Olfactory lures may bring pigs to the trap site, but placing the trap where they will easily find it (e.g., at or near feeding sites or regularly used travel corridors) will yield greater success in less time than trying to alter their movements. Orienting the door on the downwind side of the trap allows for easy detection of odor from the bait and pigs may more readily enter the trap. During drought or in arid habitats, trapping at water sites (water point trapping) can be effective.

8.3.1.1 Trapping Strategy

"Strategy" may seem like somewhat of a misnomer when it comes to wild pig trapping because simple logic dictates that you trap as many as possible, as quickly as possible. However, trapping has rarely been successful in eradicating a population of wild pigs. The reason is that most trapping programs for wild pigs do not employ a strategic approach; they are typically designed and operated to maximize the number of animals removed while minimizing costs. The problem with wild pig trapping programs that do not employ strategy is that they inevitably do not remove all wild pigs. Unfortunately, uncaptured or escaped sows have been educated about the danger of traps, and they quickly replace lost members of the sounder due to their prolific rate of reproduction (see Chapter 4). These programs inherently report sustained large numbers of wild pigs removed from the population, but rarely document a permanent reduction in density. More recently, some managers and agencies have begun to recognize that effective control of wild pigs requires a strategic management approach, and success stories are becoming more common.

There is growing recognition that removal of entire sounders, rather than individual animals or partial groups, may be a more effective strategy for controlling wild pig populations. Whole sounder removal is based upon the discovery that wild pigs in some populations exhibit high site fidelity, and are possibly even territorial (Ilse and Hellgren 1995, Gabor et al. 1999, Sparklin et al. 2009; see Chapter 3). If wild pigs show high site fidelity and an area is cleared of wild pigs (a whole sounder is removed), then recolonization of that cleared area should occur slowly. At Fort Benning, Georgia, the 5-step whole sounder removal strategy was employed, and eliminated all but 2 female wild pigs in an area that was approximately 8,000 ha. Step 1 in the whole sounder removal strategy is to survey the population using game cameras and identify where individual sounders are located. Step 2 involves identifying each unique sounder and all of the individuals in each sounder. During step 3, traps are constructed to capture individual sounders, and each sounder is allowed to habituate to the trap. Step 4 involves trapping the pigs, and step 5 consists of monitoring after the sounder is removed to ensure that all pigs in the group have been eliminated. Whole sounder removal has since been employed successfully in a number of situations where wild pigs show high site fidelity. However, data suggest that wild pigs in some areas may not use exclusive space (Boitani et al. 1994), and hence it is unknown whether whole sounder removal would be an effective strategy for eliminating wild pigs in these situations. At this time, few controlled studies have been conducted regarding whole sounder removal, and it is not completely understood how variation in spatial and social dynamics of wild pigs influences effectiveness of whole sounder removal. Nonetheless, management strategies that work to remove entire social groups of wild pigs will likely be more effective than those that do not, if eradication is the goal.

8.3.2 Hunting/Shooting

The role of recreational hunting as a means of controlling wild pig populations remains controversial, although it remains popular among recreational hunters and allowed as a method of population control in most states (see Chapter 11). To our knowledge, recreational hunting has never provided long-term control of a wild pig population. In Australia, Caley and Ottley (1995) reported little evidence to support effectiveness of recreational hunting to reduce populations. In Hawaii, Hess et al. (2006) reported no significant difference in efficiency between public hunting and agency staff hunting of wild pigs, and each method was less efficient than snaring by staff (see Chapter 17). Mayer (2014) reported that recreational hunting only removes on average 23% of a wild pig population annually, not achieving the 60–80% removal needed to reduce populations. In states with few or no established populations, legalized hunting of wild pigs encourages movement of pigs and establishment of new populations. However, in states with well-established populations, recreational hunting provides income to landowners and increases tolerance. In all cases, allowing recreational shooting establishes a "hunting culture" around wild pigs, which generally works contrary to control/eradication efforts.

Hunting wild pigs with dogs as a control method has been successful in some instances, particularly when populations are at low densities and efforts are part of a formal eradication program (Littauer 1993, Caley and Ottley 1995, Mapston 1999, Mayer et al. 2009). However, recreational hunting with dogs by itself is not generally effective at reducing or eliminating populations of wild pigs (Caley and Ottley 1995). Dog hunting usually results in capture of only 1 or 2 pigs/event, thereby limiting use for large-scale projects (Caley and Ottley 1995). Dogs are selectively bred and trained to detect pigs by scent and bring pigs to bay, where they can be shot or stabbed by pursuing hunters. Some hunters employ "catch dogs" that are released at short distance from the bayed pig specifically to grab and hold the pig until the hunter arrives. Ancillary equipment often includes protective vests for dogs to prevent injury and GPS tracking collars to facilitate following and recovering dogs during and after the hunt.

Use of a VHF-equipped wild pig, known as a "Judas pig," to guide shooters to other pigs is sometimes employed in control projects where pig populations are at low densities (Pech et al. 1992, Littauer 1993). While a recent study has suggested that not enough research has been conducted to determine which sex or age group makes the best Judas pig (West et al. 2009), some older studies have theorized that adult males, due to their solitary nature, are a poor choice (Wilcox et al. 2004). They suggested females are the most effective Judas pigs, though their sample size was small. McIlroy and Gifford (1997) explored the Judas pig technique in Australia, and suggested the technique worked best with sows captured in the same area. Because radio equipment failures may occur and some managers worry that releasing a reproductively active animal may work contrary to control, McCann and Garcelon (2008) recommended surgically sterilizing Judas pigs before releasing them. However, effects of sterilization on social behavior or effectiveness of Judas pigs has not been studied. Judas pigs were one control technique utilized in the eradication of wild pigs from Pinnacles National Monument (McCann and Garcelon 2008).

Aerial shooting, most commonly from helicopters, is an effective tool in areas with limited tree canopy and few anthropogenic structures, and has been employed by several agencies across the United States with varying success. While this technique will probably not be effective at eradicating a population of wild pigs because of increasing cost and effort required as population density decreases (Choquenot et al. 1999), it can be effective at reducing densities over a short period. Hone (1990) and Saunders (1993) reported that aerial shooting from helicopters was effective at reducing targeted populations by 65–80%, and Dexter (1996) found that movement patterns of wild pigs that survive aerial eradication programs do not change from pre-removal. Similarly, Campbell et al. (2010) reported only minor behavioral changes in pigs not removed during aerial control, the act of flying low and even shooting other pigs in a sounder did not disperse pigs from established home ranges. Bodenchuk (2014) reported helicopter aerial shooting to be the lowest cost/animal removed for the Texas USDA/APHIS/Wildlife Services program, largely because the agency focused removal efforts in areas with high densities of wild pigs. Within this program, across 9 project areas, 5,127 wild pigs were removed over 156.2 hours, at an estimated cost of US$18.27/pig. The same analysis looked at private, piston-powered helicopter take, where vendors in 7 project areas removed 523 wild pigs in 27.4 hours at a cost of US$21.11/pig removed. Fixed-wing aircraft can be employed in open habitat, and the Texas data identified 3 project areas where 256.4 hours were flown on 53 dates, and 1,495 wild pigs were removed at a cost of US$26.63/animal. Bodenchuk (2014) noted that these data were particular to Texas, where large meta-populations exist and return visits to the property are necessary. The data represent costs relative to each other rather than predictive quantities. Results of aerial control programs depend greatly on vegetation cover, crew experience, and pig densities. For example, experienced Texas Wildlife Services aerial crews flew in San Diego County, California where few pigs existed. In 28 hours of survey time in thick vegetation and mountainous habitat, no pigs were observed. In contrast, the same crew was able to remove 65 pigs in 17.5 hours of flight time in more open habitat in Arizona. In Michigan, a Wildlife Services helicopter hovered over a radio-tagged, female wild pig that was underneath conifer cover and tried to flush the individual with the rotor wash but the pig was never observed.

With increasing availability and affordability to the public of devices that incorporate military technology, thermal/night vision scopes and cameras have enabled detection and hunting/shooting of wild pigs during nighttime hours. Concurrently, many state wildlife agencies have loosened regulations associated with hunting wild pigs to legalize their take after dark and over bait. To date, little to no data are available regarding the effectiveness of night hunting with the aid of thermal/night vision technology at reducing wild pig populations. We hypothesize that while this activity might be a good tool for detecting and eliminating individual animals from the landscape, like normal recreational hunting, these activities are probably not very effective at reducing populations at moderate to high densities.

8.3.2.1 Illegalization of Hunting

The wildlife management community generally accepts that illegal transport of wild pigs for recreational hunting has been the primary cause of rapid population spread to new areas in the past few decades (Bevins et al. 2014, Lewis et al. 2017). Wild pigs in high population areas tend to exhibit high site fidelity (Ilse and Hellgren 1995, Gabor et al. 1999, Sparklin et al. 2009; see Chapter 3). Despite being in North America for over 500 years, wild pigs have been slow to increase their occupied range until recently. Only illegal transport and relocation of wild pigs can explain the spotty distribution of these animals throughout southeastern North America, as well as their sudden appearance in northeastern and north-central North America, and other states and Canadian provinces (see Chapters 12–18). While accidental escapes from domestic production facilities and high-fenced shooting operations have contributed to this range expansion, intentional relocations for hunting is the primary factor contributing to the growing spread of wild pig populations in most areas. For this reason, legality of wild pig hunting and associated increases in the spread of wild pig populations in the United States and Canada has received considerable debate.

In 2006, Kansas banned anyone but landowners from hunting wild pigs in a proactive, yet controversial, step towards eliminating their growing wild pig population (see Chapter 11). The impetus for this legislative action was spontaneous appearances of wild pigs on public hunting areas far from any established populations. Kansas authorities believed that by eliminating the incentive (hunting) to relocate wild pigs, they could reduce or eliminate establishment of new populations in the state. This would allow managers to focus their efforts on eradicating isolated subpopulations without risk of establishing new populations. The success of the Kansas model for wild pig management resulted in other states implementing similar legislative policies.

In 1999, Tennessee opened a statewide hunting season on wild pigs with no bag limit in an attempt to reduce the population. However, during the ensuing decade of unlimited hunting, the population expanded substantially, and small isolated populations of wild pigs appeared as the result of introductions for hunting. In response to this trend, Tennessee enacted legislation in 2011 that removed the "big game" status from wild pigs, and designated them as a pest species to be removed by methods other than hunting. New York successfully eradicated the few populations of wild pigs that were present in the state, and as part of their management program, they enacted legislation in 2013 that made it illegal to hunt wild pigs, thus eliminating the incentive to introduce wild pigs for the purposes of hunting. This trend continued, and other states banned hunting of wild pigs or considered legislation to that end (see Chapter 11). In contrast to the beliefs of wild pig researchers and managers, and despite the fact that hunting has never successfully eliminated wild pigs, most wild pig hunting enthusiasts consider hunting a beneficial control measure. Because of their belief and strong interest in maintaining wild pig hunting as a recreation, many hunters are opposed to legislative actions that would reduce or restrict hunting of wild pigs.

8.3.3 Snares

Snares (cable restraints) are widely used in a number of states to capture and hold wild pigs. Snares may be set under fences where pigs access crop fields, on trails, or on trees and poles used by pigs as rubs. Snares employ a flexible wire cable, usually 3 mm, with a sliding "lock" that closes to hold the pig when the animal pulls against the loop. The size of the loop varies based on setting. Under a fence, the loop must adequately fill the hole the pigs are using. Set against a rub, the loop size and height depends on the size of the pig, which is indicated by the height of mud on the rub (J. Hetzl, USDA/APHIS/Wildlife Services, personal communication). When set on a trail, a 30-cm loop is set with the bottom of the loop 5 cm above the ground. The selectivity (exclusion of non-targets) of snares depends on the skills of the trapper, but even the most experienced may occasionally capture non-target animals. When checked early each day, most non-target captures can be released unharmed. Anderson and Stone (1993) reported 7–43 hours of labor for each pig captured with snares in Hawaii, and noted that success was positively associated with density of wild pigs. Foot snares involve a cable laid flat on the ground with a spring-loaded device that elevates the snare and closes the loop when the target animal steps inside the loop. Foot snares have been used in Hawaii as a selective method of capture for individual pigs (T. J. Ohashi, USDA/APHIS/Wildlife Services, personal communication). All snares intentionally capture a single animal at a time and are not effective for wide-scale eradication or control projects. They are generally most effective for targeted removal of individuals utilizing a wallow or rub site, for those accessing crop areas, or for animals that have become trap shy.

8.3.4 Toxicants

Toxicants for wild pigs in the United States are limited, though research focused on new toxicants is ongoing. Kaput®, a warfarin-based product that interferes with blood clotting by inhibiting vitamin K-dependent clotting factors, was proposed for wild pig control and efforts are underway to register the product for legal use. In Australia, where a warfarin-based toxicant for wild pigs was used under an experimental permit, 97–99% of the targeted populations were eliminated during trials (McIlroy et al. 1989, Saunders et al. 1990). However, concerns about humaneness of the toxicant have prevented licensing. Sodium monofluoroacetate (Compound 1080) was also utilized in Australia for controlling pig populations (O'Brien et al. 1988, Twigg et al. 2006, Twigg et al. 2007), although it has not been registered in the United States for use with wild pigs. Compound 1080 interferes with the Krebs cycle, thereby halting metabolism of carbohydrates and depriving the body of energy. Wild pigs that ingest Compound 1080 typically vomit, and O'Brien et al. (1986) suggested this vomit could be hazardous to non-target species. Sodium nitrite is currently registered for wild pig removal in New Zealand, and investigations aimed at supporting US and Australian registration are underway (Snow et al. 2017a, 2019; see Chapter 9). Cowled et al. (2008) identified sodium nitrite as a potential humane toxicant for wild pigs that when metabolized causes methemoglobinemia. Wild pigs are believed to be particularly sensitive to sodium nitrite (Lapidge et al. 2012), suggesting that although it has some effects on all mammals, it may have some degree of species specificity. Sodium nitrite is a commonly used human food preservative and appears to be environmentally safe. One challenge with sodium nitrite has been formulation of baits that mask the taste of the toxicant to wild pigs and result in consumption rates sufficient for high rates of mortality (Foster et al. 2014). Additionally, strategies must be developed to effectively administer toxicants to wild pigs while restricting access to non-target species such as bears, deer, and raccoons (Snow et al 2017b; see Chapter 9).

8.3.5 Contraceptives

Contraception refers to an intentional reduction in fertility rates, and because wild pigs have a prolific rate of reproduction (Dzieciolowski et al. 1992, Ditchkoff et al. 2012, see Chapter 4), contraception has been suggested as a potentially valuable tool for management programs designed to reduce or eliminate populations of wild pigs. Additionally, contraception is a potential alternative method of population control for situations where lethal control may not be socially acceptable. Although early development of contraceptive tools for free-ranging wildlife populations began in the 1960s (Balser 1964, Linhart and Enders 1964, Greer et al. 1968), contraceptives for wild pigs did not receive much attention until the 1990s. Initial efforts focused on a GnRH (gonadotropin-releasing hormone) immunocontraceptive vaccine, GonaCon™, due to its relative success with other wildlife species (Miller et al. 2000; Killian et al. 2004, 2006b, 2007). Gonadotropin-releasing hormone, which is secreted by the hypothalamus, is responsible for initiating the cascade of reproductive hormones in both males and females, and the immunocontraceptive vaccine works by producing antibodies that neutralize GnRH (Miller et al. 2008), essentially halting the reproductive process before it can begin. Testing of GonaCon™ with captive domestic and wild pigs indicated that the immunocontraceptive is effective in controlling fertility (Killian et al. 2003, 2006a; Miller et al. 2003). Additionally, contrary to some hormone-based immunocontraceptives, GonaCon™ does not seem to cause any adverse behavioral or physiological effects (Massei et al. 2008, Quy et al. 2014). However, due to the potential for GonaCon™ to impact reproduction in non-target species, injection is currently the only method of delivery ensuring only target species are treated, and an injectable contraceptive agent is not a logistically realistic option for controlling wild pig populations.

A more recent alternative immunocontraceptive technology based on phage display (Samoylova et al. 2010, 2012, 2017) has the potential to be formulated for oral delivery, and thus eliminate logistical hurdles imposed by injectable technologies. Phage display is a laboratory technique that enables production and testing of a wide variety of novel proteins for biological activity. Early research with this technology produced 2 phage-peptide antigens that, when injected into domestic pigs, stimulated production of anti-peptide antibodies that were shown to act as anti-sperm antibodies (Samoylova et al. 2012). Recent work indicated that serum collected from immunized pigs interferes with sperm-oocyte binding and fertilization during *in vitro* fertilization systems (T. I. Samoylova, Auburn University, unpublished data). Ultimately, utility of any fertility control technology will be a function of length of effectiveness (temporary versus permanent infertility) and proportion of the population sterilized (Cowan and Massei 2008, Burton et al. 2013). Low-intensity contraceptive programs (either short duration of infertility or low proportion of population affected) show little promise for reducing populations of wild pigs, while high-intensity programs are projected to be effective. However, management programs that utilize lethal control and contraceptive technology in combination will likely be far more effective at limiting wild pig populations than programs utilizing only contraceptive technology (Burton et al. 2013). Similar to toxicants, oral delivery of contraceptives to wild pigs must be accomplished safely and effectively while excluding non-target species, and will require additional research before any oral contraceptives are registered for use in the United States (Campbell et al. 2006; see Chapter 9).

8.3.6 Fencing

Fencing has been used to reduce wild pig access to sensitive areas, with the goal of reduction or elimination of damage to those areas (Hone and Atkinson 1983, Mayer and Beasley 2018). Rattan et al. (2010) reported that a 61-cm-tall exclosure of galvanized wire effectively excluded wild pigs from deer feeders in Texas, while 51-cm fencing only reduced access by 58%. Doupé et al. (2010) indicated that fencing could significantly reduce damage to ephemeral floodplain lagoons, but suggested biological, chemical, and physical attributes of these areas were more sensitive to natural seasonal effects than pig rooting. Reidy et al. (2008) evaluated electric fencing in multiple settings and found it effective in reducing wild pig access, though not 100% effective. In a Texas field trial, 30 holes created by wild pigs were found 1 year after 37 different holes caused by wild pigs had been repaired in the same section of fence (M. J. Bodenchuk, unpublished data). While repair and wild pig removal resulted in a 19% reduction in fence damage, it was concluded that fencing alone was insufficient to restrict movement of wild pigs.

8.3.7 Diversionary Feeding

Supplemental or diversionary feeding is an accepted practice for mitigating wildlife damage that is limited in duration and extent, but this technique has not been tested with wild pigs in North America. Goulding et al. (1998) reported that strategic placement of supplemental feed reduces depredation of crops in Europe. However, supplemental feeding may lead to elevated reproductive rates and increase pig populations if conducted in dry, food-stressed environments (Wilson 2005, Ditchkoff et al. 2012). Where pigs are invasive and population reduction or eradication is desired, supplemental feeding will not meet management objectives. Ditchkoff et al. (2017) speculated that increases in availability of supplemental feed in the form of bait resulted in elevated levels of reproduction that at least partially undermined the goals of a bounty program for wild pigs. If the management goal is for wild pigs to remain on the landscape, or if removal is difficult to accomplish and resource damage is limited in both space and time, supplemental feeding may be effective in drawing wild pigs away from a vulnerable crop or resource (Goulding et al. 1998). For example, protecting sea turtle nests from damage by wild pigs in a relatively public setting may be difficult to accomplish due to political and social constraints (i.e., lethal control is not an option). In this situation, diversionary feeding may be effective at moving pigs from the beach for the short time it takes to protect nests, and result in a reduction in damage.

8.3.8 Monitoring

A variety of techniques exist for monitoring populations of wild pigs, but effectiveness of some is limited to a narrow range of conditions. Additionally, little empirical data exist that describe efficacy of monitoring techniques or programs, but is becoming an increasingly crucial area of research due to the importance of determining the success of management programs based on metrics other than just total number of pigs killed (e.g., proportion of population removed; see Chapter 9). Today, game cameras are probably the most widely used tool for monitoring populations. In their most simple and common application, cameras can be used to provide information on sounder location, size, and composition. However, cameras can also effectively estimate population density and other parameters (Hanson et al. 2008, Keiter et al. 2017). Although complex mathematical models are often used to generate reliable population estimates, simple counts of unique sounders and wild pigs using an array of cameras can be appropriate for many management applications, particularly in areas where individual sounders or wild pigs can easily be distinguished (Holtfreter et al. 2008, Keiter et al. 2017) and areas where sounders are prone to use exclusive space (Sparklin et al. 2009). Holtfreter et al. (2008), following a period of prebaiting, reported detection probabilities >0.5 for both adult and juvenile age classes when using game cameras for only 72 hours; detection probabilities in excess of 0.82 were achieved following 7 days of operation. According to MacKenzie et al. (2002), detection probabilities of 0.3 are adequate for obtaining capture histories and estimating population density. Holtfreter et al. (2008) also indicated that use of only images captured during nocturnal periods was adequate, allowing for substantial reductions in labor associated with reviewing camera images. Davis et al. (2016) focused on utilizing catch-per-unit data to estimate populations of wild pigs, and their data suggested that these models are accurate in moderate to high-density populations where >40% of the population is removed and >3 removal efforts are conducted. More recently, Davis et al. (2018) described how probability of absence of wild pigs could be estimated using data collected with cameras.

The Agricultural Research Service is currently experimenting with unmanned aerial systems (e.g., drones) with thermal imaging capability to locate wildlife in Texas. This technology, at minimum, has considerable potential for locating distinct groups of wild pigs, and may allow for accurate estimates of population density. However, research with other species suggests that replicate surveys should be conducted within several hours to ensure data accuracy, as animal detection rates, and hence density estimates, can vary considerably over time (Storm et al. 2011).

Increasingly, efforts to monitor spread of wild pigs have been initiated in areas where they are absent or have been eradicated. Recognizing that a quick response is more cost effective than allowing a population to become established, some states have implemented wild pig "hotlines" where reports may be called in by the public. The National Feral Swine Mapping System (2012) was developed in the mid-2000s to maintain records of wild pig distribution in North America (Corn and Jordan 2017). This map is regularly updated with information from professional biologists. Similarly, Alberta implemented a bounty system that in part, encourages early removal of wild pigs, but also serves as a reporting and monitoring program. More recently, monitoring of eDNA in water systems has been explored as a means to detect presence of wild pigs at low densities (Williams et al. 2018, see Chapter 9).

Monitoring damage caused by wild pigs can also be an important metric to measure the success of a management program pre- and post-control efforts. On small areas, there are a number of techniques used to document and estimate degree of damage. Field surveys by agricultural producers (however, see Chapter 7 for inherent inaccuracies potentially associated with these surveys), and acreage damaged can be totaled using ground surveys (Engeman et al. 2018) or GIS applications. Use of unmanned aerial systems with cameras can cover large areas and document visible damage for later analysis (Samiappan et al. 2018). Computer monitoring on crop harvest equipment may also provide greater insight into crop damage when the entire crop is not lost (K. C. VerCauteren, USDA/APHIS/WS/National Wildlife Research Center, personal communication).

8.3.9 Meat Markets

Development of markets for wild pig meat exists in Texas, Louisiana, and Florida, and has been discussed in several other states. In Louisiana, meat from live wild pigs delivered to meat processors is used in commercial kitchens and restaurants. In Texas, 2 processing plants coordinate with buyers that operate licensed holding facilities. Wild pigs typically over 36 kg are purchased from private trappers at the holding facility, transferred alive to the processor, and slaughtered under USDA inspection. Meat from these animals may be sold anywhere in the United States or abroad. Another processor in central Texas operates a mobile abattoir with inspection services, but markets fewer pigs because their pigs are typically shot from free-ranging populations. Yet another meat processor buys dead pigs from shooters and utilizes the meat for pet food. The number of pigs removed varies considerably between years, but estimates suggest that the entire market in Texas, Louisiana, and Florida utilizes between 30,000 and 60,000 pigs annually (N. F. Bauer, US Department of Agriculture/Food Safety and Inspection Service, unpublished data).

While removal of pigs from the landscape for food has positive impacts, meat markets generally are not effective at reducing populations of wild pigs. Because processors have specific capacities per day and the market has limits, only large-bodied pigs are removed. Additionally, the existence of a market for wild pig meat provides an incentive to maintain pig populations, and often smaller pigs are released back into the population to be "cropped" later. Landowners trapping pigs for market generally do not support or allow eradication projects that might jeopardize future financial gain. Additionally, while markets will accept pigs of either sex, a premium price is paid for large boars, the removal of which is ineffective for population control. Finally, having a live animal market provides cover for vehicles transporting pigs for illicit purposes.

8.3.10 Bounties

As with other invasive or pest species, bounties are often proposed as a potential solution for controlling populations of wild pigs. Unfortunately, and contrary to popular belief, bounties have been ineffective throughout history at effectively controlling populations of most free-ranging wildlife. Ditchkoff et al. (2017) reported results of a bounty program on wild pigs and found that despite economic incentives of US$25–40/pig for 12 months, population density increased. Total number of wild pigs euthanized during the bounty program did not substantially exceed what was normally harvested each year by hunters prior to implementation of the program, and it was believed that fraud (wild pigs harvested from areas not involved with the bounty program were submitted as part of the program) was prevalent. Ditchkoff et al. (2017) concluded that the goals of bounty participants may not have aligned with goals of the bounty program because the goal of program participants may have been to maximize their economic return rather than reduce or eliminate the population.

8.4 INTEGRATED MANAGEMENT PROGRAMS

The most successful wild pig management programs to date have incorporated a variety of control techniques, as no single method of wild pig control is suitable for all situations (see Chapter 9). While one control technique may be effective at reducing populations in naïve, high-density areas, the same technique may be ineffective at locating and eliminating wild pigs previously exposed to trapping or at lower densities. By utilizing a multi-faceted approach where numerous control techniques are implemented when and where they are most effective and efficient, wild pig management programs can maximize probability of success. Also known as integrated pest management, this broad management strategy was originally developed in the 1960s to manage agricultural insect pests, and since adopted as the preferred management strategy for dealing with wildlife pest species (Sterner 2008).

Two wild pig eradication efforts clearly demonstrate effectiveness of an integrated management approach: Pinnacles National Monument and Santa Cruz Island, both in California (see Chapter 12). However, it should be noted that both of these successful eradication programs occurred on islands or areas enclosed by pig-proof fencing. The control program at Pinnacles National Monument (McCann and Garcelon 2008) utilized several lethal control strategies to eliminate wild pigs inside a 57-km^2 fenced area. The control program began with intensive trapping that reduced the naïve population by approximately 70%. However, pigs that were less willing to enter traps, either due to natural inclination or negative conditioning, were then targeted with a combination of hunting techniques (e.g., ground hunting, hunting with dogs, Judas pigs). Without an integrated approach, it was unlikely that wild pigs would have been removed from Pinnacles National Monument.

The eradication program at Santa Cruz Island (Parkes et al. 2010) utilized a similar integrated approach. Trapping was the initial control technique utilized in each of 5 distinct fenced zones on the island, but aerial gunning, ground hunting, and Judas pigs were used thereafter to eliminate the remaining pigs in each area. Parkes et al. (2010) reported that intensity of each control technique varied depending on relative success, exemplifying an integrated pest management program conducted with an adaptive management approach. As mentioned earlier in this chapter, development of a management plan prior to implementation is an important step in the management process. However, most management plans encounter unexpected challenges along the way, and incorporation of an adaptive approach (regular monitoring of success and progress for the purpose of allowing timely modification of the plan as needed) to wild pig management is another way that probability of success can be maximized.

8.5 PROFESSIONAL VERSUS RECREATIONAL MANAGEMENT

Successful management of wild pigs in North America will ultimately require at least some degree of coordination between professional managers and the public. Whereas government agencies develop and implement control programs for wild pigs on federal and state lands, the majority of lands inhabited by wild pigs are privately owned, where management is at the discretion of the landowner. As a result, management and control programs for wild pigs across privately owned lands represent a range of effort and effectiveness, where some properties are conducting highly effective control programs, others are expending effort but lack the knowledge and/or experience to be successful, and most have no management at all. Ultimately, this patchwork of highly diverse management programs will likely hinder success at reducing populations of wild pigs at county, state, provincial, or regional scales, and at best will only maintain current populations despite significant expenditure of resources. Ultimately, this begs the question regarding the role of the public in the future of wild pig management.

Private landowners undoubtedly have a role to play in the future of wild pigs in North America due to the extent of land that is privately owned, but 3 things need to occur for the public (e.g., private landowners) to contribute substantially to wild pig management in North America. First, the public must understand that wild pigs are an invasive exotic that should be discouraged rather than encouraged, and efforts to inform landowners of the degree to which wild pigs negatively impact society and the environment must occur (see Chapter 10). Second, effectiveness of landowners at controlling or eliminating wild pigs on private lands needs to improve. In some states in southeastern North America, private lands account for 80–90% of the land area, and unless wild pigs are targeted on those lands as effectively as they might be on public lands, wild pig control will never be successful in regions where populations are well established. Finally, wild pig management will ultimately require a cooperative approach. Originally developed in an agricultural context to share equipment, reduce costs, and improve profits, cooperative management of wildlife has been successfully advocated and employed in situations where effective management will benefit from, or requires, a large land area. Probably the most common example of cooperative wildlife management has been with white-tailed deer (Guynn et al. 1983), where multiple adjoining landowners establish collective objectives and work together to achieve a level of management success that would be unattainable without the cooperation of others. In terms of wild pig management, a cooperative effort would likely either involve sharing of trapping equipment and/or labor, with the goal being complete eradication of wild pigs, or the general granting of access to professional managers for the purposes of wild pig control.

There is considerable debate among wildlife professionals concerning benefits of "recreational" wild pig management. Many landowners and outdoor enthusiasts believe, as do some wildlife professionals, that any effort, no matter how insignificant, towards reducing wild pig numbers is positive. However, there is a growing belief that "recreational" management in the form of opportunistic trapping and shooting of wild pigs may actually cause more harm than good. Several studies have demonstrated that wild boar commonly respond to hunting and human pressures by altering their behavioral patterns and space use (Maillard and Fournier 1995; Sodeikat and Pohlmeyer 2002, 2007; Keuling et al. 2008; Scillitani et al. 2010; Thurfjell et al. 2013). Wild pigs are intelligent animals (Frädrich 1974) and rapidly learn to avoid traps or situations that could be harmful. According to one study (M. D. Smith, Auburn University, unpublished data), the willingness of wild pigs to visit trap sites and amount of time spent at those sites tended to decrease following harassment. If recreational management practices are ineffective at altering population density or growth, but serve to increase avoidance of trap sites and potentially decrease trapping success, it is possible that the costs of those efforts outweigh the benefits. This situation highlights the importance of involving the public in wild pig management in a manner in which they can be successful, and the responsibility lies with professional biologists to ensure that the public is effective.

ACKNOWLEDGMENTS

Contributions from M.J.B. were supported by the USDA. Mention of commercial products or companies does not represent an endorsement by the US Government.

REFERENCES

Adams, C. E., B. J. Higginbotham, D. Rollins, R. B. Taylor, R. Skiles, M. Mapston, and S. Turman. 2005. Regional perspectives and opportunities for feral hog management in Texas. *Wildlife Society Bulletin* 33:1312–1320.

Anderson, A., C. Slootmaker, E. Harper, J. Holderieath, and S. A. Shwiff. 2016. Economic estimates of feral swine damage and control in 11 US states. *Crop Protection* 89:89–94.

Anderson, S. J., and C. P. Stone. 1993. Snaring to control feral pigs (*Sus scrofa*) in a remote Hawaiian rain forest. *Biological Conservation* 63:195–201.

Balser, D. S. 1964. Management of predator populations with antifertility agents. *Journal of Wildlife Management* 28:352–358.

Bevins, S. N., K. Pedersen, M. W. Lutman, T. Gidlewski, and T. J. Deliberto. 2014. Consequences associated with the recent range expansion of nonnative feral swine. *BioScience* 64:291–299.

Bodenchuk, M. J. 2014. Method-specific costs of feral swine removal in a large metapopulation: The Texas experience. *Proceedings of the Vertebrate Pest Conference* 26:269–271.

Boitani, L., L. Mattei, D. Nonis, and F. Corsi. 1994. Spatial and activity patterns of wild boars in Tuscany, Italy. *Journal of Mammalogy* 75:600–612.

Bomford, M., and P. O'Brien. 1995. Eradication or control for vertebrate pests? *Wildlife Society Bulletin* 23:249–255.

Burton, J. L., J. D. Westervelt, and S. S. Ditchkoff. 2013. *Simulation of Wild Pig Control via Hunting and Contraceptives*. ERDC/CERL TR-13-21. US Army Engineer Research and Development Center, Construction Engineering Research Laboratory, Champaign, IL.

Caley, P., and B. Ottley. 1995. The effectiveness of hunting dogs for removing feral pigs (*Sus scrofa*). *Wildlife Research* 22:147–154.

Campbell, T. A., S. J. Lapidge, and D. B. Long. 2006. Using baits to deliver pharmaceuticals to feral swine in southern Texas. *Wildlife Society Bulletin* 34:1184–1189.

Campbell, T. A., and D. B. Long. 2008. Mammalian visitation to candidate feral swine attractants. *Journal of Wildlife Management* 72:305–309.

Campbell, T. A., D. B. Long, and B. R. Leland. 2010. Feral swine behavior relative to aerial gunning in southern Texas. *Journal of Wildlife Management* 74:337–341.

Choquenot, D., J. Hone, and G. Saunders. 1999. Using aspects of predator-prey theory to evaluate helicopter shooting for feral pig control. *Wildlife Research* 26:251–261.

Choquenot, D., R. J. Kilgour, and B. S. Lukins. 1993. An evaluation of feral pig trapping. *Wildlife Research* 20:15–22.

Corn, J. L., and T. R. Jordan. 2017. Development of a national feral swine map, 1982–2016. *Wildlife Society Bulletin* 41:758–763.

Cowan, D. P., and G. Massei. 2008. Wildlife contraception, individuals, and populations: How much fertility control is enough? *Proceedings of the Vertebrate Pest Conference* 23:220–228.

Cowled, B. D., P. Elsworth, and S. J. Lapidge. 2008. Additional toxins for feral pig (*Sus scrofa*) control: Identifying and testing Achilles' heels. *Wildlife Research* 35:651–662.

Davis, A. J., M. B. Hooten, R. S. Miller, M. L. Farnsworth, J. Lewis, M. Moxcey, and K. M. Pepin. 2016. Inferring invasive species abundance using removal data from management actions. *Ecological Applications* 26:2339–2346.

Davis, A. J., R. McCreary, J. Psiropoulos, G. Brennan, T. Cox, A. Partin, and K. M. Pepin. 2018. Quantifying site-level usage and certainty of absence for an invasive species through occupancy analysis of camera-trap data. *Biological Invasions* 20:877–890.

Dexter, N. 1996. The effect of an intensive shooting exercise from a helicopter on the behaviour of surviving feral pigs. *Wildlife Research* 23:435–441.

Ditchkoff, S. S., D. B. Jolley, B. D. Sparklin, L. B. Hanson, M. S. Mitchell, and J. B. Grand. 2012. Reproduction in a population of wild pigs (*Sus scrofa*) subjected to lethal control. *Journal of Wildlife Management* 76:1235–1240.

Ditchkoff, S. S., and B. C. West. 2007. Ecology and management of feral hogs. *Human-Wildlife Conflicts* 1:149–151.

Ditchkoff, S. S., R. W. Holtfreter, and B. L. Williams. 2017. Effectiveness of a bounty program for reducing wild pig densities. *Wildlife Society Bulletin* 41:548–555.

Dobie, J. F. 1929. *A Vaquero of the Brush Country*. Grosset and Dunlap, New York.

Doupé, R. G., J. Mitchell, M. J. Knott, A. M. Davis, and A. J. Lymbery. 2010. Efficacy of exclusion fencing to protect ephemeral floodplain lagoon habitats from feral pigs (*Sus scrofa*). *Wetlands Ecology and Management* 18:69–78.

Dzieciolowski, R. M., C. M. H. Clarke, and C. M. Frampton. 1992. Reproductive characteristics of feral pigs in New Zealand. *Acta Theriologica* 37:259–270.

Engeman, R. M., J. Terry, L. R. Stephens, and K. S. Gruver. 2018. Prevalence and amount of feral swine damage to three row crops at planting. *Crop Protection* 112:252–256.

Foster, J. A., J. C. Martin, K. C. VerCauteren, G. E. Phillips, and J. D. Eisemann. 2014. Optimization of formulations for the lethal control of feral pigs. *Proceedings of the Vertebrate Pest Conference* 26:277–280.

Frädrich, H. 1974. A comparison of behaviour in the Suidae. Pages 133–143 *in* V. Geist and F. R. Walther, editors. *The Behavior of Ungulates and Its Relation to Management*. International Union for the Conservation of Nature and Natural Resources, Morges, Switzerland.

Gabor, T. M., E. C. Hellgren, R. A. Van Den Bussche, and N. J. Silvy. 1999. Demography, sociospatial behaviour and genetics of feral pigs (*Sus scrofa*) in a semi-arid environment. *Journal of Zoology, London* 247:311–322.

Gaskamp, J. A. 2012. Use of drop-nets for wild pig damage and disease abatement. Thesis, Texas A&M University, College Station, TX.

Gaskamp, J. A. 2015. Long-awaited BoarBuster™ deliveries begin this month. *Ag News and Views* 33:3.

Gaskamp, J. A., and K. L. Gee. 2011. Using drop-nets to capture feral hogs. *Ag News and Views* 29:6.

Goulding, M. J., G. Smith, and S. J. Baker. 1998. *Current Status and Potential Impact of Wild Boar (Sus scrofa) in the English Countryside: A Risk Assessment*. Central Science Laboratory, Ministry of Agriculture, Fisheries and Food, London, UK.

Greer, K. R., W. W. Hawkins, Jr., and J. E. Catlin. 1968. Experimental studies of controlled reproduction in elk (wapiti). *Journal of Wildlife Management* 32:368–376.

Guynn, D. C., Jr., S. P. Mott, W. D. Cotton, and H. A. Jacobson. 1983. Cooperative management of white-tailed deer on private lands in Mississippi. *Wildlife Society Bulletin* 11:211–214.

Hamrick, B., M. D. Smith, C. Jaworowski, and B. Strickland. 2011. *A Landowner's Guide for Wild Pig Management: Practical Methods for Wild Pig Control*. Publication 2659. Mississippi State University Extension Service, Mississippi State University, Mississippi State, MS.

Hansen, L. 2011. Extensive management. Pages 409–451 *in* D. G. Hewitt, editor. *Biology and Management of White-Tailed Deer*. CRC Press, Boca Raton, FL.

Hanson, L. B., J. B. Grand, M. S. Mitchell, D. B. Jolley, B. D. Sparklin, and S. S. Ditchkoff. 2008. Change-in-ratio density estimator for feral pigs is less biased than closed mark-recapture estimates. *Wildlife Research* 35:695–699.

Hess, S. C., J. J. Jeffrey, D. L. Ball, and L. Babich. 2006. Efficacy of feral pig removals at Hakalau Forest National Wildlife Refuge, Hawaii. *Transactions of the Western Section of the Wildlife Society* 42:53–67.

Higginbotham, B. 2014. The art and (some) science of trapping wild pigs: From traps to gates to triggers and more! *Proceedings of the Vertebrate Pest Conference* 26:258–268.

Holtfreter, R. W., B. L. Williams, S. S. Ditchkoff, and J. B. Grand. 2008. Feral pig detectability with game cameras. *Proceedings of the Annual Conference of the Southeastern Association of Fish and Wildlife Agencies* 62:17–21.

Hone, J. 1990. Predator-prey theory and feral pig control, with emphasis on evaluation of shooting from a helicopter. *Australian Wildlife Research* 17:123–130.

Hone, J., and B. Atkinson. 1983. Evaluation of fencing to control feral pig movement. *Australian Wildlife Research* 10:499–505.

Ilse, L. M., and E. C. Hellgren. 1995. Resource partitioning in sympatric populations of collared peccaries and feral hogs in southern Texas. *Journal of Mammalogy* 76:784–799.

Jacobson, H. A., C. A. DeYoung, R. W. DeYoung, T. E. Fulbright, and D. G. Hewitt. 2011. Management on private property. Pages 453–479 *in* D. G. Hewitt, editor. *Biology and Management of White-Tailed Deer*. CRC Press, Boca Raton, FL.

Jedrzejewski, W., and J. F. Kamler. 2004. Modified drop-net for capturing ungulates. *Wildlife Society Bulletin* 32:1305–1308.

Keiter, D. A., A. J. Davis, O. E. Rhodes, Jr., F. L. Cunningham, J. C. Kilgo, K. M. Pepin, and J. C. Beasley. 2017. Effects of scale of movement, detection probability, and true population density on common methods of estimating population density. *Scientific Reports* 7:9446.

Keuling, O., N. Stier, and M. Roth. 2008. How does hunting influence activity and spatial usage in wild boar *Sus scrofa* L.? *European Journal of Wildlife Research* 54:729–737.

Killian, G., K. Fagerstone, T. Kreeger, L. Miller, and J. Rhyan. 2007. Management strategies for addressing wildlife disease transmission: Case for fertility control. *Proceedings of the Wildlife Damage Management Conference* 12:265–271.

Killian, G., L. Miller, J. Rhyan, T. Dees, D. Perry, and H. Doten. 2003. Evaluation of GnRH contraceptive vaccine in captive feral swine in Florida. *Proceedings of the Wildlife Damage Management Conference* 10:128–133.

Killian, G., L. Miller, J. Rhyan, and H. Doten. 2006a. Immunocontraception of Florida feral swine with a single-dose GnRH vaccine. *American Journal of Reproductive Immunology* 55:378–384.

Killian, G. J., N. K. Diehl, L. A. Miller, J. C. Rhyan, and D. Thain. 2006b. Long-term efficacy of three contraceptive approaches for population control of wild horses. *Proceedings of the Vertebrate Pest Conference* 22:67–71.

Killian, G. J., L. A. Miller, N. K. Diehl, J. C. Rhyan, and D. Thain. 2004. Evaluation of three contraceptive approaches for population control of wild horses. *Vertebrate Pest Conference* 21:263–268.

Lapidge, S., J. Wishart, L. Staples, K. Fagerstone, T. Campbell, and J. Eisemann. 2012. Development of a feral swine toxic bait (Hog-Gone®) and bait hopper (Hog-Hopper™) in Australia and the USA. *Proceedings of the Wildlife Damage Management Conference* 14:19–24.

Lavelle, M. J., N. P. Snow, J. W. Fischer, J. M. Halseth, E. VanNatta, and K. C. VerCauteren. 2017. Attractants for wild pigs: Current use, availability, needs and future potential. *European Journal of Wildlife Research* 63:86.

Leopold, A. 1933. *Game Management.* University of Wisconsin Press, Madison, WI.

Levy, B., C. Collins, S. Lenhart, M. Madden, J. Corn, R. A. Salinas, and W. Stiver. 2016. A metapopulation model for feral hogs in Great Smoky Mountains National Park. *Natural Resource Modeling* 29:71–97.

Lewis, J. S., M. L. Farnsworth, C. L. Burdett, D. M. Theobald, M. Gray, and R. S. Miller. 2017. Biotic and abiotic factors predicting the global distribution and population density of an invasive large mammal. *Scientifics Reports* 7:44152.

Linhart, S. B., and R. K. Enders. 1964. Some effects of diethylstilbestrol on reproduction in captive red foxes. *Journal of Wildlife Management* 28:358–363.

Littauer, G. A. 1993. Control techniques for feral hogs. Pages 139–148 *in* L. W. Hanselka and J. F. Cadenhead, editors. *Feral Swine: A Compendium for Resource Managers.* Texas Agricultural Extension Service, College Station, TX.

Long, D. B., and T. A. Campbell. 2012. Box traps for feral swine capture: A comparison of gate styles in Texas. *Wildlife Society Bulletin* 36:741–746.

MacKenzie, D. I., J. D. Nichols, G. B. Lachman, S. Droege, J. A. Royle, and C. A. Langtimm. 2002. Estimating site occupancy rates when detection probabilities are less than one. *Ecology* 83:2248–2255.

Maillard, D., and P. Fournier. 1995. Effects of shooting with hounds on size of resting range of wild boar (*Sus scrofa* L.) groups in Mediterranean habitat. *IBEX Journal of Mountain Ecology* 3:102–107.

Mapston, M. E. 1999. Feral hog control methods. Pages 117–120 *in Proceedings of the First National Feral Swine Conference*, 2–3 June 1999, Ft. Worth, TX.

Mapston, M. E. 2010. *Feral Hogs in Texas.* Texas Cooperative Extension Publication B-6149, College Station, TX.

Massei, G., D. P. Cowan, J. Coats, F. Gladwell, J. E. Lane, and L. A. Miller. 2008. Effect of GnRH vaccine GonaCon™ on the fertility, physiology and behaviour of wild boar. *Wildlife Research* 35:540–547.

Mayer, J. J. 2014. *Estimation of the Number of Wild Pigs Found in the United States.* SRNS–STI–2014–00292, Savannah River Nuclear Solutions, LLC, Savannah River Site, Aiken, SC.

Mayer, J. J., and J. C. Beasley. 2018. Wild pigs. Pages 221–250 *in* W. C. Pitt, J. C. Beasley, and G. W. Witmer, editors. *Ecology and Management of Terrestrial Vertebrate Invasive Species in the United States.* CRC Press, Boca Raton, FL.

Mayer, J. J., R. E. Hamilton, and I. L. Brisbin, Jr. 2009. Use of trained hunting dogs to harvest or control wild pigs. Pages 275–288 *in* J. J. Mayer and I. L. Brisbin, Jr., editors. *Wild Pigs: Biology, Damage, Control Techniques and Management.* SRNL-RP-2009-00869. Savannah River National Laboratory, Aiken, SC.

McCann, B. E., and D. K. Garcelon. 2008. Eradication of feral pigs from Pinnacles National Monument. *Journal of Wildlife Management* 72:1287–1295.

McIlroy, J. C., M. Braysher, and G. R. Saunders. 1989. The effectiveness of a warfarin poisoning campaign against feral pigs, *Sus scrofa*, in Namadgi National Park, A.C.T. *Australian Wildlife Research* 16:195–202.

McIlroy, J. C., and E. J. Gifford. 1997. The 'Judas' pig technique: A method that could enhance control programmes against feral pigs, *Sus scrofa*. *Wildlife Research* 24:483–491.
McIlroy, J. C., and E. J. Gifford. 2005. Are oestrous feral pigs, *Sus scrofa*, useful as trapping lures? *Wildlife Research* 32:605–608.
McIlroy, J. C., E. J. Gifford, and R. I. Forrester. 1993. Seasonal patterns in bait consumption by feral pigs (*Sus scrofa*) in the hill country of South-eastern Australia. *Wildlife Research* 20:637–651.
Metcalf, E. M., I. D. Parker, R. R. Lopez, B. Higginbotham, D. S. Davis, and J. R. Gersbach. 2014. Impact of gate width of corral traps in potential wild pig trapping success. *Wildlife Society Bulletin* 38:892–895.
Miller, L., J. Rhyan, and G. Killian. 2003. Evaluation of GnRH contraceptive vaccine using domestic swine as a model for feral hogs. *Proceedings of the Wildlife Damage Management Conference* 10:120–127.
Miller, L. A., J. P. Gionfriddo, K. A. Fagerstone, J. C. Rhyan, and G. J. Killian. 2008. The single-shot GnRH immunocontraceptive vaccine (GonaCon™) in white-tailed deer: Comparison of several GnRH preparations. *American Journal of Reproductive Immunology* 60:214–223.
Miller, L. A., B. E. Johns, and G. J. Killian. 2000. Immunocontraception of white-tailed deer with GnRH vaccine. *American Journal of Reproductive Immunology* 44:266–274.
National Feral Swine Mapping System. 2012. NFSMS home page. <https://www.aphis.usda.gov/aphis/ourfocus/wildlifedamage/operational-activities/feraswine/sa-fs-history>. Accessed 21 Sep 2018.
North American Invasive Species Network. 2015. *The Ten Most Important Invasive Species or Invasive Species Assemblages in North America in 2015*. North American Invasive Species Network, Gainesville, FL.
O'Brien, P. H., R. E. Kleba, J. A. Beck, and P. J. Baker. 1986. Vomiting by feral pigs after 1080 intoxication: Nontarget hazard and influence of anti-emetics. *Wildlife Society Bulletin* 14:425–432.
O'Brien, P. H., B. S. Lukins, and J. A. Beck. 1988. Bait type influences the toxicity of sodium monofluoroacetate (Compound 1080) to feral pigs. *Australian Wildlife Research* 15:451–457.
Parkes, J. P., D. S. L. Ramsey, N. MacDonald, K. Walker, S. McKnight, B. S. Cohen, and S. A. Morrison. 2010. Rapid eradication of feral pigs (*Sus scrofa*) from Santa Cruz Island, California. *Biological Conservation* 143:634–641.
Pech, R. P., J. C. McIlroy, M. F. Clough, and D. G. Green. 1992. A microcomputer model for predicting the spread and control of foot and mouth disease in feral pigs. *Proceedings of the Vertebrate Pest Conference* 15:360–364.
Peine, J. D., and J. A. Farmer. 1990. Wild hog management program at Great Smoky Mountains National Park. *Proceedings of the Vertebrate Pest Conference* 14:221–227.
Pimentel, D. 2007. Environmental and economic costs of vertebrate species invasions into the United States. Pages 2-8 *in* G. W. Witmer, W. C. Pitt, and K. A. Fagerstone, editors. *Managing Vertebrate Invasive Species: Proceedings of an International Symposium*. US Department of Agriculture/Animal and Plant Health Inspection Service/Wildlife Services/National Wildlife Research Center, Ft. Collins, CO.
Quy, R. J., G. Massei, M. S. Lambert, J. Coats, L. A. Miller, and D. P. Cowan. 2014. Effects of a GnRH vaccine on the movement and activity of free-living wild boar (*Sus scrofa*). *Wildlife Research* 41:185–193.
Rattan, J. M., B. J. Higginbotham, D. B. Long, and T. A. Campbell. 2010. Exclusion fencing for feral hogs at white-tailed deer feeders. *The Texas Journal of Agriculture and Natural Resource* 23:83–89.
Reed, R. N., J. D. Wilson, G. H. Rodda, and M. E. Dorcas. 2012. Ecological correlates of invasion impact for Burmese pythons in Florida. *Integrative Zoology* 7:254–270.
Reidy, M. M, T. A. Campbell, and D. G. Hewitt. 2008. Evaluation of electric fencing to inhibit feral pig movements. *Journal of Wildlife Management* 72:1012–1018.
Saalfeld, S. T., W. C. Conway, R. Maxey, C. Gregory, and B. Ortego. 2009. Recovery of nesting bald eagles in Texas. *Southeastern Naturalist* 8:83–92.
Salinas, R. A., W. H. Stiver, J. L. Corn, S. Lenhart, C. Collins, M. Madden, K. C. VerCauteren, et al. 2015. An individual-based model for feral hogs in Great Smoky Mountains National Park. *Natural Resource Modeling* 28:18–36.
Samiappan, S., J. M. P. Czarnecki, H. Foster, B. K. Strickland, J. L. Tegt, and R. J. Moorhead. 2018. Quantifying damage from wild pigs with small unmanned aerial systems. *Wildlife Society Bulletin* 42:304–309.
Samoylova, T. I., T. D. Braden, J. A. Spencer, and F. F. Bartol. 2017. Immunocontraception: Filamentous bacteriophage as a platform for vaccine development. *Current Medicinal Chemistry* 24:3907–3920.
Samoylova, T. I., A. M. Cochran, A. M. Samoylov, B. Schemera, A. H. Breiteneicher, S. S. Ditchkoff, V. A. Petrenko, and N. R. Cox. 2012. Phage display allows identification of zona pellucida-binding peptides with species-specific properties: Novel approach for development of contraceptive vaccines for wildlife. *Journal of Biotechnology* 162:311–318.

Samoylova, T. I., N. R. Cox, A. M. Cochran, A. M. Samoylov, B. Griffin, and H. J. Baker. 2010. ZP-binding peptides identified via phage display stimulate production of sperm antibodies in dogs. *Animal Reproduction Science* 120:151–157.

Saunders, G. 1993. Observations on the effectiveness of shooting feral pigs from helicopters. *Wildlife Research* 20:771–776.

Saunders, G., B. Kay, and B. Parker. 1990. Evaluation of a warfarin poisoning programme for feral pigs (*Sus scrofa*). *Australian Wildlife Research* 17:525–533.

Scillitani, L., A. Monaco, and S. Toso. 2010. Do intensive drive hunts affect wild boar (*Sus scrofa*) spatial behaviour in Italy? Some evidences and management implications. *European Journal of Wildlife Research* 56:307–318.

Smith, T. N., M. D. Smith, D. K. Johnson, and S. S. Ditchkoff. 2014. Evaluation of continuous-catch doors for trapping wild pigs. *Wildlife Society Bulletin* 38:175–181.

Snow, N. P., J. A. Foster, J. C. Kinsey, S. T. Humphrys, L. D. Staples, D. G. Hewitt, and K. C. VerCauteren. 2017a. Development of toxic bait to control invasive wild pigs and reduce damage. *Wildlife Society Bulletin* 40:256–263.

Snow, N. P., M. J. Lavelle, J. M. Halseth, C. R. Blass, J. A. Foster, and K. C. VerCauteren. 2017b. Development of a species-specific bait station for delivering toxic bait to invasive wild pigs. *Wildlife Society Bulletin* 41:264–270.

Snow, N. P., M. J. Lavelle, J. M. Halseth, M. P. Glow, E. H. VanNatta, A. J. Davis, K. M. Pepin, et al. 2019. Exposure of a population of invasive wild pigs to simulated toxic bait containing biomarker: Implications for population reduction. *Pest Management Science* 75:1140–1149.

Sodeikat, G., and K. Pohlmeyer. 2002. Temporary home range modifications of wild boar family groups (*Sus scrofa* L.) caused by drive hunts in Lower Saxony (Germany). *Zeitschrift für Jagdwissenschaft* 48:161–166.

Sodeikat, G., and K. Pohlmeyer. 2007. Impact of drive hunts on daytime resting site areas of wild boar family groups (*Sus scrofa* L.). *Wildlife Biology in Practice* 3:28–38.

Sparklin, B. D., M. S. Mitchell, L. B. Hanson, D. B. Jolley, and S. S. Ditchkoff. 2009. Territoriality of feral pigs in a highly persecuted population on Fort Benning, Georgia. *Journal of Wildlife Management* 73:497–502.

Sterner, R. 2008. The IPM paradigm: Vertebrates, economics and uncertainty. *Proceedings of the Vertebrate Pest Conference* 23:194–200.

Stiver, W. H., and E. K. Delozier. 2009. Great Smoky Mountains National Park wild hog control program. Pages 341–352 *in* J. J. Mayer and I. L. Brisbin, Jr., editors. *Wild Pigs: Biology, Damage, Control Techniques and Management.* SRNL-RP-2009-00869. Savannah River National Laboratory, Aiken, SC.

Storm, D. J., M. D. Samuel, T. R. Van Deelen, K. D. Malcolm, R. E. Rolley, N. A. Frost, D. P. Bates, and B. J. Richards. 2011. Comparison of visual-based helicopter and fixed-wing forward-looking infrared surveys for counting white-tailed deer *Odocoileus virginianus*. *Wildlife Biology* 17:431–440.

Thurfjell, H., G. Spong, and G. Ericsson. 2013. Effects of hunting on wild boar *Sus scrofa* behaviour. *Wildlife Biology* 19:87–93.

Twigg, L. E., T. Lowe, M. Everett, and G. Martin. 2006. Feral pigs in north-western Australia: Population recovery after 1080 baiting and further control. *Wildlife Research* 33:417–425.

Twigg, L. E., T. Lowe, and G. Martin. 2007. Bait consumption by, and 1080–based control of, feral pigs in the Mediterranean climatic region of south-western Australia. *Wildlife Research* 34:125–139.

West, B. C., A. L. Cooper, and J. B. Armstrong. 2009. Managing wild pigs: A technical guide. *Human-Wildlife Interactions* Monograph 1:1–55.

Wilcox, J. T., E. T. Aschehoug, C. A. Scott, and D. H. Van Vuren. 2004. A test of the Judas technique as a method for eradicating feral pigs. *Transactions of the Western Section of the Wildlife Society* 40:120–126.

Williams, B. L., R. W. Holtfreter, S. S. Ditchkoff, and J. B. Grand. 2011a. Trap style influences wild pig behavior and trapping success. *Journal of Wildlife Management* 75:432–436.

Williams, B. L., R. W. Holtfreter, S. S. Ditchkoff, and J. B. Grand. 2011b. Efficiency of time-lapse intervals and simple baits for camera surveys of wild pigs. *Journal of Wildlife Management* 75:655–659.

Williams, K. E., K. P. Huyvaert, K. C. VerCauteren, A. J. Davis, and A. J. Piaggio. 2018. Detection and persistence of environmental DNA from an invasive, terrestrial mammal. *Ecology and Evolution* 8:688–695.

Wilson, C. J. 2005. *Feral Wild Boar in England: Status, Impact, and Management.* DEFRA, RDS National Wildlife Management Team, Exeter, UK.

9 Research Methods for Wild Pigs

James C. Beasley, Michael J. Lavelle, David A. Keiter, Kim M. Pepin, Antoinette J. Piaggio, John C. Kilgo, and Kurt C. VerCauteren

CONTENTS

9.1 INTRODUCTION

Many aspects of *Sus scrofa* biology are well described as pigs have been domesticated for nearly 10,000 years (see Chapter 2), serving as food for humans and model organisms routinely used as human surrogates in medical and forensic sciences (Swanson et al. 2004, Dekeirsschieter et al. 2009). Aspects of domestic pig biology that are well understood include their physiology, aging, genetics, and reproduction. In particular, the entire *Sus scrofa* genome was sequenced in 2012, resulting in substantial understanding of the genetic architecture and traits in this species. Despite our knowledge of domestic pigs, basic biological and ecological attributes of wild *Sus scrofa* remain understudied, particularly in their invasive ranges. Such deficiencies stem from a multitude of factors, historically reflecting a lack of interest and funding for basic ecological studies, with almost no research published on wild pigs prior to the 1950's (Figure 9.1). However, a dramatic increase in ecological studies on *Sus scrofa* has occurred over the last few decades, within both its native and invasive ranges (Figure 9.1). In North America, this increase reflects growing awareness of economic and ecological impacts that wild pigs have on both native and anthropogenic ecosystems (e.g., agriculture).

Although there is growing interest in research on wild pigs, a number of challenges exist to studying basic aspects of their ecology and management due to their social structure, physiology, and adaptability. For example, wild pigs exhibit a complex social structure, yet observational study methods commonly used to elucidate behavior and social dynamics of other social species like primates and elephants are often not feasible with wild pigs because of their secretive behavior. Similarly, while advancements in telemetry technology have revolutionized our knowledge of spatial ecology for many species, issues with rapid weight gain and body structure in wild pigs have limited the application of these technologies. However, researchers have addressed some of these issues and telemetry studies are becoming more common (e.g., Pepin et al. 2016, Kay et al. 2017). Likewise, despite its importance for modeling pig population dynamics, no studies have successfully quantified known-fate survival of neonate wild pigs, primarily because tools to conduct such assessments are limited (Keiter et al. 2017a).

Adding to challenges associated with studying wild pigs, as a heavily persecuted and in some instances regulated invasive species, research goals may be at odds with management objectives. For example, releasing captured wild pigs for ecological study may be politically unacceptable in some cases. Furthermore, there are now few landscapes where wild pigs are unmanaged, limiting

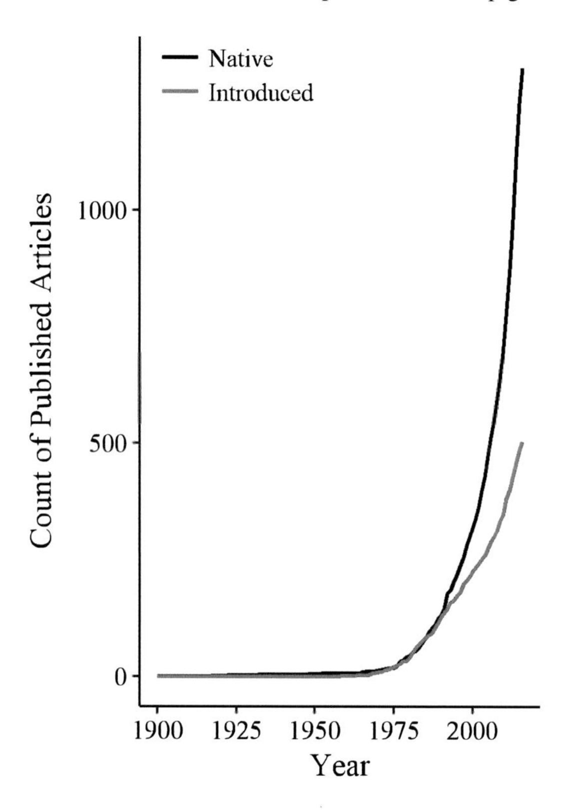

FIGURE 9.1 Counts of published articles on wild pigs within their introduced and native ranges from 1910 to November 2016. (Adapted from Web of Science; Search terms: feral pigs, feral swine, *Sus scrofa*, wild boar, wild hogs, wild pigs.)

inferences about complex aspects of pig ecology (e.g., social structure) in the absence of anthropogenic activity and how management may influence pig behavior and population dynamics. These types of data are essential to development of effective strategies for controlling this important invasive species. Wild pigs also are secretive and intelligent, and as a result can be difficult to study in areas with abundant cover and heavy management pressure.

In this chapter we highlight the current state of research tools, emphasizing advancements in technologies and methodologies used to study wild pig ecology, management, and damage assessment. While much of the chapter focuses on research in North America (invasive range), these technologies and methods are applied globally to study wild pigs. As such, we have incorporated representative literature from throughout the native and invasive ranges of this species. Topics vary from highly applied field methods such as trapping, handling, and monitoring techniques, to lab-based assessments of molecular ecology, disease, and diet, highlighting analysis techniques of these types of data where appropriate. We also include methodology associated with pig population control and damage assessment. We conclude by summarizing the current state of knowledge of wild pig ecology, pointing to future research tools and methods, and emphasizing data gaps to guide future research.

9.2 CAPTURE AND HANDLING

Capturing wild pigs for research is considerably different from capture for population control. Capture techniques vary widely and should be tailored to intended research or management outcomes. For example, if capture is for collection of biological samples for disease surveillance across all sexes and age classes, aerial gunning (if lethal sampling is acceptable) or multiple animal capture with drop nets or corral traps may be the most appropriate strategies. Conversely, if specific genders or age classes are targeted, remotely monitored and activated traps may be most valuable. Trapping is the most widely used method for live capturing wild pigs. A variety of effective traps exist, with semi-permanent corral traps and portable rigid-wire box traps being the most common (e.g., Keiter and Beasley 2017; Figure 9.2). Although corral traps require more time to construct than placement of a box trap, the benefit is that quantity of pigs captured per unit effort is typically higher. In the United States and elsewhere, commercially produced traps and gates may be available locally at farm-supply stores, can be found through Internet searches, or can be hand built (Long and Campbell 2012). In general, trap construction is relatively consistent, yet there are an array of gate designs, trigger types, and trap configurations (Long and Campbell 2012, Metcalf et al. 2014).

9.2.1 Trap Gates

There are 2 basic types of gates most commonly used for trapping wild pigs, single-catch gates that remain closed once triggered (e.g., guillotine) and continuous-catch gates (e.g., rooter, saloon) that are intended to allow additional captures once the gate is shut (Figure 9.3). Although continuous-catch doors have potential to increase capture success, there is also increased potential for escape due to intentional lack of resistance to reopen gates. Some research has found continuous-catch gates to be ineffective at increasing number of pigs captured (Smith et al. 2014). Additionally, Long and Campbell (2012) reported that capture rates of adults did not vary between box traps with side-swing or rooter gates (Figure 9.3), but rooter gates caught more juveniles. In contrast, guillotine-style gates (Figure 9.3) provide unobstructed entry into a trap and, once triggered, escape is less common. Many trappers recommend using 2 synchronized guillotine gates on opposite sides of a corral trap to decrease reluctance of wary pigs to enter traps.

Gate designs are shifting away from narrow (0.9–1.2-m wide) side-hinged styles to larger (e.g., 2.4-m-wide) overhead guillotine-style gates, although there is limited evidence to suggest a marked increase in trapping efficacy (Metcalf et al. 2014; Figure 9.3). In some circumstances, a variety of overhead capture devices including drop corral traps, drop nets, and net cannons or projectors that

FIGURE 9.2 Proven devices for the capture of wild pigs. (a) Corral traps are commonly used, while (b) air cannons with netting and (c) drop nets can also be effective. (d) Recently developed drop traps can be very effective by suspending the entire trap overhead, thus minimizing impediments to entering traps. Corral traps are typically triggered by pigs when a trip wire is contacted or by remote activation and monitoring through cameras and cell phones (e.g., (a) and (d) have become increasingly popular). Remote-activated systems are highly effective (pending remote signal coverage and strength), and can help minimize non-target captures. Drop nets and air cannon nets do not require animals to enter an enclosure and thus may improve efficacy when used on trap-wary animals and for capturing specific individuals. (Photos (a–c) by the USDA/APHIS/WS/National Wildlife Research Center and (d) by J. Beasley. With permission.)

FIGURE 9.3 Trap gate styles commonly used for capturing wild pigs. (a, f) Single-catch, guillotine-style gates rely on gravity to keep the gate closed, although some traps include mechanisms (like stops) that prohibit raising the gate without first releasing the mechanism. (b–e) Continuous-catch gates allow additional pigs to enter the trap after the gate has been triggered and closed. Gate (f) is an example of an increasingly popular design that incorporates remote monitoring and activation by cell phone and can be triggered as desired when an operator identifies target animals in the trap. (Photos (a) by the National Feral Swine Damage Management Program and (b–f) by the USDA/APHIS/WS/National Wildlife Research Center. With permission.)

fire nets over wild pigs are useful alternatives (Figure 9.2), especially for those pigs reluctant to enter more traditional cage traps (Gaskamp 2012, Bauman 2015).

Gates can be adapted for any trapping scenario (Figure 9.3). For example, remote, user-activated gate options are commercially available and include systems that incorporate cameras that enable remote identification of animals entering traps and trap activation via smart phone. These designs can be particularly useful when trying to capture specific individuals or for maximizing the proportion of a group captured. Such technology has revolutionized trapping by increasing efficiency and minimizing wasted effort associated with non-target captures. Current drawbacks are recurring costs (e.g., cellular data plans) and reliance on cellular or satellite coverage, although most of these traps can be adapted to be triggered manually via trip lines or root sticks if necessary.

9.2.2 Trap Triggers

Trap triggers typically include a gate retention mechanism that is released when a trip line is pulled. The trip line generally runs horizontally near the back of the trap, at a height that target animals will need to push through, that is linked to the gate retention mechanism. Bait is deposited behind and below the trip line so that animals will inadvertently apply pressure to it while feeding and release the gate. Another frequently used trigger mechanism is a "root stick," which is essentially a stick acting as a temporary anchor for the trip line (Figure 9.4). The stick is secured horizontally to the ground by 2 more set sticks inserted into the ground at an angle away from the direction of pull. Bait is then dispersed around the entire root stick setup; the root stick then becomes dislodged by

FIGURE 9.4 Manual trigger mechanisms for activating gates on traps for capturing wild pigs including: (a) sailing shackle and steel ring, (b) root stick, and (c) locking pliers and steel ring. Triggers (a) and (c) are activated when a wild pig applies force to a trip wire attached to the mechanism causing the ring attached to the gate to be released. Trigger (b) is activated as a wild pig consumes bait around the mechanism and dislodges any of the sticks that are acting as an anchor and holding the gate open. (Photos by the USDA/APHIS/WS/National Wildlife Research Center. With permission.)

pigs during feeding, releasing the gate. Sailing snap shackles and spring-loaded locking pliers are ideal triggering mechanisms to secure cables holding the gate open until the trip line is activated by pigs (Figure 9.4).

9.2.3 Bait

Bait used for trapping typically includes dry or fermented whole kernel corn or wheat, which performs reliably, is relatively inexpensive, and is widely accessible. Numerous alternative baits and attractants have been evaluated for attracting wild pigs (e.g., Wathen et al. 1988, Campbell and Long 2008, Williams et al. 2011), although assessment of the best bait types for any specific location or season is highly recommended. Be sure to check state game laws on baiting, as some states restrict the types, amounts, and locations of baits that can be used without special permission. Alternative baits including meat, berry flavoring, and livestock feed additives have been successful, although results are highly variable (Wathen et al. 1988, Campbell and Long 2009a). Research has yet to identify any baits specific to wild pigs (Lavelle et al. 2017, Beasley et al. 2018). Availability of natural foods during capture efforts undoubtedly influences motivation of target animals to respond to bait, and thus impacts trapping productivity (Williams et al. 2011, Long and Campbell 2012).

9.2.4 Other Capture Methods

Capture using dogs trained to locate, pursue, catch, and restrain wild pigs is a possibility given the right circumstances, including widespread landowner permission. Use of dogs is often more applicable to lethal control of wild pigs when indiscriminate captures are acceptable (Bauman 2015). Further, the outcome can be unpredictable and physically demanding on all parties involved, and efforts may prove counterproductive to other ongoing capture efforts from displacement of target pigs (Bauman 2015).

Aerial gunning from helicopters or fixed-wing aircraft can be highly productive relative to other capture methods for acquiring a large number of samples from dead pigs in a short period of time (Saunders 1993, Campbell et al. 2010). To relocate animals after aerial gunning, paper streamers and global positioning system (GPS) waypoints can be helpful for directing ground crews. Shooting animals over bait can also be effective for collecting specific individuals. Strategically placed tree stands, considering prevailing winds, and night vision or thermal optics will improve efficiency of this technique.

The aforementioned methods lend themselves to research requiring capture and release or lethal sampling. Selection of capture method(s) should include thorough consideration of available resources including labor, traps, bait, and funds, as wild pig capture typically requires significant investments. Further, capture objectives such as number of animals, targeted age class, sex, size, and spatial distribution help determine which capture strategies will be most productive and efficient to implement (Mayer 2009).

9.2.5 Handling

An important consideration when capturing wild pigs for research that will be released is ensuring they are released uninjured. A drawback of all previously mentioned capture tools (except nets) is that they are constructed of seemingly permeable, though rigid wire that has the potential to cause injury (Barasona et al. 2013, Casas-Díaz et al. 2015). To reduce damage-inflicting stimuli from outside the trap and minimize stress of captured animals, traps should be approached quietly and calmly, or shade cloth can be used to enshroud an entire trap after animals are captured (Sweitzer et al. 1997, Fowler 2011). As the shade cloth is put in place, an instant calming effect is observed with a reduction in attempted jumps and impacts with corral panels (Lavelle et al. 2019). Further, not only do pigs calm down, but they also stand still, improving the chances for a successfully placed immobilization dart or injection.

Chemical immobilization strategies for wild pigs are straightforward, although considerable variability in individual responses to drugs is common (Calle and Morris 1999). Delivery of anesthetics is typically accomplished by injection with darts or pole syringes. When anesthetizing pigs at close range (<2 m) in box traps or in nets, blow guns, or pole syringes are ideal for delivery, whereas dart guns are ideal for larger traps. Once injected, induction rates depend on drug choice, injection site, level of excitement of animals, and completeness of injection (Calle and Morris 1999). Due to the presence of fat deposits, especially on the dorsal aspect of the rump, a long needle (>3.0 cm) for intramuscular injections is required (Fowler 2011), but needle length should be adjusted for pig size. To penetrate tough outer hide and layers of fat, needles should be sturdy (e.g., 14 or 16 gauge). A combination of Telazol® (4.4 mg/kg) and xylazine (2.2 mg/kg) is commonly used to anesthetize pigs (Fenati et al. 2008, Heinonen et al. 2009). However, variability in responses among individuals has been observed with some animals exhibiting compromised thermoregulation or prolonged immobilization (Calle and Morris 1999). These variations potentially stem from variability in the quality of injections (e.g., injection location, deep fat deposits, environmental conditions affecting metabolism). This unpredictability has led to exploration of novel combinations such as ketamine/medetomidine, midazolam/detomadine/butorphanol, medetomidine/midazolam/butorphanol, butorphanol/azaperone/medetomidine, and nalbuphine/medetomidine/azaperone (Heinonen et al. 2009, Kreeger and Arnemo 2012, Ellis et al. 2019).

Although challenging, it is possible to anesthetize wild pigs with dart projectors from a blind. If darting is used for capture, we recommend use of telemetry-equipped transmitter darts to improve the likelihood of locating anesthetized animals (Walter et al. 2005, Thurfjell et al. 2009). A pig will likely travel 200–300 m into thick underbrush after darted, making recovery difficult even with transmitter darts (Thurfjell et al. 2009). One particular benefit of darting is targeting of specific individuals (e.g., for replacing batteries in collars or taking biological samples from an individual over time), although researchers should be aware darting of specific pigs can require considerable effort and is not always successful. Aerial net gunning has also been used to capture wild pigs (R. K. Brook, College of Agriculture and Bioresources, University of Saskatchewan, and D. E. Etter, Michigan Department of Natural Resources, unpublished data).

Physical handling and restraint of captured wild pigs also is possible. If traps are located near roads and are constructed with a gate or loading port to insert a chute, it is possible to move pigs from traps to a handling device or trailer for processing (Figure 9.5). Commercially available squeeze chutes designed for small livestock can be useful in restraining wild pigs (Fowler 2011). Elaborate custom-fabricated mobile handling trailers are also a possibility for extensive handling (Lavelle et al. In Review; Figure 9.5). As wild pigs are powerful and aggressive, utmost care in handling is needed.

9.3 MONITORING TECHNIQUES

Development of techniques to monitor wild pig populations and effects on the environment has recently increased. Knowledge of population density and demographic rates, such as survival and reproduction, as well as changes in these rates, is critical to understanding wild pig ecology and improving management programs that minimize impacts on natural and anthropogenic ecosystems. Here we discuss the development of methods to collect and assess these demographic data.

9.3.1 Marking and Tracking Techniques

Research on assessing vital rates, spatial ecology, and habitat use often involve marking wild pigs with devices ranging from individually identifiable ear tags to GPS collars. However, wild pigs pose a number of behavioral and physiological challenges to many commonly used methods of

FIGURE 9.5 Devices adapted or designed to facilitate handling captured wild pigs. (a, d) Squeeze chutes can be moved from a transport trailer into place after capturing wild pigs. (b, c) Custom-fabricated trailers incorporate multiple holding areas and handling stalls, and can be designed to connect to traps following capture. (Photos (a) by J. Suckow, (b) by the Range Cattle Research and Education Center - University of Florida, (c) by the USDA/APHIS/WS/National Wildlife Research Center, and (d) by R. Powers. With permission.)

tagging and tracking. For example, individual identification via direct or camera observation of tag or collar numbers is frequently hindered by mud from wallowing behavior. Additionally, fighting, particularly between males, can result in loss of external tags. Wild pigs have also been individually marked with passive integrated transponders (PIT) tags, which are relatively inexpensive and allow quantification of individual visits to sites containing monitoring stations, if the marked pigs can be coerced into passing within range of the monitoring station (Campbell et al. 2013a).

Very high frequency (VHF) or GPS collars are often used to monitor wild pig spatial ecology, habitat use, and survival. Recently, research has highlighted the need for more information on inter- and intraspecific contact rates to improve assessments of disease transmission risk by wild pigs (Pepin et al. 2016). Proximity collars, which incorporate a sensor that records contacts between collared animals at a discrete distance, provide one method to help address this knowledge gap. A challenge is that the electronic components of GPS collars can be damaged through rough treatment by wild pigs (e.g., rooting and wallowing). As such, researchers should temper their expectations of transmitter longevity and manufacturers should improve transmitter designs to withstand wear. For example, rooting behavior places considerable stress on component housings that hang from the neck and thus component housings on these types of collars require structural reinforcement.

When using GPS collars, determination of an appropriate fix schedule will depend on project objectives. For example, wild pigs tend to be less active during daylight hours in some seasons and locations (see Chapter 3), hence it may be possible to prolong battery use and minimize data upload costs by reducing fixes during the day.

Methods to attach VHF and GPS units to wild pigs differ for adult and juvenile animals. Like many ungulates, VHF and GPS units are often attached to pigs via neck collars. However, due to the circumference of the neck relative to the head, combined with the propensity of wild pigs to rub on trees and posts, collars generally must be fitted tighter than for other species or they will slip off the animal within a few weeks. Further compounding the problem of collar retention, pig body size and weight can fluctuate dramatically among seasons, resulting either in the collar becoming too loose (and falling off the animal prematurely) or too tight (creating animal welfare concerns). Sex, approximate age of the animal, and season should be taken into account prior to attaching neck collars. Researchers also must judge whether the animal is likely to experience high weight gain or loss (e.g., due to pregnancy status), and this can vary by location. Additionally, it is recommended to complement telemetry with targeted camera trapping to monitor animal welfare after collaring.

Harness transmitter attachments (see Chapter 14) encounter similar issues often to a greater degree, but were successfully employed by Fischer et al. (2016) on adult pigs for relatively short time frames (≤3 months). Researchers in Michigan also fit a GPS collar and harness on an adult female for 293 days with no visible signs of significant physical or locomotory stress. This animal had pronounced wild boar morphology and undoubtedly changed weight during this time, but perhaps not to the extent observed in southern wild pigs of more domestic origin. Ear tag transmitters on adult wild pigs are less commonly used, but may be appropriate when project goals do not require extended battery life or frequent upload of GPS points (Keuling et al. 2010).

Use of neck collars on juveniles is generally infeasible due to rapid growth and issues with neck conformation and behaviors similar to adults. Harness transmitters have been used unsuccessfully on piglets (Baubet et al. 2009, Keiter et al. 2017a), failing because of removal of transmitters by conspecifics, most likely the associated sow. Surgical implantation of VHF transmitters into the abdominal cavity of piglets has also been used with mixed efficacy, being more successful on piglets >3 kg (Keiter et al. 2017a). Ear tag transmitters may be the most effective technique currently available to monitor juvenile wild pigs, as ear tags do not require field surgery for attachment, and are typically retained on piglets >3 kg (Keiter et al. 2017a). Ear tag transmitters designed for neonate (<3 kg) pigs have been successfully deployed on wild pigs 1–2 days old, although retention times appear to be limited to several weeks (S. Chinn, Warnell School of Forestry and Natural Resources, University of Georgia, unpublished data). Further miniaturization of VHF and GPS units will likely facilitate improved monitoring of piglets in the future.

9.3.2 Aging

Knowing the age of individual wild pigs can be important for many research and population assessment purposes. Cementum annuli in molariform teeth can be used to age wild pigs, but uncertainty remains as to the age and frequency at which annular rings develop in different climates and habitats (Clarke et al. 1992, Choquenot and Saunders 1993). Patterns of premolar and molar eruption can be used to assign wild pigs to age classes: piglet (0–0.5 years); juvenile (0.5–1 years); yearling (1–1.5 years); subadult (1.5–3 years); and adult (>3 years; Mayer and Brisbin 2008). Furthermore, Halseth et al. (2018) developed a wild pig aging guide using tooth eruption and replacement with age classes categorized as 0–8 weeks, 8–20 weeks, 20–30 weeks, 30–51 weeks, 12–18 months, 18–26 months, 26–36 months, 36–48 months, and 48+ months. In addition, Mayer (2002) presented a method for using tooth wear, i.e., relative exposure of dentine on molars, to estimate ages >3 years, but criteria for assigning age likely differ by geographic location.

9.3.3 Reproduction

As with other mammals, productivity can be assessed by direct observation and examination of reproductive organs. Male reproductive condition has been determined using presence of spermatozoa and testicular measurements, including mass and volume (e.g., Sweeney et al. 1979, Johnson et al. 1982). Examination of female reproductive tracts can yield counts of corpora lutea and fetuses, which provide information on ovulation rates, prenatal litter size, and fetal sex ratio (Taylor et al.

1998, Ditchkoff et al. 2012). Estimation of fetus age from fetal crown-rump measurements can indicate conception and parturition dates (Henry 1968). Pigs do not retain placental scars, and, as such, scar counts are not useful for assessing litter size in this species.

Vaginal implant transmitters (VITs) used in conjunction with GPS collars on pregnant sows have proven effective for locating farrowing nests and neonatal piglets (Keiter et al. 2017a). Information gained can include assessment of postnatal litter size and nest and neonate characteristics. Additionally, use of VITs enable researchers to mark neonates for survival studies (see above). However, when implanting sows with VITs, accurate determination of pregnancy status (e.g., with a portable ultrasound) is essential to avoid implanting sows that are not pregnant.

9.3.4 Abundance or Density Estimation

Perhaps the information about wild pig populations most frequently sought is population abundance or density (i.e., abundance per unit area). Measured over time, this information is useful for determining success of eradication programs, determining effectiveness of management strategies, and developing relationships between numbers of pigs and impacts on resources. Physiological and behavioral characteristics of wild pigs pose challenges to accurate estimation of abundance. Future research should develop population estimation methods that address pig-specific biological complexities, as described below.

Many density estimation methods that have been applied to wild pigs rely on traditional capture-mark-recapture (CMR) methods, where unmarked individuals are uniquely marked at first capture, and recapture attempts occur. A second class of methods, based on removal sampling, estimates abundance of an initial population using the size and rate of decline during successive removal events (Davis et al. 2016). To derive density from estimates of abundance for either class of methods, the abundance estimate is divided by the area sampled, although it can be difficult to identify representative areas associated with different capture techniques.

Simple CMR methods for estimating abundance assume that: 1) marks are permanent and detected correctly, 2) individuals have equal probabilities of capture, 3) capture rate is constant throughout the study, and 4) the population is closed to demographic changes during the study. Removal methods make similar assumptions, although assumptions about marks are not relevant. These assumptions do not hold true for most wildlife studies, but severity of the violations depends on the species.

For wild pig populations consisting of matrilineal and bachelor groups or solitary adult males (Mayer and Brisbin 2009), the assumption of individual independence is violated (Keiter et al. 2017b). A simple method of estimating abundance for social species is to censor dependent offspring (i.e., individuals younger than dispersal age) from the estimate, thus providing an estimate for the adult population only (Keiter et al. 2017b). However, development of methods that can use data from all individuals would allow for more accurate estimates of demographic processes such as growth rates and density.

Due to the social structure of wild pigs, age- and sex-based variation in movement patterns (Kay et al. 2017), and social behaviors such as potential territoriality (Sparklin et al. 2009), the assumption of equal capture rates across individuals is also often violated. Mixture models (i.e., modeling sub-groups of the data using multiple probabilistic distributions in order to adequately represent a 'mixture' of different processes), such as spatially explicit capture-recapture models (SECR) can potentially improve inference by accounting for heterogeneous capture rates and spatial structure of the sampling design and population. Mixture models account for sources of heterogeneous capture rates, helping to reduce bias in abundance estimates. Incorporating spatial information provides the added benefit of being able to convert abundance estimates to density, which allows for comparison of values across habitats and risk assessment (Davis et al. 2017a). However, although incorporation of data describing heterogeneous capture rates and spatial information can improve inference, it does not completely resolve issues originating from the mechanisms driving heterogeneous capture rates and can increase uncertainty (Keiter et al. 2017b). New techniques (or pig-specific experimental designs) that are robust to low movement rates (i.e., lower probability of detection at multiple detectors) and other pig-specific biology are needed.

A variety of non-invasive methods can be used for identifying wild pigs including natural marks (Sweitzer et al. 2000, Keiter et al. 2017b), biomarkers (Reidy et al. 2011, Beasley et al. 2015), or DNA from scat or hair samples (Ebert et al. 2010, 2012; Kierepka et al. 2016). Non-invasive marks can be advantageous in management applications because they require less time and are less likely to cause behavioral changes that affect inference. Ebert et al. (2010, 2012) developed hair- and scat-based DNA sampling protocols to estimate wild pig populations within a CMR framework, and determined that scat sampling was less biased because it sampled individuals more homogenously across age classes. Scat can be collected along transects on multiple sampling occasions, using an adaptive search method to improve detection (Keiter et al. 2016). In absence of rain, DNA degradation rate in scat samples can be low over a 5-day period (Kierepka et al. 2016); however, degradation rates vary by habitat.

If CMR methods are to be used to inform effects of management (i.e., culling) on populations, data should be collected before and after control, and data collection involves procedures not necessarily conducted by management alone. Thus, when the management objective is control or eradication, an estimation method that does not require effort beyond pig removal work is most desirable. Removal sampling can be an efficient approach for estimating abundance or density in high-density populations because it relies only on documentation of removal efforts (Davis et al. 2016). Removal models estimate initial abundance by considering the number removed and search effort. Thus, data on the number removed can be divided by initial abundance to estimate the proportion removed due to control, which eliminates the need to do pre- and post-management measurements to evaluate management effects. However, as with other abundance or density estimation methods, removal models require data from multiple sampling events from the same population, which can be rare in some management contexts. Also, removal models only perform well when abundances are moderate to high (Davis et al. 2016). Thus, other metrics of population status such as site-level occupancy (which has positive relationships to abundance; Passy 2012) may be more useful for evaluating population changes once populations are at low levels (Davis et al. 2017b).

Similar to CMR models, removal models also assume a closed population. In managed areas adjacent to unmanaged areas, where immigration into the managed area may be high, use of CMR or removal models could lead to overestimation of abundance (Hanson et al. 2008). Future work on adapting CMR and removal models for open populations using pig demographic data will facilitate planning allocation of management resources by providing: 1) less biased estimates of density, 2) estimates of immigration rates from unmanaged areas, and 3) integration of long-term control and monitoring into a single framework. One method for relaxing the assumption of demographic closure is to use an integrated monitoring and management algorithm which combines a removal model with a dynamic population model, using removal sampling data as inputs (Chee and Wintle 2010).

Density is a more informative metric of population size than abundance because it can be compared across management areas of different sizes. However, to convert abundance to density an estimate of the area sampled is required. This typically requires knowledge of local animal movements, which vary regionally and seasonally due to behavior, food availability, weather, and landscape (see Chapter 3). Some analytical methods (e.g., SECR) can explicitly account for animal space use and thus estimate density, although others, including removal models, lack an inherent measurement of space use and require additional data on animal movements or other assessments of the area sampled to convert abundance to density. In the absence of local animal movement data estimating the area sampled can be challenging, and less attention has been given to methods of estimating area sampled relative to methods of estimating abundance, presenting an important research gap.

Lastly, Lewis et al. (2017) used biotic and abiotic factors to predict wild pig density on a global scale. Factors included in the model were determined based on locations of 129 wild pig populations from 5 continents, with complimentary density estimates reported in the literature. This study used generalized linear models and model selection techniques to evaluate relative importance, magnitude, and direction of the relationship for a suite of biotic (e.g., agriculture, vegetation cover, and large carnivore richness) and abiotic (e.g., precipitation and potential evapotranspiration) factors in predicting wild pig population density. In addition to predicting densities of existing populations, this information could be used to predict the potential of this invasive large mammal to expand its distribution globally.

9.3.5 Molecular Techniques

Molecular tools are providing new insights into wild pig ecology and information critical for population control, with contributions ranging from elucidating population dynamics (see above section on density estimation), describing feeding habits, delimiting populations or management units, and detecting pigs through environmental DNA (eDNA).

Much work has been devoted to documenting wild pig diet from stomach content analysis, which has recorded a seemingly endless list of plants, animals (both invertebrates and vertebrates), and fungi consumed by pigs (Ballari and Barrios-García 2013, Ditchkoff and Mayer 2009; see Chapters 3, 7). It has been documented that the diet of wild pigs is predominantly plants, but in their invasive range almost every wild pig stomach investigated has had animal material in it (Ballari and Barrios-García 2013, Ditchkoff and Mayer 2009). Readily digestible animal-based diet items may be hard to detect and likely underrepresented in traditional stomach content analysis (Ditchkoff and Mayer 2009, Valentini et al. 2009). Molecular-based metabarcoding diet analyses uses next-generation sequencing (NGS) to target a specific gene shared across taxonomic groups (e.g., plants, animals, fungi) by simultaneously sequencing all DNA extracted from a fecal sample (Valentini et al. 2009). Fecal samples typically are removed from the colon of a culled animal or collected fresh (within 24 hours of deposit) from transects (Robeson et al. 2018). Collection of fresh fecal material is critical for this method as DNA degrades rapidly once in the environment due to effects of UV radiation, bacteria, and weather (Santini et al. 2007). The goal of Robeson et al. (2018) was to develop a molecular approach for studying wild pig diets, rather than to quantify diet composition for a specific population. However, several intriguing results from this study suggest the utility of this approach for a more robust understanding of wild pig diets across the various ecosystems they occupy and for detecting diet changes throughout the year. With only 8 pigs sampled in Texas, but targeted at a time and place where quail were nesting, 1 pig's diet was largely composed of quail DNA (Robeson et al. 2018). In California, a pig was documented to have consumed a Panamint kangaroo rat (*Dipodomys panamintinus*), which has a restricted distribution in the Mojave Desert (Robeson et al. 2018). Thus the continued application of this method promises to reveal further insights into wild pig diets and the damage they impose on flora and fauna across both their native and invasive global ranges.

While metabarcoding studies offer new insights into wild pig diets, care must be taken to understand the biases and limitations of this approach and develop appropriate sampling schemes. Some environments affect degradation more than others, and across much of the invasive range of wild pigs in the United States, degradation is rapid due to hot, humid environments (Kierepka et al. 2016). It is important to understand that any set of primers used to amplify DNA of broad taxonomic groups will introduce biases and better amplify some species' DNA fragments over others (Deagle et al. 2013, 2014; Elbrecht and Leese 2015). Further, the process of NGS relies on library preparation for targeted fragments, potentially introducing additional biases (Shokralla et al. 2012). It may also be the case that DNA that is more common in a sample may be all that gets amplified, thus missing rare items in the diet sample (Vestheim and Jarman 2008). These biases can include preferential amplification of host DNA when target DNA is for other vertebrates (Deagle et al. 2006, Robeson et al 2018). These biases may be overcome by using primers that block amplification of pig DNA (Vestheim and Jarman 2008, Robeson et al. 2018), and by collecting fresh fecal samples from ground transects as exterior epithelial cells deposited from the pigs intestinal tract have begun to degrade but the prey item's DNA is still robust on the interior of the fecal sample

(Robeson et al. 2018). These issues are known and continue to be better understood and controlled (Thomas et al. 2016). In the meantime, it is critical to understand these potential biases when framing a study of wild pig diet. Even with these issues, using NGS metabarcoding to identify wild pig diet items allows for detection of animal and plant species with comparison of relative abundances of these items across geographical regions and throughout extended time periods to assess changes in diet. Further, molecular methods allow for processing of hundreds of samples in a single run, whereas stomach content analyses are extremely time intensive and sample sizes are often limited. Thus, molecular methods may allow unprecedented insight into impacts of wild pigs on native and invaded ecosystems, although costs associated with these methods are often high.

Use of DNA to study wild pigs has been more prolific in Europe than in other regions (Goedbloed et al. 2013, Podgórski et al. 2014). In Europe, wild boar are native and thus what we learn about this species in that region may not inform us about wild pigs in their invasive ranges. In North America, some studies have revealed the invasion history, dispersal patterns, and population structure of wild pigs (see Chapter 2). Mitochondrial DNA (mtDNA) sequences, which are maternally inherited, have been used to infer historical and recent movements of wild pigs in California (Sweitzer at al. 2015), across the United States (McCann et al. 2014) and Chile (Aravena et al. 2015). These studies found historical introductions from multiple sources and ongoing movement by humans. However, employing mtDNA to study wild pigs in their invasive range is likely less useful than nuclear markers that are bi-parentally inherited, due to the recent history of this invasion and ongoing long-distance movements facilitated by humans. For example, fine-scale population level studies have been conducted in Florida using 52 microsatellites (Hernandez et al. 2018), and in California using 43 microsatellites (Tabak et al. 2017). Both studies benefited from high sample sizes of pigs and molecular markers with high variability, demonstrating wild pigs have low dispersal and thus highly structured populations. Both studies also provided clear evidence that human-mediated movements of pigs is ongoing. Molecular studies that demonstrate human-mediated movements of wild pigs largely infer this because haplotypes or genotypes generally associating with a certain geographical area show up in a distant location. However, Tabak et al. (2017) used a modeling approach and conclusively demonstrated human-mediated movements and associations of movements with human economic interests that contribute to this problem.

Another method for using molecular data to explore wild pig evolution and ecology is to use high-density single nucleotide polymorphism (SNP) arrays (Ramos et al. 2009) that genotype individuals at approximately 60,000 SNP loci (e.g., PorcineSNP60 v2 BeadChip, Illumina, San Diego, California). Such high resolution molecular tools offer the capability of identifying genes that are under selection in certain populations and derive inferences about environmental influences across their range that positively or negatively affect those traits. With this level of genetic information it is unnecessary to employ other genomic reducing methods to randomly sample the genome (e.g., genotyping by sequencing, RAD Seq). However, obtaining whole genome sequences of mitochondria or ultimately sequencing whole genomes of individuals will allow understanding of differences in adaptation among individuals and populations. Although this technology is now accessible, the costs are still prohibitive for population level assessments, which require large sample sizes. Overall, with the large microsatellite panel and SNPs currently available we should be able to better understand breeding behavior, dispersal patterns, and estimate effective population sizes of wild pigs.

9.3.5.1 Environmental DNA

Understanding biological and external factors that influence population expansion and growth of wild pigs inferred from various molecular markers would help in development of management plans for controlling this species across their invasive range; it is also important to develop tools for detecting their presence. This is especially critical when wild pigs occur in low numbers in the early stages of an invasion or after implementation of population control. Detection of species in low numbers through amplification of DNA from environmental samples (eDNA; Taberlet et al. 2012) is a tool that contributes meaningfully to a suite of detection tools (e.g., cameras, tracking plots). The ability to detect pig DNA from water has been developed and deployed (Williams et al. 2017, Davis et al. 2018), allowing for detection of pigs after they have visited a water body even when there is only a single pig to detect (Williams et al. 2018). Even after just snout to water contact and with sampling occurring 15 minutes after initial contact, 3/3 samples were positive for pig eDNA (Williams et al. 2018). The utility of this approach is clear but as with all surveillance tools, there are limitations (Goldberg et al. 2016). In particular, eDNA does not distinguish between wild and domestic pigs, so it can only be effectively used in watersheds where no domestic pigs are located (Williams et al. 2018). Further field deployment and modeling have shown that sampling from smaller water bodies (wildlife guzzlers or small wallows, generally not more than 10 m diameter) produces a significantly higher probability of detection than from larger water bodies (ponds or moving streams; Davis et al. 2018). Even given these limitations, the ability to detect pigs in the environment from water samples provides a powerful surveillance tool that can be easily and inexpensively applied (Williams et al. 2016, 2018).

9.3.5.2 Pathogen Surveillance

Wild pigs carry numerous pathogens and parasites that can affect humans, livestock, and native wildlife (Bevins et al. 2014; see Chapter 5). Since wild pigs were introduced to the United States from Europe and Asia, there is concern that they may be more resistant to pathogens that do not currently exist in the United States, and thus potentially serve as a reservoir for pathogens to naïve species (Bevins et al. 2014). Given these concerns, use of rapid molecular diagnostic tools and ongoing pathogen surveillance in wild pig populations are critical to early detection of epizootics. The USDA/APHIS/Wildlife Services conducts regular and ongoing antibody surveillance of wild pig populations for a variety of pathogens including exposure to brucellosis, pseudorabies, and classical swine fever (Pedersen et al. 2014; see Chapter 5). These pathogens are tested by using various serological diagnostics. Once a pathogen is detected, then DNA sequences can be obtained through Sanger sequencing and phylogenetic or network analyses can be used to understand the origin of the pathogen. Thus molecular tools are critical to our surveillance of pathogens and our ability to understand transmission pathways.

9.4 POPULATION CONTROL

Considering the resiliency and adaptability of wild pigs, research to improve control techniques (e.g., trapping, shooting, fencing) and development of new, innovative methods (e.g., toxicants, vaccines, contraceptives) is crucial. Equally important is development and assessment of cost-effective, integrated management approaches to reduce wild pig populations and associated damage (Campbell and Long 2009b).

9.4.1 Physical Methods

Research on current lethal control methods is generally focused on ways to improve efficiency and cost-effectiveness. For example, studies have evaluated efficacy of various trap designs (Gaskamp 2012), gate dimensions (Metcalf et al. 2014), and gate styles (Long and Campbell 2012, Smith et al. 2014). Research also has evaluated numerous baits and attractants used for wild pig trapping (Wathen et al. 1988, Campbell and Long 2008, Williams et al. 2011, Lavelle et al. 2017), but further research to identify attractants that are specific to pigs would help increase trapping efficacy and cost-effectiveness. Snares have also been used as an effective control technique, particularly in rugged terrain where conditions for trap use are unsuitable (Anderson and Stone 1993, West et al. 2009), but further research focused on improving techniques to selectively snare wild pigs is needed. Similarly, thermal imaging and night vision have substantially increased shooting efficiency and effectiveness, but research has not determined cost-effectiveness and efficacy for large-scale population reduction (West et al. 2009).

The effects of commonly employed lethal control measures (e.g., controlled shooting, aerial gunning, recreational hunting, trapping) on remaining wild pigs could have negative implications if removal methods cause pigs to disperse, thereby increasing risk of disease transmission, potentially spreading the problem elsewhere, and hindering overall management (Fischer et al. 2016). Drive hunts with dogs in 2 separate studies conducted in Germany both had little impact on home range size of wild boar (Sodeikat and Pohlmeyer 2003, Keuling et al. 2008), and Campbell et al. (2012) found there was minimal effect on space use of wild pigs subjected to trapping, controlled shooting, drive shooting, and helicopter removals. Similarly, Campbell et al. (2010) reported that home range size and core area use was not altered due to aerial gunning from a helicopter, and it was therefore concluded to be a suitable control method in response to a disease outbreak. Conversely, research has also shown that a dispersal response can depend on the level of removal pressure. For example, areas of utilization and daily movement rates of wild pigs in southern Missouri were not affected following an initial simulated removal event, but increased after a second removal event was implemented (Fischer et al. 2016). Scillitani et al. (2010) also found that wild boar in Italy that were hunted multiple times per month increased their ranges and movement rates compared to those that were only hunted once per month. The dispersion likelihood of wild pigs following removal events is an important factor to consider when determining optimal management programs, but additional research is necessary for wildlife managers to make well-informed management decisions.

The Judas pig technique has been an important component of several successful control programs and may be particularly useful when trying to eliminate few remaining pigs in a population or in large, remote areas (McIlroy and Gifford 1997, Parkes et al. 2010). However, there is debate as to which sex or age class may serve as the best Judas pig candidate and whether or not inducing sows in estrus is advantageous (McIlroy and Gifford 1997, Wilcox et al. 2004, West et al. 2009, Parkes et al. 2010). Future research to ascertain the most suitable individuals, distance pigs will travel to find others, and length of time to locate new groups after a removal effort has occurred will help bolster the Judas pig method. Such research will also improve efficacy and cost-effectiveness of control programs by informing wildlife managers of the best locations to focus removal efforts (Beasley et al. 2018).

There is a continued need for research to improve and develop effective non-lethal control methods for wild pig management in addition to lethal methods. Recent research has evaluated the efficacy of different fence types under varying levels of motivation to either contain (e.g., disease outbreak scenario) or exclude (e.g., protection of agricultural crops) wild pigs from specified areas. Reidy et al. (2008) found that a 2-strand polywire electric fence effectively reduced sorghum (*Sorghum bicolor*) crop damage and provided a relatively inexpensive temporary fencing option for wild pigs. A 1.1-m high netted wire mesh fence with steel corner posts and a barbed wire strand buried at the base of the fence was reported to reduce damage to floodplain lagoons (Doupé et al. 2010). An innovative design involving low-level fencing (i.e., 86 cm in height) has been developed to exclude wild pigs from game feeders, but still allow access for white-tailed deer (Rattan et al. 2010). However, while certain fence designs (e.g., electric fencing) may prevent pigs from accessing areas of interest, such as agricultural fields, they may not be suitable when pigs are highly motivated to breach. Lavelle et al. (2011) tested the efficacy of 4.8-m × 0.86-m hog panel fencing to isolate wild pigs to a confined area during a simulated disease outbreak and found that hog panels were ~ 83% successful in containing pigs during simulated depopulation trials with paintball guns and 100% successful during aerial gunning trials from a helicopter. The authors concluded that even greater success probably would have been achieved if 1.3-m–1.5-m tall hog panels had been used, and that hog panel fencing could be successfully implemented for disease control and damage management. Baiting stations have also been explored as a means to contain wild pigs within a specified area during depopulation events, but were found to be an ineffective alternative to fencing (Campbell et al. 2012).

Repellents have tremendous potential for damage management given olfactory sensitivity of wild pigs, but research has yet to discover effective ways to implement these tools. Several odor and gustatory repellents currently marketed to prevent wild pig damage failed to reduce crop damage or tortoise (*Testudo hermanni hermanni*) nest predation (Vilardell et al. 2008, Schlageter and

Haag-Wackernagel 2012a, b). In another study, 3 commercial repellents (Morkit®, Tree Guard®, and Hot Sauce®) were found to decrease intake of seed or shelled corn by captive wild pigs, indicating that use of these products had potential to reduce depredation damage to seeded corn immediately after sowing (Santilli et al. 2003). Frightening devices also may reduce damage, but Schlageter and Haag-Wackernagel (2011) found that blinking LED lights were ineffective at deterring wild pigs from bait sites. Dakpa et al. (2009) tested a large-scale system that combined an intermittent and simultaneous shrill sound (electric horn with a frequency of 480 Hz, sound pressure of 100 dB and acoustic range of more than 274 m) with a bright light (500-watt inflorescence bulb), which proved effective in keeping wild pigs away from various types of crops over several months. However, wild pigs typically habituated to chemical repellents and frightening devices currently available, thus providing short-term solutions at best (Massei et al. 2011).

9.4.2 Pharmaceuticals

Delivery mechanisms are an essential component of pharmaceutical development for wild pig control. Baiting may be a practical and effective option for broad-scale pharmaceutical distribution, but bait consumption by wild pigs and non-target species needs to be evaluated to determine optimal baiting strategies and risk to non-target species (Campbell et al. 2006, Beasley et al. 2015). Biomarkers have been used as a cost-effective tool to measure bait uptake for a diverse range of free-ranging wildlife species, and several studies have evaluated their utility in wild pigs (Campbell et al. 2006, Massei et al. 2009, Reidy et al. 2011, Beasley et al. 2015, Baruzzi et al. 2017). Biomarkers such as tetracycline hydrochloride and iophenoxic acid can be integrated into baits and are suitable long-term markers as they can be detected for several months post-ingestion (Campbell et al. 2006, Massei et al. 2009, Reidy et al. 2011). Rhodamine B also is an effective biomarker for assessing bait uptake in wild pigs that is detectable in guard hair and whiskers for several months post-ingestion (Beasley et al. 2015, Baruzzi et al. 2017, Webster et al. 2017). In many circumstances Rhodamine B may prove to be a more favorable biomarker for assessment of large-scale pharmaceutical bait consumption and delivery due to simple collection and affordable detection from hair samples, compared to more labor intensive collection and expensive detection from blood (iophenoxic acid) or bone (tetracycline hydrochloride; Beasley et al. 2015). Preventing non-target species consumption of pharmaceuticals is also crucial both from safety and cost-effective standpoints. Results from a study in southern Texas using iophenoxic acid marked baits demonstrated that a large proportion of non-target species (i.e., raccoons (*Procyon lotor*) and opossums (*Didelphis virginiana*)) consumed baits intended for wild pigs (Campbell et al. 2006), highlighting the importance of developing wild pig-specific systems for delivering pharmaceutical baits (discussed in more detail below).

Development of a wild pig-specific bait station/feeder is also crucial for safe use and future registration of any pharmaceutical in the United States (Campbell et al. 2013b), but presents a significant challenge because it needs to adequately administer the toxicant or other pharmaceutical to substantial proportions of wild pig populations while also prohibiting access to numerous non-target species (Snow et al. 2017b, Lavelle et al. 2018). Several designs have been tested including the Boar-Operated-System™ (Massei et al. 2010) and the Hog-Hopper™ (Lapidge et al. 2012), both designed to take advantage of wild pig rooting behaviors by requiring them to lift a gate with a handle using their snout for access (Snow et al. 2017b). Additionally, Snow et al. (2017b) have designed a prototype bait station that utilizes rooting abilities of wild pigs after discovering a threshold of ~13.6 kg lid resistance prohibited access by raccoons, but still permitted access by wild pigs. Research is also ongoing to develop methods to exclude black bears (*Ursus americanus*).

9.4.2.1 Contraceptives

Interest in utility of contraceptives for wild pig management has increased amidst rising pressure from the public for humane, non-lethal control techniques. However, no contraceptives are currently registered for wild pigs in the United States. Contraceptives applicable to wild pig management would ideally be species-specific, inexpensive, cause infertility to a substantial percentage of the population, be administered orally in a single dose that causes infertility for the life of the individual, and cause no or minimal adverse side-effects (Massei et al. 2012). While a contraceptive meeting all the above requirements has yet to be developed, advancements in immunocontraceptive research, particularly with gonadotropin-releasing hormone (GnRH) injections, shows considerable potential for rendering wild pigs infertile (Killian et al. 2006, Massei et al. 2012). GnRH injections produce antibodies that inhibit hormones necessary for reproduction, thereby preventing ovulation and spermatogenesis (Massei et al. 2012). Killian et al. (2006) administered 2 different doses of GnRH to female wild pigs captured in Florida and held in outdoor pens and found that pregnancy was prevented 36 weeks after administration in all pigs receiving the higher dose, and 80% of pigs that received the lower dose. Additionally, Massei et al. (2012) reported infertility for at least 3–6 years following a single GnRH dose in 11 of 12 captive female pigs. Research has also found that GnRH injections do not appear to have an effect on behavior or activity of pigs (Massei et al. 2012, Quy et al. 2014). However, an injectable form of GnRH is impractical for broad-scale management applications, requiring additional research for alternative delivery options such as oral administration. The development of oral contraceptives, phage display, and cytotoxins is currently under investigation, but further research is needed before becoming available for wild pig management (Samoylova et al. 2012, Campbell et al. 2017, D. Eckery, USDA/APHIS/WS/National Wildlife Research Center, personal communication).

9.4.2.2 Vaccines

Wild pigs serve as disease reservoirs for several dozen known pathogens that can pose significant threats to domestic livestock, wildlife, and/or humans (Barrios-García and Ballari 2012; see Chapter 5). Particular concerns exist for the domestic swine industry, where wild pigs are an obstacle to eradication of several diseases (e.g., pseudorabies, swine brucellosis, and classical swine fever (CSF)) that can cause substantial economic losses (Hahn et al. 2010, Miller et al. 2013, Rossi et al. 2015). Although culling is a primary countermeasure for controlling diseases, vaccines can be effective for reducing disease risk and prevalence. However, vaccines have not been used extensively for wild pig management in the United States to date (West et al. 2009). Elzer (1999) successfully vaccinated wild pigs against brucellosis by administering an oral vaccine in a corn syrup, corn, and pecan mixture. Researchers in Europe also tested and utilized bait formulations with live attenuated vaccines to reduce CSF prevalence for over 15 years, but these vaccine campaigns are expensive, require dissemination of vaccines multiple times per year, and there are no species-specific delivery methods (Rossi et al. 2015). Bacillus Calmette-Guerin (BCG) and heat-killed *Mycobacterium bovis* vaccines have also been developed and tested with promising results for control of tuberculosis (Beltrán-Beck et al. 2012). Piglets are an important age class for vaccination (Ballesteros et al. 2009a), but Brauer et al. (2006) found that piglets did not consume vaccine-laden baits. Thus, researchers have attempted to develop baits specifically designed for 2–3-month old piglets (Ballesteros et al. 2009b). Marker vaccines have also been examined as an alternative to commonly used live attenuated vaccines, which allow serological differentiation between infected and vaccinated animals (Feliziani et al. 2014). Similar to contraceptives, further research is required before becoming applicable for broad-scale management in the United States (West et al. 2009).

9.4.2.3 Toxicants

Research focused on development of toxicants for use in the United States as an additional management tool shows great potential for efficient and cost-effective population control of wild pigs. Sodium fluoroacetate (1080) and yellow phosphorus have been used to control wild pigs in Australia, but have not been approved for use in the United States, primarily due to questions regarding humaneness and risks to non-target species (Cowled et al. 2008, Snow et al. 2017a). Kaput®, a warfarin-based toxicant (Poche et al. 2018), was recently registered as the first toxicant for wild pigs in the United States by the Environmental Protection Agency (EPA), but as of this writing has not been approved for operational use. One of the major challenges in developing a wild pig toxicant is creating a toxicant that is highly toxic to pigs, resulting in rapid and humane death, while also minimizing secondary hazards and risks to non-target species (Cowled et al. 2008). In an effort to address this challenge, researchers have been developing HOGGONE® (Animal Control Technologies Australia P/L, Victoria, Australia), a sodium nitrite-based toxicant for wild pigs to eventually be registered through the EPA for use in the United States (Snow et al. 2017a). Sodium nitrite induces a humane and rapid death to wild pigs upon consumption due to methemoglobinemia, reportedly within 3–4 hours, where pigs experience a loss of consciousness then death from hypoxia (Cowled et al. 2008, Snow et al. 2017a). Sodium nitrite has also been purported to pose minimal secondary hazard risks (Lapidge et al. 2012, Snow et al. 2018). A sodium nitrite-based toxicant known as BAIT-RITE Paste® has already been approved for use in New Zealand and pen trials with BAIT-RITE in New Zealand and HOGGONE in the United States achieved approximately 90% and 95% mortality, respectively (Shapiro et al. 2016, 2017a).

9.4.3 Integrated Management

Optimizing integrated management approaches that are specifically designed for individual control programs is imperative for planning cost-effective wild pig control (Campbell and Long 2009b). An informative and strategic decision-making process is useful for determining how specific integrated management approaches can be implemented within desired time frames. Recent advancements in research towards improving integrated wild pig management has been explored through modeling and take into consideration savings associated with resources protected as well as cost of control (Davis et al. 2017b). For example, bioeconomic models have been used to assess helicopter gunning and 1080 toxicant use as cost-effective control options to reduce lamb predation by wild pigs (Choquenot and Hone 2000). Models suggested annual helicopter removals were more profitable than using 1080 toxicant in terms of control costs vs. reductions in lamb predation if pasture biomass was above a certain threshold. However, once pasture biomass decreased below that threshold, toxicant use became more profitable because of increased bait uptake by wild pigs. These models demonstrate the importance of having a firm understanding of control techniques and various factors that impact efficacy, such as utilizing toxicants during periods of low natural food abundance. Similarly, Krull et al. (2016) demonstrated the most effective interval for harvesting wild pigs to consistently reduce ground disturbance damage was every 3 months. However, costs associated with a 3-month harvest interval were also considerably higher, such that the investment may not be worth the return in damage reduction. Continued modeling research similar to the above examples, which link pig population dynamics to damage and assess efficacy and cost-effectiveness of control programs are imperative for future success of wild pig control. Given the complexity of issues associated with management of invasive wild pigs, a substantial challenge to maximizing efficacy of integrative management is development of models scalable across regions that vary by environmental attributes and account for differing management objectives and stakeholder interests. In addition, there is a need to better link management efforts to both population demographics and reductions in damage, both at the local and regional scales.

9.5 DAMAGE ASSESSMENTS

It is well known wild pigs have far-reaching impacts, both directly and indirectly, on a range of ecosystems, habitats, native species, and agricultural crops (Bevins et al. 2014; see Chapter 7). Thus, to better understand the magnitude of damage and inform management decisions aimed at reducing the numerous damage types caused by wild pigs, it has become increasingly imperative to quantify these impacts with accurate techniques (Engeman 2000, Bengsen et al. 2014). Further, damage monitoring strategies should be a fundamental basis of any control program to assess efficacy of wild pig population control on natural or agricultural resources (Bengsen et al. 2014). Research on development of damage indices is no longer solely focused on the number of wild pigs removed because pig numbers do not necessarily equate to damage amounts (Melzer et al. 2009). Hone (2012) modeled effects of varying wild pig harvest rates on damage reduction and found the pig population needed to be reduced dramatically to observe changes in damage because the relationship between density and damage was saturating (damage amount increased exponentially with low pig densities and was saturated at high pig densities). Researchers are working to improve current damage assessment methods and utilize technological advancements to develop new methods that can be efficiently conducted in practical and cost-effective manners, while still producing accurate estimates of damage.

9.5.1 Agricultural Damage

Assessing agricultural crop damage by wild pigs is an important step in mitigating economic losses crop producers experience, but accurately quantifying crop damage is often difficult for numerous reasons including spatial and temporal variation of damage (Bengsen et al. 2014, Bleier et al. 2017). As a result, few practical step-by-step procedures have been developed to accurately assess damage on a broad scale (Bengsen et al. 2014, Michez et al. 2016), which has compelled researchers to develop new methods that can be easily implemented to quantify damage to a variety of crop types. For example, Engeman (2017) developed a practical damage assessment method applicable to a variety of row crops. This technique is applied shortly after planting, a particularly susceptible time for crop damage by wild pigs (Schley et al. 2008, Bleier et al. 2017), by quantifying field size, distance between rows, and cumulative length of damage along all rows of the field. A similar method has been developed for assessing damage to low-growing row crops just prior to harvest (Engeman and Ondovchik 2017). Although recent improvements to ground-based agricultural damage assessments have been explored, additional research leading to methods that can be applied more universally is needed.

Technological advancements in remote sensing, particularly with unmanned aerial systems (UAS), has promising applicability for efficient, cost-effective, and accurate agricultural damage assessments (Anderson and Gaston 2013, Michez et al. 2016, but see Gentle et al. 2011). UAS can readily provide imagery with extremely high spatial and temporal resolution capable of distinguishing crop damage caused by wild pigs (Michez et al. 2016). Considering the potential of UAS for agricultural damage assessments, Michez et al. (2016) compared a ground-based method to 2 UAS-based methods to estimate damage to corn prior to harvest. A fixed-wing UAS was flown over fields to obtain imagery, which was converted into orthophotos through photogrammetric processing. Damage was then estimated by either visually delineating areas of damage into polygons in ArcGIS or with crop height models to differentiate between damaged and undamaged plants. Both UAS methods reduced total assessment time by 75% or greater, but tended to underestimate total damage compared to the ground-based method. The authors concluded that UAS assessments were a viable method to estimate crop damage, but also recommended ground-based methods for cross-validation. Additionally, research is currently underway to develop effective UAS-based tools and techniques to assess wild pig damage for a variety of crops including corn, soybean, and peanuts. Multiple UAS platforms and sensors are being tested, as are autonomous flight software packages that systematically fly and collect data over damaged crop fields (Figure 9.6). Further developmental research is required to evaluate and optimize UAS-based crop damage assessments, but should prove to be extremely valuable for future damage assessments.

9.5.2 Damage to Ecosystems and Native Species

The extensive range of wild pigs and their destructive foraging behavior results in numerous impacts to ecosystems and native species that can be difficult to quantify (Thomas et al. 2013, Murphy et al. 2014, Keiter and Beasley 2017). Protocols to assess damage are often context specific depending

FIGURE 9.6 Aerial photos taken by UAS (a) showing extensive damage to corn fields by wild pigs, and (b) ability of UAS to detect wildlife damage to crops not visible to ground personnel walking or driving field edges. (Photos (a) by M. Lutman and (b) by S. Smith. With permission.)

on the type of impact and local environment (Fagiani et al. 2014). Damage quantification is further complicated by trying to find a balance between adequate sample size to accurately detect impacts and practical utilization of available resources, resulting in an overall lack of suitable methods to quantify the broad range of damage wild pigs cause (Thomas et al. 2013, Fagiani et al. 2014). In response to a scarcity of suitable damage assessment techniques, Fagiani et al. (2014) developed a statistically robust monitoring procedure to assess rooting effects in a lowland forest in central Italy on richness, diversity, and abundance of understory plants, invertebrates, and small mammals by comparing areas with low and high rooting damage. The authors concluded that their sampling framework could be used as an initial guide for developing assessment procedures and then tailored for specific objectives and area sampled. Additional assessments to quantify damage have recently been tested including Engeman et al. (2016), who quantified fine-scale rooting disturbance over a 5-year period to ecologically sensitive plant communities with numerous threatened and endangered species in south-central Florida. However, this technique may be impractical in many damage assessment situations considering the amount of time and resources required to complete. Thomas et al. (2013) optimized a more practical line-intercept method that was applicable to a variety of damage types through testing in Florida wetland sites. Damage was measured along evenly spaced transects, summed as a single total across all transects, and then divided by total length of all transects rather than averaging the proportion of damage for each transect line separately. The development and optimization of additional damage assessments designed to be practically implemented yet produce accurate estimates for numerous types of impacts are needed.

9.6 CONCLUSIONS AND FUTURE DIRECTIONS

Over the last several decades research has grown on the ecology and management of wild pigs throughout both their native and invasive ranges (Figure 9.1). These efforts have greatly improved our understanding of the basic biology and ecology of wild pigs, but have also led to substantial advancements in tools and technologies to capture, handle, study, and manage this important invasive species. While many of these advancements have benefited from more broadly developed technologies (e.g., molecular tools, telemetry, UAS technology, and more widespread access to cellular and satellite networks, night vision, and thermal technologies), others are a direct result of research on wild pigs and integration of a wider breadth of research expertise now focused on this species. Despite these advancements, compared to other ungulates in North America wild pigs remain a surprisingly understudied species. Thus, there remains a critical need for additional research on the ecology, management, impacts, and human dimensions associated with wild pigs (Beasley et al. 2018). Researchers also would be wise to take advantage of the growing amount of data being generated across the range of wild pigs by synthesizing these datasets (e.g., Kay et al. 2017) to develop more comprehensive and broadly applicable conclusions to aid the management of this destructive invasive species.

In most cases, management of wild pigs requires eventual euthanasia of captured individuals to reduce population size and associated damages. As a result, research on intact populations can be perceived as in direct conflict with short-term achievement of management goals. However, the ultimate goal of most wild pig research is to inform and improve management success, but research endeavors are costly and may take several years to complete, thus diminishing relevance to immediate management objectives. Nonetheless, advancements from research are essential to continued improvement of management techniques and strategies. In situations where ongoing management is occurring or necessary, we suggest researchers take advantage of such opportunities whenever possible to maximize resources and direct the application of research goals to management needs in an adaptive management framework. In particular, further development of methods and modeling approaches that allow for estimation of density and other demographic or vital parameters from culled individuals, and use of samples collected from these animals would minimize potential conflicts between research and management objectives (e.g., Davis et al. 2016). Indeed, such an approach has been applied widely among numerous wild pig studies to date and can be useful for developing an adaptive management strategy for wild pig removal programs.

Tools and techniques necessary to capture and study wild pigs have been around for centuries, yet research continues to yield novel insights into the ecology, impacts, control, and human dimensions of this species, all of which are essential to improve management success. While many future advancements will stem from further application of new or improved tools (e.g., molecular methods, tracking technologies), there also remains a need for refinement of basic tools necessary to capture and handle wild pigs. For example, advancements in trap designs to capture trap-shy pigs and development of baits and attractants that are selective for wild pigs and can be used during big game hunting seasons in areas that prohibit baiting are needed (Beasley et al. 2018). Furthermore, as new information or technology becomes available research must be amenable to the shifting needs of managers tasked with reducing wild pig populations. For example, current interests in application of toxicants or other pharmaceuticals to control wild pigs necessitates research to address the efficacy and impacts of such control options. However, until research and funding on wild pig ecology and management are prioritized by a greater number of state and federal wildlife agencies, the ability of researchers to address outstanding and future research needs will be limited.

ACKNOWLEDGMENTS

Contributions from M.J.L., K.M.P., A.J.P., and K.C.V. were supported by the USDA. Contributions from J.C.B. were partially funded by the US Department of Energy under award # DE-EM0004391 to the University of Georgia Research Foundation. Contributions from J.C.K. were funded by the US Department of Energy through the USDA under Interagency Agreement DE-AI09-00R22188. Mention of commercial products or companies does not represent an endorsement by the US Government.

REFERENCES

Anderson, K., and K. J. Gaston. 2013. Lightweight unmanned aerial vehicles will revolutionize spatial ecology. *Frontiers in Ecology and the Environment* 11:138–146.

Anderson, S. J., and C. P. Stone. 1993. Snaring to control feral pigs *Sus scrofa* in a remote Hawaiian rain forest. *Biological Conservation* 63:195–201.

Aravena, P., O. Skewes, and N. Gouin. 2015. Mitochondrial DNA diversity of feral pigs from Karukinka Natural Park, Tierra del Fuego Island, Chile. *Genetics and Molecular Research* 14:4245–4257.

Ballari, S. A., and M. N. Barrios-García. 2013. A review of wild boar *Sus scrofa* diet and factors affecting food selection in native and introduced ranges. *Mammal Review* 44:124–134.

Ballesteros, C., R. Carrasco-García, J. Vicente, J. Carrasco, A. Lasagna, J. De la Fuente, and C. Gortázar. 2009a. Selective piglet feeders improve age-related bait specificity and uptake rate in overabundant Eurasian wild boar populations. *Wildlife Research* 36:203–212.

Ballesteros, C., C. Gortazar, M. Canales, J. Vicente, A. Lasagna, J. A. Gamarra, R. Carrasco-Garcia, and J. De la Fuente. 2009b. Evaluation of baits for oral vaccination of European wild boar piglets. *Research in Veterinary Science* 86:388–393.

Barasona, J. A., J. R. López-Olvera, B. Beltrán-Beck, C. Gortázar, and J. Vicente. 2013. Trap-effectiveness and response to tiletamine-zolazepam and medetomidine anaesthesia in Eurasian wild boar captured with cage and corral traps. *BMC Veterinary Research* 9:1.

Barrios-García, M. N., and S. A. Ballari. 2012. Impact of wild boar (*Sus scrofa*) in its introduced and native range: A review. *Biological Invasions* 14:2283–2300.

Baruzzi, C., J. Coats, R. Callaby, D. P. Cowan, and G. Massei. 2017. Rhodamine B as a long-term semi-quantitative bait biomarker for wild boar. *Wildlife Society Bulletin* 41:271–277.

Baubet, E., S. Servanty, and S. Brandt. 2009. Tagging piglets at the farrowing nest in the wild: Some preliminary guidelines. *Acta Silvatica et Lignaria Hungarica* 5:159–166.

Bauman, K. D. 2015. Efficacy of feral swine removal techniques in southeast Missouri. Thesis, Southeast Missouri State University, Cape Girardeau, MO.

Beasley, J. C., S. S. Ditchkoff, J. J. Mayer, M. D. Smith, and K. C. VerCauteren. 2018. Research priorities for managing invasive wild pigs in North America. *Journal of Wildlife Management* 82:674–681.

Beasley, J. C., S. C. Webster, O. E. Rhodes, Jr., and F. L. Cunningham. 2015. Evaluation of Rhodamine B as a biomarker for assessing bait acceptance in wild pigs. *Wildlife Society Bulletin* 39:188–192.

Beltrán-Beck, B., C. Ballesteros, J. Vicente, J. De la Fuente, and C. Gortázar. 2012. Progress in oral vaccination against tuberculosis in its main wildlife reservoir in Iberia, the Eurasian wild boar. *Veterinary Medicine International* 2012:1–11.

Bengsen, A. J., M. N. Gentle, J. L. Mitchell, H. E. Pearson, and G. R. Saunders. 2014. Impacts and management of wild pigs *Sus scrofa* in Australia. *Mammal Review* 44:135–147.

Bevins, S. N., K. Pedersen, M. W. Lutman, T. Gidlewski, and T. J. Deliberto. 2014. Consequences associated with the recent range expansion of nonnative feral swine. *BioScience* 64:291–299.

Bleier, N., I. Kovács, G. Schally, L. Szemethy, and S. Csányi. 2017. Spatial and temporal characteristics of the damage caused by wild ungulates in maize (*Zea mays* L.) crops. *International Journal of Pest Management* 63:92–100.

Brauer, A., E. Lange, and V. Kaden. 2006. Oral immunisation of wild boar against classical swine fever: Uptake studies of new baits and investigations on the stability of lyophilised C-strain vaccine. *European Journal of Wildlife Research* 52:271–276.

Calle, P., and P. Morris. 1999. Anesthesia for nondomestic suids. *Zoo & Wild Animal Medicine Current Therapy* 4:639–646.

Campbell, S., C. R. Long, B. Pyzyna, M. Westhusin, C. Dyer, and D. Kraemer. 2017. Development and evaluation of an oral contraceptive bait for feral pigs. *Reproduction, Fertility and Development* 29:189–190.

Campbell, T. A., J. A. Foster, M. J. Bodenchuk, J. D. Eisemann, L. Staples, and S. J. Lapidge. 2013b. Effectiveness and target-specificity of a novel design of food dispenser to deliver a toxin to feral swine in the United States. *International Journal of Pest Management* 59:197–204.

Campbell, T. A., S. J. Lapidge, and D. B. Long. 2006. Using baits to deliver pharmaceuticals to feral swine in southern Texas. *Wildlife Society Bulletin* 34:1184–1189.

Campbell, T. A., and D. B. Long. 2008. Mammalian visitation to candidate feral swine attractants. *Journal of Wildlife Management* 72:305–309.

Campbell, T. A., and D. B. Long. 2009a. Strawberry-flavored baits for pharmaceutical delivery to feral swine. *Journal of Wildlife Management* 73:615–619.

Campbell, T. A., and D. B. Long. 2009b. Feral swine damage and damage management in forested ecosystems. *Forest Ecology and Management* 257:2319–2326.

Campbell, T. A., D. B. Long, M. J. Lavelle, B. R. Leland, T. L. Blankenship, and K. C. VerCauteren. 2012. Impact of baiting on feral swine behavior in the presence of culling activities. *Preventive Veterinary Medicine* 104:249–257.

Campbell, T. A., D. B. Long, and B. R. Leland. 2010. Feral swine behavior relative to aerial gunning in southern Texas. *Journal of Wildlife Management* 74:337–341.

Campbell, T. A., D. B. Long, and S. A. Shriner. 2013a. Wildlife contact rates at artificial feeding sites in Texas. *Environmental Management* 51:1187–1193.

Casas-Díaz, E., F. Closa-Sebastià, I. Marco, S. Lavín, E. Bach-Raich, and R. Cuenca. 2015. Hematologic and biochemical reference intervals for wild boar (*Sus scrofa*) captured by cage trap. *Veterinary Clinical Pathology* 44:215–222.

Chee, Y. E., and B. A. Wintle. 2010. Linking modelling, monitoring and management: An integrated approach to controlling overabundant wildlife. *Journal of Applied Ecology* 47:1169–1178.

Choquenot, D., and J. Hone. 2000. Using bioeconomic models to maximize benefits from vertebrate pest control: Lamb predation by feral pigs. *Human Conflicts with Wildlife: Economic Considerations* 8. <http://digitalcommons.unl.edu/nwrchumanconflicts/8>.

Choquenot, D., and G. Saunders. 1993. A comparison of three aging techniques for feral pigs from subalpine and semi-arid habitats. *Wildlife Research* 20:163–171.

Clarke, C. M. H., R. M. Dzieciolowski, D. Batcheler, and C. M. Frampton. 1992. A comparison of tooth eruption and wear and dental cementum techniques in age determination of New Zealand feral pigs. *Wildlife Research* 19:769–777.

Cowled, B. D., P. Elsworth, and S. J. Lapidge. 2008. Additional toxins for feral pig (*Sus scrofa*) control: Identifying and testing Achilles' heels. *Wildlife Research* 35:651–662.

Dakpa, P., U. Penjore, and T. Dorji. 2009. Design, fabrication and performance evaluation of wild pig repellent device. *Journal of Renewable Natural Resources. Bhutan* 5:116–126.

Davis, A. J., M. B. Hooten, R. S. Miller, M. L. Farnsworth, J. Lewis, M. Moxcey, and K. M. Pepin. 2016. Inferring invasive species abundance using removal data from management actions. *Ecological Applications* 26:2339–2346.

Davis, A. J., B. Leland, M. Bodenchuk, K. C. VerCauteren, and K. M. Pepin. 2017a. Estimating population density for disease risk assessment: The importance of understanding the area of influence of traps using wild pigs as an example. *Preventive Veterinary Medicine* 141:33–37.

Davis, A. J., R. McCreary, J. Psiropoulos, G. Brennan, T. Cox, A. Partin, and K. M. Pepin. 2017b. Quantifying site-level elimination certainty of an invasive species through occupancy analysis of camera-trap data. *Biological Invasions* 20:877–890.

Davis, A. J., K. M. Pepin, N. P. Snow, K. E. Williams, and A. J. Piaggio. 2018. Accounting for observation processes across multiple levels of uncertainty improves inference of species distributions and guides adaptive sampling of environmental DNA. *Ecology and Evolution*. doi:10.1002/ece3.4552.

Deagle, B. E., J. P. Eveson, and S. N. Jarman. 2006. Quantification of damage in DNA recovered from highly degraded samples: A case study on DNA in faeces. *Frontiers in Zoology* 16:3–11.

Deagle, B. E., S. N. Jarman, E. Coissac, F. Pompanon, and P. Taberlet. 2014. DNA metabarcoding and the cytochrome c oxidase subunit I marker: Not a perfect 485 match. *Biology Letters* 10:20140562.

Deagle, B. E., A. C. Thomas, A. K. Shaffer, A. W. Trites, and S. N. Jarman. 2013. Quantifying sequence proportions in a DNA-based diet study using Ion Torrent amplicon sequencing: Which counts count? *Molecular Ecology Resource* 13:620–633.

Dekeirsschieter, J., F. J. Verheggen, M. Gohy, F. Hubrecht, L. Bourguignon, G. Lognay, and E. Haubruge. 2009. Cadaveric volatile organic compounds released by decaying pig carcasses (*Sus domesticus* L.) in different biotopes. *Forensic Science International* 189:46–53.

Ditchkoff, S. S., D. B. Jolley, B. D. Sparklin, L. B. Hanson, M. S. Mitchell, and J. B. Grand. 2012. Reproduction in a population of wild pigs (*Sus scrofa*) subjected to lethal control. *Journal of Wildlife Management* 76:1235–1240.

Ditchkoff, S. S., and J. J. Mayer. 2009. Wild pig food habits. Pages 55–60 *in* J. J. Mayer and I. L. Brisbin, Jr., editors. *Wild Pigs: Biology, Damage, Control Techniques and Management*. SRNL–RP–2009–00869. Savannah River National Laboratory, Aiken, SC.

Doupé, R. G., J. Mitchell, M. J. Knott, A. M. Davis, and A. J. Lymbery. 2010. Efficacy of exclusion fencing to protect ephemeral floodplain lagoon habitats from feral pigs (*Sus scrofa*). *Wetlands Ecology and Management* 18:69–78.

Ebert, C., D. Huckschlag, H. K. Schulz, and U. Hohmann. 2010. Can hair traps sample wild boar (*Sus scrofa*) randomly for the purpose of non-invasive population estimation? *European Journal of Wildlife Research* 56:583–590.

Ebert, C., F. Knauer, B. Spielberger, B. Thiele, and U. Hohmann. 2012. Estimating wild boar *Sus scrofa* population size using faecal DNA and capture-recapture modeling. *Wildlife Biology* 18:142–152.

Elbrecht, V., and F. Leese. 2015. Can DNA-based ecosystem assessments quantify species abundance? Testing primer bias and biomass – sequence relationships with an innovative metabarcoding protocol. *PloS ONE* 10:e0130324.

Ellis, C. K., M. E. Wehtje, L. L. Wolfe, P. L. Wolff, C. D. Hilton, M. C. Fisher, S. Green, M. P. Glow, J. M. Halseth, M. J. Lavelle, N. P. Snow, E. H. VanNatta, J. C. Rhyan, K. C. VerCauteren, W. R. Lance, and P. Nol. 2019. Comparison of the efficacy of four drug combinations for immobilization of wild pigs. *European Journal of Wildlife Research*: 65:78.

Elzer, P. H. 1999. Vaccine delivery methods for feral swine now and in the future. Pages 94–97 *in First National Feral Swine Conference*, 2–3 June 1999. Ft. Worth, TX.

Engeman, R. M. 2000. Economic considerations of damage assessment. *USDA National Wildlife Research Center Symposia Human Conflicts with Wildlife: Economic Considerations* 3:36–41.

Engeman, R. M. 2017. Estimating feral swine damage to row crops just after planting. General procedural guide for WS-Operations to implement practical damage estimation. Wildlife Services Tech Note, Part of the WS Damage Assessment Series. USDA/APHIS/WS/National Wildlife Research Center. Ft. Collins, CO.

Engeman, R. M., and M. Ondovchik. 2017. Estimating feral swine damage to row crops at harvest. Wildlife Services Tech Note, Part of the WS Damage Assessment Series. US Department of Agriculture/Animal and Plant Health Inspection Sevice/Wildlife Services/National Wildlife Research Center. Ft. Collins, CO.

Engeman, R. M., S. L. Orzell, R. K. Felix, E. A. Tillman, G. Killian, and M. L. Avery. 2016. Feral swine damage to globally imperiled wetland plant communities in a significant biodiversity hotspot in Florida. *Biodiversity and Conservation* 25:1879–1898.

Fagiani, S., D. Fipaldini, L. Santarelli, S. Burrascano, E. Del Vico, E. Giarrizzo, M. Mei, A. V. Taglianti, L. Boitani, and A. Mortelliti. 2014. Monitoring protocols for the evaluation of the impact of wild boar (*Sus scrofa*) rooting on plants and animals in forest ecosystems. *Hystrix, the Italian Journal of Mammalogy* 25:31–38.

Feliziani, F., S. Blome, S. Petrini, M. Giammarioli, C. Iscaro, G. Severi, L. Convito, J. Pietchmann, M. Beer, and G. M. De Mia. 2014. First assessment of classical swine fever marker vaccine candidate CP7_E2alf for oral immunization of wild boar under field conditions. *Vaccine* 32:2050–2055.

Fenati, M., A. Monaco, and V. Guberti. 2008. Efficiency and safety of xylazine and tiletamine/zolazepam to immobilize captured wild boars (*Sus scrofa* L. 1758): Analysis of field results. *European Journal of Wildlife Research* 54:269–274.

Fischer, J. W., D. McMurtry, C. R. Blass, W. D. Walter, J. Beringer, and K. C. VerCauteren. 2016. Effects of simulated removal activities on movements and space use of feral swine. *European Journal of Wildlife Research* 62:285–292.

Fowler, M. E. 2011. *Restraint and Handling of Wild and Domestic Animals*. Third Edition. Wiley-Blackwell, Ames, IA.

Gaskamp, J. A. 2012. Use of drop-nets for wild pig damage and disease abatement. Thesis, Texas A&M University, College Station, TX.

Gentle, M., S. Phinn, and J. Speed. 2011. Assessing pig damage in agricultural crops with remote sensing. Final Report to the Australian Pest Animal Management Program. Department of Employment, Economic Developement and Innovation, Queensland, Australia.

Goedbloed, D. J., H. J. Megens, P. Van Hooft, J. M. Herrero-Medrano, W. Lutz, P. Alexandri, R. P. Crooijmans, M. Groenen, S. E. Van Wieren, R. C. Tdenber, and H. H. Prins. 2013. Genome-wide single nucleotide polymorphism analysis reveals recent genetic introgression from domestic pigs into northwest European wild boar populations. *Molecular Ecology* 22:856–866.

Goldberg, C. S., C. R. Turner, K. Deiner, K. E. Klymus, P. F. Thomsen, M. A. Murphy, S. F. Spear, A. McKee, S. J. Ovler-McCance, R. S. Cornman, M. B. Laramie, A. R. Mahon, R. F. Lance, D. S. Pilliod, K. M. Strickler, L. P. Waits, A. K. Fremier, T. Takahara, J. E. Herder, and P. Taberlet. 2016. Critical considerations for the application of environmental DNA methods to detect aquatic species. *Methods in Ecology and Evolution* 7:1299–1307.

Hahn, E. C., B. Fadl-Alla, and C. A. Lichtensteiger. 2010. Variation of Aujeszky's disease viruses in wild swine in USA. *Veterinary Microbiology* 143:45–51.

Halseth, J., M. J. Lavelle, N. P. Snow, and K. C. VerCauteren. 2018. Technical Note: Aging Feral Swine in the Field. USDA/APHIS/WS/National Wildlife Research Center. Ft. Collins, CO. 4 pp.

Hanson, L. B., J. B. Grand, M. S. Mitchell, D. B. Jolley, B. D. Sparklin, and S. S. Ditchkoff. 2008. Change-in-ratio density estimator for feral pigs is less biased than closed mark-recapture estimates. *Wildlife Research* 35:695–699.

Heinonen, M. L., M. R. Raekallio, C. Oliviero, S. Ahokas, and O. A. Peltoniemi. 2009. Comparison of azape rone–detomidine–butorphanol–ketamine and azaperone–tiletamine–zolazepam for anaesthesia in piglets. *Veterinary Anaesthesia and Analgesia* 36:151–157.

Henry, V. G. 1968. Length of estrous cycle and gestation in European wild hogs. *Journal of Wildlife Management* 32:406–408.

Hernandez, F. A., B. M. Parker, C. L. Pylant, T. J. Smyser, A. J. Piaggio, S. L. Lance, M. P. Milleson, J. D. Austin, and S. Wisely. 2018. Invasion ecology of wild pigs (*Sus scrofa*) in Florida, USA: The role of humans in the expansion and colonization of an invasive wild ungulate. *Biological Invasions* 19:1–16.

Hone, J. 2012. The future: Management options. Pages 121–140 *in* J. Hone, editor. *Applied Populations and Community Ecology: The Case of Feral Pigs in Australia*. John Wiley & Sons, Somerset, NJ.

Johnson, K. G., R. W. Duncan, and M. R. Pelton. 1982. Reproductive biology of European wild hogs in the Great Smoky Mountains National Park. *Proceedings of the Annual Conference of the Southeastern Association of Fish and Wildlife Agencies* 36:552–564.

Kay, S. L., J. W. Fischer, A. J. Monaghan, J. C. Beasley, R. Boughton, T. A. Campbell, S. M. Cooper, S. S. Ditchkoff, S. B. Hartley, J. C. Kilgo, S. M. Wisely, A. C. Wyckoff, K. C. VerCauteren, and K. M. Pepin. 2017. Quantifying drivers of wild pig movements across multiple spatial and temporal scales. *Movement Ecology* 5:14.

Keiter, D. A., and J. C. Beasley. 2017. Hog heaven? Challenges of managing introduced wild pigs in natural areas. *Natural Areas Journal* 37:6–16.

Keiter, D. A., F. L. Cunningham, O. E. Rhodes, Jr., B. J. Irwin, and J. C. Beasley. 2016. Optimization of scat detection methods for a social ungulate, the wild pig, and experimental factors affecting detection of scat. *PLoS ONE* 11:e0155615.

Keiter, D. A., J. C. Kilgo, M. A. Vukovich, F. L. Cunningham, and J. C. Beasley. 2017a. Development of known-fate survival monitoring techniques for juvenile wild pigs (*Sus scrofa*). *Wildlife Research* 44:165–173.

Keiter, D. A., A. J. Davis, O. E. Rhodes, Jr., F. L. Cunningham, J. C. Kilgo, K. M. Pepin, and J. C. Beasley. 2017b. Effects of scale of movement, detection probability, and true population density on common methods of estimating population density. *Scientific Reports* 7:9446.

Keuling, O., K. Lauterbach, N. Stier, and M. Roth. 2010. Hunter feedback of individually marked wild boar *Sus scrofa* L.: Dispersal and efficiency of hunting in northeastern Germany. *European Journal of Wildlife Research* 56:159–167.

Keuling, O., N. Stier, and M. Roth. 2008. How does hunting influence activity and spatial usage in wild boar *Sus scrofa* L.? *European Journal of Wildlife Research* 54:729–737.

Kierepka, E. M., S. D. Unger, D. A. Keiter, J. C. Beasley, O. E. Rhodes, Jr., F. L Cunningham, and A. J. Piaggio. 2016. Identification of robust microsatellite markers for wild pig fecal DNA. *Journal of Wildlife Management* 80:1120–1128.

Killian, G., L. Miller, J. Rhyan, and H. Doten. 2006. Immunocontraception of Florida feral swine with a single-dose GnRH vaccine. *American Journal of Reproductive Immunology* 55:378–384.

Kreeger, T. J., and J. M. Arnemo. 2012. *Handbook of Wildlife Chemical Immobilization*. Fourth Edition. Wildlife Pharmaceuticals, Inc., Ft. Collins, CO.

Krull, C. R., M. C. Stanley, B. R. Burns, D. Choquenot, and T. R. Etherington. 2016. Reducing wildlife damage with cost-effective management programmes. *PLoS ONE* 11:e0146765.

Lapidge, S., J. Wishart, L. Staples, K. Fagerstone, T. Campbell, and J. Eisemann. 2012. Development of a feral swine toxic bait (Hog-Gone®) and bait hopper (Hog-Hopper™) in Australia and the USA. *Proceedings of the Wildlife Damage Management Conference* 14:19–24.

Lavelle, M. J., K. C. VerCauteren, T. J. Hefley, G. E. Phillips, S. E. Hygnstrom, D. B. Long, J. W. Fischer, S. R. Swafford, and T. A. Campbell. 2011. Evaluation of fences for containing feral swine under simulated depopulation conditions. *Journal of Wildlife Management* 75:1200–1208.

Lavelle, M. J., N. P. Snow, C. K. Ellis, J. M. Halseth, J. W. Fischer, M. P. Glow, E. H. VanNatta, and K. C. VerCauteren. In Review. *Hog Wild: Developing a Strategy for Handling Entire Sounders of Wild Pigs.*

Lavelle, M. J., N. P. Snow, C. K. Ellis, J. M. Halseth, M. P. Glow, E. H. VanNatta, H. N. Sanders, and K. C. VerCauteren. 2019. When pigs fly: Reducing injury and flight response when capturing wild pigs. *Applied Animal Behaviour Science* 215:21–25.

Lavelle, M. J., N. P. Snow, J. W. Fischer, J. M. Halseth, E. VanNatta, and K. C. VerCauteren. 2017. Attractants for wild pigs: Current use, availability, needs and future potential. *European Journal of Wildlife Research* 63:86.

Lavelle, M. J., N. P. Snow, J. M. Halseth, J. C. Kinsey, J. A. Foster, and K. C. VerCauteren. 2018. Development and evaluation of a bait station for selectively dispensing bait to invasive wild pigs. *Wildlife Society Bulletin* 42:102–110.

Lewis, J. S., M. L. Farnsworth, C. L. Burdett, D. M. Theobald, M. Gray, and R. S. Miller. 2017. Biotic and abiotic factors predicting the global distribution and population density of an invasive large mammal. *Scientific Reports* 7:44152.

Long, D. B., and T. A. Campbell. 2012. Box traps for feral swine capture: A comparison of gate styles in Texas. *Wildlife Society Bulletin* 36:741–746.

Massei, G., J. Coats, R. Quy, K. Storer, and D. P. Cowan. 2010. The Boar-Operated-System: A novel method to deliver baits to wild pigs. *Journal of Wildlife Management* 74:333–336.

Massei, G., D. P. Cowan, J. Coats, F. Bellamy, R. Quy, S. Pietravalle, M. Brash, and L. A. Miller. 2012. Long-term effects of immunocontraception on wild boar fertility, physiology and behaviour. *Wildlife Research* 39:378–385.

Massei, G., A. Jones, T. Platt, and D. P. Cowan. 2009. Iophenoxic acid as a long-term marker for wild boar. *Journal of Wildlife Management* 73:458–461.

Massei, G., S. Roy, and R. Bunting. 2011. Too many hogs? A review of methods to mitigate impact by wild boar and feral hogs. *Human-Wildlife Interactions* 5:79–99.

Mayer, J. J. 2002. *A Simple Field Technique for Age Determination of Adult Wild Pigs.* WSRC-RP-2002-00635, Westinghouse Savannah River Company, Aiken, SC.

Mayer, J. J. 2009. Comparison of five harvest techniques for wild pigs. Pages 315–327 *in* J. J. Mayer and I. L. Brisbin, Jr., editors. *Wild Pigs: Biology, Damage, Control Techniques and Management.* SRNL-RP-2009-00869. Savannah River National Laboratory, Aiken, SC. 400 pp.

Mayer, J. J., and I. L. Brisbin, Jr. 2008. *Wild Pigs in the United States: Their History, Comparative Morphology, and Current Status.* Second Edition. The University of Georgia Press, Athens, GA.

Mayer, J. J., and I. L. Brisbin, Jr., editors. 2009. *Wild Pigs: Biology, Damage, Control Techniques and Management.* SRNL-RP-2009-00869. Savannah River National Laboratory, Aiken, SC.

McCann, B. E., M. J. Malek, R. A. Newman, B. S. Schmit, S. R. Swafford, R. A. Sweitzer, and R. B. Simmons. 2014. Mitochondrial diversity supports multiple origins for invasive pigs. *Journal of Wildlife Management* 78:202–213.

McIlroy, J. C., and E. J. Gifford. 1997. The 'Judas' pig technique: A method that could enhance control programmes against feral pigs, *Sus scrofa*. *Wildlife Research* 24:483–491.

Melzer, R. I., K. L. Twyford, C. Rowston, and J. D. Augusteyn. 2009. Pest arrest in central Queensland: Conserving biodiversity through pest management. *Australasian Journal of Environmental Management* 16:227–235.

Metcalf, E. M., I. D. Parker, R. R. Lopez, B. Higginbotham, D. S. Davis, and J. R. Gersbach. 2014. Impact of gate width of corral traps in potential wild pig trapping success. *Wildlife Society Bulletin* 38:892–895.

Michez, A., K. Morelle, F. Lehaire, J. Widar, M. Authelet, C. Vermeulen, and P. Lejeune. 2016. Use of unmanned aerial system to assess wildlife (*Sus scrofa*) damage to crops (*Zea mays*). *Journal of Unmanned Vehicle Systems* 4:266–275.

Miller, R. S., M. L. Farnsworth, and J. L. Malmberg. 2013. Diseases at the livestock-wildlife interface: Status, challenges, and opportunities in the United States. *Preventive Veterinary Medicine* 110:119–132.

Murphy, M. J., F. Inman-Narahari, R. Ostertag, and C. M. Litton. 2014. Invasive feral pigs impact native tree ferns and woody seedlings in Hawaiian forest. *Biological Invasions* 16:63–71.

Parkes, J. P., D. S. Ramsey, N. Macdonald, K. Walker, S. McKnight, B. S. Cohen, and S. A. Morrison. 2010. Rapid eradication of feral pigs (*Sus scrofa*) from Santa Cruz Island, California. *Biological Conservation* 143:634–641.

Passy, S. I. 2012. A hierarchical theory of macroecology. *Ecology Letters* 15:923–934.

Pedersen, K., C. R. Quance, S. Robbe-Austerman, A. J. Piaggio, S. N. Bevins, S. M. Goldstein, W. D. Gaston, and T. J. DeLiberto. 2014. Identification of *Brucella suis* from feral swine in select states in the United States. *Journal of Wildlife Diseases* 50:171–179.

Pepin, K. M., A. J. Davis, J. Beasley, R. Boughton, T. Campbell, S. M. Cooper, W. Gaston, S. Hartley, J. C. Kilgo, S. M. Wisely, C. Wyckoff, and K. C. VerCauteren. 2016. Contact heterogeneities in feral swine: Implications for disease management and future research. *Ecosphere* 7:e01230.

Poche, R. M., D. Poche, G. Franckowiak, D. J. Somers, L. N. Briley, B. Tseveenjav, and L. Polyakova. 2018. Field evaluation of low-dose warfarin baits to control wild pigs (*Sus scrofa*) in north Texas. *PLoS ONE* 13:e0206070.

Podgórski, T., D. Lusseau, M. Scandura, L. Sönnichsen, B. Jędrzejewska. 2014. Long-lasting, kin-directed female interactions in a spatially structured wild boar social network. *Plos ONE* 9:e99875.

Quy, R. J., G. Massei, M. S. Lambert, J. Coats, L. A. Miller, and D. P. Cowan. 2014. Effects of a GnRH vaccine on the movement and activity of free-living wild boar (*Sus scrofa*). *Wildlife Research* 41:185–193.

Ramos, A. M., R. P. M. A. Crooijmans, N. A. Affara, A. J. Amaral, A. L. Archibald, J. E. Beever, C. Bendixen, et al. 2009. Design of a high density SNP genotyping assay in the pig using SNPs identified and characterized by next generation sequencing technology. *PLoS ONE* 4:e6524.

Rattan, J. M., B. J. Higginbotham, D. B. Long, and T. A. Campbell. 2010. Exclusion fencing for feral hogs at white-tailed deer feeders. *The Texas Journal of Agriculture and Natural Resource* 23:83–89.

Reidy, M. M., T. A. Campbell, and D. G. Hewitt. 2008. Evaluation of electric fencing to inhibit feral pig movements. *Journal of Wildlife Management* 72:1012–1018.

Reidy, M. M., T. A. Campbell, and D. G. Hewitt. 2011. A mark-recapture technique for monitoring feral swine populations. *Rangeland Ecology and Management* 64:316–318.

Robeson, M. S., K. Khanipov, G. Golovko, S. M. Wisely, M. White, M. Bodenchuck, Y. Fofanov, N. Fierer, and A. J. Piaggio. 2018. Assessing the utility of metabarcoding for diet analysis of the omnivorous wild pig. *Ecology and Evolution* 8:185–196.

Rossi, S., C. Staubach, S. Blome, V. Guberti, H. H. Thulke, A. Vos, F. Koenen, and M. F. Le Potier. 2015. Controlling of CSFV in European wild boar using oral vaccination: A review. *Frontiers in Microbiology* 6:1–11.

Samoylova, T. I., A. M. Cochran, A. M. Samoylov, B. Schemera, A. H. Breiteneicher, S. S. Ditchkoff, V. A. Petrenko, and N. R. Cox. 2012. Phage display allows identification of zona pellucida-binding peptides with species-specific properties: Novel approach for development of contraceptive vaccines for wildlife. *Journal of Biotechnology* 162:311–318.

Santilli, F., L. Galardi, and M. Bagliacca. 2003. Corn appetibility reduction in wild boar (*Sus scrofa* L.) in relationship to the use of commercial repellents. *European Vertebrate Pest Management Conference* 4:213–218.

Santini, A., V. Lucchini, E. Fabbri, and E. Randi. 2007. Ageing and environmental factors affect PCR success in wolf (*Canis lupus*) excremental DNA samples. *Molecular Ecology Notes* 7:955–961.

Saunders, G. 1993. Observations on the effectiveness of shooting feral pigs from helicopters. *Wildlife Research* 20:771–776.

Schlageter, A., and D. Haag-Wackernagel. 2011. Effectiveness of solar blinkers as a means of crop protection from wild boar damage. *Crop Protection* 30:1216–1222.

Schlageter, A., and D. Haag-Wackernagel. 2012a. Evaluation of an odor repellent for protecting crops from wild boar damage. *Journal of Pest Science* 85:209–215.

Schlageter, A., and D. Haag-Wackernagel. 2012b. A gustatory repellent for protection of agricultural land from wild boar damage: An investigation on effectiveness. *Journal of Agricultural Science* 4:61–68.

Schley, L., M. Dufrêne, A. Krier, and A. C. Frantz. 2008. Patterns of crop damage by wild boar (*Sus scrofa*) in Luxembourg over a 10–year period. *European Journal of Wildlife Research* 54:589–599.

Scillitani, L., A. Monaco, and S. Toso. 2010. Do intensive drive hunts affect wild boar (*Sus scrofa*) spatial behaviour in Italy? Some evidences and management implications. *European Journal of Wildlife Research* 56:307–318.

Shapiro, L., C. Eason, C. Bunt, S. Hix, P. Aylett, and D. MacMorran. 2016. Efficacy of encapsulated sodium nitrite as a new tool for feral pig management. *Journal of Pest Science* 89:489–495.

Shokralla, S., J. L. Spall, J. F. Gibson, and M. Hajibabaei. 2012. Next-generation sequencing technologies for environmental DNA research. *Molecular Ecology* 21:1794–1805.

Smith, T. N., M. D. Smith, D. K. Johnson, and S. S. Ditchkoff. 2014. Evaluation of continuous-catch doors for trapping wild pigs. *Wildlife Society Bulletin* 38:175–181.

Snow, N. P., J. A. Foster, J. C. Kinsey, S. T. Humphrys, L. D. Staples, D. G. Hewitt, and K. C. VerCauteren. 2017a. Development of toxic bait to control invasive wild pigs and reduce damage. *Wildlife Society Bulletin* 40:256–263.

Snow, N. P., J. A. Foster, E. H. VanNatta, K. E. Horak, S. T. Humphrys, L. D. Staples, D. G. Hewitt, and K. C. VerCauteren. 2018. Potential secondary poisoning risks to non-targets from a sodium nitrite toxic bait for invasive wild pigs. *Pest Management Science* 74:181–188.

Snow, N. P., M. J. Lavelle, J. M. Halseth, C. R. Blass, J. A. Foster, and K. C. VerCauteren. 2017b. Development of a species-specific bait station for delivering toxic bait to invasive wild pigs. *Wildlife Society Bulletin* 41:264–270.

Sodeikat, G., and K. Pohlmeyer. 2003. Escape movements of family groups of wild boar *Sus scrofa* influenced by drive hunts in Lower Saxony, Germany. *Wildlife Biology* 9:43–49.

Sparklin B. D., M. S. Mitchell, L. B. Hanson, D. B. Jolley, and S. S. Ditchkoff. 2009. Territoriality of feral pigs in a highly persecuted population on Fort Benning, Georgia. *Journal of Wildlife Management* 73:497–502.

Swanson, K. S., M. J. Mazur, K. Vashisht, L. A. Rund, J. E. Beever, C. M. Counter, and L. B. Schook. 2004. Genomics and clinical medicine: Rationale for creating and effectively evaluating animal models. *Experimental Biology and Medicine* 229:866–875.

Sweeney, J. M., J. R. Sweeney, and E. E. Provost. 1979. Reproductive biology of a feral hog population. *Journal of Wildlife Management* 43:555–559.

Sweitzer, R. A., B. J. Gonzales, I. A. Gardner, D. Vuren, J. D. Waithman, and W. M. Boyce. 1997. A modified panel trap and immobilization technique for capturing multiple wild pig. *Wildlife Society Bulletin* 25:699–705.

Sweitzer, R. A., B. E. McCann, R. E. Loggins, and R. B. Simmons. 2015. Mitochondrial DNA perspectives on the introduction and spread of wild pigs in California. *California Fish and Game* 101:131–145.

Sweitzer, R. A., D. Van Vuren, I. A. Gardner, W. M. Boyce, and J. D. Waithman. 2000. Estimating sizes of wild pig populations in the north and central coast regions of California. *Journal of Wildlife Management* 64:531–543.

Tabak, M. A., A. J. Piaggio, R. S. Miller, R. Sweitzer, and H. B. Ernest. 2017. Anthropogenic factors predict movement of an invasive species. *Ecosphere* 6:e01844.

Taberlet, P., E. Coissac, M. Hajibabael, and L. H. Rieseberg. 2012. Environmental DNA. *Molecular Ecology* 21:1789–1793.

Taylor, R. B., E. C. Hellgren, T. M. Gabor, and L. M. Ilse. 1998. Reproduction of feral pigs in southern Texas. *Journal of Mammalogy* 79:1325–1331.

Thomas, A. C., B. E. Deagle, J. P. Eveson, C. H. Harsch, and A. W. Trites. 2016. Quantitative DNA metabarcoding: Improved estimates of species proportional biomass using correction factors derived from control material. *Molecular Ecology Resource* 16:714–726.

Thomas, J. F., R. M. Engeman, E. A. Tillman, J. W. Fischer, S. L. Orzell, D. H. Glueck, R. K. Felix, and M. L. Avery. 2013. Optimizing line intercept sampling and estimation for feral swine damage levels in ecologically sensitive wetland plant communities. *Environmental Science and Pollution Research* 20:1503–1510.

Thurfjell, H., J. P. Ball, P. A. Åhlén, P. Kornacher, H. Dettki, and K. Sjöberg. 2009. Habitat use and spatial patterns of wild boar *Sus scrofa* (L.): Agricultural fields and edges. *European Journal of Wildlife Research* 55:517–523.

Valentini, A., F. Pompanon, and P. Taberlet. 2009. DNA barcoding for ecologists. *Trends in Ecology & Evolution* 24:110–117.

Vestheim, H., and S. N. Jarman. 2008. Blocking primers to enhance PCR amplification of rare sequences in mixed samples – a case study on prey DNA in Antarctic krill stomachs. *Frontiers in Zoology* 5:12.

Vilardell, A., X. Capalleras, J. Budó, F. Molist, and P. Pons. 2008. Test of the efficacy of two chemical repellents in the control of Hermann's tortoise nest predation. *European Journal of Wildlife Research* 54:745–748.

Walter, W. D., D. M. Leslie, Jr., J. Herner-Thogmartin, K. G. Smith, and M. E. Cartwright. 2005. Efficacy of immobilizing free-ranging elk with Telazol® and xylazine hydrochloride using transmitter-equipped darts. *Journal of Wildlife Diseases* 41:395–400.

Wathen, G., J. Thomas, and J. Farmer. 1988. European wild hog bait enhancement study – Final report. Research/Resource Management Report Series, US Department of the Interior, National Park Service, Southeast Region, Atlanta, GA.

Webster, S. C., F. L. Cunningham, J. C. Kilgo, M. Vukovich, O. E. Rhodes, Jr., and J. C. Beasley. 2017. Effective dose and persistence of Rhodamine-B in wild pig vibrissae. *Wildlife Society Bulletin* 41:764–769.

West, B. C., A. L. Cooper, and J. B. Armstrong. 2009. Managing wild pigs: A technical guide. *Human-Wildlife Interactions Monograph* 1:1–55.

Wilcox, J. T., E. T. Aschehoug, C. A. Scott, and D. H. Van Vuren. 2004. A test of the Judas technique as a method for eradicating feral pigs. *Transactions of the Western Section of the Wildlife Society* 40:120–126.

Williams, B. L., R. W. Holtfreter, S. S. Ditchkoff, and J. B. Grand. 2011. Trap style influences wild pig behavior and trapping success. *Journal of Wildlife Management* 75:432–436.

Williams, K. E., K. P. Huyvaert, and A. J. Piaggio. 2016. No filters, no fridges: A method for preservation of environmental DNA. *BMC Research Notes* 9:298.

Williams, K. E., K. P. Huyvaert, and A. J. Piaggio. 2017. Clearing muddied waters: Capture of environmental DNA from turbid waters. *PLoS ONE* 12:e0179282.

Williams, K. E., K. P. Huyvaert, K. C. VerCauteren, A. J. Davis, and A. J. Piaggio. 2018. Detection and persistence of environmental DNA from an invasive, terrestrial mammal. *Ecology and Evolution* 8:688–695.

10 Human Dimensions and Education Associated with Wild Pigs in North America

Michael T. Mengak and Craig A. Miller

CONTENTS

10.1 INTRODUCTION

Controlling or eradicating non-native species is necessary to mitigate damage to crops, anthropogenic resources, livestock, and native plant and animal communities. Public support is often critical to success of control and eradication activities (Bremner and Park 2007). However, accurate understanding of public awareness and knowledge about a species and its impacts is necessary to effectively design an education program that ensures public support. Few studies exist on the social aspects of wild pig (*Sus scrofa*) eradication (Brook and van Beest 2014), and public support for eradication or control is unlikely if stakeholders do not sufficiently understand the problem at a level comparable to professional understanding. If stakeholders do not perceive a problem, resource managers will face greater difficulty in implementing management actions. Often, public knowledge lags behind that of resource managers and public sources of information can be suspect in the form of hearsay, anecdotes, rumors, inaccurate media accounts, and superstition. Well-designed human dimension surveys provide a baseline for understanding general levels of knowledge among the public so that managers can design effective educational materials. To be effective, control or eradication programs must engage the public, not solely resource users (e.g., hunters, farmers; Brook and van Beest 2014). Moreover, as wild pig management depends heavily on cooperation of private landowners and, in certain instances, the hunting pubic, effective management also depends on support by key stakeholder groups. Even though the current distribution of wild pigs extends into northern portions of North America, most of our information on human dimensions of wild pigs and associated education programs stem from southeastern North America and Texas, which is the primary focus of this chapter.

10.2 HUMAN DIMENSION SURVEYS

Wood and Lynn (1977) conducted one of the earliest human dimension surveys on wild pigs. They surveyed 733 natural resource managers (i.e., foresters, wildlife biologists, land managers) in 11 southern states to document the distribution of wild pigs, problems caused by wild pigs, and control methods employed by managers. Wood and Lynn (1977) surveyed individuals that represented an estimated 610,665 km^2 of land, with wild pigs reported on 109,634 km^2 (17.9% of total land area). All 11 states reported wild pigs occupying <0.01% of land (Virginia) to >33% (Texas), and sightings were reported as "frequent" on 52% of the area inhabited by wild pigs. Interestingly, wild pigs were not reported in southern Louisiana, probably due to lack of responses rather than actual absence. Overall, respondents reported wild pigs were largely confined to swamps, river bottoms, and the coastal plain (especially barrier islands; Wood and Lynn 1977).

Wild pig population trends in 1977 are of interest because survey respondents reported that populations were stable (33%) or decreasing (35%); only Texas reported increasing populations (Wood and Lynn 1977). The authors concluded that wild pig populations overall were decreasing on most areas throughout the southern United States. Widespread damage across the southern United States was not considered a problem in 1977, but perceptions were vastly different in 2018 (see below). In general, low levels of wild pig damage were reported to longleaf pine (*Pinus palustris*) regeneration (Wood and Lynn 1977), but in the 1970s, little area was devoted to regenerating longleaf pine. Some respondents reported depredation by wild pigs on food plots (primarily chufa) and agriculture row crops, but only in local situations. Fifty-six percent of respondents reported infrequent damage to timber regeneration, and 58% reported infrequent wildlife damage due to wild pigs (Wood and Lynn 1977). Wildlife damage in this survey included predation or food resource competition (primarily hard mast). Destruction of wild turkey (*Meleagris gallopavo*) or northern bobwhite quail (*Colinus virginianus*) nests was overwhelmingly considered infrequent or nonexistent (Wood and Lynn 1977). Few (38%) respondents reported employing control measures. Of those who attempted control, hunting or trapping was used most often, and 50% of respondents felt that hunting was an effective control technique (Wood and Lynn 1977). In terms of group agreement, only "national forest district rangers" seemed to reach consensus in stating that wild pig populations were not consistent with multiple use management objectives (Wood and Lynn 1977).

After Wood and Lynn (1977), few human dimension studies were conducted for the next nearly 20 years. A 1993 conference in Texas may mark the beginning of renewed interest in the wild pig problem (Hanselka and Cadenhead 1993). Rollins (1993) surveyed Texas county cooperative extension agents and reported results that differed from the earlier survey of Wood and Lynn (1977). Texas county extension agents reported wild pigs in 73% of Texas counties. County agents felt wild pigs came from adjacent counties (52%), escaped domestication (62%), or were intentionally released for hunting purposes (48%; Rollins 1993). Unlike respondents in the Wood and Lynn (1977) survey, 80% of Texas county agents reported that in counties with wild pigs, populations were increasing (Rollins 1993). Agents reported >58% of counties had wild pigs for >15 years prior to the conference (Rollins 1993). Agents further reported damage to crops (e.g., hay, small grains, corn, peanuts), non-crop property (e.g., fences, waterers, livestock, and wildlife depredation), disease transmission to livestock, and US$63,000 in losses to 1,243 sheep and goats (Rollins 1993). No other estimates of economic losses were provided, and this survey did not include an assessment of attitudes towards wild pigs.

Frederick (1998) surveyed all 58 county agricultural commissioners in California. In 1991, nearly 1,500 sheep, goats, and exotic game animals were reported killed by wild pigs (Frederick 1998). In 1996 county agricultural commissioners reported wild pigs caused over US$1.7M in damage and by 1997 pigs were reported in 78% of California counties (Frederick 1998; see Chapter 12). Impacts included damage to lawns, orchards, vineyards, pasture, ponds, row crops, and irrigation. Respondents from 21 of 29 (72%) counties reported that damage was increasing, whereas those from the remaining 8 counties reported it was the same; none reported decreasing damage. Frederick (1998) acknowledged his survey represented a small percentage of actual damage and suggested the need for a uniform system of reporting damage.

Adams et al. (2005) conducted one of the first studies to determine consequences of wild pig invasions to ranchers and farmers in Texas. Most respondents felt wild pigs came from adjacent property and were an agricultural pest. Rooting, wallowing, and crop damage were widely reported. An average of US$7,515 in crop loss and US$2,631 in control costs over the lifetime ownership of the land was reported by respondents (Adams et al. 2005). Total economic loss for 344 respondents since wild pigs first appeared on their land was US$2,585,200. Respondents expressed strong support for management and control, but scored relatively low on general knowledge questions about wild pig biology and ecology (Adams et al. 2005).

Canadian press reports during the early 2000s suggested significant numbers of wild pigs in 3 prairie provinces of central Canada (see Chapter 13). Working from survey data provided by rural municipal officials in Saskatchewan, a majority of respondents (59–70% depending on crop) reported wild pig damage was "never serious" (Brook and van Beest 2014). However, respondents chose "I don't know" 25–28% of the time (again depending on crop) when asked to rate frequency of damage. Frequency of observing wild pigs was positively correlated to perceptions of risk for wild pig induced damages. The authors concluded that despite widespread distribution of wild pigs across Saskatchewan, municipal officials were often unaware of associated risks.

Respondents to a survey of farmers in counties of southwest Georgia exhibited negative beliefs about wild pigs (Mengak 2012). The majority disagreed with statements such as "I enjoy seeing feral hogs around my property" (81.6% strongly disagree), "Feral hogs are an important part of the environment" (68.0% strongly disagree), and "Feral hogs are a welcome addition to the number of big game species I can hunt" (58.8% strongly disagree; Mengak 2012). Respondents did not support positive statements about wild pigs. Most strongly agreed with the statement "Feral hogs are a nuisance" (67.6% strongly agree), "Feral hogs should be eliminated where ever possible" (60.5% strongly agree), and "Feral hogs cause a great deal of damage to deer & turkey food plots" (52.4% strongly agree; Mengak 2012). Additionally, most respondents felt the wild pig population was higher than 2010, 2008, and 2006 (50.0%, 67.7%, and 67.6%, respectively; Mengak 2012).

A later survey of farmers and rural non-farm landowners throughout Georgia found similar results (Mengak 2016). Most respondents disagreed to some level (i.e., strongly disagree, disagree, somewhat disagree) with the statement "I enjoy seeing feral swine around my property" (78.9%). Most respondents also disagreed with statements such as "Feral swine are an important part of the environment" (67.0%), "Feral swine are not a threat to the safety of people" (68.2%), and "People should learn to live with feral swine near their homes and farms" (76.1%). Similarly, most respondents agreed to some level (i.e., somewhat agree to strongly agree) with negative statements about wild pigs: wild pigs are a nuisance (75.7%), should be eliminated (63.2%), and damage the environment (70.0%). Over two-thirds of respondents felt it should be a felony to transport and release wild pigs in Georgia and respondents again felt the wild pig population was higher than 2014, 2012, and 2010 (38.0%, 51.8%, and 53.4%, respectively; Mengak 2016).

With data from Illinois and Georgia, Harper et al. (2016) compared survey responses to 4 positive (i.e., "I enjoy seeing feral hogs around my property," "Feral hogs are an important part of the environment," "Feral hogs are not a threat to people," and "People should learn to live with feral hogs near their homes or farms") and 3 negative (i.e., "Feral hogs are a source of disease," "I worry about problems feral hogs might cause to my property," and "Feral hogs should be eliminated wherever possible") statements. This study provided an interesting comparison, as wild pig populations in Georgia have existed for centuries (see Chapter 16) while those in Illinois have been present for <2 decades (see Chapter 14). Consensus regarding wild pigs was similar among landowners in both states. No differences were found for statements addressing worry about wild pig damage and eliminating wild pigs wherever possible (Harper et al. 2016). Differences were found for wild pigs as a source of disease and the 4 positive statements; however, none of these differences were statistically meaningful due to small effect sizes (Harper et al. 2016).

Grady et al. (2019) conducted a nationwide survey sent to rural and urban residents to assess public attitudes towards wild pigs. Overall, 89% of respondents knew of wild pigs in the United States and 56% of respondents had a negative view of wild pigs. Respondents who indicated they were hunters overwhelmingly agreed with the statements such as "I believe wild pigs harm native wildlife," "I believe wild pigs degrade wildlife habitat," and "I believe wild pigs compete with other wildlife species for food" (71%) compared to non-hunters (48%). The majority of respondents also disagreed with the statement "Wild pigs should be allowed to be transported anywhere in the US without restrictions" (70%), and supported penalties for transporting wild pigs within their state (58%) and from their state to other states (61%). Hunters and respondents living in rural settings were more likely to support wild pig transportation restrictions and penalties, which is likely influenced by their increased knowledge regarding the negative risks wild pigs pose (Grady et al. 2019).

In another recent study, Mississippi residents were aware that wild pigs cause serious ecological damage and can transmit diseases to domestic livestock, but were less aware that wild pigs pose a disease threat to humans (Neal et al. 2017). A majority (>88.5%) agreed wild pigs should be hunted to reduce damage and the species should be treated as a pest animal (>67%). Similar to surveys conducted in other states, Mississippi respondents agreed wild pigs pose a threat to humans and natural resources and did not believe that wild pigs were beneficial on private property (Neal et al. 2017).

Given studies conducted to date, stakeholders hold a range of attitudes and beliefs about wildlife. However, surveys have generally targeted agricultural producers (Mengak 2012, Baldwin et al. 2014, Slootmaker et al. 2016) or land managers and rural landowners that were likely to have contact with wild pigs (Wood and Lynn 1977, Adams et al. 2005, Harper et al. 2014, Mengak 2016). Fewer surveys have specifically measured attitudes of the general public about wild pigs. Visitor perceptions of invasive alien species (IAS) on Cumberland Island National Seashore (CUIS) were the subject of an environmental attitude survey (Sharp et al. 2011). Wild pigs were perceived as the largest threat to the local ecosystem on CUIS and most visitors perceived wild pigs as a moderate or severe threat (72%; Sharp et al. 2011). Prior environmental attitudes influenced visitor perceptions. Detailed analysis of visitor surveys suggested media portrayal of IAS issues might exacerbate management problems, perhaps even fostering an anti-management subpopulation among people (Sharp et al. 2011). Results of this survey suggested increased knowledge and awareness may benefit management efforts but is not a complete solution to public resistance of management programs.

Wild boar issues in Europe are comparable to those in North America. Two populations of wild boar became established in southern England following escape from captivity (Goulding and Roper 2002). Prior to 1983, wild boar had been extirpated from Britain for several centuries (Goulding 2003). The renewed presence of wild boar attracted negative press as people reported fears related to direct danger posed by the animals, damage to crops, livestock predation, and disease transmission risk. Press coverage of wild boar identified 18 distinct public issues (Goulding and Roper 2002). Of the 18, 10 concerned negative aspects of wild boar, 5 were positive, and 3 were ambivalent to the wild boar issue. Fear of attack by wild boar was the most-frequently reported issue (Goulding and Roper 2002). The authors found no confirmed reports of unprovoked attacks on humans, and noted that wild boar occupied Europe for centuries without constituting a significant threat to humans. However, people falsely believe that wild boar were inherently dangerous to humans. Damage to agriculture and disease risk were the most frequent topics of press coverage after public safety (Goulding and Roper 2002). Notably, Mayer (2013) provided a comprehensive summary of wild pig attacks worldwide on humans and considered attacks to be a rare event.

Wild pigs were the 5th most-frequently listed of 6 common agricultural pest species in California but ranked higher when listed as both a pest and need for additional control methods (Baldwin et al. 2014). Of the most-frequently listed agricultural pests, wild pigs were responsible for 6–10% loss in estimated profit from 7 of 8 common agricultural commodities examined (Baldwin et al. 2014). Wild pigs were reported as being especially damaging to rangelands and nut orchards. The most common form of damage attributed to wild pigs was lost production (and resulting profit), disease transmission, and damage to irrigation infrastructure (Baldwin et al. 2014). Shooting and trapping were reported as the control methods used with greatest frequency and effectiveness (Baldwin et al. 2014). Respondents reported the need for further development of control methods and increased knowledge regarding economic damages and how the crop-wildland interface influences populations and distribution of wild pigs (Baldwin et al. 2014). The belief that discrepancies between the most-frequently used and most effective control methods related to both costs and lack of knowledge of the methods was also reported (Baldwin et al. 2014).

In a survey of farmers living at or below poverty level, Slootmaker et al. (2016) found the majority of respondents viewed wild pigs negatively on a semantic differential scale (e.g., good/bad, harmful/beneficial). As with other affected populations, these farmers agreed or strongly agreed with negative statements (e.g., "Feral swine are a nuisance") about wild pigs and disagreed or strongly disagreed with positive statements (e.g., "Problems related to feral swine are exaggerated"). A majority of respondents supported all 5 control options including hunting (58%), hunting with dogs (62%), aerial shooting (53%), trap and remove (80%), and poison if humane and did not affect other wildlife (42%; Slootmaker et al. 2016).

Landowner acceptability of wild pig management was examined through a mail survey of a random sample of 3,035 landowners (response rate = 58%) in Illinois counties with populations of wild pigs in 2013 (Harper et al. 2014). The survey addressed awareness of wild pigs, damage identification, and state preferences for various wild pig management practices under consideration by the Illinois Department of Natural Resources (IDNR). Even though wild pigs existed in the counties in which survey participants owned land, few (3%) reported they had pigs on their land (Harper et al. 2014). Of that minority, 56% had personally observed pigs, and 52% reported others had observed pigs on their property (Harper et al. 2014). Most landowners (65%) reported actions were taken to remove wild pigs, with allowing hunters access being the most frequent response (35%). All landowners experiencing wild pig damage reported it to IDNR (Harper et al. 2014). Of the 4 management options, landowners favored targeted sharpshooting over bait on their own property (75%) and county (81%), followed by trap and remove (78%). The least-favored option was aerial shooting by helicopter; however, this method was still favored by more than half of respondents for their land (58%) and in their county (60%; Harper et al. 2014).

Other studies that looked at damage caused by wildlife but did not specifically address wild pigs suggest public acceptance of damage control varies among stakeholder groups. For example, farmers in Utah and Wyoming, who typically own large properties and often suffer direct economic loss due to depredation of crops and livestock by varying species of wildlife, do not necessarily favor lethal control over non-lethal options (McIvor and Conover 1994). Given the choice of "lethal," "non-lethal," or "whatever works best" control options, farmers chose "lethal" for 1 of 10 species, and "whatever works best" for 9 of 10 species. Conversely, non-farmers may be non-landowners or own too small of an area to directly derive income or livelihood from land. This segment of the public never selected "lethal" control methods but instead chose "non-lethal control" as the top choice for 6 of 10 species (McIvor and Conover 1994). Farmers want efficient results but do not necessarily want to kill everything that could cause damage.

In a nationwide survey of American households, nearly two-thirds of respondents disagreed or strongly disagreed with the statement "Careful use of poisons is an acceptable method to control wildlife populations" (65%; Reiter et al. 1999). Poison baits were not rated as humane by 23% of respondents for rodents, 55% for birds, and 41% for predators (Reiter et al. 1999). For comparison, farmers in the southeastern United States with limited resources (i.e., gross farm sales <US$176,800 and household income ≤national poverty level) supported or were neutral on the humane and safe use of poison (Slootmaker et al. 2016). Thus, as use of toxicants and other tools for lethal control of wild pigs (e.g., aerial gunning, use of thermal optics, suppressors) become more available and common, resource managers will confront diverse public opinions related to their control, management, and eradication. The American public appreciates the need to control wildlife that threaten crops, livestock, and human health and safety (Reiter et al. 1999), but extending this support to wild pigs will require focused messaging and education.

10.3 EDUCATION

Results from producer and landowner surveys show wild pigs generally elicit negative attitudes, and in general, reported wanting more information on wild pig biology and control options. In response, university cooperative extension, state and federal agencies, and non-governmental organizations

TABLE 10.1
Outreach Information on Invasive Wild Pigs in North America

Outreach resource	Affiliation	Source
Website	Mississippi State University Extension Service	https://www.wildpiginfo.msstate.edu
Website	eXtension Community of Practice	https://articles.extension.org/feral_hogs
Website	Texas A&M Agrilife Extension	https://feralhogs.tamu.edu/
Website	Noble Research Institute	https://www.noble.org/news/feral-hogs
Website	USDA Feral Swine Damage Management Program	https://www.aphis.usda.gov/aphis/resources/pests-diseases/feral-swine/feral-swine-program
Website	National Wild Pig Task Force	https://www.nwptf.org
Website	University of Georgia – Cooperative Extension	https://www.georgiawildpigs.com/
YouTube video	USDA Feral Swine Damage Management Program	https://www.youtube.com/watch?v=WRQSs_PzRoA
YouTube channel	Alabama Cooperative Extension Service	https://www.youtube.com - Wild Pig Management
YouTube channel	Texas A&M Agrilife Extension	https://www.youtube.com - Feral Hogs
YouTube channel	Mississippi State University Extension Service	https://www.youtube.com - MSSTATEwfaTV
Mobile app (Apple and Android)	University of Georgia and Mississippi State University	Mobile App Store – "Feral Pig Damage"
Mobile app (Apple and Android)	University of California – Agriculture and Natural Resources	Mobile App Store – "Wild Pig Damage"
Mobile app (Apple and Android)	Texas A&M Agrilife Extension	Mobile App Store – "Feral Hog Management"

Note: Website source information was current as of this publication but is subject to change.
Courtesy of B. K. Strickland, M. T. Mengak, and M. D. Smith.

have produced education materials, smart phone apps, websites, webinars, and face-to-face field events to disseminate information to eager consumers (Table 10.1).

Synthesizing research on wild pig biology, ecology, and natural history along with reliable information on trap design, trap efficacy, and other control methods often falls to extension or outreach specialists at state land-grant universities (Hohbein and Mengak 2018). The Berryman Institute at Utah State University produced a special issue of their journal *Human-Wildlife Conflicts* to compile then current information on ecology and management of wild pigs (Conover 2007). The biennial International Wild Pig Conference and the National Wild Pig Task Force (M. D. Smith, Auburn University, personal communication) are examples of continuing research-focused forums devoted to understanding wild pig ecology and management (Beasley et al. 2018). For a complete review of past wild pig conferences in North America and internationally see Table 10.2. Online resources through individual state extension service websites are easy to locate (Table 10.1). Texas Cooperative Extension (Mapston 2004) produced one of the first websites devoted to wild pig control. Technical management guides (West at al. 2009; Hamrick et al. 2011a, b; Foster and Mengak 2015) have a similar format and target audiences. Each of these guides provides background information on history of wild pigs in the United States, wild pig biology and ecology, threats posed by wild pigs, and technical information on trapping and other control techniques. The lead federal agency addressing wild pigs is the USDA/APHIS/Wildlife Services, which has produced numerous fact sheets including a federal management protocol (USDA/Animal and Plant Health Inspection Service 2015). Similar products have been produced by non-governmental organizations (Stevens 2010) and numerous state agencies. Shorter extension publications are available from many Cooperative Extension Units (Pierce and Martensen 2000, Plasters et al. 2013, Lovallo 2014, Mengak 2015). Whereas state fish and wildlife

TABLE 10.2
List of Wild Pig Conferences in North America and Internationally

Meeting	Location	Year	Publication/Proceedings
Regional and national meetings			
Hog Subcommittee to the Forest Game Committee, Southeastern Association of Game & Fish Commissioners	Clearwater, Florida	1964	Lewis et al. (1965)
Hog Subcommittee to the Forest Game Committee, Southeastern Association of Game & Fish Commissioners	Tulsa, Oklahoma	1965	Lewis et al. (1966)
Symposium on the Research and Management of Wild Hog Populations	Georgetown, South Carolina	1977	Wood (1977)
Wild Hog Control Workshop	Gatlinburg, Tennessee	1983	Tate (1984)
Feral Pig Symposium	Orlando, Florida	1989	Black (1989)
Feral Swine Symposium	Kerrville, Texas	1993	Hanselka and Cadenhead (1993)
The wild pig in California oak woodland: Ecology and economics	Berkeley, California	1993	Tietje and Barrett (1993)
National Feral Swine Symposium	Orlando, Florida	1997	1997. Proceedings national feral swine symposium. Orlando, Florida
1999 Feral Swine Symposium	Fort Worth, Texas	1999	Anonymous (1999)
2004 Wild Pig Symposium	Augusta, Georgia	2004	Mayer and Brisbin (2009)
2006 National Conference on Wild Pigs	Mobile, Alabama	2006	Auburn University School of Forestry and Wildlife Sciences (2006)
2008 National Conference on Feral Hogs	St. Louis, Missouri	2008	Missouri Department of Conservation (2008)
2010 International Wild Pig Conference	Pensacola, Florida	2010	Berryman Institute (2010)
2012 International Wild Pig Conference	San Antonio, Texas	2012	Mississippi State University Extension Service (2012)
2014 International Wild Pig Conference	Montgomery, Alabama	2014	Auburn University and Mississippi State University (2014)
2016 International Wild Pig Conference	Myrtle Beach, South Carolina	2016	Mississippi State University Extension Service (2016)
2018 International Wild Pig Conference	Oklahoma City, Oklahoma	2018	Mississippi State University Extension Service (2018)
International meetings			
International Wild Boar Symposium	Toulouse, France	1984	Spitz and Pépin (1984)
Suid Workshop, IVth International Theriological Congress	Edmonton, Alberta	1985	Barrett and Spitz (1991)
Pigs and Peccaries Species Specialist Group, International Union for the Conservation of Nature and Natural Resources	Berlin, Germany	1990	Proceedings published in Bongo, vol. 18
Wild Boar Session and Workshop, Ongules/Ungulates 91	Toulouse, France	1991	Spitz et al. (1992)
2nd International Wild Boar Symposium and on sub-order Suiformes	Turin, Italy	1994	Proceedings published in *Ibex: Journal of Mountain Ecology*, Vol. 3
3rd European Wild Boar Research Group	Zaragoza, Spain	1998	Publication status unknown
3rd International Wild Boar Symposium	Uppsala, Sweden	2000	Proceedings published in *Wildlife Biology*, vol. 9 (Supplement 1)

(*Continued*)

TABLE 10.2 (CONTINUED)
List of Wild Pig Conferences in North America and Internationally

Meeting	Location	Year	Publication/Proceedings
4th International Wild Boar Conference	Lousa, Portugal	2002	Fonseca et al. (2002)
5th International Wild Boar and Suidae Symposium	Kraków, Poland	2004	Bobek (2004)
6th International Wild Boar Symposium	Kykkos, Cyprus	2006	Hadjisterkotis (2006)
7th Symposium on International Wild Boar (*Sus scrofa*) and on Sub-order Suiformes	Sopron, Hungary	2008	Náhlik (2008)
8th International Symposium on Wild Boar and Other Suids	York, United Kingdom	2010	FERA and DEFRA (2010)
9th International Symposium on Wild Boar and Other Suids	Hannover, Germany	2012	Institute for Terrestrial and Aquatic Wildlife Research (2012)
10th International Symposium on Wild Boar and Other Suids	Velenje, Slovenia	2014	Poličnik (2014)
11th International Symposium on Wild Boar and Other Suids	Mersch, Luxembourg	2016	Luxembourg Ministry of the Environment (2016)
12th International Symposium on Wild Boar and Other Suids	Lázně Bělohrad, Czech Republic	2018	Mendel University in Brno (2018)
Australian meetings			
Feral Pig Workshop	Perth, Australia	1980	Anonymous (1980)
Workshop on the Management of the Feral Pig in Australia	Trangie, Australia	1987	O'Brien (1987)
National Workshop on Feral Pig Management	Orange, Australia	1993	Anonymous (1993)
Feral Pig Workshop	Cairns, Australia	1999	Johnson (1999)
2003 Feral Pig Action Agenda	Cairns, Australia	2003	Lapidge (2003)

Courtesy of J. J. Mayer.

departments may participate in education efforts, they do not typically originate or organize education programs for landowners on the subject of wild pig biology or management.

A critical role of education efforts is to teach effective control methods (e.g., whole sounder removal), dispel myths, and disseminate information on disease risk (see Chapter 5), toxicants, and other control measures (Figure 10.1; see Chapters 8, 9). It must be realized that a single presentation on wild pigs is not likely to be very effective. Comprehensive educational program must have consistent messaging and the message must convey accurate information on wild pig biology. For example, at local field days in Georgia, agency and private professional trappers consistently point out that less than 5% of wild pigs trapped exceed 136 kg. This is an example of a consistent message supported by data to dispel the online myths of 181–272-kg wild pigs as common occurrences. Pigs of this size may exist but are not typical (see Chapter 2).

Texas Cooperative Extension (TCE) markets public education programs as Feral Hog Appreciation Days (FHAD; Rollins et al. 2007). TCE outreach recognizes diverse stakeholder perspectives on wild pigs by acknowledging 3 concurrent views: 1) positive views (e.g., hunters, landowners who profit from wild pig hunting), 2) negative views (e.g., individuals who suffer economic loss due to wild pig damage), and 3) "ugly" (e.g., individuals who may be more likely to confront ecological consequences of wild pigs, including competition with native species for food, degraded water quality, reduced plant and animal diversity due to activities of wild pigs). The TCE education programming includes on-site visits, field days, videos, and printed materials. A diversity of products are available to address multiple audiences. Standard programming includes a pre-test to

FIGURE 10.1 Education workshop held in Dawson County, Georgia, during December 2018. The workshop was one in a series held since 2015 intended to educate landowners and citizens about wild pig biology, wild pig diseases, Georgia wild pig hunting regulations, and effective trapping methods. Workshops were sponsored by a consortium of groups including the Georgia Association of Conservation Districts, University of Georgia - Cooperative Extension, Warnell School of Forestry & Natural Resources, Georgia Wildlife Resources Division, Georgia Department of Agriculture, and local cooperators. (Photo by M. Mengak. With permission.)

assess participant knowledge, science-based information on wild pigs including history, damage, economic impact, control options, disease and health issues, regulatory issues, and other topics. Following the program and information presentation, a customer satisfaction survey and follow up test was administered (Rollins et al. 2007). Results indicated 96% of stakeholders were satisfied with the program, 89% were satisfied with information presented, 91% rated information as valuable, and 56% reported they will take some type of action (although the specific action was not described; Rollins et al. 2007). An additional 61% of respondents reported they would benefit economically from the program and 98% said they would recommend the program to others (Rollins et al. 2007). These results are impressive (yet typical) of many extension program efforts. What is lacking, however, are studies to follow up on actions taken by program participants and directly linking educational efforts to efforts in the field to eradicate or reduce populations of wild pigs.

As mentioned briefly above, another function of a sound educational effort is to disseminate information about disease risks wild pigs pose to livestock, domestic pets, and humans. Disease risks are real (see Chapter 5) and hunters and trappers should take proper precautions when cleaning wild pigs or moving wild pigs to new locations (e.g., use of personal protective equipment). The Centers for Disease Control and USDA/APHIS/Wildlife Services produce and disseminate pamphlets and brochures discussing the disease risk associated with wild pigs (US Department of Agriculture 2011, 2013; Cunningham 2012).

Embracing current technology, smart phone apps for wild pigs include "Feral Pig Damage" from the University of Georgia and Mississippi State University, the "Wild Pig Damage" app from the University of California-Davis, and the "Feral Hog Management" app from Texas A&M Agrilife Extension (Table 10.1). Soon smart phone apps such as these will enable users to access the GPS feature on their phone and map portions of an agricultural field or other area of damage. The mapping feature will provide relatively accurate, real-time information on extent of damage when it is discovered, and data, along with GPS georeferenced coordinates, can be uploaded to a central database and sent to a university extension specialist.

Public stakeholders support wildlife control if their livelihoods are directly or indirectly threatened due to wildlife damage (Adams et al. 2005). Resource agencies must have data regarding risk and damage from wild pigs to gain trust and support for control or eradication. In general, for example, visitors to a park or natural area come with pre-existing perceptions of ecosystem health not based on local management efforts and programs. Stakeholders may oppose lethal control of wild pigs or other charismatic species (e.g., feral horses (*Equus ferus*), Sharp et al. 2011; black bear (*Ursus americanus*), Agee and Miller 2009) without fully understanding economic or ecosystem consequences resulting from exerting no control. We see this play out with wild pigs when hunters oppose eradication of wild pigs on a specific management area without any appreciation for the depredation to crops suffered by local farmers.

The majority of Mississippi residents (92%) reported some level of awareness about wild pigs (Neal et al. 2017). Respondents reported the 3 leading information sources as "word-of-mouth" (29%), "television" (22%), and "first-hand experience" (17%; Neal et al. 2017). Place of residence strongly influenced information source, and rural residents more likely obtained information from word-of-mouth and first-hand experience compared to more urban residents who relied on television (Neal et al. 2017). Knowing how citizens receive information can inform wildlife agencies seeking to reach clients in the most cost-effective manner. It must also be realized that word-of-mouth information often lacks factual credibility. Beliefs based on inaccurate information, including myths, rumors, and superstitions, are likely more common among rural populations (Neal et al. 2017). Additionally, age, gender, race, education, and income significantly affect perceptions of wild pigs by people in Mississippi (Neal et al. 2017).

Numerous websites from hunting outfitters and chat room forums and blogs devoted to wild pig hunting are readily accessible. Collectively, these sources can be grouped as generally providing stories related to wild pig hunting and they overwhelmingly promote wild pigs as a game species, often in direct opposition to state wildlife policies (see Chapter 11). Moreover, few (if any) outfitter websites and online chat forums provide science-based information on wild pig control, biology, ecology, or management (C. A. Miller, Illinois Natural History Survey, Prairie Research Institute, University of Illinois, unpublished data). They also have little impact on wild pig harvest (e.g., Todd and Mengak 2018) or damage reduction.

Many professional (Hamrick et al. 2011a), popular science (Morthland 2011, Nordrum 2014), hunting (McCombie 2013, Thomas 2016), forestry (Smith 2016) and natural history (Proctor 2012) magazines, as well as numerous newspaper articles, are raising awareness of wild pig problems and informing multiple stakeholder groups and other publics about problems posed by wild pigs. Although wild pigs were not 1 of the 10 species presented to survey participants by McIvor and Conover (1994), farmers reported getting wildlife related information from game wardens (47.7%), with newspapers (1.8%) and magazines (2.3%) ranking low as sources of information (McIvor and Conover 1994). Comparatively, non-farmers reported getting wildlife information from newspapers (96.1%) or magazines (87.9%; McIvor and Conover 1994). Given that the non-farming public may be getting information from various media outlets and online databases, Internet sources, and the worldwide web (sources that did not exist or were very limited in 1994), it is critical wildlife professionals deliver current and accurate information in formats various stakeholders use. This variety of outlets along with consistency of messaging is critical in shaping public opinion and maintaining support for wild pig management, control, and eradication if that is an agency's goal.

Minority segments of the hunting community are a powerful voice at public hearings when new hunting, trapping, or transportation regulations are proposed by state agencies. At a recent public hearing in Georgia, approximately 50 local wild pig hunters verbally overwhelmed 5 wildlife professionals and the state veterinarian when a regulation to require all wild pigs be ear-tagged prior to transport was proposed (Georgia Feral Swine Working Group, personal communication). The regulation, though supported by the farming community, state agency, and the state's wild pig working group, was tabled. Similarly, wild pig hunters and meat processors in Texas succeeded in getting a judge to issue a restraining order temporarily halting use of an EPA registered toxicant for wild pig control (Copeland 2017). At a 2010 public hearing in Tennessee, a wildlife specialist testified before the Tennessee Wildlife Resource Commission in front of approximately 300 wild pig hunters in support of Tennessee's ban on wild pig hunting. After his testimony, he was escorted out of the building with 4 armed conservation officers as members of the audience made offensive hand gestures (B. West, University of Tennessee, personal communication).

It can be difficult to dispel deeply held beliefs regarding wild pigs. The task is often made more complicated by the confusing array of laws, regulations, and jurisdiction among states (see Chapter 11). A recently completed survey of Southeastern fish and wildlife agencies illustrates this problem (Southeastern Association of Fish and Wildlife Agencies 2012). In 2 states, authority for wild pigs rests with the Department of Agriculture (DOA), while it rests with the Department of Natural Resources (DNR) in 6 states. Seven states share authority between their respective DOAs and DNRs. No states in the region have a closed season for private lands. West Virginia allows wild pig hunting in just 4 counties, has a limit of one wild pig harvested/hunter/year (West Virginia Division of Natural Resources 2018), and tries to distinguish wild boar from wild pigs by geographic location (McCoy 2017). Regulations for night hunting and hunting with dogs on private land also differ among states (Southeastern Association of Fish and Wildlife Agencies 2012). Finally, no states in the region allow release of wild pigs into the wild, but several allow release into high-fenced shooting operations. High-fenced shooting operations may or may not be permitted and inspected by state authorities and are unlikely to permanently hold wild pigs as escapes are known to occur. Laws related to high-fenced shooting operations for wild pig hunting will be difficult to change. Educational programs may help inform decision makers but the influence of local politics must never be discounted.

10.4 CONCLUSIONS

Wild pig management is a controversial topic. Some stakeholders (e.g., farmers) and professionals (e.g., wildlife managers) speak in terms of eradication, while some (e.g., hunters) speak in terms of controlling wild pigs. Other citizens and elected officials may be confused by the contrasting endpoints. Thus, involving citizens through education programs aimed at increasing awareness, disseminating factual information, and employing citizen-science networks is critical to establishing effective management programs. Further, although exposure time to damage may ameliorate perceptions of damage in the case of white-tailed deer (*Odocoileus virginianus;* Knuth et al. 1992), field activities, and educational programming by land-grant universities and state and federal agencies may be acting to continually place the wild pig issue at the forefront of landowner awareness and thus reduce complacency among affected landowners. Human dimensions studies are necessary to assess the diverse opinions of many stakeholder groups. These studies can gauge support among the public for various management options, collect information on real-time impacts such as disease spread, economic cost, and management methodologies implemented, and effectively measure knowledge of the public to ascertain if information and education efforts by state or federal agencies and extension professionals are achieving desired results. Findings from human dimensions studies can effectively shape new outreach efforts and measure impacts. As new control methodologies (e.g., contraception, toxicants, complete hunting ban, transportation bans) become available or come to be implemented by regulatory authorities, preferences among the public must be measured and considered in decision-making processes. Combining knowledge from biological and social research allows managers to develop acceptable control strategies accepted by diverse stakeholder groups.

REFERENCES

Adams, C. E., B. J. Higginbotham, D. Rollins, R. B. Taylor, R. Skiles, M. Mapston, and S. Turman. 2005. Regional perspectives and opportunities for feral hog management in Texas. *Wildlife Society Bulletin* 33:1312–1320.

Agee, J. D., and C. A. Miller. 2009. Factors contributing toward acceptance of lethal control of black bears in central Georgia, USA. *Human Dimensions of Wildlife* 14:198–205.

Anonymous. 1980. Summary of the proceedings of the feral pig workshop. Agricultural Protection Board, Department of Conservation and Land Management, Government of Western Australia, Perth, Australia.

Anonymous. 1993. *Proceedings: National Workshop on Feral Pig Management.* Orange, New South Wales, Australia

Anonymous. 1999. *Proceedings of the Feral Swine Symposium.* June 2–3. Texas Animal Health Commission, Fort Worth, TX. 149 pp.

Auburn University and Mississippi State University. 2014. *2014 International Wild Pig Conference – Science & Management.* Montgomery, Alabama, April 13–16, 2014. Conference Program with Abstracts. 41 pp.

Auburn University School of Forestry and Wildlife Sciences. 2006. *2006 National Conference on wild pigs. Mobile*, Alabama, May 21–23, 2006. Conference program with abstracts. 39 pp.

Baldwin, R. A., T. P. Salmon, R. H. Schmidt, and R. M. Timm. 2014. Perceived damage and areas of needed research for wildlife pests of California agriculture. *Integrative Zoology* 9:265–279.

Barrett, R. H., and F. Spitz, editors. 1991. *Biology of Suidae. Biologie des Suidés.* I.R.G.M., Toulouse, France. 170 pp.

Beasley, J. C., S. S. Ditchkoff, J. J. Mayer, M. D. Smith, and K. C. Vercauteren. 2018. Research priorities for managing invasive wild pigs in North America. *Journal of Wildlife Management* 82:674–681.

Berryman Institute. 2010. *2010 International Wild Pig Conference – Science & Management.* Pensacola, FL, April 11–13, 2010. Conference program with abstracts. 44 pp.

Black, N., editor. 1989. *Proceedings: Feral Pig Symposium.* April 27–29, Orlando, Florida. Livestock Conservation Institute, Madison, WI. 77 pp.

Bobek, B., editor. 2004. *5th International Wild Boar and Suidae Symposium – Abstracts.* Institute of Biology, Pedagogical University, Kraków, Poland.

Bremner, A., and K. Park. 2007. Public attitudes to the management of invasive non-native species in Scotland. *Biological Conservation* 139:306–314.

Brook, R. K., and F. M. van Beest. 2014. Feral wild boar distribution and perceptions of risk on the central Canadian prairies. *Wildlife Society Bulletin* 38:486–494.

Conover, M. R. 2007. America's first feral hog war. *Human-Wildlife Conflicts* 1:129–131.

Copeland, M. 2017. Battle raging over poisoning of feral hogs. Waco Tribune-Herald. Published 28 March 2017. Archived at <https://www.wacotrib.com/news/business/battle-raging-over-poisoning-of-feral-hogs/article_b409c020-3f48-581c-a001-a8fe5853b6a8.html>. Accessed 18 Feb 2019.

Cunningham, F. 2012. Feral swine damage control strategies. National Wildlife Research Center Fact Sheet, Ft. Collins, CO.

Fonseca, C., A. Herrero, A. Luis, and A. M. V. M. Soares, editors. 2002. Wild boar research 2002. A selection and edited papers from the 4th International Wild Boar Symposium. *Galemys* 16 (Special Edition):149–155.

Food and Environment Research Agency and Department for Environment, Food and Rural Affairs. 2010. *8th International Symposium on Wild Boar and Other Suids – Book of Abstracts.* FERA and DEFRA, York, UK. 86 pp.

Foster, M., and M. T. Mengak. 2015. *Georgia Wild Pig Management Manual.* River Valley Commission, Columbus, GA.

Frederick, J. M. 1998. Overview of wild pig damage in California. *Proceedings of the Vertebrate Pest Conference* 18:82–86.

Goulding, M. 2003. *Wild Boar in Britain.* Whittet Books Ltd., Suffolk, VA.

Goulding, M. J., and T. J. Roper. 2002. Press responses to the presence of free-living wild boar (*Sus scrofa*) in southern England. *Mammal Review* 32:272–282.

Grady, M. J., E. E. Harper, K. M. Carlisle, K. H. Ernst, and S. A. Shwiff. 2019. Assessing public support for restrictions on transport of invasive wild pigs (*Sus scrofa*) in the United States. *Journal of Environmental Management* 237:488–494.

Hadjisterkotis, E. 2006. *Sixth Symposium on International Wild Boar (Sus scrofa) and on Sub-order Suiformes – Abstracts*. Ministry of the Interior, Nicosia, Cyprus. 65 pp.

Hamrick, B., T. Campbell, B. Higginbotham, and S. Lapidge. 2011a. Managing an invasion: Effective measure to control wild pigs. *The Wildlife Professional*, Summer 2011: 41–42.

Hamrick, B., M. Smith, C. Jaworowski, and B. Strickland. 2011b. *A Landowner's Guide for Wild Pig Management*. Mississippi State University Extension Service Publication No. 2659. Mississippi State University, Mississippi State, MS.

Hanselka, C. W., and J. F. Cadenhead. 1993. *Feral Hogs: A Compendium for Resource Managers – Proceedings of a Symposium*. Texas Agriculture Extension Service, Kerrville, TX.

Harper, E. E., C. A. Miller, A. L. Stephenson, M. E. McCleary, and L. K. Campbell. 2014. Landowner attitudes and perceived risks toward wild pigs on private lands in Illinois. Job Completion Report, Federal Aid in Wildlife Restoration W-112-R-22. Human Dimensions Research Program Report HR-14-05. INHA Technical Report 2014 (16). Illinois Natural History Survey, Champaign, IL.

Harper, E. E., C. A. Miller, J. J. Vaske, M. T. Mengak, and S. Bruno. 2016. Stakeholder attitudes and beliefs toward wild pigs in Georgia and Illinois. *Wildlife Society Bulletin* 40:269–273.

Hohbein, R. R., and M. T. Mengak. 2018. Cooperative extension agents as key informants in assessing wildlife damage trends in Georgia. *Human-Wildlife Interactions* 12: 243–258.

Institute for Terrestrial and Aquatic Wildlife Research. 2012. *9th International Symposium on Wild Boar and Other Suids – Book of Abstract*. Verein der Förderer des Instituts für Wildtierforschung e.V., Hannover, Germany. 68 pp.

Johnson, C. N., editor. 1999. Feral pigs: Pest status and prospects for control. Proceedings of a feral pig workshop. James Cook University, Cairns, March. Research Report No. 13. Cooperative Research Centre for Tropical Rainforest Ecology and Management, Cairns, Australia.

Knuth, B. A., R. J. Stout, W. F. Siemer, D. J. Decker, and R. C. Stedman. 1992. Risk management concepts for improving wildlife population decisions and public communication strategies. *Transactions of the North American Wildlife and Natural Resources Conference* 57:63–74.

Lapidge, S. J., editor. 2003. *Proceedings of the Feral Pig Action Agenda*. James Cook University, Cairns, June 2003. Pest Animal Control Cooperative Research Centre, Canberra, Australia. 96 pp.

Lewis, J., J. Haygood, H. L. Holbrook, and J. McDaniel. 1966. Bibliography on European and wild hogs. Hog Subcommittee to the Forest Game Committee. Southeastern Association of Game & Fish Commissioners. 7 pp.

Lewis, J. C., R. Murray, and G. Matschke. 1965. Hog subcommittee report to the chairman of the forest game committee. Southeastern Association of Game & Fish Commissioners, 18 pp.

Lovallo, M. 2014. *Feral Swine in Pennsylvania*. Penn State Cooperative Extension Service, State College, PA.

Luxembourg Ministry of the Environment. 2016. *11th International Symposium on Wild Boar and Other Suids – Abstract Booklet*. Luxembourg Ministry of the Environment, Mersch, Luxembourg. 60 pp.

Mapston, M. E. 2004. *Feral Hogs in Texas*. Texas Cooperative Extension Bulletin B-6149. Texas A&M University, College Station, TX.

Mayer, J. J. 2013. Wild pigs attacks on humans. *Proceedings of the 15th Wildlife Damage Management Conference* 15:17–35.

Mayer, J. J., and I. L. Brisbin, Jr., editors. 2009. *Wild Pigs: Biology, Damage, Control Techniques and Management*. SRNL-RP-2009-00869. Savannah River National Laboratory, Aiken, SC. 400 pp.

McCombie, B. 2013. Swine story: A history of feral hogs in the US. *Game and Fish Magazine*, April 23, 2013. <http://www.gameandfishmag.com/editorial/history-of-feral-hogs/191026>. Accessed 18 Feb 2019.

McCoy, J. 2017. In WV, it's hard to tell a wild boar from a feral pig. Story published July 1, 2017. *West Virginia Gazette-Mail*. <www.wvgazettemail.com>. Accessed 18 Feb 2019.

McIvor, D. E., and M. R. Conover. 1994. Perceptions of farmers and non-farmers toward management of problem wildlife. *Wildlife Society Bulletin* 22:212–219.

Mendel University in Brno. 2018. *12th International Symposium on Wild Boar and Other Suids – Programme*. Lázně Bělohrad, Czech Republic. 5 pp.

Mengak, M. T. 2012. *2012 Georgia Wild Pig Survey*. Final Report. Publication WMS-12-16. Warnell School of Forestry and Natural Resources, University of Georgia, Athens, GA.

Mengak, M. T. 2015. *Wild Pigs (Sus scrofa)*. Wildlife Fact Sheet WSFNR-15-08. Warnell School of Forestry and Natural Resources, University of Georgia, Athens, GA.

Mengak, M. T. 2016. *2015 Georgia wild pig survey*. Final Report. University of Georgia, Warnell School of Forestry and Natural Resources, Publication No. WSFNR 16–23.

Mississippi State University Extension Service. 2012. *2012 International Wild Pig Conference – Science & Management*. San Antonio, TX, April 15–18, 2012. Conference program with abstracts. 33 pp.

Mississippi State University Extension Service. 2016. *2016 International Wild Pig Conference – Science, Management & Solutions*. Myrtle Beach, SC, April 17–20, 2016. Conference program with abstracts. 44 pp.

Mississippi State University Extension Service. 2018. *International Wild Pig Conference – Science, Management & Solutions*. Oklahoma City, OK, April 15–18, 2018. Conference program with abstracts. 40 pp.

Missouri Department of Conservation. 2008. *2008 National Conference on Feral Hogs – Bar the Gate*. St. Louis, MO, April 13–15, 2008. Conference program with abstracts.

Morthland, J. 2011. A plague of pigs in Texas. *Smithsonian Magazine*, January 2011. <https://www.smithsonianmag.com/science-nature/a-plague-of-pigs-in-texas-73769069/>. Accessed 19 Feb 2015.

Náhlik, A. (Chairman of Organizing Committee). 2008. *7th Symposium on International Wild Boar (Sus scrofa) and on Sub-Order Suiformes – Program*. Institute of Wildlife Management and Vertebrate Zoology, Faculty of Forestry, University of West Hungary, Sopron, Hungary. 8 pp.

Neal, D., J. Tegt, and B. Strickland. 2017. Mississippi public awareness, knowledge, and attitudes towards wild hogs. Unpublished Final Report. Mississippi State University, Mississippi State, MS.

Nordrum, A. 2014. Can wild pigs ravaging the US be stopped? *Scientific American*, October 21. 2014. <https://www.scientificamerican.com/article/can-wild-pigs-ravaging-the-u-s-be-stopped/>. Accessed 1 July 2015.

O'Brien, P., editor. 1987. *Management of the Feral Pig in Australia: Report of a Workshop*. Trangie Agricultural Research Centre, Trangie, Australia.

Pierce, R. A., and R. Martensen. 2000. *Feral Hogs in Missouri: Damage Prevention and Control*. MU Guide – Agriculture, G-9457. Cooperative Extension, University of Missouri, Columbia, MO.

Plasters, B., C. Hicks, R. Gates, and M. Titchenell. 2013. Feral swine in Ohio: Managing damage and conflicts. Fact Sheet, W-26-13. Ohio State University Extension, Agriculture and Natural Resources, Ohio State University, Columbus, MO.

Poličnik, H. (Chairman of Organizing Committee). 2014. *10th International Symposium on Wild Boar and Other Suids – Abstracts*. ERICo Velenje, Velenje, Slovenia. 130 pp.

Proctor, A. 2012. Feral hogs: Trouble on the Horizon. *Virginia Wildlife*, November/December 2012: 20–21.

Reiter, D. K., M. W. Brunson, and R. H. Schmidt. 1999. Public attitudes toward wildlife damage management and policy. *Wildlife Society Bulletin* 27:746–758.

Rollins, D. 1993. Statewide attitude survey on feral hogs in Texas. Pages 1–8 *in* C. W. Hanselka and J. F. Cadenhead, editors. *Feral Swine: A Compendium for Resource Managers*. Texas Agricultural Extension Service, Kerrville, TX.

Rollins, D., B. J. Higginbotham, K. A. Cearley, and R. N. Wilkins. 2007. Appreciating feral hogs: Extension education for diverse stakeholders in Texas. *Human-Wildlife Conflicts* 1:192–198.

Sharp, R. L., L. R. Larson, and G. T. Green. 2011. Factors influencing public preferences for invasive alien species management. *Biological Conservation* 144:2097–2104.

Slootmaker, C., E. Harper, A. Anderson, J. Holderieath, and S. Shwiff. 2016. Economic impacts of feral swine on limited resource farmers in the United States. Final Report. US Department of Agriculture/Animal and Plant Health Inspection Service/Wildlife Services/National Wildlife Research Center, Ft. Collins, CO.

Smith, M. D. 2016. Hogging the forest. *Forest Landowner*, May/June 2016. 6 pp.

Southeastern Association of Fish and Wildlife Agencies. 2012. Annual state summary report – wild hog working group. Southeastern Association of Fish and Wildlife Agencies, Jackson, MS. <https://www.dgif.virginia.gov/wp-content/uploads/seafwa-2012-annual-state-report.pdf>. Accessed 18 Feb 2019.

Spitz, F., G. Janeau, G. Gonzalez, and S. Aulagnier, editors. 1992. *Ongules/Ungulates 91: Proceedings of the International Symposium*. Toulouse, France, September 2–6, 1991. Société Francaise pour l'Etude et la Protection des Mammifères, and Toulose: Institut de Recherché sur les Grands Mammifères, Paris & Toulouse, France. 661 pp.

Spitz, F., and D. Pépin, editors. 1984. *Symposium International sur le Sanglier (International Wild Boar Symposium), Les Colloques de l'INRA 22*. INRA, Toulouse, France. 226 pp.

Stevens, R. 2010. *The Feral Hog in Oklahoma*. Second Edition. The Samuel Roberts Nobel Foundation, Ardmore, OK.

Tate, J., editor. 1984. Techniques for controlling wild hogs in the Great Smoky Mountains National Park. *Proceedings of a Workshop*, November 29–30, Research/Resources Mgmt. Rpt. SRE-72. U. S. Department of the Interior, National Park Service, Southeast Regional Office, Atlanta, GA. 87 pp.

Thomas, L., Jr. 2016. Pig problems. *Quality Whitetails*, August/September 2016. 6 pp.

Tietje, W., and R. Barrett. 1993. *The Wild Pig in California Oak Woodland: Ecology and Economics*. Integrated Hardwood Management Program, Department of Forestry and Resource Management, University of California, Berkeley, CA. 45 pp.

Todd, C. T., and M. T. Mengak. 2018. The impact of wild pig hunting outfitters on wild pig populations across the Southeast. Publication WSFNR-18-45. Warnell School of Forestry & Natural Resources, University of Georgia, Athens, GA.

US Department of Agriculture. 2011. *Feral Swine: Damage and Disease Threats*. US Department of Agriculture/Animal and Plant Health Inspection Service Program Aid No. 2086. US Department of Agriculture, Washington, DC.

US Department of Agriculture. 2013. Feral/wild pigs: Potential problems for farmers and hunters. US Department of Agriculture/Animal and Plant Health Inspection Service Program Aid No. 2086. Supersedes Agriculture Information Bulletin No. 799 first published October 2005. US Department of Agriculture, Washington, DC.

US Department of Agriculture/Animal and Plant Health Inspection Service. 2015. *Final Environmental Impact Statement – Feral Swine Damage Management: A National Approach*. US Department of Agriculture, Washington, DC.

West, B. C., A. L. Cooper, and J. B. Armstrong. 2009. Managing wild pigs: A technical guide. *Human-Wildlife Interactions Monograph* 1:1–55.

West Virginia Division of Natural Resources. 2018. *West Virginia Hunting and Trapping Regulations Summary July 2018 – June 2019*. West Virginia Division of Natural Resources, South Charleston, WV.

Wood, G. W. 1977. *Research and Management of Wild Hog Populations: Proceedings of a Symposium*. Belle Baruch Forest Science Institute of Clemson University, Georgetown, SC. 113 pp.

Wood, G. W., and T. E. Lynn. 1977. Wild hogs in southern forests. *Southern Journal of Applied Forestry* 1:12–17.

11 Wild Pig Policy and Legislation

Andrew L. Smith

CONTENTS

11.1 INTRODUCTION

The wild pig (*Sus scrofa*) invasion in North America has become increasingly dynamic over the last few decades, facilitating interactions across nearly all levels of organization (i.e., ecological, social, economic, and governmental). The situation, such as degree of impacts, management implications, and stakeholder involvement, has rapidly evolved, not only regarding proliferation of the species across the United States, but also in our comprehension of the issues associated with wild pigs (West et al. 2009). While most wild pig issues are typically associated with ecological and economic impacts, there are complex social, cultural, political, and legal implications as well. This has resulted in a highly variable and often polarized dichotomy of professional and public opinions on wild pigs; opinions that foster an equally variable spectrum of legislative and management responses. This multidimensional nature (Bevins et al. 2014) and increasing complexity (West et al. 2009), accompanied by shortfalls surrounding state management and authority, has resulted in a patchwork of laws and policies; many of which are unique to each state (West et al. 2009; Appendixes 11.1–11.5). While most laws and policies are conceived with intent to reduce or regulate wild pig populations by limiting activities such as hunting, their capture and release, and transportation (Centner and Shuman 2015), others are designed to pacify the public and other factions such as hunting communities, private industries, and special interest groups. Laws and policies occur through various hunting regulations such as license requirements, bag limits, and weapon restrictions, as well as through commercial ordinances such as relaxed private land regulations and facility operation requirements or standards (e.g., high-fenced shooting operations, slaughter or processor facilities). Some laws and policies are even miscellaneous products of legislative action passed in response to rapidly developing conflicts among state agencies, private industries, and the public.

These circumstances create a highly complex and controversial environment around the wild pig issue in North America, both from scientific and legal perspectives.

Outside of continental North America, offshore islands in the Caribbean have similar issues as those faced on the mainland, relative to biodiversity threats and economic damages. However, despite these Caribbean populations of wild pigs (as well as those in the Pacific) being some of the first introduced during the 15th and 16th centuries, and possibly earlier (Mayer and Brisbin 2008), attempted policy and regulation used to limit their proliferation has been scant. This could be a result of human populations having coexisted with wild pigs for more than 500 years, depending on these populations to supplement historical, pre-Columbian, island diets. Moreover, in the absence of large-scale agriculture and other economic drivers of policy (such as damage), wild pig issues are likely to be lower priority, perhaps resulting in less rigorous legislative processes to limit their populations.

Similar can be said for the provinces of Canada. In some areas, such as Manitoba, wild pigs are considered private property and fall under traditional labels for livestock breeds under the Animal Liability Act (C.C.S.M. c. A95; see Chapter 13). This restricts citizens from harvesting the species unless they are harassing or injuring livestock. Furthermore, since wild pigs are a non-native species to Canada, their regulation does not lie with the Crown. This prevents their listing under the Wildlife Act (C.C.S.M. c. W130), which would allow the use of hunting seasons to control individuals through hunter harvest. However, due to recent concerns for public safety and threats to natural resources (i.e., wildlife and wildlife habitat), an amendment was passed within the Exotic Animals Regulation to allow residents to kill escaped wild pigs from farms and high-fenced shooting operations.

Due to its complexity, a burgeoning involvement of stakeholders, and overall degenerative nature of the issue, a high volume of conservation-related legislation from US state governments has focused on limiting further spread of wild pigs. Resulting legal language has become diverse, and often counterintuitive. While many states (particularly in the Southeast) may share similar issues regarding wild pigs (i.e., impacts, population distribution, rate of expansion), profound differences in scientific, political, and legal opinions can be attributed to this kaleidoscope of legislative and management responses. Ultimately, and regardless of strategies that have been employed, many state-level decisions attempting to limit wild pig distribution and expansion failed to show meaningful progress (Centner and Shuman 2015), with some exceptions. In fact, some policy decisions have actually expedited range expansion of these animals (Bevins et al. 2014).

In addition to adaptability (Snow et al. 2017) and persistence of wild pigs, challenges to limiting their spread largely relates to increasing recreational values (Mayer and Brisbin 2009). Vast range expansion over the previous 15–20 years has been facilitated through introductions of wild pigs by humans from small, localized populations (Mayer and Beasley 2018) on a continental (Bevins et al. 2014) and perhaps irreparable scale. Products of this activity have been increases in wild pig hunting and national popularity, a competitive and partisan professional environment and an unparalleled threat to natural and economic resources. This chapter discusses the social and political consequences of wild pig distribution and range expansion, examines and evaluates legal provisions designed to limit this phenomenon, while briefly summarizing catalysts that facilitated it.

11.2 TRANSPORTATION LAWS GOVERNING WILD PIG MOVEMENT IN THE UNITED STATES

11.2.1 Intrastate Transport Provisions

Currently, wild pigs are recognized as the 2nd most popular big-game species in North America, second only to white-tailed deer (*Odocoileus virginianus*) in terms of numbers of animals harvested annually (Kaufman et al. 2004, Wildlife Society 2015). Having been introduced to every nonpolar continent (Long 2003, Barrios-Garcia and Ballari 2012), wild pigs are considered one of the most widely distributed mammal species globally (Massei and Genov 2004). In the United States, this popularity and subsequent expansion of wild pig distribution is a direct result of individuals wishing to create or augment hunting opportunities through introductions (Bevins et al. 2014, Snow et al. 2017), both illegally (Gipson et al. 1998, Mayer 2014) and legally. Despite being a highly complex and multidimensional issue, reasoning behind the success of wild pigs in their introduced range (aside from its adaptable biology and lack of effective predators) is fairly straightforward—human (or anthropogenic) assisted expansion (Snow et al. 2017); a characteristic that is common across many introduced taxa of non-native, invasive species.

Movement of wild pigs through human-mediated activities, such as transportation, has been occurring for well over a hundred years (Gipson et al. 1998, Mayer 2014). In fact, societies throughout the world have been transporting and introducing pigs to new ranges for thousands of years (Bevins et al. 2014). In the continental United States, wild pig introductions began during the 1500s for provisional purposes on exploratory expeditions, which eventually transitioned into recreational purposes for wealthy landowners by the late 1800s (Mayer and Brisbin 2008). Now, wild pig introductions represent an amalgamation of provisional and recreational purposes for a large demographic (i.e., hunting sector) of the United States, and has become widely popular and conventional. While movement of wild pigs through human-mediated activities has undoubtedly occurred over the last several hundred years in North America, only recently has it been attributed as a primary catalyst for the accelerated range expansion seen during the last decade or so (Bevins et al. 2014, Mayer 2014). Also, and of no coincidence, human-facilitated introductions appear to gradually increase alongside growing popularity of wild pig hunting.

These developments have prompted some state governments to adopt legal provisions designed to preclude transportation (legal or illegal) through various laws and penalties, with intent on limiting further statewide impacts. Regarding intrastate transport of wild pigs, 2 categories may be distinguished: 1) restrictive transport policies, which subjectively regulate pigs being transported within state lines through specific criteria (discussed below), and 2) prohibitive transport policies, which adopt a zero-tolerance policy on the movement and release of live individuals. The restrictive category appears to be most popular among states (especially in the Southeast), and requires that specific criteria be satisfied before a wild pig may be moved legally by vehicle. Such criteria include, but not limited to: 1) ear tags (or other form of visible identification), 2) disease testing with negative results, 3) trailer/hauler specifications, 4) terminal destination (slaughter facility or high-fenced shooting operation), and/or 5) granted permits or licenses. For example, in Mississippi and Georgia, and other states (Tables 11.1–11.5), it is legal to transport wild pigs after first obtaining a wild pig transportation permit or license, which is typically granted by the wildlife agency or agriculture commission (GA Code § 2-7-201, Miss Code Ann. § 49-7-140). The prohibitive (or zero-tolerance) category forbids transportation of live individuals under any circumstances by state law. Differences between regulatory categories are important to recognize, as many states claim transporting wild pigs within their respective boundaries is illegal, when in fact it is legal once the transportation criteria have been satisfied. States that adopted the prohibitive provision (excluding Alabama), are the only states to have achieved meaningful progress towards preventing spread and reducing populations of wild pigs. In addition to prohibiting transport, such states also compliment these policies with a prohibition on hunting supported by steep penalties for violators (discussed further below), in theory dissolving existing incentives for residents (or non-residents) to participate in recreational wild pig hunting activities.

As long as it remains legal to transport wild pigs through ambiguous, restrictive policy (within or across state lines), movements are passively encouraged, ultimately resulting in no meaningful reduction in numbers or range. Despite this fundamental and well-documented catalyst of wild pig expansion, many states have not passed legal provisions to preclude live transport. This not only supports the political and controversial nature of wild pig hunting, but also suggests that meaningful policies in the future are unlikely in these areas. Furthermore, neighboring states to those adopting lax transport regulations will continue to be vulnerable to wild pig dispersal and resultant increases

TABLE 11.1
State Agency Jurisdiction, Transportation Policy, and Violation Penalty in the Southern United States in 2015.

State/Source	Jurisdiction	Transportation Policy	Violation Penalty
Alabama AL Admin. Code Ch. 220-2; USDA/APHIS (2015)	Alabama Dept. of Conservation and Natural Resources	Illegal to sell or transport live individuals (No permits)	Class B misdemeanor – Fine ≥US$2,500 fine, mandatory
Arkansas AR code ann. §§ 2-38-501, 502, 503, 504; USDA/APHIS (2015)	Arkansas Game and Fish Commission Arkansas Livestock and Poultry Commission	Legal to transport with a license granted by Arkansas Livestock and Poultry Commission	Misdemeanor – Fine ≥US$1,000 per offense and/or ≤(30) day imprisonment; Equipment and vehicles used are subject to seizure
Florida FL Admin. Code 5C-21.015; 5C-3.007; USDA/APHIS (2015)	Florida Fish and Wildlife Conservation Commission Florida Dept. of Agriculture and Consumer Services	Legal to transport – if person(s) is registered as a Feral Swine Dealer (FSD) by the Florida Dept. of Agriculture and Consumer Services	N/A
Georgia Off. GA Code ann. §§ 2-7-201; 17-10-4; USDA/APHIS (2015)	Georgia Dept. of Natural Resources Georgia Dept. of Agriculture	Legal to transport with a Feral Hog Transport Permit issued by Dept. Of Agriculture	Misdemeanor (of high and aggravated nature) ≥US$1,500 fine
Kentucky Kt REV. STAT. §§ 150.186-150.990; USDA/APHIS (2015)	Kentucky Dept. of Fish & Wildlife Resources	Illegal to transport live individuals	Class A misdemeanor – ≤US$500 and ≥(90) day ≤ (12) mo imprisonment; Revocation of hunting, trapping, and fishing rights for (10) yr
Louisiana LA. admin. Code ch. 21 §§ 1301-1321; USDA/APHIS (2015)	Louisiana Dept. of Wildlife and Fisheries (LDWF) Louisiana Dept. of Agriculture and Forestry	Legal to transport with LDWF-issued ear tags or a Feral Swine Transporter registration permit via Board of Animal Health (5 year renewal)	N/A
Mississippi MS code ann. §§ 49-7-140; USDA/APHIS (2015)	Mississippi Dept. of Wildlife, Fisheries & Parks	Legal to transport with a Live Wild Hog Transportation Permit	Class I violation – ≥US$2,000 fine, (5) day imprisonment, and revocation of hunting, trapping, and fishing rights for (1) yr
North Carolina NC Gen. Stat. §§ 106-798, 106-798.1; USDA/APHIS (2015)	North Carolina Wildlife Resources Commission	Legal to transport with permit	Misdemeanor – Fine ≤US$5,000 per offense
South Carolina SC code §§ 50-16-20, 50-16-25; USDA/APHIS (2015)	South Carolina Dept. of Natural Resources Clemson Livestock & Poultry Health	Legal to transport – with permit and tags	Misdemeanor – Fine ≤US$1,000 and/or ≤(6) mo imprisonment; Revocation of hunting privileges for (1) yr

(Continued)

TABLE 11.1 (CONTINUED)
State Agency Jurisdiction, Transportation Policy, and Violation Penalty in the Southern United States in 2015

State/Source	Jurisdiction	Transportation Policy	Violation Penalty
Tennessee TN code ann. §§ 44-2-102; USDA/APHIS (2015)	Tennessee Wildlife Resources Agency	Illegal to transport live individuals	Class A misdemeanor – Fine ≥US$2,500 per offense
Virginia VA Admin. Code §§ 4:15-30-40; USDA/APHIS (2015)	Virginia Dept. of Game and Inland Fisheries	Legal to transport – with permit issued by VDGIF	N/A
West Virginia WV Code §§ 20-2-12; §§ 61-1-7.16; USDA/APHIS (2015)	West Virginia Dept. of Natural Resources West Virginia Dept. of Agriculture	Legal to transport and release "feral" swine but illegal to transport "wild" boar	Misdemeanor – Fine ≤US$500 and/or ≤(12) mo imprisonment with graduating offenses

Note: In the absence of available legislation, the 2015 US Department of Agriculture Feral Swine Environmental Impact statement was referenced. Entries have been standardized where possible.

in subsequent impacts. The crux with permitting or licensing transport of wild pigs, whether to a "terminal facility" or any proclaimed location, is that people who wish to introduce them illegally go through the same procedures as those who are operating within the law and with legitimate purpose. Once an applicant obtains necessary paperwork and is legal, all that remains is to make the illegal release. For instance, a person obtains a wild pig, and that individual then follows procedures to obtain a permit through the state wildlife agency or agricultural commission, proclaiming that the pig is being taken to a slaughterhouse. A background check or similar process (some states do neither) ensures the applicant does not have a history of offenses, and if not a permit is granted. It is also worth noting here that many of those engaging in illegal activity may bypass any permit or legal process all together and transport wild pigs regardless of what measures are in place. Now permitted, the individual can legally move the animal to their presumed location (e.g., slaughter facility, holding pen, high-fenced shooting operation), but movement is no longer monitored. The individual is now at liberty to move the pig via intrastate transport (legally) with potential intent of releasing it (illegally) at their discretion. This fundamental issue with wild pig movement severely inhibits meaningful enforcement actions while the pig is in transit, leaving law enforcement with only a small window of opportunity to prevent illegal releases (C. Lewis, Alabama Department of Conservation and Natural Resources, personal communication). Essentially, the small enforcement window is when the pig has reached its final destination. It is only at this point that officers can prove that the individual is no longer in compliance with their permit, and are intending to make an illegal release. It is also worth noting that only a few releases of wild pigs can result in establishment of populations and significant damage, and perhaps in some states or areas the effects are irreversible.

If entirely prohibiting transportation (zero-tolerance) is not a realistic legislative option (such as for states in the Southeast), states should develop and implement other approaches. For example, there is a fundamental need for a mechanism that allows state agencies issuing such wild pig transportation permits and licenses to ensure applicant and recipient accountability, confirming that pig(s) indeed arrived at the proclaimed destination. Furthermore, agencies should rigorously monitor and keep record of recipients, as well as denied applicants. Knowledge of counties where these permits originate and terminate would be valuable for monitoring trends and identifying key areas

TABLE 11.2
State Agency Jurisdiction, Transportation Policy, and Violation Penalty in the Northeastern United States in 2015

State/Source	Jurisdiction	Transportation Policy	Violation Penalty
Connecticut CT Gen. Stat. §§ 22-278; USDA/APHIS (2015)	Connecticut Dept. of Energy and Environmental Protection Connecticut Dept. of Agriculture	N/A for feral swine	N/A for feral swine
Delaware DE Code §§ 7201, 7202, 7203; USDA/APHIS (2015)	Delaware Dept. of Natural Resources and Environmental Control Delaware Dept. of Agriculture	Illegal to possess live or dead	Fine ≤US$500 per offense and/or ≤ (30) day imprisonment
Maine MDA Rules Ch. 223; 206; USDA/APHIS (2015)	Maine Dept. of Agriculture, Conservation and Forestry	Importation permit required and permanent identification	N/A
Maryland USDA/APHIS (2015)	Maryland Dept. of Natural Resources	N/A	N/A
Massachusetts USDA/APHIS (2015)	Massachusetts Dept. of Energy and Environmental Affairs, Division of Fisheries and Wildlife	Importation permit required	N/A
New Hampshire NH Rev. stat. ann. §§ 467:3, 467:5 USDA/APHIS (2015)	New Hampshire Fish and Game Dept. New Hampshire Dept. of Agriculture, Markets and Food	N/A	N/A
New Jersey NJ Rev. stat. §§ 23:4-63.3, 63.4; USDA/APHIS (2015)	New Jersey Dept. of Natural Resources	Legal to transport with permit	Fine ≤US$500
New York NYS ECL §§ 11-0514; §§ 71-0925; USDA/APHIS (2015)	New York Dept. of Environmental Conservation	Illegal to possess or transport	Fine ≤US$500 per animal in possession with graduating offenses
Pennsylvania 9 CFR §§ 78.30; USDA/APHIS (2015)	Pennsylvania Dept. of Agriculture Pennsylvania Game Commission	Legal to transport with permit	None found in CFR
Rhode Island RI Gen. Laws §§ 4-14-1; §§ 4-15-4; §§ 4-18-3; §§ 4-18-14; USDA/APHIS (2015)	Rhode Island Dept. of Environmental Management	Illegal to possess or import exotics (Subject to addition of Suidae family)	Fine ≤US$100 and confiscation of specimen; Fine ≤US$100 + damages
Vermont 10 Vt stat. ann. §§ 4709; USDA/APHIS (2015)	Vermont Fish and Wildlife Dept. Vermont Agency of Agriculture, Food, and Markets	Illegal to import or possess	N/A

Note: In the absence of available legislation, the 2015 US Department of Agriculture Feral Swine Environmental Impact statement was referenced. Entries have been standardized where possible.

TABLE 11.3
State Agency Jurisdiction, Transportation Policy, and Violation Penalty in the North-Central United States in 2015

State/Source	Jurisdiction	Transportation Policy	Violation Penalty
Illinois Il. Admin. Code §§ 17:700-700.30; §§ 17:2530 USDA/APHIS (2015)	Illinois Dept. of Natural Resources	Legal with permit and health certificate	Class A misdemeanor – Fine ≤US$2,500 and <(1) yr imprisonment, possible revocation of licenses and permits
Indiana In. Admin Code §§ 312:9-3-18.6; USDA/APHIS (2015)	Indiana Dept. of Natural Resources Indiana Board of Animal Health	Legal if transported for immediate euthanasia	N/A
Iowa Ia Admin. Code r. 21-77.1-14; USDA/APHIS (2015)	Iowa Dept. of Natural Resources	Illegal to transport "feral" swine, but legal if "nonferal" = w/ disease testing	Civil penalty ≤US$1,000 with continuing violation each day of ≤US$25,000
Michigan MI Comp. Laws 324 §§ 41303; USDA/APHIS (2015)	Dept. of Natural Resources Dept. of Agriculture and Rural Development	Illegal to possess and import	Felony – Fine ≥US$2,000 but ≤US$20,000 and/or ≤(2) yr imprisonment
Minnesota Mn stat. 84D.11-13; USDA/APHIS (2015)	Minnesota Dept. of Agriculture	Legal to transport with permit	Misdemeanor – Fine ≤US$1,000 and/or ≤(90) day imprisonment
Missouri MO Rev. Stat. §§ 270.400, 270.270; USDA/APHIS (2015)	Missouri Dept. of Agriculture	Legal if transporting from farm to farm, directly to slaughter, or slaughter-only market	Class A misdemeanor – Fine ≤US$2,000 and/or (1) yr imprisonment per offense
Ohio Oh Admin. Code §§ 901: 1-11-07; USDA/APHIS (2015)	Ohio Dept. of Natural Resources Ohio Dept. of Agriculture	Legal with official identification (CVI)	N/A
Wisconsin Wi NR 16.11; NR 40; Stat. 951.18; USDA/APHIS (2015)	Wisconsin Dept. of Natural Resources Wisconsin Dept. of Agriculture, Trade and Consumer Services	Legal to transport with permit	Not Found in 951.18

Note: In the absence of available legislation, the 2015 US Department of Agriculture Feral Swine Environmental Impact statement was referenced. Entries have been standardized where possible.

where pig hunting may be occurring, as well as any illegal activity. State wildlife and agriculture agencies may also consider additional screening methods or procedures which ensure that recipients of permits and licenses are vetted and trustworthy, ultimately becoming a more exclusive process.

11.2.2 Interstate Transport and Import Provisions

Under the current US federal system, regulative authority over wild pigs resides with states (West et al. 2009, USDA/Animal and Plant Health Inspection Service 2015). From a national perspective, the injurious species provision of the Lacey Act (16 U.S.C. §§ 3371–3378) protects natural and agricultural resources of the United States from interstate commerce of such species (with an emphasis on non-native and invasive species), but provides little to no assistance as wild pigs cannot be listed as injurious wildlife due to lack of federal authority to regulate subspecies (Centner and Shuman 2015). Instead, the Code of Federal Regulations allows movement of wild pigs among states if in compliance with federal and state laws where the animal originated. For compliance, wild pigs must be: 1) permanently marked by an identification tattoo or other approved swine identification tag, 2) moved directly to slaughter (or other authorized location), avoiding physical contact with other livestock while in route, 3) accompanied by a permit issued by the APHIS representative or state animal health official from where it originated, and 4) is found to be negative of disease from an official test within 30 days prior to interstate movement (9 C.F.R. §§ 78.30). Nevertheless, some states may choose to adopt these same (relaxed), or similar measures to allow import into the state (Tables 11.1–11.5).

The current regulatory framework on movement of wild pigs considerably fetters authority of the Federal Government, and thus ability to regulate inter and intrastate movement of wild pigs. With the

TABLE 11.4
State Agency Jurisdiction, Transportation Policy, and Violation Penalty in the Midwestern United States in 2015

State/Source	Jurisdiction	Transportation Policy	Violation Penalty
Kansas KS ann. Stat. §§ 47-1809; USDA/APHIS (2015)	Kansas Dept. of Agriculture	Illegal	Civil penalty – Fine ≥US$1,000 and ≤US$5,000 per offense
Nebraska NE rev. Stat. §§ 37-524; USDA/APHIS (2015)	Nebraska Game and Parks Nebraska Dept. of Agriculture	Illegal	Class IV misdemeanor – Fine ≥US$100 and ≤US$500
North Dakota ND Cent. Code §§ 36-26-03, 05; USDA/APHIS (2015)	North Dakota Board of Animal Health	Illegal	Civil penalty – Fine ≤US$5,000 per offense and total costs of control or eradication incurred from the violation
Oklahoma Ok Stat. Title 2 Ch. 1 Article 6 (FSCA); USDA/APHIS (2015)	Oklahoma Dept. of Agriculture, Food, and Forestry	Legal to transport with permit	Felony – Fine US$2,000 and/or (2) yr imprisonment
South Dakota SD admin. rule §§ 12:68:18:07.01, 12:68:18:03.01; USDA/APHIS (2015)	South Dakota Game, Fish, and Parks South Dakota Animal Industry Board	Legal with CVI and permit	N/A
Texas Tx Admin. Code Title 4 §§ 55.9; USDA/APHIS (2015)	Texas Parks and Wildlife Texas Animal Health Commission	Legal	Non-specified

Note: In the absence of available legislation, the 2015 US Department of Agriculture Feral Swine Environmental Impact statement was referenced. Entries have been standardized where possible.

TABLE 11.5
State Agency Jurisdiction, Transportation Policy, and Violation Penalty in the Western United States in 2015

State/Source	Jurisdiction	Transportation Policy	Violation Penalty
Arizona Az Admin code §§ R3-2-602, 613; §§ R12-4-401, 406 (interpreted as including "feral swine"); USDA/APHIS (2015)	Arizona Dept. of Agriculture	Legal w/ official health certificate meeting swine entry requirements	N/A regarding "feral swine"
California CA code regs. Title 14 §§ 671; Fish & Game Code §§ 2118; USDA/APHIS (2015)	California Dept. of Fish and Wildlife	Legal to transport with permit	Civil penalty ≥US$500 and ≤US$10,000; Misdemeanor – ≤(6) mo imprisonment or fine ≤US$1,000
Colorado Co rev. stat. ann. §§ 33-16-114; CO Gen. Prov. Ch. W-0 Article 2 #002(M); Article VI #008; USDA/APHIS (2015)	Colorado Parks and Wildlife Colorado Dept. of Agriculture	Illegal	Misdemeanor – Fine ≥US$250 and ≤ US$1,000; (5) license suspension points per offense
Idaho Id Code §§ 36-202(g); §§ 22-1905, 1913; USDA/APHIS (2015)	Idaho Dept. of Fish and Game Idaho State Dept. of Agriculture	Legal to transport with permit	Misdemeanor – Fine ≤US$3,000 and/or ≤(12) mo imprisonment; May receive civil penalty ≤ US$10,000 per offense
Montana Admin. Rules of MT 12.6.1540-1541; USDA/APHIS (2015)	Montana Fish, Wildlife & Parks	Illegal	N/A
Nevada Nev. Rev. Stat. 569; Nev. Admin. Code 504; USDA/APHIS (2015)	Nevada Dept. of Agriculture	Illegal	Gross misdemeanor – Fine ≤US$2,000 and/or ≤ (1) yr imprisonment
New Mexico NM Stat. §§ 77-18-6; USDA/APHIS (2015)	New Mexico Livestock Board	Illegal	Misdemeanor – Fine ≤US$1,000 and/or ≤(1) yr imprisonment
Oregon Or Admin. Rules §§ 635-056-0050; USDA/APHIS (2015)	Oregon Dept. of Fish & Wildlife Oregon Dept. of Agriculture	Illegal	N/A
Utah USDA/APHIS (2015)	N/A	N/A	N/A
Washington USDA/APHIS (2015)	Washington Dept. of Fish & Wildlife	N/A	N/A
Wyoming WGFC Ch. 10 §§ 3(b)(ii)(A); USDA/APHIS (2015)	Wyoming Game and Fish Dept. Wyoming Livestock Board Wyoming Dept. of Agriculture	Illegal	N/A

Note: In the absence of available legislation, the 2015 US Department of Agriculture Feral Swine Environmental Impact statement was referenced. Entries have been standardized where possible.

US government adopting restrictive-type policies (as opposed to prohibitive) previously discussed, wild pigs will continue to be transported, localized, and eventually naturalized (see Chapter 6), spreading their impacts and providing opportunities for additional illegal activity. Note that as new introductions of wild pigs occur, and these populations progress from small and isolated to established and connected, available policy and management responses become narrowed (Lodge et al. 2006), and the reality of meaningfully reducing their populations and resulting impacts become less likely.

11.3 LEGAL STRATEGIES FOR LIMITING WILD PIGS

11.3.1 Species Classification

Hunting regulations, transportation policies, and other laws pertaining to wild pigs vary considerably based on legal classification of the species within states. As with nominal variation used for wild pigs by laypersons (e.g., feral swine, wild hog, feral hog, and wild boar), a similar assortment of legal classifications exist throughout the United States. These include several variations of game animal, nuisance animal, invasive species, wildlife, or other species (Appendices 11.1–11.5). The means that allow individuals to participate in hunting, trapping, or transporting wild pigs depend on how states identify the species within the law. Legal language of these categories varies from strongly worded classifications such as "outlaw quadruped" (Louisiana) and "destructive species" (Tennessee) to "public nuisance" (Arkansas) and "other" (Kentucky; Wildlife Society 2015). What causes state legislatures and commissions to adopt specific language appears quite variable and obscure; however, one might conclude that encouraging or discouraging residents and non-residents from participating in various wild pig activities underlies these decisions.

11.3.2 Recreational Hunting

Several states have enacted legal provisions in attempts to enlist public assistance through hunting and other various methods of harvest on public and private lands. In addition to placing restrictions on intrastate transport, state governments have erroneously sought to reduce wild pig numbers through legally designating the animal as a state-listed game species. This method currently applies to Alabama, California, Hawaii, Ohio, and West Virginia (Mayer 2014, USDA/Animal and Plant Health Inspection Service 2015). In an effort to mobilize hunters in controlling animal densities and subsequent damages, this strategy became evident in several states during the mid-1950s (Keiter et al. 2016). As with bounties (discussed below), this approach creates demand for the species and frequently leads to population expansion rather than reduction (Bevins et al. 2014). For example, in California wild pigs were designated as a state-listed game species beginning in the 1956–1957 hunting season (Kreith 2007). During the time leading up to this legal classification, wild pig populations and their state distribution remained somewhat stable and limited. By the 1980s, the population had expanded from only a few coastal areas to 33 counties (Waithman et al. 1999). Today, wild pigs occur in 57 of 58 counties in California (Southeastern Cooperative Wildlife Disease Study 2016). However, it is worth noting that the recreational hunting approach enables states to generate much needed revenue used for indefinite control and maintenance of wild pig populations. From 2002 to 2007, California Department of Fish and Game totaled nearly US$3.2 million through wild pig tag revenues (Kreith 2007). These revenues are derived from sale of resident and non-resident tags, licenses, and/or permits, and have wide application regarding ability of the agency to purchase fuel and equipment, hire staff, trappers, and aerial gunning services to ameliorate impacts precipitated by wild pigs.

Excluding benefits from tags and license sales, a similar result was evident in Tennessee, where populations of wild pigs only existed in 6 counties over a period of 30 years (1950s–1980s). Despite being considered relatively limited in distribution by current standards, in an additional effort to reduce the distribution further, the Tennessee Wildlife Resources Agency (TWRA) installed a statewide year-long wild pig hunting season (Bevins et al. 2014). Since the institution of the liberal hunting season, wild pig populations have significantly expanded at an accelerated rate from 18 counties in 1988, to 58 counties in 2016 (Southeastern Cooperative Wildlife Disease Study 2016; Figure 11.1). Concordantly, the frequency of reports regarding crop and property damage in Tennessee has also increased (Bevins et al. 2014).

Customarily following establishment of a hunting season or classification to game status, state agencies have sought to compliment hunting regulations to assist in higher rates of harvest by: eliminating bag limits, reducing weapon restrictions, and relaxing license requirements and restrictions on the time of day when harvest of wild pigs may occur. However, manipulation of state hunting regulations, in effort to increase harvest, also occurs in areas with high wild pig abundance, whether listed as a game species or not. On public lands where pig hunting is legal, the vast consensus among states is: 1) wild pigs may only be harvested using the applicable firearm(s) and ammunition for the current hunting season in progress, 2) with no bag limit imposed, and 3) a requirement of a state-issued hunting license. While some states may create or limit hunting seasons to preserve and maintain pig populations for hunting revenues, others may also impose a limited hunting season to mitigate human related pressures and disturbance during active breeding seasons of game species such as squirrel, turkey, and deer (Alabama Department of Conservation and Natural Resources 2015, Louisiana Department of Wildlife and Fisheries 2015).

On private lands, some states eliminated the requirement for landowners to possess a state-issued hunting license for hunting and removing wild pigs on their property. Additionally, some have legalized night hunting and expanded the equipment used to harvest wild pigs, such as firearms and high-capacity magazines, suppressors, thermal or infrared optics, and lights. It is clear that

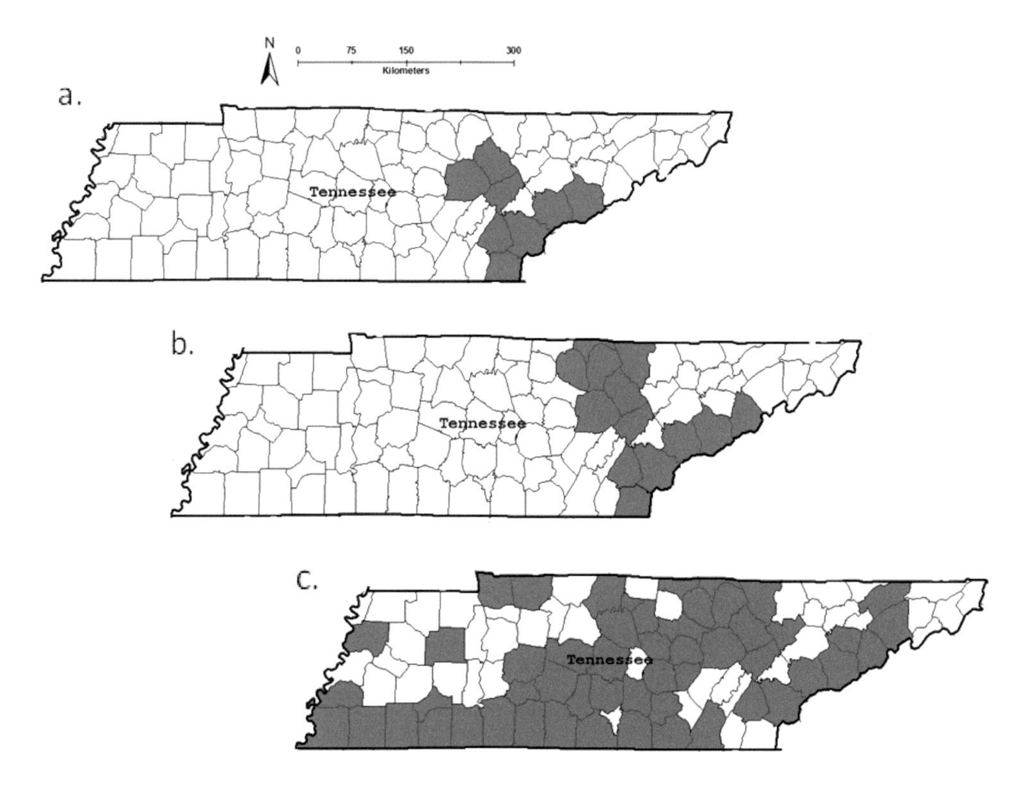

FIGURE 11.1 Wild pig distribution in Tennessee, showing known breeding populations (a) before 1950; (b) in 1988, prior to the open-season hunting program; and (c) in 2012, after the hunting program ceased in 2010. The data were provided by Daryl Ratajczak and Chuck Yoest, of the Tennessee Wildlife Resources Agency. (Reproduced from Bevins et al. 2014.)

expanding these regulations is intended to assist hunters in harvesting greater numbers of wild pigs. However, while no data currently exists to support this assumption, one might conclude that allowing such recreation may further incentivize the sport, unless this equipment is used objectively and exclusively with the goal of meaningful population reduction and not for monetary gain through outfitting. Based on social drivers such as hunter motivations (Ditchkoff et al. 2017), one might also conclude that the recent surge in popularity of wild pig hunting could be associated with methods used to harvest them.

11.3.3 Bounties

For some species perceived as hazardous to public safety or economic activity, federal and state governments have historically employed a bounty system. This system enlisted public help and support to reduce population numbers and where possible, locally eradicate deleterious species. It allows a citizen to commercially harvest a species and submit some form of proof (often the tail, ear, or other diagnostic body part) to local government, agency, or private entity for a reward. While bounties may have successfully facilitated the dispatch of large quantities of species of concern, they have largely been ineffective at reaching their designated objective of reducing populations to a manageable size (Palmer et al. 2007, Ditchkoff et al. 2017). For example, in 1952 the British government sought to offer bounties on grey squirrels that lasted 5 years, resulting in disposal of 1,000,000 squirrels (Palmer et al. 2007). However, the bounty program eventually dissolved, as squirrel populations were not reduced to manageable levels. In fact, some officials suggested that the problem had become more severe despite the bounty effort (Palmer et al. 2007).

The fundamental flaw, as evident in other cases that employed bounty programs—such as a program installed at Ft. Benning, a US military base in Georgia—is the kindling of incentives that encourage participants to exploit the system. This can occur by participants fraudulently submitting body parts from domesticated individuals or animals outside the target population, through continued reintroduction of the species to sustain income and opportunity from the bounties, or likely a combination of the 2. As noted by Ditchkoff et al. (2017), ineffective bounties result when incentives (i.e., monetary gain and recreational motivations) outweigh risks of punishment. At Ft. Benning, hunters submitted tails from each wild pig they harvested and were paid US$40.00/tail. The ultimate result of the program was failure, as participants were eventually reported to be procuring pig tails from meat processors and submitting them for payment, resulting in an expensive bill accrued by the military base (i.e., money exchanged to the participants), in addition to yielding no desired outcome (i.e., no reduction in the wild pig population; Bevins et al. 2014, Ditchkoff et al. 2017). Incentivizing production of a commodity targeted for removal generates this fundamental flaw and disables effectiveness of the bounty system at reducing target populations, and often worsens the situation (Palmer et al. 2007, Bevins et al. 2014). Ditchkoff et al. (2017) concluded that wild pig density, sounder size, and juvenile:adult female ratio had in fact increased on the military base, in some areas drastically (i.e., wild pig populations doubled in the southern area by the end of the study period). In addition to the target population increasing, estimated costs accrued by the military base were substantial (i.e., US$57,296). In review, one might conclude that in desire to harvest trophy boars, and over-incentivized reward (i.e., US$40/tail compared to other analogous programs at US$5/tail), facilitated fraudulent activity and ultimately influenced failure of the program.

Other shortcomings of the bounty system approach to control wild pigs is failure to elicit sufficient public participation to sustain the program when harvest incentives are perceived as low (Ditchkoff et al. 2017). Furthermore, maintaining ample participation once a significant amount of the population has been removed can be difficult (Bomford and O'Brien 1995); often the result of increased costs and removal efforts associated with lower population densities and educated individuals or groups within the target population (i.e., trap-shy pigs; N. Dornak, Caldwell County Feral Hog Task Force, personal communication). Additionally, hunter preferences (Ditchkoff et al.

2017) and motivations (Bartel and Brunson 2003) may also negatively affect efficacy of the bounty approach, as hunters may bias effort towards what they perceive as trophy animals—in this case large male boars. As such, hunters disregard fecund females, thereby offering little to no reduction in recruitment or reproduction (Hanson et al. 2009). A study on the coyote (*Canis latrans*) bounty program in Utah showed that program participation was unrelated to monetary gain, but instead was more aligned with ideas of a positive outdoor experience and increased big-game hunting opportunities (Bartel and Brunson 2003). Hence, wild pig hunting enthusiasts and their actions may detract from meaningful reductions in targeted pig populations, simply in an effort to sustain their recreational opportunities and trophy animal motivations. Despite the volume of well-documented cases and subsequent issues created by bounty programs, the use of bounty programs as management tools continues to be proposed as a potential solution to controlling animal densities, especially those perceived as invasive or pests (Ditchkoff et al. 2017).

Effectiveness of bounty systems employed at smaller spatial scales (i.e., county or community level) may be less obscure than at larger scales (e.g., state level). Localized programs can be measurable, transparent, and accountable. For example, the Caldwell County Feral Hog Task Force (CCFHTF) established in Texas in 2013 has illuminated benefits resulting from established, well-organized and coordinated bounty and control programs, such as community awareness. After the initial 2 years of the program, the task force reported removing up to 65% of the wild pig population in the county. Of this 65%, the CCFHTF reported that 71% resulted from the bounty program, 12% from aerial gunning services, 11% from voluntary reporting and 6% from professional trapping. Program participation not only actively removed wild pigs from the landscape through the bounty system, but also allowed landowners to qualify for cost-sharing programs to assist with purchasing trapping equipment and aerial gunning services which were procured through a contract with a private entity. However, after a 2-year period, reporting and participation in the program diminished significantly. Despite a 50% reduction in trappers and hunters actively reporting their harvest from 2015 to 2017, the task force reported that harvest declined and hunter and trapper effort was increasing (N. Dornak, personal communication). This suggests: 1) local pig populations are declining and 2) remaining pigs are becoming more difficult to remove (populations are less dense and more trap-shy). The obvious flaw is that public participation wanes after initial successes of the program. However, due to public involvement, the program established a credible relationship with the local community, providing a cornerstone for outreach, education, and thus involvement. They achieved this through numerous workshops, enhanced communication with stakeholders, and extensive media coverage. According to N. Dornak (personal communication) the latter became an essential component of the program as it facilitated mass communication and increased the outreach capabilities of the program, in addition to increasing popularity of the program and community awareness of issues associated with wild pigs.

Public participation in such efforts is critical and without public support, programs are likely to fail. In the case of Caldwell County, with public buy-in and support, the community has managed to outnumber, and thus curtail the majority of illegal activity that would lead to further proliferation of wild pigs (N. Dornak, personal communication). This fundamental accomplishment by the task force suggests and supports the concept that social toleration of wild pigs directly relates to degree of proliferation. Furthermore, if community buy-in is feeble and eventually leads to program failure, then resources necessary to implement such a program are wasted. In neighboring Hays County for example, a program similar to the one implemented in Caldwell County was introduced and after initial efforts, deemed unsuccessful. Hays County is considered to be more of a recreationally oriented community with only ~56% of land in agriculture, in contrast to the agricultural community of Caldwell County that is comprised of nearly 90% farmland (N. Dornak, personal communication). Consequently, land-use practices and community ideals in Hays County are more aligned with recreational opportunities and revenues generated from the hunting industry, leading to more interest in hunting opportunity versus income generated from crop or livestock production. This result would theoretically lead to less community support for eradication and control of a big-game commodity, despite competition with native wildlife and the ecosystem damages wild pigs cause. Therefore, implementing eradication and control programs in areas like Hays County would be remissive as they are unlikely to succeed, resulting in a waste of resources that could be more effectively applied in areas where community objectives align with wild pig population reduction rather than expansion. This concept, based on community and/or cultural demographics and ideals, should be one of the first formulaic considerations when attempting to employ bounty and control programs that involve the public, if meaningful population reduction is the goal.

11.3.4 Recreational Hunting Prohibition and Restrictions

Counterintuitive to control of wild pig populations, several states enacted laws and regulations that prohibit or strongly discourage hunting of wild pigs. Reasoning behind this strategy is that removing or reducing existing incentives to hunt pigs eliminates or severely confines hunting to designated areas. Moreover, by reducing hunting opportunities, incentives for hunting enthusiasts to translocate wild pigs (legally or illegally) are minimized. While the proximate success of this legal strategy may be evident in some states, its ultimate success from a national or regional perspective is more obscure. States that have a robust cultural history of wild pig hunting may not share the same successes as states where wild pig hunting is not culturally entrenched (Figures 11.2, 11.3). For example, Kansas and New York successfully reduced wild pig populations and prevented range expansion (Bevins et al. 2014), with assistance of this legislation. The spatial extent of the wild pig population in Kansas peaked in 2009–2010 in 21 counties, covering a geographic area of ~4,978 km^2. Subsequent to banning hunting and the implementation of other control programs, the population was reduced to only 12 counties by 2015 (~794 km^2; Southeastern Cooperative Wildlife Disease Study 2015). In New York, wild pigs occurred in 10 counties in 2012 covering a geographic area of ~850 km^2 (Southeastern Cooperative Wildlife Disease Study 2012). By 2016, New York proclaimed

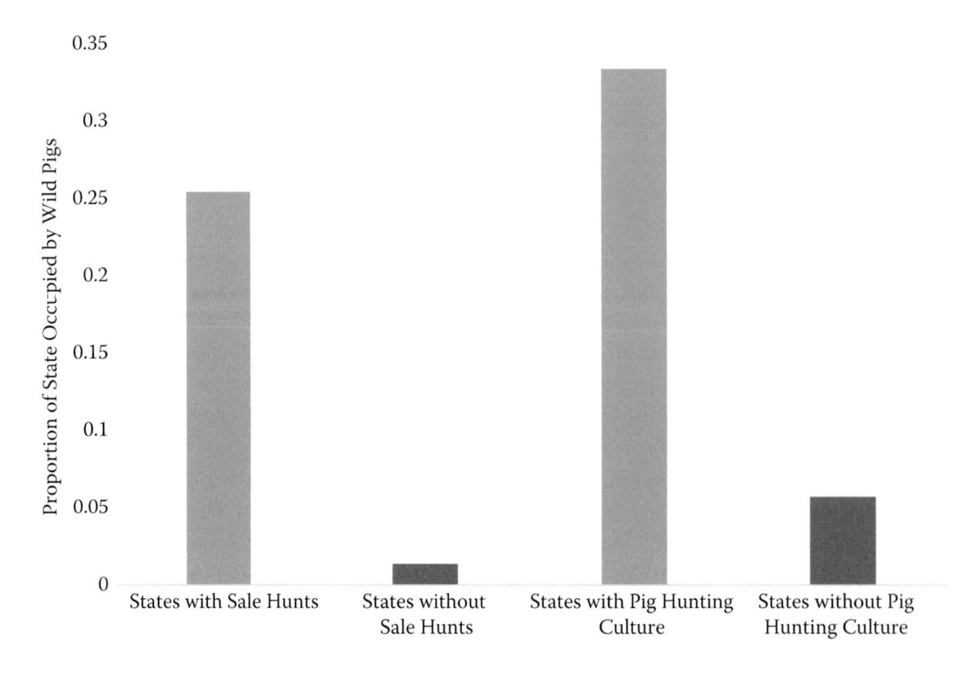

FIGURE 11.2 Wild pig distributions among states with and without pig hunting cultures and private land sale hunts.

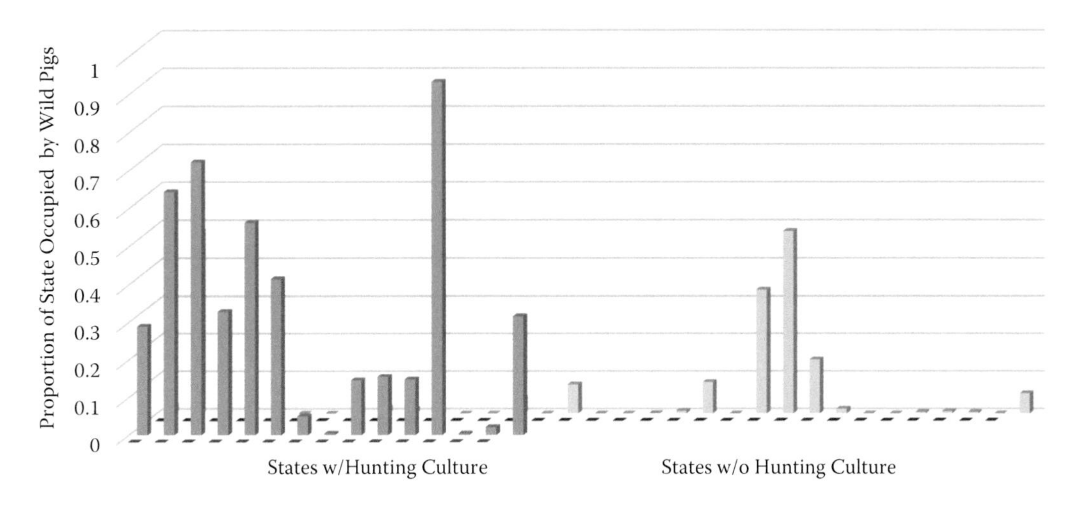

FIGURE 11.3 Wild pig distributions in states with and without historic pig hunting cultures.

complete eradication of wild pigs (Southeastern Cooperative Wildlife Disease Study 2016). Similar results were observed in Arizona where population distribution peaked in 2012 in 8 counties (~8,211 km^2) but fell to 6 counties in 2016 (Southeastern Cooperative Wildlife Disease Study 2012) after initiating a hunting ban.

In some ways analogous to banning or restricting recreational hunting, several states have partly modeled this approach by permitting only "incidental take" on public land. In Mississippi for example, hunters may take wild pigs incidentally while deer hunting during rifle season on some state wildlife management areas (Mississippi Department of Wildlife, Fisheries and Parks 2015). This opportunistic approach only permits hunters to engage in pig harvest if the opportunity presents itself while deer hunting, in contrast to focused recreational pursuit of the species. Enforcement of this incidental take provision also occurs in the remaining designated focal areas where wild pig hunting is legal in Tennessee (Tennessee Wildlife Resources Agency 2015). Success of this regulation is debatable. One could argue that it may help abbreviate the practice of hunting pigs on public lands, thus slowing illegal transport and release onto other public lands where recreational hunting of pigs is not tolerated (similar to the hunting ban provision). However, one could easily conclude that "incidental take" programs have not resulted in significant reductions in population sizes, although they may have aided in precluding establishment of new populations in some areas. However, these effects are difficult to ascertain or prove due to the saturated distribution of wild pigs in Mississippi during the past several years.

11.3.5 Commercial Enterprise Regulations (Outfitting and High-Fenced Shooting Operations)

The burgeoning popularity to hunt wild pigs has gained notable momentum over the past decade. As a result of overabundant populations, hunter motivations, and various entertainment platforms (i.e., television and social media), wild pig hunting has become a conventional practice and demand appears to be increasing. This has provided an opportunity for landowners, hunting outfitters, and other private enterprises to capitalize and supplement incomes through various services related to wild pigs. Such commercial endeavors include, but are not limited to, meat processors and transitory holding facilities, hunting guide services and property leasing, trap fabrication and manufacturing, as well as various bait and lure products, amongst others. While a diversity of ideas and subsequent attempts to exploit wild pigs as a revenue source have materialized, traditional hunting services have generated the most success, relative to income and personal gain. While some operations exist and rely on non-discriminatory harvest (i.e., either gender or age class), many specialize and charge additional fees for hunting "trophy boars," where hunters pay a premium to harvest large mature males. Such activities not only perpetuate wild pig hunting and its popularity, but also contribute further to wild pig proliferation. As a result, many states recognize issues created by this practice and are taking legal action to constrict and prevent its further establishment. One mechanism is through legal provisions, which preclude landowners from charging fees for residents and non-residents to hunt wild pigs. By doing this, states hope to eliminate incentives for landowners to sustain existing populations or introduce new individuals onto their properties. Similar steps are occurring on public lands. For example, Colorado Parks and Wildlife prohibit commercial hunting or taking of wild pigs, as well as receiving any compensation from their elimination (2 CO Code Reg. 406-0).

While intentional introductions of wild pigs occurred in the United States since the late 1800s and remained somewhat of an exclusive practice throughout the mid to late-1900s (Mayer and Brisbin 2009), recent surge in popularity of wild pigs over the past 10–15 years has facilitated establishment of more recent transitionary facilities and high-fenced shooting operations. Escapes from high-fenced shooting operations and other holding facilities, whether intentional or negligent, are a leading source of local pig populations (Mayer and Brisbin 2008, Plasters et al. 2013, Snow et al. 2017). Recent studies in California (Tabak et al. 2017) and Florida (Hernández et al. 2018) showed that proximity to such facilities was a significant indicator and predictor of wild pig immigration and genetic diversification from source populations, shedding light on how this practice contributes to range expansion in the United States. This has prompted state governments to pass legal provisions that either curtail or preclude further establishment of high-fenced shooting operations, with these regulations often intending to phase out existing practices within several years (University of Arkansas Cooperative Extension Service 2015). Dissolving the practice of high-fenced shooting operations for wild pigs is a legislative priority for several states including Arkansas, Minnesota, New York, and Tennessee (USDA/Animal and Plant Health Inspection Service 2015). To further regulate high-fence activity, some states have enacted rules requiring operators to keep track of individual wild pigs within their facility. For example, Iowa requires owners of high-fenced shooting operations to equip each of their stock with an implanted electronic identification device, which digitally contains the registration information of the owner (IA Administrative Code § 21–77.5). Such measures will prove useful in situations where pigs may escape, and help agencies monitor individuals who participate in purchasing or selling pigs (or other pig activities) by requiring their electronic registration.

Regardless of strategies used, many state agencies that attempt to limit distribution and expansion of wild pig populations frequently encounter opposition from the hunting community (Centner and Shuman 2015) and other various private enterprises and industries. For example, in 2013 the Pennsylvania Game Commission publicly announced a proposal to assist in complete elimination of wild pigs from the state (Pennsylvania Game Commission 2013). Owners and operators of high-fenced shooting operations viewed the proposal as a threat to their tradition and vocation, and before the final regulation could go into effect, the Pennsylvania legislature passed a bill stating that the Game Commission has "no authority to promulgate regulations on swine hunting preserves" (PA Statutes Annotated § 2390). As a result, high-fenced shooting operations in Pennsylvania continued business operations as usual and provided owners of rural lands with sources of income (Centner and Shuman 2015). More examples of this opposition also have occurred in other areas of the United States. In early 2017 when the Environmental Protection Agency (EPA) approved registration of the toxicant bait Kaput®, several states expressed immediate interest in using the toxicant for wild pig control. Texas, which may have the largest volume of industry associated with wild pigs, encountered immediate opposition to (and from) their legislators and lead administrative officials, the bait manufacturer (Scimetrics Ltd. Corp.), and other stakeholders from various interest groups (Statesman Media 2017). The ultimate result of this proposal was a threat of mass litigation in Texas courts from various interest groups vs. Scimetrics Ltd. Corp, and anyone who might use the bait (Dover and Marston 2017), leading to a complete withdrawal of the proposal from state legislature (Statesman Media 2017). Moreover, the Texas legislature quickly passed a bill requiring further research at an accredited state university or state agency before granting state-level approval for use of the bait. Another example of such conflict occurred in Tennessee subsequent to passage of the hunting ban in 2010. Legislative efforts of the TWRA encountered radical opposition from pig hunting communities of the Northern Cumberland Plateau region. Such focal opposition included violent threats towards TWRA personnel and events in which pig hunting enthusiasts attempted to sabotage state vehicles and equipment, and stymie management and enforcement capabilities by placing road spikes (K. Miles, Tennessee Wildlife Resources Agency, personal communication). Personnel at the Savannah River National Laboratory in South Carolina also faced similar threats by those who strongly value personal benefits associated with wild pig hunting (J. J. Mayer, Savannah River National Laboratory, personal communication).

11.4 STATE OVERSIGHT

11.4.1 Enforcement and Penalties

A wide breadth of policies and regulations regarding wild pigs exist in the United States (Tables 11.1–11.5). To complement and aid in enforcing those policies and regulations, state governments have also promulgated various fines and penalties for violators. Violations range from misdemeanors to felonies, and include penalties such as mandatory fines, license revocations, liabilities, imprisonment, or some combination. While some states may invoke more lenient penalties and judicial systems concerning wild pig violations, other states adopt more aggressive measures to punish violators and discourage illegal activity. For example, if caught and convicted of transporting wild pigs in Kentucky, the State not only issues a fine of US$500/offense (each pig represents one offense) and a 3–12-month prison sentence, Kentucky Department of Game and Fish also revokes hunting and fishing rights of the violator for 10 years (KT Revised Statutes § 150.990). This penalty is more severe than other states. In other areas of the United States, fines typically range between US$500 and US$5,000 (depending on the state) and several states enforce a revocation of hunting and fishing privileges for a duration of no more than 12 months (17 IL Administrative Code §§ 2530; MS Code Annotated §§ 49-7-140; SC Code §§ 50-16-25). While it may not influence decisions of all pig hunting enthusiasts, repercussions from severe penalties (as enforced by Kentucky) may dissuade residents or non-residents from participating in illegal transportation.

It is also worth noting that while some states choose to enforce fines and penalties, some states have yet to adopt any penalty system (at the time of this writing). This could be a result of a state not having imminent threats of wild pigs (including hunting communities that wish to translocate them), which might prompt promulgation of such rules. For example, in Montana, Wyoming, and South Dakota, where wild pig populations are either absent or isolated and limited, no penalty exists to punish those wishing to transport them illegally. However, if states wish to address this issue proactively in an effort to prevent populations from becoming established, penalty systems need to be in place before the species arrives in transit. Furthermore, if laws and penalties exist in State Code, it clarifies the state's legal tolerance and position on such activities. In states where wild pig populations are endemic and meaningful relief from damages is unlikely (e.g., southeastern United States), precluding illegal wild pig activities simply may not rank among top agency or legislature priorities. In some areas of the Southeast, members of the local populace are advocates for wild pigs, and view them as a "native" and essential component of the community (Holderieath 2017), influencing state government inaction. For example, in Louisiana and Florida, it remains legal to transport wild pigs if the aforementioned criteria (e.g., identification tags, disease testing) are satisfied and the applicant possesses a "feral swine transporter registration permit" (LA Administrative Code 21 §§ 1321) or "feral swine dealer registration license" (FL Administrative Code 5C-21.015). However, no rule or statute exists to punish those who do not operate in compliance with the law. This could also be the result of a lack in political and legislative will.

Moreover, agency ability to enforce laws with meaningful success (as in other wildlife related issues) is further handicapped by lack of personnel resources (i.e., law enforcement officials). Many state agencies, in particular but not exclusive to those in the Southeast, are vastly understaffed due to limited and intermittent nature of state budgets. This often results in 1 or 2 officers being responsible for patrolling multiple counties in some areas, limiting their abilities to respond promptly to reported violations. Not only do many of these understaffed agencies occur in regions where wild pigs tend to be most prolific, such areas also are characterized as rural and composed of large tracts of agricultural lands, national forests, wildlife management areas (WMAs), or other various public and private lands. In addition to being remote, many of these rural areas (i.e., national forests, state WMAs) are interspersed with large networks of public access such as US Forest Service and county roads. This makes it difficult for what may already be a limited number of law enforcement personnel to effectively patrol and regulate illegal activity—in addition to their everyday and more routine duties. Moreover, and considering the phenotypic variation of wild pigs, many law enforcement officials (e.g., game wardens, county sheriffs, highway patrol) may not have the training necessary to identify and distinguish between domestic or wild pigs.

In addition to the inherent difficulties surrounding state-level authority and regulation of wild pigs (Centner and Shuman 2015), another shortfall limiting the ability of a state to preclude illegal activity and spread of wild pigs is the judicial system under which violations are cited, prosecuted, and convicted. While some conservation officers are clearly equipped to recognize criminal activity related to wild pigs and are willing to apprehend and cite those who participate, successes of prosecution and conviction are more obscure and, depending on locality, uncertain. Some states have multiple citations or arrests throughout a year; however, the number of those arrests that lead to an actual conviction tend to be lower. The reason behind this is cryptic, and may likely be political in origin. For example, municipal- and county-level politics, lower courts being ill-equipped to prosecute wildlife violations, or even relaxed sentencing procedures for cases not typical of crimes that courts deal with on a normal basis (e.g., hunting violations such as possession limits, license requirements, and shooting from public roads) limit enforcement of regulations. For example, in late 2016, conservation officers from the Alabama Department of Conservation and Natural Resources (ADCNR) arrested 16 individuals from 3 states (i.e., Alabama, Florida, Mississippi) in an undercover investigation on illegal transport, release, and possession of wild pigs. According to Assistant Law Enforcement Chief C. Lewis (ADCNR), the covert operation had insurmountable evidence, including video surveillance and audio recordings, on the involved racketeering ring. Ultimately, all violators were convicted and received ~US$2,500 fines each. However, most of the crime ring received pre-trial diversions (non-adjudication) due to status as first time offenders, resulting in dropping of many of the cases and/or charges in separate county courts with penalties being negligible. Despite the amount of evidence, the prosecution was only able to sentence 1 individual (out of 16) with jail time, and those charges based on outstanding warrants unrelated to wild pig activities. It was also reported that 1 individual blatantly continued to hold captive wild pigs on his property for various pig hunting activities and was cited multiple times, after the initial arrest, stating that he had too much money invested in pigs and would continue to "feed them out and sell" (C. Lewis, personal communication).

These issues ultimately lead to question whether such regulations and policies are realistically enforceable in some regions of the United States. If so, do convictions and fines occur at rates consistent enough to preclude illegal wild pig activity and slow subsequent range expansion? With likelihood of apprehending, and most importantly convicting those who participate in illegal activity being relatively low, perhaps state agencies and legislatures should consider more severe punishments like Kentucky. While isolated populations of wild pigs have slightly expanded over the past few years (T. Brunjes, Kentucky Department of Fish and Game, personal communication), Kentucky has shown considerable success in averting a statewide wild pig proliferation. If rewards (i.e., monetary or recreational) from illegal wild pig activity continue to outweigh risks (e.g, fines, penalties) throughout much of the United States, it is unlikely that translocation and establishment of wild pig populations will cease. A cost:benefit analysis of illegal activities and legal repercussions may shed light on requirements of state governments to effectively preclude criminal activity that is facilitating rapid spread of wild pigs throughout the country.

11.4.2 Jurisdiction

Another difficulty with state-level management of wild pigs is that multiple agencies may share authority and responsibilities for management and enforcement of regulations (Tables 11.1–11.5). While one agency may benefit from license sales and other hunting related revenues, another may be protecting agricultural interests and public safety through mitigation of crop damages, impacts to infrastructure, and vehicle collisions (USDA/Animal and Plant Health Inspection Service 2015, Wildlife Society 2015). Whether wildlife and natural resources agencies or agricultural agencies have authority to enforce and regulate wild pig invasions depends on several factors. Foremost, classification of wild pigs within a state (see 11.3.1 Species Classification above) influences management and enforcement. If wild pigs are a state-listed game species, then regulative authority typically resides with state wildlife and natural resources agencies. However, other factors may play an important role in this authorization and frequently aren't as comprehensive. For instance, jurisdiction may also be associated with and influenced by management capabilities, budget and associated constraints, personnel expertise, political climate, or economic agenda, the current distribution and proliferation of wild pigs within state boundaries, as well as the degree and frequency of negative impacts in the state. These impacts can be regarded as events which disrupt economic activity (agricultural productivity), social activity (public safety), and ecological integrity. If state legislature prioritizes agricultural commodities, then it is likely that jurisdiction will fall under the agricultural agency of that state, with potential co-management with the state wildlife agency (discussed below).

In some states, wild pigs are not recognized or acknowledged as wildlife, defaulting management authority to state agriculture agencies. However, without statewide enlisted officers to regulate and enforce conservation laws, the agriculture agency must accomplish the same overarching goals (as the wildlife agency), which ultimately should aim to reduce negative impacts and consequences associated with the invasion. Thus, state agricultural agencies could slow impacts of wild pigs by appointing officials designated for wild pig related issues, in addition to creating programs such as meat and facility inspection, and other services that protect livestock, crop, infrastructure, and land-use practices. For example, some states require specifications for fenced enclosures where wild pigs may be released and held until time of slaughter or harvest (MS Code Annotated §§ 49-7-140; OK Administrative Code 35 §§ 15-34-6). In instances of escapees from such enclosures, some states adopted provisions that hold landowners or operators of the facility responsible for damages caused by escaped pigs (NH Revised Statutes Annotated §§ 467:3–5, OK Statutes 1 §§ 6–612).

Another common strategy regarding jurisdiction is a joint effort between state wildlife and agriculture agencies. For example, Arkansas adopted this strategy and the Arkansas Livestock and Poultry Commission has responsibility over establishing and enforcing regulations for wild pigs, regarding possession and transport (Wildlife Society 2015). However, the Livestock and Poultry Commission is assisted by Arkansas Game and Fish Commission that provides guidance on hunting and trapping regulations on public lands such as wildlife management areas. This arrangement allows a diverse team of professionals and officials to operate statewide with the ability to enforce laws, administer policies, and regulate wild pig invasions on multiple fronts. Other states in the southeastern United States where this approach is also evident include Florida, Georgia, Louisiana, Oklahoma, South Carolina, Tennessee, and West Virginia (Wildlife Society 2015). While many of these partnerships work well, others inherit conflicting views and objectives between agencies (and their leadership), detracting from meaningful progress (Centner and Shuman 2015). Having a diverse team of professionals from across several agencies is an invaluable resource. However, 1 agency should remain ultimately responsible for oversight and administration, so that a clear chain of command is in place. Despite the degree of national impacts it is also worth noting that 1 state (Utah) has yet to authorize any agency with jurisdiction over wild pigs (USDA/Animal and Plant Health Inspection Service 2015).

11.5 SUMMARY AND CONCLUSIONS

While efficacy of management tools used in the field to control wild pigs is well documented (see Chapter 8), the legal provisions and strategies to support and enforce these tools are less developed. Currently, no model legislation, policy, or regulation exists that applies across all states and serves interests of all stakeholders. Legislative and management success in one state may ultimately be ineffective in another. Although some states share similarities relative to wild pig distribution and expansion, other factors influence success or failure of legislative efforts, including agency or legislative political climate, agriculture, natural resources, and other economic priorities, as well as social and cultural drivers of its region. Thus, state-level decision making remains flexible and highly variable (West et al. 2009), allowing wild pig policies and regulations tailored to specific needs and demands of each state (Wildlife Society 2015). However, these decisions are not always based on scientific or economic premises, or in the interests of a majority, and instead may be political in essence. As a result, state-level legislation has shown some success, but has also been ultimately ineffective in others, and, in some instances, has functioned to accelerate proliferation of the state's wild pig population.

A primary confounding barrier to controlling wild pig populations is the proper and objective identification of the issue. We generally look for patterns and find solutions which capitalize on the species' biology and assist us in controlling animal densities. Over the last several years there has been a litany of objective research on wild pig ecology and management, aiding resource managers in reducing populations. This research has largely focused on reproductive, behavioral, and spatial aspects of the species. While these are indeed primary topics usually linked to the success of the invading species, they are not the root of the problem. Humans have continued to introduce wild pigs and other taxa on a global scale over thousands of years, for a variety of reasons (i.e., industry, recreation, sustenance). This has resulted in various institutions and governments exploring biological solutions for what in fact, is fundamentally a social problem. While it is no revelation that humans are transporting wild pigs and are an underlying cause of their proliferation, research priorities continue to focus on ecological variables which may improve model function. Resulting predictive probabilities from Snow et al. (2017) suggested wild pig populations are expanding with a northward and westward trajectory, strongly associated with a climatic predictor and winter temperature. It is also of no coincidence that observed range expansion increased simultaneously with increasing popularity of hunting for wild pigs (Keiter et al. 2016). However, these 2 regions are essentially the only remaining areas where wild pigs have yet to establish populations in superabundant numbers, and where large tracts of habitat still remain vacant, relative to conspecifics. Figuratively speaking, it is the only place left to go. Moreover, any range expansion in saturated southern states would likely be difficult to detect within existing data and under current sampling or monitoring techniques. This chapter and supporting research (Waithman et al. 1999, Hamrick et al. 2011, Bevins et al. 2014, Mayer 2014) suggests wild pig hunting popularity and demand is increasing, relative to amount of human-seeding (Snow et al. 2017) observed over the last decade. This could lead one to conclude that recreational and commercial profitable hunting incentives (i.e., creating new, or augmenting existing hunting and commercial opportunity) are also likely contributing to this bidirectional expansion estimated by Snow et al. (2017). If this is true, it strongly supports the necessity and prudency for laws which preclude private landowners (and others) from profiting from presence and abundance of wild pigs on their properties; in addition to governmental provisions which dismantle the establishment and operation of high-fenced shooting operations for wild pigs—if meaningful prevention is to occur. Further supporting this necessity are more recent results from Tabak et al. (2017) and Hernández et al. (2018), which alluded to the predictive power of high-fenced shooting operations in various wild pig occupancy/spread scenarios. By acknowledging and addressing this underlying issue, state and federal bodies of government are given the opportunity to regulate these social aspects through promulgation of objective wildlife policy and law.

States may choose to approach wild pig invasions using a variety of strategies; however, those with limited populations that proactively poised regulations in a more prohibitive (as opposed to restrictive) framework are most likely to preclude illegal activity and statewide proliferation. Other states with low or limited populations, who have yet to enact such measures, are strongly encouraged to do so, before the pig hunting fashion, culture, and subsequent industries establish and become conventional. However, likelihood of significant relief from a prohibitive framework in states or areas with high proliferation of wild pigs is unknown. This, in part, is due to the reactive nature of state governments and failure of legislatures and commissions to proactively adopt such measures in these areas. In areas where hunting is a highly valued industry and culture, whether or not such measures would be lawfully abided and meaningfully enforced is also unclear. One might conclude that those who engage in illegal activity would continue to do so regardless of governmental provisions, especially after pigs become established, abundant, and socially conventional. Nevertheless, as pig hunting enthusiasts and interest groups become more mobilized and politically astute, there remains an urgent need for greater public support in convincing legislators and other leadership that wild pigs are impairing state economies (Centner and Shuman 2015).

Although often obscure, the primary objective of a state—whether eradication, reduction, or maintenance—is typically reflected in its legislative prudency and subsequent laws and policies enacted. However, these objectives may be perceived as a focus of state wildlife and natural resources agencies, department of agriculture, or the state legislature—and generally not those of a collective majority of tax-paying citizens. Moreover, they are often decided by appointed (versus elected) officials. Because of these 2 procedures in the legislative process, decision-makers are not conveniently held accountable when passing legislation. Nonetheless, as highlighted by Centner and Shuman (2015), state legislators and government officials who fail to respond in meaningful ways to public health risks, ecological, and economic damages hosted by wild pigs, are remiss in the responsibilities of their post. If meaningful population reduction and control is in fact the clear objective of our state's governments, organizations, and citizenry, then a fundamental change is needed to catalyze our leadership to enact objective, science-based wildlife policy and law to regulate both human and wild pig populations at both the state and federal levels.

ACKNOWLEDGMENTS

The author would like to thank J.J. Mayer, J.L. Corn, and T. Jordan for the data that they provided in support of developing this project and subsequent chapter, as well as their insight and guidance regarding the substance of the data. Appreciation is also given to B.K. Strickland and B. Leopold for their support and leadership. In addition, the author would like to acknowledge the Boone & Crockett Club and Mississippi State University Extension for funding this research.

REFERENCES

Alabama Department of Conservation and Natural Resources. 2015. *Alabama Hunting and Fishing Digest 2015–2016 Requirements, Fees and Season Dates*. J. F. Griffin Publishing, LLC. Williamstown, MA.

Barrios-Garcia, M. N., and S. A. Ballari. 2012. Impact of wild boar (*Sus scrofa*) in its introduced and native range: A review. *Biological Invasions* 14:2283–2300.

Bartel, R. A., and M. W. Brunson. 2003. Effects of Utah's coyote bounty program on harvest behavior. *Wildlife Society Bulletin* 31:738–743.

Bevins, S. N., K. Pedersen, M. W. Lutman, T. Gidlewski, and T. J. Deliberto. 2014. Consequences associated with the recent range expansion of nonnative feral swine. *BioScience* 64:291–299.

Bomford, M., and P. O'Brien. 1995. Eradication or control for vertebrate pests? *Wildlife Society Bulletin* 23:249–255.

Centner, T. J., and R. M. Shuman. 2015. Governmental provisions to manage and eradicate feral swine in areas of the United States. *AMBIO* 44:121–130.

Continuing Consolidation of the Statutes of Manitoba. C.C.S.M. c. A95.

Continuing Consolidation of the Statutes of Manitoba. C.C.S.M. c.W130.

Ditchkoff, S. S., R. W. Holtfreter, and B. L. Williams. 2017. Effectiveness of a bounty program for reducing wild pig densities. *Wildlife Society Bulletin* 41:548–555.

Dover, S., and J. Marston. 2017. Joint statement of Texas Hog Hunters Association (THHA) and Environmental Defense Fund (EDF): In support of H.B. 3451 and S.B. 1454. Published Position Statement.

Gipson, P. S., B. Hlavachick, and T. Berger. 1998. Range expansion by wild hogs across the central United States. *Wildlife Society Bulletin* 26:279–286.

Hamrick, B., M. D. Smith, C. Jaworowski, and B. Strickland. 2011. *A Landowner's Guide for Wild Pig Management: Practical Methods for Wild Pig Control.* Mississippi State University Extension Service, Alabama Cooperative Extension System, Alabama A&M University, and Auburn University.

Hanson, L. B., M. S. Mitchell, J. B. Grand, D. B. Jolley, B. D. Sparklin, and S. S. Ditchkoff. 2009. Effect of experimental manipulation on survival of feral pigs. *Wildlife Research* 36:185–191.

Hernández, F. A., B. M. Parker, C. L. Pylant, T. J. Smyser, A. J. Piaggio, S. L. Lance, M. P. Milleson, et al. 2018. Invasion ecology of wild pigs (*Sus scrofa*) in Florida, USA: The role of humans in the expansion and colonization of an invasive wild ungulate. *Biological Invasions* 20:1865–1880.

Holderieath, J. 2017. Essays on feral swine: Producer welfare effects and spatiotemporal management of feral swine. Dissertation, Colorado State University, Ft. Collins, CO.

Kaufman, K., R. Bowers, and N. Bowers. 2004. *Kaufman Focus Guide to Mammals of North America.* Houghton Mifflin, New York.

Keiter, D. A., J. J. Mayer, and J. C. Beasley. 2016. What is in a "common" name? A call for consistent terminology for nonnative *Sus scrofa. Wildlife Society Bulletin* 40:384–387.

Kreith, M. 2007. *Wild Pigs in California: The Issues.* University of California Agricultural Issues Center. AIC Issues Brief 33.

Lodge, D. M., S. Williams, H. J. MacIsaac, K. R. Hayes, B. Leung, S. Reichard, R. N. Mack, et al. 2006. Biological invasions: Recommendations for US policy and management. *Ecological Applications* 16:2035–2054.

Long, J. L. 2003. *Introduced Mammals of the World: Their History, Distribution, and influence.* Commonwealth Scientific and Institutional Research Organisation. Melbourne, Victoria, Australia.

Louisiana Department of Wildlife and Fisheries. 2015. *Louisiana Hunting Regulations 2015–2016.* Baton Rouge, LA.

Massei, G., and P. V. Genov. 2004. The environmental impact of wild boar. *Galemys* 16:135–145.

Mayer, J. J. 2014. *Estimation of the Number of Wild Pigs Found in the United States.* Savannah River National Laboratory, SNRL-STI-2014-00292. Aiken, SC.

Mayer, J. J., and J. C. Beasley. 2018. Wild Pigs. Pages 219–248 *in* W. C. Pitt, J. C. Beasely, and G. W. Witmer, editors. *Ecology and Management of Terrestrial Vertebrate Invasive Species in the United States.* CRC Press, Boca Raton, FL.

Mayer, J. J., and I. L. Brisbin, Jr. 2008. *Wild Pigs in the United States: Their History, Comparative Morphology, and Current Status.* University of Georgia Press, Athens, GA.

Mayer, J. J., and I. L. Brisbin, Jr., editors. 2009. *Wild Pigs: Biology, Damage, Control Techniques and Management.* Savannah River National Laboratory, SRNL-RP-2009-00869. Aiken, SC.

Mississippi Department of Wildlife, Fisheries and Parks. 2015. *Mississippi Outdoor Digest 2015–2016.* Jackson, MS.

Palmer, G. H., J. Koprowski, and T. Pernas. 2007. Tree squirrels as invasive species: Conservation and management implications. *Managing Vertebrate Invasive Species: Proceedings of an International Symposium.* US Department of Agriculture/Animal and Plant Health Inspection Service/Wildlife Services/National Wildlife Research Center, Ft. Collins, CO.

Pennsylvania Game Commission. 2013. *Wildlife; Feral Swine and Wild Boar Eradication.* 43 Pennsylvania Bulletin 2039. Pennsylvania Game Commission, Meadville, PA.

Plasters, B. C., C. Hicks, R. Gates, and M. Titchenell. 2013. *Feral Swine in Ohio: Managing Damage and Conflicts.* Ohio State University Extension W-26-13, 1–6. Ohio State University, Columbus, OH.

Snow, N. P., M. A. Jarzyna, and K. C. VerCauteren. 2017. Interpreting and predicting the spread of invasive wild pigs. *Journal of Applied Ecology* 54:2022–2032.

Southeastern Cooperative Wildlife Disease Study. 2012. *National Feral Swine Map, 2012*. University of Georgia, Athens, GA.
Southeastern Cooperative Wildlife Disease Study. 2015. *National Feral Swine Map, 2015*. University of Georgia, Athens, GA.
Southeastern Cooperative Wildlife Disease Study. 2016. *National Feral Swine Map, 2016*. University of Georgia, Athens, GA.
Statesman Media. 2017. Texas House gives preliminary approval to feral hog poison study bill. Cox Media Group, Austin, TX. <http://www.statesman.com/news/texas-house-gives-preliminary-approval-feral-hog-poison-studybill/IJ0V1uhVf6tl1FAMK UZ46L/>. Accessed 3 Jan 2018.
Tabak, M. A., A. J. Piaggio, R. S. Miller, R. A. Sweitzer, and H. B. Enerst. 2017. Linking anthropogenic factors with the movement of an invasive species. *Ecosphere* 8:e01844.
Tennessee Wildlife Resources Agency. 2015. *Tennessee Hunting & Trapping Guide* 2015–2016. The Bingham Group, Knoxville, TN.
University of Arkansas Cooperative Extension Service. 2015. Laws and regulations governing feral hogs in Arkansas. FSA9106. University of Arkansas Division of Agriculture, Research and Extension, Little Rock, AR.
US Department of Agriculture/Animal and Plant Health Inspection Service. 2015. *Final Environmental Impact Statement. Feral Swine Damage Management: A National Approach*. Washington, DC.
Waithman, J. D., R. A. Sweitzer, D. Van Vuren, J. D. Drew, A. J. Brinkhaus, I. A. Gardner, and W. M. Boyce. 1999. Range expansion, population sizes, and management of wild pigs in California. *Journal of Wildlife Management* 63:298–308.
West, B. C., A. L. Cooper, and J. B. Armstrong. 2009. Managing wild pigs: A technical guide. *Human-Wildlife Interactions Monograph* 1:1–55.
Wildlife Society. 2015. *2015 Annual State Summary Report: Wild Hog Working Group*, 1–47. Wildlife Society, Bethesda, MD.

APPENDIX 11.1
2015–2016 State Hunting Regulations for Wild Pigs in the Southeastern United States

State	State-Listed Game Species	Listed Name	State Hunting License Required	Season	Bag Limit	Weapon Restrictions	Night Hunting Legal	Source
AL	Yes	Game Animal – Feral Swine	Yes	Yes	N/A	Private land: No Public land: Yes, subject to applicable regulations during current season in progress	Private land: Yes Public land: No	Alabama Hunting and Fishing Digest 2015–2016: Requirements, Fees, and Season Dates
AR	No	Public Nuisance – Feral Hog	Yes	Yes	N/A	Private land: No WMAs: Yes, Yes, subject to applicable regulations during current season in progress	Private land: Yes Public land: No	Laws and Regulations Governing Feral Hogs in Arkansas 2015
FL	No	Wild Hog	No	Yes	Yes	Private land: No Public land: Yes	No	Florida 2015–2016 Hunting Regulations
GA	No	Feral Hog	Yes	Yes	N/A	Private land: No Public land: Yes, subject to applicable regulations during current season in progress	Private Land: Yes Public Land: No	2015–2016 Georgia Hunting Seasons & Regulations
SC	No	Feral Hog/Wild Hog	Yes	Yes	N/A	Private land: No weapon restrictions during daylight hours Public land: Yes, subject to applicable regulations during current season in progress	Yes; on private land from the last day of February to July 1	South Carolina Department of Natural Resources Rules & Regulations Hunting & Fishing July 1, 2015–August 14, 2016
TN	No	Destructive Species - Wild Hog	Yes	Public land: Incidental take; wild hog control season on 2 WMAs (dogs only)	N/A	Yes, subject to applicable regulations during current season in progress	Yes; landowners with a granted exemption from Tennessee Wildlife Resources Agency	2016 Tennessee Hunting & Trapping Guide

(Continued)

APPENDIX 11.1 (CONTINUED)
2015–2016 State Hunting Regulations for Wild Pigs in the Southeastern United States

State	State-Listed Game Species	Listed Name	State Hunting License Required	Season	Bag Limit	Weapon Restrictions	Night Hunting Legal	Source
VA	No	Nuisance Species – Feral Hog/ Wild Hog	Yes	Yes	N/A	No	Yes	Hunting & Trapping in Virginia July 2015–June 2016
WV	Yes	Boar/Wild Boar	Yes, residents only; also requires a big-game stamp	Yes	Yes	Yes; subject to applicable regulations of current season in progress	No	West Virginia - Hunting and Trapping Regulations Summary- July 2015–June 2016

APPENDIX 11.2
2015–2016 State Hunting Regulations for Wild Pigs in the Northeastern United States

State	State-Listed Game Species	Listed Name	State Hunting License Required	Season	Bag Limit	Weapon Restrictions	Night Hunting Legal	Source
CT	No	N/A	N/A	Hunting Prohibited	N/A	N/A	N/A	2016 Connecticut Hunting & Trapping
DE	No	Feral Swine	N/A	Hunting Prohibited	N/A	N/A	N/A	2015–2016 Delaware Hunting & Trapping
ME	No	N/A	N/A	Hunting Prohibited	N/A	N/A	N/A	Maine Hunting & Trapping: The official 2015–16 State of Maine Hunting & Trapping Laws and Rules
MD	No	N/A	N/A	Hunting Prohibited	N/A	N/A	N/A	Maryland Guide to Hunting and Trapping – 2015–2016
MA	No	N/A	N/A	Hunting Prohibited	N/A	N/A	N/A	2016 Massachusetts Fish & Wildlife Guide to Hunting, Freshwater Fishing and Trapping
NH	No	Feral Boar, Feral Hog, Wild Hog	Yes	No	Yes	Yes; subject to applicable regulations of current season in progress	No	New Hampshire Fish and Game Department Hunting & Trapping Digest September 2015–August 2016
NJ	No	Feral Hog	Yes	Yes	N/A	Yes; subject to applicable regulations of current season in progress	No	2015–2016 New Jersey Hunting & Trapping Digest
NY	No	Eurasian Boar	N/A	Hunting Prohibited	N/A	No	No	New York Hunting & Trapping – 2015–2016 Official Guide to Laws & Regulations
PA	No	Feral Swine/ Wild Boar	Yes	No	N/A	Yes; subject to applicable regulations of current season in progress	No	July 1, 2015–June 30, 2016 Pennsylvania Hunting & Trapping Digest
RI	No	N/A	N/A	Hunting Prohibited	N/A	N/A	N/A	State of Rhode Island and Providence Plantations Hunting Regulations for 2015–2016 Season
VT	No	N/A	N/A	N/A	N/A	N/A	N/A	2016 Vermont Hunting, Fishing & Trapping Laws and Guide - Fish and Wildlife Regulations

APPENDIX 11.3

2015–2016 State Hunting Regulations for Wild Pigs in the Midwestern United States

State	State-Listed Game Species	Listed Name	State Hunting License Required	Season	Bag Limit	Weapon Restrictions	Night Hunting Legal	Source
IL	No	Feral Swine, Wild Pigs	Yes	No	N/A	N/A	No	Illinois Digest of Hunting and Trapping Regulations 2015–2016
IN	No	Wild Pig	No	No	N/A	N/A	N/A	Indiana Hunting & Trapping Guide – 2015–2016 Regulations Guide
IA	No	Feral Hog	Yes	No	N/A	No	Yes	2015–16 Iowa Hunting and Trapping Regulations
MI	No	Nuisance Species - Feral Swine	Yes	No	N/A	No	No	2015 Michigan Hunting and Trapping Digest
MN	No	Feral Swine	N/A	Hunting Prohibited	N/A	N/A	N/A	2015 Minnesota Hunting & Trapping Regulations Handbook
MO	No	Feral Hog	No, except during deer and turkey seasons	No	N/A	N/A, except during deer and turkey seasons	No	A Summary of Missouri Hunting and Trapping Regulations 2015–2016
OH	Yes	Feral Swine, Wild Boar	Yes	No	N/A	Yes; subject to applicable regulations of current season in progress	Yes	Ohio Hunting and Trapping Regulation 2015–2016
WI	No	Unprotected Species - Nuisance Wild Hogs, European Wild Hogs, Russian Wild boars	Yes	No	N/A	Yes; subject to applicable regulations of current season in progress	Yes	Wisconsin 2016 Small Game Hunting Regulations

APPENDIX 11.4

2015–2016 State Hunting Regulations for Wild Pigs in the North-Central United States

State	State-Listed Game Species	Listed Name	State Hunting License Required	Season	Bag Limit	Weapon Restrictions	Night Hunting Legal	Source
CO	No	N/A	N/A	N/A	N/A	N/A	N/A	Colorado Parks and Wildlife – 2016 Colorado Big Game
KS	No	Feral Hog	N/A	Hunting Prohibited	N/A	N/A	Yes	2016 Kansas Hunting & Fur Harvesting Regulations
MT	No	N/A	N/A	N/A	N/A	N/A	N/A	2016 Montana Hunting Regulations
NM	No	Unprotected Species – Feral Hog	No	No	N/A	Yes; subject to applicable regulations of current season in progress	No	2015–2016 New Mexico Hunting
NE	No	N/A	N/A	Hunting Prohibited	N/A	N/A	N/A	2015 Big Game Guide – Nebraska
ND	No	N/A	N/A	Hunting Prohibited	N/A	N/A	N/A	2015–2016 North Dakota Guides to Hunting Regulations
OK	No	Hog, Feral Swine	Yes	Yes	N/A	Private land: NoPublic land: Yes, subject to applicable regulations of current season in progress	Yes	Oklahoma Hunting: The Official 2015–2016 OklahomaHunting Guide
SD	No	N/A	N/A	Hunting Prohibited	N/A	N/A	N/A	South Dakota 2015 Hunting and Trapping Handbook
TX	No	Exotic Animal - Feral Hog	Yes	No	N/A	Yes; subject to applicable regulations of current season in progress	Yes	Texas 2015–2016 Hunting Seasons
WY	No	N/A	N/A	Hunting Prohibited	N/A	N/A	N/A	2016 Wyoming Game and Fish Department Regulations

APPENDIX 11.5
2015–2016 State Hunting Regulations for Wild Pigs in the Western United States

State	State-Listed Game Species	Listed Name	State Hunting License Required	Season	Bag Limit	Weapon Restrictions	Night Hunting Legal	Source
AZ	No	N/A	N/A	Hunting Prohibited	N/A	N/A	N/A	Arizona Game and Fish Department – 2016–17 Arizona Hunting Regulations
CA	Yes	Big Game—Wild Pig, European Wild Pigs and their hybrids (genus *Sus*)	Yes;Wild pig tags also required	No	N/A	Yes; Subject to applicable regulations of current season in progress	No	California 2015–2016 Mammal Hunting Regulations
ID	No	N/A	N/A	N/A	N/A	N/A	N/A	Idaho 2015 & 2016 Big Game Seasons & Rules
NV	No	N/A	N/A	Hunting Prohibited	N/A	N/A	N/A	Nevada Hunting Guide 2015
OR	No	Invasive Non-native, Introduced Species - Feral Swine	Yes	No	N/A	Yes; subject to applicable regulations of current season in progress	No	2015 Oregon Big Game Regulations
UT	No	N/A	N/A	N/A	N/A	N/A	N/A	2015 Utah Big Game Regulations
WA	No	Deleterious Exotic Wildlife-Feral Pigs	N/A	N/A	N/A	N/A	N/A	Washington 2015 Big Game Hunting Seasons & Regulations

12 Wild Pigs in Western North America

Michael P. Glow, John J. Mayer, Bethany A. Friesenhahn, and Kurt C. VerCauteren

CONTENTS

12.1 INTRODUCTION

The regional history and presence of wild pigs (*Sus scrofa*) in western North America collectively reflects the entire range of diversity observed in their occurrence throughout the rest of the continent. For example, like southeastern North America (see Chapter 16), wild pigs have been present in California since European colonial times. In contrast, similar to the presence of these animals in many northern tier US states and Canadian provinces/territories (see Chapters 13, 14), the occurrence of wild pigs in a number of locations in the region has been recent, sporadic, and short-lived.

For this chapter, western North America is defined as including the following US states: Alaska, Arizona, California, western Colorado, Idaho, western Montana, Nevada, New Mexico, Oregon, Utah, Washington, western Wyoming, and the Canadian provinces/territories of British Columbia and the Yukon. Ranging from 71°N down to 31°N latitude, this extensive region covers an area of 6,250,570 km^2 and encompasses 25% of the entire North American landmass. Given this extensive areal landform, western North America contains a diversity of ecological regions including portions of the tundra, taiga, northwestern forested mountains, marine West Coast forests, Great Plains, North American deserts, Mediterranean California, temperate Sierras, and southern semi-arid highlands (Commission for Environmental Cooperation 1997). Altitudes in western North America range widely from 6,194 m in elevation down to 86 m below sea level (Carpenter and Provorse 1996). Floral habitats include tundra, coniferous forest, temperate rain forest, deciduous forest, grassland, and desert. Accordingly, climatic type extremes in the region range from tundra and humid continental in the north (average annual temperature below 10°C) down to hot desert and cool steppe in the south (average annual temperature above 18°C; Peel et al. 2007). As such, the densest populations of pigs are found in Mediterranean California while colder northern portions of the region preclude potential establishment of this introduced species (Figure 12.1). Additionally, some areas in the region that were once unsuitable to wild pigs have been transformed into suitable habitat by human activities such as the introduction of irrigated crops and pastures, supplemental feeding, and establishment of man-made water resources (McClure et al. 2015).

Given the more recent appearance of wild pigs in most of this region, the presence of these animals in western North America is overwhelmingly considered to be undesirable. Even in California, where their existence has been the longest, wild pigs are increasingly being viewed as a destructive invasive species rather than a valuable big game animal. In fact, several locations in California have been the subjects of extensive and costly eradication programs for wild pigs (e.g., Trione-Annadel State Park, Channel Islands, Pinnacles National Monument, San Diego County). Further compared to some other regions (e.g., Southeast, Pacific islands), most of western North America (aside from California) does not have either the historic or societal culture engrained that would support the continued presence of these animals.

12.2 HISTORIC AND PRESENT DISTRIBUTION

Largely centered in California, wild pigs in western North America have collectively had a long history dating back to at least the late 1500s. As elsewhere on the continent, both feral/domestic pigs and wild boar x feral pig hybrids have comprised the types of wild pigs introduced into this region. With some exceptions principally associated with California, the histories and current status of these animals in the states and provinces/territories in western North America are generally localized and not connected to the adjacent governmental units. Because of this, wild pig history in this region is herein presented alphabetically for each state or province/territory in the following sections.

12.2.1 Alaska

In July of 1984, 6 female and 2 male wild pigs, purchased in California, were released onto Marmot Island in the Gulf of Alaska. Reportedly of "European wild boar" ancestry, these animals were stocked by a private citizen, pursuant to a state grazing lease that he had been granted for 1 year. Originally released on a 16.2-ha private parcel, these wild pigs soon spread throughout the island. Within a year of their introduction, the pigs were causing severe damage to the island's vegetation and soil. Resultantly, the state denied extension of the grazing permit and directed the owner to remove all of the pigs, which he was either unwilling or unable to do. In retrospect, it appears that

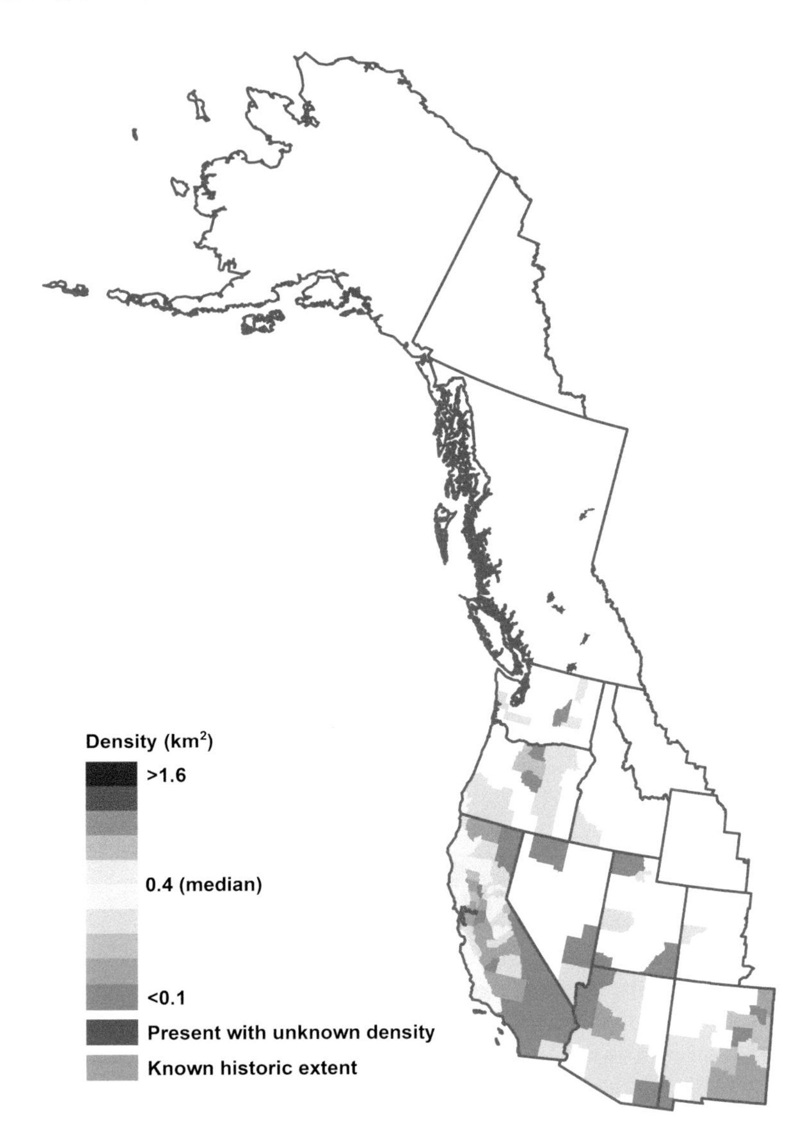

FIGURE 12.1 Relative density of wild pigs across western North America with the densest populations found along coastal California and less dense populations scattered primarily within the southern half of the region. (Courtesy of R. Miller. With permission.)

time and the elements accomplished that task. In 1998, a hunter killed what turned out to be the last wild pig seen on the island. A state survey in 2006 found no evidence of wild pigs, fresh tracks or rooting activity (Lloyd et al. 1987, Paul 2009). Other scattered reports of escaped captive-reared wild pigs in Alaska have come from the Sterling area on the Kenai Peninsula and along the Tanana/Delta drainage corridor south of Fairbanks (J. J. Mayer, Savannah River National Laboratory, unpublished data). No established populations of wild pigs currently exist in Alaska (USDA/Animal and Plant Health Inspection Service 2015).

12.2.2 Arizona

From the late 1880s to present, a population of wild pigs has been reported to exist along the Colorado River bordering Arizona and California below the city of Needles, California south to the Mexican border. This population originated from free-range domestic stock. In the 1890s, these wild pigs were estimated to number in the thousands of animals (The New York Times 1894), and by the 1950s, estimates had decreased to several hundred individuals (McKnight 1964). More recently, control activities, especially on the Havasu National Wildlife Refuge, have greatly reduced the number of wild pigs along the Colorado drainage (Neskey 2018). In the late 1950s, isolated small populations of wild pigs in the state were reported in the Dragoon Mountains, Whetstone Mountain, southeast corner of Grand Canyon National Park, San Pedro Valley of Cochise County, and on the east slope of the Chiricahua Mountains (McKnight 1964). In the 1970s, escaped wild pigs from a high-fenced shooting operation in Mesquite, Nevada established themselves along the Virgin River near Littlefield, Arizona. In the 1990s, a population of escaped wild pigs was also reported on a ranch in the San Pedro River Valley near Redington, Arizona. Other recent wild pig populations have been reported in the Mud Mountain/Black Rock Mountain area, near Cordes Junction/Cordes Lakes/Black Canyon City, southeast of Prescott, near Bowie and Willcox in northern Cochise County, and on the Buenos Aires National Wildlife Refuge southwest of Tucson (Steller 2010, Allen 2013). Wild pig populations in Arizona are estimated to number in the hundreds of animals in only a few counties at present (USDA/Animal and Plant Health Inspection Service 2015).

12.2.3 British Columbia

Similar to several Canadian provinces (see Chapter 13), the Ministry of Agriculture in British Columbia began promoting wild boar farming as part of a livestock diversification program in the 1980s. The number of these farms in British Columbia peaked in 2001 at about 40, which collectively housed less than 2,000 animals (Michel et al. 2017). As happened in other provinces, these farming ventures did not realize the necessary return on investment to stay in business. Some wild boar farming operations slaughtered their herds while others simply released their stock to survive on their own (Mayer 2018). In other instances, captive wild boar escaped confinement by breaching fences. At least 1 report of free-ranging wild boar occurred before 2000 in the Peace River region near the border with Alberta (Brook 2017). In 2007, escaped captive-reared wild pigs began to establish themselves in the wild in the area around Christina Lake. However, by 2011, all of those animals had been either recovered or killed (CBC News 2011, Shore 2016). In 2014, free-ranging wild boar were reported in the Lower Mainland, Kamloops, Okanagan, Peace and Kootenay regions of British Columbia. The province also changed its laws that year to allow hunters to kill these wild pigs (British Columbia Ministry of Forests, Lands and Natural Resource Operations 2014). In 2016, wild pigs were also reported in Cariboo-Chilcotin, Peace Country, north Okanagan regions of the province as well as in Carnation Creek on the south shores of Barkley Sound on Vancouver Island (McKinley 2016, Shore 2016). At present, wild pigs in British Columbia remain in scattered locations and are found in low numbers in the southern half of the province.

12.2.4 California

The history of wild pigs in California has been covered extensively and in great detail elsewhere (e.g., Pine and Gerdes 1973; Barrett 1977, 1978; Barrett and Pine 1980; Waithman et al. 1999; Mayer and Brisbin 2008, 2009). As such, the following constitutes an abbreviated history summary. The first introduction of pigs into California occurred in 1582, when the Spanish established a penal colony on Santa Cruz Island. Six hundred prisoners were brought to the island by ship and abandoned there with a few head of cattle, pigs, and a horse. A short time later, the prisoners killed the cattle and horse and used the hides to cover crudely made boats to escape the island; the pigs were turned loose to wander the island, where they multiplied and became free-ranging (Hogg 1920). In 1769, the Spanish introduced domestic pigs to the California mainland (Hutchinson 1946). By 1834, domestic pigs were found around 21 of the Spanish missions in the state. After 1850, as new settlements and homesteads spread throughout California in the wake of the Gold Rush, domestic pigs were commonly released to free-range in the hills and fatten during the acorn season; many of these pigs were never recovered. By the 1880s, wild pigs were being hunted as game animals in the state (Barrett 1978). In late 1925 or 1926, 12 wild pigs, which had been captured around Hooper Bald in North Carolina, were released on a private ranch in Monterey County. These animals were wild boar x feral pig hybrid stock that were descendants of wild boar that had been introduced into North Carolina in 1912 (see Chapter 16). Animals exhibiting the wild boar or hybrid phenotype have subsequently been introduced throughout many areas in the state (Mayer and Brisbin 2008, 2009). In the 1950s, wild pig range in California was restricted to just a few coastal counties, which had increased to 10 counties by the early 1960s (Waithman et al. 1999). After 1965, hunters began to show an increasing interest in wild pigs as game animals which led to numerous unregulated introductions throughout the oak woodland zone in California (Barrett 1977). In the 1970s, wild pigs were reported in 29 California counties (Mansfield 1978). This then increased to 33 counties in the 1980s, 45 counties in the 1990s (Updike and Waithman 1996), and 56 counties in the 2000s (Waithman 2001). In 2014, wild pigs were reported to be present in 57 of 58 counties in California (absent in Imperial County; USDA/Animal and Plant Health Inspection Service 2015).

12.2.5 Western Colorado

Mesa County is the only location in western Colorado where wild pigs have been reported. In the mid-1990s, escaped captive-reared wild pigs from the Little Creek Ranch near Collbran were observed in the area surrounding the ranch. The Little Creek Ranch, which had operated a commercial "wild boar" high-fenced shooting operation since 1971, was directed to repair the enclosure's fencing and recapture the escaped animals. Local hunters and wildlife officers killed several in the adjacent Grand Mesa National Forest. In 2008, the ranch was quarantined when the owner was caught trying to illegally import wild pigs that were found to be positive for pseudorabies (Lightcap 2008, J. J. Mayer, unpublished data). There are no established populations of wild pigs in western Colorado at this time (USDA/Animal and Plant Health Inspection Service 2015).

12.2.6 Idaho

In 2002, localized sightings of wild pigs in the woods near Kamiah, Idaho were reported (The Spokesman-Review 2002); however, there have been no subsequent reports of those animals persisting in the wild. Beginning in 2007, a small population of wild pigs was reported in the Bruneau Valley area in Owyhee County (Darr 2011). It was believed that hunters brought the animals in from California and released them to establish a wild population (Dumas 2009). Multi-agency eradication efforts combined with similar efforts by local landowners in 2010 and 2011 were thought to have eliminated those animals (Idaho Statesman 2012). There have been no sightings of wild pigs in the Bruneau Valley after 2011 (Zatkulak 2013). No wild pigs are thought to be currently present in Idaho (D. L. Nolte, USDA/APHIS/Wildlife Services, personal communication).

12.2.7 Western Montana

In the late 2000s, a group of domestic pigs escaped in the Grant Creek area near Missoula, but were eradicated by state agency personnel before they could become established in the wild (Stugelmayer 2010). There are currently no established populations of wild pigs in western Montana (USDA/Animal and Plant Health Inspection Service 2015).

12.2.8 Nevada

Wild pigs were recently found in 3 locations within Nevada. The earliest was in the 1980s along the Virgin River in Clark County. These pigs, reportedly of wild boar ancestry, were established when animals from an Arizona high-fenced shooting operation located near Mesquite, Nevada escaped.

This population, found mostly on public lands and estimated to number less than 50 animals in 2010, has expanded its range south along the river drainage corridor toward Lake Mead. A second population of wild pigs has been reported in Paradise Valley, along a small stretch of the Humboldt River basin about 65-km north of Winnemucca in Humboldt County. This population, initially started with released wild pigs that had been obtained in California, is primarily found on private lands and is estimated to number up to 300 animals. The last of the 3 wild pig populations in the state is located near Caliente in Lincoln County (Liakopoulos 1999; Kimak 2002, 2010). USDA/Animal and Plant Health Inspection Service (2015) noted that wild pigs in Nevada were present in low numbers in only a few counties.

12.2.9 New Mexico

Since the early 20th century, the existence of wild pigs in New Mexico has changed dramatically. The earliest known population of these animals in the state had been present in Hidalgo County since the early 1900s. These wild pigs, found in the San Luis, Animas and Peloncillo mountain ranges in the county, originated from escaped or released domestic stock from local ranches or farms in the early part of the 20th century. Mostly found on private lands, some of these animals have also expanded their range onto the Coronado National Forest (Mayer and Brisbin 2008). In the late 1950s to early 1960s, the wild pig population in the Peloncillo Mountains was estimated to number several hundred animals (McKnight 1964). By the late 1980s, these pigs had expanded their range into the immediate portions of adjacent Grant County in New Mexico and Cochise County in Arizona (Corn and Jordan 2017). Beginning in the 1970s, wild pigs started to appear in other locations in the state. Some of these animals were remnants of free-range husbandry practices or escaped livestock; however, most recently appearing wild pigs were the result of intentional releases for hunting opportunities (USDA/APHIS/Wildlife Services New Mexico 2010). In addition, still other wild pigs expanded their range from neighboring Texas along the Canadian and Pecos River drainage corridors into northeastern and southeastern New Mexico, respectively (USDA/APHIS/Wildlife Services New Mexico 2010, Moffatt 2012). By 2011, wild pigs in New Mexico had been reported in 19 of 33 counties (Moffatt 2012). Two years later, that range had expanded to 22 counties (Davis 2013). Later that year, USDA/APHIS/Wildlife Services established a pilot project to work toward eradicating wild pigs from the state (USDA/Animal and Plant Health Inspection Services 2015). USDA/Animal and Plant Health Inspection Service (2015) reported that wild pigs were found in approximately half of the counties in New Mexico in 2014. In 2017, USDA/APHIS/Wildlife Services staff estimated limited wild pig presence in Lincoln National Forest, White Mountain Wilderness, eastern Roosevelt County, southern Lea County, Eddy County, White Sands Missile Range and Hidalgo County (USDA/APHIS/Wildlife Services New Mexico 2017a).

12.2.10 Oregon

Wild pigs have been present intermittently in Oregon for over 2 centuries. The earliest known occurrence was in 1811 near present-day Astoria, Oregon, when a herd of escaped domestic pigs, described as a "troublesome pack of wild swine," were established for a short time (Rouhe and Sytsma 2007). In the first half of the 20th century, an established feral population of escaped domestic pigs existed along the Rogue River in Curry County in southwestern Oregon. In the 1950s, this population numbered fewer than 100 animals (McKnight 1964, Mayer and Brisbin 2008). In the early 1940s, wild pigs were also reported to be found in southeastern Oregon (Scheffer 1941). During the 1970s and 1980s, no wild pigs were thought to be present in the state (Mayer and Brisbin 2008); however, some appeared in the 1990s. These were the result of both illegal releases of wild pigs and escapes from commercial high-fence shooting operations (Coblentz and Bouska 2004). In addition, some wild pigs present in northern California expanded their range into portions of southern Oregon (Rouhe and Sytsma 2007). In the 1990s, wild pigs were present in Crook County, and in 2001, both Jefferson and Wasco Counties had populations of wild pigs (Coblentz and Bouska 2004). In the early 2000s, several ranchers in Oregon imported wild pigs from northern California to establish populations to hunt (Oregon Invasive Species Council 2002). In 2004, wild pigs were reported in Coos, Curry, Josephine, Jackson, Klamath, Wasco, Jefferson, Crook, and Wheeler Counties (Coblentz and Bouska 2004). By 2009, the range of Oregon's wild pigs had expanded to collectively include Benton, Coos, Crook, Curry, Deschutes, Douglas, Gilliam, Harney, Jackson, Jefferson, Josephine, Klamath, Lincoln, Malheur, Sherman, Union, Wasco, and Wheeler Counties (Mortensen 2009). In 2013–2014, wild pigs were primarily found in north-central Oregon, and to a lesser extent, in southwestern Oregon, and were estimated to number between 2,000 and 5,000 animals (USDA/Animal and Plant Health Inspection Service 2015). Successful removal efforts by USDA/APHIS/Wildlife Services have continued in Oregon through 2019 and officials postulate that the range of wild pigs may be primary limited to Wheeler County (USDA/APHIS/Wildlife Services Oregon 2019).

12.2.11 Utah

In the late 1990s, wild pigs along the Virgin River in Arizona expanded their distribution into Utah as far north as St. George (Liakopoulos 1999). In 2006, the private owners of Barrow Land and Livestock leased Fremont Island in the Great Salt Lake to use as a private hunting ranch. Among the species stocked onto the island were wild pigs. By appearance, these pigs were of wild boar x feral pig hybrid ancestry (J. J. Mayer, unpublished data). In 2010, 1 of the wild pigs escaped the island and was pursued by authorities along the Antelope Island Causeway, but then it fled back into the lake and drowned (Hollenhorst and Leonard 2011). In late 2013, another wild pig escaped the island and was observed on the Causeway during low water conditions in the lake. Later that year at the request of federal and state agencies, the island's owners began to remove the wild pigs from Fremont Island. By October 2013, all of the pigs had been removed from the island (Prettyman 2013). USDA/Animal and Plant Health Inspection Service (2015) reported that wild pigs were currently found in only a few counties in Utah.

12.2.12 Washington

Wild pigs have had a sporadic, and to date, temporary presence at 4 general locations within Washington State. In 1949, wild pigs were reported to be present on the Quinault Indian Reservation in the Olympic Peninsula (Loney 2001). There were several theories as to the origin of these animals, some of which dated back to the 1800s. During the past 50–60 years these animals were present on the Reservation in only low numbers (Associated Press 2001). In the 1990s, these wild pigs spread off of the reservation southeast to the Wynooche River valley (Loney 2001). Following that range expansion, efforts were made to eradicate the pigs both on and off of Reservation lands. By 2010, there were no more reports of wild pigs anywhere on the Olympic Peninsula (Washington Invasive Species Council 2010). In 1981, as many as 60 wild pigs were released along the Skagit River in Skagit and Whatcom counties. Between eradication efforts conducted by the Washington Department of Fish and Wildlife and animals that were killed during hunting seasons, this population of wild pigs was eliminated by mid-1982 (Mayer and Brisbin 2008). In July 2015, wild pigs, which appeared to be escaped domestics, were reported in the Columbia Basin Wildlife Area in Grant County. Following the discovery of these animals, USDA/Wildlife Services initiated an effort to eradicate them. In August 2016, the last of these was killed (Twietmeyer 2016, Nailon 2017). In the summer of 2016, a small number of what appeared to be released or escaped domestic pigs were reported in the Gifford Pinchot National Forest in eastern Lewis County. Efforts were made by multiple state agencies and some private citizens to recapture or remove these pigs. After their removal in 2016, more free-ranging pigs were reported in the same area in 2017. It is unclear if any pigs are still present in this National Forest (Nailon 2017). USDA/APHIS/Wildlife Services removed 5 wild pigs from a remote area in the northwest part of the state and Washington Department of Fish and Wildlife removed 1 wild pig from the southwestern part of the state in 2017 (USDA/APHIS/ Wildlife Services Washington 2018).

12.2.13 Western Wyoming

Wild pigs have never been reported as being present anywhere in Wyoming (USDA/Animal and Plant Health Inspection Service 2015, Mayer 2018). Because of concerns over the presence of wild pigs in Nebraska within 20 miles of the Wyoming border, the Wyoming legislature passed a law (Title 11 Chapter 26 § 11-26-101) governing feral livestock in 2009; this law further gave state officials the authority to take action against any wild pigs discovered in the state (Pelzer 2011). Lastly, the importation and possession of wild pigs is illegal in Wyoming (WGFC Chapter 10 § 3(d)(ii)(A)).

12.2.14 Yukon

In early June 2018, 7 female wild boar escaped confinement from a wild boar farm located in the Mendenhall area west of Whitehorse in southern Yukon. The Territorial government told the farmer responsible to remove the animals from the wild or face penalties (CBC News 2018a). By late August 2018, all of the escaped wild boar had been hunted down and killed (CBC News 2018b). Ultimately, the farmer was fined US$400 for allowing his animals to escape. The Yukon Department of the Environment is considering ways to ensure that no animals escape from the estimated 5–10 wild boar farms that remain in operation in the Territory (CBC News 2018c).

12.3 REGIONAL ENVIRONMENTAL ASPECTS IN WESTERN NORTH AMERICA

12.3.1 Morphology

Morphological characteristics of wild pigs in the Western region are similar to those described among other regions of North America. Given the interest in wild pigs with wild boar or hybrid phenotypic characteristics, numerous introductions of wild pigs with such characteristics have been made throughout California and into the adjacent states of Oregon, Idaho and Nevada. Because of this, Updike and Waithman (1996) stated that some wild boar characteristics were present in virtually all wild pigs in California. Interestingly, syndactylous hooves (see Chapter 2) had been reported to be present among the wild pigs found on Santa Catalina Island in California; however, that physical trait had disappeared from the island by the 1970s (Mayer and Brisbin 2008).

12.3.2 Behavior

Home ranges of wild pigs in the Western region vary and are often influenced by resource availability, particularly being tied to water sources/drainages across much of this arid region. Wide ranging reports of the average home range size of wild pigs in California have been documented. The average home range size for boars was 1.4 km^2 and 0.7 km^2 for sows on Santa Catalina Island off the southern coast of California, compared to 50+ km^2 for boars and 10–25 km^2 for sows on Dye Creek Ranch in north-central California (Barrett 1978, Baber and Coblentz 1986). Sweitzer et al. (2000) reported the average home range of wild pigs along the north and central coast of California was 3.5 km^2 and ranged from 0.5 to 15.5 km^2. Home ranges for wild pigs in other parts of the Western region have not been reported. Sows typically utilized 2–3 adjacent canyons at Dye Creek Ranch and rarely moved more than 0.5 km from their nest when their litter was <3 weeks old (Barrett 1978). Boars generally have larger home ranges, often traveling extensive distances in search of sows to breed (Barrett 1978). Considerable home range overlap of boars and sows between and within sexes has also been shown (Baber and Coblentz 1986), and they are primarily not territorial

(Barrett 1978). Varying topography does not seem to alter wild pig movements in their search of food, water, and cover (Barrett 1982).

When resources are limited in the summer, wild pigs have been reported to travel up to 10 km/day in search of food and will feed all night (Barrett 1978). Mid-slopes are utilized more often during the wet season when resources are more abundant, but wild pigs generally avoid ridgetops (Baber and Coblentz 1986). Wild pigs in the region are almost exclusively nocturnal during the summer and tend to be more active on moonlit nights, shift to crepuscular activity in the fall, and are usually diurnal during the winter, often bedding all night in thick cover for thermal protection (Barrett 1978, Waithman 2001, Mayer and Brisbin 2009).

Similar to wild pig populations in other regions, wild pigs in the Western region typically travel in matriarchal groups consisting of a sow and 1 to 3 generations of her offspring (Barrett 1978). Wild pig group size in Dye Creek, California varied seasonally, and averaged 8.4 pigs with a range of 6–10 (Barrett 1978). Although boars are generally solitary, they will travel with matriarchal groups to breed, spending up to a week or more with 1 sounder before searching for additional groups to breed, sometimes reportedly up to 11 km apart. In general, wild pig social groups over 25 are described as rare; however, large collective social units of up to 97 pigs have been reported congregated around localized resources such as irrigated pastures in arid areas (Barrett 1978).

The presence of large carnivores, particularly mountain lions (*Puma concolor*), capable of predating adult and juvenile wild pigs throughout the Western region is unique compared to most other regions of North America. Wild pigs are an important food source for mountain lions, reportedly making up 5–45% of their diet in California (Sweitzer 1998). Hopkins (1989) found that wild pigs made up 5–20% of mountain lion scat samples from the Diablo Mountains in California. Wild pigs, however, only made up 2% of the actual kills recorded from the same study. The level of mountain lion use of wild pigs as a food resource may be chiefly based on the density of these prey animals in a given area (Hopkins 1989). Craig (1986) reported that predation of wild pigs by mountain lions appeared to occur primarily during the wet season. Although research on interactions between mountain lions and wild pigs is lacking, wild pigs may exhibit behavioral responses due to the presence of mountain lions, such as increased vigilance. Increased vigilance (see Chapter 3) in response to predation risk while feeding is a trade-off between resource acquisition and reducing risk of predation and is thought to influence the fitness of individuals (Quenette and Desportes 1992, Brown et al. 1999). Research has shown that vigilance in wild pigs decreases with increasing group size (Quenette and Desportes 1992, Mayer and Brisbin 2009), but has not been tested in the presence of mountain lions. Wild pigs may also allocate more or less time in certain habitats based on predation risk, but this is also not known. Other documented and potential predators of wild pigs in the Western region include the golden eagle (*Aquila chrysaetos*), coyote (*Canis latrans*), black bear (*Ursus americanus*) and bobcat (*Lynx rufus*; Barrett 1978, Roemer et al. 2001, Christie et al. 2014).

12.3.3 Diseases and Parasites

Wild pigs in the Western region are susceptible to and serve as reservoirs for a host of disease pathogens and parasites (see Chapter 5), which can pose a significant threat to humans, livestock, and many wildlife species. Disease pathogens that are particularly threatening include brucellosis (*Brucella suis*) and pseudorabies (PRV) (Sweitzer et al. 1996), but reports of prevalence across the region vary. Brucellosis and PRV have been documented in California, specifically in Monterey, Santa Clara, Tehama, and San Luis Obispo counties and San Clemente and Santa Catalina islands (Sweitzer et al. 1996). Drew et al. (1992) found over 90% of positive brucellosis cases were in Monterey and San Luis Obispo counties in California. Pederson et al. (2013) also reported positive cases of PRV in Nevada, New Mexico, and California (Figure 12.2). Conversely, Pederson et al. (2012) reported no positive brucellosis cases in wild pigs sampled in New Mexico, Arizona, Nevada, and California, and Barrett (1978) reported similar results in north-central California. Porcine epidemic virus (PEDV), a type of coronavirus that can be up to 95% fatal in juvenile pigs,

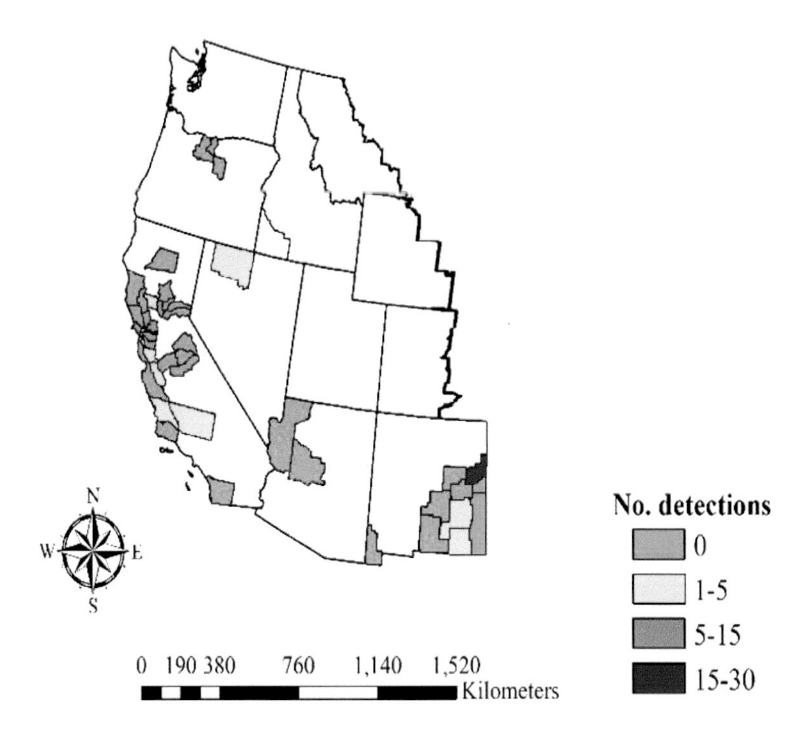

FIGURE 12.2 Pseudorabies antibody detections in wild pigs during a nationwide surveillance effort conducted 2009–2012. (Adapted from Pederson et al. 2013.)

was first documented in the United States in 2013 (Bevins et al. 2018). Since then, positive cases in wild pigs have been documented in California, but no cases were found in Arizona, New Mexico, Utah, Nevada, and Oregon (Bevins et al. 2018). Wild pigs have tested positive for another coronavirus, transmissible gastroenteritis virus, in Arizona, New Mexico, and Oregon (Bevins et al. 2018).

Hog cholera was introduced to Santa Cruz and Santa Rosa islands of California in the late 1940s and 1950s as a management tool for reducing wild pig numbers, but testing in the 1980s failed to detect this pathogen in wild pig populations on those islands as well as elsewhere in California and the Havasu National Wildlife Refuge in Arizona (Nettles et al. 1989). Since the 1990s, small numbers of wild pigs in California have tested positive for plague (*Yersinia pestis*); however, there have been no cases of transmission of this pathogen from wild pigs to humans (Smith et al. 1998, Kreith 2007). Although zoonotic parasites such as *Trichinella* spp. and *Toxoplasma gondii* in wild pigs are primarily distributed across the Southern and Midwest regions of the United States, *Trichinella* spp. has been detected in New Mexico and *Toxoplasma gondii* has been detected in California, Arizona, and New Mexico (Sweitzer et al. 1996, Hill et al. 2014).

Wild pigs in the region are also a reservoir for zoonotic parasites including *Cryptosporidium parvum* and *Giardia* sp., which pose a threat to protozoal surface water contamination given their propensity towards water sources (Atwill et al. 1997). These zoonotic parasites were significantly more prevalent in populations where densities of wild pigs exceeded 2.0 pigs/km^2 or in wild pigs <8 months of age (Atwill et al. 1997). Reports of ecto- and endoparasites on wild pigs in the region are limited, but Barrett (1978) documented a light to heavy hog louse (*Haematopinus suis*) infestation from wild pigs in north-central California.

12.3.4 Food Habits

Similar to other regions, wild pigs in the Western region are opportunistic omnivores that primarily consume herbaceous vegetation, but their diets vary seasonally depending on plant phenology and

FIGURE 12.3 (a) Green, succulent grasses and forbs are an important diet component of wild pigs in the winter and spring, but the availability of these resources is dependent upon the timing and duration of the rainy season. (b) Resource availability changes throughout the course of the year and is particularly impacted during years of drought. (Photos by E. VanNatta. With permission.)

availability (Baber and Coblentz 1987, Loggins et al. 2002, Wilcox and Van Vuren 2009). Diets during the winter and spring generally consist of succulent grasses and forbs, but the timing and duration of the rainy season largely influences the extent of time these resources remain an important component of the diet (Barrett 1978, Baber and Coblentz 1987, Loggins et al. 2002; Figure 12.3). As grasses and forbs begin to mature and decline in quality during the summer, wild pigs also consume fruits, seeds, wild oats (*Avena barbata*), and browse (Barrett 1978, Baber and Coblentz 1987). Baber and Coblentz (1987) reported that Australian saltbush (*Atriplex semibaccata*) made up 80% of all forbs consumed during the dry summer season on Santa Catalina Island, California because the plant remained green throughout the summer when other green vegetation was scarce. Additionally, wild pigs in the region have shown preference for prickly pear (*Opuntia* spp.), lemonade sumac (*Rhus integrifolia*), cholla cactus (*Opuntia* spp.), holly-leafed cherry (*Prunus ilicifolia*) and cherry plums (*Prunus* sp.; Baber and Coblentz 1987, Schauss et al. 1990, USDA/APHIS/Wildlife Services New Mexico 2010). Poison oak (*Toxicodendron diversilobum*) leaves have also been reported as an important source of browse when green vegetation is lacking during the summer (Barrett 1978). During the fall, oak mast provides an extremely important food source that can influence reproductive success, body condition, and survival (Barrett 1978, Baber and Coblentz 1987). When abundant, wild pigs will primarily consume acorns, which may be part of their diet until February (Barrett 1978). Roots, bulbs, and Manzanita berries (*Arctostaphylos* spp.; Figure 12.4) are important sources of energy when acorn crops fail (Barrett 1978, Baber and Coblentz 1987). Invertebrate consumption is generally opportunistic when rooting (Sweitzer 1998, Wilcox and Van Vuren 2009) and is not reported to vary by season (Baber and Coblentz 1987). Variation in diet between the sex or age of wild pigs has not been reported (Barrett 1978, Loggins et al. 2002).

The diet composition of wild pigs also includes various vertebrate species, but reports of the rate at which vertebrates are consumed and if they are opportunistically consumed or intentionally predated in the Western region varies. Most research suggests that vertebrate remains found in wild pig stomachs were either carrion or made up a small percentage of their diet (Barrett 1978). Barrett (1978) reported that carrion was the major source of animal matter found in wild pig stomachs, but only made up 1.6% of their total diet on an annual basis. Similarly, the only vertebrate matter Baber and Coblentz (1987) found in wild pigs was carrion and was less than 10% of their overall diet. Adult boars were reported to consume more carrion than other wild pigs because of their ability to dominate a carcass. In addition, adult boars also moved farther and more often than sows, which probably increased their chances of encountering carcasses (Barrett 1978).

Other studies, though, have shown a much larger proportion of vertebrates in wild pig diets and suggest intentional predation may be occurring. Wilcox and Van Vuren (2009) sampled 104

FIGURE 12.4 Manzanita berries (*Arctostaphylos* spp.) can be an important food source for wild pigs across portions of western North America, especially after an acorn crop failure. (Photo by M. Glow. With permission.)

stomachs over a 7-year period from wild pigs in the Diablo Range, located in central California. They found that 40.4% of stomachs contained vertebrate remains with a total of 167 individual animals including 11 mammal species, 5 birds, 3 snakes, and 1 frog. The dominant prey species were California voles (*Microtus californicus*; Figure 12.5) and Botta's pocket gophers (*Thomomys bottae*). The number of prey individuals in each stomach ranged from 2 to 18, with 1 sow's stomach containing 13 voles, 4 gophers, and 1 broad-footed mole (*Scapanus latimanus*). Wilcox (2015) similarly reported a wild pig from the same area had consumed 23 individual vertebrates consisting of 8 species including a Pacific gopher snake (*Pituophis c. catenifer*). The occurrence of multiple

FIGURE 12.5 Three California voles (*Microtus californicus*) found within 1 wild pig stomach in California. Several studies in California have documented a large proportion of small vertebrate contents in wild pig diets, suggesting intentional predation may be occurring. (Photo by J. Wilcox. With permission.)

prey individuals in each stomach more than likely characterized 1 night of foraging due to the rapid gastric emptying rate of wild pigs (Wilcox 2015). Although some predated species were primarily ground dwelling and could have been consumed opportunistically while rooting, other species were semi-arboreal or extremely agile, suggesting wild pigs may be intentionally hunting small mammals (Wilcox and Van Vuren 2009). Additionally, Loggins et al. (2002) reported that a wild pig with a nearly intact adult California ground squirrel (*Spermophilus beecheyi*) in its stomach was seen motionless over a ground squirrel burrow for more than 20 minutes before being shot, further supporting the possibility of active predation of small mammals. Interestingly, Loggins et al. (2002) and Wilcox and Van Vuren (2009) both reported a substantial increase of vertebrate remains in the diets of wild pigs in the summer and fall compared to the winter and spring. Although acorns are high in energy and provide a substantial portion of wild pig diets during this time, they are also low in protein. Thus, wild pigs may be supplementing their diet during this time by consuming small vertebrates to meet their protein demands, which is particularly important for reproducing sows (Loggins et al. 2002, Wilcox and Van Vuren 2009). Protein is also required to convert starch from acorns into stored fat (Wilcox 2015). Wild pigs that consumed vertebrate species also had significantly less rump fat compared to those that did not, and predation rates were greater in females than males (Wilcox and Van Vuren 2009).

12.3.5 Habitat Use

Preferred habitat for wild pigs in the Western region include chaparral shrublands, northern coastal sage, oak woodlands, and oak grasslands (Waithman et al. 1999, Mayer and Brisbin 2008; Figure 12.6). Habitats where wild pigs are most abundant are typically characterized by permanent water sources, extensive oak woodlands, and dense vegetation for cover (Barrett 1978, Waithman et al. 1999). Oak thickets provide suitable bedding cover, cooler temperatures, and an abundance of mast and invertebrates for foraging (Barrett 1982). During the summer, wild pigs prefer moist, cooler canyon bottoms or north-facing slopes with dense vegetation, and open mid-slopes and ridgetops

FIGURE 12.6 Preferred habitat for wild pigs in western North America includes (a) chaparral shrublands, (b, c) oak woodlands, and (d) oak grasslands. (Photos by E. VanNatta. With permission.)

are usually avoided (Barrett 1978, Baber and Coblentz 1986). Irrigated pastures also provide an important food source for wild pigs during the summer, and rooting in boulder washes for bulbs is common during years when acorn production is low (Barrett 1982).

12.3.6 Population Biology

Reported population demographics in the Western region are limited to California. The post-natal male-to-female sex ratio of over 5,500 pigs removed from Santa Catalina Island, California was 1:0.94 (Schuyler et al. 2002). Barrett (1978) similarly reported a 1:1 ratio for pigs 0–12 months in age, but also found that the adult population was biased towards females, probably due to selective hunting pressure for boars. Annual variation where populations were biased towards females one year and males the next has also been reported (Sweitzer et al. 2000). Wild pig densities in California typically average approximately 3–4 pigs/km^2 and range from 0.4 to 13.9 pigs/km^2 (Barrett 1978, Schauss et al. 1990, Waithman et al. 1999, Sweitzer et al. 2000, Teton et al. 2016). However, Baber and Coblentz (1986) reported a much larger average density of 28 pigs/km^2, ranging from 21 to 34 pigs/km^2 on Santa Catalina Island, California. Population fluctuations are primarily influenced by long-term drought, acorn production, and in some cases hunting pressure (Barrett 1978, Baber and Coblentz 1987, Waithman et al. 1999). For example, Sweitzer et al. (2000) reported wild pig densities in areas with little to no hunting pressure averaged 2.4 pigs/km^2 compared to 1.1 pigs/km^2 in moderate-heavy hunted areas. Additionally, Schauss et al. (1990) found that wild pig density estimates decreased 51% following a poor acorn year and then increased by 65% the following year after increased acorn production. The reported population age structure varies between the north and central coast regions of California: <8 months (25.5%), 8–18 months (21%), and >18 months (53.5%), compared to <8 months (55.5%), 8–18 months (5.2%), and >18 months (39.3%), respectively (Sweitzer et al. 2000). Barrett (1978) also reported an age structure of 0–12 months (48.8%), 1–2 years (22.8%), and >2 years (28.4%) in north-central California.

The most common mortality causes of wild pigs in the Western region are accidents, starvation, and predation for pigs under 6 months of age, and hunting, disease, and tooth deterioration for adult pigs (Barrett 1978). Mortality rates can be as great as 70–90% for wild pigs under 6 months of age, and the most common accidents include being crushed to death in the farrowing nest or trampled by other pigs, especially boars. However, once pigs reach 6 months, their survival rates increase (Barrett 1978). Life expectancy may be slightly greater for females than males (Schauss et al. 1990). Based on limited data, wild pigs in this region have an average life expectancy of 10–36 months but can live up to 8 years or more (Barrett 1978, Schauss et al. 1990). The effects of mountain lion predation on wild pig populations and whether mountain lion predation influences variation of wild pig densities is largely unknown (Waithman et al. 1999).

12.3.7 Reproductive Biology

Reproductive data for the Western region is limited to California and New Mexico, where breeding occurs throughout the year. Although data on reproductive parameters (e.g., pregnant female weight, body condition, percent pregnant, fetal litter size, conception/farrowing dates, social unit/group size) have been reported in Saskatchewan (Koen et al. 2018; see Chapter 13), such data has not been reported for British Columbia or the Yukon. Baber and Coblentz (1986) reported the timing of peak farrowing varies depending on availability of nutritional resources. When oak mast was abundant and fall/winter rains were substantial enough to produce ample nutritional forage, farrowing peaked during the winter and spring. Comparatively, if a mast failure occurred and fall/winter rains came late, farrowing peaked in the spring and summer. Farrowing for wild pigs at Dye Creek Ranch peaked in November, with a secondary peak in July (Barrett 1978). However, the mortality rate for piglets born in July can be extremely high due to limited nutritional resources. As a result, sows at Dye Creek Ranch averaged 2 litters/year, typically conceiving shortly after their first litter was lost (Barrett 1978), compared to 0.86 litters/year on Santa Catalina Island (Baber and Coblentz 1986). Baber and Coblentz (1986) similarly reported a mortality rate increase for piglets farrowed during the summer due to poor nutrition. Similar to other regions, pregnant sows build a farrowing nest, often lined with grass, leaves, and vegetation, shortly before parturition (Barrett 1978). Ovulation rate for wild pigs in California ranges from 6.7 to 8.5 and prenatal mortality ranges from 25 to 34% (Barrett 1978, Baber and Coblentz 1986). Average fetal litter size is 5.0–5.6 and ranges from 1 to 10 young (Barrett 1978, Baber and Coblentz 1986, Dirsmith et al. 2017). Mean fetal litter size on Santa Catalina Island was greatest in sows 2–3 years of age while litters were larger for sows 4–5 years of age at Dye Creek Ranch (Barrett 1978, Baber and Coblentz 1986). Fetal litter size was also found to increase as sow's body size increased in New Mexico (Dirsmith et al. 2017). Sows were most commonly at least 1 year in age before breeding successfully (Baber and Coblentz 1986). Dirsmith et al. (2017) found no correlation between fetal litter size and annual precipitation, but also noted limited variation in long-term annual precipitation. Limited fetal sex data suggest either a 1:1 ratio (Baber and Coblentz 1986) or a female-biased sex ratio (Koen et al. 2018).

12.3.8 Damage

Wild pigs in the Western region cause damage to a variety of ecosystems, habitats, native flora and fauna, and agricultural resources. Rooting behavior of wild pigs is particularly destructive and can cause extensive damage to numerous ecosystems (see Chapter 7) including grasslands and oak woodlands commonly found across portions of the Western region. Although rooting occurs year-round as a result of natural foraging behavior, it is especially common from mid-autumn to early spring before soils dry out and become increasingly difficult for wild pigs to excavate (Kotanen 1995, Teton et al. 2016). Damage from rooting may also vary annually (Cushman et al. 2004) and spatially (Mayer and Brisbin 2009) across different landscapes depending on resource availability. Rooting significantly increases in oak thickets when an abundance of acorns are available, but is more common in other habitats when acorns are limited (Mayer and Brisbin 2009). Rare plant species such as the Calypso orchid (*Calypso bulbosa*) in California and important archaeological and historical sites are seriously impacted by rooting (Mayer and Brisbin 2009). Wild pigs also pose a significant threat to numerous endangered species including the willow flycatcher (*Empidonax traílli extimus*) and Yuma Ridgway's rail (*Rallus obsoletus yumanesis*) on the Havasu National Wildlife Refuge (Zaun et al. 2016).

Several studies have quantified the impacts of rooting on California grasslands, which include alterations in the composition of plant species following disturbance. Although invasive plant species have slowly transformed the composition of California grasslands over the last 200 years, primarily due to land use changes (Cushman 2007, D'Antonio et al. 2007), rooting has resulted in the continued spread of invasive plant species in disturbed grasslands. For example, Cushman et al. (2004) found that recolonization of invasive plant species increased with increasing rooting disturbance. Tierney and Cushman (2006) similarly reported a substantial increase in invasive forbs and grasses 1 year following rooting disturbance and noted that those species continued to persist thereafter. However, native plant species can persist and have been shown to increase at a slow but steady rate following rooting disturbance (Kotanen 1995, Tierney and Cushman 2006). Common native perennial bunchgrass species (*Deschampsia* spp. and *Danthonia* spp.) are distinctly robust against wild pig rooting, potentially due to deep roots which are difficult for pigs to uproot (Kotanen 1995, Cushman et al. 2004). The composition and abundance of seed banks of native and invasive plant species in disturbed areas will also influence impacts (Cushman 2007). Negative impacts on soil moisture, pH, organic matter, and nitrate levels due to rooting have not been found (Moody and Jones 2000, Cushman et al. 2004, Tierney and Cushman 2006). Given the long history of wild pigs in California, negative impacts may be even more pronounced in areas across the Western region into which wild pigs have only recently expanded (Kotanen 1995).

Rooting and foraging by wild pigs have also been shown to have negative impacts in oak woodland ecosystems. Sweitzer and Van Vuren (2008) reported that the number of oak tree seedlings in exclosure plots were at least double compared to unfenced control plots, and were 4 times greater in exclosures at 1 study site in the northern coast region of California after 7 years of protection. The size of oak tree seedlings were also significantly smaller in control plots. These impacts consequently result in reduced oak regeneration in California where oak regeneration is already reduced, and impacts are especially substantial in areas where pig densities exceed 2 pigs/km^2 (Sweitzer and Van Vuren 2002). Wild pigs consume a substantial portion of acorns and thus reduce the amount of acorns available for native wildlife (Sweitzer and Van Vuren 2002). However, no studies have evaluated the direct ecological impacts of competition for acorns between wild pigs and native wildlife (Sweitzer 1998). Interestingly, it has been suggested that wild pigs may be replacing the role of grizzly bears (*Ursus arctos*), long extirpated from California, by providing a natural source of disturbance in oak woodland ecosystems (Sweitzer and Van Vuren 2002; see Chapter 6).

Wild pigs have also indirectly facilitated hyperpredation by golden eagles on the Channel Islands of California, leading to substantial reductions of native island fox (*Urocyon littoralis*) populations (Roemer et al. 2001). Golden eagles were historically transient to the islands, but the arrival of wild pigs on the island provided a stable food source that allowed the establishment of an unnaturally large breeding population of golden eagles. Following population control of wild pigs on the island, the eagles relied more heavily on foxes for prey, leading to the near extinction of island foxes (Roemer et al. 2001, 2002).

Agricultural damage by wild pigs in the Western region is also extensive and impacts numerous crops including grapes, kiwi, berries, nut orchards, artichokes, cut flowers, peas, lettuce, pumpkins, avocados, citrus fruits, and pistachios (Frederick 1998, Baldwin et al. 2014, White et al. 2018; Figure 12.7). In California, where over 400 agricultural commodities are produced (Baldwin et al. 2014), over US$1.7M in damage to agricultural resources in 40 counties was reported in 1996 (Frederick 1998), and more recent reports similarly estimated approximately US$2M in damage annually (Bigelow and Mathis 2018). In New Mexico, reported costs of damage to agriculture in cases where USDA/APHIS/Wildlife Services was contacted for assistance increased from only US$300 in 2005 to over US$218,500 in 2008 (USDA/APHIS/Wildlife Services New Mexico 2010). Wild pig activity in agricultural areas is typically greatest around harvest time when they consume ripened crops (White et al. 2018). Fruit trees and grape vines are damaged when wild pigs use them as rubbing posts or tusking/scent marking trees and damage to drip-line irrigation and nets that catch falling Feijoa (*Acca sellowiana*) fruit have also been reported (Frederick 1998). Pastures and rangeland are damaged by rooting (Baldwin et al. 2014; Figure 12.8). Additionally, 9 counties in California reported livestock predation by wild pigs in a 1996 survey (Frederick 1998). Wild pigs have also been implicated as a cause to the 2006 *E. coli* O157 spinach outbreak from ranches along the central coast of California, which resulted in 205 human illnesses. Thirty-one of these cases developed a type of kidney failure called hemolytic-uremic syndrome, and 3 individuals died (Centers for Disease

FIGURE 12.7 Wild pigs cause extensive agricultural damage in western North America to a variety of crops including grape vineyards shown here. (Photos by J. Lewis. With permission.)

FIGURE 12.8 Cattle production can be negatively impacted by wild pig rooting in pastures and rangeland. (Photo by M. White. With permission.)

Control and Prevention 2006, Jay et al. 2007, Jay-Russell et al. 2012). The proximity of agricultural fields to wildland areas, presence of fencing, and level of human activity can influence the degree of damage wild pigs cause to agricultural resources (White et al. 2018; Figure 12.9).

12.4 REGIONAL MANAGEMENT IN WESTERN NORTH AMERICA

From a historical perspective, management strategies used for wild pigs in western North America have been decidedly localized and changed dramatically over time. Following the early introductions

FIGURE 12.9 Several factors can influence the severity of damage to agricultural resources caused by wild pigs including the proximity of agricultural fields to wildland areas. (Photo by J. Lewis. With permission.)

to this region, wild pigs were being hunted as game animals in both California and along the lower Colorado River by the late 1800s (The New York Times 1894, Barrett 1978). In the early 1900s, although they were considered to be a pest, wild pigs on Santa Cruz Island were recreationally hunted (Hogg 1920). A couple of decades later, in an effort to control wild pig numbers on Santa Cruz and Santa Rosa islands, hog cholera virus (a.k.a. classical swine fever) was introduced several times based on the advice of experts (Wheeler 1944, Nettles et al. 1989). The pigs were captured live (i.e., either trapped or roped), inoculated with the virus, and released, some immediately and some after being maintained in a pen for a time. On Santa Cruz, the virus was initially introduced prior to 1944, and then reintroduced once in 1950 and again in the early to mid-1950s. On Santa Rosa, hog cholera was initially introduced in 1949, and then again in either 1952 or 1953. Following each introduction of the virus, numerous dead or dying pigs were observed on each of these islands; however, most evidence of the virus had disappeared within a few years (Nettles et al. 1989). In 1987, a survey of sera and tissues from both Santa Cruz and Santa Rosa revealed that hog cholera was no longer active within either wild pig population (Nettles et al. 1989).

Although wild pig populations expanded rapidly across numerous portions of North America in recent years, they remain absent or sparsely distributed across much of the Western region, excluding California and eastern New Mexico. Thus, management is often focused on monitoring and early detection followed by a rapid response to prevent establishment and support eradication operations when possible. Early detection of wild pigs and surveillance following an eradication effort is crucial to stop their spread and eliminate their destructive impacts. Oregon, Washington, and Idaho established a toll-free hotline as part of a "Squeal on Pigs" campaign for the public to report any wild pig sightings (Oregon Invasive Species Council 2018). Additionally, a House bill in Oregon was passed in 2009 requiring landowners to contact the Oregon Department of Fish and Wildlife immediately when wild pigs are spotted on their property (Oregon Invasive Species Council 2018). Recently developed environmental DNA (eDNA) technology is also being used for surveillance in New Mexico to detect wild pigs by sampling water sources (Williams et al. 2017; see Chapter 9) and may be particularly useful in remote portions of the state. Canada has bolstered efforts to prevent the spread of wild pigs by initiating the PigTrace Program (Canadian Food Inspection Agency 2015; see Chapter 13). Implemented by the Canadian Food Inspection Agency in 2015, the program requires all domestic and captive wild pig farmers to properly mark all individuals, keep records, and report any movements of pigs.

In addition to monitoring, Oregon, New Mexico, and Arizona all have management plans in place to eradicate established wild pig populations. Oregon originally proposed a 4-year, US$1.29M eradication plan in 2007 that included legislation to stop the release of pigs, guidelines for population estimates, public education and outreach, and monitoring following an eradication (Rouhe and Sytsma 2007). USDA/APHIS/Wildlife Services in Oregon removed 155 wild pigs using a helicopter for aerial removal in 2014 (USDA/APHIS/Wildlife Services Oregon 2015), but remnant populations still remain (US Department of Agriculture 2017). Arizona similarly developed a wild pig eradication plan for the Havasu National Wildlife Refuge in partnership with USDA/APHIS/Wildlife Services in 2016 (Zaun et al. 2016). Aerial and ground removal operations in February 2017 and February 2018 eliminated a total of 65 and 67 wild pigs, respectively, and 540 pigs have been removed from the refuge since 2006 (Neskey 2018, US Fish and Wildlife Service 2018). A partnership between numerous stakeholders in New Mexico including the US Forest Service, Bureau of Land Management, New Mexico Game and Fish, Mescalero Apache Tribe, New Mexico Cattle Growers Association, and USDA/APHIS/Wildlife Services was launched in 2013 with a goal of wild pig eradication in the state by 2018 (Avalos 2013). Efforts are currently ongoing in New Mexico where 308 wild pigs were removed by USDA/APHIS/Wildlife Services in the 2017 fiscal year (USDA/APHIS/Wildlife Services New Mexico 2018). Aerial shooting informed by Judas pigs (see Chapter 8) is one of the most productive removal techniques in New Mexico (USDA/APHIS/Wildlife Services New Mexico 2017b).

Prior to the mid-1950s, wild pigs were unclassified under California state law, and could be killed with no restrictions (Waithman et al. 1999). Initially in 1956 in Monterey County and then statewide in 1957, the management strategy for these animals in California did a complete reversal with wild pigs being classified as a big game species to recognize the valued status of these animals for hunting purposes (Updike and Waithman 1996). In Monterey County, both a season and bag limit were imposed on wild pig harvest, while the rest of the state had a year-round open season and no bag limits (Pine and Gerdes 1973, Barrett 1977). All of this directed attention to pigs as sporting animals and increased their popularity as game animals in California (Pine and Gerdes 1973). In the mid-1970s, wild pig hunting was opened year-round on a statewide basis (Mayer and Brisbin 2008). In 1992, hunters were required to purchase tags to harvest wild pigs in California (Waithman et al. 1999). There is no limit on the number of wild pig tags 1 person may purchase. In addition, successful hunters must return their completed wild pig tag report cards immediately, and non-lead ammunition is required for harvesting wild pigs in certain areas in the state (California Department of Fish and Wildlife 2018). During the 2018–19 hunting season, wild pig tags were US$22.94 for residents and US$77.76 for non-residents (California Department of Fish and Wildlife 2018). Revenue generated from wild pig hunting in California is approximately US$1.2M annually (Bigelow and Mathis 2018). A total of 53,430 tags were sold during the 2015–16 hunting season, an 8.3% decrease from the year before, and 4,223 wild pigs were harvested (7.9% success rate; Garcia and Raymond 2018). Over 60% of wild pigs harvested were in Monterey, San Luis Obispo, Kern, Mendocino, and Tehama Counties (Garcia and Raymond 2018). Rules and regulations regarding wild pig hunting in the rest of the Western region where wild pigs are present varies, but wild pig hunting in Nevada is prohibited (see Chapter 11).

Despite the big game status of wild pigs in California, eradication operations have been conducted in several locations in the state including Santa Catalina Island, Santa Rosa Island, Santa Cruz Island, San Clemente Island, Trione-Annadel State Park and Pinnacles National Monument in central California. Approximately 11,855 pigs were removed from Santa Catalina Island between 1990 and 2001 at a cost of nearly US$3.2M, primarily utilizing trapping, hunting with dogs, and ground hunting (Schuyler et al. 2002). By October 2001, it was estimated less than 300 wild pigs were left on the island (Schuyler et al. 2002). Wildlife managers successfully eradicated wild pigs from Santa Cruz Island in 14 months, removing 5,036 primarily through trapping, aerial gunning, and hunting with dogs (Parkes et al. 2010; Figure 12.10). The island was separated into 5 zones with 1-m-high fencing before eradication began and Judas pigs were used to assist removal of the last remaining pigs (Parkes et al. 2010). A similar successful eradication operation was conducted on Santa Rosa Island between July 1990 and March 1993 where 1,175 wild pigs were removed, predominately by hunting and aerial shooting (Lombardo and Faulkner 2000). In an effort to protect both the endemic and endangered species found on the island, the US Navy had eradicated all wild pigs found on San Clemente Island by 1992 (Department of the Interior 1997). Between 1985 and 1987, a program to eradicate wild pigs on Trione-Annadel State Park in Sonoma County was initiated by the California Department of Parks and Recreation and the University of California at Berkeley. Over that 3-year period, a total of 142 pigs were removed at a cost of US$165,000 by means of trapping and hunting with dogs, comprising 69% and 31% of the harvest, respectively. A fence to preclude future immigration into the park was also constructed (Barrett et al. 1988). At the Pinnacles National Monument, the National Park Service began construction of a 57-km^2 exclosure in 1983 in an effort to reduce damage caused by wild pigs and completed it in 2005 at a cost of approximately US$2M (McCann and Garcelon 2008). Additionally, removal efforts began in 2003 and wild pigs were successfully eradicated from the exclosure by 2006, removing 200 pigs for a total cost of US$623,601 (McCann and Garcelon 2008).

12.5 REGIONAL WILD PIG RESEARCH IN WESTERN NORTH AMERICA

The majority of research in the Western region has been conducted in California, mainly due to the long historical presence of wild pigs relative to other parts of the region. Comprehensive descriptive studies of wild pigs were conducted in the mid- to late 1960s and 1980s along the

FIGURE 12.10 A total of 5,036 wild pigs were removed from Santa Cruz Island, California, during a successful eradication operation. (a) The island was separated by 1-m high fencing into 5 zones before eradication efforts began that primarily utilized (b) trapping, (c) hunting with dogs, and (d) aerial gunning. (Photos by N. MacDonald. With permission.)

central California coast, in north-central California, and Santa Catalina Island, respectively (Pine and Gerdes 1973, Barrett 1978, Baber and Coblentz 1986). These studies documented numerous aspects of wild pig biology and ecology including morphology, movement rates, home ranges, reproduction, food habits, and survival. More recent studies have estimated wild pig populations in California (Waithman et al. 1999, Sweitzer et al. 2000), measured fecundity rates in New Mexico (Dirsmith et al. 2017), studied food habits (Loggins et al. 2002, Wilcox and Van Vuren 2009), and documented damage caused by wild pigs in California grasslands and oak woodlands (Kotanen 1995; Sweitzer and Van Vuren 2002, 2008; Cushman et al. 2004; Tierney and Cushman 2006). Disease monitoring has been conducted in California (Barrett 1978, Sweitzer et al. 1996, Smith et al. 1998) and other parts of the Western region as part of nationwide surveillance efforts (Nettles et al. 1989, Pederson et al. 2012, Hill et al. 2014, Bevins et al. 2018). Several studies have also documented eradication operations in California, which in some cases tested the utility of Judas pigs for removal (Lombardo and Faulkner 2000, Schuyler et al. 2002, McCann and Garcelon 2008, Parkes et al. 2010).

12.6 A CASE STUDY OF WILD PIG RESEARCH AT TEJON RANCH, CALIFORNIA

We thank Michael D. White[a], Jesse S. Lewis[b], Ben S. Teton[c], and Ryan S. Miller[d] ([a]University of California, Berkeley, [b]Arizona State University, [c]Tejon Ranch Conservancy, [d]USDA/APHIS/Veterinary Services) for providing this case study. Contributions from R.S.M. were supported by the USDA. Mention of commercial products or companies does not represent an endorsement by the US Government.

Tejon Ranch, the largest contiguous private property in California (109,265 ha), is arguably one of the most unique and biodiverse ecological landscapes in the state. The ranch straddles the

Tehachapi Mountains, includes a portion of the southern Great Central Valley to the north and western Mojave Desert to the south, and provides a landscape linkage between the Coast Ranges and Sierra Nevada. In the 1980s, captive wild pigs breached the fence of a high-fenced shooting operation near Tehachapi, California (northeast of Tejon Ranch) and colonized the Tehachapi Mountains and Tejon (Figure 12.11). Christie et al. (2014) estimated there were between 1,000 and 4,000 pigs on Tejon by 2013. The Tejon Ranch Company operates a commercial hunting program and began selling access to the ranch to harvest pigs in 2001. The Company's goal was to maximize economic returns on wild pig hunting, generating an average of US$1.2M annually between 2001 and 2008, with a maximum annual harvest of 1,200 pigs (Christie et al. 2014).

In 2008, a private conservation agreement was executed that placed 90% of the ranch under a conservation easement and created the Tejon Ranch Conservancy to steward its conservation values. The Conservancy identifies wild pigs as a primary threat to these values and is conducting management trials and research to better understand and reduce ecological impacts of wild pigs on that portion of the ranch. However, the Tejon Ranch Company does not support many of the management actions proposed by the Conservancy to control pigs because of concerns these actions would adversely affect its lucrative hunting program. Therefore, wild pig management at Tejon Ranch is an example of the challenges to wild pig management that are created when a destructive invasive species is regulated and regarded as a game animal.

In 2014, the Conservancy and USDA/APHIS/Wildlife Services entered into a research partnership to test methods to estimate wild pig populations through time, document movements of wild pigs relative to agricultural fields and other landscape variables, and better understand the ecological impacts of wild pigs. As part of this research, a 48-km^2 grid of wildlife cameras was established in the heart of the ranch, and over 100 wild pigs were captured and fit with global positioning system (GPS) collars, very high frequency (VHF) collars, or unique color bands. The Tejon Ranch Conservancy also identified 101 individual wild pigs from natural markings using wildlife

FIGURE 12.11 (a) Study area, (b) mountain lion, (c) wild pig, and (d) cattle at the Tejon Ranch in California. (Photos by J. Lewis. With permission.)

camera photos. This research will allow a direct comparison of mark-resight population estimation approaches for wild pigs using natural and artificial marks to determine what levels of population change might result from specific management actions.

Biological samples were collected from all handled animals to test for a suite of pathogens that pose potential disease risks to livestock, agriculture, and humans. Wild pig use of agricultural fields in California has been linked to widespread disease outbreaks in humans on a national scale, resulting in human illness and deaths (Jay et al. 2007, Jay-Russell et al. 2012), illustrating the importance of managing wild pigs, particularly when they may interact with the human food supply. Regular use of grape vineyards and pistachio orchards by wild pigs was documented during the growing season (White et al. 2018). Wild pigs accessed agricultural fields only at night, likely in response to sustained hunting pressure during the day, demonstrating that recreational hunting on the Ranch would not be able to lessen damage. Additionally, 1 male wild pig traveled several kilometers each day between areas of forested cover and agricultural fields, indicating the large area of potential influence that wild pigs have (Figure 12.12).

Wild pigs continue to be a management concern in California, given the extensive damages they cause and the different goals of concerned stakeholders. For example, any control measures for wild pigs currently proposed by the Conservancy to reduce ecological damages in the conserved lands of Tejon Ranch, such as culling, would not be allowed by the landowner because of the revenue derived from wild pig hunting. However, some of the economic benefits associated with hunting wild pigs in California exist only because this invasive animal is regulated as a big game species. More information to evaluate the potential economic and recreational benefits of hunting wild pigs relative to their potential ecological, agricultural, and pathological impacts is needed to develop realistic wild pig management strategies acceptable to stakeholders across the diverse landscapes of California. Ultimately, changing the regulatory status of wild pigs in California will likely prove to have the greatest influence on their control and management.

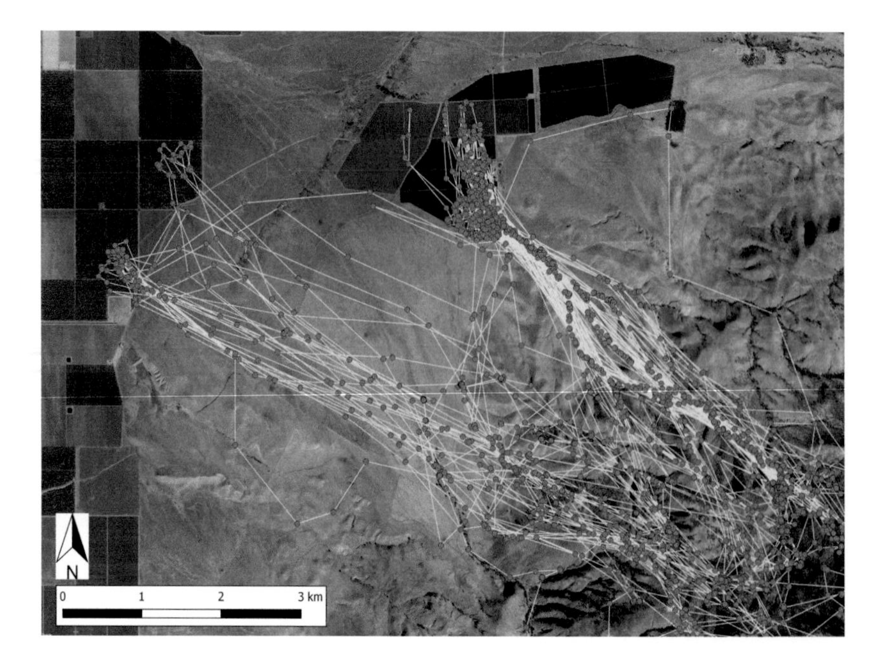

FIGURE 12.12 Movements of an adult male wild pig fit with a GPS collar (locations collected every 30 minutes) on the Tejon Ranch from August 2 to November 28, 2016. This animal made daily movements from forested habitat in the southeast to the agricultural fields in the northwest. (Courtesy of J. Lewis. With permission.)

12.7 FUTURE OF WILD PIGS IN WESTERN NORTH AMERICA

Wild pigs remain relatively scattered across much of the Western region, and population levels have yet to become established respective to other regions in North America such as the Southeast (see Chapter 16). Given the propensity of destructive ecological and agricultural impacts that wild pigs cause and a lack of historical status as a game species, excluding California, they are primarily perceived as an invasive species that should be managed as such. Thus, there is a favorable opportunity to proactively stop the further spread of wild pigs and eradicate rather than reduce populations that have become established. Widespread management efforts to eliminate wild pigs have been and continue to be implemented by state agencies in partnership with numerous collaborators. These efforts are increasingly important because predictive modeling has suggested wild pigs could be capable of inhabiting most counties in the US portion of the Western region in the next 3–5 decades (Snow et al. 2018). However, vast and remote areas common in the Western region also present a challenge to management of wild pigs. Persistent monitoring and early detection will be imperative to assist eradication. Continued efforts to improve the efficiency and effectiveness of monitoring and control techniques could aid the eradication of wild pigs within most of the region in the somewhat near future.

Despite increased awareness of the destructive nature of wild pigs, greater uncertainty exists related to their future in California, where they have expanded to 57 of 58 counties within the state (USDA/Animal and Plant Health Inspection Service 2015) and have been managed as a big game species for over 60 years. As in other regions, wild pig hunting in California presents a socio-economic challenge for management due to revenue from both pig tag sales and guided/commercial hunts, as well as an engrained culture of wild pig hunting, which creates incentives to maintain populations. However, perceptions may be shifting towards recognizing wild pigs as an invasive species, especially as damage continues to increase across the state. Evidence for this shift is supported by a new bill, AB-2805, introduced in the California Legislature in 2018. If passed, the bill would declassify wild pigs as a big game species and reclassify them as an exotic game species. Additionally, the bill would allow landowners to remove wild pigs causing damage on their property without having to obtain a depredation permit, eliminate wild pig tags for each wild pig harvested and replace it with an annual wild pig validation (US$25 for residents and US$75 for non-residents) that would allow hunters to harvest a set number of wild pigs determined by the California Department of Fish and Wildlife (CDFW) for that license year, and require the CDFW to conduct a wild pig management assessment through the 2022–23 license year to determine the effectiveness of reducing populations and associated damage (Bigelow and Mathis 2018). Hunting groups are the primary opponents to the bill, including the California Sportsman's Lobby and Safari Club International (Bigelow and Mathis 2018). While continued progression towards managing wild pigs as an invasive species will be necessary to mitigate damage, wild pigs will likely remain abundant on the California landscape until committed and concerted efforts are implemented for widespread removal of the species.

ACKNOWLEDGMENTS

The authors thank S.S. Ditchkoff for constructive editorial comments and suggestions that improved an early draft of this chapter. Contributions from J.J.M. were supported by the U. S. Department of Energy Office of Environmental Management under Contract DE-AC09-08SR22470 to Savannah River Nuclear Solutions LLC. Contributions from K.C.V. and M.P.G. were supported by the USDA. Mention of commercial products or companies does not represent an endorsement by the US Government.

REFERENCES

Allen, L. 2013. It's a boar war: Arizona's feral pigs. *Arizona Wildlife Views*, March-April: 8–12.

Associated Press. 2001. Feral pigs proliferating on Olympic Peninsula. *The Seattle Times*. <http://community.seattletimes.nwsource.com/archive/?date=20010809&slug=pigs09m>. Accessed 27 Sept 2010.

Atwill, E. R., R. A. Sweitzer, M. D. G. C. Pereira, I. A. Gardner, D. Van Vuren, and W. M. Boyce. 1997. Prevalence of and associated risk factors for shedding *Cryptosporidium parvum* oocysts and *Giardia* cysts within feral pig populations in California. *Applied and Environmental Microbiology* 63:3946–3949.

Avalos, E. 2013. Feral swine removal demonstration project. US Department of Agriculture/Animal and Plant Health Inspection Service. <https://www.usda.gov/media/blog/2013/07/24/feral-swine-removal-demonstration-project>. Accessed 26 July 2018.

Baber, D. W., and B. E. Coblentz. 1986. Density, home range, habitat use, and reproduction in feral pigs on Santa Catalina Island. *Journal of Mammology* 67:512–525.

Baber, D. W., and B. E. Coblentz. 1987. Diet, nutrition, and conception in feral pigs on Santa Catalina Island. *Journal of Wildlife Management* 51:306–317.

Baldwin, R. A., T. P. Salmon, R. H. Schmidt, and R. M. Timm. 2014. Perceived damage and areas of needed research for wildlife pests of California agriculture. *Integrative Zoology* 9:265–274.

Barrett, R. H. 1977. Wild pigs in California. Pages 111–113 *in* G. W. Wood, editor. *Research and Management of Wild Hog Populations*. Belle Baruch Forest Science Institute of Clemson University, Georgetown, SC.

Barrett, R. H. 1978. The feral hog at Dye Creek Ranch, California. *Hilgardia* 46:283–355.

Barrett, R. H. 1982. Habitat preferences of feral hogs, deer, and cattle on a Sierra Foothill range. *Journal of Range Management* 35:342–346.

Barrett, R. H., B. L Goatcher, P. J. Gogan, and E. L. Fitzhugh. 1988. Removing feral pigs from Annadel State Park. *Transactions of the Western Section of the Wildlife Society* 24:147–152.

Barrett, R. H., and D. S. Pine. 1980. History and status of wild pigs, *Sus scrofa*, in San Benito County, California. *California Fish and Game* 67:105–117.

Bevins, S. N., M. W. Lutman, K. Pederson, N. Barrett, T. Gidlewski, T. J. Deliberto, and A. B. Franklin. 2018. Spillover of swine coronaviruses, United States. *Emerging Infectious Diseases* 24:1390–1392.

Bigelow, F., and D. Mathis. 2018. *Assembly Bill 2805 Wild Pigs: Validation*. California Assembly Committee on Water, Parks and Wildlife, Rural County Representatives of California, Sacramento, CA.

British Columbia Ministry of Forests, Lands and Natural Resource Operations 2014. Province takes aim at feral pigs. *British Columbia Ministry of Forests, Lands and Natural Resource Operations*. <https://archive.news.gov.bc.ca/releases/news_releases_2013–2017/2014FLNR0021–000322.htm>. Accessed 20 Mar 2014.

Brook, R. 2017. Canada's wild pig problem. *Webinar*. <https://bcinvasives.ca/documents/Canadas_Wild_Pig_Problem_Ryan_Brook_Webinar.pdf>. Accessed 8 May 2018.

Brown, J. S., J. W. Laundre, and M. Gurung. 1999. The ecology of fear: Optimal foraging, game theory, and trophic interactions. *Journal of Mammalogy* 80:385–399.

California Department of Fish and Wildlife. 2018. *2018 California Big Game Hunting Digest*. California Department of Fish and Wildlife, Sacramento, CA.

Canadian Food Inspection Agency. 2015. Notice to industry—mandatory identification and reporting requirements for pigs and farmed wild boars. *Canadian Food Inspection Agency*. <http://www.inspection.gc.ca/animals/terrestrial-animals/traceability/notice-to-industry-2015-03-09-/eng/1425916938050/1425916938785>. Accessed 27 July 2018.

Carpenter, A., and C. Provorse. 1996. *The World Almanac of the USA*. World Almanac Books, Mahwah, NJ.

CBC News. 2011. Pigs go wild in Christina Lake, B.C. *CBC News Online*, March 19. <http://www.cbc.ca/news/canada/calgary/story/2011/03/19/bc-christina-lake-wild-pigs.html?ref=rss>. Accessed 22 June 2012.

CBC News. 2018a. Then there were 2: More wild boars on the run in Yukon shot dead. *CBC News Online*, July 17. <https://ca.news.yahoo.com/then-were-2-more-wild-002711596.html>. Accessed 3 Apr 2019.

CBC News. 2018b. All wild boars that escaped from Yukon farm now dead, says government. *CBC News Online*, July 17. <https://www.cbc.ca/news/canada/north/yukon-wild-boars-escape-dead-1.4802621>. Accessed 4 Apr 2019.

CBC News. 2018c. Big pig problem: What to do after Yukon's wild boar fiasco? *CBC News Online*, September 28. <https://ca.sports.yahoo.com/news/big-pig-problem-yukon-apos-090000529.html>. Accessed 3 Apr 2019.

Centers for Disease Control and Prevention. 2006. Multistate outbreak of E. coli O157:H7 infections linked to fresh spinach (Final Update). <https://www.cdc.gov/ecoli/2006/spinach-10-2006.html>. Accessed 14 Mar 2016.

Christie, J., E. DeMarco, E. Hiroyasu, A. Kreger, and M. Ludington. 2014. *Wild Pig Management on Tejon Ranch*. Bren School Group Project Report to the Tejon Ranch Conservancy, pp. 1–82.

Coblentz, B., and C. Bouska. 2004. *Pest Risk Assessment for Feral Pigs in Oregon*. Department of Fisheries and Wildlife, Oregon State University, Corvallis, OR.

Commission for Environmental Cooperation. 1997. *Ecological Regions of North America: Toward a Common Perspective.* NAAEC Commission for Environmental Cooperation, Montreal, Canada.

Corn, J. L., and T. R. Jordan. 2017. Development of the national feral swine map, 1982–2016. *Wildlife Society Bulletin* 41:758–763.

Craig, D. L. 1986. The seasonal food habits in sympatric populations of puma (*Puma concolor*), coyote (*Canis latrans*), and bobcat (*Lynx rufus*) in the Diablo Range of California. Thesis, San Jose State University, San Jose, CA.

Cushman, J. H. 2007. History and ecology of feral pig invasions in California grasslands. Pages 191–196 *in* M. R. Stromber, J. D. Corbin, and C. M. D'Antonio, editors. *California Grasslands Ecology and Management.* University of California Press, Berkeley and Los Angeles, CA.

Cushman, J. H., T. A. Tierney, and J. M. Hinds. 2004. Variable effects of feral pig disturbances on native and exotic plants in a California grassland. *Ecological Indicators* 14:1746–1756.

D'Antonio, C. M., C. Malmstrom, S. A. Reynolds, and J. Gerlach. 2007. Ecology of invasive non-native species in California grassland. Pages 67–83 *in* M. R. Stromberg, J. D. Corbin, and C. M. D'Antonio, editors. *California Grasslands Ecology and Management.* University of California Press, Berkeley and Los Angeles, CA.

Darr, D. 2011. Unwelcome invaders: Wild pigs pose a serious new threat to Idaho. *Boise Weekly.* <http://www.boiseweekly.com/gyrobase/unwelcome-invaders-wild-pigs-pose-a-serious-new-threat-to-idaho/Content?oid=2090820&showFullText=true>. Accessed 24 Mar 2011.

Davis, S. 2013. Drought curbs spread of feral hogs in southeastern New Mexico. *Carlsbad Current-Argus*, June 18. <http://www.currentargus.com/ci_23483408/drought-curbs-spread-feral-hogs-southeastern-new-mexico?source=most_viewed#>. Accessed 12 Nov 2013.

Department of the Interior. 1997. Endangered and threatened wildlife and plants: Determination of endangered status for three plants from the Channel Islands of southern California. *Federal Register* 62:42692–42702.

Dirsmith, K. L., M. A. Tabak, D. W. Wolfson, and R. S. Miller. 2017. New Mexico feral swine litter size and pregnancy report: 2010–2016. USDA/APHIS/VS/Center for Epidemiology and Animal Health Report.

Drew, M. L., D. A. Jessup, A. A. Burr, and C. E. Franti. 1992. Serologic survey for brucellosis in feral swine, wild ruminants, and black bear of California, 1977 to 1989. *Journal of Wildlife Diseases* 28:355–363.

Dumas, C. R. 2009. Invasive feral hogs threaten to spread livestock disease. *Capital Press.* <http://www.capitalpress.com/content/CRD-feral-hogs-update-w-B-amp-amp-W-art--p-10-121109>. Accessed 11 Jan 2010.

Frederick, J. M. 1998. Overview of wild pig damage in California. *Proceedings of the Vertebrate Pest Conference* 18:82–86.

Garcia, J., and K. Raymond. 2018. *2015–2016 Wild Pig Take Report.* California Department of Fish and Wildlife, Sacramento, CA.

Hill, D. E., J. P. Dubey, J. A. Baroch, S. R. Swafford, V. F. Fournet, D. Hawkins-Cooper, D. G. Pyburn, et al. 2014. Surveillance of feral swine for *Trichinella* spp. and *Toxoplasma gondii* in the USA and host-related factors associated with infection. *Veterinary Parasitology* 205:653–665.

Hogg, J. E. 1920. Boar hunting on Santa Cruz. *Outer's Recreation* 63:194–196, 238–240.

Hollenhorst, J., and W. Leonard. 2011. Wild pigs, sheep on remote island causing concern among wildlife officials. *KSL.com.* Utah October 28. <http://www.ksl.com/?nid=960&sid=17866097>. Accessed 4 Sept 2012.

Hopkins, R. A. 1989. *Ecology of the Puma in the Diablo Range, California.* Dissertation. University of California, Berkeley, CA.

Hutchinson, C. B., editor. 1946. *California Agriculture.* University of California Press, Berkeley, CA.

Idaho Statesman. 2012. Seen any feral pigs in Idaho? Call the N.W. swine line. *Idaho Statesman.* <http://www.idahostaesman.com/2012/07/24/2199615/seen-any-feral-pigs-in-idaho-call.html>. Accessed 31 July 2012.

Jay, M. T., M. Cooley, D. Carychao, G. W. Wiscomb, R. A. Sweitzer, L. Crawford-Miksza, J. A. Farrar, et al. 2007. *Escherichia coli* O157:H7 in feral swine near spinach fields and cattle, central California coast. *Emerging Infectious Diseases* 13:1908–1911.

Jay-Russell, M. T., A. Bates, L. Harden, W. G. Miller, and R. E. Mandrell. 2012. Isolation of *Campylobacter* from feral swine (*Sus scrofa*) on the ranch associated with the 2006 *Escherichia coli* O157:H7 spinach outbreak investigation in California. *Zoonoses and Public Health* 59:314–319.

Kimak, J. 2002. Feral pigs causing some problems along the Virgin River. *Las Vegas Review-Journal*, May 12. <http://www.reviewjournal.com/lvrj_home/2002/May-12-Sun-2002/living/18679119.html>. Accessed 15 Aug 2003.

Kimak, J. 2010. Feral pigs populate the Silver State. *Boulder City Review*, July 15. <http://bouldercityreview.com/news/feral-pigs-populate-the-silver-state/>. Accessed 23 Jan 2013.

Koen, E. L., E. Vander Wal, R. Kost, R. K. Brook. 2018. Reproductive ecology of recently established wild pigs in Canada. *The American Midland Naturalist* 179:275–286.

Kotanen, P. M. 1995. Responses of vegetation to a changing regime of disturbance: Effects of feral pigs in a California coastal prairie. *Ecography* 18:190–199.

Kreith, M. 2007. Wild Pigs in California: The Issues. AIC Issues Brief No. 33. Agricultural Issues Center, University of California, Davis, CA.

Liakopoulos, P. 1999. Outdoor adventures: Wild boar hunts on the Virgin River. *Pahrump Valley Times*, August 11. <http://www.pahrumpvalleytimes.com/1999/08/11/387289.html>. Accessed 28 Apr 2010.

Lightcap, C. 2008. *Colorado Ranch Quarantined Following Discovery of Swine Disease*. COSDA: Communication Officers of State Departments of Agriculture. <cosdablog.blogspot.com/2008/12/Colorado-ranch-quarantined-following.html>. Accessed 16 Oct 2009.

Lloyd, D. S., R. B. Smith, and K. A. Sundberg. 1987. Introduction of European wild boar to Marmot Island, Alaska. *The Murrelet* 68:57–58.

Loggins, R. E., J. T. Wilcox, D. H. Van Vuren, and R. A. Sweitzer. 2002. Seasonal diets of wild pigs in oak woodlands of the central coast region of California. *California Fish and Game* 88:28–34.

Lombardo, C. A., and K. R. Faulkner. 2000. Eradication of feral pigs (*Sus scrofa*) from Santa Rosa Island, Channel Islands National Park, California. *Proceedings of the California Islands Symposium* 5:300–306.

Loney, T. 2001. Open season on feral pigs? *The Daily World Local News*. <http://aberdeen.donrey.com/daily/2001/Aug-02-Thu-2001/news/news4.html>. Accessed 17 June 2002.

Mansfield, T. M. 1978. Wild pig management on a California public hunting area. *Transactions of the California-Nevada Section of the Wildlife Society* 25:187–201.

Mayer, J. J. 2018. Introduced wild pigs in North America: History, problems and management. Pages 299–312 *in* M. Melletti and E. Meijaard, editors. *Ecology, Evolution and Management of Wild Pigs and Peccaries: Implications for Conservation*. Cambridge University Press, Cambridge, UK.

Mayer, J. J., and I. L. Brisbin, Jr. 2008. *Wild Pigs in the United States: Their History, Comparative Morphology, and Current Status*. Second Edition. The University of Georgia Press, Athens, GA.

Mayer, J. J., and I. L. Brisbin, Jr., editors. 2009. *Wild Pigs: Biology, Damage, Control Techniques and Management*. SRNL-RP-2009-00869. Savannah River National Laboratory, Aiken, SC.

McCann, B. E., and D. K. Garcelon. 2008. Eradication of Feral Pigs from Pinnacles National Monument. *Journal of Wildlife Management* 72:1287–1295.

McClure, M. L., C. L. Burdett, M. L. Farnsworth, M. W. Lutman, D. M. Theobald, P. D. Riggs, D. A. Grear, and R. S. Miller. 2015. Modeling and mapping the probability of occurrence of invasive wild pigs across the contiguous United States. *Plos One* 10:e0133771.

McKinley, J. 2016. Officials want feral pigs stopped before they go hog wild. *Invasive Species Council of BC*, February 5. <https://bcinvasives.ca/news-events/recent-highlights/officials-want-feral-pigs-stopped-before-they-go-hog-wild/>. Accessed 4 May 2018.

McKnight, T. 1964. *Feral Livestock in Anglo-America. University of California Publications in Geology*, Vol. 16. University of California Press, Berkeley and Los Angeles, CA.

Michel, N. L., M. P. Laforge, F. M. Van Beest, and R. K. Brook. 2017. Spatiotemporal trends in Canadian domestic wild boar production and habitat predict wild pig distribution. *Landscape and Urban Planning* 165:30–38.

Moffatt, K. 2012. Hogs gone wild: Invasive pigs threaten wildlife, crops. *New Mexico Wildlife* 56:10–11.

Moody, A., and J. A. Jones. 2000. Soil response to canopy position and feral pig disturbance beneath *Quercus agrifolia* on Santa Cruz Island, California. *Applied Soil Ecology* 14:269–281.

Mortensen, C. 2009. Hog wild: Oregon's feral pig population may be about to explode. *Eugene Weekly*. <http://www.eugeneweekly.com/2009/02/05/coverstory.html>. Accessed 26 Oct 2009.

Nailon, J. 2017. Wild pigs reported in Gifford Pinchot National Forest for second year in a row. *The Chronicle*. <http://www.chronline.com/news/wild-pigs-reported-in-gifford-pinchot-national-forest-for-second/article_391c9700-e9d6-11e7-b297-07099c7b15bf.html>. Accessed 28 Dec 2017.

Neskey, J. 2018. Feral swine eradication at Havasu National Wildlife Refuge: Protecting endangered species from feral swine damage. US Department of Agriculture/Animal and Plant Health Inspection Service. <https://www.usda.gov/media/blog/2018/04/17/feral-swine-eradication-havasu-national-wildlife-refuge-protecting-endangered>. Accessed 26 July 2018.

Nettles, V. F., J. L. Corn, G. A. Erickson, and D. A. Jessup. 1989. A survey of wild swine in the United States for evidence of hog cholera. *Journal of Wildlife Diseases* 25:61–65.

Oregon Invasive Species Council. 2002. *Meeting Minutes-September 24–25, 2002*. Oregon Invasive Species Council, Astoria, OR.

Oregon Invasive Species Council. 2018. *Squeal on Pigs: Feral Swine Campaign*. <https://www.oregoninvasivespeciescouncil.org/squeal-on-pigs/>. Accessed 25 July 2018.

Parkes, J. P., D. S. L. Ramsey, N. Macdonald, K. Walker, S. McKnight, B. S. Cohen, and S. A. Morrison. 2010. Rapid eradication of feral pigs (*Sus scrofa*) from Santa Cruz Island, California. *Biological Conservation* 143:634–641.

Paul, T. W. 2009. *Game Transplants in Alaska*. Technical Bulletin No. 4, Second Edition. Alaska Department of Fish and Game, Juneau, AK.

Pederson, K., S. N. Bevins, J. A. Baroch, J. C. Cumbee, S. C. Chandler, B. S. Woodruff, T. T. Bigelow, and T. J. DeLiberto. 2013. Pseudorabies in feral swine in the United States, 2009–2012. *Journal of Wildlife Diseases* 49:709–713.

Pederson, K., S. N. Bevins, B. S. Schmit, M. W. Lutman, M. P. Milleson, C. T. Turnage, T. T. Bigelow, and T. J. DeLiberto. 2012. Apparent prevalence of swine brucellosis in feral swine in the United States. *Human-Wildlife Interactions* 61:38–47.

Peel, M. C., B. L. Finlayson, and T. A. McMahon. 2007. Updated world map of the Köppen-Geiger climate classification. *Hydrology and Earth System Sciences* 11:1633–1644.

Pelzer, J. 2011. Wyoming on guard against wild pigs. *Billings Gazette*, April 24. <https://billingsgazette.com/news/state-and-regional/wyoming/wyoming-on-guard-against-wild-pigs/article_72ebdc53-f27e-5249-97ee-d1c6d825d085.html>. Accessed 3 June 2013.

Pine, D. S., and G. L. Gerdes. 1973. Wild pigs in Monterey County, California. *California Fish and Game* 59:126–137.

Prettyman, B. 2013. Feral pigs killed after one headed for Antelope Island. *The Salt Lake Tribune*, October 31. <http://www.sltrib.com/sltrib/news/57038093-78/island-pigs-fremont-sheep.html.csp>. Accessed 12 May 2014.

Quenette, P. Y., and J. P. Desportes. 1992. Temporal and sequential structure of vigilance behavior of wild boars (*Sus scrofa*). *Journal of Mammalogy* 73:535–540.

Roemer, G. W., T. J. Coonan, D. K. Garcelon, J. Bascompte, and L. Laughrin. 2001. Feral pigs facilitate hyperpredation by golden eagles and indirectly cause the decline of the island fox. *Animal Conservation* 4:307–318.

Roemer, G. W., C. J. Donlan, and F. Courchamp. 2002. Golden eagles, feral pigs, and insular carnivores: How exotic species turn native predators into prey. *Proceedings of the National Academy of Sciences* 99:791–796.

Rouhe, A., and M. Sytsma. 2007. Feral swine action plan for Oregon. Prepared for the Oregon Invasive Species Council. Portland State University, Portland, OR.

Schauss, M. E., H. J. Coletto, and M. J. Kutilek. 1990. Population characteristics of wild pigs, *Sus scrofa*, in eastern Santa Clara County, California. *California Fish and Game* 76:68–77.

Scheffer, V. B. 1941. Management studies on transplanted beavers in the Pacific Northwest. *Transactions of the North American Wildlife Conference* 6:320–326.

Schuyler, P. T., D. K. Garcelon, and S. Escover. 2002. Eradication of feral pigs (*Sus scrofa*) on Santa Catalina island, California, USA. Pages 274–286 *in* C. R. Veitch and M. N. Clout, editors. *Turning the Tide: The Eradication of Invasive Species*. IUCN SSC Invasive Species Specialist Group, International Union for the Conservation of Nature and Natural Resources, Cambridge, UK.

Shore, R. 2016. Destructive and invasive, feral pigs spreading across B.C. *Vancouver Sun*, February 2. <http://www.vancouversun.com/technology/destructive+invasive+feral+pigs+spreading+across/11690807/story.html>. Accessed 5 May 2018.

Smith, C. R., B. A. Wilson, C. L. Fritz, C. M. Myers, J. C. Hitchcock, M. B. Madon, and S. B. Werner. 1998. Review of plague in California, 1997. *California Morbidity*, May:1–4.

Snow, N. P., M. A. Jarzyna, and K. C. VerCauteren. 2018. Interpreting and predicting the spread of invasive wild pigs. *Journal of Applied Ecology* 54:2022–2032.

Steller, T. 2010. Only the lack of water keeps wild pigs from being state menace. *Arizona Daily Star*. <http://www.azstarnet.com/news/science/environment/article_e216707d-5bdf-5779-a79f-4a4f7f4a785.html>. Accessed 22 Mar 2010.

Stugelmayer, J. 2010. Wild pigs migrate north, possibly into Montana. *Montana Kaimin: UM's Independent Campus Newspaper*. <http://www.montanakaimin.com/index,php/articles/article/wild_pigs_migrate_north_possibly_into_montana/831>. Accessed 23 Mar 2010.

Sweitzer, R. A. 1998. Conservation implications of feral pigs in island and mainland ecosystems, and a case study of feral pig expansion in California. *Proceedings of the Vertebrate Pest Conference* 18:26–34.

Sweitzer, R. A., I. A. Gardner, B. J. Gonzales, D. Van Vuren, and W. M. Boyce. 1996. Population densities and disease surveys of wild pigs in the coast ranges of Central and Northern California. *Proceedings of the Vertebrate Pest Conference* 17:75–82.

Sweitzer, R. A., and D. H. Van Vuren. 2002. Rooting and foraging effects of wild pigs on tree regeneration and acorn survival in California's oak woodland ecosystems. USDA Forest Service Gen. Tech. Report. PSW-GTR-184:219–231.

Sweitzer, R. A., and D. H. Van Vuren. 2008. Effects of wild pigs on seedling survival in California oak woodlands. *Proceedings of the California Oak Symposium* 6:267–277.

Sweitzer, R. A., D. Van Vuren, I. A. Gardner, W. M. Boyce, and J. D. Waithman. 2000. Estimating sizes of wild pig populations in the north and central coast regions of California. *Journal of Wildlife Management* 64:531–543.

Teton, B. S., K. Kunkel, and M. D. White. 2016. Grappling with wild pigs in California high country: Wild pig population and disturbance research at Tejon Ranch. *Proceedings of the Vertebrate Pest Conference* 27:124–127.

The New York Times. 1894. Wild hogs in Arizona: They roam in bands of thousands along the Colorado. July 29. <https://timesmachine.nytimes.com/timesmachine/1894/07/29/106868840.pdf>. Accessed 20 May 2016.

The Spokesman Review. 2002. It's open season on wild boars. The Spokesman-Review, Spokane, WA. <http://www.spokesmanreview.com/allstories-news-story.asp?date=120402&ID=s1267048>. Accessed 27 Sept 2009.

Tierney, T. A., and J. H. Cushman. 2006. Temporal changes in native and exotic vegetation and soil characteristics following disturbances by feral pigs in a California grassland. *Biological Invasions* 8:1073–1089.

Twietmeyer, N. 2016. Free Wilbur: The war against feral swine comes to Washington. *Seattle Weekly*. <http://www.seattleweekly.com/news/free-wilbur-the-war-against-feral-swine-comes-to-washington/>. Accessed 15 May 2018.

Updike, J., and J. Waithman. 1996. Dealing with wild pig depredations in California: The strategic plan. *Proceedings of the Vertebrate Pest Control Conference* 16:40–43.

US Department of Agriculture. 2017. Feral swine populations 2017 by county. <https://www.aphis.usda.gov/wildlife_damage/feral_swine/images/2017-feral-swine-distribution-map-county.jpg>. Accessed 25 July 2018.

US Department of Agriculture/Animal and Plant Health Inspection Service. 2015. *Final Environmental Impact Statement—Feral Swine Damage Management: A National Approach*. Washington, DC.

US Department of Agriculture/Animal and Plant Health Inspection Service/Wildlife Services New Mexico. 2010. *Feral Hog Biology, Impacts, and Eradication Techniques*. Washington, DC.

US Department of Agriculture/Animal and Plant Health Inspection Service/Wildlife Services New Mexico. 2017a. *New Mexico Fiscal Year 2017 Quarter 3 Feral Swine Quarterly Report*. Washington, DC.

US Department of Agriculture/Animal and Plant Health Inspection Service/Wildlife Services New Mexico. 2017b. *New Mexico Fiscal Year 2017 Quarter 2 Feral Swine Quarterly Report*. Washington, DC.

US Department of Agriculture/Animal and Plant Health Inspection Service/Wildlife Services New Mexico. 2018. *FY 18 WS Feral Swine State/Territory Plan*. Washington, DC.

US Department of Agriculture/Animal and Plant Health Inspection Service/Wildlife Services Oregon. 2015. *Oregon State Report*. Washington, DC.

US Department of Agriculture/Animal and Plant Health Inspection Service/Wildlife Services Oregon. 2019. *Fiscal Year 2019 Wildlife Services Feral Swine State/Territory Plan*. Washington, DC.

US Department of Agriculture/Animal and Plant Health Inspection Service/Wildlife Services Washington. 2018. *Washington Fiscal Year 2018 Quarter 1 Feral Swine Quarterly Report*. Washington, DC.

US Fish and Wildlife Service. 2018. Swine eradication efforts. US Fish and Wildlife Service. <https://www.fws.gov/refuge/havasu/swine_eradication.html>. Accessed 26 July 2018.

Waithman, J. 2001. *Guide to Hunting Wild Pigs in California*. California Department of Fish and Game, Wildlife Programs Branch, Sacramento, CA.

Waithman, J. D., R. A. Sweitzer, D. Van Vuren, J. D. Drew, A. J. Brinkhaus, I. A. Gardner, and W. M. Boyce. 1999. Range expansion, population sizes, and management of wild pigs in California. *Journal of Wildlife Management* 63:293–308.

Washington Invasive Species Council. 2010. Stop the invasion: Feral swine *Sus scrofa*. *Washington State Recreation and Conservation Office, Washington Invasive Species Council*. <https://invasivespecies.wa.gov/documents/priorities/FeralSwineFactSheet.pdf>. Accessed 14 Nov 2017.

Wheeler, S. A. 1944. California's little known Channel Islands. *United States Naval Institute Proceedings* 70:257–270.

White, M. D., K. M. Kauffman, J. S. Lewis, and R. S. Miller. 2018. Wild pigs breach farm fence through harvest time in southern San Joaquin Valley. *California Agriculture* 72:120–126.

Wilcox, J. T. 2015. Implications of predation by wild pigs on native vertebrates: A case study. *California Fish and Game* 101:72–77.

Wilcox, J. T., and D. H. Van Vuren. 2009. Wild pigs as predators in oak woodlands of California. *Journal of Mammalogy* 90:114–118.

Williams, K. E., K. P. Huyvaert, K. C. VerCauteren, A. J. Davis, and A. J. Piaggio. 2017. Detection and persistence of environmental DNA from an invasive, terrestrial mammal. *Ecology and Evolution* 8:688–695.

Zatkulak, K. 2013. No wild pigs sightings in Idaho since 2011. KTVB, Boise, ID. <http://www.ktvb.com/news/No-wild-pigs-sightings-in-Idaho-since-2011-215313451.html>. Accessed 11 Nov 2013.

Zaun, B., D. Magnuson, G. Klinger, and J. Fernandez. 2016. *Havasu National Wildlife Refuge Feral Swine Eradication Plan*. US Fish and Wildlife Service, Needles, CA.

13 Wild Pigs in North-Central North America

Ryan K. Brook and Michael P. Glow

CONTENTS

13.1 INTRODUCTION

The establishment of wild pigs (*Sus scrofa*) in north-central North America has occurred relatively recently, with most detections identified after 1990. This reflects dispersal-based spread of wild pigs northward in the United States (Snow et al. 2017), and initial introduction of the first free-ranging wild pigs from domestic sources that began during the 1980s in Canada (Brook and van Beest 2014, Michel et al. 2017). From 1990 to 2019, the most notable expansion of wild pigs in this region was a dramatic increase in the 3 Prairie Provinces of Canada (Alberta, Manitoba, and Saskatchewan; Aschim and Brook 2019). At the same time, wild pigs expanded in Kansas and Missouri. However, wild pigs were greatly reduced through intensive control efforts in Kansas (Richardson et al. 1997, Hartin et al. 2007), but still occasionally cross into Kansas along the southern border from Oklahoma. Outside of Missouri, few populations of wild pigs currently exist in the US portion of the region. Several wild pig sightings along the Canadian-US border in Manitoba, Saskatchewan, and North Dakota have generated important concerns about cross-boundary movements.

In the North-Central region, control efforts in the United States have focused on limiting natural and human-assisted movements of wild pigs northward. Control efforts in Canada have been minimal, with some localized efforts to remove wild pigs through an unsuccessful bounty program in Alberta, ground shooting and trapping in a few locations in Saskatchewan, and declaration of Manitoba as a wild pig control area in 2001 where any wild pigs could be shot on sight.

For this chapter, the North-Central region includes the Northwest Territories, Alberta, Saskatchewan, and Manitoba of Canada and all or parts of 10 US states: eastern Colorado, eastern Montana, eastern Wyoming, Iowa, Kansas, Missouri, Nebraska, North Dakota, South Dakota, and western Minnesota (Figure 13.1). This region overlaps to a high degree with the Great Plains of

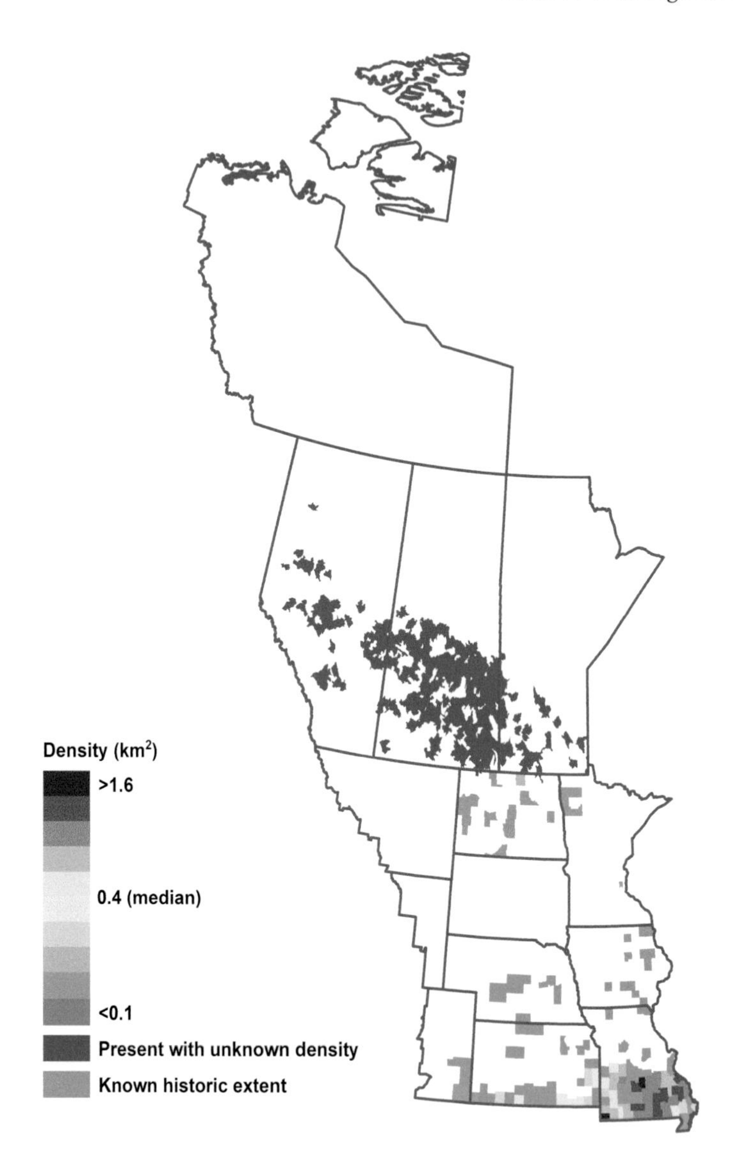

FIGURE 13.1 2019 spatial distribution of wild pig populations in north-central North America, based on observational data. Densities were generated using a hierarchical Bayesian catch-effort model that integrates removal and effort data with climatic data, habitat data, and data influencing probability of capture (e.g., tree cover, road density, topography), and represent a county average using model predictions. Shaded gray areas are where pigs were once reported but have now been eliminated as a result of control programs. Areas in dark blue include all known watersheds in Canada with wild pig detections up to 2017, based on Aschim and Brook (2019). (Courtesy of R. Miller. With permission.)

North America and includes over 4 million km², of which less than 10% is currently occupied by wild pigs.

Climate in the North-Central region varies dramatically along a 4,000-km gradient from south to north. Southern US states in the region have long, warm, summers (average July highs of 32°C) and short, warm winters (average January low of –5°C). Conversely, at northern extremes of the region in the arctic tundra, short, cool summers (average July highs of 16°C) and long cold winters (average January low of –31°C with no sunlight) are normal. Annual precipitation is similar across the region at 20–100 cm and the total proportion that falls as snow increases with latitude. Within the

Canadian portion of the region, snow lasts for at least half of the year and regularly accumulates to >30-cm deep on the landscape. Severe summer droughts are not uncommon for much of the region and are expected to reach unprecedented levels in response to climate change (Cook et al. 2015).

13.2 HISTORIC AND PRESENT DISTRIBUTION

The historical record of wild pigs is considerably different between Canada and the United States in the North-Central region, primarily due to unique histories of initial introductions. Although wild pigs were introduced to the southern United States approximately 500 years ago (see Chapter 16), they have only recently been documented in scattered locations throughout most of the North-Central region. Wild pigs were first detected in northeastern Kansas at Fort Riley Military Base (44,500 ha) in 1993 (Gipson et al. 1994). Established populations were confirmed in 18 relatively small areas across Kansas in 2004, but were notably reduced by 2016 (Gipson et al. 2006, Corn and Jordan 2017). Wild pigs were first detected in Nebraska in 2004 in 4 counties (Gipson et al. 2006); the last known wild pigs were removed from the state by 2012 (USDA/APHIS/Wildlife Services Nebraska 2015). These relatively isolated populations of wild pigs potentially originated from humans purposefully transporting and releasing animals to create new hunting opportunities, escapes from high-fenced shooting operations, dispersal from established southern populations, and/or escaped domestic swine (Gipson et al. 1997, 2006). Wild pig sightings occurred in 11 Iowa counties from 2003 to 2007 (Iowa State University Forestry Extension, 2017), but more recent statewide wild pig activity has been limited to sporadic sightings and removals (USDA/APHIS/Wildlife Services Iowa 2017, 2018).

At one time, Missouri had large populations of wild pigs, having originated from domestic stock that were released into open range during years of heavy acorn production in the Ozarks. However, by the early 1960s, wild pig populations in the state were confined to 3 small areas in southeastern Missouri (Mayer and Brisbin 2008). Then, during the 1990s, the range and abundance of wild pigs increased substantially in Missouri (Hartin 2006, Hartin et al. 2007). During this time, landowners began raising wild boar for high-fenced shooting operations or as an alternative form of agricultural livestock; in addition, large numbers of domestic pigs were released into the wild in the 1990s when pork prices dropped precipitously (Hartin et al. 2007). Additionally, the illegal release of wild pigs has been a continuous issue, resulting in the establishment of wild pig populations in over 30 Missouri counties (Davis 2018). Recent focused control efforts have successfully eradicated wild pigs from several counties in Missouri (USDA/APHIS/Wildlife Services Missouri 2015).

USDA/APHIS/Wildlife Services has reported recent, sporadic removals of wild pigs in southeastern Colorado. In 2016, surveillance began in the Arkansas River Valley in Prowers County, Colorado, an area historically thought to be absent of wild pigs, after a hunter harvested a wild pig near Granada, but no additional wild pigs have been found (USDA/APHIS/Wildlife Services Colorado 2016a). Additionally, USDA/APHIS/Wildlife Services trapped and lethally removed 5 wild pigs from Baca County, Colorado in 2016 (USDA/APHIS/Wildlife Services Colorado 2016a). Ground and aerial surveillance efforts were conducted by USDA/APHIS/Wildlife Services in southeastern Colorado and western Kansas in 2017, but no wild pigs or wild pig sign were detected (USDA/APHIS/Wildlife Services Colorado 2017).

Intermittent sightings of wild pigs have occurred in North Dakota since 1997 and these individuals were eliminated through state and federal control efforts. Similarly, there are no known established populations in South Dakota or Minnesota, with only sporadic wild pig sightings in South Dakota. While there are also no known established populations in Montana, wild pigs were reported within 5 miles of the Canadian/Montana border in 2018 (French 2019).

Captive wild boar were first introduced to Canada in the 1980s and 1990s in all provinces to diversify agricultural production and establish high-fenced shooting operations in some provinces, including Alberta, Saskatchewan, and Quebec (Michel et al. 2017). Of all of the wild boar farms in Canada, 68% were in Alberta, Saskatchewan, and Manitoba (Michel et al. 2017). Farmers were strongly encouraged to breed wild boar with domestic pigs to produce animals with larger body sizes and litters (Michel et al. 2017). A thriving international market for wild boar meat from Canada was never established, though a small niche market for wild boar meat within Canada still remains (Michel et al. 2017). Across multiple provinces and especially in the north-central region of Canada, domestic wild boar farmers have experienced chronic escapes because of insufficient fencing standards and near absence of monitoring and enforcement (Michel et al. 2017). In some instances, entire herds were released to the wild (Michel et al. 2017). Annual escape rates from domestic wild boar farms was estimated at 3% (Tokaruk 2002). The number of domestic wild boar on farms in the north-central region of Canada peaked at production of 30,000 animals in 1996, but has since dropped dramatically to 7,000 by 2011 (Michel et al. 2017). Due to even further decreases, Statistics Canada no longer records numbers of wild boar farms or animals produced in the national agriculture census every 5 years (Statistics Canada 2016).

High-fenced shooting operations were established in Alberta and Saskatchewan in the 1990s and 2000s where several continue to operate, but they have been prohibited in Manitoba since 2000 (Michel et al. 2017). Most high-fenced shooting operations claim to offer "full-blooded Russian and European wild boar" as well as "feral wild boar" on their websites. Farm-raised wild boar from Canada can be imported to high-fenced shooting operations in some US states, but are subject to permitting and disease testing (USDA/APHIS/VS/National Import Export Services 2018). Escaped animals from high-fenced shooting operations are frequently detected, often in proximity to the facility they escaped from.

From 1990 to 2000, free-ranging wild pigs were only documented across 4 watersheds in the Canadian provinces of Alberta, Saskatchewan, and Manitoba (Aschim and Brook 2019). However, by 2010, wild pigs had expanded considerably across areas of all 3 of the Prairie Provinces, especially Saskatchewan (Aschim and Brook 2019). Wild pigs continued to spread rapidly (i.e., average annual increase 41,000 km^2/year) across the majority of watersheds in the southern half of Saskatchewan and extensive areas of Alberta and Manitoba from 2011 to 2017 as pigs escaped or were released from domestic wild boar farms combined with high reproductive rates in the wild (Aschim and Brook 2019). As a result, the largest Canadian distribution of wild pigs is in the north-central region of the country. Wild pigs have not been detected in the far north of Canada (>59° North Latitude), including the Northwest Territories, likely due to extreme winter conditions that last for more than 6 months and absence of agricultural production as a high-quality food source (Aschim and Brook 2019).

Intentional inter- and intrastate movements of wild pigs by humans has been and continues to be an important concern in the United States (Beasley et al. 2018), as it can greatly increase the rate of wild pig expansion in North America. However, there is scant evidence of humans transporting wild pigs in Canada to establish new wild populations. Different states and provinces have diverse regulations regarding the transport of wild pigs and domestic wild boar (see Chapter 11) and transport of animals is poorly documented, especially where it is illegal.

13.3 REGIONAL ENVIRONMENTAL ASPECTS IN NORTH-CENTRAL NORTH AMERICA

13.3.1 Behavior

Relatively little research on wild pig behavior has been documented in the North-Central region. Wild pigs in southern Missouri traveled on average 2,700 m/day and maximum hourly movements ranged from 830 to 4,411 m and 562 to 3,991 m for pigs not subject to and subject to simulated removal activities, respectively (Fischer et al. 2016). Space-use and diurnal movement distances increased following a second simulated removal activity in southern Missouri, which, while speculative, could result in increased range expansion and disease transmission risk following repeated removal efforts (Fischer et al. 2016). Kay et al. (2017) reported an average home range size of 13.3 km^2 (range 1.9–41.6 km^2; autocorrelated kernel density estimation) and above-average movement

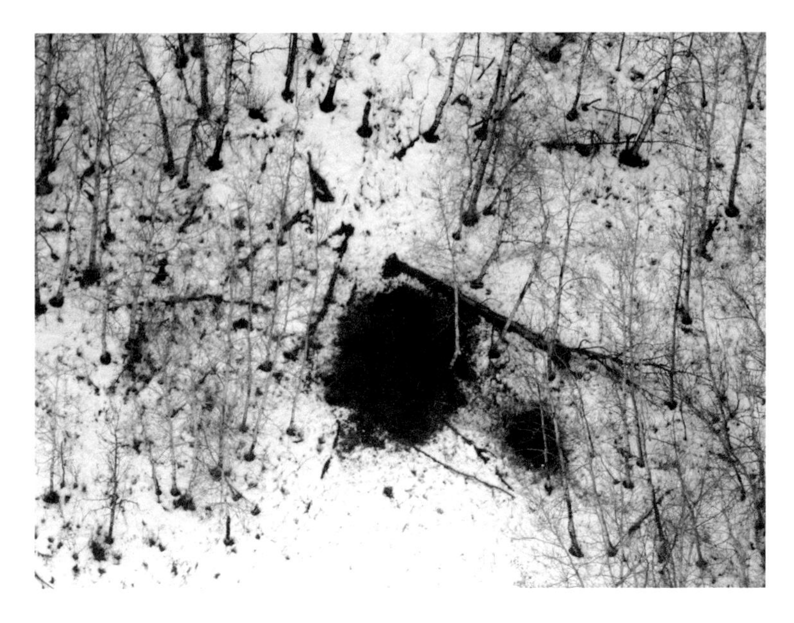

FIGURE 13.2 One large and one smaller adjacent bed occupied by 11 wild pigs that were buried under the vegetation and completely obscured from vision in central Saskatchewan during February 2017. (Photo by R. Brook. With permission.)

rates for wild pigs in the Ozark Highlands ecoregion of Missouri. In Saskatchewan, wild pigs showed no difference in activity levels at night or during the day based on trail camera data (Stolle et al. 2015). Brook and van Beest (2014) reported wild pig sightings occurred during daylight (28%), dusk (17%), dawn (16%), and night (9%). To help survive during harsh winter conditions, wild pigs make beds in wetlands consisting of vegetation such as cattails (*Typha* spp.; Figure 13.2). Biologists often refer to these winter beds as "pigloos."

O'Brien et al. (2019) used trail cameras in Saskatchewan to examine occurrence of wild pigs in winter and summer and examine if their occurrence was positively or negatively associated with other wildlife species. During winter, wild pigs were positively associated with occurrence of moose (*Alces alces*) and coyote (*Canis latrans*) and negatively associated with white-tailed deer (*Odocoileus virginianus*), mule deer (*O. hemionus*), and humans. During the summer, wild pigs were positively associated with moose and mule deer, and negatively associated with domesticated cattle (*Bos taurus*), elk (*Cervus canadensis*), and humans. The average group size of wild pigs was 10.3 and 4.7 in central and southern Saskatchewan, respectively, although sample sizes were relatively small (Koen et al. 2018).

13.3.2 Diseases and Parasites

Disease testing of wild pigs in Canada has been limited, and testing for many important diseases (e.g., swine brucellosis, bovine tuberculosis; see Chapter 5) has not occurred. This is arguably the most significant gap in understanding risks and potential impacts of wild pigs in Canada. McGregor et al. (2015) tested 22 wild pig samples from Saskatchewan for a range of parasites and diseases. Serological tests were negative for Porcine Reproductive and Respiratory Syndrome (PRRS), swine influenza (H1N1), a variant of Swine influenza (H3H2), Porcine Circovirus (PCV-2), and Transmissible Gastroenteritis Virus/Porcine Respiratory Coronavirus (TGE/PRCV). These samples also tested negative for *Toxoplasma gondii, Trichinella,* and *Mycoplasma hyopneumoniae.* However, all were positive for *Actinobacillus pleurophneumoiеae* and 16 were positive for *Lawasonia intracellularis.* Gajadhar et al. (1997) tested samples from 248 domestic wild boar for *Trichinella* from Manitoba and Saskatchewan and all samples were negative.

In Kansas, Gipson et al. (1999) used serological tests on 24 wild pigs from Fort Riley Army Base to determine they had been exposed to parvovirus, enterovirus, and swine influenza, while none tested positive for swine brucellosis, pseudorabies (PRV), or PRRS. Parasites found included lungworms (*Metastrongylus* spp.; 9/24), roundworms (*Ascaris suum*; 4/24), and whipworms (*Trichurus suis*; 2/24). In Iowa, 2 hunters were diagnosed with trichinosis after handling raw meat from a wild pig harvested at a high-fenced shooting operation (Holzbauer et al. 2014). Researchers in Missouri sampled 321 wild pigs from 1993 to 2005 for PRV, swine brucellosis, and classical swine fever (Hartin et al. 2007). The only positive was a pig from Cole County that had been exposed to swine brucellosis. Additionally, they reported a 1.3% prevalence rate (1/80) for tularemia. Wild pigs have also tested positive for PRV, swine brucellosis, *Toxoplasma gondii*, *Trichinella spiralis*, influenza A, and leptospirosis elsewhere in Missouri (USDA/APHIS/Wildlife Services Missouri 2015). Pedersen et al. (2012) tested wild pig samples for swine brucellosis in North Dakota, Iowa, Nebraska, Missouri, and Kansas in 2009–2010, and positive samples were detected in Missouri and Kansas. Five wild pigs were tested in Baca County, Colorado for PRV in 2016 and were all negative (USDA/APHIS/Wildlife Services Colorado 2016a).

In 2006, approximately 180 wild pigs were captured and illegally transported from Texas to Erikson, Nebraska, and then subsequently moved to Genoa, Nebraska before several (<10) escaped into the wild (Wilson et al. 2009). A lactating female from the escaped group and 8 piglets were lethally removed in 2007 and the adult tested positive for pseudorabies antibodies, marking the first time PRV was detected in Nebraska (Wilson et al. 2009). Seven wild pigs from Seward County, Nebraska were submitted for disease testing in 2005 and 1 tested positive for swine influenza (Wilson et al. 2009). Additionally, 31 wild pigs from Harlan County, Nebraska were tested in 2007–2008 and 1 tested positive for PRRS and another was positive for swine influenza (Wilson et al. 2009).

13.3.3 Food Habits

Very little research has been conducted on the food habits of wild pigs in the North-Central region, although it is recognized they are voracious omnivorous and do best in proximity to agriculture. Crops in agricultural regions of Alberta, Manitoba, and Saskatchewan provide an important food source on a year-round basis, including leftover crops in fields following harvest and spilled grain from storage bins during winter (Brook and van Beest 2014, Michel et al. 2017). Additionally, livestock feeding practices (i.e., swath and bale grazing) and harvested grain storage in polyethylene bags (e.g., wheat, canola) provide readily accessible high-quality food sources for wild pigs (Michel et al. 2017). Mustard seed and flaxseed have been noted as less preferred food sources (Brook and van Beest 2014).

13.3.4 Habitat Use

Wild pig distribution rapidly expanded in North America, Europe, and Asia in recent years (Brook and van Beest 2014, Massei et al. 2015, McClure et al. 2015, Vetter et al. 2015). In North America, current northern limits of wild pigs closely align with the northern extent of agricultural production, which also corresponds to the southern edge of boreal and aspen (*Populus* spp.) forests. Currently, populations occur at 58° North latitude in North America, and 60° North latitude in Europe and Asia (Brook and van Beest 2014, Lewis et al. 2017, Aschim and Brook 2019). Northern limits of wild pigs remain unclear, particularly in North America, but effects of global climate change at northern latitudes may facilitate northward expansion. An important environmental feature in the North-Central region is the Prairie Pothole Region in Canada and northern United States. This area is comprised of tens of thousands of small- to medium-sized ponds distributed across the landscape that include open water and marsh wetlands (Euliss and Mushet 1996), providing a readily available source of water and cover.

Wild pigs in Saskatchewan are typically associated with agricultural crops, deciduous forests, and low densities of paved roads (Brook and van Beest 2014). They also utilize pasture and grasslands. Seasonal variation in habitat use in Saskatchewan has been documented as wild pigs primarily use wetlands and agricultural land in the winter and grasslands during the summer (O'Brien et al. 2019). Additionally, O'Brien et al. (2019) reported wild pigs generally avoided deciduous forests in winter, likely due to a lack of food and water, or because snow tends to be deeper and softer in forests compared to open areas, making travel harder.

13.3.5 Population Biology

There are no documented predation cases in the region of wild pigs by mountain lions (*Puma concolor*), wolves (*Canis lupus*), coyotes (*Canis latrans*), or black bears (*Ursus americanus*), the only 4 mammal species likely to be natural predators. Abundant year-round food sources and few predators in agricultural regions of Canada allow wild pigs to survive in the harshest portions of the Canadian Prairie region (Michel et al. 2017). Hunter harvest has been largely undocumented in the North-Central region but appears to have little impact on populations. Richardson et al. (1997) documented that 100 sport hunters harvested 4 wild pigs in 1993 and just 3 the following year. A bounty program in Alberta resulted in 1,135 wild pigs harvested by mostly sport hunters from 2008 to 2017 during a period when wild pigs were rapidly expanding their distribution (Aschim and Brook 2019).

A number of large wild pigs have been confirmed in the North-Central region, including a 250-kg pig shot by a farmer in northwest North Dakota near the Canadian border in 2013. A 200-kg pig was also shot by a hunter in 2016 in east-central Saskatchewan. Researchers in Saskatchewan also captured a 211-kg male and a 293-kg pregnant female in 2016 (R. K. Brook and R. Powers, Department of Animal and Poultry Science, College of Agriculture and Bioresources, University of Saskatchewan, unpublished data). There are currently no reported density, postnatal male-to-female sex ratio, or age structure estimates for wild populations in the North-Central region.

13.3.6 Reproductive Biology

Reproductive data in the North-Central region is very limited. On the Canadian prairies, wild pigs reproduce continuously, with piglets born in all seasons, with the highest peak in late summer and low rates during fall and early winter (Brazeau 2019). The average pregnant sow in Saskatchewan weighed 74 kg and 54% of females ≥46 kg were pregnant in February with an average of 5.6 fetuses per pregnant female (range 4–7; Koen et al. 2018). Similarly, Brazeau (2019) found that average litter size from 52 trail camera detections of sounders with striped piglets was 5.4 (range 1–15). Fetal sex ratio of wild pigs in Saskatchewan tended to be biased towards females, but sample size was relatively small (Koen et al. 2018). Peak conception and parturition occurred November-February and February-May, respectively (Koen et al. 2018). Research on wild boar in Europe suggests litter size increases 1 piglet for every 6.6° increase in latitude (Bywater et al. 2010), but this has not been evaluated in wild pig populations in North America.

13.3.7 Damage

Wild pigs are an important and growing concern over the agricultural regions of Alberta, Saskatchewan, and Manitoba, and are associated with crop damage and harassment of livestock (Brook and van Beest 2014, Michel et al. 2017, Aschim and Brook 2019). The total number of compensation payments for crop damage by wild pigs is small (<100 across all 3 Canadian provinces) over the last decade (R. K. Brook, unpublished data). However, many rural residents are not aware that wild pigs occur locally, and it is likely that wild pig crop damage goes unreported or is reported as being caused by other species such as elk and white-tailed deer. Brook and van Beest (2014) conducted a survey of community leaders in rural municipalities in Saskatchewan that revealed that perceptions of the amount of damage to agricultural resources such as hay bales, standing hay, standing grain, pastures, and fences was relatively low (60–70% of responses indicated damage was never serious).

Agricultural damage is also a concern in the US portion of the region, but reported data is limited. Annual value of production loss to soybeans and wheat in Missouri due to wild pig damage was estimated at US$459,000 and US$27,000, respectively (Anderson et al. 2016). Bioeconomic modeling estimated that a foot-and-mouth disease outbreak contracted and spread by wild pigs in Missouri could result in over 18,600 livestock deaths and indirect and direct economic losses of US$4.4M and US$7.5M, respectively (Cozzens et al. 2010).

13.4 REGIONAL MANAGEMENT IN NORTH-CENTRAL NORTH AMERICA

Similar to wild pig management across North America, integrated approaches that account for region-specific biological/ecological (e.g., role of habitat, resources exploited by wild pigs) and social (e.g., stakeholder perception and motivation) aspects associated with wild pigs must be incorporated to better understand the issue and implement effective management programs in the North-Central region (Brook and van Beest 2014; see Chapter 8). For example, Brook and van Beest (2014) found that many rural municipalities in Saskatchewan were unaware of the risks associated with wild pigs or how to properly mitigate their impacts. Given that wild pigs are a relatively novel species to many portions of the North-Central region, increased education and public awareness of the negative issues associated with wild pigs will be important steps towards successful management of the species (Brook and van Beest 2014). Rigorous methods to prevent escape and prohibiting releases of individuals from domestic wild pig farms in Canada are also extremely important to control further spread of the species (Michel et al. 2017). Although the number of domestic wild pig farms has decreased substantially since the 1990s, wild pig populations are supplemented or new populations are established each year through releases or escapes (Michel et al. 2017).

The majority of free-ranging wild pig occurrences in Canada are in Saskatchewan where provincial government control has occurred intermittently in some areas over the last 15 years. These control activities were led by different agencies. The earliest and most effective effort has been a grassroots program in southeast Saskatchewan; the Moose Mountain Wild Boar Eradication team is comprised mostly of local farmers concerned about risks and impacts of wild pigs. Fourteen domestic wild boar escaped near Moose Mountain Provincial Park in 1998 and became established in the area. Through trial, error, and persistence, the group worked to eradicate the population. They had limited success early on and the population increased to about 59 animals, but their success increased as their methods evolved and support from the provincial government grew. The most recent government-led wild pig control effort, the Feral Wild Boar Control program, is by Saskatchewan Crop Insurance Corporation (SCIC), a provincial crown corporation affiliated with the Ministry of Agriculture. This program is largely based on passive surveillance; when a producer observes wild pigs they report them to SCIC which then deploys personnel to implement lethal control using hunting, snaring, and trapping. The general public has been encouraged to hunt wild pigs, but no evidence exists to suggest that it is an effective control measure and it may actually be serving to disperse pigs further across the landscape (Michel et al. 2017).

Since 2014, Canada has implemented a national swine tracking system for domestic pigs called PigTrace. The system requires all domestic pigs be ear tagged or tattooed and linked to a unique premise identification number. In 2015 this system also became mandatory for wild boar farms, but there has been limited compliance. Escapes and purposeful releases of individual animals and entire herds from meat farms and high-fenced shooting operations are ongoing, and because most animals do not have an ear tag or other identifying mark it is impossible to determine the farm of origin (Michel et al. 2017).

Concentrated control efforts were conducted on Fort Riley Military Installation in northeastern Kansas during 1993–1995 using traps, snares, recreational hunting, and sharpshooting (Richardson et al. 1995). Trapping was the most successful control tool (approximately 80% of pigs removed each year) and deer carcasses, deer entrails, grain, fermented corn mash, rotten fruit, and raspberry jello were used as bait. Hunting efforts by the public were relatively unsuccessful and the authors suggested that hunting should not be allowed in areas where trapping also occurs. A total of 69 wild pigs were removed and approximately 25 wild pigs remained on base in 1995, after which USDA/APHIS/Wildlife Services took over control efforts.

In 2017, the Missouri Department of Conservation, USDA/APHIS/Wildlife Services, partner agencies, and private landowners removed 6,561 wild pigs across the state and an additional 7,339 were removed from January-September, 2018 (Davis 2018). Removal efforts focused on eliminating entire sounders before moving on to new areas and recreational hunting of wild pigs was also prohibited on Missouri Department of Conservation, Corp of Engineers, and LAD Foundation lands starting in 2017 (Davis 2018). Additional management efforts include the Missouri Feral Hog Task Force which formed in response to increasing populations of wild pigs across the state with 3 primary objectives: 1) state-wide eradication of wild pigs, 2) development of a state-wide map of wild pig distribution, and 3) collection of blood samples for disease sampling (Hartin 2006).

USDA/APHIS/Wildlife Services has been an important contributor to wild pig management in the US portion of the North-Central region. Over 70 cooperators in Missouri worked with USDA/APHIS/Wildlife Services to mitigate wild pig damage in 2015 (USDA/APHIS/Wildlife Services Missouri 2015). USDA/APHIS/Wildlife Services and Nebraska Game and Parks conducted a collaborative effort to remove the last known wild pig populations in Nebraska in 2011–2012 (USDA/APHIS/Wildlife Services Nebraska 2015). In Iowa, USDA/APHIS/Wildlife Services responds to reports of wild pig sightings and provides education on wild pigs through outreach events (USDA/APHIS/Wildlife Services Iowa 2015). During the 2017 fiscal year they removed 11 wild pigs across the state (USDA/APHIS/Wildlife Services Iowa 2017). In 2018, Iowa Department of Transportation worked with USDA/APHIS/Wildlife Services to establish a system to notify personnel when domestic pigs escaped due to commercial pig trailer accidents or rollovers and have since responded to 4 incidents, 1 of which involved assisting in the capture of 190 escaped domestic pigs (USDA/APHIS/Wildlife Services Iowa 2018). Similarly, in North Dakota, USDA/APHIS/Wildlife Services works with partner agencies to investigate any wild pig reports and conducts removal operations when reports are confirmed (USDA/APHIS/Wildlife Services North Dakota 2015). In Colorado, USDA/APHIS/Wildlife Services trapped and equipped a young male wild pig with a global positioning system (GPS) collar in Baca County, Colorado to assess its movements and act as a Judas animal (USDA/APHIS/Wildlife Services Colorado 2016a). The GPS data was pivotal for gaining access from a key landowner in Baca County, which will be extremely beneficial for future removal of wild pigs from the area (USDA/APHIS/Wildlife Services Colorado 2016b).

13.5 REGIONAL WILD PIG RESEARCH IN NORTH-CENTRAL NORTH AMERICA

Relatively little scientific research has been conducted on wild pigs in the North-Central region, mainly because of recent establishment and mostly limited densities. Work in the mid-1990s focused on health and diseases of wild pigs in Kansas, and explored range expansion (Gipson et al. 1994, 1997, 1998). Research on disease and range expansion was also conducted in Missouri during the early 2000s (Hartin 2006, Hartin et al. 2007). Fischer et al. (2016) fitted 31 wild pigs with GPS harnesses in southern Missouri to determine movement and space use of wild pigs in response to simulated removal activities. Distribution of wild pigs and risk perceptions on Canadian prairies were recently studied by Brook and van Beest (2014), while Stolle et al. (2015) used trail cameras to examine diurnal and nocturnal activity patterns. Michel et al. (2017) examined spatial and temporal patterns in data on domestic production of wild boar from a Canada-wide agriculture census and documented a rapid increase in the number of farms and number of animals from 1991, peaking from 1996 to 2001, then rapidly declining in 2011. Currently, domestic wild boar farms are so rare that they are not reported in the Canada-wide census of agriculture (Statistics Canada 2016). Koen et al. (2018) determined reproductive rates for wild pigs in Saskatchewan. Recent work deploying GPS collars on wild pigs in Saskatchewan and Manitoba is ongoing (R. K. Brook, unpublished data; Figure 13.3), and national-scale mapping of wild pig distribution that tracks the spatial expansion of wild pigs from 1990 to 2017 was recently completed (Aschim and Brook 2019).

FIGURE 13.3 On the Canadian Prairies, net-gunning from a helicopter has been used effectively to capture wild pigs so that GPS collars can be deployed to address research needs. (Photo by R. Brook. With permission.)

13.6 FUTURE OF WILD PIGS IN NORTH-CENTRAL NORTH AMERICA

Given limited population control efforts on the Canadian prairies, further range expansion and increased population density of wild pigs is expected. Eradication of these populations may be unattainable as there are endemic populations in remote areas with few roads and extensive areas of forest and cropland. Expanding populations of wild pigs, composed of both free-ranging animals and those recently escaped or released from domestic wild boar farms and high-fenced shooting operations, especially in Saskatchewan, have significant potential to disperse to adjacent provinces and states. In Kansas and Missouri, where there have been or are established populations of wild pigs, there have been significant efforts to detect and eradicate them. All other US states in the North-Central region are focused on remaining free of established populations of wild pigs and have been successful so far. The ongoing northward expansion of wild pigs in the continental United States and overall expansion of populations on the Canadian prairies suggests that most of this region is at significant long-term risk of developing endemic wild pig populations over the next 20 years in the absence of effective management efforts. It remains unclear how far wild pigs may expand northward, as most releases and escapes to date have been in agricultural areas of southern Canada. White-tailed deer, which are much less winter hardy than wild pigs, have expanded their range across large areas of the North-Central region in recent years and now persist in part of the Northwest Territories, most of Alberta, and an increasing area of Saskatchewan and Manitoba (Rothley 2002, Latham et al. 2011, Dawe and Boutin 2016), suggesting wild pigs could also expand into and even beyond these areas.

ACKNOWLEDGMENTS

The authors thank R.S. Miller for generating the regional area map and G.J. Roloff and K.C. VerCauteren for editorial input. Contributions from M.P.G. were supported by the USDA. Mention of commercial products or companies does not represent an endorsement by the US Government.

REFERENCES

Anderson, A., C. Slootmaker, E. Harper, J. Holderieath, and S. A. Shwiff. 2016. Economic estimates of feral swine damage and control in 11 US states. *Crop Protection* 89:89–94.

Aschim, R., and R.K. Brook. 2019. Evaluating cost-effective methods for rapid and repeatable national scale detection and mapping of invasive species spread. *Nature Scientific Reports*. 9:7254.

Beasley, J. C., S. S. Ditchkoff, J. J. Mayer, M. D. Smith, and K. C. VerCauteren. 2018. Research priorities for managing invasive wild pigs in North America. *Journal of Wildlife Management* 82:674–681.

Brazeau, B. 2019. Reproductive ecology of invasive wild pigs (*Sus scrofa*) on the Canadian Prairies and implications for Porcine Epidemic Diarrhea Virus. Undergraduate Thesis, Department of Animal and Poultry Science, University of Saskatchewan, Saskatoon, Canada.

Brook, R. K., and F. M. van Beest. 2014. Feral wild boar distribution and rural municipal leadership perceptions of risk on the central Canadian Prairies. *Wildlife Society Bulletin* 38:486–494.

Bywater, K. A., M. Apollonio, N. Cappai, and P. A. Stephens. 2010. Litter size and latitude in a large mammal: The wild boar *Sus scrofa*. *Mammal Review* 40:212–220.

Cook, B. I., T. R. Ault, and J. E. Smerdon. 2015. Unprecedented 21st century drought risk in the American Southwest and Central Plains. *Science Advances* 1:1–7.

Corn, J. L., and T. R. Jordan. 2017. Development of the national feral swine map, 1982–2016. *Wildlife Society Bulletin* 41:758–763.

Cozzens, T., K. Gebhardt, S. A. Shwiff, M. W. Lutman, K. Pederson, and S. Swafford. 2010. Modeling the economic impact of feral swine-transmitted foot-and-mouth disease: A case study from Missouri. *Proceedings of the Vertebrate Pest Conference* 24:308–311.

Davis, C. 2018. MDC and partners eliminate more than 7,300 feral hogs from Missouri's landscape in 2018 so far. <https://mdc.mo.gov/newsroom/mdc-and-partners-eliminate-more-7300-feral-hogs-missouri%E2%80%99s-landscape-2018-so-far>. Accessed 14 Mar 2019.

Dawe, K. L., and S. Boutin. 2016. Climate change is the primary driver of white-tailed deer (*Odocoileus virginianus*) range expansion at the northern extent of its range; land use is secondary. *Ecology and Evolution* 6:6435–6451.

Euliss, N. H., and D. M. Mushet. 1996. Water-level fluctuation in wetlands as a function of landscape condition in the prairie pothole region. *Wetlands* 16:587–593.

Fischer, J. F., D. McMurty, C. R. Blass, W. D. Walter, J. Beringer, and K. C. VerCauteren. 2016. Effects of simulated removal activities on movements and space use of feral swine. *European Journal of Wildlife Research* 62:285–292.

French, B. 2019. Feral swine amass at Montana's northern border. <https://billingsgazette.com/outdoors/feral-swine-amass-at-montana-s-northern-border/article_f8122af8-16d2-55b2-bc41-710212a4343d.html>. Accessed 7 June 2019.

Gajadhar, A. A., J. R. Bisaillon, and G. D. Appleyard. 1997. Status of *Trichinella* in domestic swine and wild boar in Canada. *Canadian Journal of Veterinary Research* 61:256–259.

Gipson P. S., B. Hlavachick, and T. Berger. 1998. Range expansion by wild hogs across the central United States. *Wildlife Society Bulletin* 26:279–286.

Gipson, P. S., W. Hlavachick, T. Berger, and C. D. Lee. 1997. Explanations for recent range expansions by wild hogs into midwestern states. *Proceedings of the Great Plains Wildlife Damage Control Workshop* 13:148–150.

Gipson, P. S., C. D. Lee, S. Wilson, J. R. Thiele, and D. Hobbick. 2006. Status of feral pigs in Kansas and Nebraska. *Prairie invaders: Proceedings of the North American Prairie Conference* 20:19–24.

Gipson, P. S., R. Matlack, D. P. Jones, H. J. Abel, and A. E. Hynek. 1994. Feral pigs, *Sus scrofa*, in Kansas. *Proceedings of the North American Prairie Conference* 14:93–95.

Gipson, P. S., J. K. Veatch, R. S. Matlack, and D. P. Jones. 1999. Health status of a recently discovered population of feral swine in Kansas. *Journal of Wildlife Diseases* 35:624–627.

Hartin, R. E. 2006. Feral hogs: Status and distribution in Missouri. Thesis, University of Missouri, Columbia, MO.

Hartin, R. E., M. R. Ryan, and T. A. Campbell. 2007. Distribution and disease prevalence of feral hogs in Missouri. *Human-Wildlife Conflicts* 1:186–191.

Holzbauer, S. M., W. A. Agger, R. L. Hall, G. M. Johnson, D. Schmitt, A. Garvey, H. S. Bishop, et al. 2014. Outbreak of *Trichinella spiralis* infections associated with a wild boar hunted at a game farm in Iowa. *Clinical Infectious Diseases* 59:1750–1756.

Iowa State University Forestry Extension. 2017. Feral hog sightings in Iowa. <https://www.extension.iastate.edu/forestry/research/feral_hogs/hogsightings.html>. Accessed 19 Mar 2019.

Kay, S. L., J. W. Fischer, A. J. Monaghan, J. C. Beasley, R. Boughton, T. A. Campbell, S. M. Cooper, et al. 2017. Quantifying drivers of wild pig movement across multiple spatial and temporal scales. *Movement Ecology* 5:14.

Koen, E. L., E. Vander Wal, R. Kost, and R. K. Brook. 2018. Reproductive ecology of recently established wild pigs in Canada. *American Midland Naturalist* 179:275–286.

Latham, A. D. M., M. C. Latham, N. A. Mccutchen, and S. Boutin. 2011. Invading white-tailed deer change wolf–caribou dynamics in northeastern Alberta. *Journal of Wildlife Management* 75:204–212.

Lewis, J. S., M. L. Farnsworth, C. L. Burdett, D. M. Theobald, M. Gray, and R. S. Miller. 2017. Biotic and abiotic factors predicting the global distribution and population density of an invasive large mammal. *Scientific Reports* 7:44152.

Massei, G., J. Kindberg, A. Licoppe, D. Gačić, N. Šprem, J. Kamler, E. Baubet, et al. 2015. Wild boar populations up, numbers of hunters down? A review of trends and implications for Europe. *Pest Management Science* 71:492–500.

Mayer, J. J., and I. L. Brisbin, Jr. 2008. *Wild Pigs in the United States: Their History, Comparative Morphology, and Current Status*. Second Edition. The University of Georgia Press, Athens, GA.

McClure, M. L., C. L. Burdett, M. L. Farnsworth, M. W. Lutman, D. M. Theobald, P. D. Riggs, D. A. Grear, and R. S. Miller. 2015. Modeling and mapping the probability of occurrence of invasive wild pigs across the contiguous United States. *PloS One* 10:e0133771.

McGregor, G. F., M. Gottschalk, D. L. Godson, W. Wilkins, and T. K. Bollinger. 2015. Disease risks associated with free-ranging wild boar in Saskatchewan. *Canadian Veterinary Journal* 56:839–844.

Michel, N. L., M. P. Laforge, F. M. van Beest, and R. K. Brook. 2017. Spatiotemporal trends in Canadian domestic wild boar production predict feral wild pig distribution. *Landscape and Urban Planning* 165:30–38.

O'Brien, P. P., E. Vander Wal, E. L. Koen, C. Brown, J. Guy, F. M. van Beest, and R. K. Brook. 2019. Understanding habitat co-occurrence and the potential for competition between native mammals and invasive wild pigs at the northern edge of their range. *Canadian Journal of Zoology* 97:537-546.

Pedersen, K., S. N. Bevins, B. S. Schmit, M. W. Lutman, and M. P. Milleson. 2012. Apparent prevalence of swine brucellosis in feral swine in the United States. *Human-Wildlife Interactions* 6:38–47.

Richardson, C. D., P. S. Gipson, D. P. Jones, and J. C. Luchsinger. 1995. A long term management plan for feral pigs on Fort Riley Army Base, Kansas. *Proceedings of the Eastern Wildlife Damage Management Conference* 11:100–104.

Richardson, C. D., P. S. Gipson, D. P. Jones, and J. C. Luchsinger. 1997. Extirpation of a recently established feral pig population in Kansas. *Proceedings of the Eastern Wildlife Damage Management Conference* 7:100–103.

Rothley, K. D. 2002. Use of multiobjective optimization models to examine behavioural trade-offs of white-tailed deer habitat use in forest harvesting experiments. *Canadian Journal of Forest Research* 32:1275–1284.

Snow, N. P., M. A. Jarzyna, and K. C. VerCauteren. 2017. Interpreting and predicting the spread of invasive wild pigs. *Journal of Applied Ecology* 54:2022–2032.

Statistics Canada. 2016. Canada census of agriculture. <www.statcan.gc.ca>.

Stolle, K., F. M. van Beest, E. Vander Wal, and R. K. Brook. 2015. Diurnal and nocturnal activity patterns of invasive wild boar (*Sus scrofa*) in Saskatchewan, Canada. *Canadian Field-Naturalist* 129:76–79.

Tokaruk, B. 2002. The Status of Wild Boars at Moose Mountain Provincial Park. Internal Report, Saskatchewan Environment, Fish and Wildlife Branch. Regina, Saskatchewan, Canada.

US Department of Agriculture/Animal and Plant Health Inspection Service/Veterinary Services/National Import Export Services. 2018. Protocol for the importation of farm raised "wild" boar from Canada. Riverdale, MD.

US Department of Agriculture/Animal and Plant Health Inspection Service/Wildlife Services Colorado. 2016a. Colorado fiscal year 2016 quarter 2 feral swine quarterly report. Washington, DC.

US Department of Agriculture/Animal and Plant Health Inspection Service/Wildlife Services Colorado. 2016b. Colorado fiscal year 2016 quarter 4 feral swine quarterly report. Washington, DC.

US Department of Agriculture/Animal and Plant Health Inspection Service/Wildlife Services Colorado. 2017. Colorado fiscal year 2017 quarter 3 feral swine quarterly report. Washington, DC.

US Department of Agriculture/Animal and Plant Health Inspection Service/Wildlife Services Iowa. 2015. Iowa State Report. Washington, DC.

US Department of Agriculture/Animal and Plant Health Inspection Service/Wildlife Services Iowa. 2017. Iowa fiscal year 2017 quarter 3 feral swine quarterly report. Washington, DC.

US Department of Agriculture/Animal and Plant Health Inspection Service/Wildlife Services Iowa. 2018. Iowa fiscal year 2018 quarter 3 feral swine quarterly report. Washington, DC.

US Department of Agriculture/Animal and Plant Health Inspection Service/Wildlife Services Missouri. 2015. Missouri State Report. Washington, DC.

US Department of Agriculture/Animal and Plant Health Inspection Service/Wildlife Services Nebraska. 2015. Nebraska State Report. Washington, DC.

US Department of Agriculture/Animal and Plant Health Inspection Service/Wildlife Services North Dakota. 2015. North Dakota State Report. Washington, DC.

Vetter, S. G., T. Ruf, C. Bieber, and W. Arnold. 2015. What is a mild winter? Regional differences in within-species responses to climate change. *PloS One* 10:e0132178.

Wilson, S., A. R. Doster, J. D. Hoffman, and S. E. Hygnstrom. 2009. First record of pseudorabies in feral swine in Nebraska. *Journal of Wildlife Diseases* 45:874–876.

14 Wild Pigs in Northeastern North America

Dwayne R. Etter, Melissa Nichols, and Karmen M. Hollis-Etter

CONTENTS

14.1 INTRODUCTION

Historical records of introduced wild boar (*Sus scrofa*) in northeastern North America date to the 1890s in New Hampshire (Mayer and Brisbin 2008). More recently, escaped or intentionally released wild boar and hybrids, and in some instances domestic swine, have appeared in the region. Established breeding populations of wild pigs in the Northeast region are a relatively new phenomenon with isolated pockets becoming more common over the past 25–30 years (Corn and Jordan 2017). Most states and provinces in the Northeast region report few or no populations of these animals, and thus damage is limited to isolated areas of concentrated wild pig activity. Where they do occur, impacts to crops and natural resources and disease transmission are of primary concern; however, little is known about the extent of these impacts or the biology/ecology of wild pigs in the region.

The Northeast region varies greatly in landscape characteristics, human population density, climate, and native and introduced flora and fauna. States in this region include Connecticut, Delaware, Illinois, Indiana, Maine, Maryland, Massachusetts, Michigan, northeast Minnesota, New Hampshire, New Jersey, New York, Ohio, Pennsylvania, Rhode Island, Vermont, and Wisconsin. Canadian provinces include Labrador, New Brunswick, Newfoundland, Nova Scotia, Ontario, Prince Edward Island, and Quebec. Collectively, this region encompasses an extensive land area of over 4.5 million km^2; however, wild pig distribution is likely restricted to south of 60° latitude because of climatic and habitat limitations. The eastern seaboard from Delaware to Massachusetts, including Philadelphia and New York City, is the most densely populated area in North America with an average human density of 248 people/km^2 in 2003 (Crossett et al. 2004). Additionally, several large cities are scattered throughout the region including Chicago, Cincinnati, Cleveland, Detroit, Indianapolis, Pittsburgh, and Toronto. Conversely, in 2010 some counties in northern Maine, Michigan, and Minnesota had population densities from 1.4 to 4.3 people/km^2 (US Census Bureau 2016).

The regional climate is primarily classified as humid continental with some subarctic zones in northern areas. Humid continental climate is characterized by mild to warm summers and cold winters. Precipitation is generally spread evenly throughout the seasons, but inland areas receive greater precipitation in summer, and snowfall is usually persistent throughout winter. The subarctic region is characterized by short cool summers and long cold winters. Precipitation occurs mostly during summer and measurable snow is common in late fall through early summer (National Oceanic and Atmospheric Administration and National Climate Data Center 2017).

The climate of the Northeast region supports diverse forest types composed primarily of deciduous hardwoods and several conifers. Nearly the entire region has been historically timbered or cleared for agriculture, having a major influence on dominant tree species and distribution. Illinois, Indiana, and Ohio make up one of the most intensively farmed regions in the continental United States. Nearly 12 million ha (81%) of total land area in Illinois consists of agriculture (USDA/Natural Resources Conservation Service 2017). Conversely, nearly 84% (3.56 million ha) of the Upper Peninsula (U.P.) of Michigan is forested (Albert 1995; Figure 14.1). This region encompasses the entirety of the Great Lakes, containing 84% of the fresh water of the United States (US Environmental Protection Agency 2016). Mount Washington, New Hampshire, is the highest elevation point (1,917 m) in northeastern North America and the Appalachian Mountains run from southeastern Canada through Georgia. The Atlantic Ocean and its associated tidal backwaters border the entire northeast coast.

Comparatively little research has been published on wild pigs found in the Northeast region. However, numerous states have managed wild pigs and we summarized their knowledge and experiences as part of this chapter (see Acknowledgments). Based on survey results and personal communications with both federal and state agency professionals, we compiled a wide breadth of information on the biology and ecology of wild pigs in the Northeast region.

14.2 HISTORIC AND PRESENT DISTRIBUTION

14.2.1 Origin and Early History

Historically, one of the first introductions of wild boar into North America occurred in Sullivan County, New Hampshire in 1890 (Mayer and Brisbin 2008). Although originally contained within Corbin's Park, a privately owned game preserve behind a 3-m high wire-mesh fence buried 60 cm underground, numerous wild boar escaped through breaches in the fence over the years. A population of wild boar still exists within the park, now incorporated as the Blue Mountain Forest Association. In addition, escaped wild boar from the park continue to persist in New Hampshire and Vermont in low numbers (Mayer and Brisbin 2008, 2009). Two additional historic releases of wild boar occurred in the Adirondack Mountains: Litchfield Park, Hamilton County, New York in 1902 and around the town of Sabael, New York near Indian Lake in 1972 (Mayer and Brisbin 2008, 2009). Neither of these releases resulted in established populations of wild pigs (Appendix 14.1).

Remnant populations of wild pigs have been discovered in several northeastern states over the past few decades (Corn and Jordan 2017). A small population of escaped domestic swine became established in Gloucester County, southern New Jersey in the 1970s. Reports of wild pigs in southeast Ohio began in the 1980s and wild pigs were intentionally released into southern Indiana in the early 1990s. A few additional states have experienced escapes from fenced game ranches and intentional releases of wild boar and hybrids. Importation records and producer statements indicate that

FIGURE 14.1 (a) Intensive agricultural landscape in Illinois and (b) extensive mixed forest and scrub-shrub wetland habitat common to Michigan's Upper Peninsula (Photos (a) by B. Wilson and (b) by D. Etter. With permission.)

several fenced breeding facilities and game ranches in Michigan purchased wild boar from Canada (Saskatchewan Agriculture and Food 2001) and these are likely stock for wild pigs in Michigan. Presently, most northeastern states and provinces have little or no wild pig activity including Labrador, Maine, Maryland, Minnesota, New Brunswick, New York, Newfoundland, Nova Scotia, Prince Edward Island and Vermont (Appendix 14.1). However, both Maryland and New York report recent, sporadic sightings of wild pigs along bordering Pennsylvania. Several states also indicate that escaped or intentionally released pot-bellied pigs have become more common (e.g., Caudell et al. 2013).

14.2.2 Present Status

Presently, wild pigs are not widespread in the Northeast region and where they do exist distribution is limited to localized populations (Corn and Jordan 2017; Appendix 14.1). Although Michigan reported sightings of wild pigs in 76 of 83 counties in 2014 (Michigan Department of Natural Resources et al. 2017), more recent data indicates that free-ranging wild pigs are limited to less than 15 counties statewide. Ohio has confirmed breeding of wild pigs in 9 counties, with occurrence reports in 8 additional counties. No northeastern states or provinces have scientifically rigorous estimates of wild pig abundance or localized densities, but most indicate they are rare statewide and occur in localized clusters. In New Hampshire, 49 wild pigs were removed from 3 counties during 2009–2012 (Musante et al. 2014). On 25 March 2019, USDA/APHIS/Wildlife Services harvested 1 wild pig in Lyndonville, Vermont, but officials did not have any further evidence to suggest additional wild pigs were in the area or the rest of the state (Vermont Agency of Agriculture, Foods and Markets 2019). Indiana provided an educated deduction of 500–1,000 wild pigs statewide, and through an automated collection system, Michigan recorded over 400 have been killed statewide in the past decade. Since 2009, USDA/APHIS/Wildlife Services removed 459 wild pigs from 2 core areas in Illinois (Scott Beckerman, USDA/APHIS/Wildlife Services, personal communication; see Wild Pig Control in Illinois Case Study below).

14.3 REGIONAL ENVIRONMENTAL ASPECTS IN NORTHEASTERN NORTH AMERICA

14.3.1 Behavior

Several states in the Northeast region have radio-collared wild pigs for research or as "Judas pigs" (see Chapter 8) to provide information on wild pig movements and behavior (Figure 14.2). Home ranges of wild boar in Michigan and New Hampshire varied from 10 to 40 km^2 and wild pigs in Ohio had home ranges of 40 km^2. Except for a few studies in Texas, these home ranges are greater than most previously reported in the United States (Schlichting et al. 2016; see Chapter 3). Relative to wild pig populations in southeastern North America (see Chapter 16), populations in the Northeast region likely exist at lower densities, which may contribute to the observed differences in home range sizes. Massei et al. (1997) found that wild boar from central Italy decreased home range size during a period of high population density and short supply of food. Additional factors that could contribute to differences in reported home range sizes include the type of wild

FIGURE 14.2 Wild pig from New Hampshire fit with a radio tracking collar. (Photo by A. Musante. With permission.)

pig (i.e., wild boar or hybrid), sex, climate, season, habitat, food availability, hunting pressure, or a combination of these and other factors (Massei et al. 1997, Lemel et al. 2003, Podgórski et al. 2013, Thurfjell et al. 2014).

In Michigan, shifts in activity centers of >13 km were common for both male and female wild boar. These shifts occurred during all seasons, even during extreme winter conditions with snow depths exceeding 25 cm and temperatures below –12°C. Wild boar from New Hampshire also made frequent movements of 5–7 km, particularly in fall and winter when beech (*Fagus grandifolia*) nuts were abundant. In Europe, wild boar decreased activity in cold temperatures to conserve energy (Keuling et al. 2008, Thurfjell et al. 2014). Thurfjell et al. (2014) also concluded that wild boar activity decreased with increasing snow depth. Conversely, Lemel et al. (2003) found that an introduced population of wild boar in Sweden increased movements when daily minimum temperature was ≤–5°C and the ground was covered by snow.

Some barriers to movements by wild pigs in the Northeast region may exist, but based on studies from Indiana and Michigan, such barriers are likely limited. Wild pigs (2 adults and 9 sub-adults) were confirmed to cross (by GPS telemetry, photographs, and tracks in the snow) the East Fork White River in southern Indiana on multiple occasions. The river is approximately 300 m wide and on some days the average daily discharge exceeded 283 m^3s^{-1} (US Geological Survey National Water Data 2017). Long distance movements of wild boar were documented in Michigan, including a >60-km movement in the fall by a pair of sub-adult males before they were harvested by hunters. GPS data collected at half-hour intervals suggested that these males may have attempted multiple crossings of Interstate Highway 75, but this highway is bordered by a 1.2-m tall woven wire fence that likely prevented passage.

Based on GPS telemetry, photographs, and sign at baited sites, nocturnal activity appears common in the Northeast region during all seasons. Daily activity patterns are variable for wild pigs in North America; however, human disturbance can influence activity (Mayer and Brisbin 2009). Shooting of wild pigs is legal year-round in some northeastern states (Indiana, Michigan, New Hampshire, Ohio, Pennsylvania, and Wisconsin) and summer months can be warm and humid, potentially influencing daily activity patterns.

Although not well understood, sounder group size appears variable throughout the Northeast region. States with wild pigs that are predominantly of domestic and hybrid origin report greater maximum sounder group size (25–40 pigs) than states with primarily wild boar (12–15 pigs). However, mean sounder group size ranges from 3 to 8 individuals throughout the Northeast region. Adult males from the Northeast region tend to be loners, a characteristic common in wild pigs (Gabor et al. 1999).

In the Lower Peninsula (L.P.) of Michigan, wild boar showed high fidelity for resting (loafing) sites during all seasons. Communal resting sites were characterized by soil hummocks in dense scrub-shrub wetland habitats associated with beaver (*Castor canadensis*) ponds (Figure 14.3). Hanson and Karstad (1959) described similar loafing sites located on high ground above tidal flooding and in shaded cover for wild pigs in southeastern North America, and resting near water was common for nursing wild boar from Spain (Fernández-Llario 2004). Additionally, proximity to abundant food resources and site-specific temperatures (i.e., exposure to sun in winter) are also important characteristics for resting sites of wild pigs (Mayer et al. 2002, Fernández-Llario 2004). In February 2014, while conducting aerial searches for wild pigs in Michigan, researchers detected 3 wild boar in a nest constructed in standing corn (Figure 14.4). This was during an extended period of extreme winter conditions when snow depth exceeded 45 cm and average daily air temperatures were –10°C. A ground inspection of the site determined that the nest was constructed of corn stalks, and that in addition to abundant food, the standing corn provided a wind block with full exposure to sunlight. Standing corn fields were also used as daytime resting cover in late summer and fall in Indiana and Michigan. Additionally, while searching for wild pigs in the U.P. of Michigan in January 2017, USDA/APHIS/Wildlife Services staff found a farrowing nest containing juvenile pigs (Figure 14.5). The nest was located in a young aspen (*Populus* spp.) stand and it was roughly 1.2 m

FIGURE 14.3 Early spring wild pig resting site in dense scrub-shrub wetland in Michigan's central Lower Penninsula (Photo by D. Etter. With permission.)

FIGURE 14.4 Winter loafing nest in standing corn in Michigan's central Lower Peninsula (Photo by K. Jacobs. With permission.)

across and 40–45 cm deep. Construction materials included blackberry and raspberry (*Rubus* spp.) canes along with small aspen saplings, and the bottom was lined with coarse grasses and balsam fir (*Abies balsamea*) boughs. Silver (1957) described a wild pig bed in New Hampshire that was constructed of hemlock (*Tsuga canadensis*) limbs.

An additional behavior documented by researchers from Michigan included winter wallowing (Figure 14.6). This behavior occurred in January 2015 within 15 m of a site baited for wild pigs at

FIGURE 14.5 Wild pig farrowing nest in an early-successional aspen stand in January 2017 located in Michigan's central Upper Peninsula (Photo by B. Roell. With permission.)

FIGURE 14.6 Winter wild pig wallow discovered in Michigan's central Lower Peninsula (Photo by D. Etter. With permission.)

the edge of a beaver pond. Wallowing behavior is not fully understood, but it is believed to provide thermoregulation, relief from parasites, and disinfection of wounds (Fernández-Llario 2005, Mayer and Brisbin 2009; see Chapter 3). Stegeman (1938) also documented wild pigs breaking ice to wallow in Tennessee. Fernández-Llario (2005) determined that fall and winter wallowing was primarily limited to adult male wild boar in western Spain and he surmised it played a role in breeding behavior.

14.3.2 Diseases and Parasites

Primary disease surveillance in the Northeast region is conducted by USDA/APHIS/Wildlife Services (Figure 14.7). Wild pigs tested seropositive for classical swine fever virus or hog cholera, hepatitis E virus, *Leptospira* spp. (leptospirosis), methicillin susceptible *Staphylococcus aureus*, porcine reproductive and respiratory syndrome virus, pseudorabies virus (PRV; porcine herpesvirus), *Brucellosis suis* (brucellosis; SB), Seneca Valley virus, and swine influenza virus in the region. Parasite identification included lungworm (*Metastrongylus* sp.) in Ohio, *Toxplasma gondii* in 6 states, and *Trichinella sp.* from 3 states. Ontario, Canada documented porcine epidemic diarrhea virus (PEDV) in 2014 and Senecavirus A in 2015, potentially having a direct impact on domestic livestock. Researchers in Michigan and Wisconsin also independently tested for SB and PRV. In Ohio, free-roaming wild pigs knowingly transported from southern states tested positive for SB and PRV. In Michigan, PRV was documented in pigs from a game ranch in 2007, 2008, and 2011. A female, juvenile wild pig in Sullivan County, New Hampshire and a wild pig that was recently removed in Vermont both tested positive for PRV (Musante et al. 2014, Vermont Agency of Agriculture, Foods and Markets 2019). In 2012, 2 wild pigs in Sullivan County, New Hampshire were reportedly infested with winter ticks (*Dermacentor albipictus*), which was the first documented case of winter ticks on wild pigs in the Northeast region (Musante et al. 2014).

In the winter of 2016, several wild boar (including 2 that were radio-collared) in the L.P. of Michigan fed at a livestock carcass dump each night over a 3-week period (Figure 14.8). Additionally, an uncollared boar was photographed forming wallows in a manure pile adjacent to the dump. This

FIGURE 14.7 Anthony Musante draws a blood sample for disease testing from a harvested wild pig in New Hampshire. (Photo by A. Musante. With permission.)

FIGURE 14.8 Livestock carcass dump frequented by GPS-collared wild pigs in Michigan's central Lower Peninsula (Photo by A. Hauger. With permission.)

type of activity creates potential exposure for transmission of pathogens and parasites between livestock and wild pigs and highlights the need to continue to educate livestock producers about potential threats.

14.3.3 Food Habits

Wild pigs are opportunistic omnivores and adjust their diet seasonally based upon available food resources (Mayer and Brisbin 2009). Natural food resources in the region include forbs, grasses, tree seedlings, soft and hard mast, roots, tubers, fungi, invertebrates, and vertebrates. White-tailed deer (*Odocoileus virginianus*) carrion from winter kill can be substantial in some areas of the region and it is possible that wild pigs utilize this resource, particularly in early spring when other foods are limited (Ballari and Barrios-García 2014). In Michigan, carrion is an effective bait used for trapping wild boar in winter (Rex Schanck, USDA/APHIS/Wildlife Services, personal communication). Additionally, wild pigs in Michigan were observed consuming winter killed fish from ice covered beaver ponds (Rex Schanck, personal communication). Several states in the Northeast region allow baiting for deer and black bear (*Ursus americanus*) hunting, and hunters sometimes report observing wild pigs at bait sites. Wild pigs also forage wildlife food plots consisting of annual forbs and grasses.

When available, hard mast is a primary natural food resource used by wild pigs in fall and winter throughout the Northeast region. Some common hard mast producing species in the Northeast region include oaks (*Quercus* spp.), American beech and American hazelnut (*Corylus americana*). Hard mast, particularly oak and beech, is a common food resource for wild pigs throughout North America (Barrett 1978, Sweeney and Sweeney 1982) and for wild boar in their native range (Bieber and Ruf 2005, Ballari and Barrios-García 2014). Bieber and Ruf (2005) modeled reproduction of wild boar in Europe and determined that abundant beech and oak hard mast facilitated rapid population growth, and access to agricultural crops contributed to further population expansion. This natural affinity of wild pigs for hard mast is concerning given the present challenge of regenerating certain oak species throughout the Northeast region (Dey 2014).

14.3.4 Habitat Use

Wild pigs from the Northeast region are typically associated with mixed forest-agriculture landscapes that are abundant throughout the region. Both hard mast and agricultural crops are prevalent in these habitats and wild pigs use them extensively for feeding and loafing. Wild pigs from the region also utilize wetland habitats extensively year-round, particularly if they are in proximity to abundant food sources. Some wild pigs have also been documented to spend extended periods of time (days to weeks) in standing corn fields. In Indiana, wild pigs used standing corn for cover and feeding, but they also frequented adjacent set-aside grasslands managed for wildlife. Similar behavior has been observed in Michigan, but some also used scrub-shrub wetland habitats adjacent to standing corn for daytime resting sites while others bedded in weedy patches within corn fields. Some additional habitats occupied by wild pigs in the region include mature wooded ridges in hilly country of southern Ohio and dense spruce (*Picea* spp.) and fir forests in the U.P. of Michigan.

14.3.5 Population Biology

Given an overall scarcity of research on wild pigs in the Northeast region, several regional characteristics pertaining to the population biology of the species (e.g., density, age/sex structure) are not available. Human induced mortality, primarily from recreational hunting and lethal removal, is the primary documented source of wild pig mortality in the Northeast region. However, at least some mortality has occurred due to vehicle collisions (Conaboy 2009, USDA/APHIS/Wildlife Services 2010). The Northeast region also has several potential predators of wild pigs. Bobcat (*Lynx rufus*) have expanded their range throughout the central Midwest agricultural region including Indiana and Illinois (Roberts and Crimmins 2010), and they are abundant in most northern states and Canadian provinces. However, bobcat prey primarily on piglets (Hanson and Karstad 1959) and likely would not limit adult wild pig populations in the region (Mayer and Brisbin 2009). Additional potential predators of young wild pigs in the Northeast region include red fox (*Vulpes vulpes*), gray fox (*Urocyon cinereoargenteus*), feral dog (*Canis lupus familiaris*), Canada lynx (*Lynx canadensis*), great horned owl (*Bubo virginianus*), red-tailed hawk (*Buteo jamaicensis;* Mayer and Brisbin 2009) and fisher (*Martes pennanti*) in some northern states and Canadian provinces.

Regional predators capable of killing adult wild pigs include black bear, cougar (*Puma concolor*), eastern coyote (*Canis latrans*), gray wolf (*Canis lupus*) and eastern wolf (*Canis lycaon*). Cougar, although potentially effective predators of wild pigs (Mayer and Brisbin 2009), only appear sporadically and there is no known breeding population in the region (US Fish and Wildlife Service 2015). However, the northern portion of the region is primary range for black bears (Scheick and McCown 2014). There are documented accounts of black bears killing adult wild pigs, but it is uncertain whether they are any more than an opportunistic predator (Hanson and Karstad 1959, Mayer and Brisbin 2009). Eastern coyotes are the most abundant predator throughout the Northeast region occupying all habitats from the densely forested Upper Great Lakes to urban New York City (Gompper 2002). Many studies from southern regions of North America have documented wild pig remains in the scat of coyotes (Schrecengost et al. 2008, Etheredge et al. 2015), but the population level impacts this predator may have on wild pigs in the Northeast region is unknown (Mayer and Brisbin 2009). Coyotes from the Northeast region are known to genetically mix with wolves (Bozarth et al. 2011), are generally perceived to be physically larger than coyotes from other regions, and have replaced wolves as the top predator of adult white-tailed deer in the region (Gompper 2002). Eastern and gray wolves are an apex predator in the Upper Great Lakes states and Canadian provinces of the region. Although wild pigs are not currently abundant in these areas, there is some evidence that wolves could moderate their expansion. In their native European and Asian range, wild boar are primary prey of wolves (Davis et al. 2012). Additionally, given the recent accelerated population growth of wolves in the Upper Great Lakes Region and concurrent decline in abundance of their primary prey (white-tailed deer; Duquette et al. 2015), it is possible that wolves could prey-switch to wild pigs (Mattioli et al. 2011, Davis et al. 2012). Given the suite of potential predators of juvenile and adult wild pigs, additional study of predation in the Northeast region is warranted.

14.3.6 Reproductive Biology

Most wild pigs in the region are wild boar or hybrids, resulting in a reproductive potential that is highly variable. Although native wild boar have a high reproductive capacity when compared to other similar sized wild mammals (Bieber and Ruf 2005, Frauendorf et al. 2016), commercial breeding practices for domestic pigs have increased reproductive capacity of wild pigs with domestic pig ancestry (Mayer and Brisbin 2009). Additionally, many factors influence reproduction in domestic and wild pigs including body mass, nutrition, population density, and climate (Bieber and Ruf 2005, Mayer and Brisbin 2009, Frauendorf et al. 2016; see Chapter 4). Reproductive tracts collected from harvested wild pigs in the region indicate that most females breed by 1 year of age, and Indiana and Ohio biologists have documented fetuses in animals believed to be only 5–6 months old.

Number of litters/year is also variable throughout the region, but 2 litters every 12–13 months has been reported. Additionally, biologists from Ohio documented that some sows bred 1 month postpartum. Michigan biologists captured and attached transmitters to a group of 4 sub-adult wild boar in early August 2015 and these pigs later associated with a sounder and several juvenile pigs in October and November 2015, suggesting that the sow was the dam of both litters. Reported litter sizes for the Northeast region range from 3 to 13 with 4 to 7 being the most common (Figure 14.9). Biologists have observed a statewide average of 6.16 piglets/litter in Ohio (Plasters et al. 2013). Prior to eradication, wild pigs in 3 separate regions of New York State were producing litters averaging 4–6 piglets (USDA/APHIS/Wildlife Services 2010). Biologists in Indiana consistently documented 10 teats on harvested wild pigs and the number of teats is correlated with maximum litter size in domestic swine (Andersen et al. 2011).

Successful reproduction during the winter has been documented throughout the Northeast region. Indiana and Michigan biologists both reported capturing striped piglets in early spring. For example, a 17.2-kg partially striped wild boar was trapped and euthanized on 5 May 2017 in the U.P. of Michigan. This was in proximity to a farrowing nest that was discovered in February 2017 (Figure 14.5), and several trail camera photos of striped juvenile pigs were recorded between these dates, indicating that young were born and survived over winter. January farrowing was also confirmed for wild boar from New Hampshire on one occasion; however, most early litters were first detected in March. Additionally, biologists in Minnesota reported successful farrowing of escaped domestic pigs in winter. In Europe, wild boar typically farrow twice/year with peaks from

FIGURE 14.9 A large litter of wild pigs from Marion County, Illinois. (Photo by B. Wilson. With permission.)

January–March and again in late summer (Mauget 1982). Little is known about mortality of juvenile wild pigs from the region, but most biologists believe piglet survival and recruitment are high.

14.3.7 Damage

Localized damage to row crops from trampling and rooting by wild pigs has been reported in the Northeast region. One landowner in New York State reported US$12,000 in damage to corn, oat, and hay fields due to wild pig depredation (USDA/APHIS/Wildlife Services 2010). Damage from wild pigs feeding in standing corn can be extensive, but damage to some fields was not as severe as expected in Indiana and Michigan. Biologists from these states observed extensive shallow rooting between rows of corn in mid-summer with only minor damage to corn plants throughout the fields (Steven Gray, Michigan State University, and Steven Backs, Indiana Department of Natural Resources (INDNR), personal communication; Figure 14.10). In these situations it was unclear what resources wild pigs were seeking; however, roots or rhizomes of other plant species were not readily apparent, suggesting rooting activity may have been to access invertebrates. Rooting is also common in hay and alfalfa hay fields, but wild pigs may avoid soybean fields in the Northeast region.

Additionally, rooting in residential lawns, gardens, golf courses, cemeteries, airports, and wildlife food plots has been reported in areas with concentrations of wild pigs. Potential damage to sensitive wetland species was reported in Ohio and Ontario and wallowing in sensitive wetland areas was a concern of biologists in most states with wild pig populations. Michigan recently documented extensive rooting of regenerating young aspen (*P. tremuloides*) managed for wildlife habitat, and Illinois also experienced damage to regenerating forests. Biologists in Indiana, New Hampshire,

FIGURE 14.10 Shallow rooting between corn rows in summer in Michigan's central Lower Peninsula. (Photo by S. Gray. With permission.)

and Ohio have confirmed damage to mature trees from tusking and excavation of seedlings. In New York, wild pigs devalued timber on a tree farm after they rubbed tree bases (USDA/APHIS/Wildlife Services 2010). Competition with native wildlife for food resources (particularly hard mast) is a concern throughout the Northeast region, but specific impacts to wildlife have not been documented. Wild pigs also pose risks to human health and safety in the region primarily through vehicle collisions. Attacks on domestic livestock (e.g., swine) and pets (e.g., dogs) have also been reported in New York State (USDA/APHIS/Wildlife Services 2010).

14.4 REGIONAL MANAGEMENT IN NORTHEASTERN NORTH AMERICA

States and provinces throughout the Northeast region vary in their definitions of a wild pig (see Chapter 11), but a common thread among most jurisdictions is that wild pigs are free ranging (i.e., wild, free of human captivity or control). Some also specify that wild pigs have no man-made identifying marks (e.g., ear tags). Most states and provinces designate wild boar and their hybrids, or animals with physical characteristics of wild boar as "wild pigs."

Wild pigs are classified as an "invasive species" or an animal with no legal protection status in all states and provinces in the Northeast region except New Jersey. In defined zones within the state, New Jersey designated wild pigs as a game species and allowed hunting during deer season as part of a successful eradication effort of the only localized population within the state. Indiana, Michigan, and New York specify that *S. scrofa* and their hybrids are an invasive species and are distinct from common domestic pig breeds (*Sus domesticus*) including heritage and heirloom breeds. In June 2015, the Michigan Court of Appeals upheld Michigan Department of Natural Resources' contention that *S. scrofa* are distinct and distinguishable from *S. domesticus* (Michigan Court of Appeals, Gregory Johnson and Bear Mountain Lodge, L.L.C., v. Michigan Department of Natural Resources, 2 June 2015). Relevant Departments of Agriculture and Natural Resources or Conservation commonly possess or share jurisdiction for regulating wild pigs in most northeast states and provinces.

Laws regarding possession and transport of wild pigs vary throughout the Northeast region (see Chapter 11). In an attempt to reduce intentional and unintentional releases of wild pigs, several northeast states and provinces have made it illegal to possess or transport wild pigs (including wild boar and hybrids thereof). Reducing augmentation of existing populations or establishment of new ones through releases, whether accidental or intentional, is likely the single most important action for assuring that wild pigs do not become abundant in the Northeast region. Michigan attributes declines in statewide abundance with designating wild boar and hybrids an invasive species in 2011 (Figure 14.11). Game ranches are permitted in several states but most do not allow possession of wild boar. Minnesota has 1 grandfathered game ranch permitted to possess wild boar, but no other facilities will be permitted. Maine also has 1 facility permitted to possess wild boar, but they cannot sell hunts.

Multiple removal techniques have been employed including corral, drop and portable panel traps, neck snares, and sharpshooting. Indiana, Michigan, New Hampshire, and Ohio biologists have incorporated "Judas pigs" (see Chapter 8) to increase effectiveness of removal efforts. This technique was successful at significantly reducing a localized population of wild boar in the central L.P. of Michigan. Because of the expansive area occupied, frequent and extensive movements, and low density of wild pigs, targeting animals for removal was extremely labor intensive. "Judas pigs" allowed managers to follow movements of uncollared pigs and greatly increased removal efforts by trapping and sharpshooting.

Hunting wild pigs for population control can be controversial because the spread of wild pigs in much of North America has been facilitated by recreational hunters moving animals (Hanson and Karstad 1959, Mayer and Brisbin 2009). Additionally, hunting is generally believed to be ineffective at reducing wild pig populations and may change activity patterns and pig movements (Keuling et al. 2008, Mayer and Brisbin 2009). However, given the low densities of wild pigs throughout most of the region, opportunistic harvest by hunters could assist with localized control efforts.

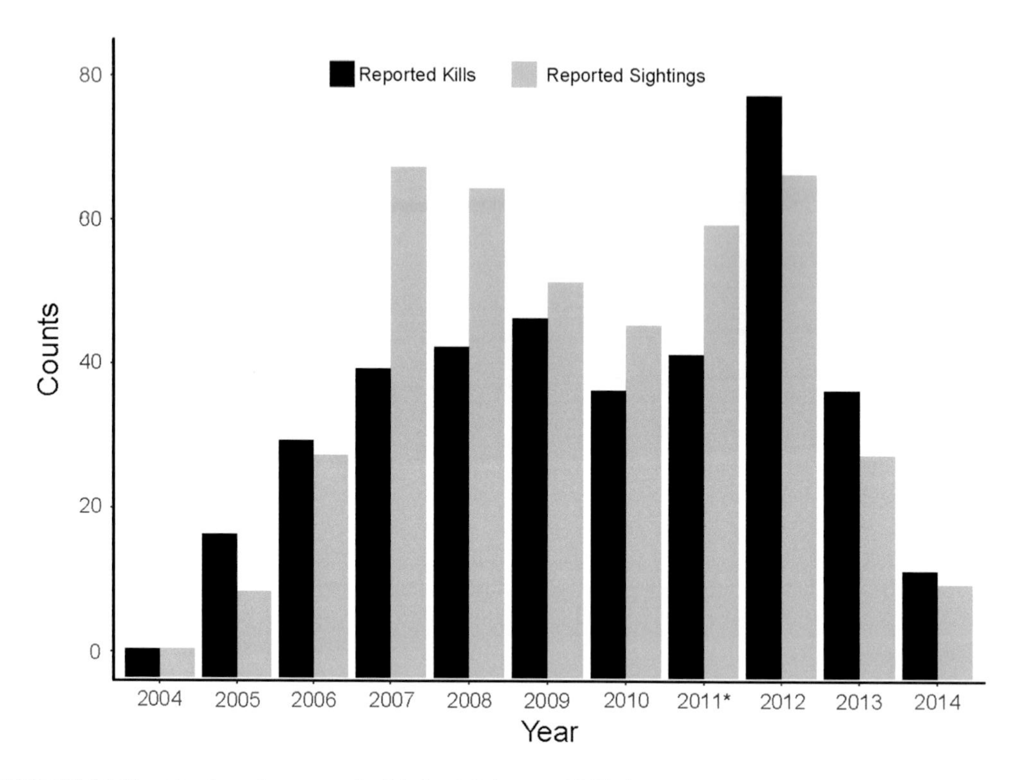

FIGURE 14.11 Number of reports of wild pig sightings and kills in Michigan, 2004–2014. An invasive species order from the Michigan Department of Natural Resources making it illegal to possess a wild boar or hybrid thereof went into effect in October 2011.

Dedicated wild pig hunters can also be knowledgeable about localized populations, which could benefit managers. Hunters using trained dogs can be highly effective at removing wild pigs. For example, hunters with trained dogs in Michigan removed more than 100 wild pigs from a core area in west-central L.P. over several years. An additional 21 wild pigs were removed by USDA/APHIS/Wildlife Services using "Judas pigs" to direct trapping and night-time shooting activities (Figure 14.12), and continued monitoring of this area failed to find additional wild pigs.

FIGURE 14.12 Wild pig with GPS tracking collar used as a "Judas pig" to find other associated wild pigs for lethal removal. (Photo by T. Wilson. With permission.)

Several states in the region have made assertive efforts to eliminate or reduce wild pig densities through active management programs. New Jersey eliminated the only localized population of wild pigs in the state using intensive lethal control combined with legal hunting during deer season. Wild pig populations in multiple areas in New York State are now believed to have been successfully eradicated through a combination of lethal removal techniques (e.g., trapping, shooting, cable snares), public outreach and education, and outlawing both recreational hunting and game ranches for this species (USDA/APHIS/Wildlife Services 2010, USDA/Animal and Plant Health Inspection Service 2015, Daniel Hojnacki, USDA/APHIS/Wildlife Services, personal communication). With coordinated public education and cooperation from private landowners, Illinois limited the spread of wild pigs from 2 core areas. Then, using a combination of lethal removal techniques including trapping, sharpshooting, and aerial control, USDA/APHIS/Wildlife Services eliminated wild pigs from Illinois (Scott Beckerman, personal communication; see Wild Pig Control in Illinois Case Study below).

14.5 A CASE STUDY OF WILD PIG CONTROL IN ILLINOIS

We thank Scott Beckerman[a], Bradley Wilson[a], and Doug Dufford[b] ([a]USDA/APHIS/Wildlife Services, [b]Illinois Department of Natural Resources) for providing this case study. Contributions from S.B. and B.W. were supported by the USDA. Mention of commercial products or companies does not represent an endorsement by the US Government.

Illinois has ideal habitat for wild pigs with adequate forest cover and abundant food sources in the form of acorns and agricultural crops. Wild pigs were first reported in 1993 along the Mississippi River in southern Illinois on the Union County Conservation Area and adjacent properties. By the late 1990s, the numbers of sightings and reports of damage resulting from these pigs had increased. By 2005, reports of wild pigs had been confirmed in central Fulton County along the Spoon River watershed. This new and distinct population consisted of domestic breeds and had once been part of a game ranch (Figure 14.13). In 2009, USDA/APHIS/Wildlife Services began working with private landowners to conduct wild pig disease surveillance in the state. In 2011, to aggressively address the wild pig problem and to protect wildlife and wildlife habitat, USDA/APHIS/Wildlife Services and

FIGURE 14.13 Mixed domestic and hybrid wild pig sounder from central Illinois. (Photo by B. Wilson. With permission.)

the Illinois Department of Natural Resources (ILDNR) signed a Cooperative Service Agreement (CSA), officially beginning collaborative wild pig management in Illinois.

After extensive field investigations, trail camera surveillance and aerial surveys conducted in the regions of Illinois where reports of wild pigs were most frequently received, it was determined that 2 self-sustaining, breeding populations were present. One population was in Fulton County while the other occurred in south-central Illinois among the adjoining counties of Fayette, Marion, Effingham, and Clay. Both populations were likely the result of former game ranches and/or intentional releases.

Illinois was divided into 3 damage management units (DMU) during wild pig elimination efforts. Two primary DMUs consisted of 1) Fulton County and 2) Fayette, Marion, Effingham, and Clay Counties. The third DMU encompassed the remainder of the state where reports of wild pigs were investigated as they were received. The ILDNR developed a database to track statewide observations of wild pigs and amended an existing hunter harvest survey to include a wild pig reporting feature. USDA/APHIS/Wildlife Services assigned a biologist for each primary DMU to work directly with private landowners requesting assistance.

The population management strategy involved the "whole sounder removal approach." This involved preparing bait sites in areas of wild pig presence and subsequently monitoring sites with motion-activated trail cameras. Once wild pigs were visiting a bait site, an open corral trap was constructed and continuously monitored with trail cameras. The drop gate was set only after the entire sounder was observed freely entering and exiting the trap. Upon successful capture, wild pigs were humanely euthanized, disease samples were taken, and carcasses were either provided to the landowner or disposed of according to the Illinois Dead Animal Disposal Act. Selective sharpshooting for trap shy and individual wild pigs was also used, especially once populations were greatly reduced. Aerial control with a helicopter was utilized on 2 occasions in 2014 to locate and remove the few remaining wild pigs.

All known breeding wild pig populations were successfully eliminated from Illinois by March 2016. A total of 376 wild pigs were removed from the Fulton County DMU and assistance was provided to 47 landowners (7,427 ha). The Fayette, Marion, Effingham, and Clay County DMU had a smaller population of wild pigs, but they were distributed over a much larger area. A total of 83 wild pigs were removed from this DMU and assistance was provided to 54 landowners (8,717 ha). Numerous reports of wild pigs have been received elsewhere in the state, though all but 1 individual has been identified as escaped domestic swine or intentionally released/escaped pet potbelly pigs; the majority of these pigs were recovered by their owners. Overall, 57% (262) of the wild pigs in Illinois were removed by trapping, 31% (145) by sharpshooting, and 12% (53) through aerial control.

Several key factors that led to successful elimination of wild pigs from Illinois included cooperation by private landowners and limiting public access to wild pigs on private property that could have led to disbursement of sounder groups. Collaboration between USDA/APHIS/Wildlife Services and ILDNR prevented a duplication of efforts, empowered multiple stakeholder groups, improved communication among agencies, and allowed for many different agencies and stakeholders to have a vested interest in the outcome. Statewide outreach efforts helped raise awareness and build support for the collaborative effort to address the growing wild pig problem. These efforts included personal consultations, informative meetings, instructional sessions, workshops and demonstrations, radio and TV appearances, brochures, flyers, posters, newspaper articles, and trade show exhibits all aimed at wild pig removal and information about reporting sightings. Utilization of aerial surveys to search for rooting damage, estimate abundance, and determine range of wild pig distribution in the primary DMUs was also important. After removing a large percentage of the population from the primary DMUs, aerial control operations proved to be important for removing the few remaining wild pigs.

In March 2014, implementation of the ILDNR administrative rule regulating the release, transportation, possession, and harvest of wild pigs in Illinois made it legal to shoot wild pigs only during the firearm deer seasons while in possession of an unfilled deer permit. This rule served to deter those who would attempt to establish populations for hunting purposes. Landowners experiencing damage were required to obtain a free nuisance wildlife removal permit in order to trap or shoot wild pigs outside of the aforementioned seasons. This helped prevent disbursement of wild pigs and encouraged a strategic, full-time approach to removal operations carried out by USDA/APHIS/Wildlife Services.

USDA/APHIS/Wildlife Services continues to work closely with ILDNR and other agencies to conduct outreach efforts and surveillance among the primary DMUs and to investigate wild pig reports throughout the remainder of the state. Detection efforts currently rely heavily upon reported damage and sightings by private landowners and hunters.

14.6 REGIONAL WILD PIG RESEARCH IN NORTHEASTERN NORTH AMERICA

Most research conducted on wild pigs in North America has been in southern US states with long-standing populations (see Chapter 16), generally focusing on wild pig ecology and effective population control. Given differences in habitats, climate, and the type of pigs present (i.e., wild boar, hybrids, or escaped domestics) in the Northeast region, compared to southern regions, there are differences in wild pig ecology and behavior. In recent years, several have initiated research projects to better understand northern wild pig ecology and inform effective management. Indiana, Michigan, New Hampshire, and Ohio have deployed VHF and GPS satellite transmitters to better understand movements, habitat use, home range, and response of pigs to targeted removal efforts (Figure 14.14).

FIGURE 14.14 Rex Schanck (USDA/APHIS/Wildlife Services) displaying a GPS satellite telemetry collar used to track movements of wild pigs in Michigan's central Lower Peninsula. (Photo by Michigan Department of Natural Resources. With permission.)

Some researchers are attempting to quantify hourly, daily, and seasonal movements, habitat use patterns, and wild pig dispersal capabilities. Researchers from Michigan are attempting to develop a resource selection function (i.e., predictive model) that portrays the likelihood of wild pig habitat use to assist in searching for and targeting wild pigs for removal from the expansive forested regions of the state. All states conducting telemetry research are interested in the efficacy of techniques for controlling wild pig populations and potential changes in activity and habitat use patterns in response to population control activities.

Effectively attaching transmitters to wild pigs can be difficult because neck-belting must be fit tight to avoid animals shedding the collar. Additionally, because of the rapid growth potential of juvenile and sub-adults, collars are typically only deployed on adults (see Chapter 9). When trapping low-density populations, this severely limits candidate pigs because younger animals are more abundant and typically easier to capture than adults (Gabor et al. 1999, Mayer and Brisbin 2009). Younger wild pigs may also display different behaviors than adults, possibly limiting valuable information for research and removal efforts. To resolve these issues, researchers in Michigan experimented with a harness constructed of rigid 2.2-mm thick and 48-mm wide belting material (Lotek Wireless, Newmarket, Ontario, Canada). This rigid belting allowed the harness and collar (constructed of the same material) to be fit loose, allowing for growth while not compromising collar shedding (Figure 14.15). Additionally, infrared reflective material was sewn to both side straps allowing sharpshooters to use infrared detecting scopes to identify "Judas pigs" and avoid removing them from targeted groups of animals. From 2014 to 2017, researchers successfully deployed this harness system on 9 wild boar ranging from 23.4 to 54.9 kg. One 29.9-kg juvenile female shed the harness within 24 hours after capture (however, she was subsequently recaptured 5 days later), but all others retained their harnesses for 52–293 days (mean = 145 days) until they were harvested. Researchers collected whole carcass weight during capture but were unable to obtain weights at time of death; however, it can be assumed that most animals would have increased mass over the time harnesses were worn. Inspection of carcasses revealed limited superficial lacerations and hair loss resulting from harnesses.

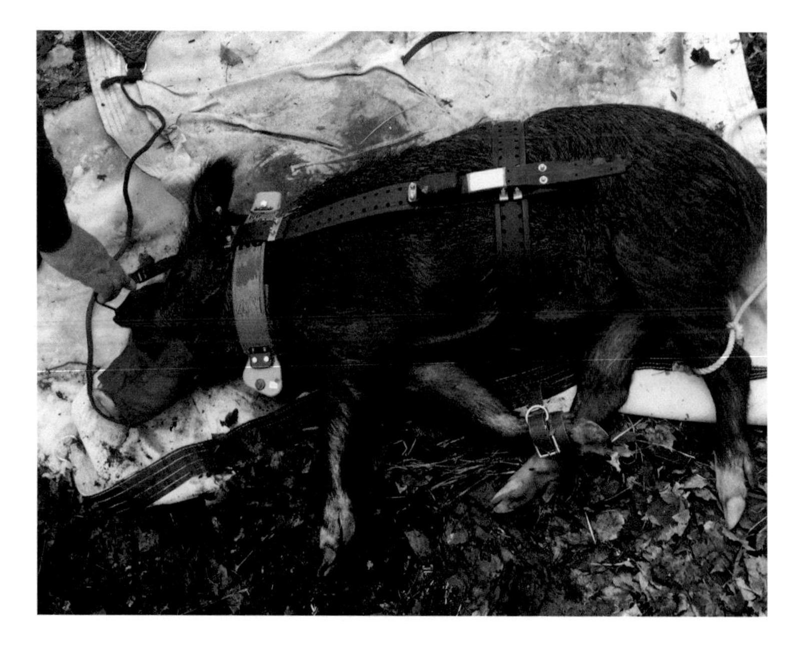

FIGURE 14.15 Sub-adult female wild pig (45.9 kg) captured in February 2017 in Michigan's Upper Peninsula A GPS satellite telemetry collar was attached with a specialized harness constructed of rigid belting material (Lotek Wireless, Newmarket, Ontario, Canada) to allow for a looser fit on juvenile and sub-adult wild pigs. (Photo by B. Roell. With permission.)

Researchers from USDA/APHIS/Wildlife Services are working with the Ohio Department of Natural Resources (ODNR) and INDNR to develop genetic profiles of wild pig populations (e.g., Caudell et al. 2013). In the future, this information could be used to explore the origin of new populations as they appear on the landscape providing information about potential dispersals or releases. This research also has implications for tracking potential disease transmission. For example, 4 wild pigs from Ohio tested positive for pseudorabies virus and 1 also tested positive for brucellosis. Genetic analysis of tissues from these pigs showed little resemblance to known wild pig populations in Ohio, but comparisons indicated similarities to pigs from Louisiana, suggesting they may have been transported into Ohio (Clint McCoy, ODNR, personal communication).

Effectively detecting wild pigs at low densities can be challenging, particularly in the Northeast region where wild pigs are known to move long-distances, terrain is sometimes remote and cover is extensive. Environmental DNA (eDNA) is gaining popularity for detecting invasive species and wild pig DNA has been successfully detected under controlled conditions (Williams et al. 2016; see Chapter 9). Researchers from the University of Michigan—Flint and Central Michigan University have develop protocols for detecting eDNA from natural watersheds, providing an additional technique for detecting wild pigs and verifying occupancy status.

14.7 FUTURE OF WILD PIGS IN NORTHEASTERN NORTH AMERICA

The future of wild pigs in the Northeast region will likely vary from historical trends of pigs observed in the southern regions of North America. Because of changes to legal status of wild pigs and assertive control efforts, the trend in wild pig abundance is declining in most states and provinces in the region. However, abundant and ideal wild pig habitat in the lower and central portions of the Northeast region could lead to eruptive population growth (Bieber and Ruf 2005) if these efforts were reduced or suspended. Information gained from other regions will be invaluable for developing effective control strategies in the Northeast region.

There are many challenges to wild pig eradication/control from the Northeast region and some challenges likely vary from those experienced in more southern regions. Presently, the public perception of wild pigs throughout the Northeast region appears to be negative and this is likely influenced by the "invasive species" designation by several government agencies. Natural resources agencies in the region have also been proactive in educating the public about negative consequences of wild pigs. Most outdoor recreationists (particularly hunters) view wild pigs as competing with native wildlife; however, there is a small fraction of hunters in the Northeast region who view wild pigs as a valued game species. If this view spreads, it could complicate management efforts. Additionally, agencies and stakeholders need to continually educate political bodies, particularly about the challenges and consequences of maintaining wild boar in captivity.

Eradicating invasive species requires effective planning and coordination among all stakeholders. The successful effort by Illinois to eradicate wild pigs is proof that it can be accomplished when multiple agencies and stakeholders work towards a common goal. Eradication also requires an understanding of the behavior of wild pig populations and the ability to use multiple lethal control techniques (Parkes et al. 2010). Common removal techniques (e.g., corral, drop and portable panel traps), which are effective for removing wild pigs when they exist at high density, may not be either time or cost effective for controlling low-density populations (see Chapter 8). Additionally, the remote and rugged terrain of many of the northern states and provinces in the region will further limit feasible lethal removal techniques.

A significant challenge for most of the northeast states and provinces is detecting wild pigs that exist at low densities and how to verify that an area is free of wild pigs following eradication efforts. Ultimately, this is a good problem to have, but additional research is needed to improve detection of wild pigs in the region. Financial resources of the primary agencies responsible for controlling wild pigs in the Northeast region may also be limited, particularly as populations are reduced and the cost/animal removed increases exponentially (Mayer and Brisbin 2009). Given potential impacts of one of the most destructive terrestrial invasive species worldwide, agencies should plan for the additional resources necessary to effectively eliminate wild pigs from northeastern North America.

ACKNOWLEDGMENTS

We thank the editors of this text for their extensive efforts to summarize the status of one of the most destructive invasive species in the world and particularly North America, wild pigs. To better summarize wild pigs within the region, the authors provided an online survey to biologists and representatives from the different states and provinces and we thank those who filled out the survey including, S. Backs (INDNR), N. Balgooyen (Wisconsin Department of Natural Resources), D. Dufford (ILDNR), C. Heydon (Ontario Ministry of Natural Resources and Forestry), M. Lovallo (Pennsylvania Game Commission), K. Marden (Maine Department of Inland Fisheries and Wildlife), C. McCoy (ODNR), J. McKnight (Maryland Department of Natural Resources), A. Musante (USDA/APHIS/Wildlife Services New Hampshire), E. Nelson (Minnesota Department of Natural Resources), F. Pogmore (USDA/APHIS/Wildlife Services Vermont), A. Randall (USDA/APHIS/Wildlife Services New Jersey), and K. Stang (New York State Department of Environmental Conservation). We also thank S. Beckerman, B. Wilson, A. Musante, and S. Backs for providing additional information and photographs. J.J. Mayer conducted extensive research to produce Appendix 14.1.

REFERENCES

Albert, D. A. 1995. Regional landscape ecosystems of Michigan, Minnesota, and Wisconsin: A working map and classification. US Department of Agriculture, US Forest Service, North Central Forest Experiment General Technical Report NC–178, St. Paul, MN.

Alison, R. 2014. Wild boars make escape. *Orillia Today*. <http://www.simcoe.com/opinion-story/4461963-wild-boars-make-escape/>. Accessed 31 Mar 2016.

Andersen, I. L., E. Naevdal, and K. E. Bøe. 2011. Maternal investment, sibling competition, and offspring survival with increasing litter size and parity in pigs (*Sus scrofa*). *Behavioral Ecology and Sociobiology* 65:1159–1167.

Ballari, S. A., and M. N. Barrios-García. 2014. A review of wild boar *Sus scrofa* diet and factors affecting food selection in native and introduced ranges. *Mammal Review* 44:124–134.

Barnes, G. 2008. Wild boar struck and killed on Rt. 2. *Telegram & Gazette*. <http://www.telegram.com/article/20081024/NEWS/810240649/1116>. Accessed 27 Oct 2008.

Barrett, R. H. 1978. The feral hog on Dye Creek Ranch, California. *Hilgardia* 46:283–355.

Berg, C. 2007. Hogs wild in Pa. *Times Argus News*. <http://timesargus.com/apps/pbcs.dl1 /article?AID=/20070408/FEATURES06/704080336/1015/FEATURES06>. Accessed 18 June 2009.

Bieber, C., and T. Ruf. 2005. Population dynamics in wild boar *Sus scrofa*: Ecology, elasticity of growth rate and implications for the management of pulsed resource consumers. *Journal of Applied Ecology* 42:1203–1213.

Bloom, B. 2004. Once a novelty, wild swine now nuisance in Crawford County. *La Cross Tribune*. <http://www.lacrossetribune.com/articles/2004/12/26/news/00lead.txt>. Accessed 25 Feb 2008.

Bostelaar, R. 2014. Shoot to kill: Wild boars are back in eastern Ontario. *Ottawa Citizen*. <http://ottawacitizen.com/news/local-news/shoot-to-kill-the-wild-boars-are-back-in-eastern-ontario>. Accessed 30 Mar 2015.

Bozarth, C. A., F. Hailer, L. L. Rockwood, C. W. Edwards, and J. E. Maldonado. 2011. Coyote colonization of northern Virginia and admixture with Great Lakes wolves. *Journal of Mammalogy* 92:1070–1080.

Campbell, L. J. 2004. Officials quiet after hydro pond draining. *Almaguin News* 119(44):1–3.

Caudell, J. N., B. E. McCann, R. A. Newman, R. B. Simmons, S. E. Backs, B. S. Schmit, and R. A. Sweitzer. 2013. Identification of putative origins of introduced pigs in Indiana using nuclear microsatellite markers and oral history. *Proceedings of the Wildlife Damage Management Conference* 15:39–41.

Cohen, L. D. 1997. Wild boars have begun to infiltrate. *Hartford Courant*. <http:/articles.courant.com/1997-06-15/news/9706150012_1_wild-boar-wild-pig-critters>. Accessed 15 Dec 2012.

Conaboy, C. 2009. Boar hits Prius on interstate. *Concord Monitor*. <http://www.concordmonitor.com/apps/pbcs.dl1/article?AID=/20091104/FRONTPAGE/911040339>. Accessed 11 Nov 2009.

Corn, J. L., and T. R. Jordan. 2017. Development of the national feral swine map, 1982–2016. *Wildlife Society Bulletin* 41:758–763.
Crossett, K. M., T. J. Culliton, P. C. Wiley, and T. R. Goodspeed. 2004. Population trends along the coastal United States: 1980–2008. NOAA, National Ocean Service, Silver Spring, MD.
Davidson, J. 2006. Wild boar in the wilds in Pennsylvania... Graybeard Outdoors. <http://www.go2gbo.com/forums/index.php?topic=92799.0>. Accessed 29 Jan 2013.
Davis, M. L., P. A. Stephens, S. G. Willis, E. Bassi, A. Marcon, E. Donaggio, C. Capitani, and M. Apollonio. 2012. Prey selection by an apex predator: The importance of sampling uncertainty. *PLoS ONE* 7:e47894. doi:10.1371/journal.pone.0047894.
Detelj, T. 2011. Wild boars spotted in NE Connecticut. *News 8 WTNH*. <http://www.wtnh.com/dpp/news/windham-cty/wild-boarsboars-spotted-in-ne-connecticut>. Accessed 13 Sep 2011.
Dey, D. C. 2014. Sustaining oak forests in eastern North America: Regeneration and recruitment, the pillars of sustainability. *Forest Science* 60:926–942.
duPont, S., and K. Wagner. 2006. Maryland Department of Health News Release: Eleven pigs discovered on quarantined Schisler farm. <http://www.mda.state.md.us/article. php?i=4325>. Accessed 23 Sep 2010.
Duquette J. F., J. L. Belant, N. J. Svoboda, D. E. Beyer, Jr., and P. E. Lederle. 2015. Scale dependence of female ungulate reproductive success in relation to nutritional condition, resource selection and multi-predator avoidance. *PLoS ONE* 10:e0140433.
Etheredge, C. R., S. E. Wiggers, O. E. Souther, L. L. Lagman, G. Yarrow, and J. Dozier. 2015. Local-scale difference of coyote food habits on two South Carolina islands. *Southeastern Naturalist* 14:281–292.
Feral Swine Federal Task Force. 2018. Feral Swine Federal Task Force News. Fiscal Year 2018, Quarter 2. 4/2/2018. APHIS National Feral Swine Damage Management Program, Ft. Collins, CO.
Fernández-Llario, P. 2004. Environmental correlates of nest site selection by wild boar *Sus scrofa*. *Acta Theriologica* 49:383–392.
Fernández-Llario, P. 2005. The sexual function of wallowing in male wild boar (*Sus scrofa*). *Journal of Ethology* 23:9–14.
Frauendorf, M., F. Gethöffer, U. Siebert, and O. Keuling. 2016. The influence of environmental and physiological factors on the litter size of wild boar (*Sus scrofa*) in an agriculture dominated area in Germany. *The Science of the Total Environment* 541:877–882.
Frye, B. 2007. Five counties in Pennsylvania are home to wild hogs. *Trib Total Media, Inc.* <http://triblive.com /x/pittsburghtrib/sports/outdoors/s_506122.html>. Accessed 2 May 2016.
Gabor, T. M., E. C. Hellgren, R. A. Van Den Bussche, and N. J. Silvy. 1999. Demography, sociospatial behaviour and genetics of feral pigs (*Sus scrofa*) in a semi-arid environment. *Journal of Zoology* 247:311–322.
Gompper, M. E. 2002. Top carnivores in the suburbs? Ecological and conservation issues raised by colonization of north eastern North America by coyotes. *Bioscience* 52:185–190.
Hanson, R., and L. Karstad. 1959. Feral swine in the southeastern United States. *The Journal of Wildlife Management* 23:64–74.
Harlow, D. 2012. State mystified by presence of wild boar that killed pig in Mercer. *Kennebec Journal*. <http://www.kjonline.com/news/wild-boar-kills-domestic-pig-raising-questions_2012-12-01.html>. Accessed 14 Dec 2012.
Hunting.net. 2010. Forum: Wild hogs. *Boars in western Maryland*. <http://www.huntingnet.com/forum/northeast/33071-wild-hogs-boars-western-maryland.html>. Accessed 18 Feb 2014.
Kelly, J. M. 2006. Roar of the boar. *The Syracure Post-Standard*. <http://www.syracuse.com/printer/printer.ssf?/base/sports0/1136540268194581.xml&coll=1>. Accessed 7 June 2006.
Keuling, O., N. Stier, and M. Roth. 2008. How does hunting influence activity and spatial usage in wild boar *Sus scrofa* L.? *European Journal of Wildlife Research* 54:729–737.
La Crosse Tribune and Leader. 1927. Hog tales. *La Crosse Tribune and Leader*. 16 January.
LaFond, K., and M. K. Salwey. 2004. Feral pig. *Wisconsin Department of Natural Resources*. <http://www.dnr.state.wi.us/org/land/wildlife/PUBL/wlnoteboo/Pig.htm>. Accessed 17 Sep 2004.
Lemel, J., J. Truvé, and B. Söderberg. 2003. Variation in ranging and activity behavior of European wild boar *Sus scrofa* in Sweden. *Wildlife Biology* 9:29–36.
Lovallo, M. 2014. Feral Swine in Pennsylvania. <https://ecosystems.psu.edu/research/centers /private-forests/news/2014/feral-swine-in-pennsylvania>. Accessed 23 Apr 2019.
MacGowan, B., and J. Caudell. 2011. Feral hogs in Indiana – what it means to you. *The Woodland Steward* 20:12–13.
Massei, G., P. V. Genov, B. W. Staines, and M. L. Gorman. 1997. Factors influencing home range and activity of wild boar (*Sus scrofa*) in a Mediterranean coastal area. *Journal of Zoology* 242:411–423.

Mattioli, L., C. Capitani, A. Gazzola, M. Scandura, and M. Apollonio. 2011. Prey selection and dietary response by wolves in a high-density multi-species ungulate community. *European Journal of Wildlife Research* 57:909–922.

Mauget, R. 1982. Seasonality of reproduction in the wild boar. Pages 509–526 *in* D. J. A. Cole and G. R. Foxcroft, editors. *Control of Pig Reproduction*. Butterworth Scientific, London.

Mayer, J. J. 2018. Introduced wild pigs in North America: History, problems and management. Pages 299–312 *in* M. Melletti and E. Meijaard, editors. *Ecology, Evolution and Management of Wild Pigs and Peccaries: Implications for Conservation*. Cambridge University Press, Cambridge, UK.

Mayer, J. J., and I. L. Brisbin, Jr. 2008. *Wild Pigs in the United States: Their History, Comparative Morphology, and Current Status*. Second Edition. The University of Georgia Press, Athens, GA.

Mayer, J. J., and I. L. Brisbin, Jr., editors. 2009. *Wild Pigs: Biology, Damage, Control Techniques and Management*. SRNL-RP-2009-00869. Savannah River National Laboratory, Aiken, SC.

Mayer, J. J., F. D. Martin, and I. L. Brisbin, Jr. 2002. Characteristics of wild pig farrowing nests and beds in the upper coastal plain of South Carolina. *Applied Animal Behaviour Science* 78:1–17.

McCann, B. 2003. The feral hog in Illinois. Thesis, Southern Illinois University, Carbondale, IL.

McNight, J. 2018. This little piggy went to Maryland? Maryland Invasive Species Council.

Michigan Department of Natural Resources. 2014. *Number of Feral Swine Reports in Michigan by County by Year* (Updated on 7 Oct 2014). Lansing, MI.

Michigan Department of Natural Resources, Department of Environmental Quality, and Michigan Department of Agriculture. 2017. Michigan invasive species program 2017 annual report. <https://www.michigan.gov/documents/invasives/Invas_annual_ 2017_single_crops-mc1_621632_7.pdf>. Accessed 14 Nov 2018.

Minnesota Department of Natural Resources. 1999. Harmful exotic species of aquatic plants and wild animals in Minnesota: Annual report 1999. Exotic Species Program of the Minnesota Department of Natural Resources, St. Paul, MN.

Minnesota Department of Natural Resources, Minnesota Department of Agriculture, and Minnesota Board of Animal Health. 2010. *Feral Swine Report to the Minnesota State Legislature*. St. Paul, MN.

Murray, B. T. 2008. Gone hog wild! Gloucester bristles as feral porker run amok. *The Star-Ledger* August 21:1–2.

Murray, M. 2015. Caught on camera: Pesky wild pigs in Delaware. *Delaware online/the News Journal*. <http://www.delawareonline.com/story/news/local/2015/04/01/photo-increases-fear-wild-pigs/70778754/>. Accessed 18 May 2015.

Musante, A. R., K. Pedersen, and P. Hall. 2014. First reports of pseudorabies and winter ticks (*Dermacentor albipictus*) with an emerging feral swine (*Sus scrofa*) population in New Hampshire. *Journal of Wildlife Diseases* 50:121–124.

National Oceanic and Atmospheric Administration, National Climate Data Center. 2017. Climate at a glance: Statewide time series. <https://www.ncdc.noaa.gov/temp-and-precip/state-temps>. Accessed 2 May 2017.

New York Times. 1880. Early days in New-York; adventures of old hunters in Sullivan County. *New York Times* 12 April.

New York Times. 1888. Beseiged by wild hogs – the terrible all-night experience of a boy. *New York Times* 28 December.

New York Times. 1898. A hunt for wild boars – Sloatsburg people much agitated about animals in the woods near them. *New York Times* February 6:9.

Parkes, J. P., D. S. L. Ramsey, N. Macdonald, K. Walker, S. McKnight, B. S. Cohen, and S. A. Morrison. 2010. Rapid eradication of feral pigs (*Sus scrofa*) from Santa Cruz Island, California. *Biological Conservation* 143:634–641.

Plasters, B. 2012. Evaluating Ohio's feral swine (*Sus scrofa*) population. Thesis, Ohio State University, Columbus, OH.

Plasters, B., C. Hicks, R. Gates, and M. Titchenell. 2013. Feral swine in Ohio: Managing damage and conflicts. *Factsheet W-26*, Ohio State University Extension. <http://ohioline.osu.edu/factsheet/W-26>. Accessed 28 Mar 2016.

Podgórski, T., G. Baś, B. Jędrzejewska, L. Sönnichsen, S. Śnieżko, W. Jędrzejewski, and H. Okarma. 2013. Spatiotemporal behavioral plasticity of wild boar (*Sus scrofa*) under contrasting conditions of human pressure: Primeval forest and metropolitan area. *Journal of Mammalogy* 94:109–119.

Roberts, N. M., and S. M. Crimmins. 2010. Bobcat population status and management in North America: Evidence of large-scale population increase. *Journal of Fish and Wildlife Management* 1:169–174.

Saskatchewan Agriculture and Food. 2001. Wild boar production: Economic and production information for Saskatchewan producers. SAF-Livestock Development Branch, Regina, Saskatchewan. ISBN: 0-88656-641 0032.

Scheick, B. K., and W. McCown. 2014. Geographic distribution of American black bears in North America. *Ursus* 25:24–33.
Schlichting, P. E., S. R. Fritts, J. J. Mayer, P. S. Gipson, and C. B. Dabbert. 2016. Determinants of variation in home range of wild pigs. *Wildlife Society Bulletin* 40:487–493.
Schrecengost, J. D., J. C. Kilgo, D. Mallard, H. S. Ray, and K. V. Miller. 2008. Seasonal food habits of the coyote in the South Carolina coastal plain. *Southeastern Naturalist* 7:135–144.
Silver, H. 1957. A history of New Hampshire game and furbearers. New Hampshire Fish and Game Department Report No. 6, Concord, NH.
Stegeman, L. C. 1938. The European wild boar in the Cherokee National Forest, Tennessee. *Journal of Mammalogy* 19:279–290.
Sweeney, J. M., and J. R. Sweeney. 1982. Feral hog (*Sus scrofa*). Pages 1164–1179 *in* J. A. Chapman and G. A. Feldhamer, editors. *Wild Mammals of North America: Biology, Management, Economics*. John Hopkins University Press, Baltimore, MD.
Thurfjell, H., G. Spong, and G. Ericsson. 2014. Effects of weather, season, and daylight on female wild boar movement. *Acta Theriologica* 59:467–472.
TidalFish.com. 2007. Feral hogs in Maryland? <http://www.tidalfish.com/forums/ showthread.php/21677 2-Feral-Hogs-in-Maryland>. Accessed 26 July 2010.
US Census Bureau. 2016. Quick facts – United States. <https://www.census.gov/quickfacts /table>. Accessed 9 May 2017.
US Department of Agriculture/Animal and Plant Health Inspection Service. 2015. *Final Environmental Impact Statement. Feral Swine Damage Management: A National Approach*. Washington, DC.
US Department of Agriculture/Animal and Plant Health Inspection Service/Wildlife Services. 2010. *2010 Status of Feral Swine in New York State*. Castleton, NY.
US Department of Agriculture/Animal and Plant Health Inspection Service/Wildlife Services. 2012. *Environmental Assessment*: *Feral Swine Damage Management in New York*. Castleton, NY.
US Department of Agriculture/Natural Resources Conservation Service. 2017. Illinois Cooperative Soil Survey. <https://www.nrcs.usda.gov/wps/portal/nrcs/detail/il/ soils/surveys>. Accessed 5 May 2017.
US Environmental Protection Agency. 2016. Great Lakes facts and figures. <https://www.epa.gov/greatlakes/great-lakes-facts-and-figures>. Accessed 5 May 2017.
US Fish and Wildlife Service. 2015. Northeast region conserving the nature of America: Eastern cougar. <http s://www.fws.gov/northeast/ecougar/index.html>. Accessed 30 June 2017.
US Geological Survey National Water Data. 2017. USGS water data for Indiana. <https://nwis.waterdata.usgs.gov/in/nwis>. Accessed 3 May 2017.
Vermont Agency of Agriculture, Foods and Markets. 2019. <https://agriculture.vermont.gov/agency-agricultur e-food-markets-news/feral-swine-captured-tests-positive-pseudorabies>. Accessed 30 May 2019.
Voorhees, S. 2006. Wild boars causing farm havoc. *R News*. <http://www.rnews.com/ print.cfm?id=35163>. Accessed 6 Mar 2009.
Weekly World News. 1986. Blazing guns save hunter's bacon. *Weekly World News* Dec. 30:11.
Werner, S. 2015. Invasive species of wild boar shot in Caledon. *Toronto Star*. <http://www.thespec.com/n ews-story/5787696-invasive-species-of-wild-boar-shot-in-caledon/>. Accessed 6 May 2016.
Wharton, J. 2011. Wild hog sighting followed by tale of hunter's kill – pigs elude Leonardtown man's camera and police. *South Maryland News online*. <http://www.somdnews.com /article/20111202/NEWS/7120298 04/1044/wild-hog-sighting-followed-by-tale-of-hunter-8217-s-kill&template=southernMaryland>. Accessed 18 Feb 2014.
Williams, K. E., K. P. Huyvaert, and A. J. Piaggio. 2016. No filters, no fridges: A method for preservation of water samples for eDNA analysis. *BMC Research Notes* 9:298.
Wisconsin Department of Natural Resources. 2005. Feral pig hunting information. *Wisconsin Department of Natural Resources*. <http://www.dnr.state.wi.us/org/land/wildlife/HUNT/ Pig/Pig_Hunting.htm>. Accessed 6 June 2006.

APPENDIX 14.1
Current Known Extent of Wild Pigs in States and Provinces of Northeastern North America

State/ Province	Earliest Reported Presence		Current Status	Reference(s)
	Year	County/Regional Municipality		
Connecticut	1997 2011	Litchfield Windham	Isolated report of escaped wild pigs; no established populations in the state	Cohen (1997), Detelj (2011)
Delaware	2014	Sussex	Single report of a wild pig; animal was suspected to be escaped domestic animal; no established populations in the state	Murray (2015)
Illinois	1993	Union	Subsequent scattered individuals or populations reported in Alexander, Edgar, Fulton, Gallatin, Hardin, Jackson, Johnson, Lawrence, Massac, Monroe, Moultrie, Pope, Pulaski, and Randolph Counties; following recent eradication program, no wild pigs remaining in the state	McCann (2003); this chapter
Indiana	Mid-1990s	Wayne	Limited distribution, primarily Jackson, Lawrence, and Wayne Counties; smaller populations and isolated sightings previously reported in Crawford, Dubois, Elkhart, Fayette, Franklin, Harrison, Henry, Jennings, Lake, Lagrange, Monroe, Orange, Pike, Randolph, Scott, Steuben, Union, and Warrick Counties; possible unknown populations	MacGowan and Caudell (2011), Caudell et al. (2013), USDA/APHIS (2015), J. J. Mayer, Savannah River National Laboratory, unpublished data
Maine	2009 2012	Hancock Somerset	Isolated reports of escaped wild pigs (e.g., from Corbin's Park); no established populations in the state	Harlow (2012), USDA/APHIS (2015), J. J. Mayer, unpublished data
Maryland	2006 2007 2007 2010 2011	Carroll Allegany Charles Garrett St. Mary's	Isolated reports of wild pigs; mostly either hunter kills or roadkills; no established populations in the state	duPont and Wagner (2006), TidalFish.com (2007), Hunting.net (2010), Wharton (2011), McNight (2018)
Massachusetts	1997 2008	Berkshire Worcester	Isolated reports of escaped wild pigs (e.g., from Corbin's Park); no established populations in the state	Cohen (1997), Barnes (2008), J. J. Mayer, unpublished data
Michigan	1986	Wexford	Scattered populations in both the Upper and Lower Peninsulas likely established by escaped wild boar from private game ranches; variously reported in 72 of 83 counties in the state between 2001 and 2014; numbers decreasing and recently only a limited distribution in primarily a few U.P. counties	Weekly World News (1986), USDA/APHIS (2015), MIDNR (2014), D. Etter, MIDNR, personal communication

(Continued)

APPENDIX 14.1 (CONTINUED)
Current Known Extent of Wild Pigs in States and Provinces of Northeastern North America

State/ Province	Earliest Reported Presence: Year	Earliest Reported Presence: County/Regional Municipality	Current Status	Reference(s)
Minnesota	2009	Pine	No established populations; isolated reports of wild pigs primarily in NE and NW; reports increasing, and breeding confirmed in Marshall County	MNDNR (1999), MNDNR et al. (2010), USDA/APHIS (2015), E. Nelson, MNDNR, personal communication
New Hampshire	1895	Sullivan, Cheshire, and Hillsborough	Low, localized numbers of escaped wild pigs from Corbin's Park, Sullivan County; an estimated 100–250 animals recently found in 4 counties, but concentrated in western part of state	Mayer and Brisbin (2008), USDA/APHIS (2015)
New Jersey	1970s	Gloucester	A localized population of wild pigs was found in Gloucester County: following recent eradication program, no wild pigs are thought to be remaining in the state; USDA/APHIS/WS recently investigated several reports of released/escaped domestic swine; all swine were captured and returned to owner or rescue facility	Murray (2008), USDA/APHIS (2015), Feral Swine Federal Task Force (2018)
New York	1825 1888 1898 1902 2004	Sullivan Ulster Orange Hamilton Cortland	Early isolated reports of wild pigs in the Shawangunk Mountains (i.e., Orange, Sullivan, and Ulster Counties) in the 1800s; localized reports of escaped wild pigs in the early 1900s and 1970s in Hamilton County; beginning in the early 2000s, escaped wild pigs from high-fenced shooting operations resulting in these animals becoming established as breeding populations in the wild in 5 counties with sightings reported from a total of at least 21 counties in the state; following a recent eradication program, no wild pigs are thought to remain in the state	New York Times (1880, 1888, 1898); Kelly (2006); Mayer and Brisbin (2009); USDA/APHIS/WS (2012); USDA/APHIS (2015)
Ohio	1980s	Vinton	Established breeding populations exist in the southeast, unglaciated portion of the state (primarily in Vinton, Jackson, Hocking, Lawrence, Scioto, and Gallia Counties), isolated emergent populations in the rest of the state	Plasters (2012), USDA/APHIS (2015), Feral Swine Federal Task Force (2018)

(Continued)

APPENDIX 14.1 (CONTINUED)
Current Known Extent of Wild Pigs in States and Provinces of Northeastern North America

State/ Province	Earliest Reported Presence		Current Status	Reference(s)
	Year	County/Regional Municipality		
Pennsylvania	2002 2002 2002 2003 2004	Wayne Cambria Bedford Wyoming Carbon	Initial sightings date back to 1993 in Pennsylvania; limited distribution in several counties primarily in the NE Central and SW Central portions of the state; the source is both escaped animals from high-fenced shooting operations and illegal releases of wild pigs; breeding populations have been reported in Bedford, Bradford, Butler, Cambria, Fulton, and Tioga Counties, but wild pigs have been reported in 9 other counties in the state	Davidson (2006), Berg (2007), Frye (2007), Lovallo (2014), USDA/ APHIS (2015)
Rhode Island	N/A	N/A	Never been reported in the state	Mayer (2018)
Vermont	1955 1956	Windsor Barton	Isolated reports of escaped animals from Corbin's Park, NH; no established populations in the state	Mayer and Brisbin (2008)
Wisconsin	1927 1999 2000 2001 2002 2004 2005	Crawford Crawford Door Ashland, Iron, Price Buffalo, Douglas, Jackson, Taylor, Trempealeau Brown, Clark, Eau Claire, Grant, Kewaunee, La Crosse, Manitowoc, Monroe, Outagamie, Polk, Richland, Sauk, Sheboygan, Vernon, Waushara Bayfield, Oneida, Portage, St. Croix, Washburn, Waupaca	Isolated reports of wild pigs in 31 of 72 counties; the source is both escaped animals from high-fenced shooting operations and illegal releases of wild pigs; no established populations in the state; depopulation of the last remaining captive wild pig farm in the state is continuing	La Crosse Tribune and Leader (1927), Bloom (2004), LaFond and Salwey (2004), WDNR (2005), Feral Swine Federal Task Force (2018)
Labrador	N/A	N/A	Never been reported in the province	Mayer (2018)
New Brunswick	N/A	N/A	Never been reported in the province	Mayer (2018)
Newfoundland	N/A	N/A	Never been reported in the province	Mayer (2018)

(Continued)

APPENDIX 14.1 (CONTINUED)
Current Known Extent of Wild Pigs in States and Provinces of Northeastern North America

State/ Province	Earliest Reported Presence		Current Status	Reference(s)
	Year	County/Regional Municipality		
Nova Scotia	N/A	N/A	Never been reported in the province	Mayer (2018)
Ontario	2004 2006 2008 2011 2011 2012 2012 2015 2017	Parry Sound Pontiac Prescott and Russell United Prince Edward Simcoe Durham Frontenac Peel Aurora and Peterborough	Isolated reports of escaped animals; no established populations in the province	Campbell (2004), Voorhees (2006), Alison (2014), Bostelaar (2014), Werner (2015), Mayer (2018), C. Heydon, Ontario Ministry of Natural Resources and Forestry, personal communication, R. Kost, College of Agriculture and Bioresources, University of Saskatchewan, personal communication
Prince Edward Island	N/A	N/A	Never been reported in the province	Mayer (2018)
Quebec	2014	Les Collines-de-l'Outaouais	Isolated report of escaped animals; no established populations in the province	Hempstead (2014) Mayer (2018); R. Kost, personal communication

Courtesy of J. J. Mayer.

15 Wild Pigs in South-Central North America

Joshua A. Gaskamp, James C. Cathey, Billy Higginbotham, and Michael J. Bodenchuk

CONTENTS

15.1 INTRODUCTION

Presence of wild pigs (*Sus scrofa*) in south-central North America dates to the early exploratory and colonization period by Europeans in North America during the 1600s. Free-ranging pigs served as a consistent protein source to early settlers, and eventually became a common form of livestock production. Since that time, wild pigs transitioned from free-range livestock with protected ownership to an invasive species with no claimed ownership, often unwanted because of associated ecological and economic damages. Currently within the South-Central region this highly adaptable species flourishes in a wide range of environments. Because of their adaptability, coupled with management aimed at increasing populations for several decades since the 1930s, wild pigs are now widespread throughout the South-Central region, which today represents one of the largest regional concentrations of these animals in North America.

For this chapter, the South-Central region includes Texas and Oklahoma. The climate in this region is humid subtropical in the east and semi-arid in the west with long, hot summers and short, mild winters. The region includes portions of the Atlantic Plain, Interior Plains, Interior Highlands, and Intermontane Plateaus physiographic regions of the United States (Fenneman and Johnson 1946). The landscape is diverse and occupied by broadleaf, pine, or mixed forests, savannahs, prairies, agricultural lands (crop, forage, and recreational), and large river drainage systems.

Wild pigs have a long history of inhabiting this region, but their populations have increased substantially since the mid-1980s. Public opinions of wild pigs in Texas and Oklahoma are mixed. Generally, agricultural commodity producers favor eradication of wild pigs to reduce damages to their industry, while many commercial hunt operators financially benefit from recreational opportunities provided by wild pigs on the landscape. Others who have niche markets, like specialty meat markets, aerial gunning businesses, and contest venues (e.g., hunting dog tournaments) often perpetuate existence of wild pigs. Differences in public opinions and associated management goals create a complex suite of challenges, strategies, and opportunities within the South-Central region.

15.2 HISTORIC AND PRESENT DISTRIBUTION

15.2.1 Origin and Early History

Wild pigs did not occur in the Americas prior to European arrival (see Chapter 2), a fact unknown to many people in the South-Central region. Wild populations established in Cuba became the source of pigs used to provision Spanish expeditions of Cortes and De Soto to Mexico and the southeastern United States. Beginning in July 1539, De Soto's expedition traveled through what is now 8 states, taking with him a large herd of pigs as a walking commissary. Throughout the trip, pigs freely grazed and many wandered off or were captured by Native Americans (Towne and Wentworth 1950). De Soto arrived in Texas in 1542 and established multiple camps before returning to Florida. At some of these camps, pigs escaped. Spanish colonists from Saltillo, Mexico attempted to move pigs north of the Rio Grande River in 1565 to avoid losing them to local Native American raids (Mayer and Brisbin 1991). By 1598, wild populations were established in Texas as Sergeant-Major Vicente de Zaldivar reported trying to buy lard rendered from pigs from the Picuris Indians (Towne and Wentworth 1950). In 1685, French explorer Rene-Robert Cavelier, Sieur La Salle established Fort St. Luis on Lavaca Bay along the Texas coast. The settlement failed, but Native Americans and explorers who came in subsequent years observed free-roaming pigs in the area. Wild pigs appeared in southeastern Oklahoma beginning in the late 19th century, potentially originating from expanding Texas populations, but more likely from escaped free-range domestic pigs.

From European settlement until the 1940s, domestic pigs in Texas were allowed to forage unconfined and were considered livestock (M. J. Bodenchuk, USDA/APHIS/Wildlife Services, personal communication). Ownership was determined by a system of ear notching and pigs were captured in late fall or winter after fattening on hard mast. Dobie (1929) cites examples of landowners in the late 1800s hunting pigs with dogs and rifles or capturing pigs in pen traps. Most captured pigs were butchered and the meat cured, but some were ear marked and released (Dobie 1929). Landowners conducted pig "drives" to take once free-roaming pigs to market or to a food processor, such as a meat packing plant (Dobie 1929). As livestock, unmarked pigs belonged to whoever could capture them, though disputes regarding ownership likely occurred. Those not captured added to a growing wild population.

Several wild boar introductions also occurred in Texas during the 1930s and early 1940s. Wild boar were first introduced along the central coastal region of Texas on the St. Charles Ranch in Aransas County by Leroy G. Denman, Sr. (Mayer 2009). Although a large population of wild pigs was already reportedly established, domestic Hampshire pigs were released onto the ranch, but did not survive in the wild. One male and 1 female wild boar were then introduced from the Brackenridge Zoo in San Antonio, Texas into a small enclosure on the ranch, but did not reproduce. An additional wild boar pair from the Houston Zoo was released in the enclosure, followed by several additional releases from zoos in St. Louis and Milwaukee (Mayer 2009). A wild sow that was caught on the ranch was then put in the enclosure and successful reproduction occurred. These cross-bred pigs were subsequently released onto the ranch before the US Department of Interior purchased the ranch that later became the Aransas National Wildlife Refuge. Following the purchase, 3,391 wild pigs were removed from the ranch between 1936 and 1939, but the remaining population increased and wild pigs still occupy the refuge today (Mayer 2009). Wild boar were next introduced by Leroy Denman in Calhoun County on the Powder Horn Ranch in 1939. Several male and female wild boar were purchased from the San Antonio, St. Louis, and San Diego zoos and an additional 2 male wild boar were purchased from Leon Springs, Texas (Mayer 2009). The wild boar were placed into an enclosure on the ranch where they eventually escaped and expanded their population. Wild boar from Powder Horn Ranch later naturally colonized Matagorda Island by crossing the Espiritu Santo Bay at low tide and an intentional introduction also occurred on San Jose Island in Aransas County. In the early 1940s, Harry Brown introduced wild boar onto his fenced property in Bexar County, Texas (Mayer 2009). Several years later, a flood washed out part of the fence and the boar escaped. Escapees established a population in this area, and through natural expansion, hybridized with populations of wild pigs already on the landscape in Medina and Bandera Counties (Mayer 2009). This hybrid population became the source for many subsequent rancher-sponsored stocking efforts, including releases in Comal, Webb, Haskell, Throckmorton, and LaSalle Counties (Mayer and Brisbin 1991).

15.2.2 Present Status

Established wild pig populations in Oklahoma and Texas expanded naturally through reproduction and dispersal and artificially through intentional trapping and translocation. Natural dispersal was generally slow and landowners managed isolated pockets of wild pigs by hunting and trapping. Though trapping and killing of free-roaming pigs was already occurring, the Oklahoma legislature changed classification of free-roaming domestic pigs from livestock to wild pigs in 2000. At the same time, the Oklahoma legislature prohibited willful release of any pig to public or private lands. Prior to 2008, intrastate movements of wild pigs was not regulated in Texas and, because pigs were considered livestock, ranchers were free to trap, transport, and release wild pigs. Although they occurred, there are no records of translocations in Oklahoma or Texas or the number of pigs moved prior to these regulations. Translocations were largely for hunting on unfenced and fenced properties. Fencing designed to restrict animal movement and hold exotic species for commercial shooting is common in this region. However, some believe that wild pigs escape from many high-fenced shooting operations, further facilitating spread of this invasive species throughout the region.

Changes in land management also favored wild pig population growth since the 1970s. Farm programs, such as the Conservation Reserve Program, reduced acreage in active agricultural production, decreasing crop damage protection in some areas. Land ownership patterns shifted with more people owning land that did not live on the property. Intensive management for deer, including food plots and supplemental feeding, further favored establishment and growth of wild pig populations. Between 2006 and 2010, wild pig populations in the region experienced marked growth. Increasing at an estimated average annual rate of 21% per year, the population essentially doubled during this period in Texas (Timmons et al. 2012). While land use changes affected this increase in abundance, range expansion has likely resulted from movements and stocking facilitated by humans. At present, wild pig populations exist in all 77 Oklahoma counties and in 230 of 254 Texas counties. Abundance within this range is variable, depending on habitat and human tolerance. In 2017, total population between the 2 states was estimated at 3.5 million wild pigs, or roughly 50% of the estimated total population of wild pigs in the United States (Stevens 2010, Timmons et al. 2012). As of 2017, there were no known, free-roaming populations of pure wild boar in the region.

15.3 REGIONAL ENVIRONMENTAL ASPECTS IN SOUTH-CENTRAL NORTH AMERICA

15.3.1 Behavior

In the South-Central region, wild pig sounders primarily consist of a sow and her litter that may combine with other sows, shoats, and piglets forming larger groups. Boars are more aggressive and often solitary, but occasionally intermix with sounders or form small bachelor groups. Ilse and Hellgren (1995) noted that sounders in south Texas were not cohesive groups, but females within sounders stayed together for a 4-month period in Oklahoma (Webb and Boyer 2018). Generally, members of a sounder remain together until natural dispersal occurs. Sounders may occupy a single area or shift among multiple areas within their home range. Seasonal or daily shifts occur when wild pigs are using habitats that provide food and cover separately. Wild pigs are capable of moving several kilometers during short time periods to avoid disturbance or predation by humans. Using global positioning system (GPS) location data, Campbell and Long (2010) estimated home range size as 5.54 +/– 2.87 km^2 (fixed kernel estimate) in San Patricio County, and 4.46 +/– 2.04 km^2 in Kleberg County in southern Texas. Using GPS location data over a 4 month period in fall, Webb and Boyer (2018) estimated home range size to be 2.22 +/– 1.52 km^2 (fixed kernel estimate) along the Red River in Oklahoma and Texas, where the river was not a barrier to movement. In 2016, 11 of 16 collared wild pigs crossed the Red River 80 times (range = 2–11 crossings). Further, Kay et al. (2017) found that sex, age, ecoregion, and distance to nearest water source or agricultural field were all significant predictors of home range size.

Wild pigs are primarily nocturnal (e.g., Australia; Saunders and Kay 1991, Caley 1997) to crepuscular (e.g., Tennessee and Texas; Singer et al. 1981, Ilse and Hellgren 1995). Campbell and Long (2010) captured wild pigs in Kleberg (9 males, 4 females) and San Patricio (5 males, 7 females) counties in Texas and used GPS collars to collect daily movements and assess influence of temperature. Kleberg County males moved greater distances per day than females, whereas females moved greater distances than males in San Patricio County. Daily activity patterns were not evident. Hourly movements in both study areas suggested primarily nocturnal activity. For both sexes, activity peaked during 2300–2400 hours in Kleberg County and at 0400 and 2100 hours in San Patricio County. Further, data pertaining to temperatures during dormant and early growing seasons (January-March) indicated wild pigs increased activity during warmer periods, although temperature was not found to influence movements in summer (May-July). Wyckoff et al. (2006) also indicated temperature could affect wild pig movements in eastern and southern Texas.

Kay et al. (2017) found that sex, age, ecoregion, atmospheric surface pressure, and distance to nearest water source were significantly associated with maximum distance (a measurement between any 2 GPS locations in a single night) across all temporal scales (daily, monthly, and overall). They also found that sex, age, ecoregion, and distance to nearest water source were significantly associated with mean hourly distance (average distance between locations in a night that are approximately 1 hour apart) across all temporal scales. These authors analyzed GPS data from wild pigs in studies across 6 states including Florida, Georgia, Louisiana, Missouri, South Carolina, and Texas. Wild pigs in the Southern Texas Plains demonstrated greatest mean hourly distance movement rates across all temporal scales.

15.3.2 Diseases and Parasites

Because wild pig populations have been increasing in Oklahoma and Texas, concerns regarding disease transmission to wildlife, livestock, and humans are growing. Disease transmission by wild pigs continues to gain regional and national interests. Prevalence of diseases within populations fluctuate over time, so multiple agencies conduct periodic surveys to monitor population health and potential for spread to other animals. In addition to continued surveillance (Figure 15.1), evidence of diseases and parasites in wild pig populations in the South-Central region is well documented.

Serological surveys for 10 wild pig populations in Texas indicated pseudorabies (PRV; 7 populations), antibodies to leptospirosis (10 populations) and swine brucellosis (*Brucella suis*; 2 populations; Corn et al. 1986). Additionally, wild pigs in Texas showed serological evidence of exposure to swine influenza (14.4% of 472 sampled in the state during 2005–2006). All showed exposure to H3N2 subtype, except in San Saba County where 4 were positive for H1N1 and 7 for H1N1 and H3N2 among the 15 samples (Hall et al. 2008). A US Department of Agriculture report indicated

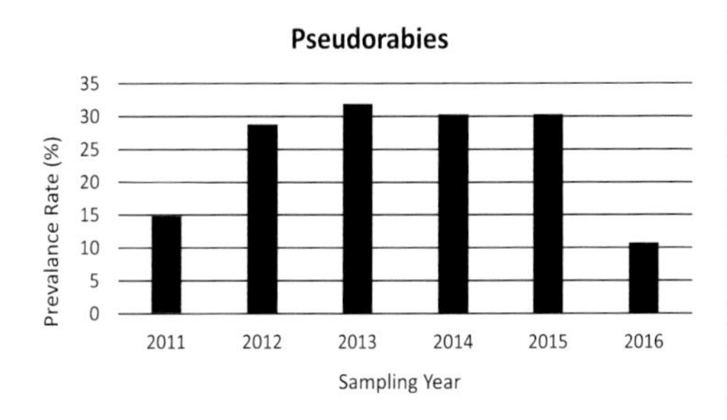

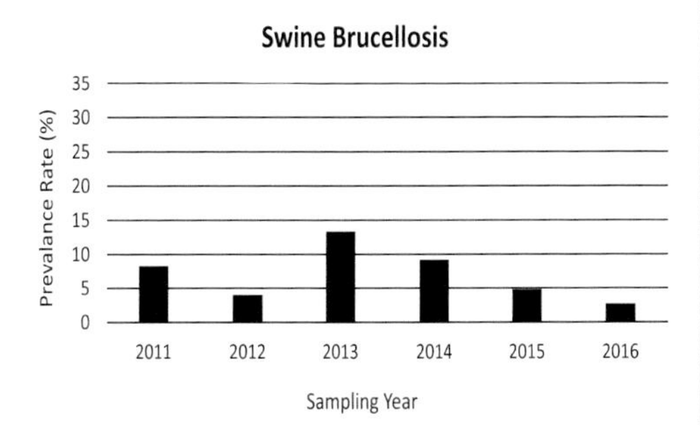

Prevalence Rate (Avg. 2011-2016)	
Classical Swine Fever	0.0%
Pseudorabies	24.4%
Swine Brucellosis	7.1%
Influenza A Virus	7.2%
Leptospirosis	46.2%
Toxoplasmosis	8.5%
Trichinosis	1.0%
Tuberculosis	0.0%
Foot and Mouth Disease	0.0%
Hepatitis E	10.9%
African Swine Fever	0.0%
Salmonella	63.6%
Porcine Reproductive and Respiratory Syndrome	0.9%
Bluetongue Virus	27.0%
Senecavirus	0.0%

FIGURE 15.1 USDA/APHIS/Wildlife Services in Oklahoma has intensively removed and tested wild pigs for various diseases from 2011 to 2016. Over 6 years, 28,253 wild pigs were removed and 5,825 were sampled for up to 20 diseases and pathogens. USDA/APHIS/Wildlife Services project personnel: Patrick Whitley, Scott Alls, and Kevin Grant. (Courtesy of P. Whitley. With permission.)

areas with relatively high incidence (40%) of avian influenza also had similar incidence of swine influenza antibodies in wild pigs. Springer (1977) reviewed literature for wild pig parasites at Aransas National Wildlife Refuge and found the population parasitized with swine kidney worm (*Stephanura dentatus*), lungworms (*Metastrongylus* spp.), roundworms (*Ascaris suum*), hookworms (*Globocephalus urosubulatus*) and several stomach worms.

Saliki et al. (1998) conducted serological surveys for viral and bacterial diseases in 120 wild pigs from 13 counties in Oklahoma (Bryan, Caddo, Choctaw, Coal, Custer, Grady, Hughes, Jefferson, Kiowa, Logan, Oklahoma, Pontotoc, and Roger Mills). Sera were examined for presence of antibodies against *Brucella suis*, porcine parvovirus, porcine reproductive and respiratory syndrome, PRV, swine influenza, transmissible gastroenteritis virus, vesicular stomatitis virus (New Jersey and Indiana serotypes), and *Leptospirosis interrogans* serovars *bratislava, canicola, grippotyphosa, hardjo, icterohernorrhagiae*, and *pomona*. No detections of antibodies to swine brucellosis, PRV, transmissible gastroententis, or vesicular stomatitis occurred. Among 44% of the samples, antibody titers were found for 1 or more *Leptospira interrogans* serovars with the 2 most frequent being *bratislava* (29%) and *pomona* (27%). Antibodies against porcine parvovirus and swine influenza virus occurred in 17% and 11% of the wild pigs, respectively. Antibodies to porcine reproductive and respiratory syndrome were found for 2 samples (2%). Though not detected in this study, later surveillance of wild pig populations in Oklahoma found antibodies to *Brucella suis* and PRV (Gaskamp et al. 2016; Figure 15.1). Hall et al. (2008) found no serological evidence of exposure to swine influenza among 50 wild pigs sampled in Oklahoma.

In addition to pathogens mentioned above, wild pigs harbor ectoparasites, including ixodid ticks capable of transferring disease to humans, wildlife, and livestock (Corn et al. 2009, Cooper et al. 2010). Sanders et al. (2013) collected 4,369 ticks from 497 wild pigs (806 sampled wild pigs including: 214 boars, 217 sows, 175 juvenile males, and 200 juvenile females) in Anderson, Bell, Bexar,

Brazos, Coryell, and San Patricio counties, Texas. They found 7 species of adult ticks including *Amblyomma americanum, A. cajennense, A. maculatum, Dermacentor albipictus, D. halli, D. variabilis*, and *Ixodes scapularis*, and nymphs of *A. americanum*, *A. cajennense*, and *D. variabilis*. These authors relayed the need for closer monitoring of tick-borne illness and wild pig exposure to *Rickettsia, Ehrlichia*, and *Borrelia*.

USDA/APHIS/Wildlife Services and the National Wildlife Research Center collaborated with USDA/Food Safety Inspection Service to determine pathogen prevalence at 2 wild pig slaughter facilities in Texas. Researchers found antibodies for a number of food borne pathogens as well as pathogens that could be transmitted to humans through contact during the slaughter process (Pedersen et al. 2017). Results of this study led to new worker safety measures for inspectors working at these facilities as well as new rules regarding record keeping for wild pigs nationwide (N. F. Bauer, Food Safety Inspection Service, personal communication).

15.3.3 Food Habits

Wild pigs in the South-Central region are opportunistic omnivores that typically range in size from 50 to 100 kg (Oliver and Brisbin 1993), with individuals weighing 100–150 kg less common. Vegetation is the primary diet of wild pigs in all seasons (Springer 1977), but dietary composition varies geographically and according to seasonal availability of foods. In Texas, Mapston (2007) found that wild pigs consume forbs and grasses, grain, bulbs, roots and tubers, soft and hard mast including grapes (*Vitis* spp.), plums (*Prunus* spp.), acorns (*Quercus* spp.), persimmons (*Diospyros* spp.), mesquite (*Prosopis* spp.), guajillo (*Acacia berlandieri*), huisache (*Vachellia farnesiana*), and prickly pear cactus (*Opuntia* spp.). Among agricultural crops, wild pigs consume corn, milo, wheat, oats, rice, soybeans, peanuts, pumpkins, watermelons, potatoes, and cantaloupes. Carrion, arthropods, amphibians, reptiles, eggs, birds, small mammals, and young of larger wild animals and livestock also occurred in wild pig diets. However, Taylor and Hellgren (1997) found that animal matter averaged only 6.7% of wild pig diets and varied seasonally. Stevens (2010) added other invertebrates like earthworms and snails to the growing list of foods consumed by wild pigs in the region.

Wild pig diets vary across the South-Central region as plant species composition is variable and changes seasonally. While prickly pear cactus and mesquite occur outside of south Texas, acorns and pecans (a high value commodity for many Oklahoma and Texas landowners) more commonly occur in fall and winter diets in the cross-timbers ecoregion from central Texas to northeast Oklahoma. Because late summer and winter are nutritionally more stressful to wild pigs and other wildlife species, interspecific competition may be important to species that rely on acorns or other mast at these times (Yarrow and Kroll 1989). Taylor and Hellgren (1997) found spring diets of wild pigs examined in south Texas to include >75% grasses, forbs, roots or tubers, or prickly pear pads. Summer diets shifted to *Opuntia* tunas (fruit) and hard mast including honey mesquite (*Prosopis glandulosa*; 27.8% of mast consumed year round) and guajillo (36.8%). Fall diets were primarily comprised of underground plant parts and corn (available in feeders), but herbaceous matter continued to occur in fall diets (32%). Nearly 60% of winter diets were grasses and corn. In the South-Central region, the most common practice used by hunters and other recreationists to attract and observe deer, wild pigs, and other exotic and introduced ungulates is to provide seasonal and year round feed stations (Figure 15.2). How feeding affects carrying capacity of wild pigs in the region is speculative, but this practice certainly increases nutrition available to wild pigs during stressful periods.

15.3.4 Habitat Use

Wild pigs tolerate most habitats in the South-Central region. Generally, the greatest densities of wild pigs occur close to consistent water sources, including riparian areas associated with rivers, lakes, and streams (Stevens 2010, Webb and Boyer 2018). Wild pigs also occur in upland areas where mast, crops, or other food sources are abundant. Both Oklahoma and Texas have large amounts of

FIGURE 15.2 Feed stations for deer commonly attract wild pigs, adults resting here. (Photo by Z. Johnson. With permission.)

agricultural land that ranges from deep soil sites suitable for farming to areas with lesser quality soils destined to remain rangeland used primarily for livestock or wildlife production. At the state level, Oklahoma and Texas have diverse ecoregions (Table 15.1), and diverse plant and animal communities among those ecoregions.

Wild pig populations expanded westward in Oklahoma (Figure 15.3) and Texas (Figure 15.4) over the last few decades, with the majority of rural lands occupied in each state (Stevens 2010, Timmons et al. 2012). Both states have rainfall gradients that offer greater precipitation in the east than west (142 cm in east Oklahoma to 43 cm in west Oklahoma and 152 cm in east Texas to

TABLE 15.1
Oklahoma and Texas Ecoregions

Oklahoma Ecoregions	Texas Ecoregions
Arkansas Valley	Piney Woods
Boston Mountains	Gulf Prairies and Marshes
Central Great Plains	Post Oak Savannah
Central Irregular Plains	Blackland Prairies
Cross Timbers	Cross Timbers
East Central Texas Plains	South Texas Plains
Flint Hills	Edwards Plateau
High Plains	Rolling Plains
Ouachita Mountains	High Plains
Ozark Mountains	Trans-Pecos
South-Central Plains	
Southwestern Tablelands	

Source: US Environmental Protection Agency (2007).

25 cm in west Texas). Drier ecosystems and climatic patterns potentially slowed westward expansion of wild pigs, but today wild pigs occur even in the drier western regions of each state. In Oklahoma, wild pigs occupy habitats from the wooded Ozark Mountains in the east to the western shortgrass prairies. In Texas, wild pigs occupy eastern pine woodlands to the Chihuahuan Desert in the west.

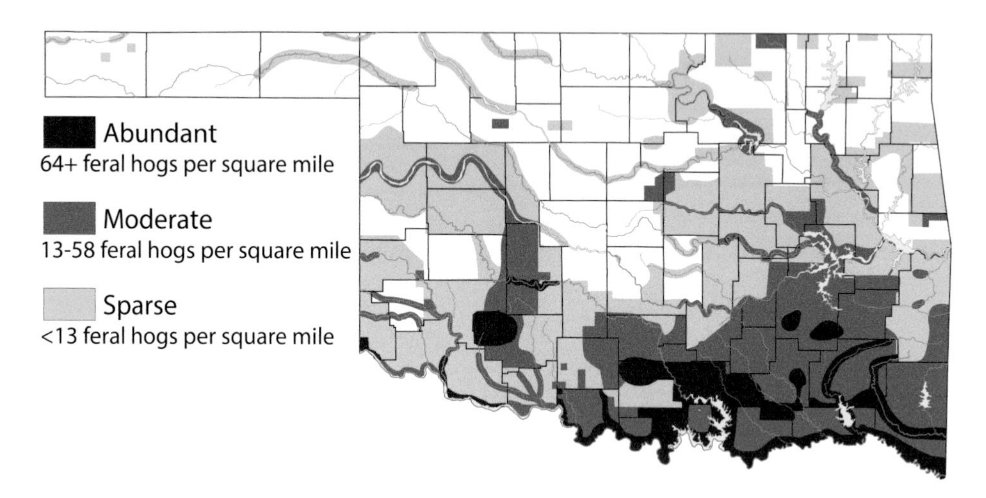

FIGURE 15.3 Estimated wild pig density in Oklahoma (2007). (Courtesy of the Noble Research Institute. With permission.)

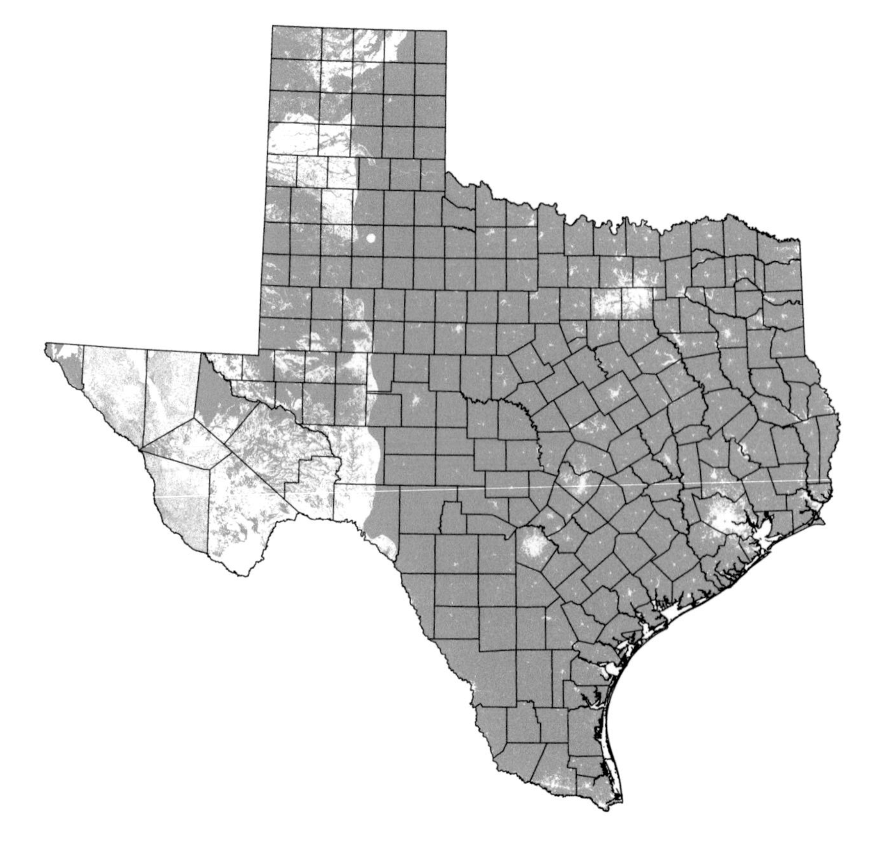

FIGURE 15.4 Suitable habitat for wild pigs, comprising 79% of the land mass in Texas. (Courtesy of Texas A&M Natural Resources Institute. With permission.)

15.3.5 Population Biology

Ilse and Hellgren (1995) reported 9.5 pigs/km^2 in southern Texas. Timmons et al. (2012) reported a range of 3.4–6.5 pigs/km^2 from 8 studies from various regions in Texas. They also used 21 studies across the United States to provide reasonable estimates of survival, litter size, and litter frequency to develop a statewide model for population growth rate in Texas, estimated at 18–21% annual growth.

Wild pig sounders can have more than 40 pigs (J. A. Gaskamp, Noble Research Institute, unpublished data), though 5–20 is more common. Sounder size varies by season, habitat, and climatic attributes, and commonly can exceed 20 individuals after multiple females farrow. Consequently, sounder demographics may also vary seasonally. In southern Texas, 111 visual observations of wild pigs across all seasons revealed the age structure consisted of 55.5% adults, 17% sub-adults and 27.5% juveniles (Ilse and Hellgren 1995).

Aside from humans, wild pigs have few routine predators in the South-Central region. Medium-sized carnivores like bobcats, foxes, and coyotes occasionally take young piglets (Mayer 2009), but likely have little effect on wild pig populations. In this region, mountain lions (*Puma concolor*) are only common along the Rio Grande River in Texas, although occasional sightings occur throughout the region. Harveson et al. (2000) examined prey use by mountain lions in La Salle, McMullen, Duval, and Webb counties in the South Texas Plains ecoregion. From examination of 75 carcasses identified as mountain lion kills, 9 taxa were represented, including wild pigs (9 carcasses). Mountain lion scats (n = 25) were also collected and contents were macroscopically and microscopically examined to determine prey items. Although this study was unable to separate scavenging from predation, frequency and volume of wild pig remains in scats were 28% and 32%, respectively.

15.3.6 Reproductive Biology

Wild pigs are the most prolific large mammal in North America and can out-compete native wildlife in the South-Central region (Taylor et al. 1998) given early age of sexual maturity (6–8 months; Barrett 1978), short gestation time (~115 days), and ability for year round breeding and farrowing. Wild pigs breed throughout the year, but Taylor et al. (1998) observed 2 seasonal peaks in parturition (January-March and June-July), and suggested nutrition and photoperiod were likely drivers of variation in seasonal reproduction. Springer (1977) reported peak farrowing season of January-May (17 of 19 observed pregnancies) for the Gulf Prairies ecoregion in Texas. He noted average litter size in this area was 4.2, and recorded 2 occurrences of 2 litters produced in 1 year. S. L. Webb (Noble Research Institute, unpublished data) and Taylor et al. (1998) found slightly larger average litter sizes (5.3) in south-central Oklahoma and south Texas (5.7), respectively, suggesting regional variation exists. Although not statistically different, average litter sizes were generally less for juveniles (4.8) than yearlings (5.0) and adult females (5.6–6.3). Litter frequency for juveniles, yearlings, and adults was 0.49, 0.85, and 1.57/year, respectively (Taylor et al. 1998). Taylor et al. (1998) also added that wild pigs had larger litter sizes and higher variance in litter size than wild boar. This result is consistent with production observed in domestic breeds that are genetically related to current wild pig populations. Annual gross fecundity of wild pigs examined in the Texas Gulf Prairies and western South Texas Plains ecoregions was greater than wild boar and over 4 times greater than native ungulates (e.g., white-tailed deer (*Odocoileus virginianus*), collared peccary (*Pecari tajacu*); Taylor et al. 1998).

15.3.7 Damage

Wild pig damage in the South-Central region is widespread and extensive, but rarely rigorously quantified. Wild pigs damage agricultural crops and infrastructure, forestry and reclamation projects, wetlands and riparian areas, and native plant communities (Campbell and Long 2009, Siemann et al. 2009), water quality, native wildlife populations, rangeland, and public and private property. Agricultural damage consists of direct removal of crops, destruction through rooting of improved pasture, damage to fences and harvest equipment, depredation on livestock, disease threats to livestock and the costs associated with lost opportunities due to threats of crop or livestock damage. Rooting by wild pigs causes physical damage to equipment, but more often leads to losses in yield due to unsuitable harvest conditions. From a study in Oklahoma, pecan harvest in areas rooted by pigs was 33.7% lower compared to unrooted areas (Boyer et al. 2018). Anderson et al. (2016) surveyed producers in 11 states, including Texas, regarding crop loss. Based on survey responses, Texas agriculture producers suffered US$89.8M in direct crop losses attributed to wild pigs. These losses included damage to grain sorghum, corn, wheat, rice, peanuts, and soybeans, but did not include losses to other crops, improved pasture, or rangeland. The 954 Texas respondents (66% had wild pigs on their land) reported spending over US$14.7M on control methods. Higginbotham et al. (2008) reported a decrease in damage following wild pig removal of US$1.5M over 2 years. M. J. Bodenchuk (personal communication) equated this to US$390 in damage averted/pig removed.

Livestock depredation is largely restricted to pasture-born lambs and goats, but is a common problem in Texas (M. J. Bodenchuk, personal communication). Exposure of livestock to diseases carried by wild pigs also poses a potential risk to agriculture (see Chapter 5). Wild pigs commonly harbor swine brucellosis and PRV (Gaskamp et al. 2016, Pedersen et al. 2017, P. Whitley, USDA/APHIS/Wildlife Services, unpublished data), both of which can be transmitted to domestic pigs, and *Brucella* from wild pigs has been identified in domestic cattle. In addition to direct losses to domestic livestock, diseases pose trade risks, potentially depressing livestock markets in affected areas. To maintain brucellosis-free status for cattle, any cow that tests positive for *Brucella* initiates an epidemiologic investigation at considerable governmental expense (M. J. Bodenchuk, personal communication).

Fence damage from wild pigs is common throughout the South-Central region. In an experimental treatment designed to evaluate efficacy of fencing to reduce crop damage and disease transmission risks, USDA/APHIS/Wildlife Services Texas repaired 37 pig-caused holes in a 12.07-km section of common "game fence" (woven wire 2.4-m high) and then removed wild pigs on both sides of the fence. One year following the removal they found 30 new, pig-caused holes in the fence.

While intensive forestry is not prevalent in the South-Central region, wild pigs pose risks to planted trees (e.g., eastern Texas; Taylor 1991). Similarly, wild pigs pose risks to reclamation projects on surface mining sites, roadside revegetation projects and wetland mitigation projects. Site preparation and newly established vegetation appear to make these sites especially attractive to wild pigs. Damage to these projects is primarily due to rooting activity and associated erosion.

Wild pigs degrade wetlands by wallowing and reducing vegetation along riparian corridors. A post-hoc analysis of over 17,000 GPS location points from wild pigs on the Welder Wildlife Refuge in Texas indicated wild pigs were within 25 m of water 24% of the time and within 100 m of water 48% of the time (T. Campbell, East Foundation, personal communication). Wild pigs also contribute pathogenic strains of *E. coli* bacteria to wetlands and streams. Six of 7 wild pigs collected in the San Saba River drainage in Texas had *E. coli* strains pathogenic to people; 4 of 7 were pathogenic to livestock (US Department of Agriculture, unpublished data).

Wild pigs affect native wildlife across much of their range by competing for food (Yarrow and Kroll 1989), through predation of birds (Tolleson et al. 1994, Synatzske 1979; Figure 15.5), amphibians and reptiles (Barron 1980), and by altering vegetative community structure and species composition which can indirectly impact many species, commonly insectivorous birds and insects. Additionally, when wild pigs utilize feed stations established to attract wildlife, potentially negative interactions with other wildlife are likely to occur. For instance, congregating animals at feed stations increases opportunity for interspecific and intraspecific disease transmission. In the South-Central region, management of wild pigs emphasizes protection of threatened or endangered species including whooping cranes (*Grus Americana*), Kemp's Ridley sea turtles (*Lepidochelys kempii*), interior least terns (*Sterna antillarum athalassos*) and Attwatter's prairie chickens (*Tympanuchus cupido attwateri*).

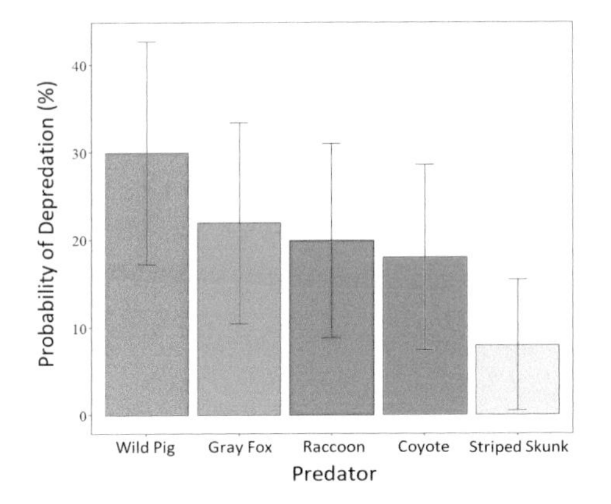

FIGURE 15.5 Impact of wild pigs on the fate of artificial turkey nests. (Courtesy of H. Sanders. With permission.)

Due to high densities of wild pigs in the South-Central region, populations are increasingly expanding into urban and suburban areas (see Chapter 19), creating additional impacts to these environments as well. Impacts include rooting damage to golf courses, green spaces, lawns, and other landscaping, vehicle collisions, and threats to human safety. From 2007 to 2017, at least 2 human fatalities have occurred in Texas due to vehicle collisions with wild pigs (M. J. Bodenchuk, personal communication).

15.4 REGIONAL MANAGEMENT IN SOUTH-CENTRAL NORTH AMERICA

15.4.1 Management Status

Prior to the 1970s, wild pig hunting in the South-Central region was popular in areas where they were abundant, but knowledge of wild pig locations was often limited to local residents. Beginning in the 1970s, wild pig hunting increased in popularity with awareness of wild pig populations becoming more widespread. Popularity of recreational hunting of wild pigs and marketability as a product for consumption or saleable hunting has likely contributed to rapid range expansion and population growth, facilitated by trap, move, and release events in Oklahoma and Texas. More recently a shift in the perception of wild pigs, from a species that provides supplemental hunting opportunities to that of an invasive species causing negative impacts to agriculture, environment, and human livelihoods and safety, has occurred. Currently, managers treat wild pigs as invasive species in the South-Central region. However, wild pigs continue to be highly sought by hunters, "hog dog" enthusiasts, and trappers. Trappers often take advantage of marketing opportunities for wild pigs in the South-Central region. For example, trappers can sell live wild pigs to high-fenced shooting operations or slaughter facilities in Oklahoma and Texas. However, buyers at these facilities either have a minimum pig weight for purchase, or pay a premium for larger animals. These restrictions or incentives may encourage illegal releases of smaller wild pigs and perpetuate existence and spread.

15.4.2 Legal Status of Wild Pigs in South-Central North America

The South-Central region primarily consists of private landholdings with no collective management goals for wild pigs. Differences in management philosophies led to a mosaic of diverse control strategies across the landscape, and absence of coordinated efforts between neighbors has made population-level control difficult. Furthermore, strong opinions and active political lobbying resulted in legislation not designed to support control efforts advised by scientific evidence or experienced resource professionals. With one of the largest concentrations of wild pigs in the United States, landowners, managers, and legislators understand that widespread eradication is not possible. With this mindset, regulations on harvest method, transportation, and marketing in the South-Central region are among the most lenient in the country (see Chapter 11). Additionally, management strategies and regulations are different in each state.

In Texas, wild pigs are legally designated as non-protected invasive livestock regardless of genetic background. Texas Parks and Wildlife Department (TPWD) requires that hunters have a valid Texas hunting license to pursue wild pigs. There is no bag limit or season and hunting of wild pigs can occur 24 hours a day. Individuals hunting pigs at night are advised to contact the local game warden to avoid potential law enforcement issues. Texas legislation allows landowners and their designated agents or lessees to kill depredating wild pigs by any legal means available, without having to possess a valid hunting license. This has caused some confusion as to who is actually required to have a license to remove wild pigs from a property. However, consensus generally exists that an individual is required to have a hunting license for recreational hunting or if they intend to keep harvested wild pigs for meat or trophy.

In Oklahoma, wild pigs are classified as an invasive species and the Oklahoma Feral Swine Control Act governs management of wild pigs. Most enforcement authority for the Act lies with the Oklahoma Department of Agriculture, Food, and Forestry (ODAFF), but Oklahoma Department of Wildlife Conservation (ODWC) regulates hunting of wild pigs during big game seasons on public lands, and authorizes night shooting exemptions. Residents and non-residents do not need a hunting license to hunt or control depredating wild pigs on private lands in Oklahoma, except during big game seasons. Hunting of wild pigs can occur at night with a free night shooting exemption administered through ODWC. To hunt wild pigs within high-fenced shooting operations licensed by ODAFF, a hunting license for captive wild pigs is required.

Initiated after the 2016 deer hunting season, ODWC included wild pig harvest questions in telephone surveys given to state hunting license holders. Though a license was not required to hunt wild pigs on private property, results of the survey indicated that 25% of surveyed Oklahoma license holders were pursuing (hunting and trapping) wild pigs with the predominant method being hunting. With an estimated 47,500 license holding pig hunters in Oklahoma, and an average 6.46 pigs harvested/hunter, statewide estimated harvest from hunting exceeds 300,000 pigs. Respondents indicated damage was one of the primary motivations to pursue wild pigs.

Some landowners and many trappers and hunters capture wild pigs in live traps or via bay and catch dogs to supply a source of protein for both human and pet consumption. Live captured wild pigs in Texas can legally be sold to one of approximately 100 buying stations in the state (permitted by Texas Animal Health Commission). More recently, processing wild pigs as a protein source in pet food has increased demand for both live and freshly killed wild pigs in Texas. The demand for wild pigs as a human food source has increased over the last 20 years. From 2004 to 2009 alone, at least 461,000 wild pigs were US Department of Agriculture-inspected and slaughtered for human consumption in Texas. Since that period, landowner awareness of selling wild pigs to buying stations and the number of buying stations has increased. Trapping has likely increased across the landscape with the introduction of these market opportunities, but opinions on impacts of this program to pig populations are mixed. There is some speculation that the payment schedule for varying size classes of wild pigs has encouraged trappers to release small piglets for re-capture and sale after growing to a marketable size.

Stocking live wild pigs on high-fenced shooting operations is also legal in Texas and Oklahoma. Boars and barrows captured in Texas can be sold to these operations permitted by TPWD, but females must go to slaughter. Live wild pigs captured in Oklahoma, regardless of age or sex, can be sold to 1 of approximately 65 licensed wild pig facilities (permitted by ODAFF). Fourteen of these are high-fenced shooting operations, and the remainder are buying stations that supply wild pigs to high-fenced shooting operations.

Awareness of damaging effects of translocations into areas with still manageable numbers of wild pigs has prompted Oklahoma to adopt regulatory practices employed in other states (e.g., Kansas). ODAFF, with the support of ODWC, has established the Oklahoma wild pig free zone in extreme northwestern Oklahoma and the panhandle (Figure 15.6). The wild pig free zone consists

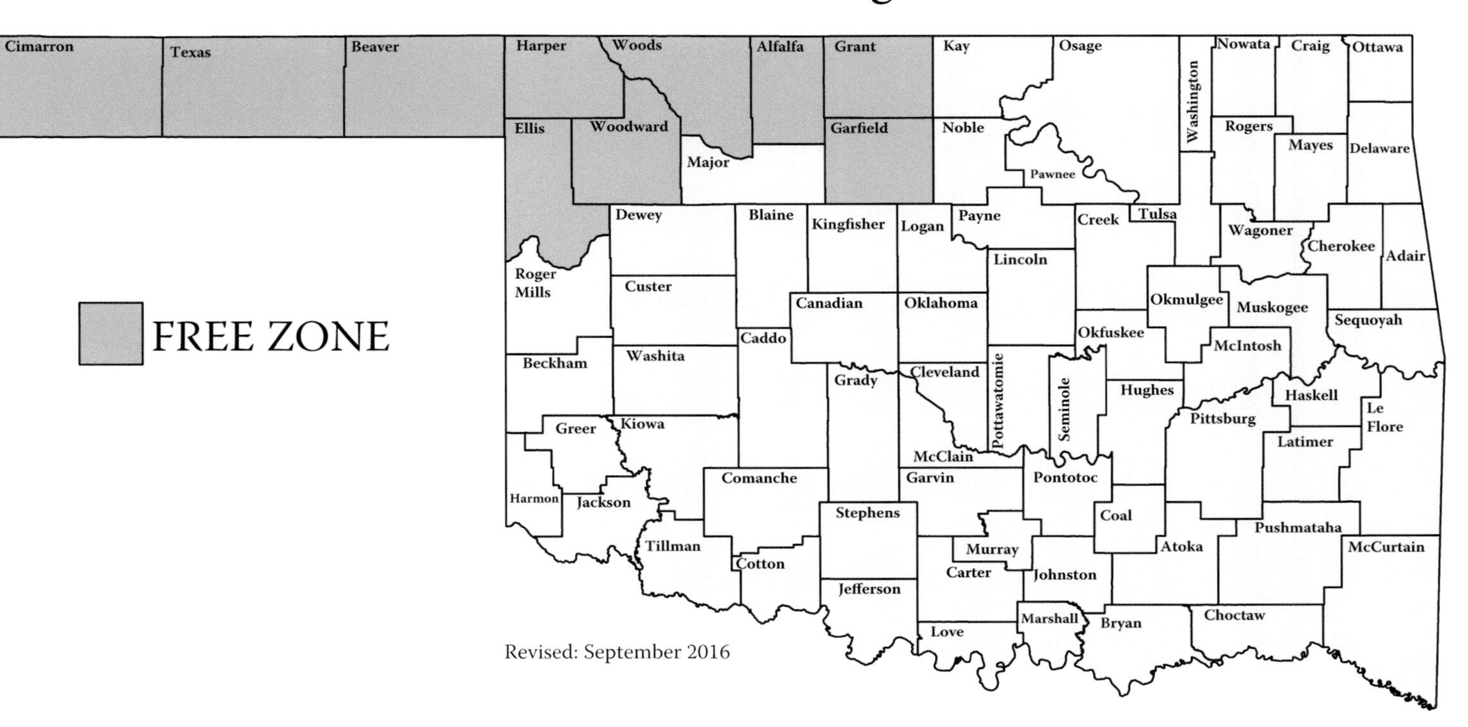

FIGURE 15.6 Oklahoma wild pig free zone, established by the Oklahoma Department of Agriculture, Food, and Forestry, with support from the Oklahoma Department of Wildlife Conservation.

of 10 counties with low wild pig densities and potential for temporary eradication. Movement of live wild pigs in or through these 10 counties is prohibited, and sightings, captures, or kills within the zone must be reported to ODAFF. The intent of this program is to keep USDA/APHIS/Wildlife Services in Oklahoma apprised of wild pig locations in the zone, focus control efforts there, and eventually grow the wild pig free zone into counties farther south and eastward.

15.4.3 Control Methods

During the 1970s and 1980s, interest in wild pigs in the South-Central region was largely restricted to recreational hunting with little regard for actual damage abatement through population control. However, as range and population size of wild pigs increased via natural reproduction and clandestine (and illegal) trapping, moving, and release events, damage caused by wild pigs became more intense and widespread (see Chapter 7). Early control methods in the region were largely limited to box traps (see Chapter 8). Many feed and hardware stores and welding shops in Texas manufactured and/or sold slight variations of the box trap. With increased availability of remote-sensing cameras in the mid-1980s, landowners began to monitor wild pig activity at trap locations. However, it soon became obvious that box traps could only capture small numbers of wild pigs per trapping event and were certainly inefficient at capturing entire sounders.

Larger, but less portable corral traps built from readily available materials (see Chapter 8) generally replaced box traps as a means to capture larger groups of wild pigs. However, even use of corral traps required adherence to a set of best management practices (BMPs) for success (Higginbotham 2013). Trappers in the region inconsistently followed these BMPs, primarily because they were time consuming, often taking weeks of preparation and pre-baiting before a trap could be set with likelihood of optimal success. Landowners in the region commonly constructed permanent corral traps for reuse later when wild pigs returned to the property. Because BMPs generally are not followed by landowners, subsequent trapping was mostly unsuccessful and unused trap structures occur on many properties across Texas and Oklahoma.

More recently, landowners and trappers in the region incorporated use of cellular technology to remotely monitor and trigger traps. In addition to remote cellular activation, innovations in trap design also appeared. A suspended trap (e.g., BoarBuster® trapping system; Figure 15.7) was a new

FIGURE 15.7 Wild pigs feeding under a BoarBuster® suspended trap. (Photo by J. Gaskamp. With permission.)

adaptation taking characteristics from drop-nets to deal with trap-wary wild pigs. Developed in the South-Central region, many landowners and trappers use this approach. Integration of human-actuated traps and automated feeders is also common and reduces landowner trips to the trap site.

In the South-Central region, shooting wild pigs with conventional firearms is popular but yields limited success in reducing damage or wild pig populations. However, shooting wild pigs was often the first practice implemented by landowners for control, because it was easy and required no additional investment if they already owned a firearm. Some landowners and hunters utilize semi-automatic rifles and spotlights at night on private lands to improve success. Consequently, this practice resulted in a law enforcement dilemma as to who might be legally spotlighting wild pigs versus illegally spotlighting game animals (e.g., white-tailed deer). Many wild pig control professionals believe that regular night hunting pressure on a property may condition pigs to associate spotlights with control efforts and further limit shooting success. In this region, investment in new technologies like night vision, infrared, and thermal scopes to eliminate spotlights has primarily been motivated by recreational value, not enhanced population control.

Likewise, hunting with assistance of specially trained bay and catch dogs is popular in the South-Central region, but is conducted more for sport or contest (e.g., hunting dog tournaments) than population control. Most wild pigs captured with dogs are hobbled, loaded, and delivered to high-fenced shooting operations or buying stations to be sold. However, it is often difficult to prevent bay dogs from following wild pigs onto neighboring properties and this practice has resulted in trespassing charges.

The Judas pig tagging and control concept is designed for low density populations where an individual pig is equipped with a GPS and/or radio telemetry collar and released to integrate with a sounder (Figure 15.8; see Chapter 8). The manager subsequently removes the sounder associated

FIGURE 15.8 Wild pigs fitted with GPS/telemetry collars can facilitate real-time tracking of sounders (e.g., Judas Technique). Pigs anesthetized for collaring in image (Photo by R. Mattson. With permission.)

with the collared pig and then allows that individual to integrate with another sounder. This technique became legal through the Oklahoma Feral Swine Control Act. However, research by S. L. Webb (unpublished data) noted several limitations of this technique in the South-Central region, including limited opportunity to harvest multiple animals in a single contact with the sounder, better suitability to daytime use, sight limitations associated with dense vegetation used by pigs in daytime, and long travel distances after each contact with the sounder. Collared individuals in the population may be more useful in refining trapping efforts (e.g., trap site location) in this region.

USDA/APHIS/Wildlife Services has long used aerial gunning from helicopters and fixed-wing aircraft to remove predators (e.g., coyotes) in the South-Central region. And, as wild pig populations continued to increase and their range expanded, this method of removal also proved cost effective against wild pigs when compared to trapping and ground shooting operations (see Chapter 8). Gunning from a helicopter with visual spotting assistance from a fixed-wing aircraft at a higher altitude has further increased efficiency of aerial gunning operations in Oklahoma. Of course, this control technique works best in more open rangeland areas of the region that have little or sparse tree cover. USDA/APHIS/Wildlife Services in Texas typically removes 28,000–30,000 wild pigs annually with aerial gunning. Private companies in Oklahoma and Texas also offer aerial gunning services to landowners interested in reducing wild pig populations. Beginning in 2012, Texas legislation dubbed the "Pork Chopper Bill" allowed these private entities to charge individuals for the experience of shooting wild pigs from aircraft. Regardless, both USDA/APHIS/Wildlife Services and private companies can only fly properties where ODAFF or TPWD provide landowner authorization permits.

One study in south Texas shed light on how aerial gunning could move animals without a barrier (fence) and how effective aerial gunning could be in depopulation scenarios. Researchers used GPS collars to collect data before, during, and after aerial gunning in southern Texas. Home range and core area sizes did not differ before and after aerial gunning. Wild pigs moved at a greater rate during the gunning phase, but this minor effect on behavior demonstrated that aerial gunning could be a valuable tool if emergency depopulation were necessary (Campbell et al. 2010).

In 2017, ODAFF contributed to the removal of 32,829 wild pigs; 17,594 by USDA/APHIS/Wildlife Services; 2,742 by private aerial hunting; 6,600 by hunting facilities; 5,289 by buying stations for slaughter, and 604 by conservation districts in corral traps acquired mid-year. Trapping (7,538 pigs) and aerial gunning (6,228 pigs) accounted for 78% of total pigs removed by USDA/APHIS/Wildlife Services in Oklahoma.

In 2003, A Texas Department of Agriculture grant funded a pilot bounty program for wild pigs in Van Zandt County, Texas. Anyone taking a wild pig by legal means within the county was eligible for US$7/animal. In 1 year, participants reported 2,062 wild pigs through the bounty program but the effort discontinued due to lack of funds. It is unlikely that the bounty program actually increased numbers of pigs removed, despite its popularity with local landowners.

15.4.4 Outreach and Education

Resource professionals from Texas and Oklahoma often collaborate on educational programs to inform the public about wild pig biology, management, and control. Educators accomplish this mostly through county presentations and publications given to landowners and more recently through videos, social media, and websites. Educational materials stemmed from research-based information are converted to language for public consumption. From 2006 to 2015, Texas A&M AgriLife Extension Service, with funding from Texas Department of Agriculture, provided relevant, research-based educational information delivered to the public through county extension agents via subject-matter specialist(s) at county, multi-county, regional, and state levels (Table 15.2).

In a separate effort, from 2012 to 2015, AgriLife Extension personnel provided 205 workshops (188 one-hour presentations and 17 four-hour workshops), 1 webinar, 4 AgriLife Extension publications, 25 YouTube videos, 2 interactive websites, and 4 social media outlets reaching 1,270,537

TABLE 15.2
Impacts of Extension Educational Programming in Texas (2006–2015)

Year	Events	# of Participants	Knowledge Increase[a]	# of Practices Adopted[b]	Economic Impact[c]	Net Promoter Score[d]
2006	27	1,995	N/A	N/A	US$919,471	50%
2007	40	3,202	68%	3.2	US$2,059,350	51%
2008	8	760	98%	3.9	US$723,165	56%
2009	11	990	89%	3.8	US$683,530	74%
2010	16	1,561	98%	3.4	US$1,676,281	67%
2011	26	2,551	98%	3.7	US$2,104,919	67%
2012	12	986	98%	3.9	US$774,025	74%
2013	11	1,099	99%	3.7	US$936,819	71%
2014	12	1,179	99%	3.1	US$499,230	69%
2015	11	841	100%	3.0	US$521,656	61%
Totals	174	15,164	–	–	US$10,898,446	–

Benefit to Cost = US$10,898,446/US$413,813 = US$26.34 benefit per US$1.00 invested in education through December 2015.

[a] Measure of the attendee's assessment of their own knowledge before and after the workshop in the areas of biology, control options, effective techniques, and damage.

[b] Measure of BMPs (from a list of 8) that the attendee plans to adopt after leaving the workshop.

[c] Based on estimates that attendees fill in on the survey with background on what is constituted as damage.

[d] Indexes program effectiveness and was measured on the question "How likely are you to recommend us to family, friends, and colleagues?"

citizens around Texas. Workshop attendees estimated knowledge received from the program, once implemented, would result in US$2,178,278 reduction in wild pig damages. Knowledge of wild pig biology, lethal control options, efficient trap/bait techniques and types/extent of pig damage by participants increased by 89%, 83%, 88%, and 74%, respectively. Additionally, participants indicated they planned to adopt an average of 2.5 BMPs/participant. Melding research and education has a strong history of delivering results leading to reduction in damage and numbers of wild pigs.

15.5 REGIONAL WILD PIG RESEARCH IN SOUTH-CENTRAL NORTH AMERICA

Much of the research conducted in the South-Central region has been synthesized and discussed throughout this chapter. Despite improved knowledge, wild pig populations continue to grow at the expense of wildlife populations and producers of agricultural commodities. Thus, a primary focus of current research revolves around methods to control wild pig populations.

Landowners generally acknowledge many reasons for reducing densities of wild pigs. However, they need viable options for wild pig control that reduce time and costs, while decreasing damage and increasing proportion of wild pigs removed from the population. Historically, landowners have employed techniques suggested by agency personnel, but scale of the current problem necessitates novel techniques that will increase effectiveness and longevity of control.

The primary objective of an agricultural commodity producer is often damage reduction to crops, typically accomplished via lethal control. One innovative non-lethal approach demonstrated that, on rangeland, electric fencing resulted in 49% fewer intrusions into bait stations with a 2-strand electric fence. In an agricultural land trial, researchers found 64% less damage to sorghum crops with a 2-strand electric fence compared to no electric fence. These authors suggest that combining electric fencing with lethal methods may be an efficient method for reducing damages to crops (Reidy et al. 2008).

Another example of research leading to novel techniques was an investigation of drop-nets, originally designed for use in trapping cervids, as a potential tool for wild pig control. In a comparison between conventional corral traps and drop-nets, drop-nets (captured 86% of identifiable wild pig populations) outperformed corral traps (captured 49% of identifiable wild pig populations) in a 2-year study in Oklahoma (Gaskamp 2012). The BoarBuster® trapping system, a suspended corral trap (Figure 15.7), was ultimately the result of melding characteristics of both systems with cellular technology.

As the South-Central region supports a high concentration of wild pigs, it is an excellent area to conduct studies focused on techniques for depopulation scenarios should disease outbreaks call for extreme biosecurity measures. For example, an evaluation of fences that could be easily constructed to quickly isolate populations demonstrated fences were 97% successful at holding wild pigs when enclosures were entered by humans for maintenance purposes, 83% effective when pursued by simulated shooting on foot, and in limited testing, 100% successful when pursued and removed by aerial gunners in a helicopter (Lavelle et al. 2011).

In early 2017, Texas became the epicenter of attention with regard to use of toxicants to kill wild pigs. For a brief period, it seemed that Texas would allow use of warfarin, an anticoagulant, as the US Environmental Protection Agency (EPA) approved a warfarin-based toxicant as a general use pesticide and Texas Department of Agriculture made it a state-restricted pesticide (Texas Department of Agriculture 2017). Use of the toxicant met opposition, given ambiguity of the label and concerns of secondary poisoning to wildlife, livestock, and humans. After threat of legal action against Texas Department of Agriculture and the toxicant manufacturer, the manufacturer voluntarily withdrew registration in Texas for the time being. Experimental use of warfarin occurred in Australia to poison wild pigs (Hone 2002). Although effective, some questioned humaneness of death (Institute of Medical and Veterinary Science 2010). Researchers discontinued experimental testing after observing apparent discomfort and pain associated with hemorrhages in internal organs, muscles, and joints, and an extended time-to-unconsciousness (Humaneness Assessment Panel 2009). Evaluation of warfarin baits in Texas (0.01% and 0.005% warfarin) tested a lower dose compared to that previously tested in Australia (0.09% warfarin) and resulted in mortality of 100% and 53% of marked wild pigs, respectively. Though raccoons were able to open feeder doors and access toxic bait (6 of 35 visitations), the authors reported no non-target wildlife fatalities (Poche et al. 2018).

In addition to warfarin, investigations on use of other toxicants have occurred in the South-Central region. In Texas, ongoing studies on use of sodium nitrite as a control measure for wild pigs intended to first understand toxin encapsulation in baits and lethality of the toxin in controlled experiments in captive wild pigs. Wild pig mortality rates reached 95% during pen trials with sodium nitrite bait (HOGGONE®; Animal Control Technologies Australia P/L, Victoria, Australia; Snow et al. 2017a). Researchers also evaluated secondary transfer to non-target species. One study suggested no risk of secondary poisoning for non-target species that consumed muscle, liver, or eyes of wild pigs poisoned with sodium nitrite. Secondary poisoning could occur for common scavengers, though, that consumed large amounts of digestive tract tissues or undigested bait from wild pig carcasses (Snow et al. 2018). To address the concern of non-target poisoning, research has also focused on development of a species-specific bait station (Snow et al. 2017b; Lavelle et al. 2018a, b; see Chapter 9). One study found that wild pigs fed more frequently from plastic compared to metal prototypes, but both feeders reduced feeding by wild pigs compared to control sites where feed was simply placed on the ground (Lavelle et al. 2018a). Optimal baiting strategies to train wild pigs to use the bait station and bait station spacing have also been explored, resulting in nearly 90% access by wild pigs and minimal non-target access (Lavelle et al. 2018b). Eventually, trials moved to larger, operational scales (e.g., ranch-level). These field trials, conducted under an EPA Experimental Use Permit, revealed unanticipated mortality to birds that consumed toxic bait dropped outside of the feeder by wild pigs. Current efforts are improving the palatability of bait to pigs and optimizing design of the bait station to reduce spillage.

15.6 FUTURE OF WILD PIGS IN SOUTH-CENTRAL NORTH AMERICA

As we look to the future, there is a need to involve landowners and managers in research on techniques to improve likelihood of adoption. Wild pig management professionals spend considerable time teaching BMPs, but many landowners and managers may not transfer those practices to the field. More specifically, engaging stakeholders as integral parts of the research, collecting data, and relaying real-life challenges may increase rate of adoption for research-based management and synchronize management efforts on larger scales. Entities like Texas Agrilife Extension, Oklahoma State University, and Noble Research Institute have a history of achieving results and adoption of practices when stakeholders are involved and have immediate access to education and research. There are a few examples of citizen-sponsored research already occurring and their potential outcomes (e.g., BoarBuster® "Recorded Captures"). Researchers and educators can use these data to illustrate distribution and intensity of trapping efforts and identify seasonal variation in trapping success (Figure 15.9), as well as measure trap efficiency (catch per unit effort) and model populations over time. Making these data available to landowners can help inform them on effective management practices in real time. Likewise, placing more attention on education, and parasite and disease issues that affect human wellness, healthy wildlife populations, livestock disease-free

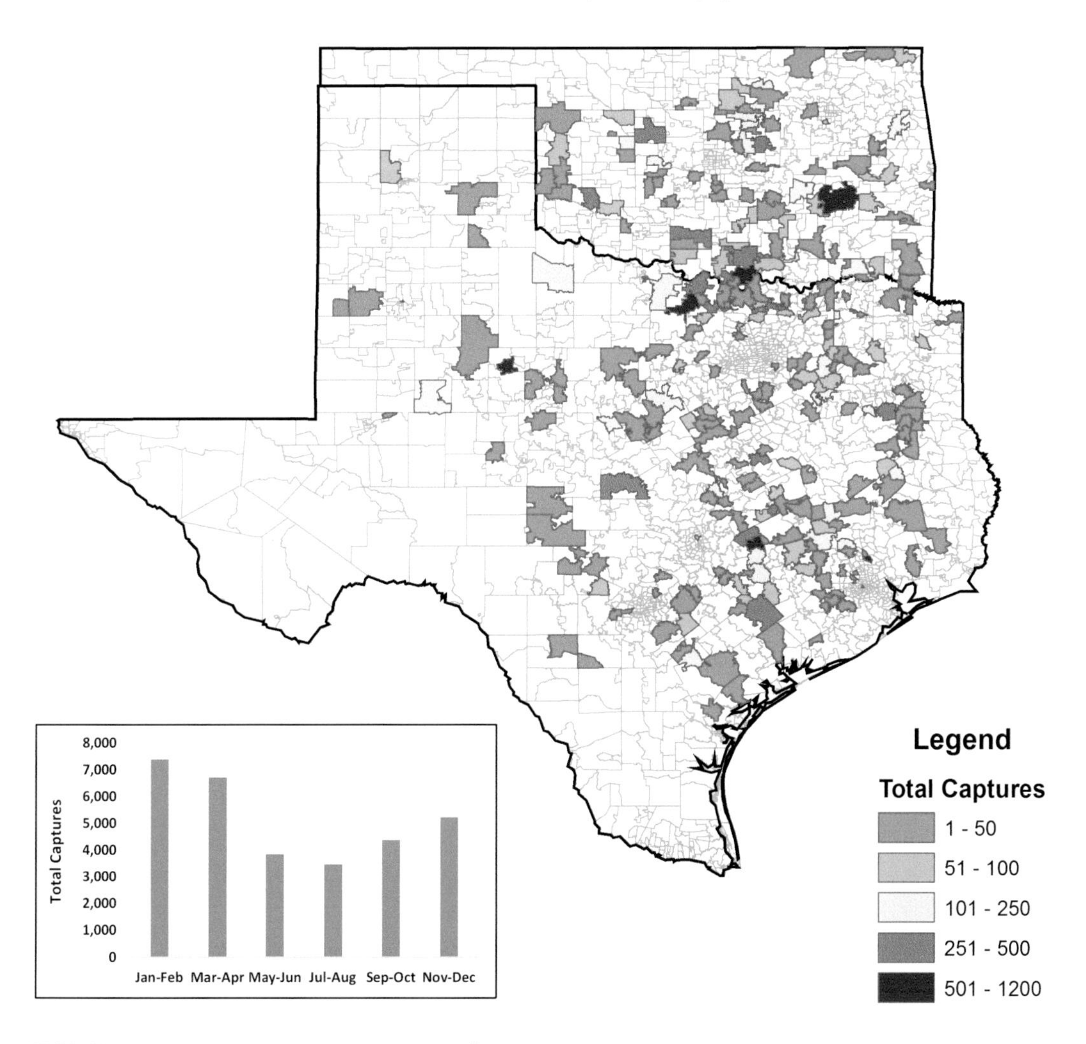

FIGURE 15.9 Capture data from BoarBuster® suspended traps in south-central North America. (Map by D. Payne. With permission.)

status, transportation and policies that surround them would improve wild pig management. Understanding long-term costs of wild pig damage to soils and native ecosystems is also an area needing investigation.

Considering estimates that suggest nearly 50% of the total wild pig population in the United States is found in the South-Central region (Stevens 2010, Timmons et al. 2012), wild pig management will be an enormous undertaking primarily focused on population reduction to reduce damage rather than eradication. Additionally, opposition from numerous stakeholder groups due to a historical tradition of wild pig hunting and incentives derived from extremely limited hunting regulations (e.g., no season, no bag limits), high-fenced shooting operations, financial earning potential associated with meat markets, and lenient legislation regarding wild pigs presents a significant challenge towards management in the region. While perceptions have started to shift as the extent and economic cost of damage caused by wild pigs is increasingly realized and has led to beneficial management advancements such as establishment of the pig free zone in the panhandle of Oklahoma, it is likely wild pigs will persist across most of the region for the foreseeable future.

REFERENCES

Anderson, A., C. Slootmaker, E. Harper, J. Holderieath, and S. A. Shwiff. 2016. Economic estimates of feral swine damage and control in 11 US states. *Crop Protection* 89:89–94.

Barrett, R. H. 1978. The feral hog at Dye Creek Ranch. California. Hilgardia, 46:283–355.

Barron, J. 1980. Vegetation impact by feral hogs: Gulf Islands National Seashore. *Mississippi Proceedings of the Second Conference on Scientific Research in the National Parks* 8:309–318.

Boyer, K. S., S. L. Webb, W. S. Fairbanks, J. A. Gaskamp, and C. Rohla. 2018. Wild pig impacts in pecan operations. Page 31 *in Proceedings of the International* Wild Pig *Conference*. 15–18 April 2018, Oklahoma City, OK.

Caley, P. 1997. Movements, activity patterns and habitat use of feral pigs (*Sus scrofa*) in tropical habitat. *Wildlife Research* 24:77–87.

Campbell, T. A., and D. B. Long. 2009. Feral swine damage and damage management in forested ecosystems. *Forest Ecology and Management* 257:2319–2326.

Campbell, T. A., and D. B. Long. 2010. Activity patterns of wild boars (*Sus scrofa*) in Southern Texas. *The Southwestern Naturalist* 55:564–600.

Campbell, T. A., D. B. Long, and B. R. Leland. 2010. Feral swine behavior relative to aerial gunning in southern Texas. *Journal of Wildlife Management* 74:337–341.

Cooper, S. M., H. M. Scott, G. R. de la Garza, A. L. Deck, and J. C. Cathey. 2010. Distribution and interspecies contact of feral swine and cattle on rangeland in south Texas: Implications for disease transmission. *Journal of Wildlife Disease* 46:152–164.

Corn, J. L., J. C. Cumbee, R. Barfoot, and G. A. Erickson. 2009. Pathogen exposure in feral swine populations geographically associated with high densities of transitional swine premises and commercial swine production. *Journal of Wildlife Diseases* 5:713–721.

Corn, J. L., P. K. Swiderek, B. O. Blackburn, G. A. Erickson, A. B. Thiermann, and V. F. Nettles. 1986. Survey of selected diseases in wild swine in Texas. *Journal of American Veterinary Medicine Association* 189:1029–1032.

Dobie, J. F.. 1929. *A Vaquero of the Brush Country*. The Southwest Press, Dallas, TX.

Fenneman, N. M., and D. W. Johnson. 1946. *Physiographic Divisions of the Conterminous US (Map)*. US Geological Survey. Reston, VA.

Gaskamp, J. A. 2012. Use of Drop-nets for Wild Pig Damage and Disease Abatement. Thesis, Texas A&M University, College Station, TX.

Gaskamp, J. A., K. L. Gee, T. A. Campbell, N. J. Silvy, and S. L. Webb. 2016. Pseudorabies virus and *Brucella abortus* from an expanding wild pig (*Sus scrofa*) population in southern Oklahoma, USA. *Journal of Wildlife Diseases* 52:383–386.

Hall, J. S., R. B. Minnis, T. A. Campbell, S. Barras, R. W. Deyoung, K. Pabilonia, M. L. Avery, et al. 2008. Influenza exposure in United States feral swine populations. *Journal of Wildlife Diseases* 44:362–368.

Harveson, L. A., M. E. Tewes, N. J. Silvy, and J. Rutledge. 2000. Prey use by mountain lions in southern Texas. *Southwestern Naturalist* 45:472–476.

Higginbotham, B. J. 2013. *Wild Pig Damage Abatement Education and Applied Research Activities*. Texas A&M Agrilife Research and Extension Center. Overton, TX.

Higginbotham, B. J., G. Clary, L. W. Hysmith, and M. J. Bodenchuk. 2008. Statewide feral hog abatement pilot project, 2006–2007. *Wildlife Damage Management, Internet Center for National Conference on Feral Hogs.*

Hone, J. 2002. Feral pigs in Namadgi National Park, Australia. *Dynamics, Impacts and Management* 105:231–242.

Humaneness Assessment Panel. 2009. Baiting of feral pigs with warfarin. <https://www. pestsmart.org.au/wp-content/uploads/2012/04/pig_baiting_warfarin.pdf>. Accessed 18 Oct 2018.

Ilse, L. M., and E. C. Hellgren. 1995. Spatial use and group dynamics of sympatric collared peccaries and feral hogs in southern Texas. *Journal of Mammalogy* 76:993–1002.

Institute of Medical and Veterinary Science. 2010. Assessing the humaneness and efficacy of a new feral pig bait in domestic pigs. Report for the Australian Government Department of the Environment, Water, Heritage and the Arts. Canberra, Australia.

Kay, S., J. Fischer, A. Monaghan, J. C. Beasley, R. Boughton, T. Campbell, S. Cooper, et al. 2017. Quantifying drivers of wild pig movement across multiple spatial and temporal scales. *Movement Ecology* 5(14):1–15.

Lavelle, M. J., N. P. Snow, J. M. Halseth, J. C. Kinsey, J. A. Foster, and K. C. VerCauteren. 2018a. Development and evaluation of a bait station for selectively dispensing bait to invasive wild pigs. *Wildlife Society Bulletin* 42:102–110.

Lavelle, M. J., N. P. Snow, J. M. Halseth, E. H. VanNatta, H. N. Sanders, and K. C. VerCauteren. 2018b. Evaluation of movement behaviors to inform toxic baiting strategies for invasive wild pigs (*Sus scrofa*). *Pest Management Science* 74:2504–2510.

Lavelle, M. J., K. C. VerCauteren, T. J. Hefley, G. E. Phillips, S. E. Hygnstrom, D. B. Long, J. W. Fischer, S. R. Swafford, and T. A. Campbell. 2011. Evaluation of fences for containing feral swine under simulated depopulation conditions. *Journal of Wildlife Management* 75:1200–1208.

Mapston, M. E. 2007. *Feral Hogs in Texas*. Agrilife Communications and Marketing, College Station, TX .

Mayer, J. J. 2009. Taxonomy and history of wild pigs in the United States. Pages 5–24 *in* J. J. Mayer and I. L. Brisbin, Jr., editors. *Biology, Damage Control Techniques and Management*. SRNL-RP-2009-00869. Savannah River National Laboratory. Aiken, South Carolina.

Mayer, J. J., and I. L. Brisbin, Jr. 1991. *Wild Pigs in the United States: Their History, Comparative Morphology, and Current Status*. The University of Georgia Press, Athens, GA.

Oliver, W. L. R., and I. L. Brisbin, Jr. 1993. Introduced and feral pigs: Problems, policy, and priorities. Pages 179–191 *in* W. L. R. Oliver, editor. *Pigs, Peccaries, and Hippos: Status Survey and Conservation Action Plan*. IUCN World Conservation Union, Gland, Switzerland.

Pedersen, K., N. E. Bauer, S. Olsen, A. M. Arenas-Gamboa, A. C. Henry, T. D. Sibley, and T. Gidlewski. 2017. Identification of *Brucella* spp. in feral swine at abattoirs in Texas, USA. *Zoonoses Public Health* 64:647–654.

Poche, R. M., D. Poche, G. Franckowiak, D. J. Somers, L. N. Briley, B. Tseveenjav, and L. Polyakova. 2018. Field evaluation of low-dose warfarin baits to control wild pigs (*Sus scrofa*) in North Texas. *PLoS ONE* 13:e0206070.

Reidy, M. M., T. A. Campbell, and D. G. Hewitt. 2008. Evaluation of electric fencing to inhibit feral pig movements. *Journal of Wildlife Management* 72:1012–1018.

Saliki, J. T., S. J. Rodgers, and G. Eskew. 1998. Serosurvey of selected viral and bacterial diseases in wild swine from Oklahoma. *Journal of Diseases* 34:834–838.

Sanders, D. M., A. L. Schuster, P. W. McCardle, O. F. Strey, T. L. Blankenship, and P. D. Teel. 2013. Ixodid ticks associated with feral swine in Texas. *Journal of Vector Ecology* 38:361–373.

Saunders, G., and B. Kay. 1991. Movements of feral pigs (*Sus scrofa*) at Sunny Corner, New South Wales. *Wildlife Research* 18:49–61.

Siemann, E., J. A. Carrillo, C. A. Gabler, R. Zipp, and W. E. Rogers. 2009. Experimental test of the impacts of feral hogs on forest dynamics and processes in the southeastern US. *Forest Ecology and Management* 258:546–553.

Singer, F. J., D. K. Otto, A. R. Tipton, and C. P. Hable. 1981. Home ranges, movements, and habitat use of European wild boar in Tennessee. *Journal of Wildlife Management* 45:343–353.

Snow, N. P., J. A. Foster, E. H. VanNatta, K. E. Horak, S. T. Humphrys, L. D. Staples, D. G. Hewitt, and K. C. VerCauteren. 2018. Potential secondary poisoning risks to non-targets from a sodium nitrite toxic bait for invasive wild pigs. *Pest Management Science* 74:181–188.

Snow, N. P., J. A. Foster, J. C. Kinsey, S. T. Humphrys, L. D. Staples, D. G. Hewitt, and K. C. VerCauteren. 2017a. Development of toxic bait to control invasive wild pigs and reduce damage. *Wildlife Society Bulletin* 41:256–263.

Snow, N. P., M. J. Lavelle, J. M. Halseth, C. R. Blass, J. A. Foster, and K. C. VerCauteren. 2017b. Strength testing of raccoons and invasive wild pigs for a species-specific bait station. *Wildlife Society Bulletin* 41:264–270.

Springer, M. D. 1977. Ecological and economic aspects of wild hogs in Texas. Pages 37–46 *in* G. W. Wood, editor. *Research and Management of Wild Hog Populations*. The Belle W. Baruch Forest Science Institute of Clemson University, Georgetown, South Carolina.

Stevens, R. L. 2010. *The Feral Hog in Oklahoma*. Samuel Roberts Noble Foundation, Ardmore, OK.

Synatzske, D. R. 1979. Status of the feral hog in Texas. Unpublished report, Texas Parks & Wildlife Department.

Taylor, R. B. 1991. The feral hog in Texas. Texas Parks and Wildlife Department Federal Aid Report Series 28, Austin, TX.

Taylor, R. B., and E. C. Hellgren. 1997. Diet of feral hogs in the western South Texas Plains. *Southwestern Naturalist* 42:33–39.

Taylor, R. B., E. C. Hellgren, T. M. Gabor, and L. M. Ilse. 1998. Reproduction of feral pigs in southern Texas. *Journal of Mammalogy* 79:1325–1331.

Texas Department of Agriculture. 2017. Commissioner Sid Miller announces TAC rule change to allow limited use of feral hog toxicant in Texas. News and Events. 21 Feb 2017.

Timmons, J. B., B. Higginbotham, R. Lopez, J. C. Cathey, J. Mellish, J. Griffin, A. Sumrall, and K. Skow. 2012. Feral hog population growth, density and harvest in Texas. Texas A&M AgriLife Extension Special Publication SP-472, College Station, TX.

Tolleson, D., W. Pinchak, D. Rollins, M. Ivy, and A. Heirman. 1994. Effect of habitat type on depredation of simulated quail nests. Page 87 *in Proceedings of the Society for Range Management Meetings*. Colorado Springs, CO.

Towne, C. W., and E. N. Wentworth. 1950. *Pigs from Cave to Corn Belt*. University of Oklahoma Press, Norman, OK.

US Environmental Protection Agency. 2007. Level III and IV ecoregions of EPA Region 6. <https://www.epa.gov/eco-research/ecoregion-download-files-region#pane-06>. Accessed 9 Dec 2018.

Webb, S. L., and K. Boyer. 2018. Wild pigs put pecans at risk, research learns more. *Noble Research Institute News and Views* 36(11):1–3.

Wyckoff, A. C., S. E. Henke, T. A. Campbell, and K. C. VerCauteren. 2006. Is trapping success of feral hogs dependent upon weather conditions? *Proceedings of the Vertebrate Pest Conference* 22:370–372.

Yarrow, G. K., and J. C. Kroll. 1989. Coexistence of white-tailed deer and feral hogs: Management and implications. *Proceeding of the Southeast Deer Study Group Meeting* 12:13–14.

16 Wild Pigs in Southeastern North America

John J. Mayer, James C. Beasley, Raoul K. Boughton, and Stephen S. Ditchkoff

CONTENTS

16.1 INTRODUCTION

The initial presence of wild pigs (*Sus scrofa*) in southeastern North America dates to the early European exploratory period of North America, and their existence in this region has continued through the present (Mayer and Brisbin 2008, 2009). These persistent populations represent some of the oldest continually existing introduced pigs in the continental United States.

Regionally, for the purposes of this chapter, southeastern North America is defined as including the following US states: Alabama, Arkansas, Florida, Georgia, Kentucky, Louisiana, Mississippi,

North Carolina, South Carolina, Tennessee, Virginia, and West Virginia. This encompasses a total land area of 1,468,304 km^2. The climate for most of this region is classified as largely humid subtropical (long summers and short, mild winters) with both the maritime temperate (cool summers and cool but not cold winters) and humid continental (warm to hot summers and cold winters) climates found in the southern Appalachians, and the tropical monsoon climate (monthly mean temperatures above 18°C in every month of the year, with a wet humid hot season (June–October) and cooler dry season (November–May)) found in extreme south Florida (Geiger 1954, Peel et al. 2007). The Southeast region includes portions of the following US physiographic regions: Atlantic Plain (Coastal Plain subregion); Appalachian Highlands (Piedmont province, Blue Ridge province, Valley and Ridge province, and Appalachian Plateaus province subregions); Interior Plains (Interior Low Plateaus subregion); and Interior Highlands (Ozark Plateaus and Ouachita province subregions; US Geological Survey 2003). Elevations range from sea level to 2,037 m above mean sea level. The landscape is occupied by broadleaf deciduous and/or conifer forest, agricultural lands, large river drainage systems, scrub/shrub lands, and swamp/marshlands (Baily 1980).

Because of their longstanding presence in the Southeast region, non-native wild pigs generally have been an accepted feature of the regional landscape for many years. In a number of areas of the Southeast region, these animals have long been considered to be an important recreational resource as a big-game animal and a prolific source of meat for human consumption in rural areas (Mayer and Brisbin 1995, Dickson et al. 2001, Smith et al. 2018). However, concurrent with a recent regional range expansion, and the associated increase in damage these animals cause to both the natural and anthropogenic environments, the opinion has shifted throughout much of the Southeast region to one of wild pigs being considered an unwanted invasive species (Bevins et al. 2014). This recent change of opinion has created polarized viewpoints across the region regarding the presence and management of these animals.

16.2 HISTORIC AND PRESENT DISTRIBUTION

16.2.1 Origin and Early History

Introduced wild pigs have had a very long history in the Southeast region spanning almost 5 centuries. The Hawaiian Archipelago is the only other part of the United States that has experienced a longer continuous presence of these animals (see Chapter 17). Given the size of the area occupied and the longevity of their occurrence, wild pig populations have had a more extensive effect (i.e., both positive and negative) on the environments in the Southeast region than in any other region of North America.

The first introductions of this species in the Southeast region were contemporaneous with the European exploration and colonization of the region in the early to mid-1500s. Many of which were Spanish expeditions that sailed from established colonies in the Caribbean to the mainland and brought livestock with them, including pigs. These pigs were the descendants of domestic stock brought to the New World from Europe, starting with Columbus's second voyage in 1493, and were subsequently dispersed throughout the Spanish colonies in the Caribbean (Mayer and Brisbin 2009; see Chapter 18).

In 1513, Juan Ponce de León led the first official Spanish expedition to the North American mainland, mapping the Atlantic and southern coastline. He returned, to what is now southwest Florida in 1521, to lead the first large-scale attempt to establish a Spanish colony on the mainland which was to serve as a base to further explore the region. On that second voyage, De León's expedition brought with them several species of domestic animals, including pigs. After arrival on the southwest coast of Florida, probably near Charlotte Harbor or Sanibel Island, the expedition offloaded its cargo and supplies and began to build the colony and a fort. After about 4 months and being unable to get the local Calusa tribe to serve the Spanish, De León decided to attack the

Native Americans. However, the Calusa fought back fiercely, killing many of the Spaniards and forcing the expedition to abandon the colony and sail back to the port of Havana, Cuba (Pickett and Pickett 2011). The fate of the expedition's pigs after the expedition departed the Florida coast is unknown.

In 1526, Lucas Vázquez de Ayllón, sailing out of the island of Santo Domingo, led a Spanish expedition that established the first European settlement on the North American mainland. Among the livestock brought along with the expedition were horses, cattle, sheep, and pigs. The expedition landed near what is now Cape Fear in North Carolina. Unable to find a suitable place for a settlement there, the expedition headed south by both land and sea, with the livestock being herded overland along the coast. Ultimately the expedition established the short-lived colony of San Miguel de Gualdape near Winyah Bay on the present-day South Carolina coast. The colony failed after a few months due to disease and starvation, and the survivors made their way back to Hispaniola (Amer 2006, Pickett and Pickett 2011). The horses were abandoned prior to the return trip to the Caribbean (Davis 2004), but the pigs' fate is unknown.

In contrast to the 2 aforementioned expeditions, Hernando de Soto's expedition is credited as the first documented source that both introduced and established pigs on the mainland. Following the examples of successful conquistadors Hernán Cortéz and Francisco Pizarro, De Soto planned to take a herd of pigs along on his journey to "La Florida" in order to supply his expeditionary force and any settlements that might be established along the group's route with meat. De Soto's expedition was large, consisting of 9 ships, over 600 men and women, 253 horses, and a great store of provisions. Among those important provisions were pigs, the original number of which is variously reported as "thirteen" up to "many" (Clayton et al. 1993, Mayer and Brisbin 2009). Whatever the number, these pigs increased to a reported total of 700. Over 3 years, De Soto and his expeditionary force traveled through what are now 14 states. Along the 4,988-km journey, pigs variously escaped into the wild and were either given to or stolen by Native Americans encountered by the Spaniards. Many of these animals were maintained in Native American villages as marked, free-ranging stock and apparently multiplied. On the return trip, the expedition found a sow that had been lost on the outward journey, and now had 13 piglets, each with notches on their ears, which apparently had been made by their Native American owners (Clayton et al. 1993).

De Soto was followed by many other European explorers and colonists that also brought pigs to the southern United States (e.g., Pedro Menéndez de Avilés, Pierre Le Moyne d'Iberville). Escaped pigs from these various expeditions and settlements went feral/wild and rapidly became established in a variety of areas. These pigs proved to be a favorite game animal for Native American hunters (Crosby 1972). In the mid-1560s, 2 short-lived French colonies in eastern Florida attributed the fact that they did not starve to the pork that was provided to them by the local Native Americans who were hunting the already established wild pigs in that area (Mayer and Brisbin 2008).

Domestic pigs were widely used in the European colonies throughout the Southeast region. The colonists, following a longstanding European husbandry practice for raising large livestock, turned these pigs out onto open range to forage for themselves, often along with cattle. However, unlike in Europe, manpower was in short supply in the colonies, so these animals often had to fend for themselves as well. When released into free-range conditions such as these, pigs will become feral/wild very quickly. During this period, these free-ranging or semi-wild pigs were typically considered to be private property in most locations, belonging to the landowner on whose land they were present. This also began the era of specific patterns of ear notching to establish ownership of free-ranging pigs. These animals were considered to be part of the landscape, with only minimal damage being reported. By the early 1900s, populations of wild pigs were widespread in the Southeast region (Mayer and Brisbin 2009).

In 1912 pure Eurasian wild boar were imported from Europe to the Southeast region. George Gordon Moore, an English financial advisor, was allowed to establish a game preserve on Hooper Bald in Graham County, North Carolina, as partial payment for his efforts in a business venture. Among the animals Moore released into 2 fenced enclosures on Hooper Bald were 13 wild boar he had purchased from a European animal dealer. From the beginning, the wild boar were able to root out of their enclosure, which was made of split-rail fencing. However, most chose to remain within the fenced preserve. After a period of about 10 years, a large organized hunt caused most of the remaining animals to break through the rail fencing and escape into the surrounding hills. These animals extended their range for a considerable distance, and even spread into Tennessee. Both wild pigs and free-ranging domestic swine were present in the area at that time, and crossbreeding occurred. All degrees of intergrades between wild boar and domestic pigs have been found in that population since then. This specific introduction had the greatest single impact on the composition of wild pigs in the United States. Numerous subsequent secondary introductions of these wild pigs from Hooper Bald were made into other locations in the Southeast region, including in Florida, Georgia, Louisiana, Mississippi, South Carolina, and Tennessee (Figure 16.1). In the mid-1920s, live-trapped wild pigs from Hooper Bald were also brought to California and released onto private ranchland in Monterey County (Mayer and Brisbin 2008, 2009).

By the late 1970s, all of the states in the Southeast region had established populations of wild pigs, with both the range and numbers only increasing minimally until 1990. Beginning in the last decade of the 20th century, all of the southern states experienced a significant increase in these animals. A major cause promoting this increase is thought to be human facilitated activities such as illegal translocations and escapees from high-fenced shooting operations, although some natural range expansion has occurred (Mayer and Beasley 2018).

FIGURE 16.1 Illustration of the subsequent introductions of wild pigs from the Hooper Bald introduction (black circle) to other locations in the United States. The dates and locations of these releases are: A) 1925–1926, Monterey County, CA; B) mid-1950s, late 1970s, Palm Beach County, FL; C) mid-1950s, Polk and Highlands Counties, FL; D) 1957, 1970, Fresno County, CA; E) early 1960s, Washington, Greene, and Unicoi Counties, TN; F) 1960s, Shasta County, CA; G) 1961, 1965, San Benito County, CA; H) 1961, Eglin Air Force Base, FL; I) 1961, Mendocino County, CA; J) 1962, 1965, 1966, Cumberland County, TN; K) 1968, Aiken County, SC; L) 1968, Tehama County, CA; M) late 1960s, Copiah County, MS; N) early 1970s, Richland and Calhoun Counties, SC; O) 1971, Boone County, WV; P) 1971, Madison County, MS; Q) 1973, near Port Gibson, MS; R) 1975, Houston County, GA; S) 1978, Allendale County, SC; and T) 1979, Lauderdale County, TN. Other subsequent translocations have also taken place from a number of these sites. Data from Mayer and Brisbin 2008. (Reproduced from Mayer and Brisbin 2009.)

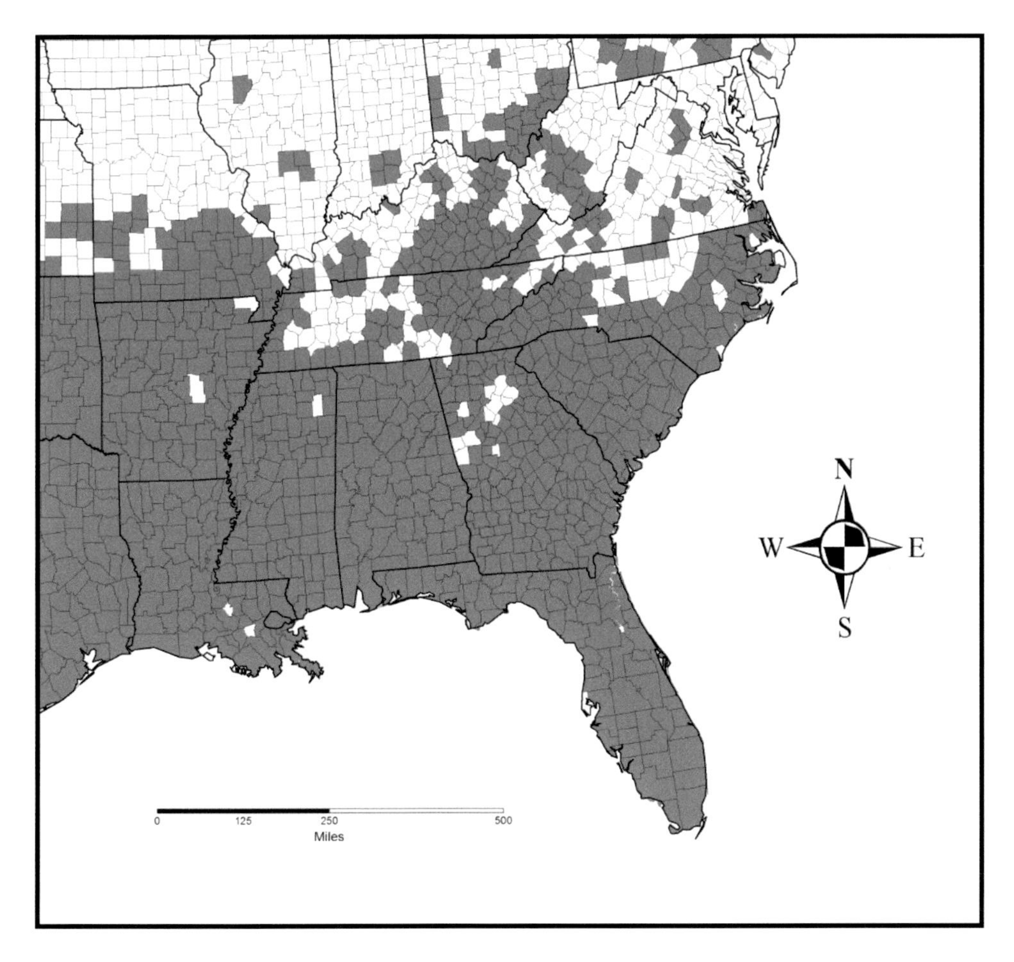

FIGURE 16.2 2015 county-by-county distribution of wild pigs in southeastern North America. (Courtesy of J. Corn. With permission.)

16.2.2 Present Status

At present, populations of wild pigs are found in every state in the Southeast region (Figure 16.2). The current occurrence varies from rare to abundant in localized areas. The area occupied in 2015 in the Southeast region encompassed 531,550 km^2 or 36% of the total land area in these 12 states (J. L. Corn, Southeastern Cooperative Wildlife Disease Study, College of Veterinary Medicine, University of Georgia, personal communication). The total estimated population in this area is about 3 million wild pigs or approximately 50% of the total US population (Mayer 2014). The current regional taxonomic status of wild pigs in the Southeast region is variable and includes populations with either only feral pigs (i.e., wild pigs solely of domestic ancestry) or wild boar x feral pig hybrids (J. J. Mayer, Savannah River National Laboratory, unpublished data).

16.3 REGIONAL ENVIRONMENTAL ASPECTS IN SOUTHEASTERN NORTH AMERICA

16.3.1 Morphology

Morphology of wild pigs in the Southeast region exhibits the same overall variation seen elsewhere in North America (Figure 16.3). One unique regional exception is the insular dwarfing documented

FIGURE 16.3 Wild pigs from the Savannah River Site in South Carolina, illustrating the variability in phenotypes present within this population. (Photos by J. Beasley. With permission.)

for the wild pigs on Ossabaw Island, Georgia, which exhibit some of the smallest adult total body mass (i.e., intact body weight including entrails) reported for the continent at 40 kg (Brisbin et al. 1977a, Mayer and Brisbin 2008). In addition, on the opposite end of the body mass spectrum, all of the exceptionally large males reported in the news media and online to weigh in excess of 350 kg were found in the Southeast region (e.g., "Hogzilla" - Georgia; "Hog Kong" - Florida; "Son of Hogzilla/Brooks Beast" - Georgia; "Monster Pig" - Alabama). Lastly, of the 8 states where syndactylous hooves and neck wattles have been reported (see Chapter 2), 6 are in the Southeast region (i.e., Arkansas, Florida, Georgia, Louisiana, Mississippi, South Carolina). More significantly, wild pig populations that exhibit both of these unique physical characteristics are only found in the Southeast region (Mayer and Brisbin 2008, 2009).

16.3.2 Behavior

The home range size of wild pigs in the Southeast region is variable, with typical home ranges covering an area of about 4–5 km^2 (Table 16.1). On average, males have larger home ranges (i.e., 5–6 km^2) compared to females (i.e., 3–4 km^2). However, pigs occupying spatially (e.g., mosaics of natural and agricultural lands) and seasonally dynamic landscapes may demonstrate much larger home ranges due to the heterogeneous arrangement of resources (Paolini et al. 2019). Periodic irregular shifts in home range occur within this species (Wood and Brenneman 1977). Kurz (1971) noted boars make excursions beyond their established home range boundaries and then return; sows typically do not exhibit this behavior. Sparklin et al. (2009) found family groups (1 or more adult sows and their young) used well-defined core areas within the social unit's home range. These core areas were in dense thickets in both pine uplands and hardwood bottoms. In the Southeast region, like elsewhere, wild pigs will utilize the path of least resistance, and travel trails made by other large mammals (e.g., white-tailed deer (*Odocoileus virginianus*), range cattle) as well as human hiking trails, foot paths, and unpaved secondary roads (Graves 1984, Mayer and Brisbin 2009, Boughton et al. 2019).

Territorial behavior in wild pigs in the Southeast region is reportedly variable. Kurz (1971) reported extensive range overlap among mature males in his study area. Similarly, spatial overlap

TABLE 16.1
Home Range Sizes Reported for Wild Pigs in Southeastern North America

Location	Home Range Size in km² (Sample Size)		Reference(s)
	Female	Male	
Congaree National Park, SC	1.9 (9)	2.2 (7)	Friebel and Jodice (2009)
Fort Benning, GA	2.84 (18)		Sparklin et al. (2009)
Great Smoky Mountains National Park, NC/TN	2.7–3.5 (4)	3.8–3.9 (9)	Singer et al. (1981)
Hobcaw Barony, SC	1.8 (3)	2.3 (3)	Wood and Brenneman (1980)
Lowndes County Wildlife Management Area, AL	4.0 (6)	4.0 (5)	Gaston et al. (2008)
Savannah River Site, SC	7.9 (4)	14.0 (3)	Crouch (1983)
Savannah River Site, SC	5.7 (5)	13.8 (6)	Hughes (1985)
Savannah River Site, SC	4.4 (1)	5.3 (4)	Kurz and Marchinton (1972)
Savannah River Site, SC	5.8 (24)	14.0 (1)	Kay et al. (2017)
Savannah River Site, SC	2.5 (2)	4.0 (2)	Mayer and Brisbin (2009)
Buck Island Ranch, FL	1.6 (9)	3.4 (9)	R.K Boughton, University of Florida, unpublished data

Source: Adapted from Mayer and Brisbin (2009).

in sounder home ranges has been observed across landscapes composed of mosaics of upland and bottomland vegetation types in South Carolina, although sounders rarely appear to use areas of overlap simultaneously (J. C. Beasley, Savannah River Ecology Laboratory, University of Georgia, unpublished data). In contrast, Conley et al. (1972) noted anecdotal evidence to support territorial behavior among large males. Sparklin et al. (2009) found family groups had nearly exclusive home ranges and had completely exclusive core areas. One of these groups even usurped and maintained exclusive control of a high-quality food resource (i.e., sunflower (*Helianthus* sp.) field), which had been used by a different group before the dominant adult females in it were removed.

Both Kurz and Marchinton (1972) and Wood and Brenneman (1977) reported wild pigs in the Southeast region were largely diurnal. However, intense hunting pressure or human activity during the day will drive pigs to become more nocturnal in their activity patterns (Stegeman 1938, Hanson and Karstad 1959). Conley et al. (1972) and Singer et al. (1981) both reported wild pigs in the southern Appalachians (i.e., an area with increased hunting pressure on this species) were more active nocturnally on a year-round basis. Kurz (1971) reported wild pigs are typically diurnal during the fall, winter, and spring, with activity peaks in early morning and late afternoon and a reduction at midday. During summer, diurnal activity is reduced and nocturnal activity is increased. This shift is likely a behavioral means of controlling their body temperature and is more evident in males than in females (Crouch 1983, Hughes 1985). Daily activity patterns also vary between sexes. Sows maintain a relatively constant activity for prolonged periods, while boars exhibit brief bursts of movement followed by lengthy periods of relative inactivity (Crouch 1983). The activity pattern of sows may vary from the norm during farrowing. Kurz (1971) stated limited data suggested sows significantly reduced their daily movements approximately one month before farrowing. In contrast to this finding, Crouch (1983) found no evidence sows reduced their daily movements during their farrowing period.

Wild pigs in the Southeast region appear to maintain home ranges but exhibit variable use of areas within their home range throughout the year. This spatial heterogeneity in home range use likely reflects food availability, population density, reproductive activity, quality and interspersion of habitat, season, climatic conditions, disturbance by humans, and social organization (Conley et al. 1972, Kurz and Marchinton 1972, Crouch 1983, Hughes 1985). For example, during poor mast years, wild pigs make extensive movements in search of alternative forage resources, with mean hourly movement increasing 5-fold (Singer et al. 1981). Intersexual differences in daily movements are apparent, with males being consistently more mobile than sows (Kurz and Marchinton 1972, Crouch 1983, Hughes 1985, Kay et al. 2017).

The social organization of wild pigs in the Southeast region is consistent with that of other populations, with the basic social unit being the sow and her litter (i.e., the family group; Figure 16.4) and sexually mature males usually being solitary (Kurz 1971, Hughes 1985, Sweeney et al. 2003). Reported wild pig group size (i.e., 2 or more animals) varies from 2 to 22 and averages 2–6 individuals (Duncan 1974, Shaffer 1976, Baber 1977, Mayer and Brisbin 2012).

Symbiotic relationships between wild pigs and 2 species of birds (i.e., Florida scrub jays (*Aphelocoma coerulescens*), common crows (*Corvus brachyrhynchos*)) in Florida have been reported whereby the birds physically forage on or groom wild pigs for ectoparasites. This behavior, which has not been reported from anywhere else in North America, has involved wild pigs being groomed while either standing or lying down. In some instances, standing wild pigs were moving about foraging with the birds "riding" on their backs. Some pigs even appeared to solicit such grooming, by walking over to the birds and lying down on their sides and waiting for the birds to begin grooming (Baber and Morris 1980, Kilham 1982).

16.3.3 Diseases and Parasites

Wild pigs in the Southeast region are susceptible to the same suite of pathogens and parasites as other regions of North America (Smith et al. 1982, Davidson and Nettles 1988; see Chapter 5). As elsewhere, both brucellosis (*Brucella suis*) and pseudorabies virus (PRV or Aujesky's Disease; Suid herpesvirus 1) are the 2 primary pathogens of concern in this region (Nettles 1989, Payeur 1989, Davis 1993). Brucellosis was first reported in wild pigs in the continental United States in South Carolina by Wood et al. (1976). Based on both confirmed infections and serological evidence, it has since been reported to also exist in wild pigs in the rest of the states in the Southeast region except for Kentucky (Zygmont et al. 1982, Nettles 1989, Pederson et al. 2012, Leiser et al. 2013). Human

FIGURE 16.4 The basic wild pig social unit (i.e., a sow and her offspring or a family group) in southeastern North America as seen here in central Florida. (Photo by Into Nature Films. With permission.)

cases of brucellosis often occur during improper exposure to wild pigs (e.g., hunters gutting animals without personal protective equipment such as plastic/surgical gloves), especially in Florida (Centers for Disease Control 2009). In 1974–1975, 22% of the human cases of brucellosis in Florida were from contact with wild pigs (Bigler et al. 1977).

Wild pigs may also play an important role in the transmission of diseases to wildlife. Eckert et al. (2019) observed extensive use of wild pig wallows by 12 native mammalian and avian species, of which 9 were observed drinking from wallows; although the role of wallows as a potential mechanism of disease transfer remains unknown. A specific regional concern with PRV is the transmission of this largely fatal pathogen to the endangered Florida panther (*Puma concolor coryi*; Glass et al. 1994). Although considered exotic to the United States, an endemic presence of vesicular stomatitis (VS) has been found and studied on Ossabaw Island, Georgia (Stallknecht et al. 1985). In the Southeast region, prior to the 1960s, VS was endemic with reported outbreaks in livestock occurring annually. Because of this, it has been suggested wild pigs represent a bridge between wild and farm environments (Figure 16.5) and it may not be coincidence the cessation of VS outbreaks corresponded to the elimination of free-ranging practices of domestic pigs in this region (Stallknecht et al. 1993).

Of 32 species of endo- and ectoparasites collected from wild pigs in the Southeast region, high prevalence rates were found for *Globocephalus urosubulatus* (78%), *Gongylonema pulchrum* (57%), *Metastrongylus apri* (52%), *Physocephalus sexalatus* (59%), *Stephanurus dentatus* (72%), and *Haematopinus suis* (64%; Smith et al. 1982). Particular interest in the Southeast region has been documentation of wild pigs' ability as a host to ticks, with *Amblyomma maculatum* and *Dermacentor variabilis* found extensively and *Amblyomma auricularium* and *Ixodes scapularis* uncommonly, in 2 different studies (Greiner et al. 1984, Merrill et al. 2018).

16.3.4 Food Habits

As determined elsewhere, wild pigs in the Southeast region are omnivorous and generally opportunistic in their food habits. Based on 6 studies conducted in the region (Henry and Conley 1972, Scott and Pelton 1975, Ackerman et al. 1978, Baron 1979, Howe et al. 1981, Wood and Roark 1980), the approximate percent composition of the wild pig diet is as follows: 90% plant material, 5% animal material, 4% fungi, and 1% other. The category "other" variously included miscellaneous debris, edible and inedible garbage, rocks/gravel, and soil/sand. Of the various plant materials consumed, mast (e.g., acorns, beechnuts, chestnuts, hickory nuts) appears to be most important and preferred

FIGURE 16.5 Comingled range cattle and a wild pig (ear-tagged for research purposes) in central Florida. (Photo by Into Nature Films. With permission.)

(Henry and Conley 1972). This is evidenced by both the preponderant usage of this food when available and the effect that abundance of this resource has on reproductive success within a population (Matschke 1964). Genetic analyses of wild pig feces collected over a full year in a rangeland setting, suggest an extremely broad choice of items consumed, with 66 plant, 68 animal, and 12 fungal species identified (Anderson et al. In Press). Concurring with stomach content studies, plant matter is usually the predominant food consumed with acorns, grasses, and many wetland species being recorded. Animals were encountered less frequently but from a broad taxonomic breadth, except for the exotic earthworm (*Pontoscolex corethurus*) which was detected in 80% of samples throughout the year, and mole crickets in 30% of samples. A variety of vertebrate DNA was present including fish, amphibian, reptiles, and mammals; surprisingly no avian species were documented (Anderson et al. In Press). Wild pigs also are efficient scavengers and routinely consume carrion when available (Turner et al. 2017).

Scott and Pelton (1975) and Wood and Roark (1980) found no difference in diet with respect to either sex or age of wild pigs analyzed. Roots also become important during fall and winter when mast crops fail (Scott and Pelton 1975, Wood and Roark 1980). In general, wild pigs tend to prefer fleshy roots or corms to woody roots (Howe et al. 1981). Consistent with food habits studies of this species elsewhere, more invertebrate than vertebrate remains are found in the stomachs of wild pigs in the Southeast region (Henry and Conley 1972, Scott and Pelton 1975). Based on estimates made using acid-insoluble ash content of wild pig scats collected in South Carolina, Beyer et al. (1994) estimated that 2.3% of the dietary volume consisted of soil.

16.3.5 Habitat Use

Wild pigs occupy and exploit a wide variety of plant communities in the Southeast region. Throughout this region, wild pigs show a documented preference for the use of riparian and wetland ecosystems (e.g., forested swamps, bottomland hardwoods, emergent marshes, sloughs, stream/river drainage corridors) as well as mesic upland hardwood forest (Kurz 1971, Graves 1984, Friebel and Jodice 2009, Sparklin 2009). Other plant communities (e.g., pine/conifer forest, shrub/scrub thickets, meadows/grassed fields) are utilized, but typically only on a temporal basis. Wild pigs have also begun to exploit developed landscapes (i.e., suburban and urban areas) in the Southeast region (see Chapter 19). Wild pigs regularly modify their space use within their home range based on climatic conditions and the availability of food and water (Kurz and Marchinton 1972, Wood and Brenneman 1980). Abrupt or seasonal changes in plant community use by wild pigs in an area are often related to food availability and dietary shifts (Sweeney 1970, Kurz and Marchinton 1972, Graves and Graves 1977). In addition, wild pigs have been reported to readily adjust to plant community changes caused by fire, logging, and natural catastrophes, except for those that result in a loss of mast (US Department of Agriculture 1981).

16.3.6 Population Biology

Based on 15 studies conducted in the Southeast region (Table 16.2), the overall postnatal male-to-female sex ratio does not typically differ from parity (i.e., 1:1). Belden and Frankenberger (1989) reported that in hunted populations, younger age classes were male biased while the adult age class was female biased. These authors also reported that in unhunted populations, the adult sex ratio was either 1:1 or slightly biased towards males. Belden and Frankenberger (1990) hypothesized such a slight male bias in the adult age class could be due to mortality in mature females resulting from pregnancy or farrowing complications. The reported population age composition in this region is 0–1 year—56%, 1–2 years—22%, and 2+ years—22% (Henry and Conley 1978, Singer and Stoneburner 1979, Mayer and Brisbin 2012). Densities of Southeastern wild pig populations average about 8 animals/km^2 and range from 0.4 to 39 pigs/km^2 (Hanson and Karstad 1959, Kight 1962, Jenkins and Provost 1964, Sweeney 1970, Kurz and Marchinton 1972, Singer and Stoneburner 1979,

TABLE 16.2
Postnatal Sex Composition of Hunter-Harvested and/or Trapped Animals Reported from Wild Pig Populations in Southeastern North America

	Percent Composition (Actual Sample Size)			
Location	**Males**	**Females**	**Difference from Parity**	**Reference(s)**
Merritt Island National Wildlife Refuge, FL	51 (90)	49 (86)	NS[a]	Baber (1977)
Brunswick Study Area, Levy County, FL	46 (56)	54 (65)	NS	Belden and Frankenberger (1979)
Fisheating Creek Wildlife Refuge, FL	55 (51)	45 (42)	NS	Belden and Frankenberger (1990)
Fisheating Creek Wildlife Refuge, FL	50 (96)	50 (96)	NS	Belden et al. (1985)
Great Smoky Mountains National Park, NC, TN	54 (89)	46 (77)	NS	Duncan (1974)
Great Smoky Mountains National Park, NC/TN	53 (276)	47 (246)	NS	Fox and Pelton (1977)
Savannah River Site, SC	52 (6,433)	48 (6,026)	<0.01	Mayer and Brisbin (2012)
Great Smoky Mountains National Park, NC/TN	44 (1,102)	56 (1,391)	<0.05	Peine and Farmer (1990)
Big Cypress National Preserve, FL	53 (258)	47 (230)	NS	Schortemeyer and McCown (1988)
Great Smoky Mountains National Park, NC/TN	54 (298)	46 (252)	<0.05	Singer and Ackerman (1981)
Cumberland Island, GA	49 (290)	51 (300)	NS	Singer and Stoneburner (1979)
Fort Benning, GA	51 (93)	49 (89)	NS	Sparklin (2009)
Savannah River Plant, SC	53 (41)	47 (36)	NS	Sweeney (1970)
Great Smoky Mountains National Park, NC/TN	53 (772)	47 (687)	<0.05	Tate (1984)
Hobcaw Barony, SC	53 (125)	47 (110)	NS	Wood and Brenneman (1977)
Buck Island Ranch, FL	48 (262)	52 (281)	NS	R.K. Boughton, University of Florida, unpublished data

[a] Not significant

Note: Each paired sex combination was tested for the difference from parity.

Source: Adapted from Mayer and Brisbin (2009).

Singer 1981, Crouch 1983, Tate 1984, Law Environmental 1988, Mayer 2005, Hanson et al. 2008, Keiter et al. 2017a). Based on computer modeling of a wild pig population over a 10-year period, Tipton (1977) reported rates of increase as r=0.30/year and λ=1.35/year. The mean lifespan for wild pigs in the Southeast region is reported to range from 8.8 to 19.6 months (Hanson 2006, Mayer and Brisbin 2012), with peak mortality typically coinciding with post-farrowing periods (Henry and Conley 1978). The maximum lifespan was estimated to be 10 years old (Henry and Conley 1978, Mayer and Brisbin 2012). Reported sources of mortality in this region include starvation, predation, disease, hunting, accidents/injuries, and vehicle collisions (Conley et al. 1972, Wood and Brenneman 1977, Mayer and Johns 2007, Mayer and Brisbin 2012, Beasley et al. 2013).

16.3.7 Reproductive Biology

Breeding occurs year-round in the Southeast region, with pregnant sows and litters being encountered during all months of the year. Conception occurs mostly in the late summer and early fall months. The primary farrowing peak is in winter, with a secondary peak in spring (Henry 1966, Sweeney et al. 1979, Belden and Frankenberger 1989, Comer and Mayer 2009, Ditchkoff et al. 2012).

Of these, summer-born pigs appear to have the best chance of survival (Belden and Frankenberger 1989). Within 24 hours prior to giving birth to their offspring, late-term pregnant sows construct farrowing nests, which were characterized for a location in western South Carolina by Mayer et al. (2002). Reported ovulation rates in southeastern wild pig populations range from 5.6 to 10.8 (Duncan 1974, Sweeney et al. 1979), with prenatal losses averaging 25.9% (Conley et al. 1972). Reported fetal litter sizes average from 3 to 8.4 and range from 1 to 16 (Hanson and Karstad 1959, Henry 1966, Conley et al. 1972, Duncan 1974, Wood and Brenneman 1977, Sweeney 1979, Comer and Mayer 2009, Ditchkoff et al. 2012). Fetal sex ratios have not differed from 1:1 (Conley et al. 1972, Comer and Mayer 2009). Females >1 year old were found to be more fecund than females <1 year old (Belden and Frankenberger 1989, Comer and Mayer 2009, Mayer and Brisbin 2012).

16.3.8 Damage

The range of damage caused by wild pigs in the Southeast region is consistent with that observed in other regions of North America (see Chapter 7). The 2 most significant areas impacted include agricultural and forestry/timber resources; however, wild pig damage is realized throughout both anthropogenic and natural environments, especially wetlands and marshes (Barrios-Garcia and Ballari 2012, Barrios-Garcia et al. 2014, Boughton and Boughton 2014, Keiter and Beasley 2017, Mayer and Beasley 2018).

Within the agricultural industry, the most important crops impacted by wild pigs in the Southeast include corn, soybeans, hay/pasture, cotton, wheat, oats, pecans, rice, peanuts, sugarcane, and sorghum; specific crop impacts vary by state (Tanger et al. 2015, Anderson et al. 2016, Mengak 2016). Other types of regional impacts in this industry include damage to stored commodities, livestock, farm equipment, fences, and stock ponds/tanks (Tanger et al. 2015, Mengak 2016). Annual economic losses to wild pig damage in the agricultural industry in individual Southeastern states are as high as US$75M (Y. Zhang, Auburn University, unpublished data). Pasture production losses in managed rangeland in central Florida alone exceeds US$2M/year (Bankovich et al. 2016).

Similarly, wild pig damage to the forestry industry includes girdling of mature pine trees (*Pinus* spp.) through rubbing, damage to the lateral roots of pine trees by rooting and chewing, and damage to the bark of trees by tusking (i.e., scent marking with the tusk glands; Conley et al. 1972, Lucas 1977). However, the most widespread and costly type of forest resource damage done by wild pigs is the depredation of planted pine seedlings. As far back as the 1950s when the Southeast region was being replanted in pine, plantings of longleaf pine (*Pinus palustris*) seedlings were documented to receive significant damage due to foraging by wild pigs (Lipscomb 1989). Wild pigs are attracted to the starchy bark of the root of longleaf seedlings, and an individual pig can destroy 200–1,000 planted longleaf seedlings in a single day (Wakeley 1954). Wood and Roark (1980) observed wild pigs typically do not completely ingest the seedling roots, but rather chew the root stock, swallow the sap and starches, and then spit out the chewed woody tissues before moving on to the next seedling. Such foraging can also include the roots of seedling slash pine (*P. elliotti*), loblolly pine (*P. taeda*), and pitch pine (*P. rigida*), although this is relatively uncommon. Wild pig damage has also been reported to hardwood/deciduous tree species. Similar to pines, wild pigs also dig up and chew/consume the root stock of certain planted hardwood seedlings (Mayer et al. 2000).

Wild pigs also impact natural areas in the Southeast region through foraging (i.e., primarily rooting; Figure 16.6), wallowing, and defecating/urinating (Keiter and Beasley 2017). Rooting by wild pigs modifies the chemistry and nutrient cycling within the soil column. It mixes the A1 and A2 soil horizons and reduces ground vegetative cover and leaf litter. Wild pig rooting in the Great Smoky Mountains National Park (GSMNP) was also found to accelerate leaching of Ca, P, Zn, Cu, and Mg from the leaf litter and upper soil horizons (Singer et al. 1984). Wild pig rooting can also severely impact aquatic and wetland habitats (Arrington et al. 1999). The destabilization of surface soils by wild pig rooting on sloped areas (e.g., along earthen dams, flood-control levies, stream banks) can result in erosion and subsequent down-gradient sedimentation. Runoff and silt-loading into

FIGURE 16.6 Wild pig rooting damage in different habitats in central Florida. (Photo by Into Nature Films. With permission.)

both standing and flowing aquatic systems can greatly degrade these habitats. Zengel and Conner (2008) reported rooting disturbance in small linear cypress-tupelo sloughs and along small creeks in Congaree National Park frequently approached 80–100%, especially during the driest months. Rooting combined with both wallowing and defecating can threaten aquatic species (e.g., endemic populations of brook trout (*Salvelinus fontinalis*) in the southern Appalachians) through siltation and contamination of stream water quality (Singer 1976, Ackerman et al. 1978, Howe et al. 1979). Kaller and Kelso (2006) found such activity increased microbial pathogens while causing a reduction in certain invertebrate communities in coastal plain streams. Because wetlands, marshes, and seepage areas attract pigs as a resource for water, wallowing, and food, they are often heavily impacted areas. Vegetation damage through rooting of mesic flatwoods dominated by slash pine and palmetto (*Seronoa repens*) was estimated as high as 6.4% of land area across 3 Floridian state parks using expenditure data from permitted wetland mitigation projects completed in the United

States (approximately US$4,600–43,300/hectare) depending upon the season and type of wetland habitat damaged (Engeman et al. 2003). Damage by wild pigs can be extensive, with 19% of exposed basin marsh and 25% of seepage slopes reported as damaged in previous studies (Engeman et al. 2004a, 2007). In the GSMNP, 49% of montane seeps and 54% of drainages had wild pig damage, with disturbance of some areas as high as 96%. Plant coverage and salamander surface density both significantly decreased under rooting disturbance (Rossell et al. 2016). The ecotones from wetland edges to upland forests are often where rooting damage is observed.

Both understory and ground-story ecosystems can be severely impacted through wild pig rooting. This is primarily due to the alteration and destructive modification this activity has on these near-surface ecosystems. For example, in intensively rooted stands on the GSMNP, 67% of all branches and logs >2.5 cm in diameter were moved by wild pigs and another 10% broken apart (Singer et al. 1984). Wild pig rooting exposed 1,400–2,800 tree roots/hectare, and 4% of exposed roots were broken (Singer et al. 1984). Bratton (1975) found wild pig rooting reduced species richness in the herbaceous understory in the GSMNP. The potential effects of rooting were found to vary among plant community types. In general, however, it was found to result in decreased plant cover, reduction in litter-layer mass, and decreased food resources for macroinvertebrates (Arrington et al. 1999).

As an omnivore, wild pigs can impact both native plant and animal species in the Southeast region. For example, a Turk's cap lily (*Lilium superbum*) population in the GSMNP was completely destroyed by wild pigs. In fact, about 1/3 of the wildflowers listed as occurring in that park have been observed to be either uprooted or eaten by wild pigs (Bratton 1974). Being an introduced species, one major concern with wild pigs has been competition with native species for available resources. Primarily associated with dietary overlap of mast, wild pigs are typically considered to compete with white-tailed deer, wild turkey (*Meleagris gallopavo*), black bear (*Ursus americanus*), raccoon (*Procyon lotor*), squirrels (*Sciurus niger* and *S. carolinensis*), and chipmunks (*Tamias striatus*; Stegeman 1938, Hanson and Karstad 1959, Conley et al. 1972, Crank 2016). Wild pigs also prey on a broad range of invertebrate and vertebrate species in the Southeast region. One of the most significant impacts of this type of wild pig foraging is the depredation of the eggs and hatchlings of reptiles and birds, although the extent to which this occurs requires further study. Examples of affected species in the Southeast region would include the American alligator (*Alligator mississippiensis*), loggerhead sea turtle (*Caretta caretta*), wild turkey, northern bobwhite (*Colinus virginianus*), and ruffed grouse (*Bonasa umbellus*; Stegeman 1938, Henry 1969, Conley et al. 1972, Hayes et al. 1996, Elsey et al. 2012).

Wild pigs also cause damage to anthropogenic environments in the Southeast region. Wild pig rooting on the side slopes of roads and rail beds can sequentially lead to soil destabilization, erosion, gully formation on those slopes, and ultimately slumping of the road/rail beds. Wild pig rooting damage on the "crawlerway" even caused problems for moving the space shuttle with the Crawler-Transporter between the Vehicle Assembly Building and launch pads 39A and 39B at the Kennedy Space Center in Florida in the 1980s (Mayer and Brisbin 2009). With the recent increase in numbers of wild pigs in developed habitats (i.e., both urban and suburban areas), incidents associated with rooting up sodded and landscaped areas (e.g., lawns, parks, greenways, sports fields, golf courses, cemeteries) has also increased (Figure 16.7; see Chapters 7, 19). In both rural and developed areas, concurrent with the recent increase in the presence of wild pigs has been an increase in the number of wild pig-vehicle collisions. Property damage and personal injury consequences of wild pig-vehicle collisions can be substantial, with the number of human fatalities increasing. Vehicle accidents with wild pigs can involve both large individual animals and multiple numbers of animals in a single collision, and occur year-round and at any time of the day (Mayer and Johns 2007, Beasley et al. 2013). Lastly, wild pigs have been known to damage other property (e.g., scientific field equipment, miscellaneous yard/patio furniture, flower boxes/pots) in the region by knocking it over either accidentally or because of their curiosity (Mayer and Brisbin 2009, 2012).

Because wild pigs are harvested for consumption by the hunting public in the Southeast region, these animals have the potential to serve as vectors for the introduction of contaminants (e.g.,

FIGURE 16.7 Wild pig rooting damage in the Vicksburg National Military Park, Vicksburg, Mississippi. (Photo by C. Nelson. With permission.)

radionuclides, metals) into the human food chain (Mayer and Brisbin 2012). The incidental ingestion of soil during foraging (Beyer et al. 1994) along with the preference of these animals for wetland areas and drainages (Crouch 1983), where many contaminants are frequently present (either through mobilization and/or sequestration), increases the potential for uptake of contaminants by wild pigs (Brisbin et al. 1977b, Stribling et al. 1986b, Oldenkamp et al. 2017).

16.4 REGIONAL MANAGEMENT IN SOUTHEASTERN NORTH AMERICA

Societal perceptions of wild pigs as well as their management in the Southeast region have evolved drastically over the last few hundred years, from that of free-range livestock, to an important big-game species, to now in many cases a destructive invasive species. Throughout the early colonization of the United States by Europeans, pigs were primarily managed as livestock, although they were allowed to freely roam the environment to forage. During this time pigs were generally considered private property, and ear notched to identify ownership (Mayer and Brisbin 2009, Mayer and Beasley 2018). This lack of regulation and management led to the widespread establishment of wild populations of pigs throughout much of the Southeast region, ultimately setting the stage for a shift in the perception to that of a potential big-game resource. This growing interest in promulgating wild pigs as a huntable game species was further catalyzed through the introduction of pure wild boar into high-fenced shooting operations throughout the United States, including the Southeast region (e.g., Hooper Bald, North Carolina; Mayer and Brisbin 2008, 2009). Escapees from these operations successfully established populations of wild boar throughout the surrounding landscapes, ultimately hybridizing with established feral/wild populations (Mayer and Brisbin 2008, McCann et al. 2014, Keiter et al. 2016a).

Ironically, many state agencies and governments seeking to reduce or eliminate populations of wild pigs today played a central role in the cultivation of the perception of pigs as a big-game resource. The distribution of wild pigs remained relatively unchanged from the early to mid-20th century, when many states began actively promoting wild pigs as a game species (Mayer and Beasley 2018). For example, a number of states, including Florida, North Carolina, Tennessee, and West Virginia, began designating wild pigs as a game animal, establishing regulations limiting seasons of harvest, methods of take, and even harvest quotas. Several southeastern states (Florida, Mississippi, North Carolina, Tennessee, and West Virginia) even invested in active stocking programs to facilitate hunter opportunities (Keiter et al. 2016a). These efforts, which were further cultivated by the promotion of wild pig hunting by various magazines and television programs, were highly successful in transforming the perception of wild pigs in the Southeast region to that of an important big-game animal. In fact, state-sponsored stocking of pigs paved the way for similar actions by both private individuals and organizations, an activity that continues (illegally) today and is largely responsible for the rapid spread of pig populations throughout the United States over the last few decades (Tabak et al. 2017). By the 1990s, all southeastern US states had established huntable wild pig populations. In response to the growing demand, hundreds of commercial high-fenced shooting operations opened throughout the United States, offering opportunities to harvest wild pigs (Mayer and Beasley 2018), with such facilities in the Southeast region having been recently identified in Florida, Louisiana, South Carolina, and Tennessee (Todd and Mengak 2018).

Their generalist behavior, high reproductive output, and lack of effective natural predators, combined with the promotion of management strategies for maintaining robust wild pig populations by many state agencies, facilitated the rapid growth of pig populations throughout the United States (Mayer and Beasley 2018). Such efforts were particularly successful throughout much of the Southeast region, where populations have been established for centuries. Wild pigs are now the second most popular big-game animal in the United States in terms of annual harvest, second only to white-tailed deer (Kaufman et al. 2004). However, an unintended consequence of the successful establishment of wild pig populations has been a concomitant increase in damages caused by pigs to both natural and anthropogenic ecosystems (Mayer and Brisbin 2009, Barrios-Garcia and Ballari 2012, Keiter and Beasley 2017). While such damages by pigs undoubtedly have always occurred, the extent of damages and number of people negatively impacted by wild pigs has risen sharply in recent decades. As a result, management has shifted towards the recognition of wild pigs as an unwanted invasive species. However, while many landowners and resource managers in the Southeast region now recognize wild pigs as a destructive invasive species in need of rigorous control, successful management of the species remains conflated by competing interests in the maintenance of wild pig populations for recreational hunting opportunities.

Throughout the early 21st century there has been a widespread transition in the management perception of wild pigs. Although variability in the management of wild pigs currently exists among southeastern states, they are now generally viewed as an unwanted invasive species by state agencies. However, the degree to which states have attempted to alter the regulation of hunting opportunities has varied. For example, North Carolina and Tennessee have removed wild pigs from their list of big-game animals. Tennessee has further outlawed all sport-hunting of wild pigs throughout the state to curb interest in promoting illegal pig translocation activities. There also is increased recognition of the role that high-fence operations play in facilitating spread of populations, and there has been increasing pressure to eliminate such opportunities in many states. In contrast, wild pigs still have legal game status in 4 West Virginia counties and can be hunted throughout most of the Southeast region. Furthermore, although not officially recognized as big game, most southeastern states are currently doing little to curb interest in pig hunting, with most states allowing unlimited bag limits with no closed seasons or weapon restrictions, and the ability to hunt pigs at night and with dogs. As a result, hunting of wild pigs remains highly popular throughout much of the Southeast region. Promotion of wild pig hunting opportunities is further exacerbated in some states such as Florida, where pigs legally can be harvested, transported live, and sold as a meat commodity.

16.5 REGIONAL WILD PIG RESEARCH IN SOUTHEASTERN NORTH AMERICA

Given the prolonged presence of wild pigs in this region, a number of studies have been conducted addressing the biology/ecology of these animals and the interactions with their host environments in the Southeast region. Such studies, conducted as early as the 1930s and 1940s (e.g., Stegeman 1938, Shaw 1941, Babero and Karstad 1945, Hamnett 1947), have continued up through the present.

Most early research focused on the game status of these introduced animals, while later studies shifted to the environmental damage caused by wild pigs. Many landmark studies on wild pigs in North America have come out of this region (e.g., Hanson and Karstad 1959, Conley et al. 1972, Wood 1977).

Central and eastern Tennessee served as an early hotbed for research on wild pigs in the Southeast region due to the abundance of animals in the area, and data from this research was published throughout the 1960s and 1970s. Much of this research was management related, and focused on topics such as reproduction (Matschke 1964; Henry 1968a, b) and basic biology and ecology (Matschke 1963, 1967; Henry et al. 1968; Henry 1970; Henry and Conley 1970, 1972, 1978). As the basic understanding of wild pig biology in the region improved, the research emphasis began to evolve to become more management oriented, where research questions were designed to understand how wild pigs might serve as a game species and how best to manage them (Matschke 1962; Henry 1966, 1969; Matschke and Hardister 1966; Henry and Matschke 1968, 1972). This early work was important as it served as some of the initial research on wild pigs in North America and is the foundation upon which wild pig research is based today.

Not long after the research in eastern and central Tennessee began, a major research program whose focus was wild pigs began at the Savannah River Site in western South Carolina. This research program is still in operation today, and arguably has contributed more to our knowledge of wild pigs than any other program across the continent (see Case Study 1 below). Early studies focused on improving our understanding of the basic biology of the animal (Sweeney et al. 1970, 1979), but later morphed into a comprehensive investigation of the morphology of wild pigs (Brisbin et al. 1977a; Mayer and Brisbin 1988, 2008; Mayer et al. 1989). Over the years, a multitude of research projects have been undertaken and completed (see review in Mayer and Brisbin 2012), and have included studies of contamination in wild pigs (Stribling and Brisbin 1978, Stribling et al. 1986b, Oldenkamp et al. 2017), behavioral studies (Crouch and Sweeney 1981, Mayer and Brisbin 1986, Mayer et al. 2002, Pepin et al. 2016, Kay et al. 2017, Turner et al. 2017, Eckert et al. 2019), development of field techniques for wild pig research (Mayer 2002, 2003; Keiter et al. 2016a, b, 2017a, b; Kierepka et al. 2016), and investigations of human-pig conflicts (Mayer et al. 2000, Mayer and Johns 2007, Beasley et al. 2013). Like earlier work conducted in central and eastern Tennessee, most research with wild pigs in North America today draws upon the information accumulated through the studies of wild pigs at the Savannah River Site.

Similar to the Savannah River Site, numerous studies have been conducted at Hobcaw Barony and other locations in South Carolina. The research conducted as part of this program has primarily been related to biology (Wood et al. 1976, Wood and Brenneman 1980, Wood and Roark 1980) and management (Wood and Brenneman 1977; Wood et al. 1977, 1992) of wild pigs.

The wild pig population on Ossabaw Island, Georgia, has also been the subject of decades of research. Beginning in the 1960s, this work included the unique aspects or features of this localized population, which were assumed to have evolved due to their isolated presence on the island since colonial times. These unique characteristics have included their morphology (Brisbin et al. 1977a), body fat physiology (Stribling et al. 1984), high physiological salt tolerance (Zervanos et al. 1983), lipid-handling enzyme and hormone systems (Martin et al. 1973, Buhlinger et al. 1978, Scott et al. 1981, Hoffman et al. 1983), and metabolic syndrome with a propensity to develop type 2 diabetes (Sturek et al. 2006, Bratz et al. 2008). Ossabaw Island wild pigs have become an important animal model for applied biomedical and basic biochemical, physiological, and genetic studies (Grundy et al. 1999, Marx 2002, Brisbin and Sturek 2009).

The GSMNP has served as a source of wild pig research for decades and much of the work in this area focused on understanding the negative impacts wild pigs have on that ecosystem. As early as the 1970s, researchers began examining the negative impacts of rooting (Belden and Pelton 1975, Howe and Bratton 1976, Bratton et el. 1982, Singer et al. 1984) and impacts to vegetation (Bratton 1974, 1975, 1979). Additional work examined blood parameters (Williamson and Pelton 1975, 1976) and general biology of the species (Scott and Pelton 1975, Howe et al. 1981, Singer et al.

1981, Johnson et al. 1982). A review of the wild pig management program in the park was provided by Stiver and Delozier (2009), and studies continue today on the impacts to seeps and salamanders (Rossell et al. 2016).

Considerable work has also been done on wild pigs in Florida. The majority of this research focused on diseases and parasites and the potential for wild pigs to serve as disease vectors (Becker et al. 1978, McVicar et al. 1981, Greiner et al. 1982, Degner et al. 1983, Gibbs and Butler 1984, Greiner et al. 1984, van der Leek et al. 1993, Romero et al. 1997). During the 1980s, research was also conducted on the recreational value and economic importance of wild pigs as game animals in Florida (Degner et al. 1982, 1983; Degner 1989). Additionally, research into control, management, and movement of wild pigs has been an important aspect of Florida research programs (Frankenberger and Belden 1976, Belden and Frankenberger 1977, Maehr et al. 1989, Hernandez et al. 2018). More recently, research has focused on studies of wild pig damage (Engeman et al. 2003; 2004a, 2004b, 2007; Boughton and Boughton 2014; Bankovich et al. 2016) as well as competition with white-tailed deer over acorn mast and feeder stations (Crank 2016) and genetic tools to analyze dietary patterns (Anderson et al. In Press). Buck Island Ranch of the Archbold Biological Station became a focal site for many of these later studies.

The Southeast Cooperative Wildlife Disease Study, based out of the College of Veterinary Medicine at the University of Georgia, has conducted research on diseases and parasites of wild pigs across the Southeast region (Nettles 1997). This group has published dozens of studies on diseases such as pseudorabies (Nettles and Erickson 1984, Pirtle et al. 1989, Nettles 1991, Corn et al. 2004), brucellosis (Zygmont et al. 1982; Nettles 1984, 1991), vesicular stomatitis (Stallnecht et al. 1986, 1987, 1993, 1999), and classical swine fever (Nettles et al. 1989). The research of this group has served as the foundation for our understanding of the potential of wild pigs to serve as vectors of disease to other wildlife species, domestic animals, and humans (see Chapter 5).

Research conducted at Fort Benning, Georgia was influential in development of control strategies for wild pigs (see Case Study 2 below). This research, based upon an initial examination of the basic biology of wild pigs on the installation (Hanson et al. 2008, 2009; Sparklin et al. 2009; Jolley et al. 2010; Ditchkoff et al. 2012), served as the foundation for the development and testing of whole sounder removal (see Chapter 8). Other research at Fort Benning examined trapping methodologies (Williams et al. 2011a, b), bounties (Ditchkoff et al. 2017), survey techniques (Holtfreter et al. 2008), and damage (Jolley et al. 2010, Eckhardt et al. 2016).

16.6 WILD PIG CASE STUDIES IN SOUTHEASTERN NORTH AMERICA

16.6.1 Case Study 1: Savannah River Site, South Carolina

Wild pigs have been present on the US Department of Energy's (DOE) Savannah River Site (SRS; prior to 1989 called the Savannah River Plant or SRP), South Carolina since before its establishment in the early 1950s (Mayer and Moore-Barnhill 2009, Mayer and Brisbin 2012). Because of their combined destructive nature and value as a game species, wild pigs have been one of the most controversial wildlife species on the SRS (Mayer 2005). In deference to that status, wild pigs of the SRS embody one of the best studied populations of this non-native species in North America (Mayer and Brisbin 2009, 2012).

The SRS is an 803-km^2 nuclear production and research facility located in Aiken, Barnwell, and Allendale Counties in the upper Coastal Plain region of western South Carolina (Figure 16.8). The entire site has been closed to general public access since its establishment. Since 1952, when the US Government acquired the SRS, forestry management practices and natural succession outside the construction and operating areas on the site have resulted in increased ecological complexity and diversity of the property. The present SRS land use comprises 96% forested habitats or undeveloped lands (Kilgo and Blake 2005).

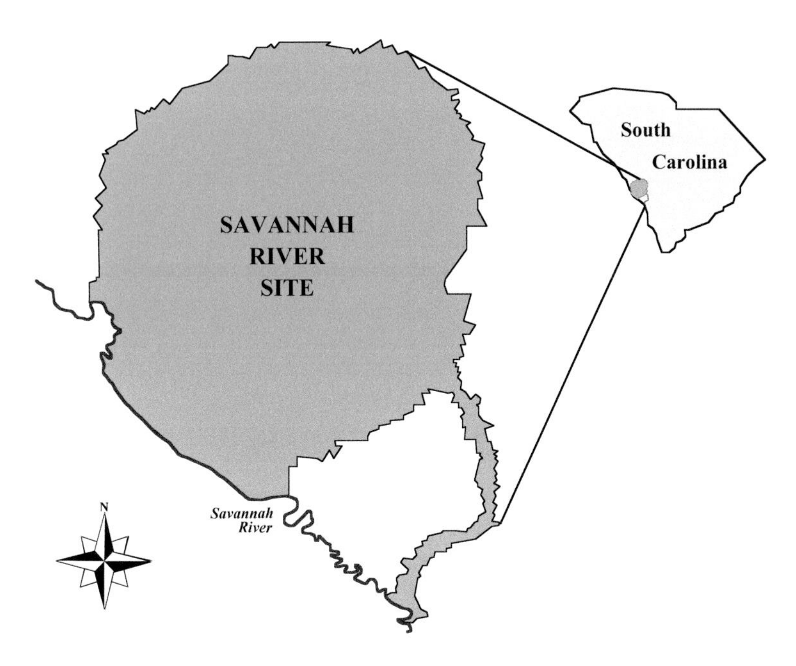

FIGURE 16.8 Location of the Savannah River Site, South Carolina. (Reproduced from Mayer and Brisbin 2009.)

16.6.1.1 History

The initial SRS wild pig population stemmed from free-ranging and semi-wild domestic pigs that were present in the area prior to the government purchase of the land. When the site was purchased by the federal government in 1950, resident farmers were given 1 year to capture and remove all free-ranging livestock. Large numbers of free-range domestic pigs could not be recovered by the end of 1951 and were left behind when the farmers were finally prohibited further access to the land in 1952. These animals thrived and multiplied in the riverswamp portion of the site (Mayer and Moore-Barnhill 2009, Mayer and Brisbin 2012).

In the mid-1970s, a small second subpopulation of wild pigs was discovered along the Upper Three Runs drainage just below US Highway 278 in the central northern portion of the site. Several theories have been posed as to the origin of these animals, but none have been verified. However, based on morphological data (Mayer and Brisbin 2008, 2012), it was evident this second subpopulation did not stem from the riverswamp subpopulation as was theorized by both Crouch (1983) and Hughes (1985). The upland wild pigs have since expanded their range toward the Savannah River, and animals exhibiting the upland phenotype have since been collected in areas adjacent to the river (Mayer and Moore-Barnhill 2009, Mayer and Brisbin 2012).

In the early 1980s, 9 wild pigs were killed on SRS public hunts near the central western boundary of the site that were found to exhibit characteristics typical of wild boar or wild boar/feral pig hybrids. The origin of these animals was traced to a release of wild pigs of mixed wild boar and feral pig ancestry onto private property, Cowden Plantation, which is adjacent to the SRS (Mayer and Brisbin 2008, 2012). These hybrid animals were released on the plantation in an effort to improve the quality of the wild pig population already present there. Wild pigs showing these hybrid characteristics have continued to increase on the SRS, and by 1989, had expanded throughout the riverswamp portion of the site (Mayer and Moore-Barnhill 2009, Mayer and Brisbin 2012; Figure 16.3).

Wild pigs have recently been observed or reported in all 46 of the wildlife management units on the SRS. The size of the SRS wild pig population has fluctuated greatly since the early 1950s. In 1952, only up to a couple of hundred pigs were estimated to exist on the site (Mayer 2005, Mayer and Moore-Barnhill 2009, Mayer and Brisbin 2012), while the current population is estimated >4,000–5,000 animals (Keiter et al. 2017a).

16.6.1.2 Damage/Impacts

Wild pigs are at present the most widespread and numerous invasive vertebrate species on the SRS. Since the late 1950s, the principle perceived role of this animal in the SRS ecosystem has been that of a nuisance or pest species due to damage caused to natural and anthropogenic systems on the site including: depredation of planted seedlings; rooting damage, vehicle collisions, depredation of wildlife food plots, radionuclide/contaminant uptake, and hazards to SRS personnel traveling on foot (Mayer and Brisbin 2012).

16.6.1.3 Management/Control

Because of the damage wild pigs cause, the SRS has used a variety of control efforts to manage the population. These activities have been conducted by several agencies operating on site with the common goal of reducing the number of wild pigs present and thus the level of damage realized. Initial control activities involved a combined trapping and opportunistic shooting program implemented by the US Forest Service-Savannah River (USFS-SR) in the mid-1950s (Mayer and Brisbin 2012). These efforts by the USFS-SR have continued up through the present with more than 974 pigs being removed between 1956 and 2018. Public hunts on the site began in the fall of 1965 in an effort to reduce the size of the white-tailed deer herd and reduce the number of deer/vehicle collisions on SRS roads. The harvest of wild pigs was also allowed during these hunts, and the annual take of pigs has increased yearly, with a total of 4,391 pigs being harvested from 1965 to 2018. In 1984, the SRFS was directed by DOE to develop a subcontracted control program to reduce the wild pig population. Implemented in 1985, this program initially only involved a subcontractor-operated trapping program, but in the early 1990s the use of trained hunting dogs to catch and remove wild pigs was incorporated. Overall, this program has removed 19,283 wild pigs from the SRS between 1985 and 2018. Other removal actions have been conducted by the University of Georgia's Savannah River Ecology Laboratory and South Carolina Department of Natural Resources for research and damage mitigation purposes, respectively; the total from these efforts has been approximately 560 wild pigs. Lastly, in 2001 and 2003, limited special public hunts for wild pigs were held to supplement other concurrent control efforts. These hunts, administered by the USFS-SR, used only teams of hunters with packs of trained hunting dogs to catch and remove wild pigs. Collectively, these 2 efforts removed a total of 173 pigs (Mayer and Moore-Barnhill 2009, Mayer and Brisbin 2012). Thus, despite having removed more than 24,980 wild pigs on the SRS between 1956 and 2018, the wild pig population on site has continued to increase (Mayer and Brisbin 2012, Beasley et al. 2013, Keiter et al. 2017a), exemplifying the challenges of controlling this destructive invasive species.

16.6.1.4 Research

The SRS wild pigs represent one of the most intensively studied populations of this non-native species in North America. Studies of this population have covered a wide spectrum of topics ranging from taxonomy and various aspects of population biology and ecology to assessments of damage/impacts and assessments of control (Table 16.3). Since the late 1960s, wild pigs have been the sole or major subject of 13 theses/dissertations, 3 book chapters, 32 published abstracts, 39 peer-reviewed journal articles, 23 conference and/or symposium proceedings chapters/articles, and 22 technical reports. In addition, animals from this population have been included in studies or data analyses reported collectively in other theses, dissertations, books, book chapters, journal articles, and technical reports. Research on the SRS wild pig population has continued to increase in the last several years, with over a dozen active projects currently. Because of the extensive historic and current research covering most aspects of the species' ecology and management, the SRS wild pig population represents a unique case study for long-term research, from which inferences can be made to other populations in the Southeast region and elsewhere.

TABLE 16.3
Examples of the Diversity of Research Conducted on Wild Pigs of the Savannah River Site, South Carolina

Research area		
General	**Specific**	**Study Reference(s)**
History of population		Jenkins and Provost (1964), Hatcher (1966), White (2004), Mayer (2005), Mayer and Brisbin (2008, 2012), Mayer and Moore-Barnhill (2009)
Taxonomy/ classification		Mayer (2005), Mayer and Brisbin (2008, 2012), Mayer and Moore-Barnhill (2009)
Biology/ecology	Age determination	Sweeney et al. (1970), Mayer (2002)
	Body fat reserves	Stribling et al. (1984)
	Diet digestibility coefficients/body condition	Miller (1979), Miller et al. (1987)
	Farrowing nest/bed characteristics	Mayer et al. (2002)
	Food habits	Sweeney (1970), Mayer and Brisbin (2012)
	Genetics	Smith et al. (1980), Kierepka et al. (2016)
	Home range/telemetry	Kurz (1971), Kurz and Marchinton (1972), Crouch and Sweeney (1981), Crouch (1983), Hughes ((1985), Pepin et al. (2016), Kay et al. (2017)
	Morphology	Sweeney (1970), Brisbin et al. (1977a), Mayer and Brisbin (1988, 1993, 2012), Mayer et al. (1989), Mayer (2003)
	Population biology	Kight (1962), Sweeney (1970), Kurz (1971), Crouch (1983), Dukes (1984), Savereno and Fendley (1989), Mayer (2009), Mayer and Brisbin (2012), Keiter et al. (2016a, 2017a, 2017b, 2017c)
	Predation of/by	Kight (1962), Sweeney (1970), Schrecengost et al. (2008), Mayer and Brisbin (2012)
	Reproduction	Sweeney (1970, 1979), Sweeney et al. (1979), Comer and Mayer (2009), Mayer and Brisbin (2012)
	Scavenging	DeVault and Rhodes (2002), Smith et al. (2017), Turner et al. (2017), Hill et al. (2018)
	Scent marking behavior	Mayer and Brisbin (1986)
	Soil ingestion	Beyer et al. (1994)
Damage/impacts	Attacks on humans	Mayer and Brisbin (2012), Mayer (2013)
	Contaminant uptake	Stribling (1978), Stribling et al. (1968a, 1986b), Mayer and Brisbin (2012), Odenkamp et al. (2017)
	Disease	Sweeney (1970), Southeastern Cooperative Wildlife Disease Study (2007), Mayer and Brisbin (2012)
	Human consumption and ecological risk assessment	Gaines (2003), Gaines et al. (2005), Smith et al. (2018)
	Hardwood seedling depredation	Mayer et al. (2000)
	Rooting/wallowing	Kirkman and Sharitz (1994), Mayer and Brisbin (2012), Eckert et al. (2019)
	Vehicle collisions	Mayer and Johns (2007), Mayer and Brisbin (2012), Beasley et al. (2013)
	Biomarkers	Beasley et al. (2015), Webster et al. (2017)

16.6.2 Case Study 2: Fort Benning, Georgia

Fort Benning is a US Army base that straddles the Alabama and Georgia border, located adjacent to Columbus, Georgia. Originally established in 1909 to provide basic training for army units in World War I, it has evolved substantially and now is known as "Home of the Infantry." Fort Benning has over 25,000 troops at any one time, and active training including mechanized, infantry, and basic training. Fort Benning is comprised of 735 km^2 on the Coastal Plain-Piedmont fall line. Frequent historic fires due to military training resulted in upland habitats predominated by longleaf pine with an herbaceous understory. This mature/climax habitat type has been maintained by prescribed fire in recent decades. Riparian bottoms bisect upland areas and are comprised of a variety of hardwood species.

16.6.2.1 History

Little information concerning wild pig populations on Fort Benning is available prior to the 1980s. More than likely, wild pigs were present on the landscape prior to its establishment due to livestock free-range practices in the region, although it is possible that any animals on the installation during its early years would have been utilized for food. The first officially documented presence of wild pigs on base was in the late 1940s and early 1950s, although no estimates of population size from that time period exist. At that time, wild pigs were being recreationally hunted on horseback by military personnel using dogs to bay up the pigs (J. A. May, Captain, US Army Retired, personal communication). It was not until much later that any efforts to monitor the base's wild pig population occurred. Most data on wild pigs on the installation come from hunter-harvested animals. Although Fort Benning hosts a substantial amount of military training activity that includes live fire exercises with tanks and other heavy military equipment, recreational hunting for active and retired military is strongly encouraged. Hunter harvest data estimated on average, >1,000 wild pigs were being harvested annually from the installation in the mid-2000s. Population studies during this same time period estimated the density of wild pigs to be between 1.07 and 2.45 pigs/km^2. However, it should be noted these estimates were conducted in a region of the base comprised primarily of upland habitats. It is likely densities of wild pigs in other portions of the base that included substantially more lowland and riparian habitats were much greater. Extrapolating these original density estimates and accounting for the quality of habitats on the base, it is likely Fort Benning has ≥3,000 wild pigs.

16.6.2.2 Damage/Impacts

Wild pigs cause similar ecological damages at Fort Benning as in other portions of the Southeast region. Their rooting in riparian areas disturbs the forest floor, leading to erosion and stream sedimentation, and they negatively impact both floral and faunal communities. A major concern with wild pigs on Fort Benning is their impact on 7 populations of relict trillium (*Trillium reliquum*), a species of conservation concern, found on base. Rooting by wild pigs severely impacts this species, and several trillium populations on Fort Benning have been fenced to eliminate impacts of wild pigs. However, even these populations are not entirely safe from wild pigs, as damage to fences (e.g., fallen trees, erosion, military activities) does occur.

Wild pigs also impact military training and equipment. Shooting ranges, targets, and other equipment are often connected into an electrical/computer infrastructure via buried cables, wires, and other devices, and rooting and other activities of wild pigs frequently damage this infrastructure causing untold costs in repair. Additionally, a portion of the base is cleared and maintained as open field for airborne (e.g., parachute) training. This area, due to its unique vegetative and resulting faunal communities, is regularly utilized by wild pigs, resulting in extensive rooting causing an uneven surface on the landing area. Numerous injuries to military personnel participating in these airborne training activities have occurred from landing on the uneven, rooted surface of this field. Finally, the cantonment area on base receives considerable damage in the form of rooting. This section of the base is well developed and contains manicured lawns that are susceptible to pig damage.

16.6.2.3 Management/Control

Wild pig control efforts on Fort Benning vary depending upon the section of the base being targeted. In the cantonment area, trapping is the primary form of control. Both corral and box traps are used during control efforts, with the objective of reducing the population size to a level where rooting is curbed. Control efforts in training areas consist primarily of hunting. Over 1,000 wild pigs are harvested annually by hunters (mostly by deer hunters), which likely only serves to maintain population levels. When damage to sensitive areas (special training areas, food plots, etc.) becomes excessive, resource managers use traps to reduce the number of wild pigs in those areas. However, these efforts do little to reduce densities base-wide. In 2007, Fort Benning instituted a bounty program in an effort to reduce the numbers of wild pigs on the installation. Although 1,138 wild pigs were removed during the program at a total cost of US$57,296 (bounties paid and program administrative costs), the population of wild pigs on Fort Benning actually increased during this period (Ditchkoff et al. 2017).

16.6.2.4 Research

Research on wild pigs at Fort Benning was conducted from 2003 to 2013 during 2 separate studies. The first study was focused on understanding the biology and ecology of wild pigs on base. Specifically, population dynamics, reproduction, food habits, and spatial ecology were examined. The second study was conducted to develop and examine control programs for wild pigs. Survey techniques, bait effectiveness, and trapping efficiency were all examined. Additionally, considerable effort was expended to understand the effectiveness of whole sounder removal in controlling pig populations. From these research efforts, 10 peer-reviewed articles, 5 theses, and 1 book chapter have been published to date.

16.7 FUTURE OF WILD PIGS IN SOUTHEASTERN NORTH AMERICA

Concomitant with increasing damages caused by wild pigs, there has been a gradual shift in the focus of pig management in the Southeast region over the last few decades, transitioning from the promotion and maintenance of wild pigs as a game species to that of an unwanted invasive species. However, these efforts are challenged by an entrenched culture of wild pig hunting, pervasive throughout much of the region. Furthermore, there are factions within the hunting realm in the Southeast region that continue to promote the management of wild pigs as an economic or game resource. As long as opportunities exist, interest in wild pig hunting and promotion of pig populations will remain a primary challenge to any control efforts. Furthermore, as with control efforts for many other invasive species, if reductions in effort or funding for control occur prior to successful eradication, populations may be capable of recovering within a relatively short time frame (Mayer 2005).

Effective management of wild pigs is particularly challenging in the Southeast region where dense forests provide reprieve from control efforts and a favorable climate facilitates maximum fecundity and population growth. Furthermore, given their widespread presence throughout much of the Southeast region over the last few hundred years, as well as historic promotion of pigs as a game species by management agencies, shifting perceptions and management of wild pigs among the general public remains a complex socioeconomic challenge. Thus, despite advances in tools and technologies for controlling wild pigs, as well as increased financial investments in pig control throughout the region, wild pigs are likely to maintain a significant presence throughout southern ecosystems for the foreseeable future. Current management throughout many areas of the Southeast region, therefore, is largely directed at reducing populations, ultimately decreasing the extent and scope of damages caused by this invasive species in particular high-impact locations. However, over the long term this paradigm will be costly and only a temporary solution to the mitigation of damages caused by pigs given the species' potential for rapid population expansion and recolonization in the absence of intensive harvest pressure.

ACKNOWLEDGMENTS

The authors thank J.L. Corn for the data provided in support of the development of this chapter. Contributions from J.J.M. were supported by the US Department of Energy Office of Environmental Management under Contract DE-AC09-08SR22470 to Savannah River Nuclear Solutions LLC. Contributions from J.C.B. were partially funded by the US Department of Energy under award # DE-EM0004391 to the University of Georgia Research Foundation.

REFERENCES

Ackerman, B. B., M. E. Harmon, and F. J. Singer. 1978. Part II. Seasonal food habits of European wild boar – 1977. Pages 94–137 *in* F. J. Singer, editor. *Studies of European Wild Boar in the Great Smoky Mountains National Park: 1st Annual Report; A Report for the Superintendent.* Uplands Field Research Laboratory, Great Smoky Mountains National Park, Gatlinburg, TN.

Amer, C. F. 2006. A survey for Lucas Vázquez de Ayllón's lost Capitana. *Legacy* 10:10–15.

Anderson, A., C. Slootmaker, E. Harper, J. Holderieath, and S. A. Shwiff. 2016. Economic estimates of feral swine damage and control in 11 US states. *Crop Protection* 89:89–94.

Anderson, W. M., R. K. Boughton, S. W. Wisely, M. M. Merrill, E. H. Boughton, M. S Robeson, II, and A. J. Piaggio. In press. Using DNA Metabarcoding to examine wild pig (*Sus scrofa*) diets in a subtropical agro-ecosystem. *Proceedings of the Vertebrate Pest Conference.*

Arrington, D. A., L. A. Toth, and J. W. Koebel, Jr. 1999. Effects of rooting by feral hogs *Sus scrofa* L. on the structure of a floodplain vegetation assemblage. *Wetlands* 19:535–544.

Baber, D. W. 1977. Social organization and behavior in the feral hog. Thesis, Florida Institute of Technology, Melborne, FL.

Baber, D. W., and J. G. Morris. 1980. Florida scrub jays foraging from feral hogs. *Auk* 97:202.

Babero, B. B., and L. H. Karstad. 1945. Studies on parasitism of feral swine in Georgia. *Journal of Parasitology* 45:44.

Baily, R. G. 1980. Descriptions of the ecoregions of the United States. Misc. Publ. No. 1391, US Department of Agriculture, Forest Service, Washington, DC.

Bankovich, B., E. Boughton, R. Boughton, M. L. Avery, and S. M. Wisely, 2016. Plant community shifts caused by feral swine rooting devalue Florida rangeland. *Agriculture, Ecosystems and Environment* 220:45–54.

Baron, J. S. 1979. Vegetation damage by feral hogs on Horn Island, Gulf Islands National Seashore, Mississippi. Thesis, University of Wisconsin, Madison, WI.

Barrios-Garcia, M. N., and S. A. Ballari. 2012. Impact of wild boar (*Sus scrofa*) in its introduced and native range: A review. *Biological Invasions* 14:2283–2300.

Barrios-Garcia, M. N., A. T. Classen, and D. Simberloff. 2014. Disparate responses of above- and below-ground properties to soil disturbance by an invasive mammal. *Ecosphere* 5(4):1–13.

Beasley, J. C., T. E. Grazia, P. E. Johns, and J. J. Mayer. 2013. Habitats associated with vehicle collisions with wild pigs. *Wildlife Research* 40:654–660.

Beasley, J. C., S. C. Webster, O. E. Rhodes, Jr., and F. L. Cunningham. 2015. Evaluation of Rhodamine B as a biomarker for assessing bait acceptance in wild pigs. *Wildlife Society Bulletin* 39:188–192.

Becker, H. N., R. C. Belden, T. Breault, M. J. Burridge, W. B. Frankenberger, and P. Nicoletti. 1978. Brucellosis in feral swine in Florida. *Journal of the American Veterinary Medical Association* 173:1181–1182.

Belden, R. C., and W. B. Frankenberger. 1977. A portable root-door hog trap. *Proceedings of the Annual Conference of the Southeastern Association of Fish & Wildlife Agencies* 31:123–125.

Belden, R. C., and W. B. Frankenberger. 1979. Brunswick hog study. Final performance report, P-R Project W-41-R, Study No. XIII-B-1. Florida Fresh Water Fish and Game Commission Wildlife Research Laboratory, Gainesville, FL.

Belden, R. C., and W. B. Frankenberger. 1989. History and biology of feral swine. Pages 3–10 *in* N. Black, editor, *Proceedings: Feral Pig Symposium.* April 27–29, Orlando, FL. Livestock Conservation Institute, Madison, WI.

Belden, R. C., and W. B. Frankenberger. 1990. Biology of a feral hog population in south central Florida. *Proceedings of the Annual Conference of the Southeastern Association of Fish & Wildlife Agencies* 44:231–249.

Belden, R. C., W. B. Frankenberger, and D. H. Austin. 1985. A simulated harvest study of feral hogs in Florida. Final performance report, P-R Project W-41-R, Study No. XIII-FEC. Florida Fresh Water Fish and Game Commission Wildlife Research Laboratory, Gainesville, FL.

Belden, R. C., and M. R. Pelton. 1975. European wild hog rooting in the mountains of east Tennessee. *Proceedings of the Annual Conference of the Southeastern Association of Game and Fish Commissioners* 29:665–671.

Bevins, S. N., K. Pedersen, M. W. Lutman, T. Gidlewski, and T. J. Deliberto. 2014. Consequences associated with the recent range expansions of nonnative feral swine. *BioScience* 64:291–299.

Beyer, W. N., E. E. Connor, and S. Gerould. 1994. Estimates of soil ingestion by wildlife. *Journal of Wildlife Management* 58:375–382.

Bigler, W. J., G. L. Hoff, W. H. Hemmert, J. A. Tomas, and H. T. Janowski. 1977. Trends in Brucellosis in Florida an epidemiological review. *American Journal of Epidemiology* 105:245–251.

Boughton, R. K., B. L. Allen., E. A. Tillman., S. M. Wisely, and R. M. Engeman. 2019. Road hogs: Implications from GPS collared feral swine in pastureland habitat on the general utility of road-based observation techniques for assessing abundance. *Ecological Indicators* 99:171–177.

Boughton, E. H., and R. K. Boughton. 2014. Modification by an invasive ecosystem engineer shifts a wet prairie to a monotypic stand. *Biological Invasions* 16:2105–2114.

Bratton, S. P. 1974. The effect of the European wild boar (*Sus scrofa*) on the high elevation vernal flora in Great Smoky Mountains National Park. *Bulletin of the Torrey Botanical Club* 101:198–206.

Bratton, S. P. 1975. The effect of the European wild boar (*Sus scrofa*) on grey beech forest in the Great Smoky Mountains. *Ecology* 56:1356–1366.

Bratton, S. P. 1979. The impact of the European wild boar on native plant communities in Great Smoky Mountains National Park. Page 216 *in Abstracts of the Second Conference on Scientific Research in National Parks*. US National Park Service, Transactions of the Proceedings series, San Francisco, CA.

Bratton, S. P., M. E. Harmon, and P. S. White. 1982. Patterns of European wild boar rooting in the western Great Smoky Mountains. *Castanea* 47:230–242.

Bratz, I. N., G. M. Dick, J. D. Tune, J. M. Edwards, Z. P. Neeb, U. D. Dincer, and M. Sturek. 2008. Impaired capsaicin-induced relaxation of coronary arteries in a porcine model of the metabolic syndrome. *American Journal of Physiology: Heart and Circulatory Physiology* 294:H2489–H2496.

Brisbin, I. L, Jr., R. A. Geiger, H. B. Graves, J. E. Pinder, III, J. M. Sweeney, and J. R. Sweeney. 1977a. Morphological characteristics of two populations of feral swine. *Acta Theriologica* 22:75–85.

Brisbin, I. L., Jr., M. W. Smith, and M. H. Smith. 1977b. Feral swine studies at the Savannah River Ecology Laboratory: An overview of program goals and design. Pages 71–90 *in* G. W. Wood, editor. *Research and Management of Wild Hog Populations*. Belle Baruch Forest Science Institute of Clemson University, Georgetown, SC.

Brisbin, I. L., Jr., and M. S. Sturek. 2009. The pigs of Ossabaw Island: A case study of the application of long-term data in management plan development. Pages 365–378 *in* J. J. Mayer and I. L. Brisbin, Jr., editors. *Wild Pigs: Biology, Damage, Control Techniques and Management*. SRNL-RP-2009-00869, Savannah River National Laboratory, Aiken, SC.

Buhlinger, C. A., P. J. Wangsness, R. J. Martin, and J. H. Ziegler. 1978. Body composition, in vitro lipid metabolism and skeletal muscle characteristics of fast-growing, lean and in slow-growing, obese pigs at equal age and weight. *Growth* 42:225–236.

Centers for Disease Control. 2009. *Brucella suis* infection associated with feral swine hunting: Three states, 2007–2008. *Morbidity and Mortality Weekly Report* 58:618–621.

Clayton, L. A., V. J. Knight, and E. C. Moore. 1993. *The De Soto Chronicles: The Expedition of Hernando De Soto to North America 1539–1543*. The University of Alabama Press, Tuscaloosa, AL.

Comer, C. E., and J. J. Mayer. 2009. Wild pig reproductive biology. Pages 51–75 *in* J. J. Mayer and I. L. Brisbin, Jr., editors. *Wild Pigs: Biology, Damage, Control Techniques and Management*. SRNL-RP-2009-00869, Savannah River National Laboratory, Aiken, SC.

Conley, R. H., V. G. Henry, and G. H. Matschke. 1972. *Final Report for the European Hog Research Project W-34*. Tennessee Game and Fish Commission, Nashville, TN.

Corn, J. L., D. E. Stallknecht, N. M. Mechlin, M. P. Luttrell, and J. R. Fischer. 2004. Persistence of pseudorabies virus in feral swine populations. *Journal of Wildlife Diseases* 40:307–310.

Crank, C. A. 2016 Potential resource competition between feral swine (*Sus scrofa*) and White-tailed deer (*Odocoileus virginianus*) on Florida rangeland. Thesis, University of Florida, Gainesville, FL.

Crosby, A. 1972. *The Columbian Exchange*. Greenwood Publishing Group, Westport, CT.

Crouch, L. C. 1983. Movements of and habitat utilization by feral hogs at the Savannah River Plant, South Carolina. Thesis, Clemson University, Clemson, SC.

Crouch, L. C., and J. R. Sweeney. 1981. Habitat utilization of feral hogs at the Savannah River Plant, South Carolina. *Bulletin of the South Carolina Academy of Sciences* 43:77.

Davidson, W. R., and V. F. Nettles. 1988. Wild swine (*Sus scrofa*). Pages 93–122 *in* W. R. Davidson and V. F. Nettles, editors. *Field Manual of Wildlife Diseases in the Southeastern United States*. Southeastern Cooperative Wildlife Disease Study, University of Georgia, Athens, GA.

Davis, B. J. 2004. The Spanish horse in the East. *The Online Museum*. <http://horseoftheamericas.com/Sphrseast.pdf>. Accessed 22 June 2017.

Davis, D. S. 1993. Feral hogs and disease: Implications for humans and livestock. Pages 84–87 *in* C. W. Hanselka and J. F. Cadenhead, editors. *Feral Swine: A Compendium for Resource Managers*. Texas Agricultural Extension Service, Kerrville, TX.

Degner, R. L. 1989. Economic importance of feral swine in Florida. Pages 39–41 *in* N. Black, editor. *Proceedings: Feral Pig Symposium*. April 27–29, Orlando, FL, Livestock Conservation Institute, Madison, WI.

Degner, R. L., L. W. Rodan, W. K. Mathis, and E. P. J. Gibbs. 1982. The recreational and commercial importance of feral swine in Florida. Industry Report No. 82–1. Florida Agricultural Market Research Center, Food and Resource Economics Department, Agricultural Experiment Station, University of Florida, Gainesville, FL.

Degner, R. L., L. W. Rodan, W. K. Mathis, and E. P. J. Gibbs. 1983. The recreational and commercial importance of feral swine in Florida: Relevance to the possible introduction of African swine fever. *Preventative Veterinary Medicine* 1:371–381.

DeVault, T. L., and O. E. Rhodes, Jr. 2002. Identification of vertebrate scavengers of small mammal carcasses in a forested landscape. *Acta Theriologica* 47:185–192.

Dickson, J. G., J. J. Mayer, and J. D. Dickson. 2001. Wild hogs. Pages 191–192, 201–208 *in* J. G. Dickson, editor. *Wildlife of the Southern Forests: Habitat & Management*. Hancock House Publishers, Blaine, WA.

Ditchkoff, S. S., R. W. Holtfreter, and B. L. Williams. 2017. Effectiveness of a bounty program for reducing wild pig densities. *Wildlife Society Bulletin* 41:548–555.

Ditchkoff, S. S., D. B. Jolley, B. D. Sparklin, L. B. Hanson, M. S. Mitchell, and J. B. Grand. 2012. Reproduction in a population of wild pigs (*Sus scrofa*) subjected to lethal control. *Journal of Wildlife Management* 76:1235–1240.

Dukes, E. K. 1984. The Savannah River Plant environment. DP-16423. E. I. du Pont de Nemours, Savannah River Laboratory, Savannah River Plant, Aiken, SC.

Duncan, R. W. 1974. Reproductive biology of the European wild hog (*Sus scrofa*) in the Great Smoky Mountains National Park. Thesis, University of Tennessee, Knoxville, TN.

Eckert, K. D., D. A. Keiter, and J. C. Beasley. 2019. Animal visitation to wild pig (*Sus scrofa*) wallows and implications for disease transmission. *Journal of Wildlife Diseases. Journal of Wildlife Diseases* 55:488–493.

Eckhardt, L. G., R. D. Menard, and S. S. Ditchkoff. 2016. Wild pigs: Inciting factor in southern pine decline? Pages 91–94 *in* C. J. Schweitzer, W. K. Clatterbuck, and C. M. Oswalt, editors. 2016. *Proceedings of the 18th Biennial Southern Silvicultural Research Conference*. e-Gen. Tech. Rep. SRS-212, US Department of Agriculture, Forest Service, Southern Research Station, Asheville, NC.

Elsey, R. M., E. C. Mouton, Jr., and N. Kinler. 2012. Effects of feral swine (*Sus scrofa*) on alligator (*Alligator mississippiensis*) nests in Louisiana. *Southeastern Naturalist* 11:205–218.

Engeman, R. M., H. T. Smith, R. Severson, M. A. Severson, J. Woolard, S. A. Shwiff, B. Constantin, and D. Griffin. 2004a. Damage reduction estimates and benefit-cost ratios for feral swine control from the last remnant of a basin marsh system in Florida. *Environmental Conservation* 31:207–211.

Engeman, R. M., H. T. Smith, R. Severson, M. A. Severson, S. A. Shwiff, B. Constantin, and D. Griffin. 2004b. The amount and economic cost of feral swine damage to the last remnant of a basin marsh system in Florida. *Journal for Nature Conservation* 12:143–147.

Engeman, R. M., H. T. Smith, S. A. Shwiff, B. Constantin, J. Woolard, M. Nelson, and D. Griffin. 2003. Prevalence and economic value of feral swine damage to native habitat in three Florida state parks. *Environmental Conservation* 30:319–324.

Engeman, R. M., A. Stevens, J. Allen, J. Dunlap, M. Dunlap, D. Teague, and B. Constantin. 2007. Feral swine management for conservation of an imperiled wetland habitat: Florida's vanishing seepage slopes. *Biological Conservation* 134:440–446.

Fox, J. R., and M. R. Pelton. 1977. An evaluation of control techniques for the European wild hog in the Great Smoky Mountains National Park. Pages 53–66 *in* G. W. Wood, editor. *Research and Management of Wild Hog Populations*. Belle Baruch Forest Science Institute of Clemson University, Georgetown, SC.

Frankenberger, W. B., and R. C. Belden. 1976. Distribution, relative abundance and management needs of feral hogs in Florida. *Proceedings of the Annual Conference of the Southeastern Association of Fish & Wildlife Agencies* 30:641–644.

Friebel, B. A., and P. G. R. Jodice. 2009. Home range and habitat use of feral hogs in Congaree National Park, South Carolina. *Human-Wildlife Conflicts* 3:49–63.

Gaines, K. F. 2003. Chapter 3: A spatially explicit model of the wild hog (*Sus scrofa*) for ecological assessment activities at the Department of Energy's Savannah River Site. Pages 91–141 *in* K. F. Gaines. *Spatial Modeling of Receptor Species for Ecological Risk Assessment Activities on the Department of Energy's Savannah River Site.* Dissertation. University of South Carolina, Columbia, SC.

Gaines, K., D. Porter, T. Punshon, and I. L. Brisbin. 2005. A spatially explicit model of the wild hog for ecological risk assessment activities at the Department of Energy's Savannah River Site. *Human and Ecological Risk Assessment* 11:567–589.

Gaston, W., J. B. Armstrong, W. Arjo, and H. L. Stribling. 2008. Home range and habitat use of feral hogs (*Sus scrofa*) on Lowndes County WMA, Alabama. Pages 1–17 *in* S. M. Vantassel, editor. National Conference on Feral Hogs. St. Louis, MO.

Geiger, R. 1954. Klassifikation der Klimate nach W. Köppen. *Landolt-Börnstein: Zahlenwerte und Funktionen aus Physik, Chemie, Astronomie, Geophysik und Technik Berlin* 3:603–607 [In German].

Gibbs, E. P., and J. F. Butler. 1984. African swine fever: An assessment of risk for Florida. *Journal of the American Veterinary Medical Association* 184:644–647.

Glass, C. M., R. G. McLean, J. B. Katz, D. S. Maehr, C. B. Cropp, L. J. Kirk, A. J. McKeiman, and J. F. Evermann. 1994. Isolation of pseudorabies (Aujeszky's disease) virus from a Florida panther. *Journal of Wildlife Diseases* 30:180–184.

Graves, H. B. 1984. Behaviour and ecology of wild and feral swine (*Sus scrofa*). *Journal of Animal Science* 58:482–492.

Graves, H. B., and K. L. Graves. 1977. Some observations on biobehavioral adaptations of swine. Pages 103–110 *in* G. W. Wood, editor. *Research and Management of Wild Hog Populations.* Belle Baruch Forest Science Institute of Clemson University, Georgetown, SC.

Greiner, E. C., P. P. Humphrey, R. C. Belden, W. B. Frankenberger, D. H. Austin, and E. P. J. Gibbs. 1984. Ixodid ticks on feral swine from Florida. *Journal of Wildlife Diseases* 20:114–119.

Greiner, E. C., C. Taylor, III, W. B. Frankenberger, and R. C. Belden. 1982. Coccidia of feral swine from Florida. *Journal of the American Veterinary Medical Association* 181:1275–1277.

Grundy, S. M., I. J. Benjamin, G. L. Burke, A. Chait, R. H. Eckel, B. V. Howard, W. Mitch, et al. 1999. Diabetes and cardiovascular disease: A statement for healthcare professionals from the American Heart Association. *Circulation* 100:1134–1146.

Hamnett, W. L. 1947. The European wild boar. *Wildlife in North Carolina* 11:12.

Hanson, L. B. 2006. Demography of feral pig populations at Fort Benning, Georgia. Thesis, Auburn University, Auburn, AL.

Hanson, L. B., J. B. Grand, M. S. Mitchell, D. B. Jolley, B. D. Sparklin, and S. S. Ditchkoff. 2008. Change-in-ratio density estimator for feral pigs is less biased than closed mark-recapture estimates. *Wildlife Research* 35:695–699.

Hanson, L. B., M. S. Mitchell, J. B. Grand, D. B. Jolley, B. D. Sparklin, and S. S. Ditchkoff. 2009. Effect of experimental manipulation on survival and recruitment of feral pigs. *Wildlife Research* 36:185–191.

Hanson, R. P., and L. Karstad. 1959. Feral swine in the southeastern United States. *Journal of Wildlife Management* 23:64–74.

Hatcher. J. 1966. History of the Savannah River Project. Atomic Energy Commission 1951–1966. Report. USDA Forest Service, SRFS, Aiken, SC.

Hayes, R. B., N. B. Marsh, and G. A. Bishop. 1996. Sea turtle nest depredation by a feral hog: A learned behavior. *Proceedings of the Annual Symposium on Sea Turtle Biology and Conservation* 15:129–134.

Henry, V. G. 1966. European wild hog hunting season recommendations based on reproductive data. *Proceedings of the Annual Conference of the Southeastern Association of Game and Fish Commissioners* 20:139–145.

Henry, V. G. 1968a. Fetal development in European wild hogs. *Journal of Wildlife Management* 32:966–970.

Henry, V. G. 1968b. Length of estrous cycle and gestation in European wild hogs. *Journal of Wildlife Management* 32:406–408.

Henry, V. G. 1969. Predation on dummy nests of ground-nesting birds in the southern Appalachians. *Journal of Wildlife Management* 33:169–172.

Henry, V. G. 1970. Weights and body measurements of European wild hogs in Tennessee. *Journal of the Tennessee Academy of Sciences* 45:20–23.

Henry, V. G., and R. H. Conley. 1970. Some parasites of European wild hogs in the southern Appalachians. *Journal of Wildlife Management* 34:913–917.

Henry, V. G., and R. H. Conley. 1972. Fall foods of the European wild hogs in the southern Appalachians. *Journal of Wildlife Management* 36:854–860.

Henry, V. G., and R. H. Conley. 1978. Survival and mortality of European wild hogs. *Proceedings of the Annual Conference of the Southeastern Association of Fish and Wildlife Agencies* 32:93–99.

Henry, V. G., J. M. Rary, G. Matschke, and R. L. Murphree. 1968. The cytogenetics of swine in the Tellico Wildlife Management Area, Tennessee. *Journal of Heredity* 59:201–204.

Henry, V. G., and G. H. Matschke. 1968. Immobilizing trapped European wild hogs with Cap-Chur-Barb. *Journal of Wildlife Management* 32:970–972.

Henry, V. G., and G. H. Matschke. 1972. Immobilizing European wild hogs with Sernylan. *Journal of the Tennessee Academy of Sciences* 47:81–84.

Hernandez, F. A., B. M. Parker, C. L. Plyant, T. J. Smyser, A. J. Piaggio, S. L. Lance, M. P. Milleson, et al. 2018. Invasion ecology of wild pigs (*Sus scrofa*) in Florida, USA: The role of humans in the expansion and colonization of an invasive wild ungulate. *Biological Invasions* 20:1865–1880.

Hill, J. E., T. L. DeVault, J. C. Beasley, O. E. Rhodes, Jr., and J. L. Belant. 2018. Effects of vulture exclusion on carrion consumption by facultative scavengers. *Ecology and Evolution* 8:2518–2526.

Hoffman, E. C., P. J. Wangsness, D. R. Hagen, and T. D. Etherton. 1983. Fetuses of lean and obese swine in late gestation: Body composition, plasma hormones and muscle development. *Journal of Animal Science* 57:609–620.

Holtfreter, R. W., B. L. Williams, S. S. Ditchkoff, and J. B. Grand. 2008. Feral pig detectability with game cameras. *Proceedings of the Annual Conference of the Southeastern Association of Fish and Wildlife Agencies* 62:17–21.

Howe, T. D., and S. P. Bratton. 1976. Winter rooting activity of the European wild boar in the Great Smoky Mountains National Park. *Castanea* 41:256–264.

Howe, T. D., F. J. Singer, and B. B. Ackerman. 1979. High elevation forage relationships of European wild boar invading Great Smoky Mountains. Report to the Superintendent. US Department of the Interior, National Park Service, Great Smoky Mountains National Park, Gatlinburg, TN.

Howe, T. D., F. J. Singer, and B. B. Ackerman. 1981. Forage relationships of European wild boar invading northern hardwood forests. *Journal of Wildlife Management* 45:748–754.

Hughes, T. W. 1985. Home range, habitat utilization, and pig survival of feral swine on the Savannah River Plant. Thesis, Clemson University, Clemson, SC.

Jenkins, J. H., and E. E. Provost. 1964. *The Population Status of the Larger Vertebrates on the Atomic Energy Commission Savannah River Plant Site*. Office of Technical Services, Department of Commerce, Washington, DC.

Johnson, K. G., R. W. Duncan, and M. R. Pelton. 1982. Reproductive biology of European wild hogs in the Great Smoky Mountains National Park. *Proceedings of the Annual Conference of the Southeastern Association of Fish & Wildlife Agencies* 36:552–564.

Jolley, D. B., S. S. Ditchkoff, B. D. Sparklin, L. B. Hanson, M. S. Mitchell, and J. B. Grand. 2010. Estimate of herpetofauna depredation by a population of wild pigs. *Journal of Mammalogy* 91:519–524.

Kaller, M. D., and W. E. Kelso. 2006. Swine activity alters invertebrate and microbial communities in a Coastal Plain watershed. *The American Midland Naturalist* 156:163–177.

Kaufman, K., R. Bowers, and N. Bowers. 2004. *Kaufman Focus Guide to Mammals of North America*. Houghton Mifflin, New York.

Kay, S. L., J. W. Fischer, A. J. Monaghan, J. C. Beasley, R. Boughton, T. A. Campbell, S. M. Cooper, et al. 2017. Quantifying drivers of wild pig movement across multiple spatial and temporal scales. *Movement Ecology* 5:1–15.

Keiter, D. A., J. J. Mayer, and J. C. Beasley. 2016a. What is in a "common" name? A call for consistent terminology for nonnative *Sus scrofa*. *Wildlife Society Bulletin* 40:384–387.

Keiter, D. A., F. L. Cunningham, O. E. Rhodes, B. J. Irwin, and J. C. Beasley. 2016b. Optimization of scat detection methods for a social ungulate, the wild pig, and experimental evaluation of factors affecting detection of scat. *PLOS ONE* 11:e0155615.

Keiter, D. A., and J. C. Beasley. 2017. Hog heaven? Challenges of managing introduced wild pigs in natural areas. *Natural Areas Journal* 37:6–16.

Keiter, D. A., A. J. Davis, O. E. Rhodes, Jr., F. L. Cunningham, J. C. Kilgo, K. M. Pepin, and J. C. Beasley. 2017a. Effects of scale of movement, detection probability, and true population density on common methods of estimating population density. *Scientific Reports* 7:9446.

Keiter, D. A., J. C. Kilgo, M. A. Vukovich, F. L. Cunningham, and J. C. Beasley. 2017b. Development of knownfate survival monitoring techniques for juvenile wild pigs (*Sus scrofa*). *Wildlife Research* 44:165–173.

Kierepka, E. M., S. D. Unger, D. A. Keiter, J. C. Beasley, O. E. Rhodes, F. L. Cunningham, and A. J. Piaggio. 2016. Identification of robust microsatellite markers for wild pig fecal DNA. *Journal of Wildlife Management* 80:1120–1128.

Kight, J. 1962. An ecological study of the bobcat, *Lynx rufus* (Schreber), in west-central South Carolina. Thesis, University of Georgia, Athens, GA.

Kilgo, J. C., and J. I. Blake, editors. 2005. *Ecology and Management of a Forested Landscape: Fifty Years on the Savannah River Site*. Island Press, Washington, DC.

Kilham, L. 1982. Cleaning/feeding symbioses of common crows with cattle and feral hogs. *Journal of Field Ornithology* 53:275–276.

Kirkman, L. K., and R. R. Sharitz. 1994. Vegetation disturbance and maintenance of diversity in intermittently flooded Carolina bays in South Carolina. *Ecological Applications* 4:177–188.

Kurz, J. C. 1971. A study of feral hog movements and ecology on the Savannah River Plant, South Carolina. Thesis, University of Georgia, Athens, GA.

Kurz, J. C., and R. L. Marchinton. 1972. Radiotelemetry studies of feral hogs in South Carolina. *Journal of Wildlife Management* 36:1240–1248.

Law Environmental. 1988. *Wildlife Assessment Study of Citrus Grove Development Tracts*. Law Environmental, Inc., Kennesaw, GA.

Leiser, O. P., J. L. Corn, B. S. Schmit, P. S. Keim, and J. T. Foster. 2013. Feral swine brucellosis in the United States and prospective genomic techniques for disease epidemiology. *Veterinary Microbiology* 166:1–10.

Lipscomb, D. J. 1989. Impacts of feral hogs on longleaf pine regeneration. *Southern Journal of Applied Forestry* 13:177–181.

Lucas, E. G. 1977. Feral hogs: Problems and control on national forest lands. Pages 17–21 *in* G. W. Wood, editor. *Research and Management of Wild Hog Populations*. Belle Baruch Forest Science Institute of Clemson University, Georgetown, SC.

Maehr, D. S., J. C. Roof, E. D. Land, J. W. McCown, R. C. Belden, and W. B. Frankenberger. 1989. Fates of wild hogs released into occupied Florida panther home range. *Florida Field Naturalist* 17:42–43.

Martin, R. J., J. L. Gobble, T. H. Hartsock, H. B. Graves, and J. H. Ziegler. 1973. Characterization of an obese syndrome in the pig. *Proceedings of the Society of Experimental Biology and Medicine* 143:198–203.

Marx, J. 2002. Unraveling the causes of diabetes. *Science* 296:686–689.

Matschke, G. H. 1962. Trapping and handling European wild hogs. *Proceedings of the Annual Conference of the Southeastern Association of Game and Fish Commissioners* 16:21–24.

Matschke, G. H. 1963. An eye lens-nutrition study of penned European wild hogs. *Proceedings of the Annual Conference of the Southeastern Association of Game and Fish Commissioners* 17:20–27.

Matschke, G. H. 1964. The influence of oak mast on European wild hog reproduction. *Proceedings of the Annual Conference of the Southeastern Association of Game and Fish Commissioners* 18:35–39.

Matschke, G. H. 1967. Aging European wild hogs by dentition. *Journal of Wildlife Management* 31:109–113.

Matschke, G. H., and J. P. Hardister. 1966. Movements of transplanted European wild boar in North Carolina and Tennessee. *Proceedings of the Annual Conference of the Southeastern Association of Fish and Wildlife Agencies* 20:74–84.

Mayer, J. J. 2002. *A Simple Field Technique for Age Determination of Adult Wild Pigs: Environmental Information Document*. WSRC-RP-2002-00635, Westinghouse Savannah River Company, Aiken, SC.

Mayer, J. J. 2003. *Total Body Mass Estimation Methodology for Wild Pigs at the Savannah River Site: Environmental Information Document*. WSRC-RP-2003-00317, Westinghouse Savannah River Company, Aiken, SC.

Mayer, J. J. 2005. Wild hog. Pages 372–377 *in* J. C. Kilgo and J. I. Blake, editors. *Ecology and Management of a Forested Landscape: Fifty Years on the Savannah River Site*. Island Press, Corvela, CA.

Mayer, J. J. 2009. Wild pig population biology. Pages 157–191 *in* J. J. Mayer and I. L. Brisbin, Jr., editors. *Wild Pigs: Biology, Damage, Control Techniques and Management*. SRNL-RP-2009-00869, Savannah River National Laboratory, Aiken, SC.

Mayer, J. J. 2013. Wild pig attacks on humans. *Proceedings of the Wildlife Damage Management Conference* 15:17–25.

Mayer, J. J. 2014. Estimation of the number of wild pigs found in the United States. SRNS-STI-2014-00292, Savannah River Nuclear Solutions, LLC, Savannah River Site, Aiken, SC.

Mayer, J. J., and J. C. Beasley. 2018. Wild pigs. Pages 219–248 *in* W. C. Pitt, J. C. Beasley, and G. W. Witmer, editors. *Ecology and Management of Terrestrial Vertebrate Invasive Species in the United States*. CRC Press, Boca Raton, FL.

Mayer, J. J., and I. L. Brisbin, Jr. 1986. A note on the scent marking behavior of two captive-reared feral boars. *Applied Animal Behaviour Science* 16:85–90.

Mayer, J. J., and I. L. Brisbin, Jr. 1988. Sex identification of *Sus scrofa* based on canine morphology. *Journal of Mammalogy* 69:408–412.

Mayer, J. J., and I. L. Brisbin, Jr. 1993. Distinguishing feral hogs from introduced wild boar and their hybrids: A review of past and present efforts. Pages 28–49 *in* C. W. Hanselka and J. F. Cadenhead, editors. *Feral Swine: A Compendium for Resource Managers*. Texas Agricultural Extension Service, Kerrville, TX.

Mayer, J. J., and I. L. Brisbin, Jr. 1995. Feral swine and their role in the conservation of global livestock genetic diversity. Pages 175–179 *in* R. D. Crawford, E. E. Lister, and J. T. Buckley, editors. *Proceedings of the Third Global Conference on Conservation of Domestic Animal Genetic Resources Rare Breeds International*. Warwickshire, England, United Kingdom.

Mayer, J. J., and I. L. Brisbin, Jr. 2008. *Wild Pigs in the United States: Their History, Comparative Morphology, and Current Status*. Second Edition. The University of Georgia Press, Athens, GA.

Mayer, J. J., and I. L. Brisbin, Jr., editors. 2009. *Wild Pigs: Biology, Damage, Control Techniques and Management*. SRNL-RP-2009-00869, Savannah River National Laboratory, Aiken, SC.

Mayer, J. J., and I. L. Brisbin, Jr. 2012. *Wild Pigs of the Savannah River Site*. SRNL-RP-2011-00295, Savannah River National Laboratory, Aiken, SC.

Mayer, J. J., I. L. Brisbin, Jr., and J. M. Sweeney. 1989. Temporal dynamics of color phenotypes in an isolated population of feral swine. *Acta Theriologica* 34:243–248.

Mayer, J. J., and P. E. Johns. 2007. Characterization of wild pig-vehicle collisions. *Proceedings of the Wildlife Damage Management Conference* 12:175–187.

Mayer, J. J., E. A. Nelson, and L. D. Wike. 2000. Selective depredation of planted hardwood seedlings by wild pigs in a wetland restoration area. *Ecological Engineering* 15:S79–S85.

Mayer, J. J., F. D. Martin, and I. L. Brisbin, Jr. 2002. Characteristics of wild pig farrowing nests and beds in the upper Coastal Plain of South Carolina. *Applied Animal Behaviour Science* 78:1–17.

Mayer, J. J., and L. A. Moore-Barnhill. 2009. Savannah River Site. Pages 331–340 *in* J. J. Mayer and I. L. Brisbin, Jr., editors. *Wild Pigs: Biology, Damage, Control Techniques and Management*. SRNL-RP-2009-00869, Savannah River National Laboratory, Aiken, SC.

McCann, B. E., M. Matthew, R. Newman, B. Schmit, S. Swaffort, R. Sweitzer, and R. Simmons. 2014. Mitochondrial diversity supports multiple origins for invasive pigs. *Journal of Wildlife Management* 78:201–2013.

McVicar, J. W., C. A. Mebus, H. N. Becker, R. C. Belden, and E. P. J. Gibbs. 1981. Induced African swine fever in feral pigs: Efficacy of diagnostic tests. *Journal of the American Veterinary Medical Association* 179:441–446.

Mengak, M. T. 2016. 2015 Georgia wild pig survey: Final report. Publ. 16–23. Warnell School of Forestry and Natural Resources, University of Georgia, Athens, GA.

Merrill, M. M., R. K. Boughton, C. C. Lord, K. A. Sayler, B. Wight, W. M. Anderson, and S. M. Wisely. 2018. Wild pigs as sentinels for hard ticks. A case study from south-central Florida. *International Journal of Parasitology: Parasites and Wildlife* 7:161–170.

Miller, S. L. 1979. Relationship of diet digestibility coefficients and body condition indices of feral swine and white-tailed deer. Thesis, University of Georgia, Athens, GA.

Miller, S. L., I. L. Brisbin, Jr., and R. W. Seerley. 1987. Indirect estimation of the digestible energy in the diets of swine by post-mortem analyses of digesta. *Acta Theriologica* 32:475–488.

Nettles, V. F. 1984. Brucellosis in wild swine. *Proceedings of the Annual Meeting of the US Animal Health Association* 88:203–204.

Nettles, V. F. 1989. Disease of wild swine. Pages 16–18 *in* N. Black, editor. *Proceedings: Feral Pig Symposium*. April 27–29, Orlando, FL. Livestock Conservation Institute, Madison, WI.

Nettles, V. F. 1991. Short- and long-term strategies for resolving problems of pseudorabies and swine brucellosis in feral swine. *Proceedings of the Annual Meeting of the US Animal Health Association* 95:551–556.

Nettles, V. F. 1997. Feral swine: Where we've been, where we're going. Pages 1.1–1.9 *in* K. L. Schmitz, editor. *Proceedings – National Feral Swine Symposium*. US Department of Agriculture/Animal and Plant Health Inspection Service, Orlando, FL.

Nettles, V. F., J. L. Corn, G. A. Erickson, and D. A. Jessup. 1989. A survey of wild swine in the United states for evidence of hog cholera. *Journal of Wildlife Diseases* 25:61–65.

Nettles, V. F., and G. A. Erickson. 1984. Psuedorabies in wild swine. *Proceedings of the Annual Meeting of the US Animal Health Association* 88:505–506.

Oldenkamp, R. E., A. L. Bryan, Jr., R. A. Kennamer, J. C. Leaphart, S. C. Webster, and J. C. Beasley. 2017. Trace elements and radiocesium in game species near contaminated sites. *Journal of Wildlife Management* 81:1338–1350.

Paolini, K., B. K. Strickland, J. Tegt, K. C. VerCauteren, and G. Street. 2019. The habitat functional response links seasonal third-order selection to second-order landscape characteristics. *Ecology and Evolution* 9:4683–4691.

Payeur, J. B. 1989. Feral swine: A potential threat to domestic cattle and swine. Pages 19–33 *in* N. Black, editor. *Proceedings: Feral Pig Symposium.* April 27–29, Orlando, FL, Livestock Conservation Institute, Madison, WI.

Pederson, K., S. N. Bevins, B. S. Schmit, M. W. Lutman, M. P. Milleson, C. T. Turnage, T. T. Bigelow, T. J. DeLiberto. 2012. Apparent prevalence of swine brucellosis in feral swine in the United States. *Human-Wildlife Interactions* 6:38–47.

Peel, M. C., B. L. Finlayson, and T. A. McMahon. 2007. Updated world map of the Köppen-Geiger climate classification. *Hydrology and Earth System Sciences* 11:1633–1644.

Peine, J. D., and J. A. Farmer. 1990. Wild hog management program at Great Smoky Mountains National Park. *Proceedings of the Vertebrate Pest Conference* 14:221–227.

Pepin, K. M., A. J. Davis, J. C. Beasley, R. Boughton, T. Campbell, S. M. Cooper, W. Gaston, et al. 2016. Contact heterogeneities in feral swine: Implications for disease management and future research. *Ecosphere* 7:e01230. doi:10.1002/ecs2.1230.

Pickett, M. F., and D. W. Pickett. 2011. *The European Struggle to Settle North America: Colonizing Attempts by England, France and Spain*, 1521–1608. McFarland and Company, Inc., Jefferson, NC.

Pirtle, E. C., J. M. Sacks, V. F. Nettles, and E. A. Rollor, III. 1989. Prevalence and transmission of psuedorabies virus in an isolated population of feral swine. *Journal of Wildlife Diseases* 25:605–607.

Romero, C. H., P. Meade, J. Santagata, K. Gillis, G. Lollis, E. C. Hahn, and E. P. J. Gibbs. 1997. Genital infection and transmission of pseudorabies virus in feral swine in Florida, USA. *Veterinary Microbiology* 55:131–139.

Rossell, C. R., Jr., H. D. Clarke, M. Schultz, E. Schwartzman, and S. C. Patch. 2016. Description of rich montane seeps and effects of wild pigs on the plant and salamander assemblages. *American Midland Naturalist* 175:139–154.

Savereno, A. J., and T. T. Fendley. 1989. *A Bibliography of Research Pertinent to the Management of Feral and Wild Hogs (Sus scrofa) in the Southeastern United States.* A report submitted to the Savannah River Ecology Laboratory, Aiken, South Carolina. Department of Aquaculture, Fisheries and Wildlife, Clemson University, Clemson, SC.

Schortemeyer, J. L., and J. W. McCown. 1988. *Big Cypress National Preserve Deer and Hog Annual Report.* Florida Game and Fresh Water Fish Commission, Naples, FL.

Schrecengost, J. D., J. C. Kilgo, D. Mallard, H. S. Ray, and K. V. Miller. 2008. Seasonal food habits of the coyote in the South Carolina coastal plain. *Southeastern Naturalist* 7:135–144.

Scott, C. D., and M. R. Pelton. 1975. Seasonal food habits of the European wild hog in the Great Smoky Mountains National Park. *Proceedings of the Annual Conference of the Southeastern Association of Fish and Wildlife Agencies* 29:585–593.

Scott, R. A., S. G. Cornelius, and H. J. Mersmann. 1981. Effects of age on lipogenesis and lipolysis in lean and obese swine. *Journal of Animal Science* 52:505–511.

Shaffer, M. L. 1976. Behavior of the European wild boar (*Sus scrofa*) in the Great Smoky Mountains National Park. Thesis, Duke University, Durham, NC.

Shaw, A. C. 1941. The European wild hog in America. *Transactions of the North American Wildlife Conference* 5:436–441.

Singer, F. J. 1976. The European wild boar in the Great Smoky Mountains National Park: Problem analysis and proposed research. NPS-SER Management Report No. 6. Uplands Field Research Laboratory, Great Smoky Mountains National Park, Gatlinburg, TN.

Singer, F. J. 1981. Wild pig populations in the national parks. *Environmental Management* 5:263–270.

Singer, F. J., and B. B. Ackerman. 1981. *Food Availability, Reproduction and Condition of European Wild Boar in Great Smoky Mountains National Park.* Research/Resource Management Report No. 43. Uplands Field Research Laboratory, Great Smoky Mountains National Park, Gatlinburg, TN.

Singer, F. J., D. K. Otto, A. R. Tipton, and C. P. Hable. 1981. Home range, movements and habitat use of European wild boar in Tennessee. *Journal of Wildlife Management* 45:343–353.

Singer, F. J., and D. Stoneburner. 1979. Feral pig management. Memorandum N1615. Uplands Field Research Laboratory, Great Smoky Mountains National Park, Gatlinburg, TN.

Singer, F. J., W. T. Swank, and E. E. C. Clebsch. 1984. The effects of wild pig rooting in a deciduous forest. *Journal of Wildlife Management* 48:464–473.

Smith, H. M., Jr., W. R. Davidson, V. F. Nettles, and R. R. Gerrish. 1982. Parasitisms among wild swine in southeastern United States. *Journal of the American Veterinary Medical Association* 181:1281–1284.

Smith, J. B., L. J. Laatsch, and J. C. Beasley. 2017. Spatial complexity of carcass location influences vertebrate scavenger efficiency and species composition. *Scientific Reports* 7:10250.

Smith, J. B., T. D. Tuberville, and J. C. Beasley. 2018. Hunting and consumption patterns of southeastern USA hunters and anglers. *Journal of Fish and Wildlife Management* 9:321–329.

Smith, M. W., M. H. Smith, and I. L. Brisbin, Jr. 1980. Genetic variability and domestication in swine. *Journal of Mammalogy* 61:39–45.

Southeastern Cooperative Wildlife Disease Study. 2007. *Diagnostic Services Section – Final Report*. Case No. CC230–07, Southeastern Cooperative Wildlife Disease Study, Athens, GA.

Sparklin, B. D., M. S. Mitchell, L. B. Hanson, D. B. Jolley, and S. S. Ditchkoff. 2009. Territoriality of feral pigs in a highly persecuted population on Fort Benning, Georgia. *Journal of Wildlife Management* 73:497–502.

Sparklin, W. D. 2009. Territoriality and habitat selection of feral pigs on Fort Benning, Georgia, USA. Thesis, University of Montana, Missoula, MT.

Stallknecht, D. E., W. O. Fletcher, G. A. Erickson, and V. F. Nettles. 1987. Antibodies to vesicular stomatitis New Jersey type virus in wild and domestic sentinel swine. *American Journal of Epidemiology* 125:1058–1065.

Stallknecht, D. E., E. W. Howerth, C. L. Reeves, and B. S. Seal. 1999. Potential for contact and mechanical vector transmission of vesicular stomatitis virus New Jersey in pigs. *American Journal of Veterinary Research* 60:43–48.

Stallknecht, D. E., D. M. Kavanaugh, J. L. Corn, K. A. Eernisse, J. A. Comer, and V. F. Nettles. 1993. Feral swine as a potential amplifying host for vesicular stomatitis virus New Jersey serotype on Ossabaw Island, Georgia. *Journal of Wildlife Diseases* 29:377–383.

Stallknecht, D. E., V. F. Nettles, G. A. Erickson, and D. A. Jessup. 1986. Antibodies to vesicular stomatitis virus in populations of feral swine in the United States. *Journal of Wildlife Diseases* 22:320–325.

Stallknecht, D. E., V. F. Nettles, W. O. Fletcher, and G. A. Erickson. 1985. Enzootic vesicular stomatitis New Jersey type in an insular feral swine population. *American Journal of Epidemiology* 122:876–883.

Stegeman, L. J. 1938. The European wild boar in the Cherokee National Forest, Tennessee. *Journal of Mammalogy* 19:279–290.

Stiver, W. H., and E. K. Delozier. 2009. Great Smoky Mountains National Park wild hog control program. Pages 341–352 *in* J. J. Mayer and I. L. Brisbin, Jr., editors. *Wild Pigs: Biology, Damage, Control Techniques and Management*. SRNL-RP-2009-00869, Savannah River National Laboratory, Aiken, SC.

Stribling, H. L. 1978. Radiocesium concentrations in two populations of naturally contaminated feral hogs (*Sus scrofa domesticus*). Thesis, Clemson University, Clemson, SC.

Stribling, H. L., and I. L. Brisbin, Jr. 1978. Fecal contamination analyses: Investigations of a new method for monitoring radiocesium levels in free-ranging feral swine. *Association of Southeastern Biologists Bulletin* 25:54.

Stribling, H. L., I. L. Brisbin, Jr., and J. R. Sweeney. 1986a. Portable counter calibration adjustments required to monitor feral swine radiocesium levels. *Health Physics* 50:663–665.

Stribling, H. L., I. L. Brisbin, Jr., and J. R. Sweeney. 1986b. Radiocesium concentrations in two populations of feral hogs. *Health Physics* 50:852–854.

Stribling, H. L., I. L. Brisbin, Jr., J. R. Sweeney, and L. A. Stribling. 1984. Body fat reserves and their prediction in two populations of feral swine. *Journal of Wildlife Management* 48:635–639.

Sturek, M., E. A. Mokelke, J. Vuchetich, J. M. Edwards, M. Alloosh, and K. L. March. 2006. In-stent neointimal hyperplasia is less fibrous in metabolic syndrome Ossabaw compared to lean Yucatan swine (abstract). *FASEB Journal* 20:A1399–A1400.

Sweeney, J. M. 1970. Preliminary investigation of a feral hog (*Sus scrofa*) population on the Savannah River Plant, South Carolina. Thesis, University of Georgia, Athens, GA.

Sweeney, J. M., E. E. Provost, and J. R. Sweeney. 1970. A comparison of eye lens weight and tooth eruption pattern in age determination of feral hogs (Sus scrofa). *Proceedings of the Annual Conference of the Southeastern Association of Game and Fish Commissioners* 24:285–291.

Sweeney, J. M., J. R. Sweeney, and E. E. Provost. 1979. Reproductive biology of a feral hog population. *Journal of Wildlife Management* 43:555–559.

Sweeney, J. M., J. R. Sweeney, and S. W. Sweeney. 2003. Feral hog, *Sus scrofa*. Pages 1164–1179 *in* G. A. Feldhammer, B. C. Thompson, and J. A. Chapman, editors. *Wild Mammals of North America: Biology, Management, and Conservation*. The Johns Hopkins University Press, Baltimore, MD.

Sweeney, J. R. 1979. Ovarian activity in feral swine (abstract only). *Bulletin of the South Carolina Academy of Sciences* 41:74.

Tabak, M. A., A. J. Piaggio, R. S. Miller, R. A. Sweitzer, and H. B. Ernest. 2017. Anthropogenic factors predict movement of an invasive species. *Ecosphere* 8:e01844. doi:10.1002/ecs2.1844.

Tanger, S. M., K. Guidry, H. Nui, C. Richard, and M. Abreu. 2015. Dollar estimates of feral hog damage to agriculture in Louisiana. Research Information Sheet 113, Agricultural Economics and Agribusiness, Louisiana State University, Baton Rouge, LA.

Tate, J., editor. 1984. Techniques for controlling wild hogs in the Great Smoky Mountains National Park. *Proceedings of a Workshop*, November 29–30, Research/Resources Mgmt. Rpt. SRE-72. US Department of the Interior, National Park Service, Southeast Regional Office, Atlanta, GA.

Tipton, A. R. 1977. The use of population models in research and management of wild hogs. Pages 91–101 *in* G. W. Wood, editor. *Research and Management of Wild Hog Populations*. Belle Baruch Forest Science Institute of Clemson University, Georgetown, SC.

Todd, C. T., and M. T. Mengak. 2018. *The Impact of Wild Pig Hunting Outfitters on Pig Populations Across the Southeast*. Publication WSFNR-18-45. Warnell School of Forestry & Natural Resources, University of Georgia, Athens, GA.

Turner, K. L., E. F. Abernethy, L. M. Conner, O. E. Rhodes, Jr., and J. C. Beasley. 2017. Abiotic and biotic factors modulate carrion fate and scavenging community dynamics. *Ecology* 98:2413–2424.

US Department of Agriculture. 1981. European boar (Section 410). Pages 1–7 *in Wildlife Habitat Management Handbook: Southern Region*. FSH 2609.23R, US Forest Service, Atlanta, GA.

US Geological Survey. 2003. A Tapestry of Time and Terrain: The Union of Two Maps – Physiographic Regions. US Department of the Interior, Reston, VA. <http://tapestry.usgs.gov/physiogr/physio.html>. Accessed 15 Mar 2017.

van der Leek, M. L., H. N. Becker, E. C. Pirtle, P. Humphrey, C. L. Adams, B. P. All, G. A. Erickson, et al. 1993. Prevalence of pseudorabies (Aujeszky's disease) virus antibodies in feral swine in Florida. *Journal of Wildlife Diseases* 29:403–409.

Wakeley, P. C. 1954. Planting the Southern pine. *Forest Service Agricultural Monograph* 18:1–233.

Webster, S. C., F. L. Cunningham, J. C. Kilgo, M. Vukovich, O. E. Rhodes, Jr., and J. C. Beasley. 2017. Minimum effective dose and persistence of Rhodamine-B in wild pig (*Sus scrofa*) vibrissae. *Wildlife Society Bulletin* 41:764–769.

White, D. L. 2004. *Deerskins and Cotton: Ecological Impacts of Historical Land Use in the Central Savannah River Area of the Southeastern US Before 1950*. Southern Research Station, USDA Forest Service, Clemson University, Clemson, SC.

Williams, B. L., R. W. Holtfreter, S. S. Ditchkoff, and J. B. Grand. 2011a. Efficiency of time-lapse intervals and simple baits for camera surveys of wild pigs. *Journal of Wildlife Management* 75:655–659.

Williams, B. L., R. W. Holtfreter, S. S. Ditchkoff, and J. B. Grand. 2011b. Trap style influences wild pig behavior and trapping success. *Journal of Wildlife Management* 75:432–436.

Williamson, M. J., and M. R. Pelton. 1975. Some biochemical parameters of serum of European wild hogs. *Proceedings of the Annual Conference of the Southeastern Association of Game and Fish Commissioners* 29:672–679.

Williamson, M. J., and M. R. Pelton. 1976. Some hematological parameters of European wild hogs. *Journal of the Tennessee Academy of Sciences* 51:25–28.

Wood, G. W., editor. 1977. *Research and Management of Wild Hog Populations: Proceedings of a Symposium*. Belle Baruch Forest Science Institute of Clemson University, Georgetown, SC.

Wood, G. W., and R. E. Brenneman. 1977. Research and management of feral hogs on Hobcaw Barony. Pages 23–35 *in* G. W. Wood, editor. *Research and Management of Wild Hog Populations*. Belle Baruch Forest Science Institute of Clemson University, Georgetown, SC.

Wood, G. W., and R. E. Brenneman. 1980. Feral hog movements and habitat use in South Carolina. *Journal of Wildlife Management* 44:420–427.

Wood, G. W., J. B. Hendricks, and D. E. Goodman. 1976. Brucellosis in feral swine. *Journal of Wildlife Diseases* 12:579–582.

Wood, G. W., E. E. Johnson, Jr., and R. E. Brenneman. 1977. Observations on the use of succinylcholine chloride to immobilize feral hogs. *Journal of Wildlife Management* 41:798–800.

Wood, G. W., and D. N. Roark. 1980. Food habits of feral hogs in coastal South Carolina. *Journal of Wildlife Management* 44:506–511.

Wood, G. W., L. A. Woodward, D. C. Matthews, and J. R. Sweeney. 1992. Feral hog control efforts on a coastal South Carolina plantation. *Proceedings of the Annual Conference of the Southeastern Association of Fish and Wildlife Agencies* 46:167–178.

Zengel, S. A., and W. H Conner. 2008. Could wild pigs impact water quality and aquatic biota in floodplain wetland and stream habitats at Congaree National Park, South Carolina? *Proceedings of the 2008 South Carolina Water Resources Conference*. <http://works.bepress.com/william_conner1/7/>. Accessed 15 Oct 2010.

Zervanos, S. M., W. D. McCort, and H. B. Graves. 1983. Salt and water balance of feral vs. domestic Hampshire hogs. *Physiological Zoology* 56:67–77.

Zygmont, S. M., V. F. Nettles, E. B. Shotts, Jr., W. A. Carmen, and B. O. Blackburn. 1982. Brucellosis in wild swine: A serologic and bacteriologic survey in the southeastern United States and Hawaii. *Journal of the American Veterinary Medical Association* 181:1285–1287.

17 Wild Pigs in the Pacific Islands

Steven C. Hess, Nathaniel H. Wehr, and Creighton M. Litton

CONTENTS

17.1 INTRODUCTION

Wild pigs (*Sus scrofa*) are perhaps the most abundant, widespread, and ecologically significant introduced large vertebrate currently found on oceanic islands in the Pacific basin. They have been most studied in the Hawaiian Islands, where they are most commonly known as feral pigs, and are representative of many other Pacific islands. The spatial and temporal distribution of wild pigs in the Pacific islands coincides almost entirely with the prehistoric discovery and contemporary human occupation of these islands. Wild pigs in Hawaii today represent a mixture of several strains of domestic swine, Asiatic wild boar, and European wild boar (Diong 1982, Tomich 1986). Pigs now occupy all but the most arid regions of many Pacific islands and are especially abundant in densely forested landscapes.

The Pacific islands include more than 2,000 islands that lie within the Pacific Ocean with a diversity of geographic, climatic, and ecological features, including incredible biodiversity (Finucane et al. 2012). The region encompasses approximately 155°W to 130°E and 15°S to 25°N (Keener et al. 2013) and is typically broken into 3 specific regions: Polynesia, Micronesia, and Melanesia. Micronesia and Melanesia are situated along the western-most extent of the Pacific adjacent to the Philippines to the north and Australia to the south. Polynesia is the largest of the 3 regions forming an expansive triangle extending from New Zealand in the southwest to Hawaii in the north to Easter Island in the southeast. The climate is generally characterized as tropical with distinct wet and dry seasons where monthly temperatures only vary approximately 2°C between the warmest and coldest months (Finucane et al. 2012, Keener et al. 2013). Atmospheric and oceanic circulation patterns have a substantial effect on the timing, duration, and intensity of the wet and dry seasons, and given the geographic area encompassed by the Pacific islands, these seasons vary across the region (Finucane et al. 2012). Accordingly, the wet season in the northern (including Hawaii) and southern (including Samoa) portions of the region is during November-April, whereas the western portion of the Pacific islands (including the Mariana and Marshall Islands) is during May-October (Finucane et al. 2012, Keener et al. 2013). Variation in elevation between islands also influences annual rainfall. Islands characterized by mountainous terrain often receive much greater rainfall totals compared to islands near sea level that lack mountainous terrain (Finucane et al. 2012). Several volcanic mountains characterize the islands of Hawaii (Mauna Kea: 4,205 m, Mauna Loa: 4,170 m) and Maui (Haleakalā: 3,055 m), and can have seasonal snow and temperatures as cold as –15°C (Finucane et al. 2012, Keener et al. 2013). The geologic origin of the Pacific islands has resulted in a variety of island types including volcanic islands, atolls, and limestone islands (Finucane et al. 2012), leading to a diversity of ecosystems throughout the region. Ecosystems found within Pacific islands include montane forests, alpine shrublands, coastal plains and wetlands, mangroves, coastal dunes, and lowland grasslands (Finucane et al. 2012).

Pigs are culturally and socially important for at least 2 reasons: 1) the contemporary popularity of subsistence and recreational hunting on many Pacific islands, and 2) the traditional mythological, religious, and ceremonial symbolism of pigs throughout Polynesia, particularly the Hawaiian legend of Kamapua'a, the hog child demigod. However, an extensive body of scientific literature has documented severe ecological alteration caused by wild pigs on Pacific islands (Nogueira-Filho et al. 2009, Leopold and Hess 2016), most of which had no native ground-dwelling mammals prior to the arrival of humans. Thus, a large number of wild pigs are routinely removed to reduce damage to natural areas, agriculture, ornamental plants in residential areas, and resort destinations such as golf courses. Fencing has been used to protect areas from wild pigs, and a multitude of methods have been developed and employed to eradicate pigs inside these exclosures once fences have been constructed: trapping, snaring, hunting (often with the assistance of dogs), and the use of toxicants in a few cases (Nogueira et al. 2007). Pigs have been successfully eradicated from a few small central Pacific islands and large natural areas of other islands totaling approximately 750 km^2 (Hess and Jacobi 2011), which has resulted in considerable ecosystem recovery. However, some fundamental aspects of the life history and ecology of wild pigs in the Pacific islands are still poorly known due to the inaccessible environments they inhabit and the lack of available methods to efficiently and accurately estimate abundance in this region. Research and management of wild pigs on Pacific islands are also poorly integrated and understood with respect to many continental areas. Policy and management would benefit from addressing both the protection of natural resources through the eradication of wild pigs in some areas, and the production and maintenance of sustained-yield populations for subsistence and recreational hunting in other, sometimes adjacent areas.

17.2 HISTORIC AND PRESENT DISTRIBUTION

17.2.1 Origin and Early History: Neolithic Voyagers of the Pacific

The remarkable history of wild pigs in the central Pacific islands originated with the banded pig (*Sus scrofa vittautus*; Boie, 1828) from Southeast Asia, which were brought for subsistence on oceanic voyages by the Neolithic Lapita culture, ancestors of the Polynesians (Larson et al. 2005). All 8 *Sus* species endemic to Southeast Asia (*S. barbatus*, *S. celebensis*, *S. philippensis*, *S. scrofa*, *S. cebifrons*, *S. oliveri*, *S. ahoenobarbus*, and *S. verrucosus*) originated from the region that includes the Malay Peninsula, Sumatra, Java, Borneo, and the Philippines (Melletti and Meijaard 2018). Two separate human-mediated dispersals of domestic pigs from this region into the Pacific islands have been documented using mitochondrial DNA examinations of ancient and contemporary pig specimens (Larson et al. 2007). A Pacific clade of *S. scrofa*, including those that were brought to the Hawaiian Islands, dispersed from mainland Southeast Asia to Java, Sumatra, and the islands of Wallacea and Oceania, while a separate dispersal carried pigs from mainland East Asia to western Micronesia, Taiwan, and the Philippines. The first appearance of *S. scrofa* in Wallacea was associated with the arrival of the Lapita culture between 5,000 and 1,500 B.C.E.

Archaeological evidence suggests pigs were brought to the Hawaiian Islands by Polynesian voyagers as early as ~1,220 C.E. (Wilmshurst et al. 2011). Supporting this, skeletal remains of pigs were documented from some of the earliest known Polynesian settlements at Bellows, O'ahu (Pearson et al. 1971), and Halawa, Moloka'i (Kirch and Kelly 1975, Kirch 1982). Additionally, Burney et al. (2001) documented pig bones from prehistoric habitation near a sinkhole and cave system on Kaua'i. Polynesian ancestry in most (70%) of 57 contemporary pigs from 4 Hawaiian Islands has recently been confirmed by the presence of a mutation in the *MC1R* gene that results in black pelage coloration (Linderholm et al. 2016). However, future work may result in a more complete picture of the ancestry of Hawaiian wild pigs.

Both skeletal remains of pigs and early historic observations indicate that the Polynesian pig (pua'a) was a smaller animal than contemporary Hawaiian wild pigs, weighing only 27–45 kg (Ziegler 2002). Polynesian pigs were abundant small animals with a limited distribution, possibly due to lack of earthworms or other sufficiently nutritious foods beyond areas of human influence. It was historically noted that, not long before European contact, well over a thousand pigs would be cooked and eaten at the consecration of an important temple or *heiau* (Ziegler 2002). Ships of Captain James Cook's voyage were provisioned with about 600 pigs during 4 months in 1778 and 1779 (Tomich 1986). Cook traded for pigs in Hawaii in 1779 and noted, "We could seldom get any above fifty or sixty pounds [23–27 kg] weight." However, another historical account describes Cook trading iron or knives for "A pig one fathom [1.8 m] long..." (Westervelt 1923). Ellis (1827) also observed in 1823 that natives possessed, "a small species of hogs, with long heads and small erect ears," that were sometimes found in the mountains.

Pua'a repeatedly interbred with multiple varieties of domestic swine, Asiatic wild boar, and European wild boar introduced by explorers and colonists beginning in the 1770s. The first of these introductions was a boar and a sow brought from England to Ni'ihau in 1778 by Cook (Tomich 1986). Pigs have subsequently become the most abundant large mammal throughout many Pacific islands. The larger size of contemporary Hawaiian pigs and admixture of Polynesian and other domestic and wild swine, as well as their great abundance, broad distribution, extraordinary reproductive capability, and variable pelage coloration contributes to the hybrid vigor of these animals.

17.2.2 Cultural Aspects of Pigs in the Pacific Islands

Unlike most other invasive species and swine throughout the world, pigs have a culturally important history in the Hawaiian Islands. Domesticated pua'a were associated with strong cultural and supernatural values in Hawaii prior to Western contact and were used in ritual slaughter during religious and ceremonial occasions and as a substitute for human sacrifice in other Polynesian cultures (Giovas 2006). In addition to being a prestigious possession and source of food primarily limited to those of high status (Giovas 2006), the adventures of the pig child, Kamapua'a, a powerful demigod who ranged over the islands, were described in oral tradition (Charlot 1987; Figure 17.1). The namesake of the traditional Hawaiian land management system, ahupua'a, refers directly to the pua'a, and its importance among resources that were collected and offered during the annual tributes in honor of the god Lono, associated with fertility, agriculture, rainfall, music, and peace. The legend of Kamapua'a recounts his pursuit by man to be punished for mischievous actions, rather than for recreation or subsistence (Charlot 1987). According to oral tradition and ethnohistory on other Pacific islands, pigs may have been exterminated for uprooting crops and gardens (Giovas 2006). The contemporary social importance of pigs has shifted nearly completely from domestication and mythology to the most pursued large wild game species for subsistence and recreation.

Notwithstanding small populations of loosely controlled and free-ranging animals that existed in ancient times, traditional and historical evidence suggests these animals remained largely semi-domesticated, depending mainly on the food and shelter provided by the periphery of the *kauhale* (homestead), and extending into adjacent lowland forests (Maly 1998). Although the starchy cores of native Hāpu'u tree ferns (*Cibotium* spp.) are seasonally important sources of food (Diong 1982), early roaming populations were limited nutritionally and anthropogenic sources of food were extremely important. Before the arrival of Europeans, native Hawaiian forests were devoid of abundant non-native fruits such as strawberry guava (*Psidium cattleyanum*), but more importantly, major

FIGURE 17.1 Image of *Kamapua'a*, the Hawaiian hog child demigod among rainforest, flowing springs, and volcanic cinder cones. (Photo by S. M. ʻOhukaniʻōhiʻa Gon III. With permission.)

sources of protein. While wild pigs were undoubtedly capable of surviving on Pacific islands, the introduction of non-native fruit producing plants and non-native earthworms substantially increased the availability of carbohydrates and protein to wild pigs, which enabled populations to increase precipitously and help to support large wild populations of pigs today.

17.2.3 Present Status

In the Pacific, wild pigs inhabit nearly every island throughout the region (Wehr et al. 2018; Table 17.1). In the Hawaiian Islands, wild pigs are present on the islands of Kauaʻi, Oʻahu, Molokaʻi, Maui, and Hawaii, but were eradicated from the island Lānaʻi by the mid-1930s (Tomich 1986, Hess and Jacobi 2011). Additionally, wild pigs no longer occur on Kahoʻolawe, Laysan, or Niʻihau for various reasons (Tomich 1986). Tomich (1986) stated the presence of wild pigs on Maui was limited to the mountains of west Maui and Kīpahulu Valley prior to the 1960s. Other tropical Pacific islands have notably large populations of wild pigs, including Guam, the Northern Mariana Islands, and Samoa (National Park Service 1997; Kessler 2002, 2011). Spanish colonizers introduced domestic pigs to the Marianas between 1672 and 1685 (Intoh 1986). Pigs were eradicated from the Mariana island of Sarigan in 1998 (Kessler 2002), but an attempt to eradicate pigs from Anatahan was discontinued after a major volcanic eruption in 2003 (Kessler 2011). Wild pigs are found on all the major islands of Samoa except Aunuʻu, and are generally regarded as a resource management problem in national parks (National Park Service 1997).

17.3 REGIONAL ENVIRONMENTAL ASPECTS IN THE PACIFIC ISLANDS

17.3.1 Behavior

Rigorous information from direct study of home range and dispersal of wild pigs on Pacific islands is limited. Giffin (1978) reported wild pig home ranges were 5.2–10.4 km^2 in mountain-pasture

TABLE 17.1
Presence of Wild Pigs on Select Pacific Islands

Pacific Island	Present	Absent
American Samoa	X[a]	
Bonin Islands	X[c]	
Cocos Islands		X[d]
Easter Island		X[e]
Fiji	X[a,b]	
French Polynesia	X[a,b]	
Galapagos Islands	X[a,b]	
Guam	X[a,b]	
Hawaiian Islands	X[a,b]	
Indonesia	X[a,b]	
Juan Fernandez	X[a,b]	
Kiribati	X[a]	
Marshall Islands	X[a]	
Micronesia	X[a]	
Nauru	X[a]	
New Caledonia	X[a,b]	
Niue	X[a]	
Norfolk Island	X[b]	
Northern Mariana Islands	X[a,b]	
Palau	X[a,b]	
Papua New Guinea	X[a,b]	
Pitcairn Island		X[e]
Solomon Islands	X[a,b]	
Tokelau Islands	X[b]	
Tonga	X[a,b]	
Tuvalu	X[f]	
Vanuatu	X[b]	
Wake Island		X[g]
Wallis and Futuna	X[a]	

[a] Barrios-Garcia and Ballari (2012).
[b] Long (2003)
[c] Kawakami and Okochi (2010)
[d] Woodroffe and Berry (1994), New World Encyclopedia Contributors (2017)
[e] Allen et al. (2001), Giovas (2006)
[f] McKinnon (2009)
[g] Griffiths et al. (2014)
Source: Reproduced from Wehr et al. (2018).

habitat and 1.3 to 5.2 km^2 in rainforest habitat on Hawaii Island. Diong (1982) found the mean home range of 5 boars and 4 sows in Kīpahulu Valley, Maui to be 2.03 km^2 and 1.57 km^2, respectively. Salbosa (2009) reported home ranges of 4 pigs (2 boars and 2 sows) in the Pelekunu Valley of Moloka‘i varied from 0.051 to 0.928 km^2. Although wild pigs make small-scale movements in response to food availability, there is no evidence to suggest that they migrate seasonally (Giffin 1978, Mitchell et al. 2009). Limited movements and home ranges are supported by the fact that today, pigs persist and can be quite difficult to eradicate from enclosed areas as small as 8 km^2 in Hawaii. Boars tend to be solitary while sows travel in matriarchal groups, and movements are generally during crepuscular hours (Diong 1982, Nogueira et al. 2007). Sows construct farrowing nests in tree holes, caves, or open habitat (Diong 1982). In the Kīpahulu Valley, Maui, the average size of matriarchal groups is 2.6 (range 1–9 pigs), and groups greater than 6 pigs are rare (Diong 1982). Movements vary depending on habitat. For example, Nogueira et al. (2007) reported that wild pigs traveled greater distances in mountain-pasture habitat compared to rain forests, possibly due to less water and cover in mountain-pasture habitats.

17.3.2 Diseases and Parasites

Zoonotic diseases and other pathogens (see Chapter 5) in tropical environments of the Pacific islands have negatively affected public health and animal husbandry. Although there is limited information linking many of these pathogens directly to wild pigs, outbreaks of trichinosis after human consumption of barbecued wild pig have been documented (Barrett-Connor et al. 1976), and the annual incidence of leptospirosis in Hawaii has consistently been among the greatest in the United States (Katz et al. 2011). Additionally, 2 common serovars in pigs and humans suggest a link between these infections, despite the presence of multiple *Leptospira* serovars and multiple mammalian hosts for this disease (Buchholz et al. 2016). Wild pigs in Hawaii have tested seropositive for pseudorabies and brucellosis, where the prevalence rate of brucellosis was the greatest detected among 35 states sampled (14.4%; Pederson et al. 2012, 2013). Porcine epidemic diarrhea virus (PEDV) has spilled over from domestic pigs into wild pig populations, including in Hawaii where seropositive samples have been detected on O‘ahu and Kaua‘i (Bevins et al. 2018). The first case of PEDV in domestic pigs was confirmed on Oahu in November 2014, but a sample from a wild pig on O‘ahu tested positive for PEDV in April 2014, suggesting that PEDV may have been present longer (Bevins et al. 2018). Bovine tuberculosis (*Mycobacterium bovis*) has been isolated from wild pigs on the island Moloka‘i. A multispecies outbreak was controlled by culling infected swine and axis deer (*Axis axis*), depopulating infected cattle, and restricting cattle production to areas without pigs (Essey et al. 1983, Miller and Sweeney 2013). Other human health threats associated with wild pigs are poorly understood in Hawaii and warrant further study. For example, fecal contamination of soil and water with *Enterococcus* bacteria is a major problem resulting in runoff into nearshore waters that are popular for recreation. However, exclusion of pigs did not affect *Enterococcus* bacteria levels in soil runoff, which was more closely related to seasonality in rainfall, but further research is warranted (Dunkell et al. 2011a).

Diong (1982) reported that the most prevalent ectoparasite for wild pigs in Maui was the pig louse (*Haematopinus suis*; 85.4% of animals tested), but mange mites (*Sarcoptes scabiei*) were also found. Common internal parasites are lungworms (*Metastrongylus elongates*) and kidney worms (*Stephanurus dentatus*) at prevalence rates of 89.4% and 81.5%, respectively (Diong 1982). Additional parasites include roundworms (*Ascaris lumbricoides*), nodular worms (*Oesophagostomum* sp.), stomach worms (*Physocephalus* sp.), and whipworm (*Trichuris suis*; Diong 1982).

17.3.3 Food Habits

Similar to other regions, wild pigs are omnivorous, although their diet consists predominantly of plant matter (Nogueira et al. 2007; see Chapter 3). Wild pigs frequently consume hāpu‘u (Hawaiian tree fern; *Cibotium chamissoi* and *C. splendens*), which constitutes a major portion of their diet because its core is rich in starches and sugars (Diong 1982, Nogueira et al. 2007). Other fern species (*Sadleria* spp., *Marattia* spp., and *Pteridium* spp.), flowering plants (*Astelia* spp. and *Freycinetia arborea*) and fruits (strawberry guava; passion fruit, *Passiflora edulis*; banana poka, *Passiflora mollissima*; and papaya, *Carica papaya*) are also preferred food items (Stone 1985, Nogueira et al. 2007, Nogueira-Filho et al. 2009). Gosmore (*Hypochoeris radicata*) is a primary food item for wild pigs in alpine grassland habitat of Maui and mountain-pasture habitat in Mauna Kea, whereas ‘Ōhelo berry (*Vaccinium reticulatum*) is important in mountain-scrub forests of Mauna Kea (Diong 1982).

Seasonal dietary shifts, which are common throughout the range of wild pigs, have also been noted. For example, Diong (1982) found that tree ferns are an abundant food source for wild pigs from January to August. However, when preferred strawberry guava fruits become available in September, wild pigs switch their feeding habits and strawberry guava fruits constitute a major portion of their diet until their availability declines. An analogous shift was observed on the island of Hawaii where wild pigs shifted from consuming primarily tree ferns to banana poka fruits when they became available (Nogueira-Filho et al. 2009). Non-native earthworms are the most important source of animal protein for wild pigs in Hawaii, but carrion is also an important source of animal protein for wild pigs in Hawaii Volcanoes National Park (Diong 1982, Nogueira et al. 2007, Abernethy et al. 2016). Diong (1982) reported 90% of wild pig stomachs sampled for food contents contained earthworms. The most common species was *Pontoscolex corethrurus*, but wild pigs also consumed *Amynthas cortices*, *Amynthas gracilis*, and *Eisenia eiseni*. Small traces of snails, ants, beetle larvae, and insects were also found within stomach contents. Additional fruits that wild pigs consumed on a lesser basis included thimble berry (*Rubus rosifolius*), ʻakala berry (*Rubus hawaiiensis*), kukui nut (*Aleurites moluccanus*), and roseapple (*Eugenia jambos*; Diong 1982).

17.3.4 Habitat Use

Wild pigs utilize a variety of habitats across the Pacific islands. Rainforests are an important habitat type for wild pigs because of the abundance of water, food, and shelter all within a small area (Diong 1982). Additional habitats include open arid areas with kiawe (mesquite; *Prosopis pallida*) trees on the island of Niʻihau (Stone 1985), kiawe flats in Maui (Diong 1982), near sugarcane (*Saccharum officinarum*) fields (Diong 1982), and coconut forests and areas with standing fresh water on Anatahan Island (part of the Mariana Islands; Kessler 2011). Wild pigs also utilize habitats at higher elevations, although these are primarily found in Hawaii. For instance, most suitable wild pig habitat at Hawaii Volcanoes National Park is between 2,500 and 3,100 m and is characterized by mountain-pasture vegetation (Nogueira et al. 2007). Wild pigs also occupy māmane (*Sophora chrysophylla*) forests of Mauna Kea at 3,000 m (Stone 1985). However, rainforests are reportedly able to support greater densities of wild pigs than these elevated habitats. Habitats where wild pigs are reportedly absent include alpine stone deserts, barren lava flows, and cinder fields (Diong 1982).

17.3.5 Population Biology

Hawaii is one of the few places in the Pacific islands where population dynamics have been studied, and these data have usually been collected during the course of eradicating wild pigs from relatively large areas. Capture and necropsy data from 711 wild pigs at Hakalau Forest National Wildlife Refuge (Hakalau Forest NWR) on Hawaii Island revealed 352 sows and 359 boars, suggesting an essentially even population sex ratio (Hess et al. 2006). There was also no difference in sex ratios within 6 age categories determined by dentition (Hess et al. 2006). Mature boars regularly weigh up to 70 kg, though much larger ones have been recorded (Tomich 1986). At Hakalau Forest NWR, the estimated mean mass of 170 sows and 160 boars was 40.8 kg and 46.9 kg, respectively, while the maximum was 88.0 kg (Hess et al. 2006). Density of wild pigs varies substantially between locations, but in general, densities are greater on oceanic islands than mainland habitats (Lewis et al. 2017). Pig density ranged from 1 to 6.5 pigs/km^2 in the ʻŌlaʻa Forest of Hawaii Volcanoes National Park and the adjoining Puʻu Makaʻala Natural Area Reserve (Anderson and Stone 1994). Hess et al. (2006) reported a density of 12.1 pigs/km^2 within one managed area at Hakalau Forest NWR; however, densities in unmanaged adjacent areas were estimated to be over 2.5 times greater. Anderson and Stone (1993) reported a maximum density of 30.7 pigs/km^2 from Maui, and Giffin (1978) stated pig densities ranged from 38.6 to 123.6 pigs/km^2 on Hawaii Island, but did not indicate the source of these data. Few rigorous density estimates are available for wild pig populations outside of the Hawaiian Islands, but Conry (1988) estimated as many as 110 pigs/km^2 in secondary forests of Guam.

Diong (1982) reported average age of a wild pig population in Kīpahulu Valley, Maui was 19.1 months with the following age structure: 0–12 months (41.7%), 13–23 months (22.9%), 24–35 months (~19%), 36–48 months (~10%), and >48 months (6.4%). The oldest pig they captured was a toothless sow estimated to be 6 years old. Hess et al. (2006) reported a maximum age of 48 months for sows and 60 months for boars among 711 captured pigs. High mortality rates (60%) for juveniles (0–6 months of age) have been reported and contributed by flooding of farrowing nests, insufficient shelter from rain (leading to cold exposure), accidental abandonment, and entrapment in mud (Diong 1982). Kidney worm parasitism and dentition failure due to dentoalveolar disease are the most common causes of mortality among adult pigs. An analysis of 757 pigs removed from a closed management unit at Hakalau Forest NWR showed that >43% of the population would need to be removed on an annual basis to reduce density, and approximately 70% of the population would have to be removed to reduce the population by half in each successive year (Hess et al. 2006). Simulation modeling by Barrett and Stone (1983) suggested 30–40% removal on a semiannual basis was needed to maintain pigs at half their equilibrium density.

17.3.6 Reproductive Biology

Wild pigs in Hawaii have a reproductive capacity comparable to other regions across the globe. Diong (1982) found that 21.9% of 41 mature sows (5 months old or greater) in Kīpahulu Valley, Maui were pregnant and 29.3% were lactating. Mean ovulation rate was 7.7 (range 2–14), mean fetal litter size was 5.9 (range 3–10), and estimated prenatal mortality rate was 23.3%. At Hakalau Forest NWR, of 327 sows of known reproductive status, 77 (23.5%) were pregnant. The number of embryos/pregnancy averaged 6.7 (range 2–12; Hess et al. 2006). The overall rate of lactation was 8.3%, but lactation exhibited marked seasonality. While sows were pregnant year-round, pregnancy rate was least during July-September and greatest during January-March. Lactation followed this same general pattern with a lag such that the peak in lactation occurred in April-June. Conversely, Diong (1982) reported 2 farrowing periods in Kīpahulu Valley, Maui during November-March and June-September. Sexual maturity has been reported to be as early as 5 and 7 months for sows and boars, respectively (Diong 1982).

17.3.7 Damage

The effects of wild pigs on natural environments and ecosystems in the Pacific islands have been widely studied. The scientific literature on wild pigs in Hawaii contained 36 peer-reviewed publications, of which 31 addressed direct effects of pigs on native vegetation (Leopold and Hess 2016). Thirteen studies addressed indirect effects on ecosystem processes and 7 addressed both direct and indirect effects of pigs. Wild pigs differ fundamentally from other ungulate species because, in addition to herbivory and trampling, pigs also wallow, dig, and root in soil (Engeman et al. 2006; see Chapter 3), primarily in mesic and wet forests. Furthermore, plants endemic to Hawaii may be particularly vulnerable to damage due to the lack of coexistence with herbivorous large mammals prior to wild pigs (Nogueira-Filho et al. 2009). Physical disturbance by rooting or digging was the most frequently described effect of wild pigs on vegetation (Ralph and Maxwell 1984, Katahira et al. 1993, Pratt et al. 1999). In an experiment on the role of small-scale physical disturbance in seedling mortality in a Hawaiian rain forest, a significantly greater proportion of terrestrial seedlings than epiphytic seedlings were damaged, and more were damaged in locations with pigs than in the absence of pigs (Drake and Pratt 2001). Wild pigs also reduced growth and survival of tree ferns, thereby reducing substrate for epiphytic establishment of native forest species (Busby et al. 2010, Murphy et al. 2013). However, Sweetapple and Nugent (2004) reported little effect on tree ferns at the population level from pig browsing. The number and diversity of sprouting seedlings increased in wet forests with wild pig activity, but the relative density and basal area of seedlings was associated negatively with pig disturbance; although pigs maintained open soil for seedling establishment in mineral soil, they also trampled small seedlings (Busby et al. 2010).

Numerous studies found that non-native plants were associated with the presence and activity of pigs (e.g., rooting). For example, Aplet et al. (1991) examined associations between rooting by wild pigs and 25 non-native plant species. Their results suggested a strong relationship between pig-induced soil disturbance and composition of the non-native portion of the plant community, particularly meadow rice grass (*Ehrhata stipoides*), firetree (*Morella faya*), Kāhili ginger (*Hedychium gardnerianum*), and palmgrass (*Setaria palmifolia*), indicating that these species may both encourage and benefit from pig activity. Rooting by pigs increased non-native species in grassland communities of Mauna Loa and Haleakalā formerly dominated by native species (Spatz and Mueller-Dombois 1975, Jacobi 1981), and was also found to facilitate non-native plant seedlings more than endozoochorous dispersal in a continental environment (Barrios-Garcia and Simberloff 2013). Wild pigs are also thought to preferentially consume and disperse seeds of invasive common guava (*Psidium guajava*), strawberry guava, banana poka, firetree, and java plum (*Eugenia cumini*) in Hawaii (Giffin 1978, Diong 1982, Stone 1985, Stone and Loope 1987, Stone and Holt 1990, Williams 1990, Stone et al. 1991, LaRosa 1992). Ground disturbance and sub-canopy openings created by pigs, coupled with the spread of seeds, likely enhances the establishment of shade-tolerant non-native plants such as strawberry guava and soapbush (*Clidemia hirta*) in closed-canopy forests (Diong 1982, Smith 1992, Martin et al. 2009). However, research in Guam suggests that seed dispersal by wild pigs provides an important ecosystem function that is otherwise lacking on the island (Gawel et al. 2018). Pigs also spread the fungal pathogen *Phytopthora cinnamomi*, which has been associated with canopy dieback of native ‘ōhi‘a (*Metrosideros polymorpha*) trees in the 1970s (Kliejunas and Ko 1976).

Several studies have examined the recovery of native plant communities after landscape-scale removal of pigs. Loope et al. (1991) found removal of wild pigs from a montane bog on Maui reversed damage to vegetation, and subsequent invasion of non-native plant species was minimal. Loh and Tunison (1999) found that native understory cover increased 48%, while non-native understory cover increased 190% in the first 2 years following the removal of pigs from a tropical montane wet forest. Cover of non-native banana poka, however, was reduced from 81% to 40% with pig removal. Hess et al. (2010) found a 2.4-fold increase in understory cover of native ferns and slight decreases in bryophyte cover and exposed soil in a tropical montane forest over a 16-year period concurrent with pig removal at Hakalau Forest NWR. Widespread invasion by non-native grasses and herbs did not occur after wild pig removal at Hakalau Forest NWR, which may have been due to shade under dense canopy cover or lack of non-native propagules (Katahira 1980). In similar montane wet forest communities, nearly all measurements of native plant density and cover were significantly greater where pigs had been removed for at least 6.5 years compared to areas inhabited by pigs (Cole et al. 2012, Cole and Litton 2014). While species richness of ground-rooted plants increased after pig removal, the unassisted recovery of species of conservation interest was limited outside of existing populations, and invasive strawberry guava increased several fold at sites where it had become established prior to pig removal, most likely as a result of the removal of top-down control by pigs on understory communities (Cole and Litton 2014).

Wild pigs have long been known to have indirect and interactive effects on ecosystem processes. These effects include accelerated soil erosion (Stone and Loope 1987), altered soil microarthropod communities (Vtorov 1993, Wehr et al., In Press), decreased soil bacterial diversity (Wehr et al., 2019), altered nutrient cycling (Singer 1981, Coblentz and Baber 1987, Long et al. 2017), and increased nitrogen mineralization and subsequent decreased soil depth of forest floors in Hawaiian montane wet forests (Vitousek 1986). The only study that has examined the effect of wild pigs on soil erosion in Hawaii was inconclusive (Dunkell et al. 2011b). They found that, although pig exclusion did not reduce total suspended solids in runoff, there was reduction in runoff volume after pigs had been excluded. Long et al. (2017) found that pig removal positively affected several critical soil physical and chemical properties, including increased stable soil aggregates and porosity, decreased bulk density, increased cycling and availability of soil nitrogen, and increased concentrations of a suite of macronutrients.

Wild pigs in Hawaii also have complex effects that extend far beyond plant communities into invertebrate and vertebrate species. Foraging and habitat degradation by wild pigs has negatively impacted numerous native Hawaiian birds, many of which are threatened or endangered. For instance, destruction of nectar-producing plants such as the Hawaiian raspberry (*Rubus hawaiiensis*) and Hawaiian lobelioids by wild pigs can have adverse effects on many nectar feeding birds including the ʻIʻiwi (scarlet honeycreeper; *Drepanis coccinea*), ʻAkohekohe (crested honeycreeper; *Palmeria dolei*) and ʻAlawī (Hawaii creeper; *Oreomystis mana*; Stone 1985, Nogueira-Filho et al. 2009). Changes in understory composition related to wild pig damage may also impact the endangered Kiwikiu (Maui parrotbill; *Pseudonestor xanthophyrys*) and Poʻouli (black-faced honeycreeper; *Melamprosops phaeosoma*; Stone 1985). Egg and gosling predation by wild pigs has also led to the decline of several ground-nesting species including the Nēnē (Hawaiian goose; *Branta sandvicensis*; Diong 1982). Additionally, wild pigs create nutrient-rich wallows and troughs in tree fern trunks when they eat the inner core of the trunk, which create breeding sites for alien mosquitoes (*Culex quinquefasciatus*; Stone and Loope 1987, Nogueira-Filho et al. 2009). These mosquitoes then vector avian malaria (*Plasmodium relictum*) and avian poxvirus (*Poxvirus avium*), which are often lethal to native Hawaiian forest birds and is considered to be one of the most important factors in the decline of Hawaiian avifauna (LaPointe et al. 2009, Nogueira-Filho et al. 2009). The endangered ʻAlae ʻUla (Hawaiian moorhen; *Gallinula chloropus sanvicensis*) is also negatively impacted by wild pigs due to nest predation and flooding of wetlands where moorhens breed and nest due to habitat damage upstream (US Department of Agriculture 2016).

Complex relationships between multiple invasive species may represent a strong potential influence on ecosystem processes consistent with invasional meltdown (Simberloff and Von Holle 1999). Non-native earthworms are an important dietary source of animal protein consumed by pigs in Hawaii and may also be indirectly propagated by wild pigs (Giffin 1978, Barrett and Stone 1983, Anderson 1995). While studies of reciprocal benefits between pigs and worms have not yet been completed, the hypothesis of mutualism between pigs and non-native earthworms has been suggested by other studies. Both canopy cover and rooting frequency by pigs were positively associated with worm abundance in Hawaii (Lincoln 2014), and other agricultural practices that promote earthworms may also contribute to pig abundance (Diong 1982). Nitrogen-fixing invasive trees whose spread is often facilitated by wild pigs, have been found to further enhance earthworm populations, exhibiting 2–8 times more biomass under the introduced firetree than the dominant native ʻōhiʻa tree (Aplet 1990). Diong (1982) also suggested earthworms aggregate under pig feces to exploit enhanced nutrient availability. Earthworms may in turn accelerate organic decomposition and soil aeration (González et al. 2006). The associations between these non-native species suggests that successful reductions in one species could have a trickle-down impact on the others, thereby positively impacting the native species of the ecosystem more substantially than believed.

A diversity of agricultural crops including bananas, coconut, sweet potatoes, yams, coffee, macadamia nuts, and sugarcane are grown in the Pacific islands region and are subject to damage by wild pigs (Bittenbender and Hirae 1980, Pacific Agricultural Plant Genetic Resource Network 2006, USDA/National Agricultural Statistics Service 2017). However, there are very few reports and limited research quantifying the impact wild pigs have on agricultural resources in this region. Wild pigs reportedly cause hundreds of thousands of dollars of damage to agricultural commodities in Hawaii where USDA/APHIS/Wildlife Services help provides protection to farmers against crop damage in Maui, Molokaʻi, and Oahu (USDA/APHIS/Wildlife Services 2015).

17.4 REGIONAL MANAGEMENT IN THE PACIFIC ISLANDS

The recognition of damage to forests and agriculture caused by wild pigs in the Hawaiian Islands led to some early control efforts. For example, from 1910 to 1958 the Hawaii Territorial Board of Agriculture and Forestry removed 169,592 wild pigs on all of the major islands in the archipelago (Diong 1982). More recently, the National Park Service was among the first to eradicate pigs from

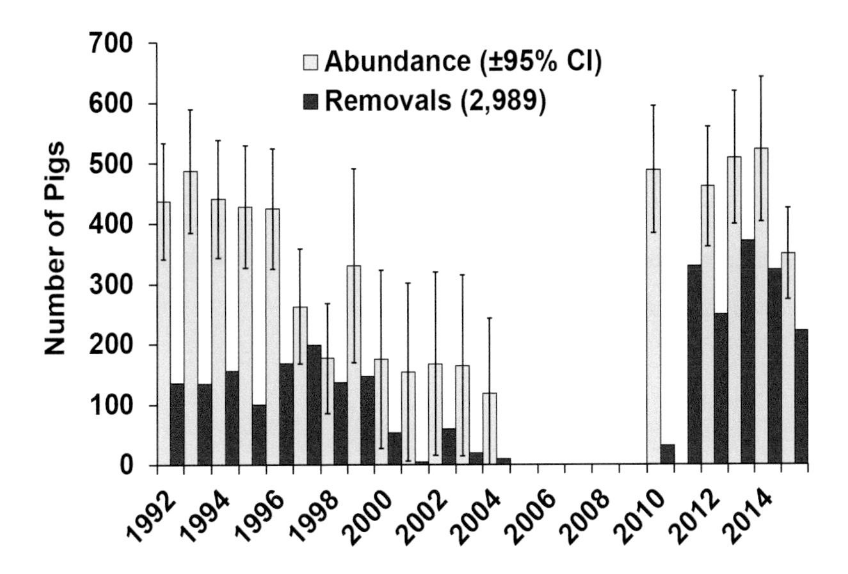

FIGURE 17.2 Estimated abundance and number of wild pigs removed from Hakalau Forest National Wildlife Refuge, Hawaii (1992–2016).

large areas of the Hawaiian Islands (Figure 17.2). Due to the steep terrain of Maui, wild pigs did not begin to invade remote parts of Kīpahulu Valley until the 1970s (Anderson and Stone 1993). Conventional control methods such as trapping and hunting dogs were precluded because the valley could not be accessed without helicopters. Snaring was used to eradicate pigs from a 1,400-ha area of Kīpahulu during a 45-month period beginning in 1986 (Anderson and Stone 1993). Hunting dogs, shooting, and snaring were also used to remove pigs from 7,800 ha of Hawaii Volcanoes National Park from 1980 to 1989 (Katahira et al. 1993), which increased to 16,200 ha by 2007. In Hakalau Forest NWR, a dense, wet montane forest, similar methods were used to remove pigs from a 4,500-ha area in 1988–2004. The long period of time to complete removal was due in part to the large size of one management unit (>2,000 ha), interspersed game production areas, high densities of pigs and reluctance to use snares (Hess et al. 2006).

In conjunction with federal landowners, the Nature Conservancy of Hawaii, the Natural Area Reserve System of the Hawaii Division of Forestry and Wildlife, East Maui Watershed Partnership, and the Three-Mountain Alliance of Hawaii Island have all adopted and refined techniques for managing ungulates (e.g., wild pigs, feral goats, feral sheep) across larger landscapes. Many of these lands adjoin each other, which has created buffers and large blocks of ungulate-free areas with high conservation value, thereby eliminating ingress between enclosed units. The success of this collaborative effort across contiguous lands of varied ownership demonstrates potential power of coordinated efforts with similar goals. Wild pigs have now been removed from >700 km^2 in Hawaii, allowing the gradual recovery of forest ecosystems (Hess and Jacobi 2011; Table 17.2). The largest pig-free areas occur on federally owned lands, primarily in national parks, which are associated with geological features such as volcanoes. Consequently, forestlands, which are the most attractive habitat to wild pigs, but also contain the greatest native biodiversity, have been protected to a lesser extent from degradation by wild pigs, at least until recently (Tummons 2010). The largest invasive species mitigation expenditure in Hawaii during fiscal year 2014 was for feral ungulates, including wild pigs, which are the most widespread species geographically (feral goats and sheep, axis deer, and other species were also included in this expenditure). Funds were used primarily for the construction of exclusionary fences, totaling more than US$5.2M (Rago and Sugano 2015).

Wild pigs have been successfully eradicated from some insular ecosystems on the periphery of the Pacific region. Among the California Channel Islands (see Chapter 12), pigs have been

TABLE 17.2
Wild Pig Eradications from the Hawaiian Islands

Location	Area (km^2)	Year Eradicated	Method
Lāna'i Island	361.3	mid-1930's	Shooting
Kīpahulu Valley, Maui	14	1988	Snaring
Hakalau Forest National Wildlife Refuge, Hawaii	44.5	2004	Dogs, shooting, snaring
Hawaii Volcanoes National Park, Hawaii	161.8	2007	Dogs, shooting, snaring
Ōla'a-Kīlauea, Hawaii	141.2	2010	Driving, trapping, shooting, snaring

Source: Reproduced from Hess and Jacobi (2011).

eradicated from Santa Rosa (Lombardo and Faulkner 2000), Santa Cruz (Morrison et al. 2007, Parkes et al. 2010), Santa Catalina (Schuyler et al. 2002), and San Clemente islands (Knowlton et al 2007). Eradication of pigs from 21,450-ha Santa Rosa Island required 33 months to remove 1,175 pigs (Lombardo and Faulkner 2000). On 25,000-ha Santa Cruz Island, the eradication of 5,036 pigs required 14 months (Parkes et al. 2010). In comparison, eradication of 18,000 pigs from the 58,465-ha Galápagos island of Santiago took 360 months (Cruz et al. 2005). Eradication of wild pigs from most Pacific islands will be a challenging prospect because of relative inaccessibility, logistical difficulties associated with native habitats and topography, and/or cultural opposition. However, some other regions with similar issues have reported some successes. Wild pigs have been eradicated from 20 islands of New Zealand, primarily by hunting, but also with the aid of toxicants in at least 2 cases (Database of Island Invasive Species Eradications 2015). Also, Giovas (2006) contended that prehistoric Polynesians either allowed swine herds to die out or intentionally exterminated them on at least 10 central Pacific islands due to insufficient or fluctuating resources required for husbandry.

In addition to trapping, a variety of techniques have been utilized or implemented in management programs designed to control wild pigs in the Pacific islands. Snaring, although not legal in much of North America, has been used successfully for eliminating wild pigs in the Pacific islands. However, it is considered controversial by animal rights advocates, leading wildlife managers in the region to prefer other methods (Maguire et al. 1997). More recently, baiting with locally abundant fruit such as papaya and macadamia nuts has proven successful and allowed trapping to replace snaring in some cases. A common practice to initiate eradication of wild pigs from newly enclosed natural areas such as Hakalau Forest NWR has been unsupervised public hunting, although it is inefficient in comparison to staff hunting and snaring (Hess et al. 2006). However, it is believed that when practical, inclusion of public hunting in control efforts may foster a more positive relationship with the public. Future control methods which may be highly efficacious, but have not yet been legalized for use in Hawaii, may include toxicants (Shapiro et al. 2015). Not only is it believed that toxicants will be more effective at controlling wild pigs than more traditional techniques, it is also believed that they will be more cost efficient. A previous comparison of eradication costs using a combination of fencing, hunting, snaring, and trapping in Hawaii Volcanoes National Park were >33 times more expensive than control efforts using the toxicant warfarin in Namadgi National Park, Australia (Hone and Stone 1989).

Control of non-native ungulates is the single most expensive natural resource management activity in natural areas in Hawaii, requiring construction, continuous maintenance, and cyclical replacement of fences, as well as a major concurrent effort in removing ungulates through hunting, trapping, and snaring (Anderson and Stone 1993). However, reinvasion from the reservoir of animals in surrounding areas is a constant problem due to treefalls enabling access through downed fences. The Hakalau Forest NWR has intensively managed wild pigs and monitored pig presence and distribution during surveys of all managed areas since 1988, providing a rare opportunity to examine long-term population level responses of pigs to management actions (Figure 17.2). Surveys have been conducted at Hakalau Forest NWR for the presence, distribution, and age of wild pig sign (e.g., scat, rooting, tracks, browsed vegetation) from 1987 to 2004 and from 2010 to 2015 using field methods developed by Stone et al. (1991). A calibrated model based on these sign surveys and the concurrent number of pigs removed from 1 large management unit was applied to estimate pig abundance in other management units (Hess et al. 2006, Leopold et al. 2016). The resulting time series of pig abundance provided managers with a means to evaluate and refine control efforts in an adaptive management framework. Hakalau Forest NWR has removed nearly 3,000 pigs since 1988, resulting in their eradication from approximately 45 km^2 by 2004, but deteriorating fences due to lack of maintenance (e.g., limited resources) led to the reinvasion of nearly every management unit by 2010. Moreover, active restoration and unassisted recovery of dense tropical forest vegetation after pigs were initially removed has apparently made the task of pig removal more difficult (Leopold et al. 2016). Setbacks such as these highlight the necessity of continued investment in these programs, if only to ensure that ground is not lost during periods of inactivity.

Presently, wild pigs are likely the most widespread and popular game species for subsistence and recreational hunting in the Hawaiian Islands (Hawaii Department of Land and Natural Resources, Division of Forestry and Wildlife 2018; Figure 17.3). Maintenance of populations of introduced ungulates for subsistence and hunting, which is considered an important management objective, is also generally considered to be incompatible with conservation of native biota and watershed functions in Hawaii. Therefore, management actions for conservation are generally based upon construction and maintenance of expensive barriers for exclusion of large non-native mammals from important natural areas (Hess and Jacobi 2011). However, these management actions reduce the land area available for maintenance of sustainable game populations and hunting activities. Despite the importance of wild pigs for subsistence and hunting, planning for continued sustained-yield populations of wild pigs has been notably lacking. Bag limits are generally set at 1 or 2 pigs/hunter/day without regard for abundance, recruitment, or any other vital information on population dynamics in the wild (Ikagawa 2013). In the only study to predict land area necessary to provide sustained-yield hunting of wild pigs in Hawaii, Hess and Jacobi (2014) used a simplistic analysis to estimate that 1,300–1,700 km^2, representing 12–16% of the land area of Hawaii Island, would be

FIGURE 17.3 The Hawaii wildlife conservation stamp, depicting a Central European boar in a wallow, is required annually for a state hunting license, illustrating the value of wild pigs to hunters in Hawaii. The nominal fee is used to support wildlife populations and habitat and manage hunting in the state.

needed to produce the total estimated weight of wild game meat harvested annually on that island (187,334 kg; Delparte and Melrose 2012). This analysis was based on the range of dressed weight/whole pig, the proportion of a pig population that can be sustainably removed annually, and the density of pig populations in the wild. These projections suggest that wild pigs could be eradicated from considerably more area in Hawaii before subsistence and hunting would be negatively impacted.

17.5 FUTURE OF WILD PIGS IN THE PACIFIC ISLANDS

The detrimental effects of wild pigs on natural ecosystems have been well documented (Nogueira-Filho et al. 2009, Leopold and Hess 2016), and continued expansion of pig-free areas would benefit conservation of native biota, watersheds, hydrological function in natural ecosystems, and potentially human health. However, the cultural importance of subsistence and recreational hunting precludes complete eradication of pigs from many large islands of the Pacific (Hess and Jacobi 2014). This conflict requires efforts to identify and prioritize areas of conservation concern for protection from populations of non-native wild pigs and establishment of adjacent areas for production and maintenance of sustained-yield populations for subsistence and recreational hunting. Additional research regarding public attitudes and beliefs associated with wild pigs on Pacific islands would facilitate the reconciliation of some conflicts between conservation and hunting.

There are also gaps in scientific information on methodology that could efficiently estimate abundance and provide an understanding of population dynamics of wild pigs (see Chapter 9). These data are needed to inform policies on management and sustained-yield hunting programs. Additionally, these data would enable managers to establish daily and seasonal bag limits that would help manage pig populations at desired levels of abundance, thereby reducing conflicts arising from damage to agriculture and native ecosystems. Integrating research findings and management practices from other locations where wild pigs are considered both pests and resources would be particularly beneficial for the Pacific islands. While major impacts of wild pigs on Pacific island ecosystems have been studied, some critical gaps in scientific information remain. Specifically, we need to improve our understanding of the association between wild pigs and avian malaria, coral reef sedimentation, ecosystem processes (e.g., nutrient cycling) and zoonotic diseases (e.g., cryptosporidiosis, leptospirosis), as well as how wild pigs affect water quality, watershed function, and threatened and endangered plants. Finally, it will be crucial to monitor long-term, landscape-scale ecological responses to the removal of wild pigs, so that impacts of later eradications can be better predicted.

ACKNOWLEDGMENTS

We thank editors and reviewers S.S. Ditchkoff, J.D. Jacobi, J.J. Mayer, B.K. Strickland, and K.C. VerCauteren for many helpful comments. Support for this work was provided by the Invasive Species program of the US Geological Survey, the University of Hawaii at Mānoa via the US Department of Agriculture/National Institute of Food and Agriculture Hatch (HAW01127H) and McIntyre Stennis (HAW01123M) Programs, and the Department of Defense Strategic Environmental Research and Development Program (RC-2433). Any use of trade, firm, or product names is for descriptive purposes only and does not imply endorsement by the US Government.

REFERENCES

Abernethy, E., K. Turner, J. Beasley, T. DeVault, W. Pitt, and O. Rhodes, Jr. 2016. Carcasses of invasive species are primarily utilized by invasive scavengers in an island ecosystem. *Ecosphere* 7(10):e01496.

Allen, M. S., E. Matisoo-Smith, and A. Horsburgh. 2001. Pacific 'babes': Issues in the origins and dispersal of Pacific pigs and the potential of mitochondrial DNA analysis. *International Journal of Osteoarchaeology* 11:4–13.

Anderson, S. J., and C. P. Stone. 1993. Snaring to remove feral pigs *Sus Scrofa* in a remote Hawaiian rain forest. *Biological Conservation* 63:195–201.

Anderson, S. J., and C. P. Stone. 1994. Indexing sizes of feral pig populations in a variety of Hawaiian natural areas. *Transactions of the Western Section of the Wildlife Society* 30:26–39.

Anderson, S. P. 1995. Some environmental indicators related to feral pig activity in a Hawaiian rain forest. Thesis, University of Hawaii at Mānoa, Honolulu, HI.

Aplet, G. H. 1990. Alteration of earthworm community biomass by the alien *Myrica faya* in Hawaii. *Oecologia* 82:414–416.

Aplet, G. H., S. J. Anderson, and C. P. Stone. 1991. Association between feral pig disturbance and the composition of some alien plant assemblages in Hawaii Volcanoes National Park. *Vegetatio* 95:55–62.

Barrett, R. H., and C. P. Stone. 1983. *Hunting as a Control Method for Wild Pigs in Hawaii Volcanoes National Park*. Unpublished report, Hawaii Volcanoes National Park, HI.

Barrett-Connor, E., C. F. Davis, R. N. Hamburger, and I. Kagan. 1976. An epidemic of trichinosis after ingestion of wild pig in Hawaii. *The Journal of Infectious Diseases* 133:473–477.

Barrios-Garcia, M. N., and S. Ballari. 2012. Impact of wild boar (*Sus scrofa*) in its introduced and native range: A review. *Biological Invasions* 14:2283–2300.

Barrios-Garcia, M. N., and D. Simberloff. 2013. Linking the pattern to the mechanism: How an introduced mammal facilitates plant invasions. *Austral Ecology* 38:884–890.

Bevins, S. N., M. W. Lutman, K. Pederson, N. Barrett, T. Gidlewski, T. J. Deliberto, and A. B. Franklin. 2018. Spillover of swine coronaviruses, United States. *Emerging Infectious Diseases* 24:1390–1392.

Bittenbender, H. C., and H. H. Hirae. 1980. *Common Problems of Macadamia Nut in Hawaii.* College of Tropical, Agriculture and Human Resources, University of Hawaii, Honolulu, HI.

Boie, H. 1828. Auszüge aus Briefen von Heinr.Boie zu Java an Hn. Schlegel, Conservator anim. vertebr. am Königl. niederl. *Museum. Isis van Oken, Jena* 21:1.02–1.035.

Buchholz, A. E., A. R. Katz, R. Galloway, R. A. Stoddard, and S. M. Goldstein. 2016. Feral swine *Leptospira* seroprevalence survey in Hawaii, USA, 2007–2009. *Zoonoses and Public Health* 63:584–587.

Burney, D. A., H. F. James, L. P. Burney, S. L. Olson, W. Kikuchi, W. L. Wagner, M. Burney, et al. 2001. Fossil evidence for a diverse biota from Kaua'i and its transformation since human arrival. *Ecological Monographs* 71:615–641.

Busby, P. E., P. Vitousek, and R. Dirzo. 2010. Prevalence of tree regeneration by sprouting and seeding along a rainfall gradient in Hawaii. *Biotropica* 42:80–86.

Charlot, J. 1987. The Kamapua'a literature: The classical traditions of the Hawaiian pig god as a body of literature. Brigham Young University–Hawaii, Honolulu, HI.

Coblentz, B. E., and D. W. Baber. 1987. Biology and control of feral pigs on Isla Santiago, Galapagos, Ecuador. *Journal of Applied Ecology* 24:403–418.

Cole, R. J., and C. M. Litton. 2014. Vegetation response to removal of non-native feral pigs from Hawaiian tropical montane wet forest. *Biological Invasions* 16:125–140.

Cole, R. J., C. M. Litton, M. J. Koontz, and R. K. Loh. 2012. Vegetation recovery 16 years after feral pig removal from a wet Hawaiian forest. *Biotropica* 44:463–471.

Conry, P. J. 1988. Management of feral and exotic game on Guam. *Transactions of the Western Section of the Wildlife Society* 24:26–30.

Cruz, F., C. J. Donlan, K. Campbell, and V. Carrion. 2005. Conservation action in the Galàpagos: Feral pig (*Sus scrofa*) eradication from Santiago Island. *Biological Conservation* 121:473–478.

Database of Island Invasive Species Eradications. 2015. Island Conservation, Coastal Conservation Action Laboratory UCSC, IUCN SSC Invasive Species Specialist Group, University of Auckland and Landcare Research New Zealand. <http://diise. islandconservation.org>. Accessed 2 Nov 2018.

Delparte, D., and J. Melrose. 2012. Hawaii County food self-sufficiency baseline 2012. University of Hawaii at Hilo Geography and Environmental Studies Department for the Hawaii County Department of Research and Development, Hilo, HI.

Diong, C. H. 1982. Population biology and management of the feral pig (*Sus scofa* L.) in Kipahulu Valley, Maui. Dissertation, University of Hawaii at Mānoa, Honolulu, HI.

Drake, D. R., and L. W. Pratt. 2001. Seedling mortality in Hawaiian rain forest: The role of small-scale physical disturbance. *Biotropica* 33:319–323.

Dunkell, D. O., G. L. Bruland, C. I. Evensen, and M. J. Walker. 2011a. Effects of feral pig (*Sus scrofa*) exclusion on enterococci in runoff from the forested headwaters of a Hawaiian watershed. *Water, Air, and Soil Pollution* 221:313–326.

Dunkell, D. O., G. L. Bruland, C. I. Evensen, and C. M. Litton. 2011b. Runoff, sediment transport, and effects of feral pig (*Sus scrofa*) exclusion in a forested Hawaiian watershed. *Pacific Science* 65:175–194.

Ellis, W. 1827. *Narrative of a tour through Hawaii.* Second Edition. H. Fisher, Son, and P. Jackson, London, UK.

Engeman, R. M., A. Stevens, J. Allen, J. Dunlap, M. Daniel, D. Teague, and B. Constantin. 2006. Feral swine management for conservation of an imperiled wetland habitat: Florida's vanishing seepage slopes. *Biological Conservation* 134:440–446.

Essey, M. A., D. E. Stallknecht, E. M. Himes, and S. K. Harris. 1983. Follow-up survey of feral swine for *Mycobacterium bovis* infection on the Hawaiian island of Moloka'i. *Proceedings of the United States Animal Health Association* 87:589–595.

Finucane, M. L., J. J. Marra, V. W. Keener, and M. H. Smith. 2012. Pacific islands region overview. Pages 1–34 *in* V. W. Keener, J. J. Marra, M. L. Finucane, D. Spooner, and M. H. Smith, editors. *Climate Change and Pacific Islands: Indicators and Impacts. Report for the 2012 Pacific Islands Regional Climate Assessment (PIRCA).* Island Press, Washington, DC.

Gawel, A. M., H. S. Rogers, R. H. Miller, and A. M. Kerr. 2018. Contrasting ecological roles of non-native ungulates in a novel ecosystem. *Royal Society Open Science* 5:170151.

Giffin, J. 1978. Ecology of the feral pig on the island of Hawaii. Pittman-Robertson Project W-15-3, Study II. Hawaii Department of Land and Natural Resources, Division of Fish and Game, Honolulu, HI.

Giovas, C. M. 2006. No pig atoll: Island biogeography and the extirpation of a Polynesian domesticate. *Asian Perspectives* 45:69–95.

González, G., C. Y. Huang, X. Zou, and C. Rodríguez. 2006. Earthworm invasions in the tropics. *Biological Invasions* 8:1247–1256.

Griffiths, R., A. Wegmann, C. Hanson, B. Keitt, G. Howland, D. Brown, B. Tershy, et al. 2014. The Wake Island rodent eradication: Part success, part failure, but wholly instructive. *Proceedings of the Vertebrate Pest Conference* 26:101–111.

Hawaii Department of Land and Natural Resources, Division of Forestry and Wildlife. 2018. Public lands hunting information survey report. <http://dlnr.hawaii.gov/recreation/files/ 2019/01/SurveyReport_2017.pdf>. Accessed 6 Mar 2019.

Hess, S. C., and J. D. Jacobi. 2011. The history of mammal eradications in Hawaii and the United States associated islands of the Central Pacific. Pages 67–73 *in* C. R. Veitch, M. N. Clout, and D. R. Towns, editors. *Island Invasives: Eradication and Management. Proceedings of the International Conference on Island Invasives.* IUCN, Gland, Switzerland.

Hess, S. C., and J. D. Jacobi. 2014. How much land is needed for feral pig hunting in Hawaii? *Pacific Conservation Biology* 20:54–56.

Hess, S. C., J. J. Jeffrey, D. L. Ball, and L. Babich. 2006. Efficacy of feral pig removals at Hakalau Forest National Wildlife Refuge. *Transactions of the Western Section of the Wildlife Society* 42:53–67.

Hess, S. C., J. J. Jeffrey, L. W. Pratt, and D. L. Ball. 2010. Effects of ungulate management on vegetation at Hakalau Forest National Wildlife Refuge, Hawaii Island. *Pacific Conservation Biology* 16:144–150.

Hone, J., and C. P. Stone. 1989. A comparison and evaluation of feral pig management in two national parks. *Wildlife Society Bulletin* 17:419–425.

Ikagawa, M. 2013. Invasive ungulate policy and conservation in Hawaii. *Pacific Conservation Biology* 19:270–283.

Intoh, M. 1986. Pigs in Micronesia: Introduction or reintroduction by the Europeans? *Man and Culture in Oceania* 2:1–26.

Jacobi, J. D. 1981. Vegetation changes in a subalpine grassland in Hawaii following disturbance by feral pigs. Technical Report 41, Cooperative National Park Resources Studies Unit, University of Hawaii at Mānoa, Honolulu, HI.

Katahira, L. 1980. The effects of feral pigs on a montane rain forest in Hawaii National Park. *Proceedings: Conference in Natural Science, Hawaii Volcanoes National Park* 3:173–178.

Katahira, L. K., P. Finnegan, and C. P. Stone. 1993. Eradicating feral pigs in montane mesic habitat at Hawaii Volcanoes National Park. *Wildlife Society Bulletin* 21:269–274.

Katz, A. R., A. E. Buchholz, K. Hinson, S. Y. Park, and P. V. Effler. 2011. Leptospirosis in Hawaii, USA, 1999–2008. *Emerging Infectious Diseases* 17:221–226.

Kawakami, K., and I. Okochi. 2010. *Restoring the Oceanic Island Ecosystem: Impact and Management of Invasive Alien Species in the Bonin Islands.* Springer, Tokyo.

Keener, V. W., K. Hamilton, S. K. Izuka, K. E. Kunkel, L. E. Stevens, and L. Sun. 2013. Regional climate trends and scenarios for the US national climate assessment: Part 8. Climate of the Pacific islands. NOAA Technical Report NESDIS 142–8, Washington, DC.

Kessler, C. C. 2002. Eradication of feral goats and pigs and consequences for other biota on Sarigan Island, Commonwealth of the Northern Mariana Islands. Pages 132–140 *in* C. R. Veitch, editor. *Turning the Tide: The Eradication of Invasive Species, Proceedings of the International Conference on Eradication of Island Invasives.* IUCN, Gland, Switzerland.

Kessler, C. C. 2011. Invasive species removal and ecosystem recovery in the Mariana Islands; challenges and outcomes on Sarigan and Anatahan. Pages 320–324 *in* C. R. Veitch, M. N. Clout, and D. R. Towns, editors. *Island Invasives: Eradication and Management*. IUCN, Gland, Switzerland.

Kirch, P. V. 1982. The impact of prehistoric Polynesians on the Hawaiian ecosystem. *Pacific Science* 36:1–14.

Kirch, P. V., and M. Kelly, editors. 1975. Prehistory and human ecology in a windward Hawaiian valley: Halawa Valley, Moloka'i. Pacific Anthropological Records 24. Department of Anthropology, Bernice P. Bishop Museum, Honolulu, HI.

Kliejunas, J. T., and W. H. Ko. 1976. Dispersal of *Phytophthora cinnamomi* on the island of Hawaii. *Phytopathology* 66:457–460.

Knowlton, J. L., C. J. Donlan, G. W. Roemer, A. Samaniego-Herrera, B. S. Keitt, B. Wood, A. Aguirre-Munoz, et al. 2007. Eradication of non-native mammals and the status of insular mammals on the California Channel Islands, USA, and Pacific Baja California Peninsula Islands, Mexico. *Southwestern Naturalist* 52:528–540.

LaPointe, D. A., C. T. Atkinson, and S. I. Jarvi. 2009. Managing disease. Pages 405–424 *in* T. Pratt, P. Banko, C. Atkinson J. Jacobi, and B. Woodworth, editors. *Conservation Biology of Hawaiian Forest Birds: Implications for Island Avifauna*. Yale University Press, New Haven, CT.

LaRosa, A. M. 1992. The status of banana poka in Hawaii. Pages 271–299 *in* C. P. Stone, C. W. Smith, and J. T. Tunison, editors. *Alien Plant Invasions in Native Ecosystems of Hawaii: Management and Research*. University of Hawaii Cooperative National Park Resources Studies Unit, Honolulu, HI.

Larson, G., T. Cucchi, M. Fujita, E. Matisoo-Smith, J. Robins, A. Anderson, B. Rolett, et al. 2007. Phylogeny and ancient DNA of *Sus* provides insights into neolithic expansion in Island Southeast Asia and Oceania. *Proceedings of the National Academy of Sciences* 104:4834–4839.

Larson, G., K. Dobney, U. Albarella, M. Fang, E. Matisoo-Smith, J. Robins, S. Lowden, et al. 2005. Worldwide phylogeography of wild boar reveals multiple centers of pig domestication. *Science* 307:1618–1621.

Leopold, C. R., and S. C. Hess. 2016. Conversion of native terrestrial ecosystems in Hawaii to novel grazing systems: A review. *Biological Invasions* 19:161–177.

Leopold, C. R., S. C. Hess, S. J. Kendall, and S. W. Judge. 2016. Abundance, distribution, and removals of feral pigs at Big Island National Wildlife Refuge Complex 2010–2015. Hawaii Cooperative Studies Unit Technical Report HCSU-075, Hilo, HI.

Lewis J. S., M. L. Farnsworth, C. L. Burdett, D. M. Theobald, M. Gray, and R. S. Miller. 2017. Biotic and abiotic factors predicting the global distribution and population density of an invasive large mammal. *Nature Scientific Reports* 7:44152.

Lincoln, N. K. 2014. Effect of various monotypic forest canopies on earthworm biomass and feral pig rooting in Hawaiian wet forests. *Forest Ecology and Management* 331:79–84.

Linderholm, A., D. Spencer, V. Battista, L. Frantz, R. Barnett, R. C. Fleischer, H. F. James, et al. 2016. A novel MC1R allele for black coat colour reveals the Polynesian ancestry and hybridization patterns of Hawaiian feral pigs. *Royal Society Open Science* 3:160304.

Loh, R. K., and J. T. Tunison. 1999. Vegetation recovery following pig removal in 'Ola'a-koa rainforest unit, Hawaii Volcanoes National Park. Technical Report 123, University of Hawaii Cooperative National Park Resources Studies Unit, Honolulu, HI.

Lombardo, C. A., and K. R. Faulkner. 2000. Eradication of feral pigs (*Sus scrofa*) from Santa Rosa Island, Channel Islands National Park, California. Pages 300–306 *in* D. H. Browne, H. Chaney, and K. Mitchell, editors. *Proceedings of the Fifth California Islands Symposium*. Santa Barbara Museum of Natural History, Santa Barbara, CA.

Long, J. L. 2003. *Introduced Mammals of the World: Their History, Distribution and Influence*. CSIRO, Clayton, Australia.

Long, M. S., C. M. Litton, C. P. Giardina, J. Deenik, R. J. Cole, and J. P. Sparks. 2017. Impact of nonnative feral pig removal on soil structure and nutrient availability in Hawaiian tropical montane wet forests. *Biological Invasions* 19:749–763.

Loope, L. L., A. C. Medeiros, and B. H. Gagné. 1991. Recovery of vegetation of a montane bog in Haleakalā National Park following protection from feral pig rooting. Technical Report 77, University of Hawaii Cooperative National Park Resources Studies Unit, Honolulu, HI.

Maguire, L., P. Jenkins, and G. Nugent. 1997. Research as a route to consensus? Feral ungulate control in Hawaii. *North American Wildlife and Natural Resources Conference* 62:135–145.

Maly, K. 1998. Nā Ulu Lā'au Hawaii (Hawaiian Forests). A briefing document prepared for the State of Hawaii Natural Area Reserves System by Kumu Pono Associates.

Martin, P. H., C. D. Canham, and P. L. Marks. 2009. Why forests appear resistant to exotic plant invasions: Intentional introductions, stand dynamics, and the role of shade tolerance. *Frontiers in Ecology and the Environment* 7:142–149.

McKinnon, R. 2009. *South Pacific*. Lonely Planet, Melbourne, Australia.

Melletti, M., and E. Meijaard, editors. 2018. *Ecology, Conservation and Management of Wild Pigs and Peccaries*. Cambridge University Press, Cambridge, UK.

Miller, R. S., and S. J. Sweeney. 2013. *Mycobacterium bovis* (bovine tuberculosis) infection in North American wildlife: Current status and opportunities for mitigation of risks of further infection in wildlife populations. *Epidemiology & Infection* 141:1357–1370.

Mitchell, J., W. Dorney, R. Mayer, and J. McIlroy. 2009. Migration of feral pigs (*Sus scrofa*) in rainforest of north Queensland: Fact or fiction? *Wildlife Research* 36:110–116.

Morrison, S. A., N. Macdonald, K. Walker, L. Lozier, and M. R. Shaw. 2007. Facing the dilemma at eradication's end: Uncertainty of absence and the Lazarus effect. *Frontiers in Ecology and the Environment* 5:271–276.

Murphy, M. J., F. Inman-Narahari, R. Ostertag, and C. M. Litton. 2013. Invasive feral pigs impact native tree ferns and woody seedlings in Hawaiian forest. *Biological Invasions* 16:63–71.

National Park Service. 1997. *National Park of American Samoa General Management Plan*. US Department of the Interior/National Park Service, Pago Pago, AS.

New World Encyclopedia Contributors. 2017. *Bibliographic details for Cocos (Keeling) Islands*. Paragon House, St. Paul, MN.

Nogueira, S. S. C., S. L. G. Nogueira-Filho, M. Bassford, K. Silvius, and J. M. V. Fragoso. 2007. Feral pigs in Hawaii: Using behavior and ecology to refine control techniques. *Applied Animal Behaviour Science* 108:1–11.

Nogueira-Filho, S. L. G., S. S. C. Nogueira, and J. M. V. Fragoso. 2009. Ecological impacts of feral pigs in the Hawaiian Islands. *Biodiversity and Conservation* 18:3677–3683.

Pacific Agricultural Plant Genetic Resource Network. 2006. Regional strategy of the *ex situ* conservation and utilization of crop diversity in the Pacific islands region. <https://www.croptrust.org/wp/wp-content/uploads/2014/12/Pacific-FINAL-29Jan07.pdf>. Accessed 6 Dec 2018.

Parkes, J. P., D. S. L. Ramsey, N. Macdonald, K. Walker, S. McKnight, B. S. Cohen, and S. A. Morrison. 2010. Rapid eradication of feral pigs (*Sus scrofa*) from Santa Cruz Island, California. *Biological Conservation* 143:634–641.

Pearson, R. J., P. V. Kirch, and M. Piettruewsky. 1971. An early site at Bellows Beach, Waimanalo, Oahu, Hawaiian Islands. *Archaeology and Physical Anthropology in Oceania* 6:204–234.

Pederson, K., S. N. Bevins, J. A. Baroch, J. C. Cumbee, S. C. Chandler, B. S. Woodruff, T. T. Bigelow, and T. J. DeLiberto. 2013. Pseudorabies in feral swine in the United States, 2009–2012. *Journal of Wildlife Diseases* 49:709–713.

Pederson, K., S. N. Bevins, B. S. Schmit, M. W. Lutman, M. P. Milleson, C. T. Turnage, T. T. Bigelow, and T. J. DeLiberto. 2012. Apparent prevalence of swine brucellosis in feral swine in the United States. *Human-Wildlife Interactions* 61:38–47.

Pratt, L. W., L. L. Abbott, and D. K. Palumbo. 1999. Vegetation above a feral pig barrier fence in rain forests of Kilauea's East Rift, Hawaii Volcanoes National Park. Technical Report 124, University of Hawaii Cooperative National Park Resources Studies Unit, Honolulu, HI.

Rago, F., and D. Sugano. 2015. Can't see the forest for the (albizia) trees: An invasive species update. Report No. 3. Legislative Reference Bureau, Honolulu, HI.

Ralph, C. J., and B. D. Maxwell. 1984. Relative effects of human and feral hog disturbance on a wet forest in Hawaii. *Biological Conservation* 30:291–303.

Salbosa, L. L. H. 2009. An analysis of feral pig (*Sus scrofa*) home ranges in a Hawaiian forest using GPS satellite collars. Thesis, University of Hawaii at Mānoa, Honolulu, HI.

Schuyler, P. T., D. K. Garcelon, and S. Escover. 2002. Eradication of feral pigs (*Sus scrofa*) on Santa Catalina Island, California, USA. Pages 274–286 *in* C. R. Veitch and M. N. Clout, editors. *Turning the Tide: The Eradication of Invasive Species*. IUCN SSC Invasive Species Specialist Group. IUCN, Gland, Switzerland.

Shapiro, L., C. Eason, C. Bunt, S. Hix, P. Aylett, and D. MacMorran. 2015. Efficacy of encapsulated sodium nitrite as a new tool for feral pig management. *Journal of Pest Science* 89:489–495.

Simberloff, D., and B. Von Holle. 1999. Positive interactions of nonindigenous species: Invasional meltdown? *Biological Invasions* 1:21–32.

Singer, F. J. 1981. Wild pig populations in the national parks. *Environmental Management* 5:263–270.

Smith, C. W. 1992. Distribution, status, phenology, rate of spread, and management of *Clidemia* in Hawaii. Pages 241–253 *in* C. P. Stone, C. W. Smith, and J. T. Tunison, editors. *Alien Plant Invasions in Native Ecosystems of Hawaii: Management and Research*. University of Hawaii Cooperative National Park Resources Studies Unit, Honolulu, HI.

Spatz, G., and D. Mueller-Dombois. 1975. Succession patterns after pig digging in grassland communities on Mauna Loa, Hawaii. *Phytocoenologia* 3:346–373.

Stone, C. P. 1985. Alien animals in Hawaii's native ecosystems: Towards controlling the adverse effects of introduced vertebrates. Pages 251–297 *in* C. P. Stone and J. M. Scott, editors. *Hawaii's Terrestrial Ecosystems: Preservation and Management*. University of Hawaii Cooperative National Park Resources Studies Unit, Honolulu, HI.

Stone, C. P., P. K. Higashino, L. W. Cuddihy, and S. J. Anderson. 1991. Preliminary survey of feral ungulate and alien and rare plant occurrence on Hakalau Forest National Wildlife Refuge. Technical Report 81, University of Hawaii Cooperative National Park Resources Studies Unit, Honolulu, HI.

Stone, C. P., and R. A. Holt. 1990. Managing the invasions of alien ungulates and plants in Hawaii's natural areas. *Monographs in Systematic Botany from the Missouri Botanical Garden* 32:211–221.

Stone, C. P., and L. L. Loope. 1987. Reducing negative effects of introduced animals on native biotas in Hawaii: What is being done, what needs doing, and the role of national parks. *Environmental Conservation* 14:245–258.

Sweetapple, P. J., and G. Nugent. 2004. Seedling ratios: A simple method for assessing ungulate impacts on forest understories. *Wildlife Society Bulletin* 32:137–147.

Tomich, P. Q. 1986. *Mammals in Hawaii: A Synopsis and Notational Bibliography*. Second Edition. Bishop Museum Press, Honolulu, HI.

Tummons, P. 2010. Just 3 percent of forest bird habitat in Hawaii is protected from ungulates. *Environment Hawaii* 21:1–8.

US Department of Agriculture. 2016. Feral swine: Impacts on threatened and endangered species. Program Aid No. 20195f.

US Department of Agriculture/Animal and Plant Inspection Service/Wildlife Services. 2015. Hawaii State Report. Washington, DC.

US Department of Agriculture/National Agricultural Statistics Service. 2017. *2017 State Agriculture Overview: Hawaii*. Hawaii Field Office, Honolulu, HI.

Vitousek, P. M. 1986. Biological invasions and ecosystem properties: Can species make a difference? Pages 163–178 *in* H. A. Mooney and J. A. Drake, editors. *Ecology of Biological Invasions of North America and Hawaii*. Springer-Verlag, New York.

Vtorov, I. P. 1993. Feral pig removal: Effects on soil microarthropods in a Hawaiian rain forest. *Journal of Wildlife Management* 57:875–880.

Wehr, N. H., S. C. Hess, and C. M. Litton. 2018. Biology and impacts of Pacific Island invasive species. 14. *Sus scrofa*, the feral pig (Artiodactyla: Suidae). *Pacific Science* 74:177–198.

Wehr, N. H., K. M. Kinney, N. H. Nguyen, C. P. Giardina, and C. M. Litton. 2019. Changes in soil bacterial community diversity following the removal of invasive feral pigs from a Hawaiian tropical montane wet forest. *Scientific Reports*. 9:14681.

Wehr, N. H., C. M. Litton, N. K. Lincoln, and S. C. Hess. In Press. Relationships between soil macroinvertebrates and nonnative feral pigs (*Sus scrofa*) in Hawaiian tropical montane wet forests. Biological Invasions.

Westervelt, W. D. 1923. *Hawaiian Historical Legends*. Fleming H. Revell Co., New York.

Williams, J. 1990. The coastal woodland of Hawaii Volcanoes National Park: Vegetation recovery in a stressed ecosystem. Technical Report 72, University of Hawaii Cooperative National Park Resources Studies Unit, Honolulu, HI.

Wilmshurst, J. M., T. L. Hunt, C. P. Lipoc, and A. J. Anderson. 2011. High-precision radiocarbon dating shows recent and rapid initial human colonization of East Polynesia. *Proceedings of the National Academy of Sciences* 108:1815–1820.

Woodroffe, C., and P. Berry. 1994. Scientific studies in the Cocos (Keeling) Islands: An introduction. *Atoll Research Bulletin* 399:1–16.

Ziegler, A. C. 2002. *Hawaiian Natural History, Ecology, and Evolution*. University of Hawaii Press, Honolulu, HI.

18 Wild Pigs in Mexico and the Caribbean

J. Alfonso Ortega-S, Johanna Delgado-Acevedo, Jorge G. Villarreal-González, Rafael Borroto-Páez, and Roberto Tamez-González

CONTENTS

In memory of Jorge G. Villarreal-González, passionate about wildlife conservation and management, may God grant you peace.

18.1 INTRODUCTION

Invasive wild pigs (*Sus scrofa*) have been present in Mexico and the insular Caribbean for more than 500 years. These animals arrived with the Spaniard colonization of the Americas and constitute the oldest populations of wild pigs anywhere in North America. Mexico and the Caribbean also sustained the earliest impacts from this invasive species in North America. Following their escape/release into the wild around new Spanish colonies, wild pigs quickly became a pest, killing the colonists' cattle and ravaging cultivated crops of maize and sugarcane (Ensminger 1961, Donkin 1985). In 1505, as an effort to reduce damage, an official proclamation was issued by the Spanish Crown directing colonists to reduce the numbers of wild pigs found around the Caribbean colonies (Zadik 2005). In 1506, 13 years after the first introduction of domestic pigs to the West Indies, Spanish colonists began hunting wild pigs due to extensive damages incurred (Ensminger 1961, Donkin 1985). Damage by these non-native animals in this region remains an issue (Borroto-Páez and Mancina 2017, Mayer 2018).

For this chapter, we define Mexico and the Caribbean as the insular area including the Greater, Lesser, and Leeward Antilles, the Lucayan Archipelago, including the Bahamas and Turks and Caicos Islands, and Bermuda. Numerous landforms across the islands create a variety of ecosystems. Precipitation in the region is produced by northeast trade winds. Tropical forests and crops thrive due to the region's moisture, temperature, and altitude, but tropical forests have diminished due to anthropogenic activities including urban development, intensive agriculture, and habitat fragmentation.

18.2 HISTORIC AND PRESENT STATUS

18.2.1 Origin and Early History

The introduction of wild pigs to this region began on September 25, 1493, when Christopher Columbus' second trip to the "Occidental Indias" departed with 18 ships and 1,500 men from the Port of Cádiz, Spain (Jimeno 1880, Crosby 1972). During Columbus' journey, several species of livestock from La Gomera in the Canary Islands were brought on board, including 8 domestic pigs which are considered to be the initial ancestors of wild pigs in the New World (Jimeno 1880, Crosby 1972). Spanish supply ships that followed Columbus' second voyage brought more pigs from the Canary Islands (Mayer and Beasley 2018). Garcia (1994) and Del Rio (1996) further reported that domestic Iberian hogs were part of cargo brought to the New World on subsequent voyages. Pigs were introduced as a source of food (Elliot 2007) and soon became a mainstay of the Spanish, indigenous peoples', and slaves' diets on the islands of the Spanish West Indies (Crosby 1972). These animals were free-ranging in an environment that had abundant food, negligible competition, and almost no predators, which enabled them to multiply rapidly (Mayer and Beasley 2018). The arrival of pigs to other Caribbean Islands closely match the beginning of Spaniard, English, French, and Dutch colonization. Free-ranging populations were soon established in Cuba, Puerto Rico, Haití, Dominican Republic and other Caribbean Islands (Garcia 1994, Del Rio 1996).

In 1519, Hernán Cortéz departed the Port of Santiago de Cuba with an expedition that included a herd of pigs, purchased in Cuba and Jamaica, to sustain his troops on their journey to the mainland (Zadik 2005). After arriving on the Yucatan Peninsula, Cortéz took control of La Villa Rica de la Vera Cruz, the present-day city of Veracruz, in July of 1519 (Hassig 2006). Cortéz ordered Diego de Ordás to begin raising pigs in Veracruz before he set out for the final conquest of the Aztec capital Tenochtitlán. Once the conquest was completed and he was (more or less) in control of Mexico, Cortéz sent for more livestock, including pigs, to be brought from Santo Domingo, Jamaica, and Cuba. There is even evidence that he sent for pigs from as far away as Genoa, Lombardia, and Barcelona. Of the pigs that were brought to the New World by Cortéz, not all of them sustained his expeditionary force; those that were not eaten escaped and reverted to the wild (Zadik 2005). Pig production as a source of animal protein was vital to support colonization with the establishment of new rural population centers and early mining operations in Mexico. These animals also became very important in the culture of indigenous people and mestizos, both of whom adopted the production of pigs in their backyards.

The first chroniclers of American conquest (e.g., Oviedo, Gomara, Las Casas, Colón, Herrera) were the first to provide data for biological introductions and impacts of wild pigs. Several fundamentally referred to how pigs became abundant and dangerous to the native fauna after being introduced to the Antilles Islands. In these descriptions, they generally referred to the island of Hispaniola, although not explicitly. Esquemelin (1678) referenced how abundant and dangerous wild pigs were during the first years of the Spanish colonization. He also made reference to what could have been the first strategy to control an invasive species population in the Americas, stressing the need to bring poison from the metropolis, kill horses, and introduce the poison in horse's viscera and place them in the rural areas to decrease high densities of wild pigs threatening livestock and humans.

Of the various expeditions that headed to Cuba during the first years of the Spanish conquest of the New World, several ended in shipwrecks that could have accidentally introduced new animal species. It is not until the early 1500s that Diego Velázquez de Cuéllar initiated Cuba's colonization and with it the introduction of various domestic animals. Pigs were introduced to Cuba as a food source in 1509 from the Canary Islands (Borroto-Páez and Mancina 2017). One of the first references of the abundance of wild pigs in the Antilles is found in a letter from Diego Velázquez to the King of Spain in 1514. He explains the fertility of the soil and the abundance of wild pigs, which he estimated had grown since their introduction to number 30,000 individuals (Jimeno 1880). De Ribera (1757) made reference to the abundance of cows, pigs, and horses in Cuba, and that many of them were free in the fields without owners. He also commented that 30 years after the colonization of the island, there were more cows, pigs, and horses in Cuba than were needed. He specifically pointed out that "pigs were infinite and tasted better than the Spaniard pigs." Initially, pig farming was practiced under free-range conditions, where pigs were kept in natural/unfenced areas and able to move, disperse, and feed at will. Pigs raised under these conditions quickly became wild and successfully increased in numbers due to their adaptability to the environment, high reproductive rate, and no predators except for feral dogs, which were also introduced. In addition, raising pigs was one of the most important activities in the Antilles in the 16th century. On Mona Island, for example, wild pigs have been present since at least 1593, with estimates of the population on the island numbering 300–700 pigs in 1975 (Wiewandt 1977). Pigs were reportedly released on the island of Barbados during the 16th century by Portuguese sailors for a future source of meat and were subsequently noted as abundant when the English colonized the island in 1625 (Giovas et al. 2019). However, recent evidence suggests that wild pig reports on the island actually represented the presence of wild pigs and an introduced peccary species (*Tayassu pecari* or *Pecari tajacu*; Giovas et al. 2019).

18.2.2 Present Status

After nearly 500 years since the arrival of domestic pigs to Mexico, pig production still occupies a very important place in Mexican husbandry. Pork accounts for about 25% of the meat produced in Mexico and is highly demanded in rural communities (Secretaría de Agricultura, Ganadería, Desarrollo Rural, Pesca y Alimentación 1999). For the Mexican gastronomic culture, the arrival of the domestic pigs caused a true culinary revolution, allowing indigenous people and mestizos to create many new rural dishes. The role of domestic pigs in the rural economy in Mexico is very important. Rural families often raise 1 or more domestic pigs as a way to supplement the family's income (Figure 18.1), because pigs can be readily sold locally. Pork is also commonly used in religious celebrations and the preparation of the meat, skin, blood, and fat in varied ways is highly valued.

Wild pigs are reportedly present in 5 main geographic areas of Mexico: 1) the border strip of the Rio Grande including the northern portions of the states of Chihuahua, Coahuila, Nuevo Léon, and Tamaulipas, 2) the protected area of the "Laguna de Términos" in the state of Campeche in southern Mexico, 3) the biosphere reserve of "Sierra La Laguna" in the state of Baja California Sur, 4) the biosphere reserve "La Michilia" in the state of Durango, and 5) the biosphere reserve "Sierra de Huautla"

FIGURE 18.1 In rural Mexico, raising domestic pigs in backyards is a common practice to provide fresh meat and supplemental income for families. (Photo by J. G. Villarreal-González. With permission.)

in the state of Morelos in central Mexico (Weber 1995, Breceda et al. 2009, Hidalgo-Mihart et al. 2014, Villarreal-González and Flores 2015; Figure 18.2). The most important area is the border strip of the Rio Grande which extends over 1,600 km from Cd. Juarez, Chihuahua just across El Paso, Texas to Matamoros, Tamaulipas, bordering Brownsville, Texas, where wildlife crossings from the United States to Mexico and vice versa have been common for many years. The presence of wild pigs along the Rio Grande River was first reported in the early 1900s by Edgar Alexander Mearns (Mayer and Brisbin 2008). Wild pig distribution in this area covers a total of 29,000 km^2, which includes 3,000 km^2 in Chihuahua, 3,500 km^2 in Coahuila, 11,500 km^2 in Nuevo Léon, and 11,000 km^2 in Tamaulipas. The presence of wild pigs in Mexico has been documented for the last 10 years based on trail cameras, hunting harvests, direct observation, and signs (e.g., tracks, feces, hairs present on the ground or on barbed wire fences) in wildlife conservation/management units (UMAs) and cattle-wildlife operations throughout the states of Coahuila, Nuevo Léon, and Tamaulipas (Villarreal-González and Flores 2015; Figure 18.3). Wild boar and hybrids have been introduced in north and central Mexico for hunting purposes. These animals are now found in 16 UMAs located in the Mexican states of Sonora (1), Chihuahua (1), Coahuila (1), Nuevo León (6), Tamaulipas (1), Aguascalientes (1), Guanajuato (1), Hidalgo (2) and Estado de México (2; Álvarez-Romero et al. 2008).

Populations of wild pigs in the Caribbean have been reported recently on the islands of Abaco, Andros, Big Major Cay, Great Inagua, and Stranger's Cay in the Bahamas, Hispaniola, Jamaica, Mona, and Puerto Rico in the Greater Antilles; Anegada, Barbuda, Dominica, Montserrat, St. Croix, St. Eustatius, St. John, and St. Thomas in the Lesser Antilles; Bonaire and Curacao in the Leeward Antilles; and Bermuda (Borroto-Páez and Woods 2012, van Buurt and Debrot 2012, Mayer 2018). In Cuba, wild pigs are reported in 32 protected areas and 6 offshore islands (Cayo Coco, Cayo Guajaba, Isla de la Juventud, Cayo Romano, Cayo Paredón Grande, and Cayo Sabinal; Figure 18.4), where frequent conflicts of interest occur between conservation authorities, protected area managers, agriculture authorities, and individuals hunting illegally (Borroto-Páez 2011, Borroto-Páez and Mancina 2017).

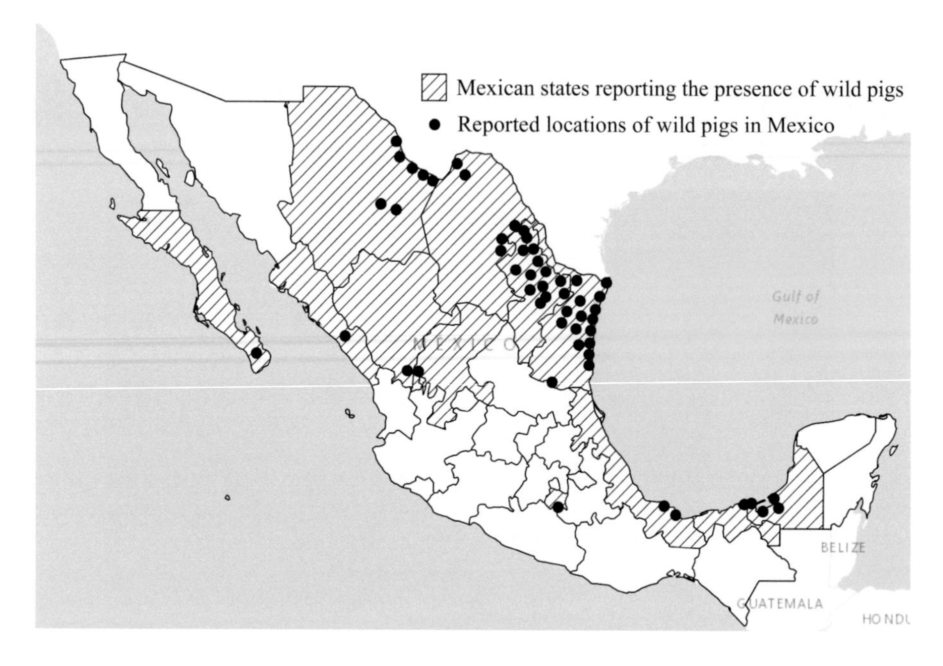

FIGURE 18.2 Recent distribution of introduced wild pigs in Mexico. Based on reports in Weber (1995), Breceda et al. (2009), Hidalgo-Mihart et al. (2014), Villarreal-González and Flores (2015), J. G. Villarreal-González (Consejo Estatal de Flora y Fauna de Neuvo León, unpublished data), and L. Lecuona (USDA/APHIS/WS/International Services Mexico, unpublished data). (Courtesy of J.J. Mayer. With permission.)

FIGURE 18.3 Group of wild pigs in a cattle operation in Anáhuac County, Nuevo León in northern Mexico. (Photo by M. Cagnasso. With permission.)

FIGURE 18.4 Wild pigs in a rural area of Cuba eating sea grape (*Coccoloba uvifera*) fruit. (Reproduced from Borroto-Páez 2011. With permission.)

18.3 REGIONAL ENVIRONMENTAL ASPECTS IN MEXICO AND THE CARIBBEAN

Despite a long historical presence in the region, studies documenting any aspects of wild pig population and reproductive biology, behavior, habitat use, and food habits in Mexico and the Caribbean are scarce.

18.3.1 Diseases and Parasites

Wild pigs in Mexico and the Caribbean represent a significant threat to the dissemination and transmission of parasites and pathogens to humans, domestic livestock, and wildlife (see Chapter 5). Preliminary studies in the state of Nuevo Léon in northern Mexico reported the presence of *Pasteurella* sp., *Entamoeba* sp., and *Eimeria* sp. in lung samples from 6 wild pigs. Four species of helminths were identified in wild pigs' gastrointestinal tracts sampled in Jamaica, which pose a transmission risk to livestock and hunters (Okoro et al. 2016). Ticks, fleas, and lice have been reported and are commonly found on wild pigs in northern Mexico (Ruiz-Fons et al. 2006).

Occurrences of African swine fever (ASF) have been reported in the Caribbean, including outbreaks in Cuba in 1971 and 1980, the Dominican Republic in 1978, and Haití in 1979 (Gallardo et al. 2015). The origin in each of these outbreaks appeared to have been contaminated food waste, which originated in either Europe or Africa and was fed to domestic pigs (Diaz 1980). All of these outbreaks were followed by the depopulation of each country's domestic pig herds, resulting in the slaughter of hundreds of thousands of animals (Ebert 1985, Zilinskas 1999). ASF outbreaks in these island nations were all successfully eliminated (Gallardo et al. 2015) and domestic pigs have been repopulated (Ebert 1985). In the early 1980s, it was determined that wild pigs in Haití also harbored ASF virus, primarily in the area north of Gonaives where wild pigs tested positive for the virus (Mulhern 1983). Through combined efforts of the US Department of Agriculture and the University of Georgia's Southeastern Cooperative Wildlife Disease Study, ASF was eradicated in the Haitian wild pig population.

18.3.2 Damage

Most available information of wild pigs in Mexico and the Caribbean is related to their effects on native and endemic flora and fauna. Wild pigs are a notorious conservation problem on islands worldwide (Cruz et al. 2005, Nogueira-Filho et al. 2009, Hilton and Cuthbert 2010). Mexico has 10–15% of the terrestrial species of the world, constituting the 4th most megadiverse country in the world (Mittermeier and Mittermeier 1992). Additionally, Mexico is ranked 3rd in mammal diversity in the world with 550 species, of which 170 species (30.9%) are endemic to the country (Ceballos and Arroyo-Cabrales 2012). At the same time, 61 species of exotic mammals, including wild pigs, have been reported in Mexico, which are considered a real threat to biodiversity conservation (Lowe et al. 2000). There are approximately 7,000 endemic species in the insular Caribbean including approximately 700 terrestrial vertebrates (Myers et al. 2000). This region is especially threatened and vulnerable to invasive species because endemic species do not share an ecological and evolutionary history with wild pigs (Wiewandt 1977, van Buurt and Debrot 2012).

The Antilles have the highest extinction rate in the world. Impacts of Spanish colonization and the introduction of plants and animals, especially mammalian predators including pigs, are to blame for many of these extinctions as the result of predation and competitive disadvantage of insular species with a long evolutionary history without predators (Morgan and Woods 1986; MacPhee and Flemming 1999; Borroto-Páez 2009, 2011; Borroto-Páez and Woods 2012; Borroto-Páez and Mancina 2017). For example, some species of island shrews (*Nesophontes* spp.), solenodon (*Solenodon* spp.), cave rat (*Boromys* spp.), edible rat (*Brotomys* spp.), and hutia (*Geocapromys* spp., *Mesocapromys* spp., *Mysateles* spp., and *Capromys* spp.) disappeared between the 16th and 19th centuries (Borroto-Páez and Mancina 2017).

A number of impacts from wild pigs have been documented in the Virgin Islands National Park located on St. John in the US Virgin Islands. Wild pigs in the park predate several native and protected species found on the island including the slipperyback skink (*Mabuya inabouia*), Puerto Rican racer (*Borikenophis portoricensis*), blind snake (*Typhlops richardii*), ground lizard (*Ameiva exsul*), and legless lizard (*Amphisbaena fenestrata*; Department of the Interior National Park Service 2003). Wild pigs also depredate eggs and chicks of the bridled quail-dove (*Geotrygon mystacea*), Bahama pintail duck (*Anas bahamensis*), and West Indian Nighthawk (*Chordeiles gundlachii*; Department of the Interior National Park Service 2003).

In the Caribbean, wild pig populations on Montserrat Islands have increased substantially after volcanic eruptions on the southern portion of the island resulted in the establishment of an exclusion zone, which allowed wild pigs to proliferate in the uninhabited area (Peh et al. 2015). Consequently, there has been a decline of the critically endangered mountain chicken frog (*Leptpdactylus fallax*) due to predation by wild pigs (Peh et al. 2015). In Cuba, wild pigs feed on crabs and eggs of reptiles

FIGURE 18.5 Wild pig in Guanahacabibes, Cuba depredating a land blue crab (*Cardisoma guanhumi*). (Reproduced from Borroto-Páez 2011. With permission.)

and birds (Figure 18.5). Borroto-Páez (2009) suggested that wild pig rooting could have a negative effect on the burrows of solenodons (*Solenodon cubanus*) in Cuba. Additional impacts of wild pigs in Cuba are summarized in Tables 18.1 and 18.2. On Cayo Coco, the high density of wild pigs (i.e., over 4 pigs/km^2) is a threat to a breeding colony of Waterhouse's leaf-nosed bat (*Macrotus waterhousei*; Borroto-Páez and Woods 2012). Extirpation of the northern Bahamas boa (*Chilabothrus exsul*) from Stranger's Cay in the Bahamas occurred after the introduction of pigs, which are anecdotally assumed to be the cause of extinction (Borroto-Páez and Woods 2012). In Andros, Bahamas, C. R. Knapp (Shedd Aquarium and San Diego Zoo Institute for Conservation Research, unpublished data) found a negative correlation between the number of species of amphibians and reptiles and the presence of wild pigs. He also observed the absence of rock iguanas (*Cyclura* sp.) from sites where wild pigs were present. Rock iguanas are endemic to the Caribbean and effects of wild pigs on these species are well documented. For example, wild pigs are a major predator of juvenile rock iguanas (Alberts 2000, 2004). Wild pigs also threaten the Mona Island iguana (*Cyclura stejnegeri*) due to egg predation (Pérez-Buitrago et al. 2008, 2016).

The most dramatic ecosystem impacts caused by wild pigs is habitat destruction by excessive rooting and wallowing (Taylor 2003; Figure 18.6). Rooting behavior of wild pigs causes extensive damage to natural plant assemblages, accelerating soil erosion, inhibiting succession, and accelerating the spread of exotic floral species (Bratton 1975, Howe et al. 1976, Wood and Barrett 1979, Stone and Keith 1987, Mungall 2001; see Chapter 7). For example, the Puerto Rico applecactus (*Harrisia portoricensis*), a columnar cactus endemic to Puerto Rico, is currently extinct from Puerto Rico and only a few individuals remain in Mona, Monito, and Desecheo (US Fish and Wildlife Service 1990, Liogier 1994). Among the causes of its extinction are changes in vegetation due to wild pigs (US Fish and Wildlife Service 1990, Rojas-Sandoval and Meléndez-Ackerman 2012).

Rooting in Virgin Islands National Park leaves large areas at risk to erosion by water and wind where rooting up to 1 m in depth has been reported (Department of the Interior National Park Service 2003). Erosion and sedimentation from wild pig rooting negatively impacts mangroves found within the park and damage to archeological and historical sites from wild pig rooting, such as those found at Cinnamon and Reef Bays, has also been reported (Department of the Interior National Park Service 2003). In the mid-1990s, wild pigs also caused substantial rooting damage on a commonly used hiking trail totaling US$50,000 in repair costs (Department of the Interior National Park Service 2003).

TABLE 18.1
Wild Pig Impacts on Cuba's Vegetation (V), Agricultural Products (AP), and Fauna (F)

Common name	Scientific Name	Impacts	Reference(s)
Flora			
Sea grape	*Coccoloba uvifera*	V	R. Borroto-Páez, Sociedad Cubana de Zoologia, unpublished data
Royal palm	*Roystonea regia*	V, AP	Le Riverend (1992), Frías (1844)
Guava	*Psidium* sp.	V, AP	Le Riverend (1992), Frías (1844)
Sweet potato vine	*Ipomoea batatas*	AP	Frías (1844)
Sugar cane	*Saccharum officinarum*	AP	R. Borroto-Páez, unpublished data
Corn	*Zea mays*	AP	Frías (1844)
Mango	*Mangifera indica*	AP	Frías (1844)
Breadfruit tree	*Artocarpus altilis*	AP	Frías (1844)
Avocado tree	*Persea americana*	V, AP	Frías (1844)
Summer squash	*Cucurbita pepo*	AP	Frías (1844)
Cassava	*Manihot esculenta*	AP	Frías (1844)
Banana	*Musa* sp.	AP	
Clammy cherry	*Cordia obliqua*	V, AP	Anónimo (1841), Frías (1844)
West Indian elm	*Guazuma ulmifolia*	V, AP	Anónimo (1841), Frías (1844)
Mahoe	*Hibiscus elatus*	V, AP	Anónimo (1841)
Bullytree	*Pouteria multiflora*	V	Anónimo (1841)
Egg-fruit	*Pouteria dominigensis*	V	Anónimo (1841)
Black lancewood	*Oxandra lanceolate*	V	Anónimo (1841)
Genipap	*Genipa Americana*	V	Anónimo (1841)
Haya	*Oxandra laurifolia*	V	Anónimo (1841)
False breadnut	*Pseudolmedia spuria*	V	Anónimo (1841)
Fauna			
Common blue crab	*Callinectes sapidus*	F	R. Borroto-Páez, unpublished data
Crabs	Several species	F	Escobar (1995)
Land blue crab	*Cardisoma guanhumi*	F	R. Borroto-Páez, unpublished data
Hermit crabs	Paguroidea		
Ground-nesting bird eggs	Several species	F	ACC & ICGC (1990), Escobar (1995)
Ground-nesting reptiles eggs	Several species	F	ACC & ICGC (1990), Escobar (1995)
Cuban land snails	*Polymita muscarum*	F	Fernández Velázquez (1990)Maceira Filgueira et al. (2010)Maceira Filgueira et al. (2010)Espinosa et al. (2004)Espinosa et al. (2004)Espinosa et al. (2004)
	Liggus sp.	F	
	Zachrysia auricoma	F	
	Farcimen ventricosum	F	
	Veronicela sp.	F	
	Zachrysia rangelina	F	
Waterhouse's leaf-nosed bat	*Macrotus waterhousei*	F	ACC & ICGC (1990)
Earthworm	Several species	F	R. Borroto-Páez, unpublished data
Solenodon cubanus burrows	*Solenodon cubanus*	F	Borroto-Páez (2009, 2011), Borroto-Páez and Woods (2012)
Vines used by Carabalí hutia	*Mysateles prehensilis*	F	Borroto-Páez (2009, 2011), Borroto-Páez and Woods (2012)

TABLE 18.2
Ecological (E), Public Health (PH), and Economic (EC) Impacts of Wild Pigs in Guanahacabibes National Park, Pinar del Río Province, Western Cuba

Impacts	Types
Disrupt natural succession and regeneration of natural vegetation	E
Spread invasive plants	E, EC
Disease vectors	PH
Soil erosion	E, EC
Sources of mosquito breeding in soil roots	PH
Fencepost deterioration	EC
Water source contamination	PH, EC
Wild pig–vehicle collisions	PH, EC
Subsistence agriculture damage	EC

Source: Borroto-Paez (2011).

FIGURE 18.6 One of the greatest impacts of wild pigs is rooting disturbance. (Photo by A. Treviño. With permission.)

Seed dispersal by wild pigs in Montserrat may lead to widespread replacement of native vegetation by invasive non-native trees (e.g., Java plum (*Syzygium cumini*), guava (*Psidium* sp.); Peh et al. 2015). Researchers have also suggested that composition and regeneration of native plant communities on Mona Island are strongly impacted by wild pigs (Cintrón 1991, Rojas-Sandoval et al. 2016). In Dominica, researchers have found thousands of seedlings of native palms, but only a small portion of them survive, possibly due to predation by wild pigs (Zona et al. 2003). On Barbuda, wild pigs are causing a continuous ecological decline in ecosystems and are responsible for the destruction of dry forest understories (Francis et al. 1994, Lindsay 2014).

Though not exactly wild pigs, we would be remiss to not mention the famous swimming pigs that have recently become a popular tourist attraction in the Bahamas and Caribbean Islands (Schwartz 2016, Todd 2018). The first and primary island popular for tourists to swim with and feed pigs was Major Big Cay, though domestic pigs have been placed on Exumas, Abaco, Eleuthera, Grand Bahama, and probably a few other small islands for tourism purposes. Pigs were originally brought to Major Big Cay from nearby Staniel Cay about 50 years ago so tourists and residents did not have to smell backyard pigpens. The sound of an outboard motor meant supplemental food for these pigs and they responded by heading to the beach and incoming boats. Touristic interests took note of this and by the late 2000s swimming with the pigs was the most popular tourist attraction in the Bahamas according to the Bahamian Tourist Bureau. These islands, though, likely do not have the resources necessary to sustain populations of free-ranging pigs. Thus, native flora and fauna will be consumed and supplemental feeding is needed. Evidence of starvation and pigs dying with nothing but sand in their stomachs have generated animal welfare concerns (Ross 2017) and aggressive encounters with people are on the rise. Should pigs be maintained on these islands for any length of time they would very likely cause negative short and long-term impacts, as the islands support very vulnerable ecosystems. Conservation authorities must assess current impacts of these pigs and work with those with touristic interests to eliminate or at least minimize impact of this known invasive species that is being used for entertainment of tourists and economic benefit of tour companies.

18.4 REGIONAL MANAGEMENT IN MEXICO AND THE CARIBBEAN

One of the most common reasons for the expansion of wild pigs is the intentional release in areas where the species were not present. Landowners in many cases think having wild pigs adds value to the hunting experience ranches offer. However, Article 27 in the Mexican general wildlife law specifically prohibits the release of exotic species in natural habitats, including wild pigs (Diario Oficial de la Federación 2016).

Traditional control methods to manage wild pig populations include complete eradication and erection of exclusion areas. As an emergency measure, headstart projects have been established on several Caribbean Islands in an attempt to increase recruitment of struggling rock iguana populations (Alberts 2004, Hudson and Alberts 2004, Pérez-Buitrago et al. 2008). Fences have been established on Mona Island to create predator exclusion zones to protect iguana and turtle nests from wild pig predation (Pérez-Buitrago 2008, 2016; Rojas-Sandoval et al. 2016). Jamaica also started an eradication program for wild pigs and other predators in 1997, resulting in a 6-fold increase in nesting females and hatchlings rock iguanas from 1991 to 2013. Recreational and subsistence hunting of wild pigs are additional management tools that may aid in conserving rock iguanas in the Caribbean (Knapp and Owens 2015, García and Gerber 2016).

A control program for wild pigs was established in Virgin Islands National Park to accomplish several goals, including: 1) substantial population reduction, 2) reduction of range expansion, 3) preservation of natural resources, 4) disturbance mitigation to archeological sites, and 5) increasing public safety for visitors (Department of the Interior National Park Service 2003). The National Park Service and USDA/APHIS/Wildlife Services utilized fencing, monitoring, and lethal removal (i.e., trapping, shooting) in coordination with the Virgin Islands Department of Planning and Natural Resources, Virgin Islands Division of Fish and Wildlife, and Virgin Islands Department of Economic Development and Agriculture to accomplish these goals and efforts are ongoing (Department of the Interior National Park Service 2003, 2017). On Montserrat Island, the Montserrat Department of the Environment permitted trapping and hunting in 2009 as part of a management program to reduce wild pigs on the island that has been successful within portions of the island (Peh et al. 2015). However, removal efforts in the southern portion of the island, where the majority of pigs are, is challenging because recent volcanic activity and risk of further eruptions has made the area relatively inaccessible (Peh et al. 2015). Although challenging and expensive, Peh et al. (2015) found that economic losses (e.g., nature-based tourism) in the absence of management would be 5 times greater than the current cost of management efforts.

While the problem of exotic invasive species in Mexico is well known, no official eradication or management programs have been established for wild pigs. Along the US-Mexico border region, the main management practice used to control wild pigs is hunting (Figure 18.7). Some of the requirements for wild pig hunting according to the Mexican general wildlife law include the

FIGURE 18.7 Wild pig harvested in a cattle operation in Los Aldama County, Nuevo León, in northern Mexico. (Photo by S. Montemayor. With permission.)

following: 1) hunting must be within an officially registered UMA and 2) the owner or manager of the UMA needs to submit an application and obtain tags for legal transportation of harvested animals (Diario Oficial de la Federación 2000). State offices responsible for issuing harvest permits in the states of Coahuila, Nuevo Léon, and Tamaulipas have simplified the procedure to encourage hunters to harvest as many wild pigs as possible in an effort to manage populations. Box traps are also used in northern Mexico to capture wild pigs (Figure 18.8). However, it is important to point out these 2 management techniques are not going to solve the wild pig problem in Mexico, but are the only efforts presently implemented.

One of the most urgent actions related to the wild pig problem in Mexico is to establish an education and outreach program to educate people about the risk of pathogen and parasite transmission to humans consuming or handling wild pigs. In a survey conducted by Villarreal-González (2017) that was sent to hunters, ranch owners, and field workers from 100 different ranches, 100% of the participants responded: 1) no protective equipment (latex gloves, glasses, butcher aprons) were used to avoid direct contact with blood, other fluids, viscera, or meat (Figure 18.9), 2) animal residues were not properly disposed of to avoid consumption by other animals, and 3) safety procedures for the consumption of wild pig meat were being ignored.

18.5 FUTURE OF WILD PIGS IN MEXICO AND THE CARIBBEAN

The perception of Mexican ranchers, hunters, and managers that wild pigs are an asset for their hunting operations is erroneous in many cases. Ninety-five percent of participants in a recent survey indicated their desire to maintain wild pigs in their operations to provide hunting opportunity and were even interested in introducing more to increase populations (Villarreal-González 2017). Based on this, it is possible to predict that the wild pig population in Mexico could increase exponentially

FIGURE 18.8 Wild pigs captured in a box trap near Falcon Lake in the state of Tamaulipas in northern Mexico. (Photo by A. Flores. With permission.)

FIGURE 18.9 Most people in Mexico do not use protective equipment to avoid contact with blood, viscera, and other internal contents when processing harvested wild pigs. (Photo by J. G. Villarreal-González. With permission.)

in the near future if further education and control actions are not implemented. Wild pigs are similarly expected to continue to persist and expand on many islands of the Caribbean where management is minimal or non-existent. Urgent actions from governments, private land owners, and landowner organizations are needed to stop range expansion of wild pigs, reduce or eliminate their populations, and mitigate their destructive impacts across Mexico and the Caribbean.

REFERENCES

ACC e ICGC (Academia de Ciencias de Cuba e Instituto Cubano de Geodesia y Cartografía). 1990. Estudio de los grupos insulares y zonas litorales del archipiélago cubano con fines turísticos: Cayos Sabinal, Guajaba y Romano. Ed. Científico-Técnica, La Habana.

Alberts, A. C. 2000. *West Indian Iguanas: Status Survey and Conservation Action Plan.* IUCN-the World Conservation Union, Gland, Switzerland.

Alberts, A. C. 2004. Conservation strategies for West Indian rock iguanas (genus *Cyclura*): Current efforts and future directions. *Iguana* 11:212–223.

Álvarez-Romero, J. G., R. A. Medellín, A. Oliveras de Ita, H. Gómez de Silva and O. Sánchez. 2008. Animales exóticos en México: Una amenaza para la biodiversidad. Comisión Nacional para el Conocimiento y Uso de la Biodiversidad, Ciudad de México, México, d.f [In Spanish].

Anónimo. 1841. Diálogo entre un labrador y su hijo. Lección IV. Crianza de ganado, haciendas o hatos y corrales. Memorias de la Sociedad Económica de La Habana. *Agricultura. Cartilla rústica* 171–194 [In Spanish].

Borroto-Páez, R. 2009. Invasive mammals in Cuba: An overview. *Biological Invasions* 11:2279–2290.

Borroto-Páez, R. 2011. Los mamíferos invasores o introducidos. Pages 220–241 *in* R. Borroto-Páez and C. A. Mancina, editors. *Mamíferos en Cuba.* UPC Print, Vaasa, Finland.

Borroto Páez, R. and C. A. Mancina. 2017. Biodiversity and conservation of Cuban mammals: Past, present, and invasive species. *Journal of Mammalogy (Special issue)* 98:964–985.

Borroto-Páez, R., and C. A. Woods. 2012. Status and impact of introduced mammals in the West Indies. Pages 241–258 *in* R. Borroto-Páez, C. A. Woods, and F. E. Sergile, editors. *Terrestrial Mammals of the West Indies. Contributions.* Wocahoota Press and Florida Museum of Natural History.

Bratton, S. P. 1975. The effects of European wild boar, *Sus scrofa*, on gray beech forest in the Great Smoky Mountains. *Ecology* 56:1356–1366.

Breceda, A., A. Arnaud-Franco, S. Álvarez-Cárdenas, P. Galina Tessaro and J. Montes-Sánchez. 2009. Evaluación de la Población de Cerdos Asilvestrados (*Sus scrofa*) y su Impacto en la Reserva de la Biosfera Sierra La Laguna, Baja California Sur, México. *Tropical Conservation Science* 2:173–188 [In Spanish].

Ceballos, G., and J. Arroyo-Cabrales. 2012. Lista Actualizada de los Mamíferos de México 2012. *Revista Mexicana de Mastozoología Nueva época* 2:27–80 [In Spanish].

Cintrón, B. 1991. Introduction to Mona Island. *Acta Científica* 5:6–9 [In Spanish].

Crosby, A. W. 1972. *The Columbian Exchanges. Biological and Cultural Consequences of 1492.* Greenwood Press, Westport, CT.

Cruz, F., C. J. Donlan, K. Campbell, and V. Carrion. 2005. Conservation action in the Galapagos: Feral pig (*Sus scrofa*) eradication from Santiago Island. *Biological Conservation* 121:473–478.

de Ribera, N. J. 1757. *Descripción de la Isla de Cuba.* Editorial de Ciencias Sociales, La Habana, Cuba [In Spanish].

del Río-Moreno, J. L. 1996. El Cerdo. Historia de un Elemento Esencial de la Cultura Castellana en la Conquista y Colonización de América (siglo XVI). *Anuario de Estudios Americanos* 53:13–35 [In Spanish].

Department of the Interior National Park Service. 2003. *Final Environmental Assessment: Sustained Reduction Plan for Non-Native Wild Hogs within Virgin Islands National Park.* National Park Service, Virgin Islands National Park, St. John, VI.

Department of the Interior National Park Service. 2017. Virgin Islands National Park: Nonnative species. National Park Service, Virgin Islands National Park, St. John, VI. <https://www.nps.gov/viis/learn/nature/nonnativespecies.htm>. Accessed 17 May 2019.

Diario Oficial de la Federación. 2000. *Ley General de Vida Silvestre. Publicado el 03 de Julio del año 2000.* México, D.F. [In Spanish].

Diario Oficial de la Federación. 2016. *Ley General de Vida Silvestre. Publicado el 12 de Julio del año 2016.* México, D.F. [In Spanish].

Diaz, O. S. 1980. Report on the status of African swine fever in the Dominican Republic. *Proceedings of the Inter-American Meeting on Foot-and-Mouth Disease and Zzoonoses Control* 12:55–59.

Donkin, R. A. 1985. The peccary:- With observations on the introduction of pigs to the New World. *Transactions of the American Philosophical Society* 75:1–152.

Ebert, A. 1985. Porkbarreling pigs in Haiti: North American "swine aid" an economic disaster for Haitian peasants. *The Multinational Monitor* 6:1–6.

Elliot, J. H. 2007. *Empires of the Atlantic World: Britain and Spain in America* 1492–1830. Yale University Press, New Haven, CT.

Ensminger, M. E. 1961. *Swine Science*. The Interstate Printers and Publishers, Danville, IL.

Escobar, T. R. 1995. *Isla de la Juventud. Vertebrados introducidos por causas deliberadas*. Editorial Científico Técnica, Pinos Nuevos, Ciudad de la Habana, Cuba [In Spanish].

Espinosa, J., J. J. Ortea, J. Fernández Milera, W. Oliva. 2004. Catálogo ilustrado de los moluscos terrestres y fluviales del Pan de Guajaibón, Área Protegida Mil Cumbres, Pinar del Río, Cuba. *Revista de la Academia Canaria de Ciencias* 16:179–220.

Esquemelin, J. O. 1678. *Piratas de América*. Comisión Nacional Cubana de la Unesco, La Habana, Cuba [In Spanish].

Fernández Velázquez, A. 1990. Ecologia de *Polymita muscarum* (Gastropoda: Fruticicollidae) en la provincia de Holguín. *Revista Biología* 4:3–13.

Francis, J., C. Rivera, and J. Figueroa. 1994. Toward a woody plant list for Antigua and Barbuda: Past and present. General Technical Report SO-102, US Department of Agriculture, Forest Service, Southern Forest Experiment Station, New Orleans, LA.

Frías, J. J de. 1844. *Ensayo sobre la crianza de ganados en la Isla de Cuba*. Oficina del Faro Industrial, La Habana, Cuba [In Spanish].

Gallardo, C., A. de la Torre Reoyo, J. Fernández-Pinero, I. Iglesias, J. Muñoz and L. Arias. 2015. African swine fever: A global view of the current challenge. *Porcine Health Management* 1:21.

García, M. A., and G. P. Gerber. 2016. Conservation and management of *Cyclura* iguanas in Puerto Rico. *Herpetological Conservation and Biology* 11:61–67.

García M. B., 1994. Los Primeros Pasos del Ganado en México. Relaciones Estudios de Historia y Sociedad. *Colegio de Michoacán* XV 59:11–44 [In Spanish].

Giovas, C. M., G. D. Kamenov, and J. Krigbaum. 2019. $^{87}Sr/e^{86}Sr$ and ^{14}C evidence for peccary (Tayassuidae) introduction challenges accepted historical interpretation of the 1657 Ligon map of Barbados. *PLoS ONE* 14:e0216458.

Hassig, R. 2006. *Mexico and the Spanish Conquest*. University of Oklahoma Press, Norman, OK.

Hidalgo-Mihart, M. G., D. Pérez-Hernández, L. A. Pérez-Solano, F. Contreras-Moreno, J. Angulo-Morales and J. Hernández-Nava. 2014. Primer registro de una población de cerdos asilvestrados en el área de la Laguna de Términos, Campeche, México. *Revista Mexicana de Biodiversidad* 85:990–994 [In Spanish].

Hilton, G. M., and R. J. Cuthbert. 2010. The catastrophic impact of invasive mammalian predators on birds of the UK Overseas Territories: A review and synthesis. *Ibis* 152: 443–458.

Howe, T. D., F. J. Singer, and B. B. Ackerman. 1976. Foraging relationships of European wild boar invading northern hardwood forest. *Journal of Wildlife Management* 45:748–754.

Hudson, R. D., and A. C. Alberts. 2004. The role of zoos in the conservation of West Indian iguanas. Pages 274–289 *in* A. C. Alberts, R. L. Carter, W. K. Hayes, and E. P. Martins, editors. *Iguanas: Biology and Conservation*. University of California Press, Berkeley, CA.

Jimeno. F. 1880. Mamíferos indígenas y animales domésticos. *Revista de Cuba (Periódico Mensual)* 8:421–430 [In Spanish].

Knapp, C. R., and A. K. Owens. 2015. Home range and habitat associations of a Bahamian iguana: Implications for conservation. *Animal Conservation* 8:269–278.

Le Riverend, J. 1992. *Problemas de la formación agraria de Cuba. Siglos XVI–XVII*. Editorial Ciencias Sociales, La Habana, Cuba [In Spanish].

Lindsay, K. C. 2014. The ferns of Antigua and Barbuda: A case of resurgence and resilience. *Pteridologist* 6:14–18.

Liogier, H. A. 1994. Descriptive flora of Puerto Rico and adjacent islands. Editorial de la Universidad de Puerto Rico, San Juan, PR [In Spanish].

Lowe S., M. Browne, S. Boudjelas, and M. De Poorter. 2000. 100 of the world's worst invasive alien species: A selection from the Global Invasive Species Database. Invasive Species Specialist Group, Species Survival Commission, World Conservation Union (IUCN), Auckland, New Zealand.

Maceira Filgueira D, R. Pascual Pérez and J. Reyes Brea. 2010. Land molluscs of the Silla de Romano protected area, north coast of Cuba, and their conservation problems. *Tentacle* 18:22–25.

MacPhee R. D. E, and C. Flemming. 1999. Requiem æternam: The last five hundred years of mammalian species extinctions. Pages 333–371 *in* R. D. E. MacPhee, editor. *Extinctions in Near Time: Causes, Contexts, and Consequences*. Kluwer Academic and Plenum, New York.

Mayer, J. J. 2018. Introduced wild pigs in North America: History, problems and management. Pages 299–312 *in* M. Melletti and E. Meijaard, editors. *Ecology, Evolution and Management of Wild Pigs and Peccaries: Implications for Conservation*. Cambridge University Press, Cambridge, UK.

Mayer, J. J., and J. C. Beasley. 2018. Wild pigs. Pages 219–248 *in* W. C. Pitt, J. C. Beasley, and G. W. Witmer, editors. *Ecology and Management of Terrestrial Vertebrate Invasive Species in the United States*. CRC Press, LLC, Taylor & Francis Group, Boca Raton, FL.

Mayer, J. J., and I. L. Brisbin, Jr. 2008. *Wild Pigs in the United States: Their History, Comparative Morphology and Current Status*. University of Georgia Press, Athens, GA.

Mittermeier R., and C. G. Mittermeier. 1992. La Importancia de la Diversidad Biológica de México. Pages 63–73 *in* J. Sarukhan and R. Dirzo, editors. *México Ante los Retos de la Biodiversidad*. Comisión Nacional para el Conocimiento y Uso de la Biodiversidad, México, D.F. [In Spanish].

Morgan, G. S., and C. A. Woods. 1986. Extinction and the zoogeography of West Indian land mammals. *Biological Journal of the Linnaean Society* 28:167–203.

Mulhern, F. 1983. 10th Coordinating Committee Meeting Held at PAPPADEP-Delmas, 14–16 Nov. 1983. Programme pour l'éradication de la peste porcine africaine et pour le développement de l'élevage porcin, Interamericano Institute de Ciencias Agricola, Port-au-Prince, Haiti.

Mungall, E. C. 2001. Exotics. Pages 736–764 *in* S. Demarais and P. R. Krausman, editors. *Ecology and Management of Large Mammals in North America*. Prentice Hall, Upper Saddle River, NJ.

Myers, N., R. A. Mittermeier, C. G. Mittermeier, G. A. B. da Fonseca, and J. Kent. 2000. Biodivesity hotspots for conservation priorities. *Nature* 403:853–858.

Nogueira-Filho, S. L. G., S. S. C. Nogueira, and J. M. V. Fragoso. 2009. Ecological impacts of feral pigs in the Hawaiian Islands. *Biodiversity and Conservation* 18:3677–3683.

Okoro, C. K., B. S. Wilson, J. Lorenzo-Morales, and R. D. Robinson. 2016. Gastrointestinal helminths of wild hogs and their potential livestock and public health significance in Jamaica. *Journal of Helminthology* 90:139–143.

Peh, K. S. H., A. Balmford, J. C. Birch, C. Brown, S. H. M. Butchart, J. Daley, J. Dawson, et al. 2015. Potential impact of invasive alien species on ecosystem services provided by a tropical forested ecosystem: A case study from Montserrat. *Biological Invasions* 17:461–475.

Pérez-Buitrago, N., M. A. García, A. Sabat, J. Delgado, A. Álvarez, O. McMillan, and S. M. Funk. 2008. Do headstart programs work? Survival and body condition in headstartered Mona Island iguanas *Cyclura cornuta stejnegeri*. *Endangered Species Research* 6:55–65.

Pérez-Buitrago, N., A. M. Sabat, and W. O. McMillan. 2016. Nesting migrations and reproductive biology of the Mona rhinoceros iguana, *Cyclura stejnegeri*. *Herpetological Conservation and Biology* 11(Monograph 6):197–213.

Rojas-Sandoval,. J., and E. Meléndez-Ackerman. 2012. Factors affecting establishment success of the endangered Caribbean cactus *Harrisisa portoricensis* (Cactaceae). *International Journal of Tropical Biology* 60:867–879.

Rojas-Sandoval, J., E. J. Meléndez-Ackerman, J. Fumero-Cabán, M. García-Bermúdez, J. Sustache, S. Aragón, M. Morales-Vargas, G. Olivieri, and D. S. Fernández. 2016. Long-term understory vegetation dynamics and responses to ungulate exclusion in the dry forest of Mona Island. *Caribbean Naturalist Special Issue* 1:138–156.

Ross, D. 2017. This is what really killed the famous swimming pigs. *National Geographic*. <https://news.nationalgeographic.com/2017/03/swimming-pigs-bahamas-death.html>. Accessed 4 Mar 2019.

Ruiz-Fons F., I. G. Fernández de Mera, P. Acevedo, U. Höfle, J. Vicente, J. de la Fuente, C. Gortazár 2006. Ixodid ticks parasitizing Iberian red deer (*Cervus elaphus hispanicus*) and European wild boar (*Sus scrofa*) from Spain: Geographical and temporal distribution. *Veterinary Parasitology* 140:133–142.

Schwartz, R. 2016. An island in the Bahamas where pigs swim free. *The New York Times*. <https://www.nytimes.com/interactive/2016/03/30/magazine/voyages-big-major-cay-pigs>. Accessed 31 Mar 2019.

Secretaría de Agricultura, Ganadería, Desarrollo Rural, Pesca y Alimentación. 1999. *Situación Actual y Perspectiva de la Producción de Carne de Porcino en México 1990–1998*. México, D.F. [In Spanish].

Stone, C. P., and J. O. Keith. 1987. Control of feral ungulates and small mammals in Hawaii's National Parks: Research and management strategies. Pages 277–287 *in* C. G. H. Richards and T. Y. Ku, editors. *Control of Mammal Pests*. Taylor and Francis, London, UK.

Taylor, R. 2003. The feral hog in Texas. *Texas Parks and Wildlife*. <https://tpwd.texas.gov/publications/pwdpubs/media/pwd_bk_w7000_0195.pdf>. Accessed 27 June 2017.

Todd, T. R. 2018. *Pigs of Paradise: The Story of the World-Famous Swimming Pigs*. Skyhorse Publishing, New York.

US Fish and Wildlife Service. 1990. Endangered and threatened wildlife and plant: Rules and Regulations. US Fish and Wildlife Service Report 50 CRF: 32252–32255, Federal Register, Atlanta, GA.

van Buurt, G., and A. O. Debrot. 2012. Exotic and invasive terrestrial and freshwater animal species in the Dutch Caribbean. Institute for Marine Resources & Ecosystem Studies. Report No. C001/12, The Ministry of Economic Affairs, Agriculture and Innovation, Netherlands.

Villarreal-González, J. G. 2017. Informe de Resultados de la Encuesta: "Presencia de Marranos Alzados y/o Jabalíes Euroasiáticos *Sus scrofa* en Predios Rurales y UMAS Extensivas del Noreste de México: Coahuila, Nuevo León y Tamaulipas." Consejo Estatal de Flora y Fauna Silvestre de Nuevo León, A.C. Monterrey, Nuevo León, México. Informe Inédito. 1–30 [In Spanish].

Villarreal-González, J. G., and G. J. A. Flores. 2015. Impacto del marrano alzado y el jabalí europeo en hábitats del matorral espinoso tamaulipeco en el noreste de México. *Ciencia Uanl* 18:1–9 [In Spanish].

Weber, M. 1995. La Introduccion del jabalí Europeo a la Reserva de la Biosfera la Michilia, Durango: Implicaciones ecologicas y epidimiologicas. *Revista Mexicana de Mastozoologia* 1:69–73 [In Spanish].

Wiewandt, T. A. 1977. Ecology, behavior, and management of the Mona Island Ground Iguana, *Cyclura stejnegeri*. Dissertation, Cornell University, Ithaca, NY.

Wood, G. W., and R. H. Barrett. 1979. Status of the wild pigs in the United States. *Wildlife Society Bulletin* 7:237–246.

Zadik, B. J. 2005. The Iberian pig in Spain and the Americas in the time of Columbus. Thesis, University of California, Berkeley, CA.

Zilinskas, R. A. 1999. Cuban allegations of biological warfare by the United States: Assessing the evidence. *Critical Reviews in Microbiology* 25:173–227.

Zona, S., James, A., Maidman, K. 2003. Native palms of Dominica. *Palms* 47:151–157.

19 Wild Pig Populations along the Urban Gradient

Jesse S. Lewis, Kurt C. VerCauteren, Robert M. Denkhaus, and John J. Mayer

CONTENTS

19.1 INTRODUCTION

Wild pigs' (*Sus scrofa*) association with humans and their residences dates back centuries, with a corresponding well-documented history of conflict (Mayer and Brisbin 2009). Recently, with expanding populations of wild pigs and humans in North America, conflicts have rapidly increased, particularly in urbanized areas. The adaptability and behavioral plasticity of wild pigs, in addition to land-use change and human tolerance, have contributed to opportunities for wild pigs to occupy and, in many cases, thrive in urban settings. Indeed, wild pigs widely occur in urbanized environments across both their native and non-native ranges (Gaston 2010, Forman 2014, Adams 2016). As a result, the colonization of urbanized areas by wild pigs leads to diverse problems for people, property, and ecological communities. Solutions for mitigating these conflicts are increasingly being explored and implemented by urban land managers. As interactions with wild pigs in urbanized areas are expected to continue increasing, it is critical that the public, government and nongovernment organizations, and wildlife managers work together to control, mitigate, and prevent the invasion of urbanized areas by wild pigs. They can be destructive and dangerous animals in urbanized areas, and there are unique challenges with managing this invasive species in developed settings.

To address this growing issue, we review 3 central topics of wild pig invasion into urbanized areas. First, we evaluate wild pigs' exploitation of the urban environment in recent decades. We describe the urban gradient, how wild pigs have invaded different types of urbanization, and changes in wild pig behavior and population characteristics in relation to urbanized habitat. Second, we review the myriad impacts of wild pigs in urbanized environments, including property damage, human health, and ecological damage. And third, we explore how wild pig populations can be controlled and managed in urban environments considering diverse human values and government regulations. We also present 2 case studies of high profile urban areas (Dallas/Fort Worth, Texas and Berlin, Germany) where wild pigs have invaded and been relatively well studied. Importantly, not all urbanization is created equal and the characteristics of different forms of urbanization can have varying effects on wild pig populations and options for their management.

19.2 EXPANDING HUMAN AND WILD PIG POPULATIONS

Urbanization is one of the fastest growing forms of land-use change across the world and covers hundreds of thousands of square kilometers globally (Schneider et al. 2009, Theobald 2014). Further, it is projected to increase by hundreds of thousands of square kilometers over the next several decades (Seto et al. 2011). Urbanization can dramatically change landscape characteristics, which can influence animal populations and the behavior of individuals (Gaston 2010, Forman 2014). Although urbanization can negatively impact many native plants and animals, other species can thrive in urbanized landscapes. In particular, native and non-native and generalist species can be especially successful in urban settings. Urbanized areas can be attractive to species for a variety of factors, including novel and supplemental food resources associated with humans (e.g., trash, lawns, direct feeding), reduced predation and competition, and novel habitat features (Gaston 2010).

How species interact with areas of human residences depends on the form of urbanization, which can be viewed across a gradient, from low to high-density urbanization (Forman 2014). At one end of the spectrum, wildland habitat is characterized by the absence of urban residences and development. Low-density residential development (i.e., rural and exurban development) is often intermixed with natural, undeveloped areas and can be permeable to animal movement (Hansen et al. 2005). In addition, rural development can be associated with agriculture, which can increase wild pig populations (Lewis et al. 2017). High-density residential development (i.e., suburban and urban development) might be impermeable to movement for many species. However, these forms of high-density urbanization are often associated with open space and greenways (e.g., parks, lawns, and natural habitat) that are intermixed with developed areas and can act as travel corridors for animals. The term "peri-urban" is commonly applied outside of the United States to describe areas of urbanization that are adjacent to high-density urbanization; peri-urban can include low-density development (Forman 2014) and high-density development (Licoppe et al. 2013). All forms of urbanization can be associated with nearby wildland areas, which can act as sources for wildlife populations that utilize urbanized landscapes (Forman 2014). The contact zone between natural areas and urbanized areas is referred to as the wildland–urban interface (WUI), which can experience increased conflicts between people and wildlife (Radeloff et al. 2005).

In addition to humans expanding areas of urbanization across the United States and globally, wild pig populations are correspondingly expanding their range at an accelerated rate and invading new areas. In Europe, wild pig populations have expanded in response to reduced hunting pressure, mild winters, reforestation, agriculture, and supplemental feeding (Massei et al. 2015). In addition, wild pigs have expanded their distribution across their non-native range of sub-Saharan Africa, Australia, North America, South America, and many islands. Particularly in the United States, wild pigs have substantially expanded their distribution over the last several decades (Southeastern Cooperative Wildlife Disease Study 2016, Snow et al. 2017). Based on available habitat, there is tremendous potential for wild pigs to invade additional areas in the United States and globally (Lewis et al. 2017). Ultimately, expanding human and wild pig populations are expected to result in increased conflicts across the gradient of urbanization in the future.

19.3 HISTORY OF OCCURRENCE ACROSS THE GRADIENT OF URBANIZATION

Although wild pigs have a long history of associating with humans and their residences, the degree of that association varies depending upon the type of urbanization. People living in low density, rural communities, especially those associated with agriculture, have a relationship with wild pigs dating back thousands of years (Larson et al. 2005). In North America, wild pigs have been a part of the rural landscape for centuries, particularly in the southern United States. Demonstrating their association with early Americans, the presence of wild pigs in rural communities has been captured in classic literature, such as in "The Adventures of Huckleberry Finn" by Mark Twain (published in 1885). Living along the banks of the Mississippi River, Huckleberry Finn states: "I stood on the bank and looked out over the river. All safe. So I took the gun and went up a piece into the woods, and was hunting around for some birds when I see a wild pig; hogs soon went wild in them bottoms after they had got away from the prairie farms. I shot this fellow and took him into camp." Later on in the story, Huck Finn reports "There was generly [sic] hogs in the garden, and people driving them out." Although fictional accounts of wild pigs, they were a common occurrence in many rural landscapes during early American history. This is supported by accounts documenting the presence of wild pigs since at least the 1700s and 1800s (and dating back into the 1500s in some cases) in the southern and eastern United States, including Alabama, Arkansas, Florida, Georgia, Mississippi, South Carolina, Texas, and Virginia, in addition to populations on the west coast in California (Mayer and Brisbin 2008; see Chapter 2).

More recently, however, wild pigs are increasingly occurring in areas of high-density urbanization. At a global scale, wild pigs inhabiting and accessing urbanized environments have dramatically increased during the last several decades (Licoppe et al. 2013; Figure 19.1), both within their native and non-native ranges (see case studies 1 and 2 below). Although little data are available in the literature about wild pig conflicts in urbanized habitat, there are many reports documented in news articles. Globally, wild pigs were reported as causing conflict in at least 44 cities or towns in 25 countries (Cahill et al. 2012, Licoppe et al. 2013). In the United States alone (as of 2017), wild pigs have invaded urban habitat associated with 150 towns and cities across the country since 2000 (Figure 19.2), which demonstrates a growing problem (Figure 19.3). In addition, based on news reports, wild pig conflicts tend to occur most in suburban habitat (Figure 19.4). However, conflicts are likely greater in areas of low-density urbanization, but these conflicts are probably underreported in news articles. Overall, the increasing trend of wild pig conflict in urbanized habitat is expected to continue.

19.4 WILD PIG POPULATIONS AND BEHAVIOR IN URBANIZED AREAS

For wild animals to access areas of urbanization, 2 features are commonly necessary: 1) populations occupying wildlands adjacent to areas of urbanization (e.g., WUI), and 2) habitat corridors that allow animals access into urbanized areas (Forman 2014). Indeed, natural areas adjacent to urbanized areas can provide the source population for wild pigs accessing urbanized landscapes (Stillfried et al. 2017a). Further, as with many species, especially large mammals, wild pigs are

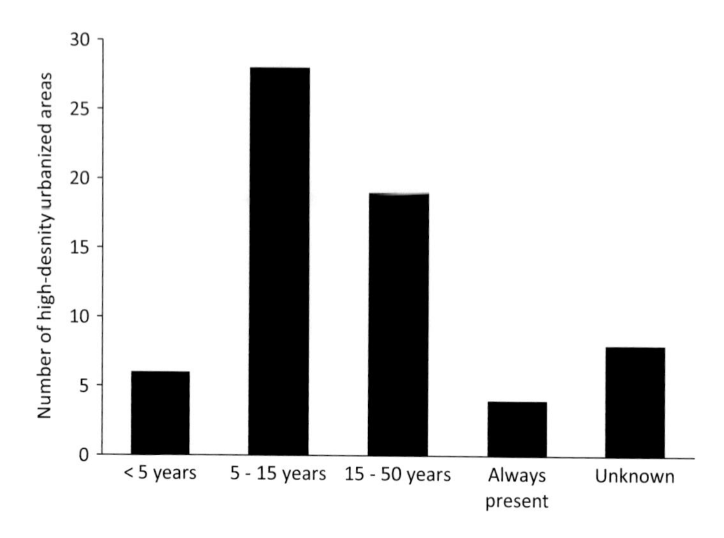

FIGURE 19.1 Number of high-density urbanized areas invaded by wild pigs globally. The x-axis presents the time since 2013 that wild pigs were believed to colonize an urbanized area. (Adapted from Licoppe et al. 2013.)

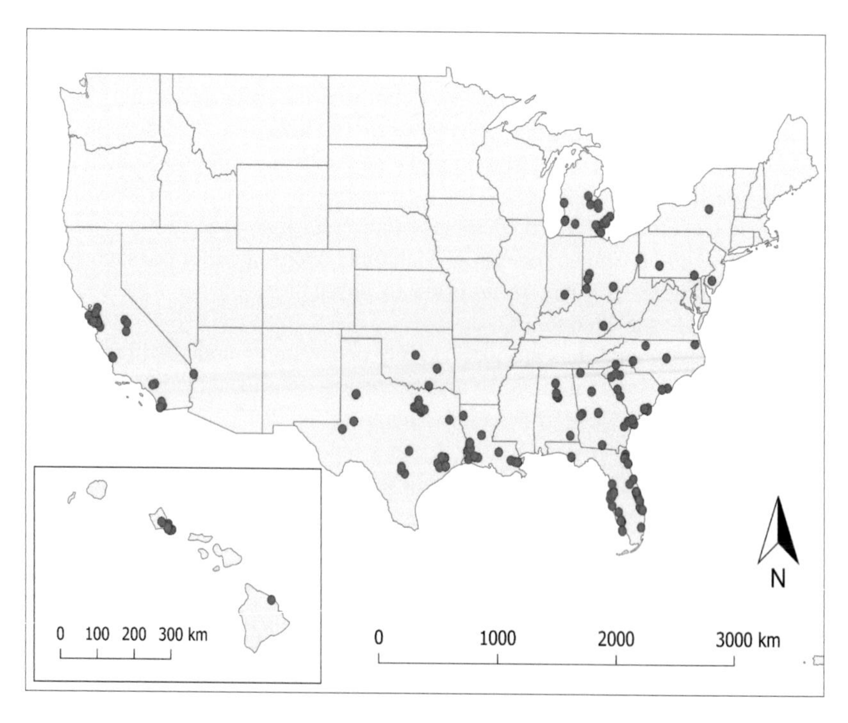

FIGURE 19.2 Locations of municipalities in the United States that have reported problems with wild pigs in urbanized areas since 2000. Hawaii is included in the inset.

more likely to occur in large patches of habitat (Virgós 2002); thus urbanized areas associated with large areas of adjacent habitat are more likely to experience invading wild pigs. Travel corridors for animals can occur along greenways, vegetated highways or utility right-of-ways, riparian pathways, or other forms of cover (Forman 2014, Castillo-Contreras et al. 2018).

The primary reasons why wild pigs colonize urbanized areas is to access food and water, a reduction of hunting or predation pressure, and expansion of urbanization into forest and other

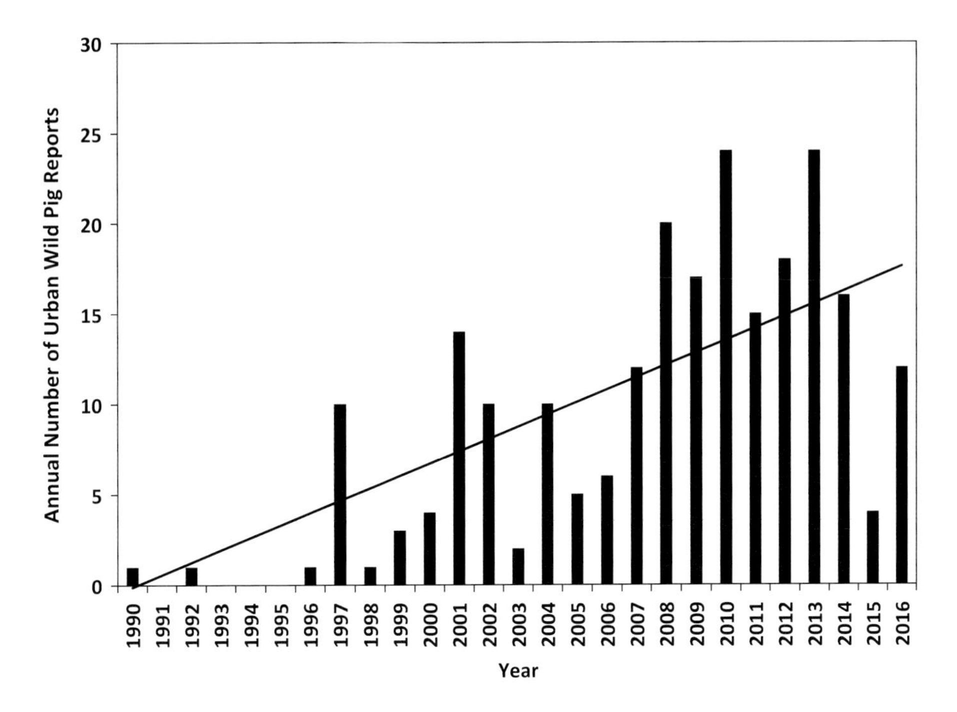

FIGURE 19.3 The number of wild pig reports in urbanized habitat in the United States has increased from 1990 to 2016.

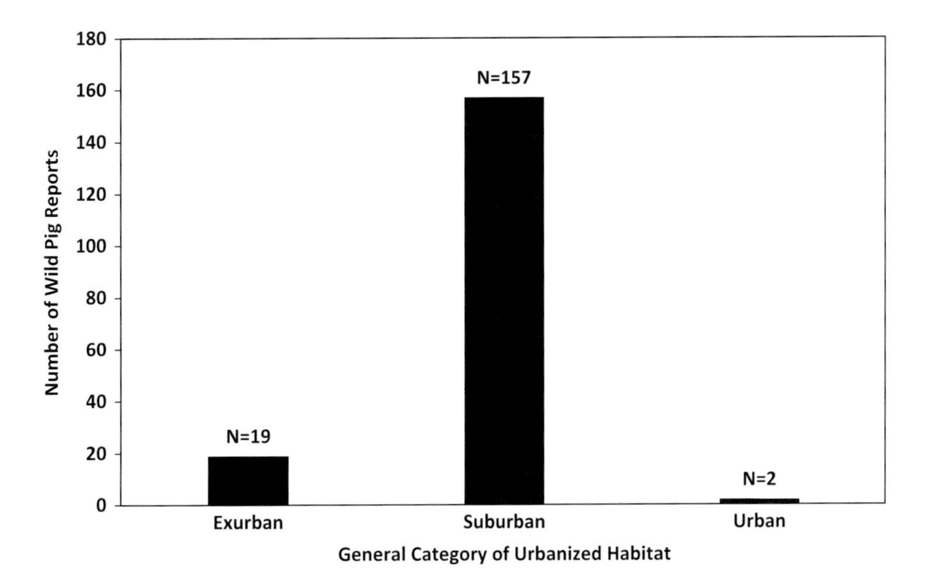

FIGURE 19.4 Wild pig reports in the United States across different levels of urbanization since 2000.

undeveloped vegetation types preferred by wild pigs (Cahill et al. 2012, Licoppe et al. 2013, Toger et al. 2018; Figure 19.5). Human-related food sources in urbanized areas can be available through people's vegetable gardens, domestic trash and other food resources, and the direct feeding of wild pigs by people (Figure 19.6). Intentional or supplemental feeding of wild pigs by the public is common in many countries (Massei et al. 2015) and an important driver of wild pig persistence in some urbanized landscapes (Cahill et al. 2012). Perhaps surprisingly, it is believed that wild pigs most often

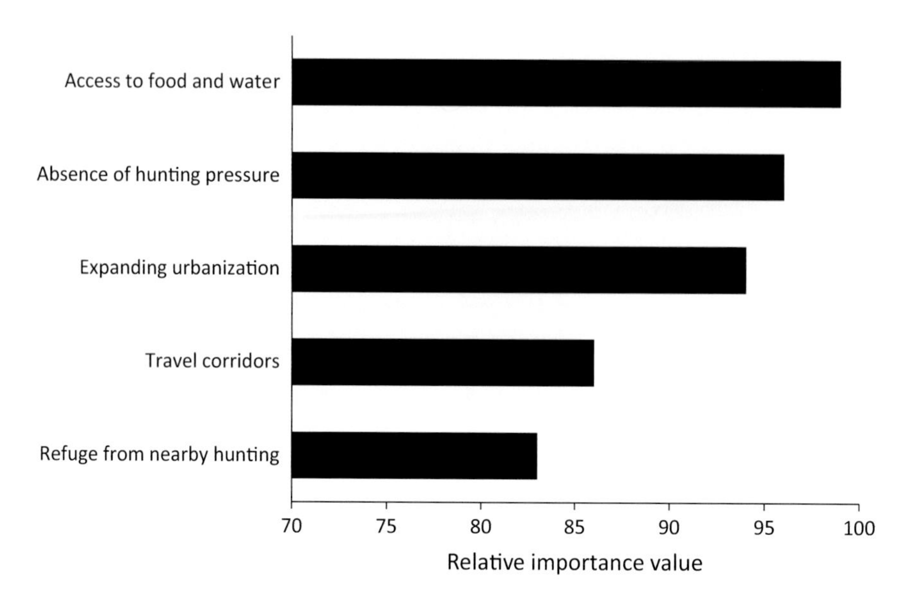

FIGURE 19.5 The top 5 reasons that wild pigs invade high-density urbanized habitat. The x-axis (relative importance value) represents a score based on responses from survey participants. (Adapted from Licoppe et al. 2013.)

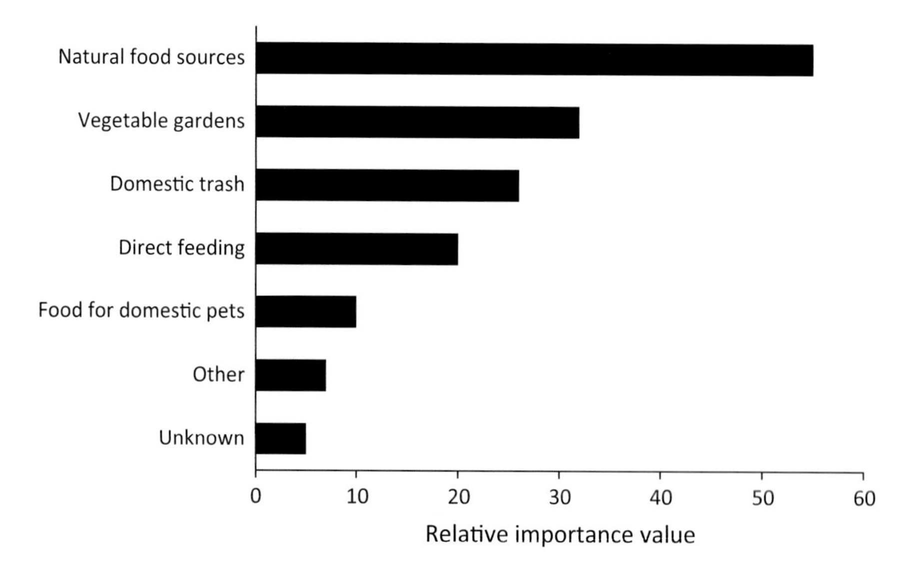

FIGURE 19.6 Food resources that wild pigs access when invading high-density urbanized habitat. The x-axis (relative importance value) represents a score based on responses from survey participants. (Reproduced from Licoppe et al. 2013.)

are accessing natural food resources in urbanized areas, which can be present in open space, public parks, and other undeveloped areas (Licoppe et al. 2013; Figure 19.6). Indeed, field studies demonstrated the diet of wild pigs in highly urbanized landscapes can be primarily comprised of natural food items, with anthropogenic food items (e.g., trash) infrequently consumed (Stillfried et al. 2017b). Further, the amount of energy consumed by wild pigs in urban areas can be greater than their rural counterparts, demonstrating energetic benefits of exploiting food in urbanized habitat (Stillfried et al. 2017b). However, anthropogenic food items can be more common in the diet of wild pigs in other regions (Cahill et al. 2012) or as a secondary choice when natural foods are unavailable (Stillfried et al. 2017b).

The behavior of wild pigs accessing urbanized landscapes is strongly influenced by the trade-off of (1) accessing available resources and (2) danger or stress (e.g., risk or disturbance, mortality, and predation), which is related to foraging ecology and behavior (Luniak 2004, Stephens et al. 2007, Toger et al. 2018). Because areas of urbanization exhibit habitat characteristics altered by anthropogenic activities, animals will often modify their behavior to access resources associated with human disturbance. These behavioral modifications can be expressed by changes in activity patterns, daily and seasonal movements, space use, habitat selection, and diet. Although urbanization is a pervasive form of development across the world, relatively little research has been conducted with wild pigs in urbanized environments.

Other ungulates are influenced by urbanization, and studies of these species can inform predictions for how wild pigs might respond to urbanized landscapes. For example, white-tailed deer (*Odocoileus virginianus*) have experienced a recent population boom and have successfully exploited habitat influenced by urbanization (DeNicola et al. 2000). In some areas, deer have become more urbanized over the last several decades, and animals using urbanized areas have exhibited increased survival and body mass (Harveson et al. 2007). It is believed that the deer's behavioral plasticity has enabled them to exploit this rapidly expanding environmental change (Harveson et al. 2007). It is unknown how long it might take ungulate populations in proximity to urbanized landscapes to use those areas, but generalist species with plastic behaviors appear likely to take advantage of these areas.

19.4.1 Behavioral Strategies in Urbanized Habitat

Wild pigs can exhibit high behavioral plasticity in response to human disturbance (Keuling et al. 2008, Podgórski et al. 2013). Further, animals within the same population can exhibit varying behaviors in response to human activities. For example, wild pigs accessing agriculture in rural landscapes can demonstrate varying strategies depending on age, sex, social affiliation, and prior experiences. Wild pigs accessing agricultural crops can be classified as "field sows" who use agriculture exclusively, "commuters" who move between agriculture and forest, and "forest sows" who remain in forested areas (Keuling 2009). Sounders either were associated with agriculture or forest, whereas yearlings were primarily commuters. Overall, most animals were field sows, thus indicating that wild pigs will be attracted to high quality food resources when available. This has important implications for how wild pigs might exploit food resources in urbanized areas as wild pigs should exhibit similar strategies when accessing areas of high-density urbanization as well, where food resources are often concentrated and nearby cover may be present. Within a population associated with urbanization, although animals might exhibit different behavioral strategies, it is expected that if urbanization contained high quality food that most animals would access this resource. Thus, if food was inaccessible to wild pigs in urbanized landscapes, wild pigs would be expected to use those areas less frequently. Indeed, wild pigs use landscape resources based on availability both spatially and seasonally (Barrios-Garcia and Ballari 2012; see Chapter 3). Although available food is thought to be the primary determinant for wild pigs accessing urbanized areas, there are other potential reasons animals use these areas as well (Figure 19.5).

As a result of accessing urbanized landscapes, wild pigs can change their behavior in at least 3 ways. First, wild pigs can exhibit spatial avoidance and not use areas associated with urbanization. This might only be observed for specific populations or age and sex classes. Second, wild pigs might exhibit temporal avoidance and minimize the effects of human disturbance by shifting their behavior to access urbanization when such disturbance is reduced. And third, wild pigs can habituate to humans and thus not avoid some types of human disturbance when they do not feel threatened. Wild pigs have displayed both avoidance and habituation behaviors depending upon characteristics of the urban environment. We will thus consider these 3 behavioral strategies when describing wild pig ecology in urbanized landscapes.

19.4.1.1 Spatial Avoidance

Many wildlife species or individuals within populations (e.g., particular age or sex classes) will avoid urbanization (Hansen et al. 2005, Radeloff et al. 2005). In some highly urbanized landscapes, wild pigs are reported to be associated with forest and are less likely to occur in areas of urbanization (Saito and Koike 2013, Saito et al. 2016). However, in other systems, wild pigs use habitat associated with urbanization and human activities (e.g., Cahill et al. 2012, Frantz et al. 2012, Stillfried et al. 2017a). When wild pigs access areas modified by human activities, animals will often concentrate their activities along edges of high-value forage areas (Linkie et al. 2007, Thurfjell et al. 2009). In addition, wild pigs will avoid large open agricultural fields, although animals might be more likely to use these areas when abundant food is available (Thurfjell et al. 2009). In other forms of urbanization, wild pigs might exhibit similar patterns of edge use and interior avoidance. For example, in some systems, areas along the WUI might be most likely to experience wild pig invasions, whereas urban centers would be avoided.

Although a growing number of urbanized areas report issues with wild pigs, other cities and towns in proximity to wild pig populations have not reported conflicts (Gaston 2010). Many instances of wild pigs using urbanized landscapes are likely undetected or unreported. In addition, for some systems, wild pigs might avoid urbanization and human disturbance. When wild pigs access urbanized areas, animals can modify their behavior to avoid human disturbance while still accessing resources.

19.4.1.2 Temporal Avoidance

Animals will shape their daily activity patterns to maximize fitness, which includes avoiding human disturbance, stress, and mortality factors. Indeed, wild pigs will shift their activity patterns in response to diverse human activities. For example, wild pigs will become more nocturnal to avoid disturbance associated with hunting pressure (Keuling et al. 2008, Ohashi et al. 2013) and to access agricultural crops in rural landscapes (White et al. 2018, Lewis unpublished data). Thus, we would predict that if wild pigs were disturbed by humans in more urbanized landscapes, they would increase nocturnal activity. As expected, wild pigs can be mostly active at night when using habitat associated with urbanization (Cahill et al. 2003, Podgórski et al. 2013). During nocturnal periods, human activity and disturbance is reduced, which provides opportunities for wild pigs to access available resources. During diurnal periods, animals can seek refuge in areas of cover and associated undeveloped areas. In comparison, in wildland adjacent to urbanization where human disturbance is reduced, wild pigs can be more active throughout the entire daily time period (Podgórski et al. 2013), indicating that animals will be more active during the middle of the day in the absence of human disturbance. Thus, wild pigs exhibit substantial plasticity in their behaviors, where adjacent populations can exhibit contrasting daily activity patterns depending on the degree to which they are disturbed by humans and have access to areas of urbanization (Podgórski et al. 2013). In other cases, wild pigs can also be active during the daytime in urban areas (Cahill et al. 2012; see next section below).

19.4.1.3 Habituation

Animals can become habituated to human activities, which erodes their avoidance behavior to anthropogenic influences. Habituation is the process whereby animals demonstrate a reduced response to a stimulus over time due to learning that human activities do not produce a sufficient negative experience (Bejder et al. 2009). For example, in places where wild pigs are not persecuted, humans reportedly can safely walk very close to these animals (Galhano-Alves 2004). In urban areas, humans can facilitate habituation by wild pigs either intentionally or unintentionally. We follow the definition of Cahill et al. (2012), who defined habituated wild pigs as animals accustomed to people and did not flee from humans or their activities. Wild pigs that are active during diurnal time periods in urban areas (Cahill et al. 2012) can indicate a population that is habituated to human activities because they are not avoiding time periods when people are most active and human disturbance is likely greatest.

Wild pigs can become habituated to humans in areas where they are fed by residents or they might not perceive people as a threat (Cahill et al. 2012, Licoppe et al. 2013). The direct feeding of wild pigs in urban environments can be a common practice in many areas (Cahill et al. 2012), which predictably leads to animals accessing urbanized landscapes more frequently to take advantage of available food. In addition, human values, attitudes, and acceptance towards wild pigs reduces negative associations that are experienced by wild pigs in urbanized landscapes, and encourages their use near people.

As observed in rural landscapes (Keuling 2009), wild pigs exploiting more urbanized landscapes might exhibit varying behaviors within a population. Within the same urbanized region, wild pigs can be nocturnal (Cahill et al. 2003), indicating avoidance of human disturbance, as well as active during daytime (Cahill et al. 2012), indicating habituation to people. In addition, females with young are more commonly classified as habituated to human activities compared to males (Cahill et al. 2012), which might indicate that age and sex classes use urbanized areas differently depending on life history characteristics. It is unclear, however, what mechanisms lead to potential differences in wild pig behavior and how these patterns might change among age and sex classes or within and between years.

19.4.2 Seasonal and Annual Use

Due to seasonal and annual variation in climate, available resources, and other factors, animal use of urbanized landscapes can increase or decrease across time periods. Wild pigs might increase activity in urbanized areas during time periods when resources are limited in adjacent wildlands or resources are elevated in urbanized areas. Because naturally occurring food can be seasonally variable and anthropogenic-associated food in urbanized areas can be more constant year-round, wild pigs might access urbanized areas more during periods of natural food scarcity (Podgórski et al. 2013, Castillo-Contreras et al. 2018). For example, food for wild pigs can be scarcer during the summer in some natural systems due to low precipitation and hard soil substrates; thus wild pigs can seek out food in urbanized areas more during this time of year (Cahill et al. 2012). During periods of high food availability in undeveloped areas (e.g., mast production in forests), wild pigs might concentrate activities away from urbanization (Podgórski et al. 2013). In addition, other climate conditions, such as deep winter snows, might influence wild pig use of wildlands or urban areas due to limited access. Thus, across the gradient of urbanization, we would predict increased wild pig use of urbanized areas during summer and winter and reduced use of these areas during spring and fall in many systems. However, when crops are seasonally available in low-density rural landscapes, wild pigs will access these areas accordingly (Thurfjell et al. 2009). In addition, wild pig access and damage in urbanized areas might be most common during drought years because irrigated lawns and green spaces may be the only areas of moisture widely available (Adams 2016).

19.4.3 Space Use, Habitat Selection, and Movements

Limited research indicates that home ranges of wild pigs can vary depending on the type of urbanization that wild pigs are accessing. Home ranges of wild pigs can be larger in rural areas with agriculture because wild pigs will travel several kilometers between areas of cover and agricultural food resources (Keuling 2009). In contrast, wild pigs may exhibit smaller home ranges in highly urbanized areas compared to adjacent natural areas, which could be related to more concentrated areas of food and cover (Fischer et al. 2004, Podgórski et al. 2013). In addition, less habitat might be available in highly urbanized areas and boundary effects could restrict animal movements. Although wild pigs may exhibit smaller extents of space use in urbanized areas, animals may spend more time moving in these areas (Podgórski et al. 2013), although movement distances can vary widely (Pepin et al. 2016). Few studies have evaluated how wild pigs select and use habitat in relation to urbanization; however, it is predicted that wild pigs will focus on using areas of cover across the gradient of urbanization. In rural landscapes associated with agriculture, wild pigs reportedly use 1-km buffers of cover, which can produce a "refuge effect," to access food resources (Amici et al. 2012). Indeed, wild pigs select for edge vegetation and avoid exposed agricultural fields (Thurfjell et al. 2009), with most crop damage occurring along forest edges (Linkie et al. 2007). It is predicted that wild pig access to other forms of urbanization would also most likely occur along edges of security cover, such as the WUI. In urban areas, wild pigs select for natural areas, which can be close to roads and houses (Stillfried et al. 2017c). However, no studies have explicitly evaluated and compared fine-scale patterns of habitat selection of wild pigs across the gradient of urbanization. With the increasing use of satellite telemetry and remote cameras to evaluate wild pig populations, researchers can evaluate fine-scale habitat selection of wild pigs in relation to multiple levels of urbanization to better understand these patterns.

19.4.4 Population Characteristics

Urbanization can influence population characteristics, including survival, reproduction, demographics, population density, body condition, disease, and genetics (Gaston 2010, Gehrt et al. 2010). Some ungulates, including white-tailed deer, have become more urban adapted over the last several decades and demonstrate increased survival and body condition (DeNicola et al. 2000, Harveson et al. 2007). Little information, however, is available for how population characteristics of wild pigs change across the gradient of urbanization. Increased mortality might be observed through vehicle collisions and management activities to reduce wild pig populations, but it is largely unknown how these factors impact population characteristics. Limited data indicates that in some systems wild pigs might exhibit lower population density in urban areas compared to adjacent natural areas, although the daily distance moved for wild pigs can be twice as long in urban areas (Podgórski et al. 2013) and can vary widely (Pepin et al. 2016).

Further work is necessary to understand how wild pig abundance and other population characteristics changes across the gradient of urbanization. Across broad spatial scales, we would predict that population abundance of wild pigs would be greatest in wildland and rural areas, due to the presence of agricultural crops, and decrease as the density of urbanization increases due to management of populations and smaller areas of habitat (Figure 19.7). In addition, at fine spatial and/or temporal scales, wild pigs might exhibit high population densities in suburban and urban habitat due

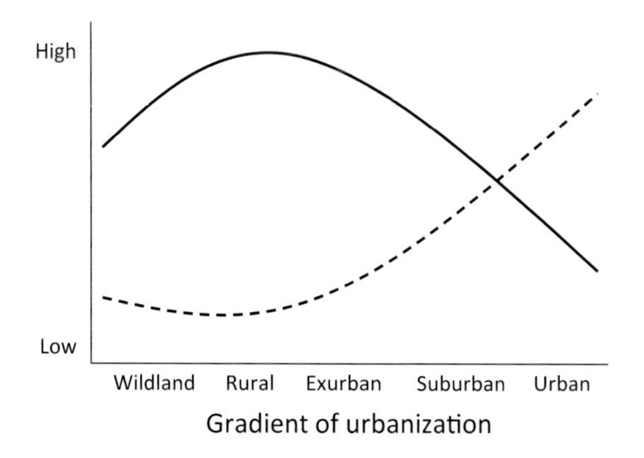

FIGURE 19.7 Predicted relationship of relative population abundance of wild pigs (solid line) and management restrictions and challenges (dashed line) along the gradient of urbanization.

to availability of food and cover, restricted movements and dispersal, and challenges in implementing management activities to suppress population density.

In some areas, body mass of wild pigs using urbanized areas can be correlated with fluctuations in oak mast (Cahill and Llimona 2004), which indicates that natural food resources are important determinants for why populations access urbanized landscapes. Further, habituated female wild pigs in urban areas can exhibit greater body weight compared to non-habituated animals (Cahill et al. 2012), indicating that food resources in urban environments can increase body condition. Although untested for wild pigs using urbanized areas, this could potentially influence survival, litter size, and population density. In some systems, females with juveniles and piglets more commonly use urbanized areas compared to males (Cahill and Llimona 2004, Cahill et al. 2012), whereas in other systems, both sexes use areas about equally (Brown 1985).

Urbanization can influence dispersal and population genetic structure of wild pigs. Dispersing wild pigs appear more likely to travel through human modified landscapes compared to other species (e.g., red deer *Cervus elaphus*; Prévot and Licoppe 2013), which influences colonization of habitat and population connectivity. Although wild pigs can travel through human modified landscapes, including areas of urbanization, animal movement can be restricted in some systems due to high levels of urbanization and major roads (Wyckoff et al. 2012, Tadano et al. 2016). Thus, urbanization can be an important landscape feature that influences wild pig movements and population structure.

19.5 DAMAGE AND IMPACTS

There are myriad impacts of wildlife accessing urbanized landscapes, and these impacts often center around conflicts with people and their property (Gaston 2010, Licoppe et al. 2013, Adams 2016). Wild pigs are an especially impactful species, given their large body size, foraging behavior, and population characteristics (e.g., adaptability, often traveling in large groups, rapid population growth). The full spectrum of damage caused by invasive wild pigs has long been recognized everywhere as being both extensive and varied (Mayer and Brisbin 2009, Barrios-Garcia and Ballari 2012; see Chapter 7). Predictively, as these animals expand their distribution into urbanized areas, similar damages occurred. The primary impacts of wild pigs in urbanized habitat includes damage to property, human safety and health, and the natural environment (Licoppe et al. 2013; Figure 19.8).

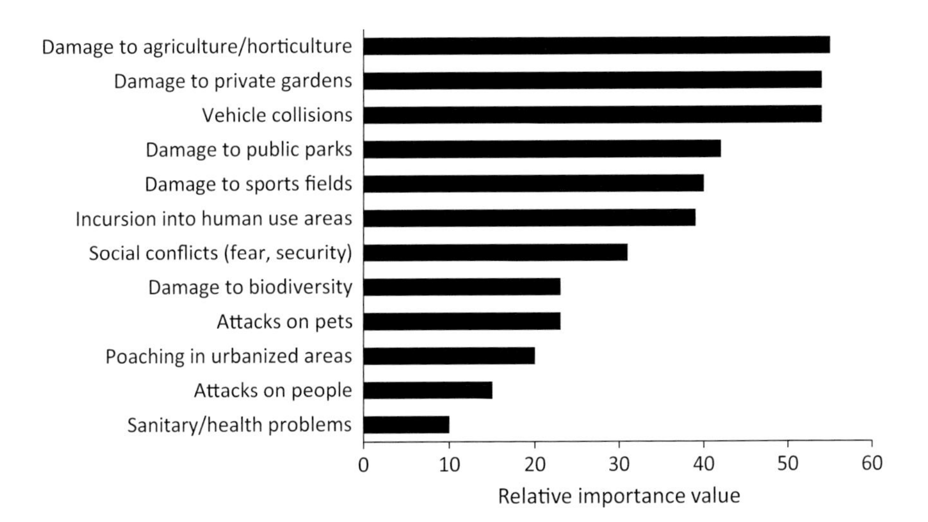

FIGURE 19.8 Impacts of wild pigs when invading high-density urbanized habitat. The x-axis (relative importance value) represents a score based on responses from survey participants. (Adapted from Licoppe et al. 2013.)

19.5.1 Damage from Rooting and Foraging

Much of the wild pig damage and impacts in developed areas results from their foraging activities. The food (e.g., naturally occurring plant/animal food resources, vegetable gardens, household garbage, direct feeding, and domestic pet foods) and water (e.g., ponds, lakes, streams, engineered drainage features, potable water supply reservoirs) resources potentially available and often easily accessible to wild pigs venturing into the urban landscape can be extensive (Jansen et al. 2007, Cahill et al. 2012, Licoppe et al. 2013). These prevalent food and water resources are a sufficient incentive for these animals to return to developed areas to forage (Mayer 2013). This both encourages and supports the occupation of developed areas by wild pigs. The damage associated with the use of these resources by wild pigs depends upon foraging style, including rooting, foraging on the ground, and browsing/grazing (Thomson and Challies 1988).

Rooting behavior (i.e., using the animal's snout to excavate soil in a forward pushing/scooping motion in search of food) of wild pigs is commonly observed in urbanized areas. Typically rooting occurs in grassed or sodded areas, where wild pigs are feeding on plant roots or invertebrates (Mayer and Brisbin 2009). In rural landscapes, rooting can be common in areas of agriculture, which can lead to extensive damage and economic loss (Barrios-Garcia and Ballari 2012). Within more developed areas, wild pig rooting damage to sodded or grassed areas has been observed in a variety of settings (e.g., residential lawns, public parks, greenways/greenbelts, golf courses, sports fields, cemeteries, and levees/dikes; Mayer and Brisbin 2009, Licoppe et al. 2013, Stillfried et al. 2017a; Figures 19.9, 19.10). Wild pigs can cause considerable rooting damage in a single night (Frederick 1998). With the aforementioned targeted resources, most of the rooting seen in developed areas is shallow, less than 10 cm in depth (Cahill et al. 2003). Wild pigs will also do shallow rooting in the leaf litter in forested areas (e.g., in private lands and public parks for mast resources) and around mulched landscaping (e.g., planted ornamental shrubs and bushes). Aside from the direct damage to the turf, such foraging activities can also result in erosion and down-gradient sedimentation (Mayer and Brisbin 2009).

Wild pigs will also forage in agricultural fields, vegetable gardens, fruit trees/orchards and ornamental floral gardens associated with developed areas. This results in damage due to direct feeding on these resources as well as trampling impacts and uprooting of plants (Mayer and Brisbin 2009, Podgórski et al. 2013, Adams 2016). Although wild pigs often feed on natural foods

FIGURE 19.9 Rooting damage from wild pigs at a house in Mississippi. (Photo by S. Alls. With permission.)

FIGURE 19.10 Rooting damage from wild pigs in a cemetery in Texas. (Photo by B. Higginbotham. With permission.)

in urbanized areas (Licoppe et al. 2013, Stillfried et al. 2017b), the consumption of garbage by wild pigs in some municipalities can be quite large (e.g., 58% by mass in stomachs: Hafeez et al. 2008). Wild pigs feeding on these materials can result in garbage/refuse dispersal through turned over waste containers. Such impacts result in both the potentially extensive scattering of litter and unsanitary conditions in localized areas (Cahill et al. 2012). In addition, wild pigs might make garbage resources more accessible to other urban exploiters (e.g., raccoons (*Procyon lotor*), coyotes (*Canis latrans*), crows (*Corvus* spp.)). Foraging can also occur at garbage dumps or landfills in developed areas (Podgórski et al. 2013). The nocturnal foraging of wild pigs in landfills, which are required by law to be covered with fill dirt at the end of daily operations, can uncover household waste. This further exposes uncovered garbage, which can lead to consumption and scattering by both avian and mammalian scavengers (J. J. Mayer, Savannah River National Laboratory, unpublished data).

Wild pigs cause a variety of property damage, most of which results from activities associated with foraging. There has been a recent increase in observed property damage by wild pigs in suburban or developed areas (Mayer and Brisbin 2009). This damage most often includes fencing (e.g., around gardens, parks, sports fields) resulting from the pigs breaching these barriers (Licoppe et al. 2013). Wild pigs have also been reported to dig up and break buried sprinkler and irrigation system pipes in agricultural fields and in yards and fields in more urbanized areas to access the water contained in those lines. Larger irrigation/sprinkler systems in parks, golf courses, and sport fields also are damaged by these animals (Frederick 1998, Licoppe et al. 2013). Lastly, wild pigs damage other property (e.g., flower pots, yard furniture, and other outdoor items) when accessing human residences (Mayer and Brisbin 2009).

19.5.2 Vehicle Collisions

In both rural and developed areas, concurrent with the recent increase in the presence of wild pigs has been an increase in the numbers of wild pig-vehicle collisions (Mayer and Johns 2007). Given greater traffic volume in developed areas, the frequency of such accidents has become a regular phenomenon in some urban settings (Stillfried et al. 2017a). Both the property damage and personal injury consequences of wild pig-vehicle collisions can be substantial, with the number of human fatalities increasing. Vehicle accidents with wild pigs can involve both large individual animals and multiple numbers of animals in a single collision, and occur year-round and throughout the 24-hour daily time period (Mayer and Johns 2007, Beasley et al. 2014). As such, these collisions represent a growing safety hazard within the urban gradient. There have also been collisions with wild pigs during the landing and taking off of aircraft in metropolitan airports, as well as trains on rail lines in developed areas (Mayer and Brisbin 2009).

19.5.3 Human Safety and Aggressive Interactions

Wild pigs can be aggressive, tenaciously defending themselves against both natural predators and conspecifics. This aggressive behavior by wild pigs also has been documented to include attacks on humans and their pets. Such attacks are rare, but have been on the increase in developed areas globally (Figure 19.11), and the consequences of these attacks can be serious, including fatalities in some cases. Some attacks have included solitary wild pigs entering occupied buildings in developed areas prior to attacking human victims. Very often these animals will mount a defensive attack when they feel cornered or threatened. In the late 1990s, the East Bay Regional Park District in central California had a number of incidents involving wild pigs charging district employees and groups of school children (Frederick 1998). Behaviors leading to wild pig attacks on humans in developed areas include: walking with a dog (leashed or un-leashed); threatening or chasing a wild pig (e.g., out of a garden); approaching an obviously wounded or injured wild pig; approaching or attempting to feed or pet/touch a wild pig; and blocking the path of a moving/escaping wild pig (Mayer 2013). In addition to attacks on people, as potentially dangerous animals, wild pigs frightening people has also been reported as a frequent occurrence in urban areas (Licoppe et al. 2013, Tari et al. 2016).

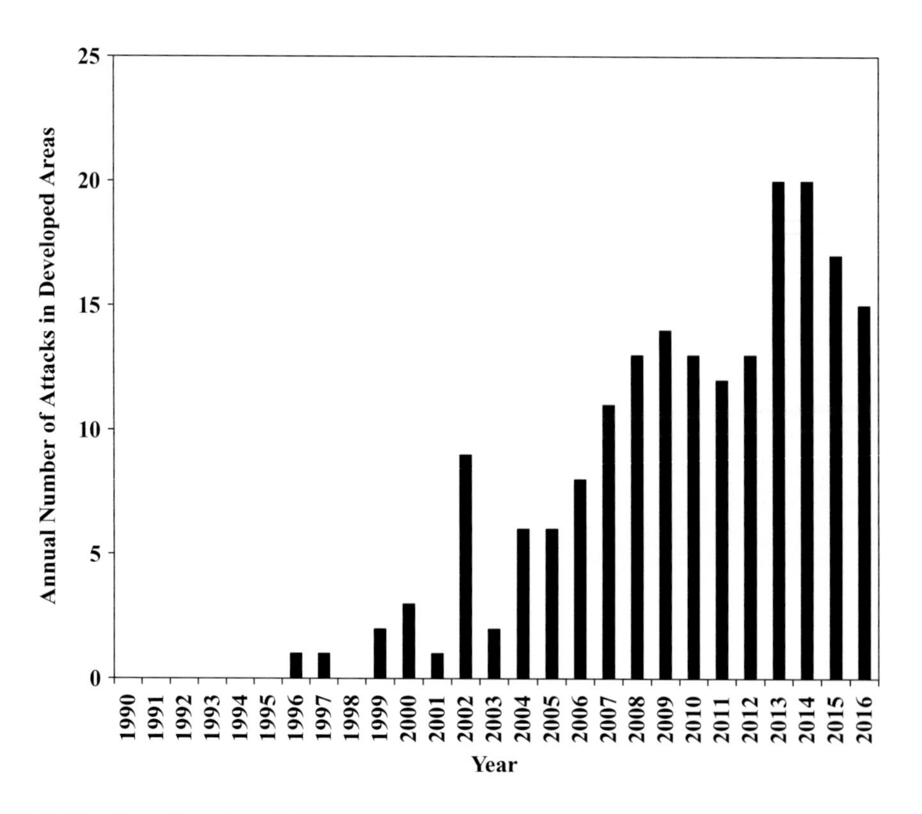

FIGURE 19.11 Annual number of wild pig attacks on humans in developed (i.e., suburban and urban) areas (N=187) globally between 1990 and 2016. (Adapted from Mayer 2013.)

19.5.4 Pathogens, Disease, and Water Contamination

Wild pigs are susceptible and can serve as disease reservoirs/vectors for a number of bacterial and viral pathogens, many of which are zoonotic in nature and can impact humans, as well as pets, livestock, and wildlife species (see Chapter 5). Thus, wild pigs can represent a potentially severe public health issue within the urban gradient (Jansen et al. 2007, Licoppe et al. 2013). As wild pig numbers increase in urbanized areas, so should the frequency of their contact with humans, domestic animals, and pathogen transmission pathways (e.g., water bodies or drainage corridors). Diseases of concern associated with wild pigs in urbanized areas include brucellosis, bovine tuberculosis, pseudorabies virus, hepatitis E virus, leptospirosis, *Toxoplasma gondii*, salmonella, and *E. coli* (Jansen et al. 2007, Mayer and Brisbin 2009, Schielke et al. 2009, Licoppe et al. 2013). In rural landscapes characterized by agriculture, wild pigs can exhibit high exposure rates to pathogens such as *Toxoplasma gondii* (e.g., 33 to 55% seroprevalance; Richomme et al. 2010), which can increase the opportunity for the spread of this pathogen in areas of low-density urbanization. In landscapes characterized by higher levels of urbanization, wild pigs can harbor pathogens that are potentially fatal to humans. For example, approximately 20% of a wild pig population (estimated at approximately 5,000 animals) tested positive for leptospirosis in Berlin, Germany (Jansen et al. 2007).

In part related to the aforementioned disease impacts, wild pigs also have a great potential to impact surface water resources within the urban gradient (Jansen et al. 2007, Licoppe et al. 2013). These aquatic systems can include both natural (e.g., rivers, streams, lakes, and ponds) and man-made (e.g., dammed impoundments, water supply reservoirs, constructed ponds and waterways, irrigation ditches) bodies of water. Wild pigs foul these water resources by wallowing, rooting/foraging, defecating, and urinating. These activities can degrade water chemistry parameters (e.g., pH, dissolved oxygen, turbidity, sedimentation/silting) as well as introduce pathogens into these flowing and standing aquatic systems. Some of these watersheds are potable water supply systems, which could enable the transmission of zoonotic pathogens into down-gradient populations of human and domestic animals. The water quality degradation and pathogen introductions could also impact local wildlife and fish populations using those ecosystems.

It is surprising that surveys of wild pig managers indicate that health and sanitary issues are considered a relatively small impact of wild pigs in urbanized habitat (Licoppe et al. 2013; Figure 19.8). Given that wild pigs can harbor a suite of pathogens that can affect people, pets, livestock, and wildlife, this impact is likely more important than perceived by people in many areas. Although recent research highlights this threat (as described above), this is an area in need of further evaluation to understand the risks to health resulting from wild pigs in urbanized habitat.

19.5.5 Economic Costs

Unlike agricultural areas impacted by wild pigs (agricultural production, facilities, and operations; Pimentel 2007, Anderson et al. 2016), the economic cost of wild pig damage within higher density urbanization has not been well quantified to date. It is estimated that wild pigs cause US$1.5 billion in damage to crops and other resources in the United States annually (Pimentel 2007). In addition, across 10 US states, wild pigs caused US$190 million in damage and loss in agricultural fields associated with rural landscapes (Anderson et al. 2016). Although largely unquantified across broad scales, it is expected that the economic cost of wild pig damage is substantial in higher density urbanization. One case study reported that 4 residential subdivisions and 1 golf course in California sustained US$64,000 in rooting damage to turf and ornamental plants by wild pigs in 1 year (Frederick 1998). Further, that same study reported that San Benito County spent US$14,500 to replace and repair irrigation systems damaged by wild pigs. In addition, the East Bay Regional Park Division in the San Francisco Bay area spent in excess of US$60,000 in control costs to reduce wild pig damage to turf and irrigation systems. Wild pigs are reportedly causing hundreds of thousands of dollars' worth of damage annually to golf courses in California (Pearson 2004). Further research is necessary to estimate the economic, social, and ecological cost of wild pigs across the urban gradient.

19.5.6 Ecological Impacts

In contrast to wildland habitat (Barrios-Garcia and Ballari 2012), wild pig damage is often not associated with impacts to either wildlife or sensitive species in urbanized habitat (Licoppe et al. 2013; Figure 19.8). This is perhaps not surprising because in many cases the wildlife species associated with human residences are often adaptable, invasive, or common. However, it is possible that urbanized habitat provides wild pigs with resources that help maintain or increase populations in nearby wildland habitat, where wild pigs could have a greater impact on sensitive species and landscape features. The impacts of urbanization acting as a subsidy to support wild pig populations, which in turn harms nearby native species, has yet to be explored and warrants further consideration. Indeed, there are likely several parallels to agriculture providing supplemental food to wild pigs that persist in nearby natural areas, which can potentially lead to impacts to biodiversity in adjacent wildlands.

19.6 CONTROL AND MANAGEMENT OPTIONS ALONG THE URBAN GRADIENT

19.6.1 Legal, Political, and Public Considerations for Management Options

Human values and government regulations influence management activities of wild pigs across the gradient of urbanization. In landscapes with low levels of urbanization (e.g., rural), human values tend to be more utilitarian, where people are more open to hunting, trapping, and controlling wildlife populations (Manfredo et al. 2003). In contrast, in more urbanized landscapes (e.g., suburban and urban), human values tend to be more protectionist, where people are less supportive of killing animals and more supportive of protecting wildlife (Manfredo et al. 2003). For example, euthanization of wild pigs in urban areas can be unpopular with the public (Cahill et al. 2012). In addition, government regulations restrict the use of some wildlife control measures in more urbanized areas (e.g., use of firearms) to promote human safety (Adams 2016). Ultimately, management options to control wild pig populations are most numerous in wildland and rural landscapes and they typically decrease with increasing density of urbanization (Figure 19.7). Further, the increased diversity of values in more urbanized landscapes can increase the likelihood for social conflict when managing wildlife populations (Patterson et al. 2003). To reduce the potential for social conflicts when managing wild pigs, it is essential to include diverse stakeholders early in the process of crafting management plans and to openly address concerns and ideas (Schuett et al. 2009, Adams 2016, Crowley et al. 2017). Overall, these are important considerations when evaluating and implementing control options to manage wild pig populations across the gradient of urbanization.

19.6.2 Management Action Options

As in rural areas, an integrated management approach, employing multiple methods for control, may provide best results. The level of control the general public will accept is often related to a threshold of damage or degree of risk to human health and safety (Loker et al. 1999, Licoppe et al. 2013). Lethal control techniques (i.e., trapping followed by euthanasia and various forms of shooting) are predominately more effective at reducing or eliminating wild pig populations compared to non-lethal methods; but as stated, lethal means are generally less accepted in suburban and urban areas. For example, a survey conducted in Berlin, Germany found that while 44% of participants agreed that the urban wild pig population should be substantially reduced, only 26% thought the population should be removed lethally (Kotulski and König 2008). In urban areas, control methods are often conducted in view of the public and lethal removal may be perceived as inhumane or dangerous due to high-density human populations (Wittmann et al. 1998, Ferretti et al. 2014). One strategy may be to trap wild pigs and transport them alive to another location where they can be humanely euthanized and properly disposed of or processed for human consumption. If lethal methods are not an option or can only be used sparingly in an urban area, lethal removal can be used in more rural areas immediately surrounding that area to help reduce the further encroachment by wild pigs.

Non-lethal options are available for wild pig control, although effectiveness varies by method and circumstance. Wire mesh or electric fencing can be constructed around certain areas to reduce damage caused by wild pigs to areas such as golf courses, cemeteries, schools, parks, and gardens, but costs must be considered (Reidy et al. 2008, Campbell et al. 2011, Cahill et al. 2012). Reducing or eliminating access to anthropogenic food sources such as garbage bins and prohibiting residents from feeding wildlife, including birds, can also be effective (Cahill et al. 2012, Licoppe et al. 2013). Trapping and relocating wild pigs may be considered humane by certain members of the public, but relocation should not be promoted in the wild pigs' non-native range and, in many states, is illegal for this species (Miller et al. 2012, USDA/Animal and Plant Health Inspection Service 2015). Relocation may spread disease, can cause stress or mortality for the animals, is expensive, and perpetuates the problem elsewhere, and pigs may return to the same area if relocation is in proximity to where they were captured (Cahill et al. 2012, Ferretti et al. 2014, Adams 2016). Chemical repellents or frightening devices may serve as a short-term solution to reduce damage, but wild pigs quickly become habituated to the products currently available (Schlageter and Haag-Wackernagel 2011, 2012). Recent advancements in contraceptive research have shown potential as a viable control tool (Cowan and Massei 2008, Ferretti et al. 2014), but no contraceptives have been registered for use in wild pigs in the United States. Managers in some municipalities intentionally provide supplemental food to wild pigs with the goal of reducing wild pig conflicts in highly urbanized areas (Fischer et al. 2004). However, managers and researchers widely recommend not providing supplemental food because this often only increases the problem of wild pigs in urbanized habitat (Cahill et al. 2012).

Successful urban wild pig management will often be a challenge and depend on finding a balance between utilizing effective yet socially acceptable control methods (Hadidian 2009, Ferretti et al. 2014). Further, even when management actions are successful in controlling wild pigs, populations can quickly return. For example, although wild pigs can be eliminated from urbanized landscapes through sustained management actions (Brown 1985), extirpation of wild pigs is often short lived and populations can quickly reinvade areas where they were recently removed (J. J. Mayer, unpublished data). Thus, active management and monitoring is essential to maintain long-term control and management of wild pig populations.

19.6.3 Working with the Public, Outreach, and Education

To minimize conflicts between humans and wild pigs, managers can engage the public and implement outreach and education strategies that focus on: 1) prevention, 2) monitoring, and 3) communication (Licoppe et al. 2013; see Chapter 10). Prevention strategies often involve the removal of attractants to wild pigs and constructing fencing to protect gardens and prevent vehicle collisions. In addition, banning direct feeding, and in some cases implementing fines for violations, can be necessary to reduce habitation of wild pigs to people (Cahill et al. 2012, Licoppe et al. 2013). Monitoring of wild pigs (e.g., public reporting, surveys of damage and conflict, surveys with remote cameras) can be effective for understanding where wild pigs are using areas of urbanization and places in need of public outreach. Often, citizen science can assist with monitoring efforts. Communicating the issues and topics of wild pigs in urbanized areas is critical to engage and inform the public. This can occur through written communication (reports, online websites, fliers, mailings, newspaper articles), visual and personal communication (public presentations, news reports, meetings), and radio and audio communication (radio broadcasts and stories).

19.7 URBAN WILD PIG CASE STUDIES

19.7.1 Case Study 1: Dallas/Fort Worth, Texas

Wild pigs have inhabited Texas for over 400 years before becoming an urban wildlife issue in the 1990s. Utilizing riparian habitat corridors, which resemble the network of vehicular highways crossing the region in all directions, leading from undeveloped rural agricultural lands through residential suburban areas and finally, into extensively developed urban centers, wild pigs have taken advantage of unguarded green spaces to establish residency alongside humans in most of the state's major cities, including the densely populated Dallas/Fort Worth Metroplex.

Texas' first formal urban wild pig control program was developed and implemented by staff of the Fort Worth Nature Center & Refuge (FWNC&R), a 1450+ hectare urban green space owned by the City of Fort Worth and managed by the City's Park and Recreation Department as a natural, native landscape. Wild pigs, having traveled along the West Fork of the Trinity River, which bisects the refuge, were first observed on the FWNC&R in July 1999, whereupon staff members embarked on a 3.5-year effort to devise a control plan that was acceptable to all relevant stakeholder groups.

An array of both internal (i.e., City of Fort Worth Departments including Park and Recreation, Police, Legal, Animal Control, Risk Management, and City Council) and external (i.e., conservation groups, state and federal resource management agencies, community groups, and animal welfare groups) stakeholders were identified and included in the control plan development process from the onset. While time-consuming and sometimes cumbersome, this provided all concerned groups the opportunity to be educated on the wild pig issue and to have all of their priorities noted. Internal stakeholders were primarily concerned with the health and safety of citizens, the economic impact of the wild pig population and who would conduct control efforts. External stakeholder priorities varied among the groups. Conservation groups and the state and federal agencies prioritized resource protection and the use of lethal control methods. Community groups were most concerned with visitor safety, and the priority of the animal welfare community was in the welfare of the individual pigs.

Through education, all stakeholders came to appreciate the need for wild pig control on the FWNC&R. Suggestions to use exclusion or trap and relocate as non-lethal control measures were judged unfeasible. Wild pig relocation required a permit from the Texas Animal Health Commission (Texas Animal Health Commission 2003) and the local agent stated that no permit would be granted to relocate FWNC&R wild pigs. Exclusion, in the form of installing a pig-proof fence surrounding the FWNC&R, was beyond the financial resources of the City of Fort Worth and did nothing to control the wild pig population already on the property. Immunocontraception use was proposed but was, and is, still in the development stage (Killian et al. 2008). Stakeholder concerns for lethal control measures were categorized into 4 areas: 1) control method, 2) euthanasia method, 3) carcass disposal, and 4) who would be responsible for conducting control operations.

Concerns regarding the specific lethal control method employed centered on human safety, efficiency, and humane treatment of individual pigs. Controlled shooting, either by a trained marksman over bait or in the form of a public hunt, was deemed unacceptable by many internal and external groups. Aerial hunting, legal and often used in the state, was impractical due to heavy forest cover over much of the FWNC&R. The use of dogs and snares was deemed inhumane and not considered despite being effectively used in other areas (Mapson 2004). Live trapping using portable cage traps and semi-permanent corral traps was accepted as the safest, most efficient and most humane means of capturing wild pigs. These traps could also be operated during the FWNC&R's nonoperational hours and closed when the facility was open to the public and allowed for the safe release of non-target captures.

Most stakeholders initially promoted chemicals as the euthanasia method of choice; however, the American Veterinary Medical Association Panel on Euthanasia recommends a properly placed gunshot as a humane method for pig euthanasia in the field (Beaver 2001). Gunshot was selected as the appropriate euthanasia method because of reduced cost, easier access and handling, and the efficiency with which it could be employed. This efficiency leads to reduced time in traps, especially in proximity to humans, and reduced stress for trapped wild pigs.

Carcass disposal proved to be the most controversial facet of the control plan as stakeholder groups debated using the meat by providing food to charities versus moving the carcasses to discreet disposal areas on the FWNC&R where natural nutrient cycling could take place. Potential disease transmission, either through processing the carcass or consuming the meat, potentially left the City of Fort Worth open to liability and led to the decision to dispose of carcasses onsite. This decision also demonstrated that the control program was designed to protect the resource and not to provide benefits for human populations, thereby solidifying animal welfare community support for lethal control.

The City of Fort Worth's Animal Control Department, charged with coping with the City's domestic animal problems, lacked the resources and experience to implement a live trap and euthanize program for wild pigs. Texas Parks and Wildlife, the state agency charged with managing wildlife populations in the state claimed no responsibility for a non-native species on municipal property, and USDA/APHIS/Wildlife Services required a prohibitive monthly fee and no limitations on method, timing, or location used for wild pig control. Professional trappers could not be authorized to discharge firearms on FWNC&R property. FWNC&R staff biologists, by training, had the expertise and experience to conduct all facets of the proposed control program and, as City of Fort Worth employees, could be authorized to use firearms under strictly controlled guidelines.

In September 2002, the final control program, approved by all stakeholders and authorized by the City Council, included live trap and euthanize via gunshot methodology with carcass disposal to occur onsite. All activities were to be conducted by FWNC&R biologists. Immediately upon authorization, local media began a campaign to build opposition to the program (Tinsley 2002). Previous involvement from all relevant stakeholders led to only 6% of comments (emails, letters, telephone calls) being in opposition to the program as described.

Active control measures began in January 2003 and have resulted in the removal of 298 wild pigs from the FWNC&R through spring 2017. Program effectiveness has been measured through a reduction in/elimination of wild pig damage, sightings, and confrontations/attacks as well as continued acceptance of the lethal control program by the general public and the media. Control activities have been so successful that during the 41-month period between August 2009 and December 2012, only 1 month (June 2011) showed wild pig activity on the FWNC&R. The removal of a sounder of 9 pigs on June 28, 2011 eliminated any activity.

The success of the FWNC&R wild pig control program has led to the development of other urban wild pig control programs in the Dallas/Fort Worth Metroplex using the FWNC&R model of stakeholder involvement to reduce or eliminate stakeholder opposition and to gain public support. Lacking some of the resources available on the FWNC&R, each community has modified the control methodology by using other city personnel (e.g., police officers, animal control officers, non-biologist park and recreation employees) or contracting with a professional trapping service to conduct the program. Some communities chose to utilize chemical euthanasia within highly developed residential areas with carcasses being incinerated to avoid secondary poisoning. Other communities, particularly those employing outside trapping services, avoid euthanasia issues by requiring trapped pigs to be taken directly to a slaughterhouse with part of the trapper's fees being covered by the revenue generated. These coordinated efforts are the result of 3 regional conferences sponsored by a coalition of communities and organizations in 2010, 2013, and 2016. These conferences focused on educating municipal decision-makers on the scope of the wild pig problem in the Metroplex region to facilitate future support of those forced to implement or manage wild pig control programs in their communities.

19.7.2 Case Study 2: Wild Boar in Berlin, Germany

We thank Milena Stillfried[a], Stephanie Kramer-Schadt[a], and Sylva Ortmann[b] ([a]Department of Ecological Dynamics, [b]Department of Evolutionary Ecology, Leibniz Institute for Zoo and Wildlife Research (IZW), Berlin, Germany) for providing this case study of wild pigs in Germany.

As in the United States, the phenomena of urban sprawl and increasing human-wildlife conflict is occurring in urban areas in Germany. In fact, Berlin is often called the "capital of wild boar." Twenty percent of Berlin is covered by forest, creating optimal conditions for wildlife. In addition to Eurasian wild boar (hereafter referred to as wild boar) occurring at high density in urban forests, they also regularly occur in urban parks and private gardens. In urban areas, wild boar strongly selected for natural landscapes, which often were close to roads and houses (Stillfried et al. 2017c). Wild boar from urban forests and within the city of Berlin exhibit different behaviors than their rural counterparts, where urban wild boar are habituated to human presence and are regularly observed during the day (Figure 19.12). In contrast, rural wild boar are shy and difficult to detect. Across private and public properties in urbanized areas, wild boar cause rooting damage, are potential vectors of zoonotic disease, and threaten people. An effective management strategy is necessary to reduce conflicts between humans and wild boar in Berlin.

Scientific findings about population dynamics contribute to an understanding of urbanization processes in wild pigs and can therefore play an important role in optimizing management efforts. The most common hypothesis for the origin of Berlin's wild boar assumed that animals occurring in developed areas originated from neighboring urban forests, since wild boar in urban forests are already habituated to humans. But surprisingly there are 3 genetically isolated urban forest populations of wild boar and only 1 large population from the rural areas around Berlin (Stillfried et al. 2017a). Another surprising finding was that individuals which originated from urban areas did not originate from neighboring urban forests but from the rural population. Habitat in developed areas of Berlin might represent attractive population sinks (where population growth is declining) that are colonized from an increasing rural population.

We assume the developed areas of Berlin were attractive to wild boar due to high energetic and easily accessible anthropogenic food. But a macroscopic stomach content analysis revealed that urban wild boar in Berlin foraged almost exclusively on natural food sources (Stillfried et al. 2017b). Berlin has more mast producing trees due to mixed, deciduous species in urban forests and residential forest, which is in contrast to rural forests that are mostly dominated by pine

FIGURE 19.12 Eurasian wild boar in an urban area of Berlin, Germany. (Photo by M. Stillfried. With permission.)

monocultures. Thus, the amount of energy was greater in stomachs of urban wild boar than rural animals. While wild boar in rural areas avoid being in proximity to humans, wild boar in urban areas have learned to tolerate human presence and are able to find food within human-dominated landscapes as shown by a movement analysis of GPS-tagged wild boar (Stillfried et al. 2017c). It is unclear if they will continue to use only natural food sources or if they will change their behavior and adapt to anthropogenic sources. Therefore, it is important to secure garbage bins and prevent direct feeding, because it is likely that wild boar would shift their diet to include anthropogenic food sources if these become easily accessible or when natural resources are limited.

19.8 FUTURE RESEARCH AND MANAGEMENT NEEDS

Nearly all research on wild pigs in urbanized landscapes has been conducted in their native range of Eurasia. Further, this research has primarily occurred in only a handful of cities or rural landscapes over a relatively restricted geographic distribution. Given the rapidly expanding populations of wild pigs in their native and non-native ranges (Massei et al. 2015, Snow et al. 2017), potential for future population expansion (Lewis et al. 2017), and well-documented conflicts of wild pigs in urbanized habitat globally (Licoppe et al. 2013), it is critical that research is completed across the full extent of their global distribution to better understand patterns of wild pig invasion of urbanized habitat. Particularly in North America, research should focus on current strong-holds of wild pigs in the southern and western United States, in addition to recently established populations in the northern United States and Canada.

Given the multiple types of urbanization that wild pigs can inhabit, we propose that researchers and managers evaluate wild pig populations across a gradient of urbanization, from low- to high-density urbanization. Three key components of wild pig research to focus on include: 1) wild pig behavior and the factors leading to avoidance and attraction to urbanized habitat, 2) population characteristics (i.e., population size and density, reproduction, growth rates, body condition, space use, habitat use, dispersal, colonization, and movements) and genetic characteristics (i.e., structuring, migration, and adaptations), and 3) best management strategies and actions to control wild pig populations and reduce conflicts with people, property, and natural resources associated with urbanized habitat. The field of urban ecology is a relatively new discipline (Forman 2014), and wild pigs are an ideal species to study to not only reduce growing conflicts with people, but also to evaluate as a model system to better understand how invasive species interact with urbanized habitat.

19.9 CONCLUSIONS

The adaptability and rapidly expanding populations of both wild pigs and humans across local, national, and global scales will predictably lead to increased interactions and conflict. These conflicts are likely to be especially strong in urbanized landscapes. Because the invasion of wild pigs into urbanized habitat is ultimately a result of human activities, people can address this issue in at least 3 ways: 1) stop transporting wild pigs to new areas, 2) manage wild pig populations associated with urbanized habitat to reduce their population size, and 3) manage human residences and property that are associated with habitat where wild pigs occur to discourage wild pig use of urbanized areas. By managing resources associated with human residences, wild pigs will have less incentive to access urbanized areas. When wild pigs access urbanized areas, the suite of options available to manage those populations will be determined by human values, resources, and government regulations. Bringing together diverse stakeholders to proactively develop management plans enables land managers to rapidly respond to wild pig invasions, control invading populations, and reduce human conflict. Ultimately, conflicts with wild pigs in urbanized habitat are expected to intensify in the future unless management plans are implemented that explicitly address this growing issue. Management and leadership across all levels of government and the public are essential to further reduce wild pig conflicts in urbanized habitat. Continued outreach, resources, and actions are necessary across all sectors to reduce conflict with wild pigs in urbanized habitat and prevent problems in new areas.

ACKNOWLEDGMENTS

We were supported by the Center of Epidemiology and Animal Health of US Department of Agriculture (USDA)/ Animal and Plant Health Inspection Service (APHIS)/Veterinary Services and the National Wildlife Research Center and National Feral Swine Damage Management Program of USDA/APHIS/Wildlife Services, Conservation Science Partners, and Arizona State University. J.J. Mayer's work on this chapter was supported by the US Department of Energy Office of Environmental Management under Contract DE-AC09-08SR22470 to Savannah River Nuclear Solutions LLC. Mention of commercial products or companies does not represent an endorsement by the US Government. We thank B. Strickland and reviewers for providing thoughtful feedback that improved earlier versions of this chapter.

REFERENCES

Adams, C. E. 2016. *Urban Wildlife Management*. CRC Press, Boca Raton, FL.

Amici, A., F. Serrani, C. M. Rossi, and R. Primi. 2012. Increase in crop damage caused by wild boar (*Sus scrofa* L.): The "refuge effect." *Agronomy for Sustainable Development* 32:683–692.

Anderson, A., C. Slootmaker, E. Harper, J. Holderieath, and S. A. Shwiff. 2016. Economic estimates of feral swine damage and control in 11 US states. *Crop Protection* 89:89–94.

Barrios-Garcia, M. N., and S. A. Ballari. 2012. Impact of wild boar (*Sus scrofa*) in its introduced and native range: A review. *Biological Invasions* 14:2283–2300.

Beasley, J. C., T. E. Grazia, P. E. Johns, and J. J. Mayer. 2014. Habitats associated with vehicle collisions with wild pigs. *Wildlife Research* 40:654–660.

Beaver, B. V. 2001. Report of the AVMA panel on Euthanasia. *Journal of the American Veterinary Medical Association* 218:669–696.

Bejder, L., A. Samuels, H. Whitehead, H. Finn, and S. Allen. 2009. Impact assessment research: Use and misuse of habituation, sensitisation and tolerance in describing wildlife responses to anthropogenic stimuli. *Marine Ecology Progress Series* 395:177–185.

Brown, L. N. 1985. Elimination of a small feral swine population in an urbanizing section of central Florida. *Florida Scientist* 2:120–123.

Cahill, S., and F. Llimona. 2004. Demographics of a wild boar *Sus scrofa* L., 1758 population in a metropolitan park in Barcelona. *Galemys* 16:37–52.

Cahill, S., F. Llimona, L. Cabañeros, and F. Calomardo. 2012. Characteristics of wild boar (*Sus scrofa*) habituation to urban areas in the Collserola Natural Park (Barcelona) and comparison with other locations. *Animal Biodiversity and Conservation* 35:221–233.

Cahill, S., F. Llimona, and J. Gràcia. 2003. Spacing and nocturnal activity of wild boar *Sus scrofa* in a Mediterranean metropolitan park. *Wildlife Biology* 9:3–13.

Campbell, T. A., D. B. Long, and G. Massei. 2011. Efficacy of the Boar-Operated-System to deliver baits to feral swine. *Preventive Veterinary Medicine* 98:243–249.

Castillo-Contreras, R., J. Carvalho, E. Serrano, G. Mentaberre, X. Fernández-Aguilar, A. Colom, C. González-Crespo, et al. 2018. Urban wild boars prefer fragmented areas with food resources near natural corridors. *Science of the Total Environment* 615:282–288.

Cowan, D. P., and G. Massei. 2008. Wildlife contraception, individuals and populations: How much fertility control is enough? *Proceedings of the Vertebrate Pest Conference* 23:220–228.

Crowley, S. L., S. Hinchliffe, and R. A. McDonald. 2017. Conflict in invasive species management. *Frontiers in Ecology and the Environment* 15:133–141.

DeNicola, A. J., K. C. VerCauteren, P. D. Curtis, and S. E. Hyngstrom. 2000. *Managing White-Tailed Deer in Suburban Environments*. Cornell Cooperative Extension, Ithaca, NY.

Ferretti, F., A. Sforzi, J. Coats, and G. Massei. 2014. The BOS™ as a species-specific method to deliver baits to wild boar in a Mediterranean area. *European Journal of Wildlife Research* 60:555–558.

Fischer, C., H. Gourdin, and M. Obermann. 2004. Spatial behaviour of the wild boar in Geneva, Switzerland: Testing methods and first results. *Galemys* 16:149–155.

Forman, R. T. 2014. *Urban Ecology: Science of Cities*. Cambridge University Press, New York.

Frantz, A., S. Bertouille, M. Eloy, A. Licoppe, F. Chaumont, and M. Flamand. 2012. Comparative landscape genetic analyses show a Belgian motorway to be a gene flow barrier for red deer (*Cervus elaphus*), but not wild boars (*Sus scrofa*). *Molecular Ecology* 21:3445–3457.

Frederick, J. M. 1998. Overview of wild pig damage in California. *Proceedings of the Vertebrate Pest Conference* 18:82–86.

Galhano-Alves, J. P. 2004. Man and wild boar: A study in Montesinho Natural Park, Portugal. *Galemys* 16:223–230.

Gaston, K. J. 2010. *Urban Ecology*. Cambridge University Press, New York.

Gehrt, S. D., S. P. Riley, and B. L. Cypher. 2010. *Urban Carnivores: Ecology, Conflict, and Conservation*. John Hopkins University Press, Baltimore, MD.

Hadidian, J. 2009. The socioecology of urban wildlife management. Pages 202–213 *in* M. J. Manfredo, J. J. Vaske, P. J. Brown, D. J. Decker, and E. A. Duke, editors. *Wildlife and Society: The Science of Human Dimensions*. Island Press, Washington, DC.

Hafeez, S., M. Abbas, A. H. Khan, and E. U. Rehman. 2008. Analysis of the diet of wild boar in Islamabad City, Pakistan. *Journal of Applied Biosciences* 9:403–406.

Hansen, A. J., R. L. Knight, J. M. Marzluff, S. Powell, K. Brown, P. H. Gude, and K. Jones. 2005. Effects of exurban development on biodiversity: Patterns, mechanisms, and research needs. *Ecological Applications* 15:1893–1905.

Harveson, P. M., R. R. Lopez, B. A. Collier, and N. J. Silvy. 2007. Impacts of urbanization on Florida Key deer behavior and population dynamics. *Biological Conservation* 134:321–331.

Jansen, A., E. Luge, B. Guerra, P. Wittschen, A. D. Gruber, C. Loddenkemper, T. Schneider, M. Lierz, D. Ehlert, and B. Appel. 2007. Leptospirosis in urban wild boars, Berlin, Germany. *Emerging Infectious Diseases* 13:739–742.

Keuling, O. 2009. Commuting, shifting or remaining?: Different spatial utilisation patterns of wild boar *Sus scrofa L.* in forest and field crops during summer. *Mammalian Biology-Zeitschrift* für *Säugetierkunde* 74:145–152.

Keuling, O., N. Stier, and M. Roth. 2008. How does hunting influence activity and spatial usage in wild boar *Sus scrofa L.*? *European Journal of Wildlife Research* 54:729–737.

Killian, G., L. Miller, J. Rhyan, T. Dees, D. Perry, and H. Doten. 2008. Evaluation of GNRH contraceptive vaccine in captive feral swine in Florida. *Proceedings of Wildlife Damage Management Conference* 10:128–133.

Kotulski, Y., and A. König. 2008. Conflicts, crises and challenges: Wild boar in the Berlin City—a social empirical and statistical survey. *Natura Croatica* 17:233–246.

Larson, G., K. Dobney, U. Albarella, M. Fang, E. Matisoo-Smith, J. Robins, S. Lowden, et al. 2005. Worldwide phylogeography of wild boar reveals multiple centers of pig domestication. *Science* 307:1618–1621.

Lewis, J. S., M. L. Farnsworth, C. L. Burdett, D. M. Theobald, M. Gray, and R. S. Miller. 2017. Biotic and abiotic factors predicting the global distribution and population density of an invasive large mammal. *Nature Scientific Reports* 7:4415.

Licoppe, A., C. Prévot, M. Heymans, C. Bovy, J. Casaer, and S. Cahill. 2013. Wild boar/feral pigs in (peri-) urban areas. International survey report as an introduction to the workshop: Managing wild boar in human-dominated landscapes. International Union of Game Biologists, Congress IUGB:1–31.

Linkie, M., Y. Dinata, A. Nofrianto, and N. Leader-Williams. 2007. Patterns and perceptions of wildlife crop raiding in and around Kerinci Seblat National Park, Sumatra. *Animal Conservation* 10:127–135.

Loker, C. A., D. J. Decker, and S. J. Schwager. 1999. Social acceptability of wildlife management actions in suburban areas: 3 cases from New York. *Wildlife Society Bulletin* 27:152–159.

Luniak, M. 2004. Synurbization—adaptation of animal wildlife to urban development. *Proceedings of the International Urban Wildlife Symposium* 4:50–55.

Manfredo, M., T. Teel, and A. Bright. 2003. Why are public values toward wildlife changing? *Human Dimensions of Wildlife* 8:287–306.

Mapson, M. E. 2004. *Feral Hogs in Texas*. Texas Cooperative Extension Publication B-6149. College Station, TX.

Massei, G., J. Kindberg, A. Licoppe, D. Gačić, N. Šprem, J. Kamler, E. Baubet, U. Hohmann, A. Monaco, and J. Ozoliņš. 2015. Wild boar populations up, numbers of hunters down? A review of trends and implications for Europe. *Pest Management Science* 71:492–500.

Mayer, J. J. 2013. Wild pig attacks on humans. *Proceedings of the Wildlife Damage Management Conference* 15:17–25.

Mayer, J. J., and I. L. Brisbin. 2008. *Wild Pigs in the United States: Their History, Comparative Morphology, and Current Status*. University of Georgia Press, Athens, GA.

Mayer, J. J., and I. L. Brisbin. 2009. Wild pigs: Biology, damage, control techniques and management. SRNL-RP-2009-00869. Savannah River National Laboratory.

Mayer, J. J., and P. E. Johns. 2007. Characterization of wild pig-vehicle collisions. *Proceedings of the Wildlife Damage Management Conference* 12:175–187.

Miller, B., J. LaCour, C. Yoest, T. G. Bowman, D. Coyner, S. Dobey, M. Dye, et al. 2012. Annual State Summary Report: Wild Hog Working Group. Southeastern Association of Fish and Wildlife Agencies, Hot Springs National Park, AR.

Ohashi, H., M. Saito, R. Horie, H. Tsunoda, H. Noba, H. Ishii, T. Kuwabara, et al. 2013. Differences in the activity pattern of the wild boar *Sus scrofa* related to human disturbance. *European Journal of Wildlife Research* 59:167–177.

Patterson, M. E., J. M. Montag, and D. R. Williams. 2003. The urbanization of wildlife management: Social science, conflict, and decision making. *Urban Forestry & Urban Greening* 1:171–183.

Pearson, H. 2004. Fore! Wild pigs on worldwide attack. *The Guardian* October 29. <https://www.theguardian.com/sport/2004/oct/30/comment.harrypearson>. Accessed 18 Apr 2017.

Pepin, K. M., A. J. Davis, J. Beasley, R. Boughton, T. Campbell, S. M. Cooper, W. Gaston, S. Hartley, J. C. Kilgo, and S. M. Wisely. 2016. Contact heterogeneities in feral swine: Implications for disease management and future research. *Ecosphere* 7:1–11.

Pimentel, D. 2007. Environmental and economic costs of vertebrate species invasions into the United States *in* G. W. Witmer, W. C. Pitt, and K. A. Fagerstone, editors. *Managing Vertebrate Invasive Species: Proceedings of an International Symposium*. US Department of Agriculture/Animal and Plant Health Inspection Service/Wildlife Services/National Wildlife Research Center, Ft. Collins, CO.

Podgórski, T., G. Baś, B. Jędrzejewska, L. Sönnichsen, S. Śnieżko, W. Jędrzejewski, and H. Okarma. 2013. Spatiotemporal behavioral plasticity of wild boar (*Sus scrofa*) under contrasting conditions of human pressure: Primeval forest and metropolitan area. *Journal of Mammalogy* 94:109–119.

Prévot, C., and A. Licoppe. 2013. Comparing red deer (*Cervus elaphus* L.) and wild boar (*Sus scrofa* L.) dispersal patterns in southern Belgium. *European Journal of Wildlife Research* 59:795–803.

Radeloff, V. C., R. B. Hammer, S. I. Stewart, J. S. Fried, S. S. Holcomb, and J. F. McKeefry. 2005. The wildland–urban interface in the United States. *Ecological Applications* 15:799–805.

Reidy, M. M., T. A. Campbell, and D. G. Hewitt. 2008. Evaluation of electric fencing to inhibit feral pig movements. *The Journal of Wildlife Management* 72:1012–1018.

Richomme, C., E. Afonso, V. Tolon, C. Ducrot, L. Halos, A. Alliot, C. Perret, M. Thomas, P. Boireau, and E. Gilot-Fromont. 2010. Seroprevalence and factors associated with *Toxoplasma gondii* infection in wild boar (*Sus scrofa*) in a Mediterranean island. *Epidemiology and Infection* 138:1257–1266.

Saito, M., and F. Koike. 2013. Distribution of wild mammal assemblages along an urban–rural–forest landscape gradient in warm-temperate East Asia. *PLoS ONE* 8:e65464.

Saito, M. U., H. Momose, S. Inoue, O. Kurashima, and H. Matsuda. 2016. Range-expanding wildlife: Modelling the distribution of large mammals in Japan, with management implications. *International Journal of Geographical Information Science* 30:20–35.

Schielke, A., K. Sachs, M. Lierz, B. Appel, A. Jansen, and R. Johne. 2009. Detection of hepatitis E virus in wild boars of rural and urban regions in Germany and whole genome characterization of an endemic strain. *Virology Journal* 6:58.

Schlageter, A., and D. Haag-Wackernagel. 2011. Effectiveness of solar blinkers as a means of crop protection from wild boar damage. *Crop Protection* 30:1216–1222.

Schlageter, A., and D. Haag-Wackernagel. 2012. Evaluation of an odor repellent for protecting crops from wild boar damage. *Journal of Pest Science* 85:209–215.

Schneider, A., M. A. Friedl, and D. Potere. 2009. A new map of global urban extent from MODIS satellite data. *Environmental Research Letters* 4:1–11.

Schuett, M. A., D. Scott, and J. O'Leary. 2009. Social and demographic trends affecting fish and wildlife management. Pages 18–30 *in* M. J. Manfredo, J. J. Vaske, P. J. Brown, D. J. Decker, and E. A. Duke, editors. *Wildlife and Society: The Science of Human Dimensions*. Island Press, Washington, DC.

Seto, K. C., M. Fragkias, B. Güneralp, and M. K. Reilly. 2011. A meta-analysis of global urban land expansion. *PLoS ONE* 6:e23777.

Snow, N. P., M. A. Jarzyna, and K. C. VerCauteren. 2017. Interpreting and predicting the spread of invasive wild pigs. *Journal of Applied Ecology* 54:2022–2032.

Southeastern Cooperative Wildlife Disease Study. 2016. *National Feral Swine Map, 2016*. University of Georgia, Athens, GA.

Stephens, D. W., J. S. Brown, and R. C. Ydenberg. 2007. *Foraging: Behavior and Ecology*. University of Chicago Press, Chicago, IL.

Stillfried, M., J. Fickel, K. Börner, U. Wittstatt, M. Heddergott, S. Ortmann, S. Kramer-Schadt, and A. C. Frantz. 2017a. Do cities represent sources, sinks or isolated islands for urban wild boar population structure? *Journal of Applied Ecology* 54:272–281.

Stillfried, M., P. Gras, M. Busch, K. Börner, S. Kramer-Schadt, and S. Ortmann. 2017b. Wild inside: Urban wild boar select natural, not anthropogenic food resources. *PLoS ONE* 12:e0175127.

Stillfried, M., P. Gras, K. Boerner, F. Goeritz, J. Painer, K. Roellig, M. Wenzler, H. Hofer, S. Ortmann, and S. Kramer-Schadt. 2017c. Secrets of success in a landscape of fear: Urban wild boar adjust risk perception and tolerate disturbance. *Frontiers in Ecology and Evolution* 5:157.

Tadano, R., A. Nagai, and J. Moribe. 2016. Local-scale genetic structure in the Japanese wild boar (*Sus scrofa leucomystax*): Insights from autosomal microsatellites. *Conservation Genetics* 17:1125–1135.

Tari, T., G. Sándor, G. Heffentraeger, and A. Náhlik. 2016. Wild boar habituation to urban areas in Hungary, in the light of Web presence. *Proceedings of the International Hunting and Game Management Symposium* 5:26.

Texas Animal Health Commission. 2003. *Regulations for Trapping or Moving Feral (Wild) Swine*. Austin, TX.

Theobald, D. M. 2014. Development and applications of a comprehensive land use classification and map for the US. *PLoS ONE* 9:e94628.

Thomson, C., and C. N. Challies. 1988. Diet of feral pigs in the podocarp-tawa forests of the Urewera Ranges. *New Zealand Journal of Ecology* 11:73–78.

Thurfjell, H., J. P. Ball, P. A. Åhlén, P. Kornacher, H. Dettki, and K. Sjöberg. 2009. Habitat use and spatial patterns of wild boar *Sus scrofa* (L.): Agricultural fields and edges. *European Journal of Wildlife Research* 55:517–523.

Tinsley, A. M. 2002. City will trap, kill wild hogs. *Fort Worth Star-Telegram*, 16 September 2002 section B:1.

Toger, M., I. Benenson, Y. Wang, D. Czamanski, and D. Malkinson. 2018. Pigs in space: An agent-based model of wild boar (Sus scrofa) movement into cities. *Landscape and Urban Planning* 173:70–80.

US Department of Agriculture/Animal and Plant Health Inspection Service. 2015. *Final Environmental Impact Statement. Feral Swine Damage Management: A National Approach*. Washington, DC.

Virgós, E. 2002. Factors affecting wild boar (*Sus scrofa*) occurrence in highly fragmented Mediterranean landscapes. *Canadian Journal of Zoology* 80:430–435.

White, M., K. Kauffman, J. S. Lewis, and R. Miller. 2018. Wild pigs breach farm fence through harvest time in southern San Joaquin Valley. *California Agriculture* 72:120–126.

Wittmann, K., J. J. Vaske, M. J. Manfredo, and H. C. Zinn. 1998. Standards for lethal response to problem urban wildlife. *Human Dimensions of Wildlife* 3:29–48.

Wyckoff, A. C., S. E. Henke, T. A. Campbell, D. G. Hewitt, and K. C. VerCauteren. 2012. Movement and habitat use of feral swine near domestic swine facilities. *Wildlife Society Bulletin* 36:130–138.

20 The Future of Wild Pigs in North America

Stephen S. Ditchkoff, James C. Beasley, John J. Mayer, Gary J. Roloff, Bronson K. Strickland, and Kurt C. VerCauteren

CONTENTS

20.1 INTRODUCTION

As described and reiterated throughout this book, wild pigs (*Sus scrofa*) are a well-established ecological and economic issue in North America, causing considerable damage to both anthropogenic and ecological resources. They are considered 1 of the top 2 most destructive, invasive terrestrial vertebrates in North America (the other is the invasive Burmese python (*Python bivittatus*) in Florida; North American Invasive Species Network 2015), and estimates of their damages are in the billions of dollars. Although most news and reports concerning impacts of wild pigs are discouraging, more recently there has been a growing number of success stories concerning their management. Within only a few years of the US federal government's investment in wild pig control in 2014 (USDA/Animal and Plant Health Inspection Service 2015), several states with small populations of wild pigs reported successful eradication. Idaho, Maryland, New Jersey, and New York have all reported successful elimination of wild pig populations within their borders, and all wild pigs are believed to have been removed from Illinois, Iowa, Maine, Minnesota, Wisconsin, and Washington as well. While surveillance in these states must be maintained as the potential for reinvasion (misguided anthropogenic or natural) exists, these success stories are cause for optimism and suggest wild pig control programs in other areas have the potential to also be successful. However, other parts of the country have much greater densities of wild pigs, as well as cultural differences that will factor into the success of any wild pig management program. For this reason, management strategies and tools, public support, and governmental support and legislative actions will need to evolve if we can expect to continue to reduce populations and impacts of wild pigs across the continent.

This brief chapter focuses on the future of wild pig management in North America, and describes some of the steps that will be required for continued reduction in wild pig populations and damages. Specifically, we discuss the development of unified goals, educating the public, the need to improve our scientific knowledge, how this knowledge is necessary to inform decision-makers, and the importance of increasing resources available for wild pig management. Finally, we pull out our collective crystal ball and provide our predictions of the future of wild pigs in North America. Hopefully, our consolidated thoughts on these subjects will positively influence wild pig management as we move forward.

20.2 DEVELOPING A MORE UNIFIED GOAL

One of the common obstacles to the implementation of virtually any wildlife management program is the variety of opinions and preferences regarding the program's ultimate goals. Oftentimes, some interest groups desire more of a species, while others desire less. Additionally, there are always a diversity of opinions on the best way to achieve management goals, and many of these opinions are founded on emotional, moral, ethical, or cultural beliefs. Wild pig management is no different in this regard. Some groups feel strongly that the existence of wild pigs in North America is desirable, while other groups contend that the damages caused by wild pigs far outweigh any positive attributes they provide. This disparity in opinions is further complicated by the large percentage of the public, as well as decision-makers, that are generally uninformed regarding the facts on wild pigs in North America. This situation highlights the need for more unified goals among stakeholders regarding management of this species. Until a greater percentage of landowners, legislators, and the general public are in agreement on common goals for wild pig management, efforts will not be as successful as they could be otherwise.

Relative to concerns of the impacts wild pigs can have on native species and ecosystems, influential conservation groups have spoken out in favor of increased management of this species. For example, The Wildlife Society, the professional society of the wildlife profession, has put forth a Final Position Statement (The Wildlife Society 2011) and the Boone and Crockett Club, a well-respected conservation organization, has made efforts to inform their membership and other outdoor enthusiasts about the ills of wild pigs. In addition, in recent years there have been numerous articles in popular hunting magazines educating readers about the negative impacts wild pigs are having on game species and other natural and agricultural resources. Many of these magazines are doing a commendable job of not glamorizing wild pig hunting and rather, are advocating for recreational hunters to be part of the solution.

To this end, we contend the establishment of local, state, regional, and national task forces whose objectives are to support and promote responsible wild pig management, control, or eradication is crucial. Task forces are usually comprised of representatives from most of the main stakeholder groups with interests in wild pigs and generally provide a unification of ideas and objectives due to their transparency, reliance upon empirical data for the generation of their opinions, and objectivity. Additionally, because they operate outside the restrictions of the individual groups of which they are comprised, they are able to generate thoughtful opinion statements, as well as work to inform and influence decision-makers. In recent years a number of these task forces have arisen and been influential in legislative decisions, improvements in economic support for control efforts, and the generation of unified management goals and objectives. If we hope for continued management success with wild pigs, active and influential task forces focused on wild pig management will have an important role to play. At the continental scale, the formation of a quad-lateral task force including Canada, the United States, Mexico, and Tribal Nations must be considered, and communication and limited coordination is already ongoing.

As a case in point, the National Wild Pig Task Force (NWPTF) in the United States has been in place since 2016 and is serving multiple valuable roles. Examples of activities include providing information to state and local task forces and, through a variety of committees, assimilating and disseminating information on topics of special interest like toxicants, knowledge gaps requiring research, and best management practices (Beasley et al. 2018).

20.3 PUBLIC EDUCATION

As thoroughly discussed in Chapter 10, public education will be critical to the successful management of wild pigs in North America. The diversity of opinions on wild pigs and agreement on the extent to which they should be managed will be a continual challenge, but the level of knowledge about wild pig biology, management, and impacts that is possessed by the public can be increased.

Wildlife professionals must work to ensure the public, and ultimately legislators, understand the facts concerning wild pigs. Too much of their information is gleaned from non-professional sources readily available on the Internet, and much of this information is not science-based. The continued work of university extension professionals, natural resource agencies, and other wildlife biologists in educating the public with factual information on wild pig biology and management will be essential for continued progress in the management of this invasive species. Local, regional, and national task forces will have an important role in driving this educational effort, but the leadership of extension and outreach professionals will be vital to ensure this knowledge and information reaches the affected stakeholders in the community. Traditional communication strategies like workshops, seminars, and print media will have to be employed, as will contemporary strategies leveraging the outreach potential of social media and other digital technologies.

20.4 IMPROVING OUR KNOWLEDGE

As part of educating the public, we must work to improve our knowledge base on the biology, impacts, and management of wild pigs. Surprisingly, we currently find ourselves, even as professionals, somewhat uneducated on numerous aspects of this destructive species. In general, we understand the impacts of wild pigs are far reaching, and possibly more impactful than any other terrestrial vertebrate in North America, due to the myriad of anecdotal accounts of wild pig damage. We also have a strong understanding of some aspects of wild pig biology (e.g., physiology, nutrition, reproduction) due to their role as a surrogate for human health experimentation and their importance as a domestic food source. However, despite our depth of knowledge in these areas, surprisingly little empirical data are available in other extremely important areas of wild pig biology and management. Beasley et al. (2018) described the current state of our knowledge on wild pigs, and pointed to specific areas where more data are needed to advance the goal of reducing the impacts of wild pigs in North America. They argued that an improved understanding of basic biological and behavioral aspects of wild pigs (e.g., movement patterns, social dynamics, and resource selection) are necessary in the development of effective control strategies for populations of wild pigs. We know wild pigs damage a wide array of ecological and anthropogenic resources (see Chapter 7), but data on the extent and economic impact of these damages is generally not available. These data are going to be critical to inform both the public and decision-makers, and hopefully to influence legislation that will contribute to the future control of wild pigs (Beasley et al. 2018). Development and evaluation of control techniques have been at the forefront of wild pig research for some time, and there have been some documented successes in eradicating wild pigs (e.g., McCann and Garcelon 2008, Parkes et al. 2010). However, we are only beginning to understand that improved tools and technologies (see Chapter 8) will be required to overcome many of the management challenges that wild pigs will present down the road.

20.5 INCREASE RESOURCES AVAILABLE FOR CONTROL

Ultimately, the future success of wild pig management in North America will be a function of increased availability of economic resources for population management and damage mitigation efforts. Currently, the economic burden of wild pig management is being collectively shouldered by private landowners and governmental agencies. Though the North American Model of Wildlife Conservation (NAMWC) is founded on public ownership of wildlife (Organ et al. 2012), where they occur on private land, the access to and thus management of game species has been largely driven by private landowners. State, provincial, and other governmental agencies establish management boundaries (e.g., seasons, bag limits, etc.), but the decisions of individual landowners have great influence on populations and habitat. To confuse the issue, the classification of wild pigs varies and is in flux among many states and provinces. For example, some states still designate wild pigs as a game species while others classify them as unprotected pests or somewhere in between. We argue that, as a non-native and invasive species, the tenets of the NAMWC do not apply to wild pigs (see Bodenchuk and VerCauteren 2016 for a complete discussion of this topic). Management of wild pigs will always consist primarily of population reduction or elimination because of their population growth potential and negative impacts, and governmental (state, provincial, and federal) investment in management programs will determine the degree of success achieved. However, it should be noted that the legal classification of wild pigs (game, non-game, pest, etc.) in each individual state in the United States has implications regarding the eligibility of Pittman-Robertson funds for use in management of the species. If Pittman-Robertson funds, which constitute the majority of economic resources available in most state agencies for management of wildlife, are not available for use in managing wild pigs, it could significantly hamper a state's ability to invest in wild pig control programs.

Governmental investment in wild pig control has increased significantly since 2014, and at least in the short term, allocation of resources for this purpose seems secure. However, the extent to which governmental support will continue will ultimately be a function of how well policy makers are educated on the impacts of wild pigs and how efficiently progress to reduce these impacts is made. As described earlier (also see Chapter 10), most people that are not strongly tied to the outdoors (e.g., agricultural producers, hunters, outdoor recreationists) are relatively ignorant of the impact wild pigs have on anthropogenic and ecological resources. Continued effort to understand and, more importantly, to economically quantify losses due to the presence of wild pigs on the landscape will be crucial in garnering support of the public and policy makers for increased investments in wild pig management programs in the coming years. The level of investment available today has resulted in measurable strides in some parts of the continent. However, continued success will only become more difficult as the low-hanging fruit (i.e., small isolated populations have been eradicated) disappears and control efforts become primarily focused in areas where control of wild pigs is more challenging. Relatedly, as our understanding of foreign animal diseases (e.g., classical swine fever, African swine fever) and how outbreaks in wild pigs could lead to national and global crisis increases, the pressure to reduce these impacts through reductions in free-ranging pigs could also drive improvements in funding. Increased economic support of wild pig management will ultimately determine the degree of success that is achieved.

20.6 THE FUTURE

Fifteen to 20 years ago, many of us were in agreement that wild pigs would eventually be found in all 50 states and north into Canada to the highest latitude where weather and climate would allow them to survive. Refreshingly, we don't believe that to be the case anymore. Given our recent success stories, we believe wild pigs will likely be eradicated from the northern tier of the United States in the near future. As part of this trend, we also predict an ever-growing percentage of states and governmental lands will continue to outlaw recreational hunting of wild pigs to aid control efforts (see Chapter 11). In southern and western North America (including Hawaii), where wild pigs are firmly entrenched as part of the landscape and/or culture, these populations will likely shrink in size and distribution as resources and tools available for control increase, but will endure because of the logistical, cultural, and economic challenges associated with their eradication. Similarly, we predict that most Canadian provinces will reduce their wild pig populations in size and distribution, but will struggle to completely eradicate them due to the expansive and wilderness nature of the areas in which wild pigs are entrenched.

As part of this elevated effort in wild pig management, we predict we will see significant advancements in the tools and technologies available for control. Toxicants and contraceptives, which are currently unavailable (i.e., due primarily to incomplete permitting processes or the lack of data on environmental or non-target impacts) for wild pig management in North America, will be developed soon and should allow for more effective and efficient control. For these reasons and others, as we stare into our collective crystal ball, we feel optimistic. Not too long ago, we had little hope in North

America for success in wild pig management. However, in a few short years we have seen considerable success, and all signs point towards continued advances and improvements in the manner with which we manage wild pigs. While we may not be around to witness the eventual eradication of wild pigs from all, or even most, of North America, we are confident we will be witness to significant reductions in their range, and are proud to be playing a small part in this future success story.

ACKNOWLEDGMENTS

Contributions from K.C.V. were supported by the USDA. Contributions from J.J.M. were supported by the US Department of Energy Office of Environmental Management under Contract DE-AC09-08SR22470 to Savannah River Nuclear Solutions LLC. Contributions from J.C.B. were partially funded by the US Department of Energy under award # DE-EM0004391 to the University of Georgia Research Foundation. Mention of commercial products or companies does not represent an endorsement by the US Government.

REFERENCES

Beasley, J. C., S. S. Ditchkoff, J. J. Mayer, M. D. Smith, and K. C. VerCauteren. 2018. Research priorities for managing invasive wild pigs in North America. *Journal of Wildlife Management* 82:674–681.

Bodenchuk, M., and K. C. VerCauteren. 2016. Management of feral swine. In *Proceedings of the 27th Vertebrate Pest Conference*, R. M. Timm and R. A. Baldwin, editors. Davis, CA.

McCann, B. E., and D. K. Garcelon. 2008. Eradication of feral pigs from Pinnacles National Monument. *Journal of Wildlife Management* 72:1287–1295.

North American Invasive Species Network. 2015. *The Ten Most Important Invasive Species or Invasive Species Assemblages in North America in 2015.* North American Invasive Species Network, Gainesville, FL.

Organ, J. F., V. Geist, S. P. Mahoney, S. Williams, P. R. Krausman, G. R. Batcheller, T. A. Decker, et al. 2012. The North American Model of Wildlife Conservation. The Wildlife Society Technical Review 12–04. Bethesda, MD.

Parkes, J. P., D. S. L. Ramsey, N. MacDonald, K. Walker, S. McKnight, B. S. Cohen, and S. A. Morrison. 2010. Rapid eradication of feral pigs (*Sus scrofa*) from Santa Cruz Island, California. *Biological Conservation* 143:634–641.

The Wildlife Society. 2011. Final Position Statement: Feral Swine in North America. <https://wildlife.org/wp-content/uploads/2014/05/feral_swine_080211.pdf>.

US Department of Agriculture/Animal and Plant Health Inspection Service. 2015. *Final Environmental Impact Statement. Feral Swine Damage Management: A National Approach.* Washington, DC.

Index

R

S

V

W

X

Y

Z